Legumes in India

Applications in Food, Medicine and Industry

Legumes in India

Applications in Food, Medicine and Industry

mamatha rao
Department of Botany
Bangalore University, Bangalore

Ane Books India
New Delhi ♦ Chennai ♦ Mumbai
Bangalore ♦ Kolkata ♦ Thiruvananthapuram ♦ Lucknow

Legumes in India : Applications in Food, Medicine and Industry

Published by

Ane Books India

Head Office	4821, Parwana Bhawan, 1st Floor, 24 Ansari Road, Darya Ganj, New Delhi - 110 002, INDIA Tel.: +91(011) 23276843-44, Fax: +91(011) 23276863 e-mail: kapoor@anebooks.com, Website: www.anebooks.com
Chennai	Avantika Niwas, 1st Floor, 19 Doraiswamy Road, T. Nagar, Chennai - 600 017 Tel.: +91(044) 28141554, 28141209 e-mail: anebooks_tn@airtelbroadband.in
Mumbai	G-012, Ground Floor, Jui Nagar Railway Station Building, Jui Nagar (East), Navi Mumbai - 400 706 Tel.: +91(022) 64522840, 64522841 e-mail: anebooksmum@gmail.com
Bangalore	38/1, 1st Floor, Model House, First Street, Opp. Shamanna Park, Basavannagudi, Bangalore - 560 004 Tel.: +91(080) 41681432 e-mail: anebang@airtelbroadband.in
Kolkata	Flat No. 16A, 220 Vivekananda Road, Maniktalla, Kolkata - 700 006 Tel.: +91(033) 23547119 e-mail: anekol@vsnl.net
Thiruvananthapuram	# 6, TC 25/2710, Kohinoor Flats, Lukes Lane, Ambujavilasam Road, Thiruvananthapuram - 695 001, Kerala Tel.: +91(0471) 4068777, 4068333 e-mail: anebookstvm@airtelbroadband.in
Lucknow	C-26, Sector-A, Mahanagar, Lucknow - 226 006 Mobile - +91 93352 29971

ISBN (10) : 81-8052-135-4
ISBN (13) : 978-81-8052-135-5

Printed at Rashtriya Printers, Delhi

with able assistance from

Dr. Shalini, B.U.

Dedicated
to
Prakruthi,
my Daughter

Foreword

The Legumes as a bioresource group are vital for human welfare, yielding important pulses as well as tannins, dyes, forage and fodder, green manure, timber, fibers, beverages, oil, etc. Besides various industrial and other uses these plants play a very significant role in human nutrition and medicine, particularly in their area of natural occurrence, i.e., tropical and the subtropical parts of the world.

There are about 179 genera and 1,152 species of legumes recorded from India. The enormous economic potential of a very large number of species of the Legumes in India has not been fully realized. There is immense scope for bioprospecting the diverse chemical compounds in Indian legumes. Vast information on various aspects of Legumes such as taxonomy, biology, chemistry and their application in medicine, food and industry, exists by way of publications all over the world. In addition to the information available on public domain there is extensive ethnobotanical information on the legumes in unpublished manuscripts and reports. There is need to collect this vast data. It is necessary to develop a comprehensive database of the Legumes occurring in India to facilitate systematic evaluation and sustainable utilization in the Indian context. The present work addresses these important needs.

This book by Dr Mamatha Rao primarily deals with a survey of economically important legumes in India, identifying the important uses of 719 legume species, giving the current valid botanical names and synonyms; basic scientific information needed to understand the chemistry and the implications of legume species in various contexts; information on i) the distribution of various classes of chemical compounds and their uses, ii) nutritive value of edible legumes, and iii) the anti-nutritional and toxic factors in them; and information on legumes used in different traditional and modern systems of medicine. An interesting aspect of the book is the information profiles of 289 important legume species. The extensive 'glossary' of medical terms along with a comprehensive 'bibliography' at the end of the book is both very useful.

This excellent compilation of diverse information would be of immense use to administrators, policy makers and conservationists, besides students, teachers and researchers in botany, environmental and economic botany, biochemistry, pharmaceutical sciences, agricultural sciences and biotechnology. With its emphasis on traditional systems of medicine this work should also be of interest to practitioners of Indian medicine systems.

The onerous undertaking of Dr Mamatha Rao in bringing this vast repository of utilitarian information on the Indian Legumes into public domain is indeed commendable.

DR M SANJAPPA
Director

Botanical Survey of India
Kolkatta

Preface

Legumes, the members of the botanical family Leguminosae, to which the pulse crops belong, are an important component of the vegetation all over the world, more particularly in the tropical countries.

Legumes have been used by humans in a multitude of ways for millennia and are arguably more varied and important than the grass family that provides the cereals and millets. Legume products contribute to the world economy through food, feed, drink, pharmaceuticals, medicine, biofuel, biotechnology, building construction, textiles, furniture, crafts, paper industry, mining, fuel wood, manufacturing processes, chemicals, fertilizers, waste recycling, horticulture, forestry, pest control, ecological restoration and so many others.

Grain and forage legumes are grown on 350 million hectares, about 12.15 per cent of arable land. Twenty seven per cent of world's primary crop grain legumes alone form 35 per cent of dietary proteins and more than 35 per cent of world's processed vegetable oil. Forage legumes provide protein, fibre and energy for the dairy and meat products.

Forty to 60 million metric tons of nitrogen are fixed in agriculture by legumes annually, of which three to five million metric tons in the natural ecosystems. Many species are used as green manure, soil improvers and stabilizers.

Legumes provide substitutes for coffee, tea, tobacco, betel nut, hops, garlic, ebony wood, cocaine, and soap.

Legume dyes, gums, resins and other products are used in processed foods, cosmetics, inks, paints and polishes.

Legume chemical defence compounds are used in many ways to protect crops from insect and nematodal pests, and viral, bacterial and fungal diseases. They are used in medicine as purgatives, abortifacients, aphrodiasiacs, hallucinogens, anti-inflammatory, antioxidant, antiseptics, purgatives, arrow and ordeal poisons, and antidotes to poison. They also are insecticides, molliscides and piscicides.

Legumes paint the surface of the earth in a myriad of colours and contribute largely to horticulture and cutflower industry.

Vast information on various aspects of the Legumes such as taxonomy, biology, chemistry and their application in medicine, food and industry, exists in diverse publications all over the world. The more important sources are Pollhill and Raven (1981), Lewis *et al.*, (2005) and the ILDIS website.

In spite of voluminous literature, the importance of the Legumes as a crucial bioresource has not been appreciated adequately, particularly in the developing countries. It requires quite an effort to obtain information of any significance on the biology, chemistry and applications of the Legumes. The main reason is, though specific areas are well known to special interest groups, much of the information is beyond the reach of the general interest groups, as it is spread in several thousand diverse publications in different languages.

There are about 179 genera and 1,152 species of legumes recorded from India. There is immense scope for bioprospecting, to vastly enhance the resource potential of the legumes in India.

In addition to the scientific and other information available on the Legumes in India, there is extensive ethnobotanical information on the legumes in unpublished manuscripts and reports. There is a long felt need to collect and collate all the information on the Legumes in India to develop a comprehensive database to facilitate systematic evaluation, identify areas for further research and sustainable utilization. This cannot be realized in one go. There are some specialized legume database systems but they are circumscribed by their specialist objectives and scope. The present book is aimed at being a comprehensive information source as a prelude to a legume utilitarian information service system, by providing as much and as diverse information as possible.

The book presents a survey of the economic potential for 18 important uses for 719 legume species in India. Various Chapters deal with the basic scientific information needed to understand the chemistry and the implications of legume species in various contexts. Specific Chapters provide information on i) the distribution of various classes of chemical compounds and their uses, ii) nutritive value of edible legumes, and iii) the anti-nutritional and toxic factors in them, and the Legumes used in different traditional and modern systems of medicine. Information profiles of 289 important legume species, giving current valid botanical names and synonyms, and the economic use of these species in different contexts, more particularly as medicine, constitute an important component of this book. The extensive glossary of medical terms, a comprehensive bibliography and an elaborate index lend user support.

It is hoped that the vast and diverse information presented in this book would be of use to administrators, policy makers and conservationists, besides students, teachers and researchers in botany, environmental and economic botany, biochemistry, pharmaceutical sciences, agricultural sciences and biotechnology. The emphasis on traditional systems of medicine in the book should also be of interest to practitioners of Indian medicine systems.

I have had the good fortune of receiving very kind help from the following:

The Authorities of the Libraries of the Bangalore University, University of Agricultural Sciences, Bangalore, Government Central Pharmacy, Bangalore, Regional Institute of Ayurveda, Bangalore.

Dr B U Shalini, who has been with me in the lab for a very long time, extended unconditional support and help for which I express immense gratitude; The book is hers as much as mine.

I gratefully acknowledge the ever willing help of my friends Mr. Rohit, Dr. Kiran, Mrs. Anita, Ms. Vani and Mr. Narasimha Murthy. My special thanks to Dr. Diwakar and Mr. Purushotham.

Ms. Savitha Murthy and Mr. Guruprasad who helped in various stages of writing and proof reading, making the compilation of the book an easier task. You helped me realize my dream.

Mr. Hemanth Gowda for being a continuous source of emotional and moral support.

My publisher, Mr. Sunil Saxena for his unstinting enthusiasm, uncompromising and unconditional faith in my abilities. I will be ever grateful to him for giving me this opportunity.

I would also like to thank the staff of Ane Books especially Mrs. Deepa and Mr. Praveen, for their untiring efforts in the publication of the book.

Prof. C. Kameswara Rao for imbibing the skill of technical writing, for his sharp editing skills, for his unfaltering support, motivation and patience during the long strenous hours of compilation and writing.

I am grateful to Dr M Sanjappa, Director, Botanical Survey of India, Kolkata, for his advice and for kindly writing the Foreword

I thank Mr. Mohan who spent tireless long hours in typing the manuscript.

The writing of the book was a difficult task as decisions to include or delete information was hard. I humbly accept responsibility for any inadequacies and mistakes that I have made.

Suggestions and constructive criticism is most welcome.

mamatha rao

Contents

PART-I

Bioresource Potential of Legumes in India

CHAPTER-1

Introduction

1.1 THE LEGUMES

A 'legume' is the seedpod (fruit), such as that of pea or bean (see Box 1.1). Members of the botanical family Leguminosae are called Legumes, as the predominant fruit type in this family is the legume. Legume type fruits exhibit enormous diversity in size, shape, number of seeds contained and several other features. Other types of fruit, as for example a lomentum that breaks into single-seeded indehiscent segments, also occur in the Leguminosae.

Legumes are a very important component of the vegetation of most countries, more particularly so in the tropics and the subtropics. All of us are familiar with a number of leguminous plants, and use them in our day-to-day life (see Box 1.2).

In terms of biological diversity and diverse uses to mankind, the legumes occupy a predominant position, among all the plant families, comparable only to the Poaceae, the family of grasses, cereals and millets, also known as the Graminae. Apart from the various industrial and other uses, such as sources of tannins, dyes, forage and fodder, green manure, timber, fibres, beverages, oil, etc., the legumes play a significant role in human nutrition and medicine, is of far reaching importance.

Pulses (see Box 1.3) are a major component of the vegetarian diet, more particularly in Asia. India is the largest pulse growing country in the world accounting for 20 per cent of the acreage and nine per cent of the production under the food grains. Pulses are the richest sources of proteins providing 80 per cent of the dietary protein for vast populations in the developing countries.

Legumes play an important role in crop mixtures and rotation, as they enhance soil fertility by fixing atmospheric nitrogen in nodules on their roots, through a symbiotic association with bacteria of the *Rhizobium* group.

Though protein rich, and high in nutritive value, the pulses may also contain toxic substances that have adverse effects on human health when consumed either raw or improperly cooked. Many such anti-nutritional factors in the seeds are reported to cause various health hazards in humans. Several of such toxic factors also provide therapeutic relief and help in combating several diseases.

Box 1.1

Legumes and Pods

The term legume is derived from the Latin word *legere* meaning 'to gather'. The legume is a fruit type that splits shedding the seed, which has to be gathered, in contrast to the cereals, which have to be thrashed, to harvest the grain.

A pod is the seed case, of a two-valved fruit, which splits, usually on both the longitudinal seams (sutures), to shed the seed. The term pod is also used to describe similar fruits of non-leguminous plants, such as mustard.

Box 1.2

Common Leguminous Plants

Sweet pea (*Pisum sativum*), several kinds of vegetable beans (*Phaseolus vulgaris, Phaseolus lunatus, Vicia faba, Lablab purpureus*), soybean (*Glycine max*), winged bean (*Psophocarpus tetragonolobus*), cluster bean (*Cyamopsis tetragonoloba*), chickpea (*Cicer arietinum*), cowpea (*Vigna unguiculata*), red gram (*Cajanus cajan*), black gram (*Vigna mungo*), green gram (*Vigna radiata*), horse gram (*Macrotyloma uniflorum*), lentil (*Lens culinaris*), groundnut (*Arachis hypogaea*), tamarind (*Tamarindus indica*), and fenugreek (*Trigonella foenum-graecum*).

Rain tree (*Samanea saman*), copper pod tree (*Peltophorum pterocarpum*), gulmohr (*Delonix regia*), Ashoka tree (*Saraca asoca*), gliricidia (*Gliricidia sepium*), peacock bush (*Caesalpinia pulcherrima*), species and varieties of *Erythrina*, *Cassia*, *Bauhinia*, *Acacia* and *Mimosa*.

Rosewood (*Dalbergia latifolia*), red sandalwood (*Pterocarpus santalinus*), and liquorice (*Glycyrrhiza glabra*).

1.2 THE NEED FOR A DATABASE OF PLANTS

There is a vast amount of diverse kinds of information on the legumes, spread over thousands of different kinds of publications. There is an urgent need to bring this massive information together in one place, in order to help the interest groups to evaluate and utilize the existing knowledge, and use this as a basis for new knowledge and applications.

An effective utilization and conservation of natural resources requires information on all aspects of the resources and such information should be evaluated systematically. This can be achieved only when all available information on the subject is organised in the form of Databases, which are an orderly and structured packets of information, facilitating storage, modification and retrieval of information, in any of several forms or combinations.

> **Box 1.3 Grams and Pulses**
>
> The word gram refers to the whole edible legume seed (bengal gram, black gram, green gram), while the word *dhal* is the vernacular name for the split gram (cotyledons) (*channa dhal, urad dhal, mung dhal*).
>
> The term pulse is collectively applied to the legume seeds or the plants that yield them, and is used as food (peas, beans, lentils).
>
> Tender legume seeds and pods, cooked and eaten, are considered as vegetables.

The objectives behind the compilation of plant databases are:

a) It is of utmost importance to provide information, to a variety of users, on the botany and taxonomy of economically useful species. Correct information on the identity of such species is of great significance in several issues such as conservation and Intellectual Property Rights.

b) The prime objective of database of plants is to provide all available information on them, in one place. This will bring to light the little known uses of plants. Such information will help to introduce new plant resources into other areas, and also to bring to light new uses of existing species. These measures will help in reducing the pressure on the over stretched conventional plant resources.

c) Information on the chemistry, nutritional and therapeutic properties of plants helps us to choose foods rich in a particular compound of nutritional value or those with an appropriate therapeutic use. Large sections of urban population spend millions of rupees on nutriceuticals, the dietary supplements, like vitamins, minerals (particularly iron and zinc), antioxidants, etc., most of which can be obtained from edible parts of plants easily and inexpensively. The public should be advised on food items that are rich in nutritional supplements.

d) Almost every plant has its own therapeutic properties, in addition to its other uses. A number of minor ailments have been managed for millennia, through using plants in folk and home medicine, in most parts of the world, more particularly in the developing countries. With acculturation and consequent neglect, this fund of knowledge is fast disappearing. Plant databases help to keep this information alive.

e) The industry is forever looking for new plant based chemicals, as structural modification of such compounds is far easier and cheaper than *de novo* design and total synthesis of compounds (see Chapter 5). Edible oils, essential oils, alkaloids, saponins, tannin and a variety of other chemical compounds are used in various industries. A database will highlight the sources of phytochemical compounds with economic potential and provides an easier access to information and source material.

1.3 THE NEED FOR A DATABASE OF INDIAN LEGUMES

There are about 179 genera and 1,152 species of legumes recorded from India. In addition to native species, several economically useful species such as soybean, winged bean, etc., have been introduced.

Legumes, being one of the most economically important groups of plants, have attracted the attention of scientists from various disciplines all over the world all the time. A voluminous amount of information is available on the taxonomy, biology, chemistry of legumes and their application in medicine, food and industry. It is very much necessary to gather and compile all such information together, in order to facilitate its evaluation and utilization. A Database is the only means of achieving this objective.

The enormous economic potential of a very large number of species of the legumes in India has not been fully realized. This hampers a complete or sustainable utilization of these species. It is of utmost importance that all the available information is compiled and put in one place to facilitate an easy access to it. This will help in a periodical evaluation of the potential of the species involved.

In addition to published information, there is a lot more of valuable unrecorded information on the uses of plants with people in the remote and tribal areas. This culture based information, constitutes Ethnobotany, Ethnomedicine and Ethnopharmacology (Kameswara Rao, 2000), will be lost forever, if not recorded the soonest (see Chapter 17).

There is an unlimited scope for bioprospecting diverse chemical compounds in legumes. For such efforts it is necessary to know what is already known, in order to save time and effort, and to prevent duplication of efforts.

1.4 LEGUME DATABASES

There is no utility-based comprehensive database of legumes, particularly for the Indian legumes. There are several plant databases that provide contextual information on legumes, such as, a) The Bean Bag (www.huh.Harvard.edu), a newsletter, b) NAPRALERT, DEREP and MEDFLOR (http://pcog8.pmmp.uic.edu/mep/MCP.html) that deal with the chemistry and properties of natural products; and c) The Database of Medicinal Plants (Kameswara Rao, 2000; www.medicinalplants-kr.org) deals with medicinal uses.

The following databases provide different types of important information on the legumes:

a) **International Legume Database and Information Service:**

The International Legume Database and Information Service (ILDIS) is an international project that aims to document and catalogue the world's legume species diversity in a readily accessible form. Research groups in many countries are participating on a co-operative basis to pool information in the ILDIS World Database of Legumes, which is being compiled to provide a worldwide information service through publications, electronic access and enquiry services.

ILDIS, basically a taxonomic database, is a collaborative organization involving legume experts and institutions from all over the world. The activities of ILDIS are managed from the ILDIS Co-ordinating Centre based at the Centre for Plant Diversity and Systematics, School of Plant Sciences, The University of Reading, UK. ILDIS is accessible at http://www.ildis.org/.

Much of the information for the ILDIS World Database of Legumes has been compiled at the ILDIS Regional Centres, located throughout the world. Information has been collected from local herbaria, national botanists, and from literature written in many different languages, making accessible a wealth of information that would otherwise remain hidden.

The partner research institutions of ILDIS include the Royal Botanic Gardens Kew, UK, Missouri Botanical Garden, USA, University of Southampton, UK, The University of Reading, UK, BIOSIS, UK, Komarov Botanical Institute, Leningrad, Russia, Agricultural University, Wageningen, The Netherlands, National Botanical Research Institute, Lucknow, India, South China Institute of Botany, China and Jardim Botanico do Rio de Janerio, Brazil.

b) **The Legume Information System:**

The Legume Information System (LIS) is a publicly accessible legume resource that integrates genetic and molecular data from multiple legume species and enables cross-species genomic, transcript and map comparisons. The intent of the LIS is to help researchers leverage data-rich model plants to fill knowledge gaps across crop plant species and provide the ability to traverse between interrelated data types. LIS, a component of the Model Plant Initiative (MPI), which is being developed as part of a cooperative research agreement between the National Centre for Genome Resources (NCGR) and the United States Department of Agriculture (USDA) Agricultural Research Service (ARS). This site can be accessed at http://www.comparative-legumes.org.

c) **INARIS, an agro-forestry database:**

The Integrated National Agricultural Resources Information System (INARIS) is a cooperative effort of different organizations to provide information on agriculturally relevant species. The National Research Centre for Agroforestry, Jhansi, and the Indian Agricultural Statistical Research Institute, Pusa, are mainly involved. This database, that provides information on agriculturally important legumes, among other species, can be reached at http://www.mirror.inaris.gen.in/net/.

d) **FAO agricultural database:**

The Food and Agricultural Organization of the United Nations developed a database that provides information on agriculturally important legumes, among other species relevant to agriculture. This can be reached at http://www.fao.org/ag/AAGP/AGPC/.

e) **Dr. Duke's Phytochemical and Ethnobotanical Databases:**

Accessed at http://www.ars-grin.gov/duke/, this database provides information on the phytochemistry, bioactivity and ethnobotany of legumes, along with other economically important plant species. The Agricultural Research Service of the US is involved in the development and maintenance of this database.

None of the existing databases serve the objectives of the present volume.

1.5 OBJECTIVE AND SCOPE OF THE PRESENT VOLUME

The major effort of this book is to provide:

a) a survey of economically important legumes in India, identifying the more important uses of 719 legume species (Caesalpiniaceae 124, Fabaceae 504, and Mimosaceae 91);
b) information on (i) the distribution of various classes of chemical compounds and their uses, (ii) nutritive value of edible legumes, and (iii) the anti-nutritional and toxic factors in them;
c) information on legumes used in different systems of medicine;
d) profiles of 289 species of legumes (Caesalpiniaceae 45, Fabaceae 198 and Mimosaceae 46), providing diverse information on their uses;
e) basic scientific information needed to understand the chemistry and the implications of use of the legume species in various contexts, to make the book more self-contained; and
f) to use currently valid names of species.

The names of chemical compounds and disease symptoms are given in the alphabetical order to provide for an easy reference.

The material in this book may not be everything that is available on Legumes in India, but provides the foundation for further development of a utility centered database on this important group of plants.

CHAPTER-2

The Legumes

2.1 INTRODUCTION

The Leguminosae are one of the three largest families of flowering plants, the other two being Asteraceae (Compositae) and Orchidaceae (Airy Shaw, 1973). The members of the Leguminosae are among the world's economically most important species.

As per earlier estimates, the Leguminosae include 650 genera and 18,000 species (Airy-Shaw, 1973), but more recent estimates vary from 690 to 800 genera, and 14,000 to 20,000 species (Cronquist, 1988, 1992). Legumes exhibit an enormous amount of diversity in their habitats, form, structure, functions, chemistry and uses to man.

Legumes are nearly cosmopolitan in distribution, though more predominant in the tropics and subtropics than in the temperate regions. The species are found in temperate, humid and arid zones, highlands, savannas, low lands and there are even a few aquatic species. Legume habitats range from wetland to dry and cold deserts, from tropical forests to alpine habitat, from sea level to over 7,000 feet in the mighty Himalaya.

Though legumes are recognised as a natural group of species, their classification is varied (see Box 2.1).

Box 2.1

Families of the Legumes

The earlier taxonomists placed the legumes under one family, the Leguminosae, and subdivided the family into three subfamilies, the Papilionoideae, Caesalpinioideae, and Mimosoideae. Some later taxonomists regarded these subfamilies as families, the Papilionaceae, Caesalpiniaceae and Mimosaceae. This aspect is detailed later in this Chapter (2.6.6).

The application of International Code of Botanical Nomenclature (ICBN), which guides the naming of plants, necessitates the use of the name Fabaceae, in place of the family names Leguminosae or Papilionaceae, and the subfamily name Faboideae in place of Papilionoideae. However, the older names may be used as alternate names.

2.2 THE ANTIQUITY OF THE LEGUMES

It is not easy to determine the exact time of origin of legumes, as their fossil record is rather sparse. Fossil woods (Muller-Stoll and Madel, 1967) and pollen (Muller, 1981) ascribable to the Caesalpiniaceae and the Mimosaceae are known from the Maastrichtian, in the Northern Hemisphere, dating back to 65 to 70 million years before present (mybp). From the fossil record, it is also evident that both the range of distribution and diversity of these two families increased until the Eocene (55 to 38 mybp). Fossils ascribed to the Fabaceae were discovered from the early Eocene (Berry, 1937) and Oligocene, about 30 to 24 mybp (Bachofen-Scht, 1949; Heywood, 1978; Anonymous, 1981a).

2.3 CENTRES OF ORIGIN OF CULTIVATED LEGUMES

The time and place of origin of cultivated plants are determined on the basis of different kinds of botanical, phytogeographical and paleobotanical evidence. Two important criteria for determining the Centres of Origin of cultivated plants are, a) the existence of a large number of wild relatives of a cultivated species, and/or b) the existence of a large number of cultivated varieties (within each species) in the wild, in a defined area. However, this is by no means an easy task and the origin of cultivated plants is often regarded as an 'impenetrable secret'.

Box 2.2

Centres of Origin

A) Twelve Centres of Origin of Cultivated Plants

a) Chinese-Japanese Centre, b) Indo-Chinese-Indonesian Centre, c) Australian Centre, d) Hindustani Centre, e) Central Asian Centre, f) Near Eastern Centre, g) Mediterranean Centre, h) African Centre, i) European-Siberian Centre, j) South American Centre, k) Central American and Mexican Centre and l) North American Centre, (Zeven and Zhukovsky, 1975).

B) Eight Centres of Origin of Cultivated Legumes

a) Chinese Centre, b) Indian Centre, c) Asiatic Centre, d) Near Eastern Centre, e) Mediterranean Centre, f) Ethiopian Centre, g) South Mexican and Central American Centre and h) South American Centre (Vavilov, 1932; Zeven and Zhukovsky, 1975; Kochaar, 1981; Yamaguchi, 1983).

Box 2.3

Seeds/Seed Fragments of Edible Legumes Discovered from Various Archaeological Sites

8,000 BCE: *Pisum sativum* (pea), *Vicia faba* (faba bean) and *Lens culinaris* (lentil) from the Neolithic sites in Greece

7,500–6,500 BCE: *Pisum sativum, Vicia faba, Phaseolus vulgaris* (common bean) and *Lens culinaris* from archaeological sites at Hacilar, Beidha and Jarmo (Kurdistan, South West Asia)

6,000 BCE: *Phaseolus vulgaris* and *Phaseolus lunatus* (Lima bean) in intermontane deposits in Peruvian Valley

5,000 BCE: *Phaseolus vulgaris, Lens culinaris* and *Vicia faba* from the Bronze Age of Austria and Switzerland

4,975 BCE: *Phaseolus vulgaris* from Tehuacan Valley (Mexico)

3,000 BCE: *Phaseolus vulgaris, Phaseolus lunatus* and Canavalia beans in Coprolites (dried and often 'fossilized' human excreta) in coastal Peru

2,400 BCE: Carbonized beans in Troy and in Tombs in Egypt

2,000 BCE: *Phaseolus vulgaris, Pisum sativum* and *Vicia faba* from the Iron Age Germany

Sources: Smartt, 1976; Kochaar, 1981; Narayanan and Dubadghao, 1982.

Central Asia is generally regarded as the birthplace of all legumes and Asia Minor as the secondary Centre of Origin. In his classic work 'Origin of Cultivated Species', de Candolle (1882) wrote that the legume species existed in Western Asia long before they were cultivated. The Centres of Origin of legumes were also the Centres of Origin of many other crop plants. Different Centres of Origin of the present day cultivated plants and of legumes are given in Box 2.2.

2.4 DOMESTICATION OF LEGUMES

Early man was nomadic, but became semi-nomadic 10,000 years ago, gathering different species of plants to supplement food obtained from hunting and fishing. Man then started domesticating plants and animals, marking the beginning of agriculture and animal husbandry. All domesticated animals and cultivated species/varieties of plants have been selected from wild populations. Domestication of plants and animals for food and other uses became essential when food sources in the vicinity were either seasonal or became scarce, causing frequent shortages. Additionally, searching for food in alternative distant places became very strenuous and time consuming. Domestication and long periods of repeated selection for traits of agronomical and nutritional significance, have drastically and simultaneously reduced the toxic effects of the wild plant populations. The efforts of agricultural scientists and conscious farmers during the past two centuries have resulted in tapping the enormous diversity inherent in the natural populations of different species of plants and animals. This necessitated the recognition of several botanical varieties, cultivated varieties (cultivars) and local favourites called 'land races', within each of several widely domesticated animal and crop plant species. For example, there are over 1,00,000 varieties and land races of rice (*Oryza sativa)* and some 15,000 varieties and land races of *Phaseolus vulgaris*, the common bean. Though the cultivars co-existed with their naturally occurring progenitors and wild relatives for a long time, the wild forms of many present day cultivated plants, including legumes, have declined and are hard to be found today.

There are several compelling reasons for the domestication of legumes, among which the following are more significant:

a) the protein content of pulses is far higher than that of any cereal,
b) they mature rapidly and can be easily cultivated even in climates and soils where other crops may fail, and
c) legume seeds have features invaluable for safe transportation and storage, such as low water content and seed coats impervious to water, and they are important in crop rotation, as they enrich the soil by fixing nitrogen in their root nodules.

2.5 HISTORY OF CULTIVATION OF LEGUMES

The history of cultivation of pulses is intricately connected to the history of agriculture, in both the New and the Old World countries, including India. The cultivated legumes have been important crops since the beginning of domestication and have been a part of our cultural heritage.

Legumes have been cultivated for over 8,000 years in various parts of the world. Egyptian dynasties featured soybean in cropping systems as early as 2000-3000 BCE. The first written record of any pulse is of soybean, in the writings of the Chinese Emperor Shen-Nung in 2800 BCE. Pulse crops were mentioned, along with the cereals, in *Yajurveda* (*Anuvaka* 244) and *Atharvaveda* (*Anuvaka* 475, Hymn 53), the oldest and the most revered Hindu religious documents in *Samskrith*, the oldest versions dated around 1,300 to 1,000 BCE and the currently known versions from around 300 BCE. Pulses were also mentioned in other Hindu religious texts such as *Purana*s, written about 300 BCE to 300 CE. The native people of both the North and South America have grown beans from very early times. Evidence of pulse crops was found from the very early times in archaeological sites of both the Old and New worlds. Archaeological studies indicate the association of man with peas from the Stone Age (see Box 2.3).

Medicinal importance of legumes was known from ancient times. Pythogoras, the Greek philosopher, forbade his disciples in 582 BCE, from eating beans, as he believed that beans came from the putrid matter of humans. Another belief among the ancients was that *Pisum sativum* and Rosemary were a good remedy for edema and jaundice. Roman physicians used peas boiled in seawater to cure erysipelas. As detailed in the later chapters, legumes were also extensively employed in the indigenous systems of medicine in China and India, dating back to 1,000 BCE.

(Box 2.3 Contd.)

1. The dates in years given in all the cases are an estimate mostly basing on circumstantial evidence and do not represent absolute calculations.
2. BCE (Before Common/Current Era) and CE (Common/Current Era) are the current terms replacing BC (Before Christ) and AD (Anno Domini, the year of the Lord), to eliminate the religious overtones in the replaced terms.
3. The Stone Age, Bronze Age and Iron Age are a chronological sequence but were not concurrent in all civilizations/countries. The Stone Age considered to have begun two million years ago ended around 6,000 BCE in the Middle East, but lingered till 4,000 BCE in Europe and till 2,500 BCE in the Americas. The Bronze Age in South East Asia was dated around 4,500 BCE, in Greece around 3,000 BCE, in China around 1,800 BCE and in Americas around 1,000 CE.

In the olden times, beans were used in elections, as votes cast by people. In a Court of law, a white bean signified acquittal and a black bean signified condemnation (Yamaguchi, 1983).

2.6 SYSTEMATIC BOTANY OF LEGUMES

Systematic Botany (or Taxonomy) is that area of botany, which deals with the naming, relationships and classification of plants. The legumes being diverse and complex, their systematic issues are also quite complex. The basic and essential issues are discussed in the following sections.

Brief descriptions of the three families, based on Willis (1973), Heywood (1978) and Purseglove (1991) are given here. For more details, standard books on plant taxonomy need to be consulted.

2.6.1 Caesalpiniaceae

The family includes about 180 genera and 2,500-3,000 species (Heywood, 1978). They are tropical and subtropical in distribution, although some genera such as *Cercis, Gleditsia* and *Ceratonia*, occur in more temperate climates. With few exceptions, the Caesalpiniaceae are trees or shrubs.

Most species produce bacterial root nodules. The stem is erect; but some species are climbers, climbing by tendrils (*Bauhinia vahlii*), hooks (species of *Caesalpinia*) or thorns that are modified branches (species of *Gleditsia*). The genus *Cassia* is among the remarkable angiosperm genera, whose species are herbs, shrubs or trees. Leaves are nearly always alternate, pinnate, occasionally bipinnate, rarely simple; stipules paired, mostly deciduous. Flowers are zygomorphic, often showy, usually hermaphrodite; sepals five or four by union of two upper sepals, mostly free, sometimes much reduced when two large and calyx like bracteoles cover the bud, usually imbricate; petals five or fewer, ascendingly imbricate (the odd petal innermost; (Fig. 2.1a); stamens ten or fewer, free to variously connate, dehiscing lengthwise or by terminal pores (species of *Cassia*); ovary superior, unilocular, ovules one to many, style simple. Fruit is a legume, indehiscent or drupaceous. Seeds are arillate in some species, rarely with endosperm.

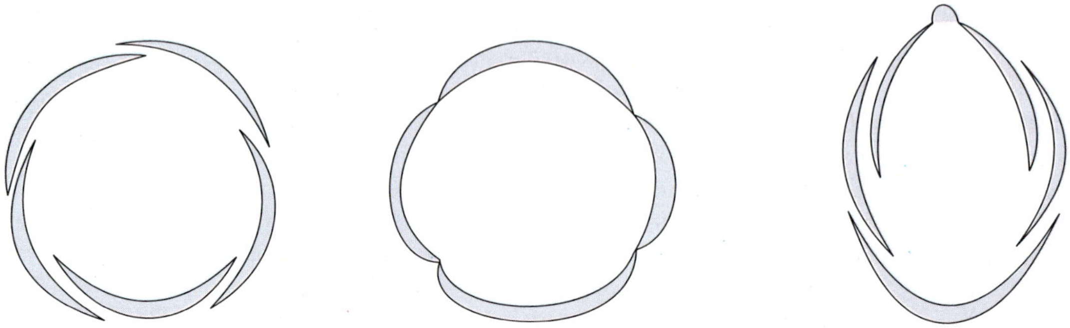

(a) (b) (c)

Fig. 2.1: Typical aestivation of the corollas in the legume families. a: Caesalpiniaceae; b: Mimosaceae; c: Fabaceae

2.6.2 Mimosaceae

The Mimosaceae include about 56 genera and 2800-3000 species that are tropical and sub-tropical, concentrated in the Southern Hemisphere (Heywood, 1978; Purseglove, 1991). Most are trees, shrubs or climbers, although some species are herbaceous (*Mimosa pudica)* and one is an aquatic species (*Neptunia oleracea).*

Roots of most species develop bacterial nodules. The stem is commonly erect; the climbers climb by emergences (*Acacia* species). Leaves usually bipinnate, rarely once pinnate, sometimes reduced to phyllodes (species of *Acacia*); stipules sometimes spinescent or emergences from stem (species of *Acacia*). Flowers actinomorphic, small, usually sessile and massed into cylindrical spikes or globose heads; sepals usually five, valvate, united into a toothed or lobed calyx; petals are the same number as sepals, valvate (petals just touching each other by the margins; (Fig. 2.1b), free or connate; stamens numerous, usually free; anthers small, versatile, often with an apical gland, dehiscing longitudinally; ovary one locular, superior; ovules few, style usually filiform; stigma small and terminal. Fruit dehiscent or indehiscent, sometimes a lomentum.

(a) (b) (c)

Fig. 2.2: Flowers typical of the legume families. a: *Delonix regia (Caesalpiniaceae);* b: *Mimosa pudica (Mimosaceae)*; c: *Clitoria ternatea (Fabaceae)*

2.6.3 Fabaceae

The Fabaceae have about 480 genera and 12,000 species constituting the largest family/ subfamily of legumes (Heywood, 1978). The species are trees, shrubs, herbs or climbers. The family is worldwide in distribution, with more species in the tropics, while the more advanced herbaceous genera are common in the temperate regions.

Root of most species bear nodules. Stem commonly erect, many species are climbers (species of *Abrus, Clitoria, Vicia).* Leaves usually alternate and mostly compound, pinnate, trifoliolate or digitate; stipulate, stipules spinescent in *Robinia,* and large and leaf like in *Pisum sativum*; stipels often present at the base of individual leaflets; flowers zygomorphic and typically papilionaceous (see Box 2.4) mostly hermaphrodite; calyx tubular and usually five toothed, imbricate; petals five, descendingly imbricate (the odd petal outermost; (Fig. 2.1c) enclosing stamens and ovary; stamens ten, monadelphous or diadelphous (with nine united + one free or 5 + 5) or even free (*Sophora);* anthers dehisce length wise; ovary superior, one locular, ovules one to many on ventral suture. Fruit usually a legume, splitting along dorsal or ventral suture, sometimes indehiscent, occasionally jointed and breaking into one seeded segments. Seeds are usually without endosperm.

Box 2.4

Papilionaceous Flowers

The papilionaceous flower is asymmetrical and typical of the Fabaceae (Papilionaceae), the name alluding to a butterfly (*Papilio, Papilion:* butterfly) (Photograph 2.3). Of the five petals constituting the corolla, the odd member is the largest, upper most and outermost, and is called the standard or *vexillum.* The two lateral petals below the standard, are the *alae (*wings) or wing petals. The *alae* overlap the *carina* (keel, boat), formed by the fusion of two petals, which encloses the stamens (variously fused by their filaments), the ovary and the style. In the bud condition, the standard encloses the wings and the keel, all of which may be basally fused variously with each other. Only a small group of species of the Tribe Sophoreae differs from the papilionaceous floral architecture.

2.6.4 Distinguishing the Legume Families

Commonly, the three legume families are easy to distinguish from each other and other plant families, basing on a combination of characters (see Box 2.5), though some confusing situations arise.

2.6.5 Relative Evolutionary Status of the Legume Families

Corner (1949) considered that the early legumes lived in conditions resembling present day tropical rain forests, which provided the right conditions for a rapid evolutionary progression. The exclusively tropical Caesalpiniaceae, which retained many floral and other characteristics resembling those of the ancestral groups of plants, from which they evolved, are regarded the least evolved type among the three legume families. The Mimosaceae, higher in evolutionary advancement than the Caesalpiniaceae, have also retained many ancestral features, but changed considerably from the ancestral type, and have migrated into the subtropics and warm temperate areas. On the other hand, the Fabaceae, the most evolved of the three families, acquired great morphological variability and adaptations, spanned far and wide and are now found in all parts of the world.

2.6.6 Systematic Position of the Legume Families in Plant Classification

The position of a particular plant group, with reference to the other related groups, in a system of classification, is called the systematic position, which reflects the relative evolutionary advancement of the group.

The legumes are a fairly evolved group, occupying an above the middle position, among the flowering plants. It is generally agreed among taxonomists that the legumes are closely related to the families in the Order Rosales, having co-evolved from the so-called Rosalean stock, a large complex group of families with a similar set of shared characters. Whether treated as a single family (with three subfamilies) or three different families, the legumes were included in the Order

Box 2.5

Distinguishing Legume Families

While identifying plant groups, one should go by a combination of features and not overly depend upon any single feature, since almost any single feature may be found in more than one, sometimes even unrelated, plant groups.

The combination of characters, such as the presence of root nodules, stipulate compound leaves, perigynous flowers with a hypanthium, the odd member of the corolla is upper most unlike in the other dicotyledonous families, monocarpellary unilocular ovary with ovules on one suture, fruit a legume or lomentum, generally sets the legumes apart from other families.

In the Mimosaceae, the flowers are actinomorphic and petals are valvate (Fig. 2.1b), while in both the Caesalpiniaceae and the Fabaceae, the flowers are zygomorphic and petals are imbricate (Fig. 2.1a, 2.1c).

In the Caesalpiniaceae the petals are ascendingly imbricate (Fig. 2.1a).

In the Fabaceae, the flowers are characteristically papilionaceous and stamens are variously united, which distinguishes them from the other legume families. Even when the flowers are nearly actinomorphic, corolla not papilionaceous and the stamens are all free (as in the tribe Sophoreae), the aestivation (arrangement in bud condition) of the petals in all members of the Fabaceae is decendingly imbricate (Fig. 2.1c).

Box 2.6

Classification of Legumes

Bentham (1865) and Taubert (1894): Leguminosae with three subfamilies, the Caesalpinioideae, Mimosoideae and Papilionoideae.

Wettstein (1933-35): Mimosaceae and Papilionaceae as separate families, the latter with two subfamilies, Caesalpinioideae and Papilionatae.

Soo (1963), Hutchinson (1964) and Cronquist (1988, 1992): Three families Caesalpiniaceae, Mimosaceae and Papilionaceae (Fabaceae).

Takhtajan (1997): Fabaceae (with three subfamilies)

Rosales by the early taxonomists. Hutchinson (1964) placed the three legume families in an Order of their own, the Leguminales, within the Rosalean group, while Cronquist (1988, 1992) placed the three families in the Order Fabales. Takhtajan (1997) treated legumes as a single family Fabaceae in the Order Fabales.

The classification of the legumes by different taxonomists is given in Box 2.6. There is still a lack of a general agreement regarding the treatment of the major divisions of legumes, which are recognised either as separate families, Caesalpiniaceae, Mimosaceae and Fabaceae (Papilionaceae) or as subfamilies of a single family, the Fabaceae (Leguminosae), namely Caesalpinioideae, Mimosoideae and Papilinoideae (Faboideae, Papilionatae). Hutchinson (1964) remarked that it is "merely a matter of opinion or conscience (or even of prejudice!) whether Caesalpiniaceae should be treated as a separate family from Mimosaceae and Fabaceae or all the three regarded as subfamilies of Leguminosae (Box 2.6).

2.6.7 Detailed Taxonomic Treatment of the Legumes

In most taxonomic situations, Families of plants contain Genera and Genera include Species. Large taxonomic groups that show wide ranges of diversity cannot be treated in this simple manner. In order to formulate more homogenous taxonomic units, bringing more clarity and purpose to classification and to reflect phenetic and phylogenetic relationships, taxonomists subdivide larger groups into subgroups. Every taxonomic unit from the Species to the Kingdom may require such subdivision, though not always. Orders are divided into Suborders, Families into Subfamilies, Subfamilies into Tribes and Subtribes, to include groups of related Genera. Genera may be divided into Sections and Subsections. Similarly, Species are subdivided into Subspecies, Varieties and Forms. Varieties, Cultivars and Land Races are recognized in many cultivated species. Such an extensively hierarchical and elaborate classification is needed to meet scientific objectives, only in certain groups of plants.

Only Bentham (1865), Taubert (1894) and Hutchinson (1964) have studied the entire group of legumes and proposed an elaborate hierarchical classification, below the level of the subfamily. However, there are serious differences among the three. Most taxonomic treatments, even many Floras (books that deal with vegetation of a specified area) do not give such details, as non-taxonomists such as the general readers and even most botanists, have no significant use for such a detailed taxonomic treatment. Those interested should consult publications referred above, or monographs dealing with the botany and taxonomy of legumes, to obtain an insight to these highly technical taxonomic details.

2.6.8 Scientific Names of Plants

Scientific naming of organisms and their groups is an important part of taxonomy and is essential for uniform and unambiguous international communication. The name of a species contains two parts: the generic name *(Pisum)* and a specific epithet *(sativum),* together forming the binomial. Thus, *Pisum sativum* is the full name of the species, the pea plant. Binomials are written in *italics*, indicating that they originated from Latin and/or Greek and not from the modern languages. Linnaeus stabilized the system of binomial nomenclature for organisms by 1753.

In order to facilitate authentication of scientific names and literature tracking, it is also a mandatory practice to attach to the binomial, the name of the taxonomist who first gave the species its scientific name (called the Authority). This practice is author citation, without which the formal citation of a name of a plant group is incomplete. In the case of the pea plant, Linnaeus named the species in 1753 and so the correct and complete citation of the name is *Pisum sativum* Linnaeus. The author's name may be abbreviated (Linnaeus as L.). Since it is cumbersome to use the author's name every time the taxon is mentioned, author citations are given only once in a particular publication, usually when the name was first mentioned. Complete citations of currently valid names of species and synonyms dealt in this book are given in the Tables 3.1, 3.2 and 3.3 and in the species profiles.

Now plants are invariably named as per the mandatory guidelines of the International Code of Botanical Nomenclature (ICBN) for wild species and as per the International Code of Nomenclature of Cultivated Plants (ICNCP), for the cultivated species. For various scientific reasons and on account of attempts to stabilize nomenclature basing on the provisions of the ICBN and ICNCP, it becomes sometimes necessary to change the scientific names of plants. In some cases, species of one genus may have to be moved to different genera for scientific reasons, as for examples species of the genus *Cassia* to the genus *Senna*, when binomials have to be changed. The names rejected on account of these reasons are called synonyms and using them is scientifically incorrect. The changed names of some common legumes are given in Box 2.7. In order to facilitate an easier reference to the current names and their synonyms, nomenclatural equivalents are given in Table 3.4.

Changes in plant names may cause some inconvenience or even confusion to non-taxonomists, particularly when the species is in much common use with its earlier name (like that of horse gram), but one must use the current correct names, in the interests of science. As far as possible, the current correct names are used in this book.

Box 2.7

Changed names of some Common Legumes

Common name	*Current name*	*Synonym*
Ashoka tree	*Saraca asoca*	*Saraca indica*
Babchi	*Cullen corylifolium*	*Psoralea corylifolia*
Black gram	*Vigna mungo*	*Phaseolus mungo*
Cluster bean	*Cyamopsis tetragonoloba*	*Cyamopsis psoralioides*
Copper pod tree	*Peltophorum pterocarpum*	*Peltophorum ferrugineum*
Green gram	*Vigna radiata*	*Phaseolus radiatus*
Horse gram	*Macrotyloma uniflorum*	*Dolichos biflorus*
Red gram	*Cajanus cajan*	*Cajanus indicus*

The generic names are often abbreviated (*P. sativum)* to save printing space, but now with electronic/digital search facilities into databases, this practice causes serious problems. Searching in databases with '*P.*', even among the legumes, would retrieve all the genera with this abbreviation, such as *Pisum, Peltophorum, Pterocarpus, Psophocarpus, Psoralea*, etc., and when a full generic name is specified, the search engine may miss reference to all the abbreviated names. In order to avoid this problem, abbreviation of generic names is avoided.

2.7 TWO HALLMARK FEATURES OF THE LEGUMES

In biology there is hardly any characteristic that entirely and wholly characterizes any species or groups. Nevertheless, certain combinations of characters stand out making the groups possessing them more or less unique. Often, these are not single characters, but syndromes that are integrals or consequence of complex genetics and physiology, resulting from long evolutionary progression and fixation.

In addition to the combination of external features of the legumes that make them stand apart from other plant groups, as detailed earlier in this Chapter, the legumes possess two unrelated hallmark features, a) characteristic seed storage proteins, and b) the ability to fix nitrogen, that are not found in other plant groups.

2.7.1 Seed Storage Proteins

Most of the legume species contain two globulin proteins, vicilin and or legumin, as the major components of storage proteins in their seeds. Though only a handful of species were analysed for molecular characterization of vicilin and legumin, these two proteins are among the most elaborately characterized plant proteins. Future research may show that a) vicilin and legumin do not occur in all legume species and/or b) that other plant species related to the legumes also contain seed storage proteins similar to these two. Nevertheless, vicilin and legumin distinguish legumes from other plant groups.

2.7.2 Nitrogen Fixation

Almost all species of the legumes studied produce tubercular structures on their roots called the root nodules. Species of the bacterial genus *Rhizobium* are aerobic, and are present in the soil. They attach to the root hairs of legume plants and invade the developing roots of the legumes. An excessive multiplication of the bacteria in the cortical cells of the roots triggers development of nodules.

The association between the legume roots and the bacteria is symbiotic. The bacteria help in the fixation of atmospheric nitrogen for the host plants, and in turn depend on their host for space and carbohydrates for their metabolism. The process of conversion of atmospheric nitrogen to ammonia by these bacteria, called nitrogen fixation, is quite complex. When the bacterial nodules decompose, their protein content becomes available in the soil, as a source of nitrogen for plants.

There are organisms other than these bacteria, such as species of *Aulosira, Nostoc* and *Anabaena* (Blue-green algae), that can fix nitrogen or plants that produce bacterial nodules in the aerial parts (species of *Ardisia*, Myrsinaceae), but the syndrome of physical root nodules and the physiological process of nitrogen fixation are almost exclusive to the legumes.

2.8 THREATENED SPECIES OF LEGUMES IN INDIA

An uncontrolled increase in human population and/or overexploitation of land area for agricultural, industrial and other uses, place an enormous pressure on our natural resources. In course of time, natural habitats are destroyed and the resources dwindle. This, along with other activities detrimental to natural resources, threatens their continued availability, particularly of plants and animals. Conservation is a scientific programme to ensure that the natural resources are used in a controlled and sustainable manner, so that they last longer, and possibly are never totally lost. Several international organizations and international treaties serve the very important cause of conservation and sustainable utilization of our natural resources.

2.8.1 IUCN and the Species Survival Commission

The International Union for Conservation of Nature and Natural Resources (IUCN) is the most important world body of 74 sovereign states, 105 government agencies, 674 non-governmental organizations and 32 affiliates concerned with the conservation of nature world wide. The Species Survival Commission (SSC) is one of the six important organs of the IUCN.

2.8.2 Red Lists of Threatened Species

The IUCN separately publishes Red lists of Threatened Animals and Plants to disseminate information on the threat or conservation status of species, to guide naturalists and policy makers.

The Red Lists are the most comprehensive and authoritative global surveys of threatened species. They identify species that are most threatened, there by serving as tools to help to set priorities for conservation action and providing baseline information for monitoring the status of different groups of plants and animals. The Red Lists provide a framework for more specific information about species under threat, such as national or regional lists, conservation strategies, alert regions, nations and communities to species of international conservation concern. The Red Lists are the technical resources used to create and strengthen national species protection Laws. These lists are periodically revised, to reflect changed situations.

The first ever IUCN Red List of Threatened plants of 1997, containing 33,798 species names in different threat categories, was released in April 1998 (Walter and Gillett, 1998). The listed species belong to 369 plant families and 200 countries. They constitute 12.5 per cent of known species of vascular plants. The publication of this list is a turning point in global efforts in plant conservation.

The IUCN Red List of Threatened Species is an authentic document. The threat status of species in this list has to be periodically reviewed, and basing on the revised status, species are to be re-categorised.

Species are assigned to different categories of threat basing on the degree of threat of extinction they individually face. The commonly used terms 'Endangered' and 'Threatened' have come to be used interchangeably, leading to a wide spread confusion. Having been aware of this situation, the Species Survival Commission has reviewed the terminology and published the IUCN Red List Categories in 1994 (Table 2.1), the use of which is mandatory. Species (or subspecific taxa) facing any threat of extinction should be assigned to one of these categories on the basis of detailed and specific criteria, given by the IUCN, which need to be consulted. The Red List of Threatened Animals of 1996 is totally according to these guidelines.

2.8.3 Problems with the IUCN 1997 Red List of Plants

The 1997 IUCN Red List of plants suffers from the following:

a) The Red Lists of Plants is based on the pre-1994 categories. At a time when the IUCN is revising even the 1994 categorization, using discarded categories like Extinct (Ex), Extinct/Endangered (Ex/E), Endangered (E), Vulnerable (V), Rare (R), and Indeterminate (I), is considered inadequate.

b) The Indian species in the 1997 IUCN Red List of plants are based on publications whose premises are not scientifically sound and have been criticized. The more important deficiencies are that the data are not quantitative, does not cover the entire range of distribution of the taxa and that the information did not come from field evaluation.

For these reasons the Indian species in the 1997 IUCN Red list of plants have to be re-evaluated by experts in different groups of plants.

2.8.4 Red List of Threatened Species of Vascular Plants in India

In order to publicise the species of vascular plants in India under different categories of threat, as listed in the 1997 IUCN Red List, and to make the information readily available to all those interested in the subject, the Red List of threatened vascular plant species in India was compiled from the IUCN 1997 Red List (Kameswara Rao *et al.*, 2003). Among the 1,236 species of vascular plants under threat in India, there are 78 species of Legumes, as given in Table 2.2.

2.8.5 CITES Appendices

The Convention on International Trade in Endangered Species of Wild Flora and Fauna (CITES) is an international treaty organization that regulates international trade in endangered species. Periodically CITES publishes lists of plants and animals whose trade is restricted. These are in the form of Appendices of three categories, I, II and III, indicating the degree of threat. Species under Appendix I are under severe trade control, those in Appendix II are subjected to a less severe control, while those in Appendix III should be carefully watched. These appendices are also periodically reviewed, increasing or relaxing international trade control.

The IUCN 1997 Red List indicates the status of species in CITES appendices, to control trade in them. Only *Pterocarpus santalinus*, (the red sandal wood), a legume tree species, appears in Appendix II of CITES.

Table 2.1

IUCN Red List Categories of 1994

I. Evaluated

A. If adequate data are available:

a) **Extinct (EX):** when there is no reasonable doubt that the last individual of the taxon has died.

b) **Extinct in the Wild (EW):** when a taxon is known only to survive in cultivation cell outside the past range.

c) **Threatened:**

- i) **Critically Endangered (CE):** when a taxon is facing an extremely high risk of extinction in the wild in the immediate future.
- ii) **Endangered (EN):** when a taxon is not critically endangered but is facing a very high risk of extinction in the wild in the near future.
- iii) **Vulnerable (VN):** when a taxon is facing a high risk in the wild in the medium term future.

d) **Lower Risk (LR):** when it has been evaluated, does not satisfy the criteria of the threat under the above mentioned categories.

- i) **Conservation Dependent (CD):** when a taxon is the focus of conservation programme, the cessation of the programme would place the taxon in the threatened categories.
- ii) **Near Threatened (NT):** when a taxon does not qualify for conservation dependent, but which is close to qualifying for Vulnerable.
- iii) **Least Concern (LC):** when a taxon does not qualify for Conservation Dependent, or Near Threatened.

B. Data Deficient (DD): when data are deficient to make an assessment of the risk of extinction a taxon faces.

II. Not Evaluated (NE): when a taxon has not yet been assessed against the criteria.

Note: These threat status categories are being revised by the IUCN. A taxon is any taxonomic group, such as the species.

Table 2.2

Red List of Legume Species in India

IUCN Red List treated all the Legumes in a single family. The cross references cited for the species, after each binomial or locality, are not relevant to the purpose of this book, and so are excluded. They can be found in the IUCN Red List (Walter and Gillett, 1998), or in the compilation of the Indian species (Kameswara Rao *et al.*, 2003). The nomenclature in this table is as per the IUCN publication.

Abbreviations: E: endangered; I: indeterminate; R: rare; V: vulnerable

Leguminosae (Fabaceae)

R *Acacia campbellii* Arn. 14782

R 14782 India – Andhra Pradesh 14782

I *Acacia hohenackeri* Craib 4988

I India – Tamil Nadu 4988

I *Acacia wightii* Baker

I India – Kerala (Travancore)

I India – Tamil Nadu (Tirunelveli Hills)

R *Albizia thompsonii* Brandis 14782

R 14782 India – Andhra Pradesh (Cuddapah) 14782

R 14782 India – Orissa (Ganjam) 14782

R 14782 India – Tamil Nadu (Coimbatore) 14782

I *Alysicarpus beddomei* Schindler 2268

I India – Tamil Nadu (Nilgiri Hills) 2268

I *Alysicarpus meeboldii* Schindler

I India – Jammu & Kashmir

I *Alysicarpus monilifer* (L.) DC. var. *venosus* Blatter & Hallberg

I India – Rajasthan 7958

I *Amphicarpaea anomalus* Bunge

I India – Jammu & Kashmir (Ladakh)

I *Amphicarpaea edgeworthii* Benth.

I India (W. Himalaya)

I *Amphicarpaea flemingii* Ali

I India – Punjab

I *Astragalus anomalus* Bunge

I India – Jammu & Kashmir (Ladakh)

I *Astragalus bakerii* Ali

I India – Jammu & Kashmir

I *Cajanus sericus* (Benth.) Maesen
- I India – Andhra Pradesh (Vizakapatanam Hills)
- I India – Karnataka (Concan)

I *Crotalaria bidiei* Gamble
- I India – Tamil Nadu (Wynaad, Nilgiri Hills)

I *Crotalaria bourneae* Fyson
- I India – Tamil Nadu

I *Crotalaria clarkei* Gamble
- I India – Tamil Nadu

E *Crotalaria clavata* Wight & Arn. 11494
- E 11494 India – Tamil Nadu 11494

I *Crotalaria conferta* Fyson
- I India – Tamil Nadu (Palani Hills, Madurai Dist)

R *Crotalaria digitata* Hook.
- R 13883 India – Karnataka 13883
- R 13883 India – Tamil Nadu 13883

I *Crotalaria filipes* Benth.
- I India – Karnataka (South Canara)

E *Crotalaria fysonii* Dunn var. *glabra* Gamble
- E 13883 India – Tamil Nadu (Palani Hills, Madurai Dist) 13883

R *Crotalaria globosa* Wight & Arn.
- R 13883 India – Karnataka (Mysore) 13883
- R 13883 India – Tamil Nadu 13883

E *Crotalaria kodaiensis* Debberm. & Biswas 11494
- E 11494 India – Tamil Nadu (Kodaikanal hills) 11494

E *Crotalaria longipes* Wight & Arn. 11494
- E 11494 India – Tamil Nadu (Kolli & Nilgiri hills) 11494

R *Crotalaria lutescens* Dalz.
- R 13883 India – Karnataka (South Canara) 13883
- R 13883 India – Maharashtra 13883

I *Crotalaria noveoides* Griff. 13883
- I 13883 India – Meghalaya 13883

R *Crotalaria peduncularis* Graham ex Wight & Arn. 11494
- R 11494 India – Kerala (Travancore) 11494
- R 11494 India – Tamil Nadu 11494

R *Crotalaria priestleyoides* Benth. ex Baker 11494
- R 11494 India – Karnataka 11494
- R 11494 India – Tamil Nadu 11494

R *Crotalaria rigida* Heyne ex Roth 11494

R 11494 India – Andhra Pradesh (Krishna) 11494

R 11494 India – Karnataka 11494

R 11494 India – Tamil Nadu 11494

E *Crotalaria sandoorensis* Beddome ex Gamble 11494

E 11494 India – Karnataka (Sandoor Hills, Bellary Dt) 11494

R *Crotalaria scabra* Gamble 13883

R 13883 India – Tamil Nadu 13883

V *Crotalaria shevaroyensis* Gamble

V 18228 India – Tamil Nadu 18228

V *Cynometra beddomei* Prain 14782

V 18228 India – Kerala (Wynaad) 14782

V *Cynometra bourdillonii* Gamble 11494

V 11494 India – Karnataka 11494

R *Cynometra travancorica* Beddome

R 13883 India – Karnataka 13883

R 13883 India – Kerala (Travancore Hills) 13883

R 13883 India – Tamil Nadu (Tirunelveli Hills) 13883

R *Desmodium wightii* Graham ex Wight & Arn.

R 18228 India – Tamil Nadu (Nilgiri & Palani Hills) 18228

R *Desmodium wynaadense* Beddome ex Gamble

R 18228 India – Kerala (Travancore Hills) 18228

R 18228 India – Tamil Nadu 18228

I *Dialium travancoricum* Bourd. 14782

I 14782 India – Kerala 14782

I *Dolichos ciliatus* Klein

I India – Tamil Nadu (East coast)

R *Eleiotis trifoliolata* Cooke 14782

R 14782 India – Karnataka (Eastern) 14782

R *Flemingia gracilis* (Mukerjee) Ali 14782

R 14782 India – Karnataka 14782

R 14782 India – Maharashtra 14782

I *Gleditsia assamica* Bor 14782

I 14782 India (Northeast) 14782

R *Hedysarum astragaloides* Benth. ex Baker

R 13883 India – Himachal Pradesh 13883

R 13883 India – Jammu & Kashmir 13883

E *Humboldtia bourdillonii* Prain 11494

E 11494 India – Kerala (Peermade Ghats) 11494

E 11494 India – Tamil Nadu (Courtallum Ghats) 11494

R *Humboldtia decurrens* Beddome ex Oliver 11494

R 11494 India – Kerala (South Western Ghats) 11494

E *Humboldtia laurifolia* Vahl

E 13883 India – Kerala (Malabar) 13883

E *Humboldtia unijuga* Beddome 9250

E 11494 India – Kerala 11494

E 11494 India – Tamil Nadu (Tirunelveli Hills) 11494

R *Indigofera barberi* Gamble 11494

R 11494 India – Andhra Pradesh (East Ghats) 11494

R 11494 India – Tamil Nadu (East Ghats) 11494

R *Indigofera caerulea* Roxb. var. *monosperma* (Santapau) Santapau

R 13883 India – Gujarat 13883

R 13883 India – Rajasthan 13883

I *Indigofera cedrorum* Dunn

I India – Himachal Pradesh

I India – Jammu & Kashmir

I *Indigofera coerulea* Roxb. var. *monosperma* Sant. 7771

I India – Gujarat 7771

R *Indigofera constricta* (Thwaites) Trimen

R 13883 India – Karnataka (Nikund Ghat) 13883

R 13883 India – Kerala (Cochin) 13883

I *Indigofera simlensis* Ali

I India – Himachal Pradesh

I *Inga cynometroides* (Beddome) Baker 14782

I 14782 India – Kerala (south Travancore Hills) 14782

R *Kingiodendron pinnatum* (DC.) Harms 14782, 21025

R 14782 India – Karnataka 14782

R 14782 India – Kerala 14782

R 14782 India – Tamil Nadu 14782

I *Lasiobema scandens* (Linn.) de Wit. var. *horsefieldii* (Wallich ex Prain) de Wit

I India – Assam

I India – Meghalaya

I *Mucuna minima* Haines 7771

I 7771 India – Orissa (Sambalpur) 7771

V *Nogra dalzellii* (Baker) Merr. 14782

V 14782 India – Karnataka 14782

V 14782 India – Maharashtra 14782

E *Nogra filicaulis* (Kurz) Merr. 14782

E 14782 India – Madhya Pradesh 14782

I *Phanera khasiana* (Baker) Thoth.
I India – Arunachal Pradesh
I India – Meghalaya
I *Phanera nicobarica* Balakr. & Thoth. 7771
I India – Nicobar Islands 7771
I *Phanera wallichii* (Macbr.) Thoth.
I India – Assam
I India – Meghalaya
E *II Pterocarpus santalinus* Linn.f. 19218, 21025
E 20263 India 20263
R *Pueraria bella* Prain 14782
R 14782 India – Arunachal Pradesh 14782
R *Rhynchosia beddomei* Baker
R 13883 India – Andhra Pradesh (Cuddapah) 13883
R 13883 India – Karnataka (Bellary) 13883
V *Rhynchosia velutina* Wight & Arn.
V 13883 India – Tamil Nadu 13883
I *Sesbania paludosa* (Roxb.) Prain
I India – Kerala (Travancore)
R *Smithia agharkarii* Hemadri 13883
R 13883 India – Maharashtra 13883
I *Smithia venkobarowii* Gamble
I India – Kerala (Peermade, Travancore Hills)
R *Tephrosia barberi* Drumm. 14782
R 14782 India – Tamil Nadu 14782
R *Tephrosia calophylla* Beddome 14782
R 14782 India – Tamil Nadu 14782
I *Tephrosia collina* V. Sharma var. *lanuginocarpa* V. Sharma
I India – Gujarat (Kutch)
I India – Rajasthan
R *Tephrosia jamnagerensis* Santapau
R 13883 India – Gujarat 13883
I *Tephrosia pentaphylla* (Roxb.) Sweet
I India – Karnataka (Bellary)
R *Tephrosia wynaadensis* Drumm. 14782
R 14782 India – Kerala 14782
I *Thermopsis inflata* Cambess.
I India – Himachal Pradesh
I India – Jammu & Kashmir
I *Vicia benthamiana* Ali
I India – Jammu & Kashmir
R *Vigna khandalensis* (Sant.) Raghavan & Wadhwa 14782
R 14782 India – Maharashtra 14782

2.9 SPECIES OF LEGUMES VENERATED IN INDIA

Veneration of nature was advocated in the oldest of the sacred texts of the Hindus, such as the four *Vaedaah*s (dated 1,300 to 1,000 BCE; later versions c300 BCE), the two principle epics *Raamaayana* (c300 BCE), *Mahaabhaaratha* (c300 BCE to 300 CE) and the *Puraana*s (4th to 16th century CE), the most revered among several hundreds of religious texts.

A large number of plants and animals are venerated in different religions of the world but Hinduism has made this an important part of rituals, projecting many species of plants and animals as especially dear to certain Gods and Godesses.

The Buddhists and Jains adopted tree worship from the older prevailing religions. So a number of venerated species are common to the Buddhist, Jainist and the Hindu traditions. The *Ashoka* tree *(Saraca asoca),* which is sacred to the Hindus, is sacred to the Buddhists too, as it is associated with the birth of Gauthama Buddha. This species is sacred to the Jainists as well, as Lord Mahaveera was considered to have attained enlightenment under the *Ashoka* tree. *Bauhinia variegata*, *Albizia lebbeck* and *Butea monosperma* are also sacred to the Buddhists (Table 2.3).

Venerated species of plants have been grown for over three millennia, with reverence and sentiment, in sacred groves, in and around temples, religious centres and villages, and even dwellings, in formal or informal gardens.

Bennet *et al.,* (1992) compiled extensive information on plants venerated in India, citing about 250 species of plants as sacred to the Hindus, in different contexts. Only the leguminous species are discussed here.

Some legume species are regarded to have originated from the body parts of Hindu Gods. For example, *Vamana Puraana* states that *Butea monosperma* (*Palasha*) originated from the right side of *Yama* (the God of death) and *Acacia catechu* (*Khair*) from the mid region of *Brahma*'s (the Creator) body.

Different venerated species of plants are used ceremoniously on the ascribed days of significance, to propitiate the respective Gods and Goddesses. Whole plants or their parts such as leaves, flowers, fruits or seeds are used in religious worship, each part with a different import. Flowers of *Saraca asoca, Erythrina variegata,* leaves of *Prosopis spicigera*, seeds of *Vigna mungo* and *Vigna radiata* are used in the worship of Lord *Vishnu*. The flowers of *Erythrina variegata* and incense of *Albizia odoratissima* are prescribed in the worship of Lord *Ishwara* (*Mahadeva*).

In addition to offering plants or their parts in prayers, there are other ways in which plants are used during religious ceremonies. The branches of *Butea monosperm* are prescribed as *samidhas* (fire wood in fire worship, *homa or havan)*, while propitiating the Moon, and the branches of *Acacia catechu,* to propitiate Mars. *Mimosa pudica* is used in bath water to propitiate the Sun and *Albizia lebbeck* to propitiate Venus. The roots of *Pterocarpus santalinus* are used as a talisman to protect the wearer from many evil influences.

Plants are also associated with the *saptharishis,* the seven principal sages of Hinduism. Among these, *Prosopis spicigera* (*Banni*) symbolizes *Rishi* Vashishta.

The following 34 species of Legumes are frequently used in religious ceremonies and grown in sacred groves, and in religious and temple gardens in India.

Caesalpiniaceae

Bauhinia acuminata, Bauhinia purpurea, Bauhinia racemosa, Bauhinia scandens var. *anguina, Bauhinia tomentosa, Bauhinia vahlii, Bauhinia variegata, Cassia fistula, Ceratonia siliqua, Saraca asoca, Senna auriculata* and *Tamarindus indica.*

Fabaceae:

Aschynomene aspera, Butea monosperma, Cicer arietinum, Clitoria ternatea, Dalbergia sissoo, Erythrina variegata, Pterocarpus marsupium, Pterocarpus santalinus, Sesbania grandiflora, Sesbania sesban, Uraria lagopodioides, Vigna mungo and *Vigna radiata.*

Mimosaceae:

Acacia farnesiana, Acacia ferruginea, Acacia nilotica ssp. *indica, Acacia suma, Adenanthera pavonina, Albizia lebbek, Albizia odoratissima, Mimosa pudica* and *Prosopis cineraria.*

The religious significance attributed to legume species is detailed in Table 2.3. Importantly, many of these species are of great medicinal importance, besides several being edible.

Table 2.3

Venerated Legume Species

As it is not possible to include the enormous information available on venerated legumes, only the salient points are given here. The interested should refer Bennett *et al.,* (1992), who have compiled vast details cited from religious texts and folklore, on the venerated plants..

Sl. No.	Species	Part used	Religious significance
1.	**VENERATED TREES**		
I	**Caesalpiniaceae**		
1	*Bauhinia purpurea*	Leaves, flowers	The leaves and flowers are offered to the Sun God on *Nirashanarka vratha* and to the Goddess *Saraswati* on *Shree Mahasaraswati vratha.*
2	*Bauhinia racemosa*	Tree	Tree is sacred to Hindus, worshipped during the *Dusserah* festival.
3	*Bauhinia variegata*	Tree, flowers	The flowers are offered to Lord *Vishnu* on *Vaikuntah chaturdhasi vratha;* often depicted in Buddhist sculptures.
4	*Cassia fistula*	Twigs, flowers	Flowers are used in religious ceremonies; stakes driven into the ground are worshipped in Mysore.
5	*Ceratonia siliqua*	Leaves	The leaves are used to worship during *Vinayaka pooja.*
6	*Saraca asoca*	Tree, flowers	This tree, 'The Sorrowless tree' or 'Heart's ease', mentioned in *Raamaayana* and *Mahaabhaaratha,* is one of the most important medicinal and venerated trees in India, both among Hindus and Buddhists. The handsome orange-red flowers are used in temple decoration. *Vamana Puraana* suggests the use of these flowers in the worship of Lord *Vishnu.* The tree is also associated with *Budha* and *Indra. Sasthi,* a minor cult in West Bengal, considers this tree as the guardian deity of the institution of the family, associated with fertility and is worshipped particularly by the women folk. The tree is a symbol of love and is dedicated to '*Kamadeva*', the God of Love, who uses its blossoms as one of the five arrows in his quiver. *Ashoka* trees are always planted in Buddhist monasteries. As per a Buddhist legend, when Buddha's mother Mahamaya became aware that she had conceived him, she retired to a grove of *Ashoka* trees. Gautama Buddha was born under one of these trees in the *Lumbini* garden. Hiuen-Tsang, the Chinese traveller, who came to India in 630 CE, mentions seeing the *Ashoka* tree under which Buddha was born. A sapling of the tree was taken by Prince Mahendra, the son of Emperor Ashoka, to Ceylon (Sri Lanka) in about 250 BCE, which was planted in Anuradhapura. Pillars now support its great branches and it is one of the oldest trees of historical importance in the world. According to a legend of western Uttar Pradesh, an *Ashoka* tree was the place of the Jain Leader, Mahavir's enlightenment.
7	*Tamarindus indica*	Twigs, flowers	Leaves and tender branches are used to propitiate Lord *Shiva.* The *Gadaba*s of the hills in Visakhapatnam and Srikakulam districts of Andhra Pradesh worship the tamarind tree as a personification of the Goddess *Talupulamma,* during the festival *Advirajula panduga,*

Sl. No.	Species	Part used	Religious significance
			celebrated for the safety of the people in forests. Tamarind tree is also used in the worship of Lord *Varuna*, the rain God, for a good rainy season, during the festival called *Pasaru panduga*.
II	**Fabaceae**		
1	*Butea monosperma*	Tree, twigs, flowers	This beautiful tree is sacred to *Soma* (Moon). The flowers are offered to other Gods also. This tree is supposed to have sprung from the feather of a falcon imbued with *Soma*. The plant is much used by the Hindus in religious ceremonies, particularly that connected with blessing calves to ensure large quantities of milk when mature. The middle of the three leaflets is believed to represent Lord *Vishnu*, the one on the left Lord *Brahma* and the one on the right Lord *Shiva;* hence the leaves are used in worship in *Chaturmas*. The Leaves are used as platters on the occasion of investiture of the sacred thread. The dry twigs are used as *samidha*s for *homa*s, the sacred fires, to invoke benevolence of Moon. When a boy becomes a *saadhu,* he would eat a leaf of this plant and his staff is made of its wood. According to *Vamana Puraana* the plant has originated from the right side of Yama's body. The leaves are offered to the Lord *Vishnu* on *Shree Narasimha Jayanti Vratha.* Astrologers believe that this plant should be used in sacred fires. In Himachal Pradesh this tree is considered sacred and is used in funeral pyres, as sandal wood is used elsewhere. Buddhists also consider the plant sacred, as it is believed that Queen Mahamaya clutched a branch of this tree at the very moment of the birth of Gautam Buddha.
2	*Dalbergia sissoo*	Tree	The tree is regarded as sacred and planted by the Hindus.
3	*Erythrina variegata*	Flowers	The flowers are used to worship Lord *Mahadeva* and Lord *Vishnu* according to *Vamana Puraana* and *Brahma Vaivarata Puraana*.
4	*Pterocarpus marsupium*	Tree	The tree is worshipped by *Bhumi*s of Purulia district in West Bengal.
5	*Pterocarpus santalinus*	Heart wood	Hindus use the red paste of the wood to mark their foreheads (*tilak*) after bathing. Wood is also used for carving idols. According to astrologers the root, when used as talisman, will have the benevolent effect of the celestial body *Raahu.*
6	*Sesbania grandiflora*	Flowers	The flowers are offered to *Shiva.* The leaves are offered in the name of the celebrated Sage *Athreya* on *Saptarshi Vratha.*
III	**Mimosaceae**		
1	*Acacia farnesiana*	Flowers	Used in worship on several different occasions.
2	*Acacia ferruginea*	Leaves, flowers	Offered to the Lord *Samba* during the *Shree Shanipradosha Vratha* and to *Rishi Vasishta* on *Saptarshi Vratha* and *Rishi Panchami Vratha.*
3	*Acacia nilotica*	Flowers, pods	In Bundelkand, girls decorate the spiny green branches of Babul with colourful flowers at the time of the festival *Mamulian*. They offer edible fruits to this tree for the fulfillment of their wishes.
4	*Acacia polyacantha*	Tree	The tree is associated with the birth star *Dhanishta,* whose presiding deity is *Vaasu.*
5	*Adenanthera pavonina*	Heart wood	The red paste of the heartwood is used as a substitute for red sandalwood.
6	*Albizia lebbeck*	Tree	Astrologers recommend this tree for water baths to receive benevolence of Venus. The tree is sacred especially to Buddhists.
7	*Albizia odoratissima*	Fragrant exudate	As per *Vamana Puraana* the fragrant material '*kaleya chandana*' from this species is used in the worship of *Mahadeva.*

Sl. No.	Species	Part used	Religious significance
8	*Prosopis cineraria*	Tree	The tree is mentioned in *Atharvavaeda* and the later *Samhitha*s. The *Vaedah*s praised the qualities of the *Shami* tree. According to *Vamana Puraana,* the plant has originated from *Katyayana* and leaves are used to worship *Vishnu.* This tree is associated with the birth star *Dhanishta* whose presiding deity is *Vaasu.* It is considered as one of the abodes of *Shiva* and one of S*hivas*'s names is '*Shamiroha*', the one who ascends the s*hami* tree. In *Mahaabhaaratha,* The *Paandavaas* hid their weapons in the *shami* tree during their *agnathavaasa* for one year. The dry twigs are used as s*amidha*s. The leaves are offered to *Ganapathi.* The *Bhil*s and *Naika*s of Gujarat worship only the s*hami* tree during *Dusserah* festival and nothing else.
2.	**VENERATED SHRUBS**		
I	**Caesalpiniaceae**		
1	*Bauhinia acuminata*	Flowers	The flowers are offered to Lord *Shiva.*
2	*Bauhinia tomentosa*	Shrub, flowers	The flowers are offered to Lord *Shiva.* The plant is often cultivated near Hindu temples.
3	*Senna auriculata*	Plant, flowers	According to folklore, this species '*Aonla*', was first born from *Brahma*'s tears of love. The other trees and men were created later. Seeing or smelling of flowers of *Aonla* is auspicious.
II	**Fabaceae**		
1	*Sesbania sesban*	Leaves, flowers	The leaves and flowers are particularly sought for *Karthika pooja* during September–October.
3.	**VENERATED HERBS**		
I	**Fabaceae**		
1	*Aeschynomene aspera*	Pith	Beads made of the pith are dyed in different colours and made into garlands to decorate idols.
2	*Cicer arietinum*	Seeds	Roasted seeds are offered to Lord *Hanumaan* and Goddess *Santhoshi Maatha.*
3	*Uraria lagopodioides*	Plant	During Vaedic times this species was invoked as a Goddess.
4	*Vigna mungo*	Seeds	Bread made from the finely ground pulse is used in many of the religious ceremonies by the Hindus. As per *Vamana Puraana* the seeds of this plant are used to worship Lord *Vishnu.*
5	*Vigna radiata*	Seeds	As per *Vamana Puraana* the seeds of this plant are used in worship of Lord *Vishnu.*
II	**Mimosaceae**		
1	*Mimosa pudica*	Plant	Water baths with this species invoke the benevolence from the Sun.
4.	**VENERATED CLIMBERS**		
I	**Caesalpiniaceae**		
1	*Bauhinia scandens*	Stems	The serpentine branches are used in snake worship and carried about by mendicants for protection against snakes.
2	*Bauhinia vahlii*	Leaves	The leaves are offered especially at *Durga pooja* and *Dusserah pooja.*
II	**Fabaceae**		
1	*Clitoria ternatea*	Flowers	The beautiful blue flowers are held sacred to the Goddess *Durga.* They are offered to Lord *Vishnu* at the time of *Shree Narasimha Jayanthi Vratha, Shree Anantapadmanabha Vratha* and *Shree Satyanarayana* on *Shree Satyanarayana Vratha* and to *Ganapathi* at *Sankastha Chauturthi Vratha.* The blue flowers are widely used in *pooja*s in West Bengal.

Box 2.8

Legume Species that Symbolize Planets, Zodiac Signs and Birth Stars

Planets

Soma	*Butea monosperma* (Muthuga, Palasha)
Mangala	*Acacia catechu* (Khair)
Shani	*Acacia ferruginea*, *Prosopis spicigera* (Shami, Banni)
Kuja	*Cajanus cajan* (Pigeon pea)
Budha	*Vigna radiata* (Green gram)
Guru	*Cicer arietinum* (Chickpea)
Shukra	*Lablab purpureus* (Country bean)
Rahu	*Vigna mungo* (Green gram)
Ketu	*Macrotyloma uniflorum* (Horse gram)

Zodiac Signs

Mesha/Aries	*Pterocarpus santalinus* (Rakhtha chandana)
Karkataka/Cancer	*Butea monosperma* (Muthuga)
Vrishchika/Scorpio	*Acacia catechu* (Khair)
Makara/Capricorn	*Dalbergia sissoo* (Bheete)
Kumbha/Aquarius	*Prosopis spicigera* (*Shami*)

Birth Stars

Mrigasira	*Acacia catechu* (Khair)
Pubba	*Butea monosperma* (Muthuga, Palasha)
Dhanishta	*Prosopis spicigera*/ *Acacia suma* (Shami)

2.10 LEGUME SPECIES IN ASTROLOGY

The Hindu almanac is based on the movement of sun through 27 constellations. Each constellation has a reigning deity and a sacred plant is associated with it for worship, in order to ward off the evil effects of celestial bodies.

Astrologers also claim that the celestial bodies influence the functions and diseases of the human body. For example, Sun is believed to influence the digestive system and diseases associated with him are inflammation, fever and bile disorders. The root of *Mimosa pudica* symbolizes the Sun, and so its decoction is used to treat fistula. *Albizia lebbeck* symbolizes Venus, and diseases of the head, gums and eyes are treated with the root extract of this species.

Nine grains and grams have become important for our survival, and have become a part of our daily fare. These nine species collectively called *Navadhanya*, were projected to symbolize the nine planets and are offered to Gods on important occasions. Significantly, six species of legumes, *Cajanus cajan* (red gram), *Cicer arietinum* (chickpea), *Lablab purpureus* (lablab bean), *Macrotyloma uniflorum* (horse gram), *Vigna mungo* (green gram) and *Vigna radiata* (black gram), figure in this group.

Hindu astrology has ascribed species of plants to planets, zodiac signs and birth stars. The nine planets and the 27 birth stars are associated with certain plant species. Each of the 12 Indian zodiac signs has long been equated with each of the 12 western zodiac signs and symbolic plants were determined for these many religious gardens in India grow all the species that signify the nine planets '*Navagraha vana*' and the 27 birth stars in '*Nakshathra vanas*' and (star gardens), providing a single place of tree worship for everyone.

The Legumes that signify the planets, zodiac signs and birth stars are given in Box 2.8. In some cases, the same species may be significant in more than one context, as for example, *Butea monosperma* and *Acacia catechu.*

Deep-rooted and large-scale veneration of plants in Hinduism immensely reflects the importance the ancient Indians have ascribed to Nature, realizing the benefits from individual species of plants and animals and the need to protect them for human survival. Hinduism, often on the fear of God's wrath, has successfully conserved abundant plant and animal resources, recognizing their importance as sources of food, medicine, and other necessities, and the consequences of their unrestricted exploitation. The cultivation of these species, considered sacred, in different kinds of gardens, provides a new dimension to gardening, fulfilling the two primary objectives of gardening, that of aesthetic appeal and conservation.

CHAPTER-3

The Economic Potential of Legumes in India

3.1 INTRODUCTION

Legumes provide many of man's basic necessities such as food, shelter, medicine and clothing, in some measure or the other. They also provide very diverse and invaluable raw materials for the industry. This chapter gives an insight into the very broad range of economic uses of legume species occurring in India.

3.2 ECONOMIC POTENTIAL OF LEGUMES IN INDIA

About 1,152 species of legumes belonging to 179 genera were recorded in India (Sanjappa, 1992). Of these, 721 species (125 of the Caesalpiniaceae, 504 of the Fabaceae, and 92 of the Mimosaceae), were chosen for evaluating their economic potential. Among the many economic uses, the legumes have in our day-to-day life, 18 uses were identified as more important for the purpose of this evaluation. The various uses of species in the three legume families are given in Tables 3.1, 3.2 and 3.3. Significantly, several legume species are put to diverse uses. The sources of information for these tables are given in the introduction to the species profiles.

Species in Tables 3.1, 3.2 and 3.3 constitute working lists, which can be modified as time and need demand. From the literature available, the correct names of the species are used, giving along, the relevant and important synonyms.

The names of several legume species have changed. In order to facilitate an easy reference to the new names, new equivalents of the synonyms are given in Table 3.4.

In the following text, the number and percentage of species with a particular use are given, for the three families and also for all legumes, in order to highlight their relative importance. The family with the largest percentage of species under each category is indicated in bold. A summary of this information is given in Table 3.5.

The following account provides an overall picture of the economic uses of legume species in India, to reflect their potential. Issues such as the nutritional value, anti-nutritional factors, toxicity, uses in different systems of alternative medicine, and uses in the management of cancer, are discussed in separate chapters.

3.2.1 Edible Legumes

It was realized centuries ago, that pulses should be an important component of our diet. The combinations of pulse and rice (*dal-chawal*) or pulse and unleavened wheat bread (*dal-roti*), have long become the standard fare of an average Indian diet.

Legume seeds are the most important and very rich source of protein, particularly for the vegetarians. They contain essential amino acids like lysine but may be deficient in methionine. Legume seeds are also rich in vitamins and minerals. From the nutritional point of view the most valuable legume seed is the soybean, which is considered a meat substitute, as it contains proteins close to those in egg, both in quality and quantity. The nutitive potential of the legumes is discussed in Chapter 6.

Table 3.6 provides information on the legume species and their edible parts. In addition to seeds, tender roots, root tubers, tender shoots, bark, leaves, flowers and tender pods of several legumes are also edible. They are consumed in the raw condition as *chutney,* or cooked as vegetables. Tender shoots, leaves and flowers of species of *Bauhinia, Caesalpinia, Cassia, Saraca, Senna* and *Tamarindus,* from the Caesalpiniaceae; species of *Astragalus, Butea, Caragana, Clitoria, Crotalaria, Cytisus, Gliricidia, Indigofera, Medicago, Mellilotus, Ougenia, Pterocarpus, Pueraria, Sesbania, Trifolium* and *Wisteria,* from the Fabaceae; and species of *Acacia, Adenanthera, Albizia, Leucaena, Neptunia* and *Prosopis,* from the Mimosaceae are consumed. Tuberous roots of *Pachyrrhiza erosus* of the Fabaceae, roots of species of *Albizia, Leucaena* and *Prosopis*, and the bark of species of *Acacia* and *Prosopis* are often used as vegetable.

Caesalpiniaceae: 42/125, 33.6%, Fabaceae: 200/504, 39.5%; **Mimosaceae: 38/92, 41.7%;** Total; 280/721, 38.9%.

3.2.2 Legumes in Medicine

A very large number of legume species are used in medicine. Legumes are projected as the medicine chest of the future (Adams, 2000), as more legume species with medicinal uses are being identified. Several species that contain therapeutically important phytochemicals were suggested for conservation as they are overexploited and are endangered (Morris, 1997). The medicinal uses of the legumes are discussed in several different Chapters in this book.

Legumes have been used in such Alternative Systems of Medicine as *Ayurveda, Siddha, Unani,* Chinese, Tibetan and Homoepathy, for treatment of various ailments for centuries (Anonymous, 1972; 1988, 1992a, b; 1995; Kent, 1975; Kirthikar and Basu, 1975; Ray and Gupta, 1980; Beal and Reinhard, 1981; Clarke, 1982; Ray *et al.*, 1980 Stuart, 1985; Yeshidanden, 1986; Hasan, 1989; Boericke, 1991; Thakur *et. al.*, 1991; Sharma, 1993; Fathima, 1994; Sivarajan and Balachandran, 1994; Keys, 1995; Yoganarasimhan, 1996; 2000; Kameswara Rao, 2000). Legume species used in different Alternative Systems of Medicine in India are discussed in the Chapters 18–21.

In view of the very large number of leguminous species used in medicine, it is not possible to provide a comprehensive review of the different uses of all of them in this Chapter, and so only the salient uses are mentioned here.

Abrus precatorius and *Abrus pulchellus* are used for digestive disorders (Birkhill, 1935; Irvin, 1961) and as a general sweetening agent (Fulla *et al.*, 1990). *Abrus precatorius* has platelet aggregation, anti-inflammatory, antiallergic and anticancer activity and is used in Chinese medicine for treating laryngitis, hepatitis and bronchitis (Sheng-Chukuo *et al.*, 1995).

Astragalus onobrychis is useful in folk medicine as diuretic, diaphoretic, antihypoxic and against rheumatic diseases (Hoppe, 1975; Benbessat, 1993), while *Astragalus membranaceus* and *Astragalus mongolicus* are used as antiperspirant, diuretic and tonic.

Caesalpinia bonduc has antimicrobial activity (Weenic *et al.*, 1990); *Caesalpinia ferrea* is used in gastroduodenal diseases (Cumbria, 1942; Cruz, 1965a, b) and as antiulcer agent in folk medicine (Bacchi *et al.*, 1995); *Cassia grandis* is used to treat anaemia and is an astringent, expectrorant and a sedative (Morton, 1981).

Cajanus cajan is useful in treatment of sickle cell disease (Ekeke and Shode, 1985; Iwu, 1986; Ekeke and Shode, 1990).

Caragana sinica is used in treating hyper proliferative or inflammatory diseases, cancer, and injury; used in folk medicine to treat hypertension and arthritis (Kulanthaivel *et al.*, 1995).

Consumption of *Castanospermum australe* (black beans), which is a rich source of genistein, lowers risk of breast and prostate cancer as well as the risk of heart disease, and fights leukaemia that strikes bone marrow and other blood forming tissues (Adams, 2000).

Crotalaria sessiliflora is used in Chinese folk medicine as antitumour agent (Anonymous, 1975 a,b; 1977).

Cullen corylifolium (Psoralea corylifolia) seeds are used in Chinese and Indian folk medicine to treat skin diseases (Yang *et al.*, 1996), leucoderma, leprosy, psoriasis and inflammatory diseases of the skin and as an inhibitor of lipid peroxidation (Haraguchi *et al.*, 2000). The seed has cytotoxic activity against stomach carcinoma cell lines (Yang *et al.*, 1996). It is also useful as a tonic, aphrodisiac and as a remedy for menstrual disorders, bone fracture, osteomalacia, osteoporosis and related conditions (Minura *et al.*, 1996).

Erythrina variegata has been used in folk medicine in Japan and China (Deshpande *et al.*, 1977; Hedge *et al.*, 1997; Tanaka *et al.*, 2000) for its antimicrobial (Telikepalli *et al*, 1990) and antiinflammatory action (Hedge *et al.*, 1997).

Glycine max has anticholesterolemic (Lovati, 1992), antioxidant (Ohminami *et al.*, 1984), antitumour (Konoshima *et al.*, 1991, 1992) and HIV infection inhibitory properties (Nakashima *et al.*, 1989). It protects cells from damage by hydrogen peroxide (Yoshikoshi *et al.*, 1996), reduces plasma cholesterol level both in experimental animals (Terpstra *et al.*, 1984) and in man (Sitori *et al.*, 1977). It has inhibitory effect on human cytomegalo virus, influenza virus and HIV Type I (Nakashima *et al.*, 1989; Hayashi *et al.*, 1997). Genistein, abundant in soybean lowers the risk of breast and prostate cancers, heart diseases, fights leukaemia that strikes the bone marrow and other blood forming tissues (Adams, 2000).

Glycyrrhiza echinata and *Glycyrrhiza glabra* have been in use for hundreds of years as the source of drug *liquiritiae radix* (Donas *et al.*, 1993), which has immunomodulatory (Kroes *et al.*, 1993) and desmutagenic activity (Zani *et al.*, 1993). These two species were used in ancient Chinese medicine as a tonic, expectorant, antiphlogistic and antitussive agents (Tang and Eisenbrand, 1992). They are also antimicrobial (Mitscher *et al.*, 1980; Li *et al.*, 1998), antimutagenic, (Mitscher *et al.*, 1986; Sakai *et al.*, 1986), antiulcer (Doll *et al.*, 1962), antiviral (Pompei *et al.*, 1979; Baba and Shigeta, 1987), antiplatelet aggregation (Tawata *et al.*, 1990), anti-inflammatory (Finney and Sommers, 1958; Sotomatsu *et al.*, 1959; Yamamoto *et al.*, 1963; Kuroyangi and Sato, 1966; Ohuchi *et al.*, 1981; Amagaya *et al*, 1984; Akamatsu *et al.*, 1991), antiallergic (Sotomatsu *et al.*, 1959; Kuroyanagi *et al.*, 1966; Kuroyanagi and Sato, 1966; Amagaya *et al.*, 1984; Inoue *et al.*, 1987), antihepatitic (Karayanagi and Sato, 1966; Suzuki *et al.*, 1977; Amagaya *et al.*, 1984), antifungal and antibacterial (Saleh *et al.*, 1990). The two species are used as detoxicants of xenobiotics in rat liver (Moon and Kim, 1997), and in traditional oriental herbal prescription for treatment of diabetic neuropathy. They also inhibit sorbitol accumulation in human red blood cells (Aida *et al.*, 1990). *Glycyrrhiza inflata* is anti-inflammatory (Shibata *et al.*, 1991), inhibits human immunodeficiency virus (HIV) (Hatano, 1988), and has radical scavenging activity (Hatano *et al.*, 1988), inhibits leucokotriene synthesis in human polymorphonuclear neutrophils (Kimura, 1988) and xanthine oxidase (Hatano, 1989). *Glycyrrhiza uralensis* has potent aldose reductase inhibiting activity (Aida *et al.*, 1989).

Lathyrus sativus is antimycotic (Schenk *et al.*, 1991) and neurotoxic. *Lespedeza capitata* is used as diuretic and antihypertensive (Wagner and Elbl, 1992).

Mucuna species (velvet bean) inhibit cholinesterase and serotonin activity, which is a brain neurotrasmitter that may be involved in learning, sleep and control of moods (Adams, 2000).

Pachyrrhizus erosus has antitumour activity; used in treatment of skin eruptions (Kardono *et al.*, 1990). *Parkia speciosa* is antidiabetic, laxative (Suvachittanont and Pothirakit, 1988) and mitogenic (Suvachittanont and Jarancharanapet, 2000). *Phaseolus vulgaris* has antifertility activity (Sun *et al.*, 1983). *Prosopis alba* is used to treat rheumatisum and 'rain pain' (Munoz *et al.*, 1981). *Psophocarpus tetragonolobus* is antitumour and used to treat acne and eczema (Adams, 2000).

Pueraria lobata has hepatoprotective action *in vitro* (Arao *et al.*, 1998), useful as antipyretic and antispasmodic agent (Arao *et al.*, 1995; 1997). It is antidiarrhoel, diaphoretic and antiemetic (Xie *et al.*, 1994; Keung *et al.*, 1996), antioxidative (Wang 1996), estrogenic (Rong *et al.*, 1998) and anticomplementary (Oh *et al.*, 2000); used against angina pectoris, hypertension, deafness, optic nerve atrophy or retinitis. (Oshima *et al.*, 1988; Huang, 1993). Species of *Pueraria* (Kudzu) are rich in a number of chemicals including daidzein, which is antiinflammatory, antimicrobial and cancer preventive, and genistein is antileukaemic (Adams, 2000).

Senna alexandrina is an age old purgative. *Senna occidentalis* has antibiotic activity (Gaind *et al.*, 1966) and antimalarial activity (Weenen *et al.*, 1990); *Senna tora* is used in Chinese herbal medicine to improve vision and is aperient, antiasthenic, diuretic (Namba, 1980), antimutagenic (Choi *et al.*,1997) and fungitoxic (Pandey *et al.*, 1982).

Sophora flavescens is used in folk medicine as an antipyretic, analgesic, anthelmintic, stomachic and antitumour (Lee *et al.*, 1997) and decreases airway hypersensitivity in several animal models with bronchial allergic inflammation (Oosterhout *et al.*, 1993). The root of *Sophora flavescens* var. *angustifolia* is used as a traditional medicine in China and Japan for centuries (Saito *et al.*, 1990). *Styphnolobium japonicum* has radioprotective, antihaemorrhagic and antimicrobial activities (Ishida *et al.* 1989; Sokolchik *et al.*, 1992).

Tamarindus indica has antimicrobial activity; seeds of *Trigonella foenum-graecum* were clinically demonstrated to be hypoglycaemic (Ali *et al.*, 1995).

Vicia faba is used in treating conditions such as hypertension, heart impairment, renal failure and liver cirrhosis (Vered *et al.*, 1997).

Species of *Bauhinia, Caesalpinia, Humboldtia, Saraca* of Caesalpiniaceae; *Abrus, Alysicarpus, Astragalus, Cajanus, Caragana, Crotolaria, Desmodium, Indigofera, Myroxylon, Sophora, Tephrosia, Uraria, Vigna* of Fabaceae; *Acacia, Albizia, Entada, Leucaena, Mimosa, Pithecelobium* and *Prosopis* of Mimosaceae are used commonly in different systems of medicine (Kameswara Rao, 2000).

Caesalpiniaceae: 87/125, 69.6%; **Fabaceae: 363/504, 71.7%;** Mimosaceae: 64/92, 70.3%; Total: 514/721, 71.5%.

3.2.3 Legumes as Green Manure

Restoration and maintenance of soil fertility is a basic and critical problem in agriculture. It is especially serious in developing countries, many of which suffer from poor productivity, on account of soils deficient in plant nutrients.

The high cost and long term negative effects of synthetic fertilisers make it increasingly important to develop strategies for the renewed use of green manure crops, especially for fragile farming ecosystems in the tropics.

Species of the bacterial genus *Rhizobium* colonising the nodules on the roots of many legumes fix nitrogen, making them a good source of organic manure. Their fresh foliage usually contains 0.5 to 1.0 per cent nitrogen. The foliage of perennial legumes may accumulate from 100 to 600 kg of nitrogen per hectare. When incorporated into the soil, legume organic manure greatly contributes to soil fertility, moisture, nutrient retention, and general cultivability (tilth) of the land, and this is the favourite method adopted in organic farming.

Used with discretion, many legumes are valuable for initial planting on barren grounds, wastelands or even sand dunes. Once established, they can create conditions for the growth of other species. Legume species can be used as pioneers in preparing difficult sites for farming and forestry (Kochaar, 1981; Hill, 1989).

Species of *Acacia, Albizia, Alysicarpus, Cassia, Canavalia, Centrosema, Crotalaria, Dalbergia, Desmanthus, Desmodium, Gliricidia, Indigofera, Lathyrus, Leucaena, Medicago, Phaseolus, Pongamia, Prosopis, Rhynchosia, Senna, Sesbania, Stylosanthes, Tamarindus, Tephrosia, Trifolium, Vicia* and *Vigna* are used as green manure.

Caesalpiniaceae: 15/125, 12.0%; **Fabaceae: 146/504, 28.9%;** Mimosaceae: 16/92, 17.6%; Total: 177/721, 24.6%.

3.2.4 Legumes as Fodder

Fodder (or forage) is the food (feed) of domestic and farm animals. Fodder is lopped from trees and shrubs and chopped. Arable lands with herbaceous or semi-herbaceous plants, left for animals to graze on, are pastures. Several pulse crops such as groundnut (*Arachis hypogaea*) are the most common pasture and fodder sources among legumes. The tender branches and foliage of a number of leguminous trees and shrubs, which are relished by animals, form another source of forage especially in dry areas and during periods of scarcity. Forage and fodder trees and large shrubs provide shelter and shade to the livestock and also yield firewood and charcoal.

Uncertain supplies and increasing costs of both conventional and processed animal feed, fertilisers and food coupled with population increase, have now placed leguminous trees and shrubs as important sources of fodder, more than ever before.

Species of *Acacia, Alysicarpus, Cassia, Senna, Centrosema, Dalbergia, Desmodium, Gleditsia, Gliricidia, Indigofera, Lathyrus, Medicago, Phaseolus, Prosopis, Pterocarpus, Sesbania, Tamarindus, Trifolium, Vicia* and *Vigna,* are used as sources of fodder, pasture and green manure (Kochaar, 1981; Hill, 1989).

Caesalpiniaceae: 19/125, 15.2%; **Fabaceae: 224/504, 44.3%;** Mimosaceae: 38/92, 41.8%; Total: 281/721, 39.0%.

3.2.5 Oils from Legumes

Chinese extracted oil from plants since pre-historic times. Two types of oils, namely fatty oils (lipids) and essential oils occur in plants. Lipids in legumes are discussed in Chapter 10.

Fatty oils are polymers of fatty acids. They are fixed, non-volatile and insoluble in water. The non-drying fatty oils are used as food and lubricants. The process of hydrogenation of vegetable oils yields solid fats that do not go rancid rapidly. Fatty oils are usually concentrated in the seeds in majority of the legume seeds.

Groundnut (*Arachis hypogaea*) oil is a globally important cooking medium. Soybean seed oil is fast gaining an equal importance. Seed oil of *Pongamia pinnata* has been used medicinally for centuries and is now being promoted as 'vegetable diesel'. Species of *Acacia, Aeschynomene, Albizia, Butea, Dichrostachys, Erythrina, Gliricidia, Myroxylon, Psophocarpus* and *Spartium* are rich in seed oil (Kochaar, 1981; Hill, 1989).

Essential oils are volatile and highly fragrant. They are generally the derivatives of terpenes and benzenes, mostly found in fragrant plant parts such as flowers. They are used for perfuming soaps, tobacco, as deodorants, in toilet preparations and for flavouring various foods and beverage products. The essential oil from the flowers of *Acacia farnesiana* is used to impart a pleasant odour to perfumes, soaps, deodorants and cosmetics. Essential oils also serve as antiseptics, stimulants, ingredients in medicines, as laboratory reagents, solvents in paint industry, components of polishes, pastes, and ink, and as insecticide.

Caesalpiniaceae: 25/125, 20.2%; Fabaceae: 69/504, 13.6%; **Mimosaceae: 19/92, 20.9%;** Total: 113/721, 15.7%.

3.2.6 Fibre from Legumes

Fibre is composed of vastly elongated lignified cells from the leaves, stems, or bark of plants. Fibre is one of the basic necessities of man. As civilisation advanced, the demand for vegetable fibre also increased greatly. In general, fibre yielding plants occupy the second place in economic importance, next to food plants.

Legumes are among the important fibre yielding species. Fibre from legumes is used in making cloth and cordage and as stuffing material. Soft fibres are used for weaving, making mats, chairs, furniture, stuffing for mattresses, upholstery, life belts, thatches, fishing nets, screens, curtains, ropes and paper. Hard fibres are used in making brooms, brushes and baskets.

Several species of *Acacia, Bauhinia, Butea, Crotalaria, Erythrina, Sesbania* and *Spartium* are the usual sources of fibre (Kochaar, 1981; Hill, 1989).

Caesalpiniaceae: 17/125, 13.6%; Fabaceae: 46/504, 9.0%; **Mimosaceae: 15/92, 16.5%;** Total: 78/721, 10.8%.

3.2.7 Legume Gums and Resins

Gums are widely distributed degradation products of cellulose or other carbohydrates in plants, and are basically a pigment suspending medium. Gums are produced either in the bark or in the seeds. They have a wide variety of uses. They impart lustre to silk and fine textiles and used for sizing paper, as granulating agent in fertilisers, and as a binder in explosives and polishes. Gums are also used as a component of adhesives, as emulsifying and smoothening agent in bakery products, confectionaries, soft drinks, cosmetics, lotions, ointments and detergents as a stiffening agent in ice creams, as binders in medicinal pills, cough syrup, foam stabiliser in beer, as purifier in wine making. They are used as an ingredient in assaying minerals, as ceramic glazes and in printing inks. Gums are medicinally used for soothing inflammation of intestinal mucosa or inflamed or burnt skin and as astringents.

The legumes are the most important source of gums. Copious gum exudes from tree bark of species of *Acacia, Albizia, Astragalus, Bauhinia* and *Caesalpinia* where as species of *Acacia, Cassia, Crotalaria, Cyamopsis, Indigofera, Senna, Sesbania* and *Trigonella* are rich in seed gums. Seed gum from *Cyamopsis tetragonoloba* and *Trigonella foenum-graecum* is an important commodity in international trade.

Resins are exudates of certain tree species of Leguminosae. They differ from gums from being insoluble in water and soluble in organic solvents, such as alcohol. Resins are an oxidation product of different essential oils and are a mixture of terpenoids and flavonoids. The main constituents of resins are resin esters. Resins are secreted in plant tissues in special cavities or resin ducts. Resins are odouriferous and so are used as perfume fixatives and ingredients of incense. They are used for caulking boats and canoes, and in the manufacture of varnishes, paints, lacquers, paper, soap, boot polish, lubricating agents, linoleum, oilcloth, printing ink and plastic paints. They are also used to flavour tobacco (Kochaar, 1981; Hill, 1989). Rural people use resin-bearing branches as torches.

Species of *Copaifera* and *Myroxylon* are a very rich source of resins.

Caesalpiniaceae: 15/125, 12.0%; Fabaceae: 32/504, 6.3%; **Mimosaceae: 38/92, 41.7%;** Total: 85/721, 11.8%.

3.2.8 Tannins from Legumes

Man has used vegetable tannins since ancient times to colour fabric and to process hides.

Tannins are complex phenolic compounds produced by plants. They are organic, acidic, non-nitrogenous, glucosidal products with astringent properties and are bitter to taste. They accumulate in the bark, wood, fruits and seeds and in some cases the leaves. Economically, tannins are of special interest because of their ability to chemically bind with certain types of proteins, such as those present in animal skin rendering it into leather, the strong, flexible, resistant and water insoluble product of day-to-day use. Tannins are also used in the textile industry, in the manufacture of ink, paints, varnishes and paper, and to colour food, beverages and as medicine. Tannins in legumes are further discussed in Chapter 11.

Common tannin yielding legumes include *Acacia nilotica, Acacia dealbata, Acacia decurrens, Cassia fistula, Ceratonia siliqua, Hardwickia binata, Lespedeza juncea, Pongamia pinnata, Pterocarpus marsupium, Schotia brachypetala, Sesbania grandiflora, Spatholobus parviflorus* and *Tamarindus indica,* and several species of *Acacia, Albizia, Bauhinia, Caesalpinia, Flemingia, Indigofera* and *Parkia* (Kochaar, 1981; Hill, 1989).

Caesalpiniaceae: 24/125, 19.4%; Fabaceae: 20/504, 3.9%; **Mimosaceae: 43/92, 47.2%;** Total: 87/721, 12.1%.

3.2.9 Dyes from Legumes

The dye industry in India dates back to 2nd century CE, as evidenced from the paintings of Ajanta caves. Hundreds of species of plants are still used for dyeing especially in the tribal and rural societies.

Most natural dyes are water soluble plant products. They have been used to colour food, yarns, textiles, leather, and wood, as well as to decorate houses, since the earliest times, in various parts of the world. They are also used in cosmetic dyeing of the body, nails and hair. Egyptian, Greek, Roman and other early civilisations have used them extensively. Indigo, the blue dye, from species of *Indigofera* grown in India and Africa, is one of the oldest of plant dyes. Legume pigments and dyes are discussed in Chapter 12.

The following legumes are among the common sources of dye: *Butea monosperma, Caesalpinia pulcherimma, Ceratonia siliqua, Clitoria ternatea, Dichrostachys cineraria, Haematoxylum campechianum, Mucuna pruriens, Pterocarpus santalinus* and *Tamarindus indica,* and species of *Acacia, Albizia, Bauhinia,* and *Indigofera* (Kochaar, 1981; Hill, 1989).

Caesalpiniaceae: 19/125, 15.3%; Fabaceae: 47/504, 9.3%; **Mimosaceae: 21/92, 23.0%;** Total: 87/721, 12.1%.

3.2.10 Timber from Legumes

Wood is the secondary xylem composed of lignin and cellulose, which constitutes the hardest part of the trunk and branches of trees and shrubs. Timber is wood suitable for building or carpentry. Wood is one of the most important of our day-to-day necessities such as house building, furniture, material for interior decoration, panelling, fuel, medicine, paper, rayon, and various natural chemical compounds. *Acacia catechu, Acacia nilotica, Albizia lebbeck, Cassia fistula, Dalbergia latifolia, Dalbergia sissoo, Desmodium oojeinense, Pongamia pinnata, Prosopis spicigera, Pterocarpus dalbergioides, Pterocarpus indicus, Pterocarpus marsupium, Pterocarpus santalinus* and *Xylia xylocarpa* are among the more commonly used legume timbers (Kochaar, 1981; Hill, 1989).

The quality of the wood depends upon its composition, which is assessed by studying its anatomy, along with the physical and chemical properties. Detailed information on the wood of 87 species in 40 genera of Legumes was compiled by Chauhan and Vijendra Rao (2003).

Caesalpiniaceae: 44/125, 35.5%; Fabaceae: 52/504, 10.3%; **Mimosaceae: 53/92, 58.2%;** Total: 149/721, 20.7%.

3.2.11 Legumes as Ornamentals, Avenue Trees, Hedges, Shade or Support Plants

Legumes have largely contributed to the beauty of our natural and man made environments. Some of the most glorious ornamental and flowering plants are legumes. *Amherstia nobilis* (Caesalpiniaceae), endemic to Burma, grown in many Indian gardens, is regarded as the Queen of Flowers. Economically important or endangered or showy legumes were chosen as the national flower/emblem by some countries/States. For example, *Acacia pycnantha* (the Golden Wattle, Mimosaceae) is the national floral emblem of Australia, while the State floral emblem of South Australia is *Swainsona formosa* (the Sturt's desert pea, also known as *Clianthus formosa,* Fabaceae, a threatened species).

Legumes that are commonly used as ornamentals, hedge plants or avenue trees include *Butea monosperma, Delonix regia, Peltophorum pterocarpum, Pongamia glabra, Samanea saman, Saraca asoca, Sesbania grandiflora* and species of *Acacia, Bauhinia, Brownea, Cassia, Erythrina* and *Senna* (Kochaar, 1981; Hill, 1989). The list is endless.

Many of the ornamentals have also other functional values such as providing fodder, floral nectar for honeybees, in stabilising and improving soil and providing shelter for insects and humans.

Coffee and tea plantations need species that provide shade and weak stemmed crop plants such as pepper need a live support, both of which are provided by many legumes such as species of *Butea, Calpurnia, Dalbergia, Desmodium, Pterocarpus, Sesbania* of Fabaceae, *Acacia, Albizia, Parkia, Pithecellobium* of Mimosaceae, *Bauhinia, Caesalpinia, Cassia, Senna, Peltophorum, Tamarindus* of Caesalpiniaceae.

Caesalpiniaceae: 68/125, 54.4%; Fabaceae: 108/504, 21.3%; Mimosaceae: 44/92, 48.4%; Total: 220/721, 30.6%.

3.2.12 Legumes as Cover Crops, Windbreaks or Soil Binders

Growing leguminous plants is a very effective way to bind soil to prevent its erosion. Some of these species are useful also as windbreaks. *Arachis hypogaea, Chamaecrista pumila, Calpurnia lasiogyne, Calopogonium mucunoides, Cytisus scoparius, Lysiloma latisiliquum, Parkinsonia aculeata* and *Senna hirsuta,* and species of *Canavalia, Clitoria, Desmodium, Indigofera, Lespedeza, Lupinus, Medicago, Melilotus, Sesbania, Vigna* and *Vicia* are among the legumes used as cover crops/windbreak/soil binders (Kochaar, 1981; Hill, 1989).

Caesalpiniaceae: 05/124, 4.0%, **Fabaceae: 102/504, 20.2%;** Mimosaceae: 18/91, 19.8%; Total: 125/721, 17.4%.

3.2.13 Beverages from Legumes

The leaves, the flowers or the seeds of the species of *Cicer arietinum, Gymnocladus dioica, Hymenaea courbaril, Tamarindus indica, Glycyrrhiza glabra, Trigonella foenum-graecum, Entada rheedii* and *Lysiloma latisiliquum,* and species of *Acacia, Albizia, Cassia, Senna, Canavalia, Myroxylon* and *Parkia* are used in preparing beverages, used as substitute for tea, coffee and as nutritive or refreshing drinks (Kochaar, 1981; Hill, 1989).

Caesalpiniaceae: 14/125, 11.29%; Fabaceae: 23/504, 4.5%; **Mimosaceae: 12/92, 13.1%;** Total: 49/721, 6.8%.

3.2.14 Legume Hosts for the Lac Insect

Lac, used as sealing wax, in electrical switches, polishes, decorative and ornamental work, is the secretion of species of the insect genus *Laccifer*, the most important being *Laccifer lacca* Kerr., which yields most of the commercial lac. The insect larvae settle on the branches of the host plants and draw nutrition. In spite of synthetic material similar to lac, natural lac is still in great demand.

Legumes such as *Albizia lucida, Butea monosperma, Caesalpinia enneaphylla, Dalbergia lanceolaria* var. *paniculata, Desmodium oojeinense, Dichrostachys cineraria, Peltophorum pterocarpum, Pithecellobium dulce, Pongamia pinnata,* and *Xylia xylocarpa,* and species of *Albizia* and *Acacia* are important host plants on which lac insects are cultivated for commercial production of lac (Kochaar, 1981; Hill, 1989).

Caesalpiniaceae: 04/125, 3.2%; Fabaceae: 04/504, 0.8%; **Mimosaceae: 17/92, 18.7%;** Total: 25/721, 3.4%.

3.2.15 Legumes in Paper Industry

The basic raw material in the manufacture of paper is cellulose fibre, which is available from nearly every species of plants. The quantity, quality and economic considerations limit the actual choice of raw material. A variety of hard woods with long fibres are mixed in varying proportions in the manufacture of paper.

Several legume species like *Acacia auriculiformis, Acacia mearnsii, Albizia lebbeck, Butea monosperma, Dalbergia paniculata, Dalbergia sissoo, Pongamia pinnata, Prosopis juliflora, Pterocarpus marsupium* and *Xylia xylocarpa* provide the pulp for paper manufacture (Kochaar, 1981; Hill, 1989).

Caesalpiniaceae: 04/125, 3.2%; Fabaceae: 22/504, 4.3%; **Mimosaceae: 16/92, 17.6%;** Total: 42/721, 5.8%.

3.2.16 Legumes as Substitute for Soap

Saponins are triterpenoidal or steroidal compounds characterised by their bitter taste, foaming in water and lysing biological membranes. Saponins have the same properties as soaps and detergents, and are present in different parts of a large number of species of plants and a few animals. Legume saponins are further discussed in Chapter 14. Fruits of *Acacia sinuata (shikakai)* are the most commonly used vegetable soap. *Albizia amara, Albizia lebbeck, Entada rheedii, Cicer arietinum, Cyamopsis tetragonoloba, Psophocarpus tetragonolobus, Samanea saman, Trifolium repens, Trigonella foenum-graecum, Vigna angularis, Vigna mungo* and *Vigna radiata* are good sources of legume saponins (Kochaar, 1981; Hill, 1989; Kameswara Rao, 2000).

Caesalpiniaceae: 04/125, 3.2%; Fabaceae: 09/504, 1.8%; **Mimosaceae: 12/92, 13.1%;** Total: 25/721, 3.4%.

3.2.17 Pesticides from Legumes

Plants synthesise several secondary metabolites of the chemical groups of glycosides, tannins, terpenoids, saponins, lignans, lactones, alkaloids and even proteins like lectins, which protect them from pests. Species containing such compounds have been used as sources of natural pesticides for a long time. In recent times, the demand for natural pesticides has grown enormously with the global enthusiasm for organic farming. Two of the most potent insecticides of plant origin are rotenone from the legumes and pyrethrum from the members of the Asteraceae. Rotenone, a potent insecticidal flavonoid, is usually extracted from *Derris elliptica* and *Tephrosia purpurea.* Pesticidal legumes are further discussed in Chapter 15.

Species of *Albizia chinensis, Amorpha fruticosa, Astracantha gummifera,* and *Castanospermum australe,* and species of *Derris, Delonix, Lonchocarpus, Sophora* and *Tephrosia* are important legume sources of insecticides and pesticides (Kochaar, 1981; Hill, 1989).

Caesalpiniaceae: 04/125, 3.2%; **Fabaceae: 50/504, 9.9%;** Mimosaceae: 05/92, 5.5%; Total: 59/721, 8.2 7%.

3.2.18 Legumes as Molluscicides and Piscicides

A number of legumes, including some edible legumes, contain chemical compounds such as non-protein amino acids, lectins, alkaloids, saponins, cyanogenic glycosides, and others which are toxic to man and other mammals. They occur in small quantities. When present in the water, such compounds as saponins are toxic to aquatic animals, as they lyse and destabilise membranes of the respiratory tract. These compounds occur in low concentrations in the plants and so are not so toxic to man and cattle, even when ingested through normal food volumes. Several legumes that contain saponins are frequently used to catch fish by stupefying them, though this practice is not generally encouraged. Such plants are cut or crushed and mixed, in required quantities, with the water in which the fish live. Such a method is not so useful in running water but is effective in ponds and lakes. The fish do not die, but are stupefied and float on the surface of water, so that they are easily gathered into baskets. *Abrus precatorius, Andira inermis, Butea monosperma*, and *Gymnocladus dioica*, and species of *Albizia, Cassia, Crotolaria, Lathyrus, Senna* and *Tephrosia* are used as pesticides (Kochaar, 1981; Hill, 1989).

Caesalpiniaceae: 11/125, 8.8%; Fabaceae: 824/504, 16.2%; **Mimosaceae: 24/92, 26.4%**; Total: 117/721, 16.2%.

3.3 RELATIVE ECONOMIC IMPORTANCE OF LEGUME FAMILIES AND SPECIES

Among the 18 categories of use, the largest percentage of all legumes is used as medicine, second largest are the edible species and those used as fodder and the third largest are the ornamentals (Table 3.5).

Among the three families, the Mimosaceae include the largest percentage of species for 12 of the 18 uses, namely, edibility, oil, fibre, gum and resin, tannin, dye, timber, beverage, host for Lac insect, paper industry, soap and piscicide. The Fabaceae have the largest percentage of species for five uses, namely, medicine, manure, fodder and forage, cover crops, and Piscicide, while the Caesalpiniaceae have the largest percentage of species used as ornamentals (Table 3.5).

Each of the three families has the largest percentage of species used in medicine, but the Fabaceae has the largest percentage of species under this category, among the three families (Table 3.5). In the Caesalpiniaceae, the second largest use is as ornamental plants and third largest use is as timber. In the Fabaceae, the second largest use is as fodder and the third largest use is as food. The second largest use in Mimosaceae is as timber and the third largest is as ornamentals. Thus the top three positions for all the three families are shared among only five of the 18 uses, namely, edibility, medicinal, fodder, timber and ornamental uses (Table 3.5). The importance of these uses makes the relevant species valuable.

Contrary to the common belief that the Fabaceae are economically the most useful family, the Mimosaceae have turned out to have the largest number of uses (12 of 18 uses). Legumes are known to be important as sources of edible seeds and timber, but their far greater use in medicine was not well recognized.

3.4 FREQUENCY INDEX OF USES OF LEGUME SPECIES

Many species have multiple uses. The larger the number of uses of a species, the greater is its economic importance. This does not mean that species with fewer uses are not important. For example, *Senna alexandrina* (*Cassia anguistifolia*) has only three uses but it is medicinally a very important species widely used as a purgative, even in modern medicine.

Frequency index is a measure of the number of uses of a species against the total number of identified uses. This is expressed as percentage or a similar value. Since there are large differences in the number of species in each family, actual numbers would be misleading. Legume species with seven or more uses are listed in Table 3.7 to indicate their relative importance.

Fifteen out of 125 species (12.0 per cent) in the Caesalpiniaceae, 42 out of 504 species (8.3 per cent) in the Fabaceae and 31 out of 92 species (34.0 per cent) in the Mimosaceae, have seven or more uses (Table 3.7). Thus the species of the Mimosaceae are economically more important than the others.

Albizia lebbeck (15 of the 18 uses), *Butea monosperma, Sesbania grandiflora, Acacia nilotica* ssp., *indica, Lysiloma latisiliquum* and *Prosopis juliflora* (14 uses), *Tamarindus indica, Erythrina variegata* and *Pongamia pinnata* (13 uses), *Cerationia siliqua, Senna siamea, Robinia pseudoacacia, Acacia farnesiana, Albizia amara* and *Albizia procera* (12 uses), are species worth looking into (Table 3.7).

3.5 FUTURE PROSPECTS

The distribution of a variety of chemical compounds and their application in different contexts is discussed in the later chapters. The species profiles in the later part of the book contains diverse uses of the legumes in India. All this information read along with the information in this chapter provides an insight into the amazing diversity of the economic potential of the Indian legumes. A comparison of the uses of the legume species occurring in India, with those uses for the same species outside India, reflects the possibility that the Indian species could be put to more uses than at present. The overall conclusion is that a well-focussed and systematic research on the Indian legumes is likely to greatly enlarge the spectrum of their utility.

Table 3.1
Economic Potential of Legumes in India: Caesalpiniaceae

Eighteen known uses, identified as important for Legumes, are indicated for each species by '+'; and '-' indicates that, particular use is not known for that species. The correct binomials are used and the important synonyms, if any, are given parenthesis preceded by '=', for each species. A wider synonymy is given in the information profiles for those species for which profiles were compiled.

The genera and species are in alphabetical order.

Sl. No.	*Botanical Name*	Edible	Medicinal	Manure	Fodder/Forage	Oil	Fibre	Gum/Resin	Tannin	Dye	Timber	Ornamental/Avenue/Hedge/ Shade/Support	Cover/Soil Binder/Windbreak	Beverage	Lac insect host	Paper industry	Soap substitute	Insecticide/Pesticide	Molluscicide/piscicide
		1	2	3	4	5	6	7	8	9	10	11	12	13	14	15	16	17	18
1.	***Acrocarpus fraxinifolius*** **Arn.**	-	+	-	-	-	-	-	-	-	+	-	-	-	-	-	-	-	-
2.	***Afzelia africana*** **Pers.**	+	+	-	-	-	-	-	-	-	+	-	-	-	-	-	-	-	+
3.	***Amherstia nobilis*** **Wall.**	-	-	-	-	-	-	-	-	-	-	+	-	-	-	-	-	-	-
4.	***Bauhinia acuminata*** **L.**	+	+	-	-	+	-	-	-	-	-	+	-	-	-	-	-	-	-
5.	***Bauhinia corymbosa*** **Roxb.**	-	-	-	-	-	-	-	-	-	-	+	-	-	-	-	-	-	-
6.	***Bauhinia diphylla*** **Buch-Ham.**	-	+	-	-	-	+	-	-	-	-	+	-	-	-	-	-	-	-
7.	***Bauhinia divaricata*** **L.**	-	+	-	+	-	+	-	-	-	-	+	-	-	-	-	-	-	-
8.	***Bauhinia foveolata*** **Dalz.** (=*Bauhinia lawii* Baker., *Piliostigma foveolatum* (Dalz.) Thoth.)	-	+	-	-	-	-	-	-	-	+	-	-	-	-	-	-	-	-
9.	***Bauhinia galpinii*** **N.E. Brown**	-	-	-	-	-	-	-	-	-	-	+	-	-	-	-	-	-	-
10.	***Bauhinia malabarica*** **Roxb.**	+	+	-	+	-	-	-	+	-	+	-	-	-	-	-	-	-	-

Sl. No.	*Botanical Name*	1	2	3	4	5	6	7	8	9	10	11	12	13	14	15	16	17	18
11.	***Bauhinia monandra* Kurz**	-	+	-	+	-	-	-	-	-	-	+	-	-	-	-	-	-	-
12.	***Bauhinia parvifolia* Tejism. & Binn.**	-	+	-	-	-	+	-	-	-	-	-	-	-	-	-	-	-	-
13.	***Bauhinia phoenicea* Wt. & Arn.**	-	+	-	-	-	-	-	-	-	+	-	-	-	-	-	-	-	-
14.	***Bauhinia purpurea* L.** (=*Bauhinia triandra* Roxb., *Phanera purpurea* (L.) Benth.)	+	+	+	+	+	+	+	+	+	+	+	-	-	-	-	-	+	+
15.	***Bauhinia racemosa* Lam.** (=*Piliostigma racemosa* (Lam.) Benth.)	+	+	+	+	+	+	+	-	-	+	+	-	-	-	-	-	-	-
16.	***Bauhinia rufescens* Lam.**	-	+	-	-	-	+	-	+	-	+	+	-	-	-	-	-	-	-
17.	***Bauhinia scandens* L.** (=*Bauhinia anguina* Roxb.)	-	-	-	-	-	+	-	-	-	+	+	-	-	-	-	-	-	-
18.	***Bauhinia semla* Wunderlin** (=*Bauhinia retusa* Roxb.)	+	+	-	-	-	-	+	-	-	+	-	-	-	-	-	-	-	-
19.	***Bauhinia tomentosa* L.** (=*Bauhinia pubescens* DC.)	+	+	-	-	+	+	-	-	+	+	+	-	-	-	-	-	-	-
20.	***Bauhinia vahlii* Wt. & Arn.** (= *Bauhinia racemosa* Vahl., *Phanera vahlii* (Wt. & Arn.) Benth.)	+	+	-	+	-	+	+	+	-	+	+	-	-	-	-	-	-	-
21.	***Bauhinia variegata* L.** (=*Bauhinia variegata* L., var. *chinensis* DC., *Phanera variegata* (L.) Benth.)	+	+	-	+	+	+	+	+	+	+	+	-	-	-	-	-	-	-
22.	***Bauhinia ungulata* L.** (=*Bauhinia macrostachya* Benth.)	-	-	-	-	-	+	-	-	-	-	-	-	-	-	-	-	-	-
23.	***Bauhinia wallichii* J.F. Macbr.**	-	+	-	-	-	+	-	-	-	-	-	-	-	-	-	-	-	-
24.	***Brownea ariza* Benth.** (= *Brownea princeps* Linden)	-	-	-	-	-	-	-	-	-	-	+	-	-	-	-	-	-	-
25.	***Brownea coccinea* Jacq.**	-	+	-	-	-	-	-	-	-	-	+	-	-	-	-	-	-	-
26.	***Brownea grandiceps* Jacq.**	-	-	-	-	-	-	-	-	-	-	+	-	-	-	-	-	-	-
27.	***Caesalpinia bonduc* (L.) Roxb.** (= *Caesalpinia crista sensu auct.*, *Caesalpinia bonducella* (L.) Fleming)	-	+	-	-	+	-	-	-	-	-	+	-	-	-	-	-	-	+
28.	***Caesalpinia cacalaco* Humb. & Bonpl.**	-	+	-	-	-	-	-	-	-	-	+	-	-	-	-	-	-	-
29.	***Caesalpinia coriaria* (Jacq.) Willd.** (= *Caesalpinia thomaea* Spreng., *Libidibia coriaria* (Jacq.) Schltdl.)	-	+	+	-	-	-	-	+	+	-	+	-	-	-	-	-	-	-
30.	***Caesalpinia crista* L.** (=*Caesalpinia chinensis* Roxb., *Caesalpinia paniculata* (Lam.) Roxb.)	-	+	-	+	+	+	-	-	-	-	+	-	-	-	-	-	-	+
31.	***Caesalpinia cucullata* Roxb.** (=*Mezoneuron cucullatum* (Roxb.) Wt. & Arn.)	-	+	-	-	+	-	-	+	-	-	-	-	-	+	-	-	-	-
32.	***Caesalpinia decapetala* (Roth) Alston**	-	+	-	-	-	-	-	+	-	-	+	-	-	-	-	-	-	-

Sl. No.	Botanical Name	1	2	3	4	5	6	7	8	9	10	11	12	13	14	15	16	17	18
33.	***Caesalpinia digyna* Rottler**	-	+	-	+	+	-	-	+	-	-	+	-	-	-	-	-	-	-
34.	***Caesalpinia enneaphylla* Roxb.**	-	+	-	-	+	-	-	+	-	-	+	-	-	+	-	-	-	-
35.	***Caesalpinia ferrea* Mart.**	-	+	-	-	-	-	-	-	-	-	-	-	-	-	-	-	-	-
36.	***Caesalpinia gilliesii* (Hook.) Dietr.**	-	-	-	-	-	-	-	-	-	-	+	-	-	-	-	-	-	-
37.	***Caesalpinia globulorum* Bakh. f. & Royen** (=*Caesalpinia major* (Medik.) Dandy & Exell.)	-	+	-	-	-	-	-	-	-	-	-	-	-	-	-	-	-	-
38.	***Caesalpinia jayabo* M. Gomez**	-	+	-	-	-	-	-	-	-	-	-	-	-	-	-	-	-	-
39.	***Caesalpinia mimosoides* Lam.**	-	+	-	-	-	-	-	-	-	-	-	-	+	-	-	-	-	-
40.	***Caesalpinia pulcherrima* (L.) Sw.,** (=*Poinciana pulcherrima* L.)	+	+	-	-	-	-	-	+	+	-	+	-	-	-	-	-	-	+
41.	***Caesalpinia sappan* L.**	-	+	-	-	+	-	-	+	+	+	+	-	-	-	-	-	-	-
42.	***Caesalpinia tortuosa* Roxb.**	-	+	-	-	-	-	-	-	-	-	-	-	-	-	-	-	-	-
43.	***Cassia artemisioides* DC.**	-	-	-	-	-	-	-	-	-	-	+	-	-	-	-	-	-	-
44.	***Cassia eremophila* Vogel**	-	-	-	-	-	-	-	-	-	-	+	-	-	-	-	-	-	-
45.	***Cassia ferruginea* Schrad.**	-	-	-	-	-	-	-	-	-	-	+	-	-	-	-	-	-	-
46.	***Cassia fistula* L.**	+	+	-	-	-	-	+	+	+	+	+	-	-	-	-	-	-	-
47.	***Cassia grandis* L.f.**	+	+	-	-	-	-	-	-	-	-	+	-	-	-	-	-	-	-
48.	***Cassia javanica* L.**	-	+	-	-	-	-	-	-	-	-	+	-	-	-	-	-	+	-
49.	***Cassia javanica* L., ssp. *nodosa* (Roxb.) K. Larsen & S.S. Larsen** (=*Cassia nodosa* Roxb.)	+	-	+	-	-	-	-	-	-	+	+	-	-	-	-	-	-	-
50.	***Cassia obovata* Colla.**	-	+	-	-	-	-	-	-	-	-	-	-	-	-	-	-	-	-
51.	***Cassia renigera* Benth.**	-	-	-	-	-	-	-	-	-	-	+	-	-	-	-	-	-	-
52.	***Cassia roxburghii* DC.**	-	+	-	-	-	-	-	-	-	+	+	-	-	-	-	-	-	-
53.	***Ceratonia siliqua* L.**	+	+	-	+	+	-	+	+	+	+	+	-	+	-	+	-	+	-
54.	***Chamaecrista absus* (L.) H.S. Irwin & Barneby** (=*Cassia absus* L.)	-	+	-	-	+	-	-	-	-	-	-	-	-	-	-	-	-	+
55.	***Chamaecrista mimosoides* (L.) E. Greene** (=*Cassia mimosoides* L.)	+	+	+	-	-	-	-	-	-	-	-	-	+	-	-	-	-	-
56.	***Chamaecrista nigricans* (Vahl) E. Greene** (=*Cassia nigricans* Vahl)	-	+	-	-	-	-	-	-	-	-	-	-	+	-	-	-	-	-
57.	***Chamaecrista pumila* (Lam.) K. Larsen** (=*Cassia pumila* Lam., *Senna prostrata* Roxb.)	-	+	+	-	-	-	-	-	-	-	-	+	-	-	-	-	-	-
58.	***Colvillea racemosa* Bojer**	-	-	-	-	-	-	-	-	-	-	+	-	-	-	-	-	-	-
59.	***Copaifera gorskiana* Benth.**	-	-	-	-	-	-	-	-	-	-	+	-	-	-	-	-	-	-
60.	***Copaifera mopane* (Benth.) Léonard**	-	+	-	-	-	-	-	-	-	-	-	-	-	-	-	-	-	-
61.	***Copaifera officinalis* L.**	-	-	-	-	-	-	+	-	-	-	-	-	-	-	-	-	-	-

Sl. No.	*Botanical Name*	1	2	3	4	5	6	7	8	9	10	11	12	13	14	15	16	17	18
62.	***Cynometra cauliflora* L.**	+	+	-	-	+	-	-	-	-	-	-	-	-	-	-	-	-	-
63. 64.	***Cynometra iripa* Kostel** ***Cynometra ramiflora* L.**	- -	+ +	- -	- -	- +	- -	- -	- -	- +	- +	- -	- -	- -	- -	- -	- -	- -	- -
65.	***Delonix elata* (L.) Gamble** (=*Poinciana elata* L.)	-	+	+	-	-	-	-	-	-	+	+	-	-	-	-	-	+	-
66.	***Delonix regia* (Hook.) Raf.** (=*Poinciana regia* Hook.)	+	+	-	-	-	-	+	-	-	+	+	-	-	-	-	-	+	-
67.	***Dialium guianense* (Aubl.) Sandwith**	+	+	-	-	-	-	-	-	-	+	-	-	-	-	-	-	-	-
68.	***Dialium indicum* L.**	+	-	-	-	-	-	-	-	-	-	-	-	-	-	-	-	-	-
69.	***Dialium travancoricum* Bourd.**	-	-	-	+	-	-	-	-	-	+	-	-	-	-	-	-	-	-
70.	***Gleditsia assamica* Bor.**	-	-	-	-	-	-	-	-	-	+	-	-	-	-	-	-	-	-
71.	***Gleditsia sinensis* Lam.**	-	+	-	-	-	-	-	-	-	-	-	-	-	+	-	-	-	-
72.	***Gleditsia aquatica* Marshall** (= *Gleditsia triacantha* Gaertn., *Gleditsia inermis* Mill.)	+	+	-	+	+	-	-	+	-	+	+	-	-	+	-	-	-	-
73.	***Gymnocladus assamicus* P.C.Kanjilal**	-	-	-	-	-	-	-	-	-	-	-	-	-	-	-	+	-	-
74.	***Gymnocladus chinensis* Baill.**	-	+	-	-	-	-	-	-	-	-	-	-	-	-	-	-	-	-
75.	***Gymnocladus dioica* (L.) K.Koch**	+	-	-	-	-	-	-	-	-	+	-	-	+	-	-	-	-	+
76.	***Haematoxylum campechianum* L.** (=*Haematoxylon campechianum* L.)	+	+	-	-	-	-	-	-	+	+	+	-	-	-	-	-	-	-
77.	***Hardwickia binata* Roxb.**	-	+	+	+	+	+	+	+	+	+	+	-	-	-	+	-	-	-
78.	***Humboldtia bourdillonii* Prain**	+	-	-	-	-	-	-	-	-	-	-	-	-	-	-	-	-	-
79.	***Humboldtia brunonis* Wall.**	-	+	-	-	-	-	-	-	-	-	-	-	-	-	-	-	-	-
80.	***Humboldtia decurrens* Oliv.**	-	-	-	-	-	-	-	-	-	+	-	-	-	-	-	-	-	-
81.	***Humboldtia laurifolia* Vahl.**	-	+	-	-	-	-	-	-	-	-	-	-	-	-	-	-	-	-
82.	***Humboldtia vahliana* Wt.**	-	-	-	-	-	-	-	-	-	+	-	-	-	-	-	-	-	-
83.	***Hymenaea courbaril* L.** (=*Hymenaea retusa* Hayne., *Inga megacarpa* M.E. Jones)	+	+	-	-	-	-	+	-	-	-	-	-	+	-	-	-	-	-
84.	***Hymenaea verrucosa* Gaertn.**	-	+	-	-	-	-	+	-	-	+	-	-	-	-	-	-	-	-
85.	***Intsia bijuga* (Colebr.) O. Kuntze**	+	+	-	-	-	-	-	+	-	+	-	-	-	-	-	-	-	-
86.	***Intsia hookeri* Prain**	-	-	-	-	-	-	-	-	+	+	-	-	-	-	-	-	-	-
87.	***Intsia palembanica* Miq.**	-	-	-	-	-	-	-	-	+	+	-	-	-	-	-	-	-	-
88.	***Kingiodendron pinnatum* (DC.) Harms** (=*Hardwickia pinnata* Roxb.)	-	+	-	-	+	-	+	-	-	+	-	-	-	-	-	+	-	-
89.	***Lysidice rhodostegia* Hance**	+	-	-	-	-	-	-	-	-	+	+	-	-	-	-	-	-	-
90.	***Maniltoa polyandra* (Roxb.) Harms** (=*Cynometra polyandra* Roxb.)	-	+	-	-	+	-	-	-	-	+	+	-	-	-	-	-	-	-
91.	***Moullava spicata* (Dalz.) Nicolson** (=*Wagatea spicata* (Dalz.) Wight)	-	+	-	-	+	-	-	+	-	-	-	-	-	-	-	-	-	-
92.	***Parkinsonia aculeata* L.**	+	+	-	+	+	+	-	-	-	-	+	+	-	-	+	-	-	-
93.	***Peltophorum africanum* Sond.**	-	-	-	-	-	-	-	-	-	-	+	-	-	-	-	-	-	-

Sl. No.	*Botanical Name*	1	2	3	4	5	6	7	8	9	10	11	12	13	14	15	16	17	18
94.	***Peltophorum dasyrrhachis* (Miq.) Kurz**	-	+	-	-	-	-	-	-	-	-	-	-	-	-	-	-	-	-
95.	***Peltophorum linnaei* Griseb.**	-	-	-	-	-	-	-	-	-	-	+	-	-	-	-	-	-	-
96.	***Peltophorum pterocarpum* (DC.) K. Heyne** (=*Peltophorum ferrugineum* (Decne.) Benth.)	-	+	-	+	-	-	-	+	+	+	+	+	-	+	-	-	-	-
97.	***Pterolobium hexapetalum* (Roth) Sant. & Wagh**	-	+	-	-	-	-	-	-	-	-	-	-	-	-	-	-	-	-
98.	***Saraca asoca* (Roxb.) Willd.** (=*Saraca indica sensu* Bedd.)	+	+	+	-	-	-	-	-	-	+	+	-	-	-	-	-	-	-
99.	***Saraca declinata* Miq.** (=*Saraca cauliflora* Baker)	+	-	-	-	-	-	-	-	-	+	+	-	-	-	-	-	-	-
100.	***Schizolobium parahyba* (Vell.) Blake** (=*Cassia parahyba* Vell.)	-	-	-	-	-	-	-	-	-	+	+	-	-	-	+	-	-	-
101.	***Schotia brachypetala* Sonder**	+	+	-	-	-	-	-	+	-	+	+	-	-	-	-	-	-	-
102.	***Senna alata* (L.) Roxb.** (=*Cassia alata* L.)	+	+	-	-	-	-	-	+	-	-	+	-	+	-	-	-	-	+
103.	***Senna alexandrina* Miller** (=*Cassia senna* L.,) *Cassia angustifolia* Vahl., *Senna angustifolia* (Vahl.) Batka)	+	+	-	-	-	-	-	-	+	-	-	-	-	-	-	-	-	-
104.	***Senna auriculata* (L.) Roxb.** (=*Cassia auriculata* L.)	+	+	+	+	-	+	+	+	-	-	+	-	+	-	-	-	-	+
105.	***Senna bicapsularis* (L.) Roxb.** (=*Cassia bicapsularis* L.)	+	-	-	-	-	-	-	-	-	-	+	-	-	-	-	-	-	-
106.	***Senna corymbosa* (Lam.) H.S. Irwin & Barneby** (=*Cassia corymbosa* Lam.)	-	-	-	-	-	-	-	-	-	-	+	-	-	-	-	-	-	-
107.	***Senna didymobotrya* (Fresen.) H.S. Irwin & Barneby)** (=*Cassia didymobotrya* Fresen)	-	-	+	-	-	-	-	-	-	-	+	-	-	-	-	-	-	-
108.	***Senna divaricata* (Nees & Blume) Lock.**	-	-	-	-	-	-	-	-	-	-	+	-	-	-	-	-	-	-
	(=*Cassia divaricata* Nees & Blume)																		
109.	***Senna floribunda* (Cav.) H.S. Irwin & Barneby** (=*Cassia floribunda* Cav., *Cassia laevigata* Willd.)	+	+	-	-	-	-	-	-	-	-	+	-	+	-	-	-	-	-
110.	***Senna hirsuta* (L.) H.S. Irwin & Barneby** (=*Cassia hirsuta* L.)	+	+	-	-	-	-	-	-	-	-	+	+	-	-	-	-	-	-
111.	***Senna italica* Mill.** (=*Cassia italica* (Mill.) Spreng., *Cassia obtusa* Roxb.)	-	+	-	-	-	-	-	-	-	-	-	-	-	-	-	-	-	-
112.	***Senna marilandica* (L.) Link.** (=*Cassia marilandica* L.)	-	+	-	-	-	-	-	-	-	+	-	-	-	-	-	-	-	-

Sl. No.	*Botanical Name*	1	2	3	4	5	6	7	8	9	10	11	12	13	14	15	16	17	18
113.	***Senna multiglandulosa* (Jacq.) H.S. Irwin & Barneby** (=*Cassia tomentosa* L.f.)	-	+	-	-	-	-	-	-	-	-	-	-	-	-	-	-	-	-
114.	***Senna multijuga* (Rich.) H. S. Irwin & Barneby** (=*Cassia multijuga* Rich.)	-	+	-	-	-	-	-	-	-	-	+	-	-	-	-	-	-	-
115.	***Senna obtusifolia* (L.) H.S. Irwin & Barneby** (=*Cassia obtusifolia* L.)	+	+	-	-	-	-	-	-	+	-	-	-	+	-	-	-	-	-
116.	***Senna occidentalis* (L.) Link** (=*Cassia occidentalis* L.)	+	+	+	-	+	-	-	-	-	-	-	-	+	-	-	-	-	+
117.	***Senna odorata* (R. Morris) Randall** (=*Cassia australis sensu* Sims., *Cassia odorata* R. Morris)	-	-	-	-	-	-	-	-	-	-	+	-	-	-	-	-	-	-
118.	***Senna podocarpa* (Guill. & Perr.) Lock** (=*Cassia podocarpa* Guill. & Perr.)	-	-	-	-	-	-	-	-	-	-	+	-	-	-	-	-	-	-
119.	***Senna siamea* (Lam.) H.S. Irwin & Barneby** (=*Cassia siamea* Lam.)	+	+	+	+	-	-	-	+	+	+	+	-	+	+	+	-	-	-
120.	***Senna spectabilis* (DC.) H.S. Irwin & Barneby** (=*Cassia spectabilis* DC.)	-	+	-	-	-	-	-	-	-	-	+	-	-	-	-	-	-	-
121.	***Senna sophera* (L.) Roxb.** (*Cassia sophera* L.)	+	+	-	+	+	-	-	-	-	-	+	-	-	-	-	-	-	-
122.	***Senna timoriensis* (DC.) H.S. Irwin & Barneby** (=*Cassia montana* Naves & Villar, *Cassia timorensis* DC.)	-	-	-	-	-	-	-	-	-	-	-	-	-	-	-	-	-	-
123.	***Senna tora* (L.) Roxb.** (=*Cassia tora* L.)	+	+	+	+	+	-	-	+	+	-	-	-	+	-	+	+	-	-
124.	***Senna uniflora* (Mill.) H.S. Irwin & Barneby** (=*Cassia uniflora* Mill.)	+	+	-	-	-	-	-	-	-	-	-	-	+	-	-	-	-	-
125.	***Tamarindus indica* L.**	+	+	+	+	+	+	+	+	+	+	+	+	+	-	-	-	-	-

Table 3.2
Economic Potential of Legumes in India: Fabaceae

Eighteen known uses, identified as important for Legumes, are indicated for each species by '+'; and '-' indicates that particular use is not known for that species. The correct binomials are used and the important synonyms, if any, are given parenthesis preceded by '=', for each species. A wider synonymy is given in the information profiles for those species for which profiles were compiled. The genera and species are in alphabetical order.

Sl. No.	Botanical Name	Edible	Medicinal	Manure	Fodder/Forage	Oil	Fibre	Gum/Resin	Tannin	Dye	Timber	Ornamental/Avenue/Hedge/ Shade/Support	Cover/Soil Binder/Windbreak	Beverage	Lac insect host	Paper industry	Soap substitute	Insecticide/Pesticide	Molluscicide/piscicide
		1	2	3	4	5	6	7	8	9	10	11	12	13	14	15	16	17	18
1.	***Abrus precatorius*** **L.**	+	+	-	-	+	+	-	-	-	-	-	-	+	-	-	-	-	+
2.	***Abrus pulchellus*** **Thw.** (=*Abrus fruticulosus non* Wt. & Arn.)	+	+	-	-	-	+	-	-	-	-	-	-	+	-	-	-	-	+
3.	***Aeschynomene americana*** **L.**	-	-	+	+	-	-	-	-	-	-	-	+	-	-	-	-	-	-
4.	***Aeschynomene aspera*** **L.**	+	+	-	+	-	+	-	-	-	-	-	-	-	-	+	-	-	-
5.	***Aeschynomene indica*** **L.**	-	+	+	+	-	-	-	-	-	-	-	-	-	-	-	-	-	-
6.	***Aeschynomene uniflora*** **E. Mey.**	-	-	-	-	-	-	-	-	-	+	-	-	-	-	-	-	-	-
7.	***Alhagi maurorum*** **Medik. & Boiss.** (=*Alhagi pseudoalhagi* Desv.)	+	+	-	+	-	+	-	+	-	-	-	-	-	-	-	-	-	-
8.	***Alysicarpus belgaumensis*** **Wt.**	-	-	-	+	-	-	-	-	-	-	-	-	-	-	-	-	-	-
9.	***Alysicarpus bupleurifolius*** **(L.) DC.**	-	+	-	-	-	-	-	-	-	-	-	-	-	-	-	-	-	-
10.	***Alysicarpus glumaceus*** **(Vahl.) DC.**	+	+	-	+	-	-	-	-	-	-	-	+	-	-	-	-	-	-
11.	***Alysicarpus longifolius*** **(Spreng.)** ***Wt. & Arn.***	-	+	-	+	-	-	-	-	-	-	-	-	-	-	-	-	-	-
12.	***Alysicarpus monilifer*** **(L.) DC.**	+	+-	-	+	-	-	-	-	-	-	-	-	-	-	-	-	-	+
13.	***Alysicarpus ovalifolius*** **(Schumach.)** ***J. Leon.***	-	-	-	+	-	-	-	-	-	-	-	-	-	-	-	-	-	-
14.	***Alysicarpus pubescens*** **Law.**	-	+	-	+	-	-	-	-	-	-	-	-	-	-	-	-	-	-
15.	***Alysicarpus scariosus*** **(Spreng.) Thw.** (=*Hedysarum scariosum* Rotter ex Spreng.)	-	+	-	+	-	-	-	-	-	-	-	-	-	-	-	-	-	-
16.	***Alysicarpus vaginalis*** **(L.) DC.**	+	+	+	+	-	-	+	-	-	-	-	+	-	-	-	-	-	-
17.	***Amorpha fruticosa*** **L.**	-	-	-	-	-	-	-	-	-	-	-	-	-	-	-	-	+	-
18.	***Amphicarpaea bracteata*** **(L.) Fernald**	+	+	+	-	-	-	-	-	-	-	-	-	-	-	-	-	-	-
19.	***Andira inermis*** **(Wt.) DC.**	-	+	-	-	-	-	-	-	-	+	+	+	-	-	-	-	+	+
20.	***Apios americana*** **Medik.**	+	+	-	-	-	-	-	-	-	-	-	-	-	-	-	-	-	-
21.	***Arachis hypogaea*** **L.**	+	+	+	+	+	-	-	-	-	-	-	+	+	-	-	-	-	-
22.	***Astracantha gummifera*** **(Labill.) Podl.** (=*Astragalus gummifer* Labill.)	+	+	-	+	-	-	+	-	-	-	-	-	-	-	-	-	+	-

Sl. No.	*Botanical Name*	1	2	3	4	5	6	7	8	9	10	11	12	13	14	15	16	17	18
23.	***Astragalus aegacanthoides* Parker**	-	+	-	-	-	-	-	-	-	-	-	-	-	-	-	-	-	-
24.	***Astragalus candolleanus* Benth.**	-	+	-	+	-	-	-	-	-	-	-	-	-	-	-	-	-	-
25.	***Astragalus chlorostachys* Lindl.**	-	+	-	-	-	-	-	-	-	-	-	-	-	-	-	-	-	-
26.	***Astragalus christianus* L.**	+	-	-	-	-	-	-	-	-	-	-	-	-	-	-	-	-	-
27.	***Astragalus corrugatus* Bertol.**	-	-	-	+	-	-	-	-	-	-	-	-	-	-	-	-	-	+
28.	***Astragalus fatimensis* Chiov.**	+	-	-	-	-	-	+	-	-	-	-	-	-	-	-	-	-	-
29.	***Astragalus graveolens* Benth.**	-	+	-	-	-	-	-	-	-	-	-	-	-	-	-	-	-	-
30.	***Astragalus hamosus* L.**	+	+	-	-	-	-	-	-	+	-	-	-	-	-	-	-	-	-
31.	***Astragalus heratensis* (Bunge) Podl.**	-	-	-	-	-	-	+	-	-	-	-	-	-	-	-	-	-	-
32.	***Astragalus himalayanus* Klotz.**	-	+	-	-	-	-	-	-	-	-	-	-	-	-	-	-	-	-
33.	***Astragalus multiceps* Benth.**	+	+	-	+	-	-	-	-	-	-	-	-	-	-	-	-	-	-
34.	***Astragalus sarcocolla* Dymock**	-	+	-	-	-	-	+	-	-	-	-	-	-	-	-	-	-	-
35.	***Astragalus strobiliferus* (Benth.) Podl.**	+	+	-	-	-	-	+	-	-	-	-	-	-	-	-	-	-	-
36.	***Astragalus tribuloides* Delile**	-	+	-	-	-	-	-	-	-	-	-	-	-	-	-	-	-	-
37.	***Astragalus verus* Olivier**	-	+	-	-	-	-	+	-	-	-	-	-	-	-	-	-	-	-
38.	***Astragalus zanskarensis* Bunge ssp., *zanskarensis*** (=*Astragalus oplites* Parker)	-	+	-	-	-	-	-	-	-	-	-	-	-	-	-	-	-	-
39.	***Baphia nitida* Lodd.**	+	+	-	+	-	-	-	-	+	+	+	-	-	-	-	-	-	-
40.	***Bolusanthus speciosus* (Bolus) Harms** (=*Lonchocarpus speciosus* Bolus)	-	+	-	-	-	-	-	-	-	+	+	-	-	-	-	-	-	-
41.	***Brya ebenus* (L.) DC.**	-	-	-	-	-	-	-	-	-	+	+	-	-	-	-	-	-	-
42.	***Butea monosperma* (Lam.) Taub.** (=*Butea frondosa* Roxb.)	+	+	+	+	+	+	+	+	+	+	+	-	-	+	+	-	-	+
43.	***Butea superba* Roxb.**	-	+	-	+	+	+	+	-	+	-	+	-	-	-	-	-	-	+
44.	***Cajanus albicans* (Wt. & Arn.) Maesen**	-	+	-	-	-	-	-	-	-	-	-	-	-	-	-	-	-	-
45.	***Cajanus cajan* (L.) Millsp.** (=*Cajanus indicus* Spreng.)	+	+	+	+	+	-	-	-	-	-	-	+	+	+	-	-	-	-
46.	***Cajanus crassus* (King) Maesen** (= *Atylosia crassa* King)	-	+	-	-	-	-	-	-	-	-	-	-	-	-	-	-	-	-
47.	***Cajanus goensis* Dalz.** (=*Atylosia barbata* (Benth.) Baker, *Atylosia goensis* (Dalz.) Dalz.)	+	+	-	-	-	-	-	-	-	-	-	-	-	-	-	-	-	-
48.	***Cajanus heynei* (Wt. & Arn.) Maesen** (=*Dunbaria heynei* Wt. & Arn.)	-	-	+	+	-	-	-	-	-	-	-	+	-	-	-	-	-	-
49.	***Cajanus lineatus* (Wt. & Arn.) Maesen** (=*Atylosia lineata* Wt. & Arn.)	+	+	-	-	-	-	-	-	-	-	-	-	-	-	-	-	-	-
50.	***Cajanus mollis* (Benth.) Maesen** (=*Atylosia mollis* Benth.)	-	+	-	-	-	-	-	-	-	-	-	-	-	-	-	-	-	-
51.	***Cajanus platycarpus* (Benth.) Maesen** (=*Atylosia platycarpa* Benth.)	-	+	-	-	-	-	-	-	-	-	-	-	-	-	-	-	-	-
52.	***Cajanus scarabaeoides* (L.) Thou.** (= *Dolichos scarabaeoides* L., *Atylosia scarabaeoides* (L.) Benth., *Rhynchosia scarabaeoides* (L.) DC.)	+	+	+	-	-	-	-	-	-	-	-	-	-	-	-	-	-	-

Sl. No.	*Botanical Name*	1	2	3	4	5	6	7	8	9	10	11	12	13	14	15	16	17	18
53.	***Cajanus trinervius* (DC.) Maesen** (=*Collea trinervia* DC.)	-	+	-	-	-	-	-	-	-	-	-	-	-	-	-	-	-	-
54.	***Callerya atropurpurea* (Wall.) Schot** (=*Adinobotrys atropurpureus* (Wall.) Dunn., *Millettia atropurpurea* (Wall.) Benth)	+	-	-	-	-	-	-	-	-	-	-	-	-	-	-	-	-	+
55.	***Calopogonium mucunoides* Desv.**	+	-	-	-	-	-	-	-	-	-	-	-	-	-	-	-	-	+
56.	***Calpurnia aurea* (Ait.) Benth.**	-	-	-	-	-	-	-	-	-	-	+	-	-	-	-	-	-	-
57.	***Calpurnia lasiogyne* E. Mey.**	-	-	-	-	-	-	-	-	-	-	-	+	-	-	-	-	-	-
58.	***Canavalia africana* Dunn**	+	+	-	-	-	-	-	-	-	-	-	-	-	-	-	-	-	+
59.	***Canavalia cathartica* Thou.**	+	+	-	-	-	-	-	-	-	-	-	-	-	-	-	-	-	-
60.	***Canavalia ensiformis* (L.) DC.** (=*Dolichos ensiformis* L.)	+	+	+	+	-	-	-	-	-	-	-	+	+	-	-	-	-	-
61.	***Canavalia gladiata* (Jacq.) DC.** (=*Dolichos gladiata* Jacq., *Canavalia ensiformis sensu auct.*)	+	+	+	+	-	-	-	-	-	-	+	+	-	-	-	-	-	-
62.	***Canavalia rosea* (Sw.) DC.** (=*Canavalia maritime* (Aubl.) Thou.)	-	+	-	-	-	-	-	-	-	-	-	+	-	-	-	-	-	-
63.	***Caragana ambigua* Stocks**	+	-	-	+	-	-	-	-	-	-	-	-	-	-	-	-	-	-
64.	***Caragana arborescens* Lam.**	+	-	-	-	-	-	-	-	-	-	-	-	-	-	-	-	-	-
65.	***Caragana brevifolia* Komarov**	-	+	-	-	-	-	-	-	-	-	-	-	-	-	-	-	-	-
66.	***Caragana brevispina* Benth.**	-	+	-	+	-	-	-	-	-	+	-	-	-	-	-	-	-	-
67.	***Caragana gerardiana* Benth.**	-	+	-	-	-	-	-	-	-	+	-	-	-	-	-	-	-	-
68.	***Caragana jubata* (Pallas) Poir.** (=*Robinia jubata* Pallas)	-	+	-	-	-	-	-	-	-	-	-	-	-	-	-	-	-	-
69.	***Caragana pygmaea* (L.) DC.**	-	-	-	+	-	-	-	-	-	-	-	-	-	-	-	-	-	-
70.	***Caragana sinica* (Buc'hoz) Rehder**	+	+	-	-	-	-	-	-	-	+	-	-	-	-	-	-	-	-
71.	***Caragana ulicina* Stocks**	-	-	-	+	-	-	-	-	-	-	-	-	-	-	-	-	-	-
72.	***Caragana versicolor* Benth.**	-	+	-	+	-	-	-	-	-	+	-	-	-	-	-	-	-	-
73.	***Castanospermum australe* A. Cunn. & C. Fraser**	+	+	-	-	-	-	-	-	-	+	+	-	-	-	-	-	+	-
74.	***Centrosema brasilianum* (L.) Benth.**	-	-	-	+	-	-	-	-	-	-	-	-	-	-	-	-	-	-
75.	***Centrosema pubescens* Benth.**	-	-	+	+	-	-	-	-	-	-	-	+	-	-	-	-	-	-
76.	***Centrosema virginianum* (L.) Benth.** **(=*Clitoria virginiana* L.)**	-	+	-	+	-	-	-	-	-	-	-	-	-	-	-	-	-	-
77.	***Cicer arietinum* L.**	+	+	+	+	-	-	-	-	+	-	-	-	+	-	-	-	-	-
78.	***Cicer microphyllum* Benth.**	+	-	-	+	-	-	-	-	-	-	-	-	-	-	-	-	-	-
79.	***Cicer songaricum* DC.**	-	+	-	-	-	-	-	-	-	-	-	-	-	-	-	-	-	-
80.	***Clianthus dampieri* Lindl.**	-	-	-	-	-	-	-	-	-	-	+	-	-	-	-	-	-	-
81.	***Clianthus puniceus* (G. Don) Lindl.**	-	-	-	-	-	-	-	-	-	-	+	-	-	-	-	-	-	-
82.	***Clitoria annua* Grah.** (=*Clitoria biflora* Dalz.)	+	-	-	-	-	-	-	-	-	-	-	-	-	-	-	-	-	-

Sl. No.	*Botanical Name*	1	2	3	4	5	6	7	8	9	10	11	12	13	14	15	16	17	18
83.	***Clitoria mariana* L.**	-	+	-	-	-	-	-	-	-	-	-	-	-	-	-	-	-	-
84.	***Clitoria ternatea* L.**	+	+	-	+	+	-	-	+	+	-	+	+	-	-	-	-	-	-
85.	***Codariocalyx gyroides* (Link) Hassk.** (=*Desmodium gyroides* (Link) DC.)	-	+	+	+	-	-	-	-	-	-	-	+	-	-	-	-	-	-
86.	***Codariocalyx motorius* (Houtt.) H. Ohashi** (= *Desmodium gyrans* (L.) DC., *Desmodium motorium* (Houtt.) Merr.)	-	+	+	+	-	-	-	-	-	-	-	+	-	-	-	-	-	-
87.	***Colutea arborescens* L.**	-	+	-	-	-	-	-	-	-	-	+	-	-	-	-	-	-	-
88.	***Colutea nepalensis* Sims**	-	+	-	-	-	-	-	-	-	-	+	-	-	-	-	-	-	-
89.	***Crotalaria acicularis* Benth.**	-	+	-	-	-	-	-	-	-	-	-	-	-	-	-	-	-	-
90.	***Crotalaria agatiflora* Schweinf.**	-	+	-	-	-	-	-	-	-	-	-	-	-	-	-	-	-	-
91.	***Crotalaria alata* D. Don.** (=*Crotalaria bialata* Roxb.)	-	+	+	+	-	-	-	-	-	-	+	-	-	-	-	-	+	-
92.	***Crotalaria albida* Roth**	-	+	-	-	-	-	-	-	-	-	-	-	-	-	-	-	-	-
93.	***Crotalaria anagyroides* Kunth**	-	-	+	+	-	-	-	-	-	-	+	+	-	-	-	-	-	-
94.	***Crotalaria assamica* Benth.** (=*Crotalaria sericea* Burm. f.)	-	+	-	-	-	+	-	-	-	-	-	-	-	-	-	-	-	-
95.	***Crotalaria berteroana* DC.**	-	+	-	-	-	-	-	-	-	-	-	-	-	-	-	-	-	-
96.	***Crotalaria bracteata* DC.**	-	+	-	-	-	-	-	-	-	-	-	-	-	-	-	-	-	-
97.	***Crotalaria breviflora* DC.**	-	+	-	-	-	-	-	-	-	-	-	-	-	-	-	-	-	-
98.	***Crotalaria burhia* Benth.**	-	+	-	+	-	+	-	-	-	-	-	-	-	-	-	-	-	-
99.	***Crotalaria calycina* Schrank.**	-	+	-	-	-	-	-	-	-	-	-	-	-	-	-	-	-	-
100.	***Crotalaria cytisoides* DC.**																		
	(=*Priotropis cytisoides* (DC.) Wt. & Arn.)	-	-	+	-	-	-	-	-	-	-	-	-	-	-	-	-	-	-
101.	***Crotalaria ferruginea* Benth.**	-	+	+	+	-	-	-	-	-	-	-	-	-	-	-	-	-	-
102.	***Crotalaria hebecarpa* (DC.) Rudd.** (=*Goniogyna hebecarpa* DC.)	-	+	-	-	-	-	-	-	-	-	-	-	-	-	-	-	-	-
103.	***Crotalaria incana* L.** (=*Crotalaria affinis* DC.)	-	-	-	+	-	-	-	-	-	-	-	-	-	-	-	-	-	+
104.	***Crotalaria juncea* L.** (=*Crotalaria sericea* Willd.)	+	+	+	+	-	+	+	-	-	-	-	-	-	-	+	-	-	-
105.	***Crotalaria laburnifolia* L.**	+	+	+	-	-	-	-	-	-	-	+	-	-	-	+	-	-	-
106.	***Crotalaria laevigata* Lam.**	-	+	-	-	-	-	-	-	-	-	-	-	-	-	-	-	-	-
107.	***Crotalaria linifolia* L.f.**	-	+	-	-	-	-	-	-	-	-	-	-	-	-	-	-	-	-
108.	***Crotalaria medicaginea* Lam.**	-	+	-	+	-	-	-	-	-	-	-	-	-	-	-	-	-	-
109.	***Crotalaria mysorensis* Roth**	-	+	-	-	-	-	-	-	-	-	-	-	-	-	-	-	-	-
110.	***Crotalaria nana* Burm. f.**	-	+	-	-	-	-	-	-	-	-	-	-	-	-	-	-	-	-
111.	***Crotalaria orixensis* Willd.**	-	+	-	-	-	-	-	-	-	-	-	-	-	-	-	-	-	-
112.	***Crotalaria pallida* Ait.** (=*Crotalaria mucronata* Desv., *Crotalaria striata* DC.)	-	+	+	+	-	+	+	-	+	-	-	+	+	-	-	-	-	-

Sl. No.	Botanical Name	1	2	3	4	5	6	7	8	9	10	11	12	13	14	15	16	17	18
113.	***Crotalaria prostrata* Willd.**	-	+	-	-	-	-	-	-	-	-	-	-	-	-	-	-	-	-
114.	***Crotalaria retusa* L.**	-	+	+	-	-	+	-	-	+	-	+	-	-	-	-	-	+	-
115.	***Crotalaria rubiginosa* Willd.**	-	+	-	-	-	-	-	-	-	-	-	-	-	-	-	-	-	-
116.	***Crotalaria saltiana* Andr.**	-	+	-	-	-	-	-	-	-	-	-	-	-	-	-	-	-	-
117.	***Crotalaria sessiflora* L.**	-	+	-	-	-	-	-	-	-	-	-	-	-	-	-	-	-	-
118.	***Crotalaria spectabilis* Roth** (=*Crotalaria sericea* Retz.)	-	+	+	+	-	+	-	-	-	-	+	+	-	-	+	-	-	-
119.	***Crotalaria trichotoma* Bojer** (=*Crotalaria usaramoensis* Bak. f.)	-	-	+	-	-	+	-	-	-	-	-	+	-	-	-	-	-	-
120.	***Crotalaria trifoliastrum* Willd.**	-	+	-	-	-	-	-	-	-	-	-	-	-	-	-	-	-	-
121.	***Crotalaria verrucosa* L.**	-	+	-	-	-	-	-	-	-	-	-	-	-	-	-	-	-	+
122.	***Cullen corylifolium* (L.) Medik.** (=*Psoralea corylifolia* L.)	-	+	+	+	+	-	+	-	-	-	-	-	-	-	-	-	-	-
123.	***Cullen plicatum* (Delile) Stirt.** (=*Cullen plicata* (Delile) Stirt., *Psoralea plicata* Delile)	-	-	-	+	-	-	-	-	+	-	-	-	-	-	-	-	-	-
124.	***Cyamopsis tetragonoloba* (L.) Taub.** (=*Cyamopsis psoralioides* (Lam.) DC.)	+	+	+	+	+	-	+	-	-	-	+	-	-	-	+	-	-	-
125.	***Cytisus proliferus* L. f.**	-	-	-	+	-	-	-	-	-	-	-	-	-	-	-	-	-	-
126.	***Cytisus scoparius* (L.) Link** (=*Sarothamnus scoparius* (L.) Koch)	+	+	-	-	+	+	-	-	+	-	-	+	+	-	+	-	-	+
127.	***Dalbergia assamica* Benth.** (=*Dalbergia lanceolaria auct non* L.f.)	-	-	-	-	-	-	-	-	-	+	+	-	-	-	-	-	-	-
128.	***Dalbergia congesta* Wt. & Arn.**	-	+	-	-	-	-	-	-	-	-	-	-	-	-	-	-	-	-
129.	***Dalbergia coromandeliana* Prain**	-	+	-	-	-	-	-	-	-	-	-	-	-	-	-	-	-	-
130.	***Dalbergia cultrata* Benth.**	-	+	-	-	+	-	+	-	-	+	-	-	-	-	-	-	-	-
131.	***Dalbergia horrida* (Dennst.) Mabberly** (=*Dalbergia multiflora* Heyne)	-	+	-	+	-	-	-	-	-	-	-	-	-	-	-	-	-	-
132.	***Dalbergia junghuhnii* Benth.**	-	-	-	-	+	-	-	-	-	-	-	-	-	-	-	-	-	-
133.	***Dalbergia lanceolaria* L.f.**	+	+	-	-	+	-	-	+	-	+	+	-	-	-	-	-	-	-
134.	***Dalbergia latifolia* Roxb.** (=*Dalbergia emarginata* Roxb.)	-	+	-	+	+	-	+	+	-	+	+	-	-	-	-	-	-	-
135.	***Dalbergia melanoxylon* Guill. & Perr.**	-	+	-	-	-	-	-	+	-	+	-	-	-	-	-	-	-	-
136.	***Dalbergia oliveri* Prain**	-	-	-	-	-	-	-	-	-	+	-	-	-	-	-	-	-	-
137.	***Dalbergia parviflora* Roxb.**	-	-	-	-	+	-	-	-	-	+	-	-	-	-	-	-	-	-
138.	***Dalbergia pinnata* (Lour.) Prain**	+	+	-	-	-	-	-	-	-	-	-	-	-	-	-	-	-	-
139.	***Dalbergia pseudo-sissoo* Miq.**	-	-	-	+	-	-	-	-	-	+	-	-	-	-	-	-	-	-
140.	***Dalbergia reniformis* Roxb.**	-	-	-	-	-	-	-	-	-	+	-	-	-	-	-	-	-	-
141.	***Dalbergia rubiginosa* Roxb.**	-	+	-	-	-	-	-	-	-	-	-	-	-	-	-	-	-	-
142.	***Dalbergia sissoides* Wt. & Arn.**	-	-	-	+	-	-	-	-	-	+	-	-	-	-	-	-	-	-
143.	***Dalbergia sissoo* DC.**	-	+	-	+	+	-	-	+	-	+	+	-	-	-	+	-	-	-
144.	***Dalbergia spinosa* Roxb.**	-	+	-	-	-	-	-	-	-	-	-	-	-	-	-	-	-	-
145.	***Dalbergia stipulacea* Roxb.** (=*Dalbergia ferruginea* Roxb.)	+	+	-	-	-	-	-	-	-	+	-	-	-	-	-	-	-	+

Sl. No.	*Botanical Name*	1	2	3	4	5	6	7	8	9	10	11	12	13	14	15	16	17	18
146.	***Dalbergia sympathetica* Nimmo**	-	+	-	-	-	-	-	-	-	-	-	-	-	-	-	-	-	-
147.	***Dalbergia volubilis* Roxb.**	-	+	-	+	-	-	-	-	-	-	-	-	-	-	-	-	-	-
148.	***Dalhousiea bracteata* (Roxb.) Benth.**	-	+	-	-	-	-	-	-	-	-	-	-	-	-	-	-	-	-
149.	***Dendrolobium triangulare* (Retz.) Schindl.** (=*Desmodium triangulare* (Retz.) Merr.)	+	+	+	+	-	-	-	-	-	-	-	+	-	-	-	-	-	-
150.	***Dendrolobium umbellatum* (L.) Benth.** (=*Desmodium umbellatum* (L.) DC.)	+	+	-	+	-	-	-	-	-	-	-	-	-	-	-	-	-	-
151.	***Derris cuneifolia* Benth.**	-	+	-	-	-	-	-	-	-	-	-	-	-	-	-	-	+	+
152.	***Derris elegans* Benth.**	-	+	-	-	-	-	-	-	-	-	-	-	-	-	-	-	+	+
153.	***Derris elliptica* (Wall.) Benth.** (=*Pongamia elliptica* Wall.)	+	+	-	-	-	-	-	-	-	-	-	-	-	-	-	-	+	+
154.	***Derris ferruginea* (Roxb.) Benth.**	-	-	-	-	-	-	-	-	-	-	-	-	-	-	-	-	+	+
155.	***Derris malaccensis* Prain**	-	-	+	-	-	-	-	-	-	-	-	-	-	-	-	-	+	+
156.	***Derris marginata* (Roxb.) Benth.** (=*Dalbergia marginata* Roxb.)	-	-	-	-	-	-	-	-	-	+	-	-	-	-	-	-	-	-
157.	***Derris robusta* (DC.) Benth.** (=*Dalbergia robusta* DC.)	-	+	-	+	-	-	-	-	+	+	-	-	-	-	-	-	+	-
158.	***Derris scandens* (Roxb.) Benth.** (=*Dalbergia scandens* Roxb.)	-	+	-	-	+	+	-	-	-	-	+	-	-	-	-	-	+	+
159.	***Derris trifoliata* Lour.**	-	+	-	+	-	+	-	-	-	-	-	-	-	-	-	-	+	+
160.	***Desmodium adscendens* (Swartz) DC.**	-	+	+	+	-	-	-	-	-	-	-	+	-	-	-	-	-	-
161.	***Desmodium alysicarpoides* Meeuwen**	-	+	-	-	-	-	-	-	-	-	-	-	-	-	-	-	-	-
162.	***Desmodium barbatum* (L.) Benth.**	-	+	-	+	-	-	-	-	-	-	-	-	-	-	-	-	-	-
163.	***Desmodium caudatum* (Thunb.) DC.**	-	+	-	-	-	-	-	-	-	-	-	-	-	-	-	-	-	-
164.	***Desmodium dichotomum* (Willd.) DC.**	-	+	+	+	-	-	-	-	-	-	-	-	-	-	-	-	-	-
165.	***Desmodium discolor* Vogel**	-	-	-	+	-	-	-	-	-	-	-	-	-	-	-	-	-	-
166.	***Desmodium elegans* DC.** (=*Desmodium tiliaefolium* G. Don)	-	+	-	+	-	+	-	-	-	+	-	-	-	-	+	-	-	-
167.	***Desmodium gangeticum* (L.) DC.**	-	+	+	+	+	+	-	-	-	-	-	+	-	-	+	-	-	-
168.	***Desmodium heterocarpon* (L.) DC.**	-	+	+	+	-	-	-	-	-	-	-	+	-	-	-	-	-	-
169.	***Desmodium heterophyllum* (Willd.) DC.** (=*Desmodium triflorum* (L.) DC. var., *majus* Wt. & Arn.)	-	+	+	+	-	-	-	-	-	-	-	+	-	-	-	-	-	-
170.	***Desmodium laxiflorum* DC.** (=*Desmodium diffusum* DC.)	-	+	+	+	-	-	-	-	-	-	-	+	-	-	-	-	-	-
171.	***Desmodium microphyllum* (Thunb.) DC.**	+	+	-	+	-	-	-	-	-	-	-	-	-	-	-	-	-	-
172.	***Desmodium multiflorum* DC.**	-	+	-	-	-	-	-	-	-	-	-	-	-	-	-	-	-	-
173.	***Desmodium oblongum* Benth.**	-	-	-	-	-	-	-	-	+	-	-	-	-	-	-	-	-	-
174.	***Desmodium oojeinense* (Roxb.) H. Ohashi** (=*Ougeinia oojeinensis* (Roxb.) Hochr.)	+	+	-	+	-	+	-	+	-	-	+	-	-	+	-	-	+	+

Sl. No.	*Botanical Name*	1	2	3	4	5	6	7	8	9	10	11	12	13	14	15	16	17	18
175.	***Desmodium podocarpum* Hook & Arn.**	+	+	-	-	-	-	-	-	-	-	-	-	-	-	-	-	-	-
176.	***Desmodium pulchellum* (L.) Benth.**	-	+	-	+	-	-	-	-	-	-	-	-	-	-	-	-	-	-
177.	***Desmodium renifolium* (L.) Schindl.**	-	+	-	-	-	-	-	-	-	-	-	-	-	-	-	-	-	-
178.	***Desmodium repandum* (Vahl) DC.**	+	+	+	-	-	-	-	-	-	-	-	+	-	-	-	-	-	-
179.	***Desmodium salicifolium* (Poir.) DC.** (=*Desmodium diversifolium* (Poir.) DC.	-	-	-	+	-	-	-	-	-	-	-	-	-	-	-	-	-	-
180.	***Desmodium sequax* Wall.**	-	+	-	-	-	-	-	-	-	-	-	-	-	-	-	-	-	-
181.	***Desmodium styracifolium* (Osb.) Merr.** (=*Desmodium capitatum* (Burm. f.) DC., *Desmodium celebicum* Schindl.)	-	+	+	-	-	-	-	-	-	-	-	+	-	-	-	-	-	-
182.	***Desmodium tortuosum* (Swartz) DC.**	-	+	+	+	-	-	-	-	-	-	-	-	-	-	-	-	-	-
183.	***Desmodium triflorum* (L.) DC.** (=*Desmodium triflorum* (L.) DC. var. *minus* Wight & Arn)	+	+	+	+	-	-	-	-	-	-	+	+	-	-	-	-	-	-
184.	***Desmodium velutinum* (Willd.) DC.** (=*Desmodium latifolium* (Roxb.) DC.)	-	+	+	+	-	+	-	-	-	-	-	+	-	-	+	-	-	-
185.	***Desmodium zonatum* Miq.**	-	+	-	-	-	-	-	-	-	-	-	-	-	-	-	-	-	-
186.	***Dioclea hexandra* (Ralph) Mabberly** (=*Mucuna hexandra* Ralph)	-	+	-	-	-	-	-	-	-	-	-	-	-	-	-	-	+	-
187.	***Dioclea javanica* Benth.**	-	+	-	-	-	-	-	-	-	-	-	-	-	-	-	-	+	-
188.	***Dolichos tenuicaulis* (Baker) Craib.**	-	+	+	-	-	-	-	-	-	-	-	-	-	-	-	-	-	-
189.	***Dolichos trilobus* L.**	+	+	+	+	-	-	-	-	-	-	-	-	-	-	-	-	-	-
190.	***Dumasia cordifolia* Baker**	+	-	-	+	-	-	-	-	-	-	-	-	-	-	-	-	-	-
191.	***Dysolobium pilosum* (Willd.) Marech.** (=*Vigna pilosa* (Willd.) Baker)	+	+	-	+	-	-	-	-	-	-	-	-	-	-	-	-	-	-
192.	***Endosamara racemosa* (Roxb.) Gees** (=*Millettia racemosa* (Roxb.) Benth.)	-	+	-	+	-	-	-	-	-	-	-	-	-	-	-	-	-	+
193.	***Eriosema chinense* Vogel**	+	+	-	-	-	-	-	-	-	-	-	-	-	-	-	-	-	-
194.	***Eriosema himalaicum* H. Ohashi**	+	+	-	-	-	-	-	-	-	-	-	-	-	-	-	-	-	-
195.	***Erythrina addisoniae* Hutch & Dalz.**	-	-	-	-	-	-	-	+	-	-	-	-	-	-	-	-	-	-
196.	***Erythrina arborescens* Roxb.**	-	+	-	-	+	-	-	-	-	+	+	-	-	-	-	-	-	-
197.	***Erythrina blakei* Parker**	-	-	-	-	-	-	-	-	-	-	+	-	-	-	-	-	-	-
198.	***Erythrina caffra* Thunb.**	-	+	-	-	-	-	-	-	-	-	+	-	-	-	-	-	-	-
199.	***Erythrina corallodendron* L.**	-	+	-	-	-	-	-	-	-	-	+	-	-	-	-	-	-	-
200.	***Erythrina crista-galli* L.**	-	+	-	-	+	-	-	-	-	-	+	-	-	-	-	-	-	-
201.	***Erythrina fusca* Lour.**	+	+	-	-	-	-	-	-	-	-	+	+	-	-	-	-	-	-
202.	***Erythrina humeana* Spreng.**	-	+	-	-	-	-	-	-	-	-	+	-	-	-	-	-	-	-
203.	***Erythrina poeppigiana* (Walp.) O.F. Cook**	-	-	-	-	-	-	-	-	-	-	+	-	-	-	-	-	-	-
204.	***Erythrina resupinata* Roxb.**	-	+	-	-	-	-	-	-	-	-	+	-	-	-	-	-	-	-
205.	***Erythrina speciosa* Andr.**	-	-	-	-	-	-	-	-	-	-	+	-	-	-	-	-	-	-
206.	***Erythrina stricta* Roxb.**	+	+	-	-	-	-	-	-	-	+	+	-	-	-	-	-	-	-
207.	***Erythrina suberosa* Roxb.** (=*Erythrina sublobata* Roxb.)	-	+	-	-	-	+	-	-	+	+	+	-	-	-	-	-	-	-

Sl. No.	*Botanical Name*	1	2	3	4	5	6	7	8	9	10	11	12	13	14	15	16	17	18
208.	***Erythrina subumbrans* (Hassk.) Merr.**	-	+	+	-	-	-	-	-	-	+	+	-	-	-	-	-	-	-
209.	***Erythrina variegata* L.** (=*Erythrina corallodendron* Lour., *Erythrina orientalis* (L.) Merr.)	+	+	+	+	+	+	+	+	+	+	+	-	-	-	+	-	-	+
210.	***Euchresta horsfieldii* (Lesch.) Benn.**	-	+	-	-	-	-	-	-	-	-	-	-	-	-	-	-	+	+
211.	***Euchresta japonica* Regel.**	+	-	-	-	-	-	-	-	-	-	-	-	-	-	-	-	-	-
212.	***Flemingia chappar* Benth.** (=*Moghania chappar* (Benth.) O. Kuntze)	+	+	-	-	-	-	-	-	-	-	-	-	-	-	-	-	-	-
213.	***Flemingia grahamiana* Wt. & Arn.**	-	+	-	-	+	-	+	-	+	-	-	-	-	-	-	-	-	-
214.	***Flemingia lineata* (L.) Ait.** (=*Moghania lineata* (L.) Kuntze.)	-	+	-	-	-	-	-	-	-	-	-	-	-	-	-	-	-	-
215.	***Flemingia macrophylla* (Willd.) Merr.** (=*Crotalaria macrophylla* Willd., *Flemingia semialata* Ait.), *Flemingia wallichii* Wt. & Arn.)	+	+	-	-	-	-	-	-	+	-	+	-	-	+	-	-	-	-
216.	***Flemingia procumbens* Roxb.**	+	+	-	-	-	-	-	-	-	-	-	-	-	-	-	-	-	-
217.	***Flemingia stricta* Roxb.**	-	+	-	-	-	-	-	-	-	-	-	-	-	-	-	-	-	-
218.	***Flemingia strobilifera* (L.) Ait.** (=*Moghania bracteata* (Roxb.) H.L. Li)	-	+	-	+	-	-	-	-	-	-	+	-	-	-	-	-	+	-
219.	***Flemingia tuberosa* Dalz.** (=*Moghania tuberosa* (Dalz.) O. Kuntze)	+	+	-	-	-	-	+	+	-	-	-	-	-	-	-	-	-	-
220.	***Flemingia wightiana* Wt. & Arn.**	-	+	-	-	-	-	-	-	-	-	-	-	-	-	-	-	-	-
221.	***Geissaspis cristata* Wt. & Arn.**	-	-	-	-	-	-	-	-	-	-	+	-	-	-	-	-	-	-
222.	***Genista monspessulana* (L.) L. John** (=*Cytisus monspessulanus* L., *Teline monspessulana* (L.) Koch)	-	+	-	-	-	-	-	-	-	-	-	-	-	-	-	-	-	-
223.	***Genista tinctoria* L.**	+	+	-	+	+	-	-	-	+	-	-	-	+	-	-	-	-	-
224.	***Gliricidia sepium* (Jacq.) Walp.** (=*Gliricidia maculata* (Kunth) Walp.)	+	+	+	+	+	-	-	-	-	+	+	+	-	-	-	-	+	+
225.	***Glycine javanica* L.**	-	-	+	-	-	-	-	-	-	-	-	-	-	-	-	-	-	-
226.	***Glycine max* (L.) Merr.** (=*Glycine soja sensu auct.)*	+	+	+	+	+	+	-	-	-	-	-	-	+	-	+	-	-	-
227.	***Glycyrrhiza glabra* L.**	+	+	-	-	-	-	-	-	+	-	-	-	-	-	-	-	+	-
228.	***Gueldenstaedtia multiflora* Bunge**	-	+	-	-	-	-	-	-	-	-	-	-	-	-	-	-	-	-
229.	***Hedysarum coronarium* L.**	-	-	+	+	-	-	-	-	-	-	+	-	-	-	-	-	-	-
230.	***Hymenaea verrucosa* Gaertn.** (=*Trachylobium hornemannianum* Hayne)	-	+	-	-	-	-	+	-	-	-	-	-	-	-	-	-	-	-
231.	***Indigastrum parviflorum* (Wt. & Arn.) Schrire** (=*Indigofera parviflora* Wt. & Arn.)	-	-	-	-	-	-	-	-	+	-	-	-	-	-	-	-	-	-
232.	***Indigofera angustifolia* L.**	-	+	-	-	-	-	-	-	+	-	-	-	-	-	-	-	-	-
233.	***Indigofera arborea* Roxb.**	-	+	-	-	-	-	-	-	-	-	-	-	-	-	-	-	-	-
234.	***Indigofera argentea* Burm. f.**	-	+	-	-	-	-	-	-	+	-	-	-	-	-	-	-	-	-
235.	***Indigofera arrecta* A. Rich.**	-	+	+	-	-	-	-	-	+	-	+	+	-	-	-	-	-	-

Sl. No.	Botanical Name	1	2	3	4	5	6	7	8	9	10	11	12	13	14	15	16	17	18
236.	***Indigofera aspalathoides* DC.** (=*Indigofera aspalathifolia* Roxb.)	-	+	-	-	-	+	-	-	-	-	-	-	-	-	-	-	-	-
237.	***Indigofera astragalina* DC.**	-	+	-	-	-	-	-	-	-	-	-	-	-	-	-	-	-	-
238.	***Indigofera atropurpurea* Horn.**	+	-	-	-	-	+	-	-	-	-	+	-	-	-	-	-	-	-
239.	***Indigofera australis* Willd.**	-	-	-	+	-	-	-	-	-	-	-	-	-	-	-	-	-	-
240.	***Indigofera cassioides* DC.** (=*Indigofera pulchella* Roxb.)	+	+	-	+	-	-	-	-	-	-	-	-	-	-	-	-	-	-
241.	***Indigofera coerulea* Roxb.**	+	+	+	+	-	-	-	-	+	-	+	-	-	-	-	-	-	-
242.	***Indigofera colutea* (Burm. F.) Merr.** (=*Galega colutea* Burm. f.)	-	+	-	-	-	-	-	-	-	-	-	-	-	-	-	-	-	-
243.	***Indigofera cordifolia* Roth**	+	+	+	+	-	-	-	-	+	-	-	-	-	-	-	-	-	-
244.	***Indigofera dosua* D. Don**	+	+	-	+	-	-	-	-	+	-	-	-	-	-	-	-	-	-
245.	***Indigofera frutescens* L.f.**	-	+	-	-	-	-	-	-	-	-	-	-	-	-	-	-	-	-
246.	***Indigofera galegoides* DC.**	-	-	-	+	+	-	-	-	-	-	-	+	-	-	-	-	-	+
247.	***Indigofera glabra* L.**	+	+	-	-	-	-	-	-	-	-	-	-	-	-	-	-	-	-
248.	***Indigofera glandulosa* Wendl.**	+	+	+	+	-	-	-	-	-	-	-	-	-	-	-	-	-	+
249.	***Indigofera glomerata* E. Mey.**	-	+	-	-	-	-	-	-	-	-	-	-	-	-	-	-	-	-
250.	***Indigofera hebepetala* Baker**	-	-	-	-	-	+	-	-	-	-	-	-	-	-	-	-	-	-
251.	***Indigofera hendecaphylla* Jacq.** (=*Indigofera endecaphylla* Jacq.)	-	+	+	+	-	-	-	-	-	-	-	+	-	-	-	-	-	+
252.	***Indigofera heterantha* Brandis**	+	+	-	+	-	+	-	-	-	-	+	-	-	-	-	-	-	-
253.	***Indigofera hirsuta* L.**	-	+	+	+	-	-	-	-	+	-	-	+	-	-	-	-	-	-
254.	***Indigofera hochstetteri* Baker** (=*Indigofera anabaptista* Baker)	-	-	+	-	-	-	-	-	-	-	-	-	-	-	-	-	-	-
255.	***Indigofera linifolia* (L.f.) Retz.**	+	+	-	+	-	-	-	+	-	-	-	-	-	-	-	-	-	-
256.	***Indigofera linnaei* Ali**	+	+	+	+	-	-	-	-	-	-	-	-	-	-	-	-	-	+
257.	***Indigofera longiracemosa* Baill.**	-	-	-	-	-	-	-	-	+	-	-	-	-	-	-	-	-	-
258.	***Indigofera mysorensis* DC.**	-	+	-	-	-	-	-	-	-	-	-	-	-	-	-	-	-	-
259.	***Indigofera nummularifolia* (L.) Alston** (=*Indigofera echinata* Willd.)	+	+	-	+	-	-	-	-	-	-	-	-	-	-	-	-	-	-
260.	***Indigofera oblongifolia* Forssk.**	-	+	+	+	-	-	-	-	+	-	+	-	-	-	-	-	-	-
261.	***Indigofera prostrata* Willd.**	-	+	-	+	-	-	-	-	-	-	-	-	-	-	-	-	-	-
262.	***Indigofera spicata* Forssk.**	-	-	+	-	-	-	-	-	+	-	-	+	-	-	-	-	-	-
263.	***Indigofera suffruticosa* Mill.**	-	+	+	-	-	-	-	-	-	-	-	+	-	-	-	-	-	-
264.	***Indigofera tinctoria* L.** (=*Indigofera sumatra* Gaertn.)	-	+	+	-	-	+	-	-	+	-	+	-	-	-	-	-	-	-
265.	***Indigofera trifoliata* L.** (=*Indigofera glandulifera* Hayata)	-	+	+	+	-	-	-	-	-	-	+	-	-	-	-	-	-	-
266.	***Indigofera trita* L.f.** (=*Indigofera carinata* De Wild.)	-	+	+	+	+	-	-	-	-	-	+	-	-	-	-	-	-	-
267.	***Indigofera uniflora* Roxb.**	-	+	-	+	-	-	-	-	-	-	-	-	-	-	-	-	-	-
268.	***Indigofera wightii* Wt. & Arn.**	-	+	-	-	-	-	-	-	-	-	-	-	-	-	-	-	-	-
269.	***Indigofera zollingeriana* Miq.** (=*Indigofera teysmannii* Miq.)	-	-	+	-	-	-	-	-	-	-	-	+	-	-	-	-	-	+

Sl. No.	Botanical Name	1	2	3	4	5	6	7	8	9	10	11	12	13	14	15	16	17	18
270.	***Indopiptadenia oudhensis* (Brandis) Brenan**	-	-	-	+	-	-	-	-	-	-	-	-	-	-	-	-	-	-
271.	***Kummerovia stipulacea* (Maxim) Makino** (=*Lespedeza stipulacea* Maxim, *Lespedeza striata* (Thunb.) Hook & Arn. var. *stipulacea* (Maxim.) Debeaux)	-	+	+	+	+	-	-	-	-	-	+	+	-	-	-	-	-	+
272.	***Lablab purpureus* (L.) Sweet** (= *Dolichos lablab L.*)	+	+	-	+	-	-	-	-	-	-	-	+	-	-	-	-	-	-
273.	***Lablab purpureus* (L.) Sweet var., lignosus**	+	+	-	+	-	-	-	-	-	-	-	-	-	-	-	-	-	-
274.	***Laburnum anagyroides* Medik.** (=*Cytisus laburnum* L.)	+	+	-	-	-	-	-	-	-	-	-	-	-	-	-	-	-	-
275.	***Lathyrus aphaca* L.**	+	+	-	+	-	-	-	-	-	-	-	-	-	-	-	-	-	-
276.	***Lathyrus hirsutus* L.**	-	-	+	+	-	-	-	-	-	-	-	+	-	-	-	-	-	-
277.	***Lathyrus imphalensis* Wall.**	-	-	-	+	-	-	-	-	-	-	-	-	-	-	-	-	-	-
278.	***Lathyrus odoratus* L.**	+	-	-	+	+	-	-	-	-	-	+	-	-	-	-	-	-	+
279.	***Lathyrus pratensis* L.**	-	+	-	-	-	-	-	-	-	-	-	-	-	-	-	-	-	-
280.	***Lathyrus sativus* L.**	+	+	+	+	+	-	-	-	-	-	-	-	-	-	-	-	-	+
281.	***Lathyrus sphaericus* Retz.**	-	-	-	-	-	-	-	-	-	-	-	-	-	-	-	-	-	+
282.	***Lathyrus tingitanus* L.**	-	-	+	+	-	-	-	-	-	-	+	-	-	-	-	-	-	+
283.	***Lens culinaris* Medik.**	+	+	+	+	-	-	-	-	-	-	-	-	-	-	-	-	-	-
284.	***Lespedeza juncea* (L.f.) Pers.**	+	+	+	+	+	-	-	+	-	-	-	+	-	-	+	-	-	-
285.	***Lonchocarpus sericeus* (Poir.) DC.**	-	+	-	-	-	-	-	-	-	-	-	-	-	-	-	-	+	+
286.	***Lotus corniculatus* L.**	+	+	+	+	+	-	-	-	+	-	-	+	-	-	-	-	-	-
287.	***Lotus hispidus* DC.**	-	-	-	+	-	-	-	-	-	-	-	-	-	-	-	-	-	-
288.	***Lotus pedunculatus* Cav.** (=*Lotus uliginosus* Schkuhr.)	-	-	-	+	-	-	-	-	-	-	-	-	-	-	-	-	-	-
289.	***Lotus tetragonolobus* L.**	+	-	-	+	-	-	-	-	-	-	-	-	+	-	-	-	-	-
290.	***Lupinus albus* L.**	+	+	+	+	+	+	-	-	-	-	-	+	+	-	+	+	-	-
291.	***Lupinus angustifolius* L.**	+	+	+	+	+	-	-	-	-	-	+	+	-	-	-	-	-	-
292.	***Lupinus luteus* L.**	+	-	+	+	+	-	-	-	-	-	+	+	-	-	-	-	-	-
293.	***Lupinus mexicanus* Lag.** (=*Lupinus bilineatus* Benth.)	-	-	-	-	-	-	-	-	-	-	+	-	-	-	-	-	-	-
294.	***Lupinus mutabilis* Sweet**	+	-	-	-	-	-	-	-	-	-	+	-	-	-	-	-	-	+
295.	***Lupinus pilosus* L.** (=*Lupinus hirsutus* L.)	+	+	-	-	-	-	-	-	-	-	+	-	-	-	-	-	-	-
296.	***Macroptilium atropurpureum* (DC.) Urban** (=*Phaseolus atropurpureus* DC.)	+	-	-	+	-	-	-	-	-	-	+	+	-	-	-	-	-	-
297.	***Macroptilium lathyroides* (L.) Urban** (=*Phaseolus lathyroides* L.)	+	-	+	+	-	-	-	-	-	-	-	-	-	-	-	-	-	-
298.	***Macrotyloma geocarpum* (Harms) Marechal & Baud.** (=*Kerstingiella geocarpa* Harms)	+	-	-	-	-	-	-	-	-	-	-	-	-	-	-	-	-	-

Sl. No.	Botanical Name	1	2	3	4	5	6	7	8	9	10	11	12	13	14	15	16	17	18
299.	***Macrotyloma uniflorum* (Lam.) Verdc.** (*=Dolichos biflorus sensu auct.*)	+	+	+	+	+	-	-	-	-	-	-	-	-	-	-	-	-	-
300.	***Mastersia assamica* Benth.** (=*Mastersia cleistocarpa* Baker)	-	+	-	-	-	-	-	-	-	-	-	-	-	-	-	-	-	-
301.	***Medicago edgeworthii* Sirj** (=*Trigonella pubescens* Baker)	-	+	-	-	-	-	-	-	-	-	-	-	-	-	-	-	-	-
302.	***Medicago falcata* L.**	+	-	-	+	-	-	-	-	-	-	+	+	-	-	-	-	-	-
303.	***Medicago laciniata* (L.) Miller**	-	-	-	+	-	-	-	-	-	-	-	-	-	-	-	-	-	-
304.	***Medicago lupulina* L.**	+	+	+	+	-	-	-	-	-	-	+	-	-	-	-	-	-	-
305.	***Medicago minima* (L.) L.** (=*Medicago polymorpha* L. var, *minima* L.)	-	-	-	+	-	-	-	-	-	-	-	-	-	-	-	-	-	-
306.	***Medicago monantha* (C.Mey) Trautv.** (=*Trigonella incisa* Benth.)	+	+	-	+	-	-	-	-	-	-	-	-	-	-	-	-	-	-
307.	***Medicago orbicularis* (L.) Bart.** (=*Medicago polymorpha* L. var., *orbicularis* L.)	-	-	+	+	-	-	-	-	-	-	-	-	-	-	-	-	-	-
308.	***Medicago orthoceras* (Kar. & Kir.) Trautv.** (=*Trigonella polycerata* M. Bieb.)	+	+	-	+	-	-	-	-	-	-	-	-	-	-	-	-	-	-
309.	***Medicago polymorpha* L.** (=*Medicago hispida* Gaertn.)	+	+	+	+	-	-	-	-	-	-	+	+	-	-	-	-	-	-
310.	***Medicago sativa* L.**	+	+	+	+	+	-	-	-	+	-	+	+	+	-	-	-	-	-
311.	***Medicago tornata* (L.) Miller. ssp.,** helix ***(Willd.)* Ooststr. & Reichg.** (=*Medicago obscura* Retz.)	-	-	-	+	-	-	-	-	-	-	-	-	-	-	-	-	-	-
312.	***Medicago turbinata* (L.) All.**	-	-	-	+	-	-	-	-	-	-	-	-	-	-	-	-	-	-
313.	***Meizotropis buteiformis* Voigt**	-	+	-	-	-	-	-	-	-	-	-	-	-	-	-	-	-	-
314.	***Melilotus albus* Medik.**	+	+	+	+	+	-	-	-	-	-	-	+	-	-	-	-	-	-
315.	***Melilotus indicus* (L.) All.** (=*Trifolium indicum* L.)	+	+	+	+	-	-	-	-	-	-	-	+	-	-	-	-	-	-
316.	***Melilotus officinalis* (L.) Pallas**	+	+	+	+	+	-	-	-	-	-	+	+	-	-	-	-	-	-
317.	***Melilotus suaveolens* Ledeb.**	+	+	-	+	-	-	-	-	-	-	-	-	-	-	-	-	-	-
318.	***Millettia caudata* (Benth.)Baker**	-	+	-	-	-	-	-	-	-	-	-	-	-	-	-	-	-	-
319.	***Millettia extensa* (Benth.) Baker** (=*Millettia auriculata* Brandis)	-	+	-	+	-	+	-	-	-	-	-	-	-	-	-	-	+	+
320.	***Millettia glaucescens* Kurz**	-	-	-	-	-	-	-	-	-	+	-	-	-	-	-	-	-	-
321.	***Millettia leucantha* Kurz**	-	-	-	-	-	-	-	-	-	+	-	-	-	-	-	-	-	-
322.	***Millettia pachycarpa* Benth.**	-	+	-	-	+	-	-	-	-	-	-	-	-	-	-	-	+	+
323.	***Millettia peguensis* Ali**	-	-	-	-	-	-	-	-	-	-	+	-	-	-	-	-	-	-
324.	***Millettia piscidia (Roxb.)* Wt.**	-	-	-	-	-	-	-	-	-	-	-	-	-	-	-	-	+	+
325.	***Millettia pulchra* Kurz**	-	-	-	-	-	-	-	-	-	+	-	-	-	-	-	-	-	-
326.	***Millettia thonningii* (Schum. & Thonn.) Baker**	-	+	-	-	-	-	-	-	-	-	+	-	-	-	-	-	+	-

Sl. No.	Botanical Name	1	2	3	4	5	6	7	8	9	10	11	12	13	14	15	16	17	18
327.	***Mucuna bracteata*** **DC.**	+	+	-	-	+	-	-	-	-	-	-	-	-	-	-	-	-	-
328.	***Mucuna gigantea*** **(Willd.) DC.**	+	+	-	-	-	-	-	-	-	-	-	-	-	-	-	-	-	+
329.	***Mucuna hainanensis*** **Hyata. ssp., *multilamellata* Wilmot** (=*Mucuna nigricans sensu auct.)*	+	+	-	-	-	-	-	-	-	-	-	-	-	-	-	-	-	-
330.	***Mucuna monosperma*** **Wight** (=*Mucuna anguina* Wall.)	+	+	-	-	-	-	-	-	-	-	-	-	-	-	-	-	-	-
331.	***Mucuna pachylobia*** **(Piper & Tracy) Rock**	+	-	-	-	-	-	-	-	-	-	-	-	-	-	-	-	-	-
332.	***Mucuna pruriens*** **(L.) DC. var., *hirsuta* (Wt. & Arn.) Wilmot** (=*Mucuna hirsuta* Wt. & Arn.)	+	-	-	-	-	-	-	-	-	-	-	-	-	-	-	-	-	-
333.	***Mucuna pruriens*** **(L.) DC. var., *pruriens*** (=*Dolichos pruriens* L., *Mucuna cochinchinensis* (Lour.) A. Cheval., *Mucuna atropurpurea (*Roxb.) Wt.)	+	+	+	+	+	-	-	-	-	-	+	+	+	-	-	-	-	+
334.	***Mucuna pruriens*** **(L.) DC. var., *utilis* (Wt.) Bruck** (=*Mucuna deeringiana* (Bort.) Merr., *Mucuna utilis* Wt.)	+	+	+	+	+	-	-	-	-	-	-	+	-	-	-	-	-	-
335.	***Mucuna urens*** **L. Medik.** (=*Dolichos urens* L.)	+	+	-	-	-	-	-	-	-	-	-	-	-	-	-	-	-	-
336.	***Mundulea sericea*** **(Willd.) A. Chev.** (=*Cytisus sericeus* Willd.)	-	+	-	-	-	-	+	-	-	-	-	-	-	-	-	-	+	+
337.	***Myroxylon balsamum*** **(L.) Harms** (=*Toluifera balsamum* L.)	+	+	-	-	+	-	+	-	-	+	+	-	-	-	-	+	+	-
338.	***Myroxylon peruiferum*** **L.f.**	+	+	-	-	-	-	-	-	-	-	-	-	-	-	-	-	-	-
339.	***Neonotonia wightii*** **(Wt. & Arn.) Lackey**	+	+	+	+	-	-	-	-	-	-	-	+	-	-	-	-	-	-
340.	***Ononis spinosa*** **L., ssp., *antiquorum* (L.) Arcang.** (=*Ononis antiquorum* L.)	-	+	-	-	-	-	-	-	-	-	-	-	-	-	-	-	-	-
341.	***Ononis spinosa*** **L., ssp., *hircina* (Jacq.) Gams.** (=*Ononis arvensis* L., *Ononis spinosa* L.)	+	+	-	-	-	-	-	-	-	-	-	-	-	-	-	-	-	-
342.	***Ophrestia pentaphylla*** **(Dalz.) Verdc.** (=*Glycine pentaphylla* Dalz.)	-	+	-	-	-	-	-	-	-	-	-	-	-	-	-	-	-	-
343.	***Ormocarpum cochinchinense*** **(Lour.) Merr.** (=*Ormocarpum orientale* (Spreng.) Merr.)	-	+	-	-	-	-	-	-	-	-	-	-	-	-	-	-	-	-
344.	***Ormocarpum sennoides*** **(Willd.) DC.**	-	+	-	-	-	-	-	-	-	-	-	-	-	-	-	-	-	-
345.	***Ormosia robusta*** **Baker** (=*Arillaria robusta* Kurz.)	-	-	-	-	-	-	-	-	-	+	-	-	-	-	-	-	-	-

Sl. No.	Botanical Name	1	2	3	4	5	6	7	8	9	10	11	12	13	14	15	16	17	18
346.	***Ormosia travancorica*** **Bedd.** (=*Placolobium travancoricum* (Bedd.) Yakovlev)	-	+	-	-	-	-	-	-	-	+	-	-	-	-	-	-	-	-
347.	***Oxytropis microphylla*** **(Pallas) DC.**	-	-	-	+	-	-	-	-	-	-	-	-	-	-	-	-	-	-
348.	***Oxytropis mollis*** **Benth.**	-	+	+	+	-	-	-	-	-	-	-	-	-	-	-	-	-	-
349.	***Oxytropis tatarica*** **Baker**	-	+	-	-	-	-	-	-	-	-	-	-	-	-	-	-	-	-
350.	***Pachyrhizus erosus*** **(L.) Urb.**	+	+	+	+	+	+	+	-	-	-	-	-	-	-	-	-	+	+
351.	***Pachyrhizus tuberosus*** **(Lam.) Spreng.**	+	+	-	-	-	-	-	-	-	-	-	-	-	-	-	-	+	+
352.	***Paracalyx scariosus*** **(Roxb.) Ali**	-	+	-	-	-	-	-	+	-	-	-	-	-	-	-	-	-	-
353.	***Parochetus communis*** **D. Don**	-	+	-	-	-	-	-	-	-	-	-	-	-	-	-	-	-	-
354.	***Phaseolus acutifolius*** **A. Gray**	+	+	-	+	-	-	-	-	-	-	-	+	-	-	-	-	-	-
355.	***Phaseolus coccineus*** **L.** (=*Phaseolus vulgaris* L. var., *coccineus* (L.) *Phaseolus multiflorus* Lam.)	+	+	-	+	+	-	-	-	-	-	+	-	-	-	-	-	-	-
356.	***Phaseolus lunatus*** **L.**	+	+	+	+	+	-	-	-	-	-	-	+	-	-	-	-	-	-
357.	***Phaseolus vulgaris*** **L.**	+	+	-	+	-	-	-	-	-	-	-	-	-	-	-	-	-	-
358.	***Physostigma venenosum*** **Balf.**	-	+	-	-	-	+	-	-	+	-	-	-	-	-	-	-	+	+
359.	***Piscidia piscipula*** **(L.) Sarg.**	-	+	-	-	-	-	-	-	-	+	-	-	-	-	-	-	-	+
360.	***Pisum sativum*** **L.** (=*Pisum arvense* L.)	+	+	+	+	+	-	-	-	-	-	-	-	-	-	-	-	-	-
361.	***Pongamia pinnata*** **(L.) Pierre** (=*Derris indica* (Lam.) Bennett, *Pongamia glabra* Vent.)	+	+	+	+	+	+	+	+	-	+	+	-	-	-	-	+	+	+
362.	***Pseudarthria viscida*** **(L.) Wt. & Arn.**	-	+	-	-	-	-	-	-	-	-	-	-	-	-	-	-	-	-
363.	***Psophocarpus palustris*** **Desv.**	+	-	-	+	+	-	-	-	-	-	-	+	-	-	-	+	-	-
364.	***Psophocarpus scandens*** **(Endl.) Verdc.**	+	-	-	+	+	-	-	-	-	-	-	+	-	-	-	+	-	-
365.	***Psophocarpus tetragonolobus*** **(L.) DC.**	+	+	+	+	+	-	-	-	-	-	-	+	-	-	-	+	-	-
366.	***Pterocarpus dalbergioides*** **DC.**	-	+	-	-	-	-	-	-	+	+	-	-	-	-	-	-	-	-
367.	***Pterocarpus echinatus*** **Pers.**	-	-	-	-	-	-	-	-	-	+	+	-	-	-	-	-	-	-
368.	***Pterocarpus indicus*** **Willd.**	+	+	-	-	-	-	+	-	+	+	+	-	-	-	-	+	-	-
369.	***Pterocarpus marsupium*** **Roxb.**	+	+	+	+	+	-	+	+	+	+	+	-	-	-	+	-	-	-
370.	***Pterocarpus santalinus*** **L. f.**	+	+	-	+	-	-	-	-	+	+	-	-	-	-	-	-	-	-
371.	***Pueraria montana*** **(Lour.) Merr., var., *chinensis* (Ohwi) Sanjappa & Pradeep** (=*Pueraria lobata* (Willd.) Ohwi var., *thomsonii* (Benth.) Maesen,)	+	+	-	+	-	+	-	-	-	-	-	+	-	-	-	-	-	+
372.	***Pueraria montana*** **(Lour.) Merr. var., *lobata* (Willd.) Sanjappa & Pradeep** (=*Pueria lobota* (Willd.) Ohwi)	+	+	+	+	-	+	-	-	-	-	+	+	-	-	-	-	-	+
373.	***Pueraria peduncularis*** **(Benth.) Benth.** (=*Pueraria yunnanensis* Franch.)	+	-	-	-	-	-	-	-	-	-	-	-	-	-	-	-	-	-
374.	***Pueraria phaseoloides*** **(Roxb.) Benth.** (=*Neustanthus phaseoloides* Benth.)	+	-	-	+	-	-	-	-	-	-	-	-	-	-	-	-	-	-

Sl. No.	Botanical Name	1	2	3	4	5	6	7	8	9	10	11	12	13	14	15	16	17	18
375.	***Pueraria tuberosa*** **Willd.** (=*Hedysarum tuberosum* Willd.)	+	+	+	+	-	-	+	-	-	-	-	+	-	-	-	-	-	-
376.	***Rhynchosia bracteata*** **Baker** (=*Rhynchosia mollissima* Dalzell)	+	-	-	-	-	-	-	-	-	-	-	-	-	-	-	-	-	-
377.	***Rhynchosia cana*** **(Willd.) DC.**	-	-	+	-	-	-	-	-	-	-	-	-	-	-	-	-	-	-
378.	***Rhynchosia capitata*** **(Roth) DC.**	-	-	+	+	-	-	-	-	-	-	-	-	-	-	-	-	-	-
379.	***Rhynchosia minima*** **(L.) DC.**	-	+	+	+	-	-	-	-	-	-	-	-	-	-	-	-	-	+
380.	***Rhynchosia rufescens*** **(Willd.) DC.**	-	+	+	-	-	-	-	-	-	-	-	-	-	-	-	-	-	-
381.	***Rhynchosia suaveolens*** **(L.f) DC.**	-	+	-	-	-	-	-	-	-	-	-	-	-	-	-	-	-	-
382.	***Rhynchosia volubilis*** **Lour.**	+	+	-	-	-	-	-	-	-	-	-	-	-	-	-	-	-	-
383.	***Robinia pseudoacacia*** **L.**	+	+	+	+	+	-	-	+	+	+	+	+	-	-	+	-	+	-
384.	***Rothia indica*** **(L.) Druce**	+	+	+	-	-	-	-	-	-	-	-	-	-	-	-	-	-	-
385.	***Securigera securidaca*** **(L.) Degen & Dorfl.** (=*Securigera coronilla* DC.)	-	+	-	-	+	-	-	-	-	-	-	-	-	-	-	-	-	-
386.	***Sesbania bispinosa*** **(Jacq.) W.Wight** (=*Sesbania aculeata* (Willd.) Pers., *Sesbania aegyptiaca sensu* Bojer)	+	+	+	+	-	+	+	-	-	-	-	-	-	-	-	-	+	+
387.	***Sesbania cannabina*** **(Retz.) Pers.**	+	-	+	+	-	+	+	-	-	-	+	+	-	-	+	-	-	-
388.	***Sesbania grandiflora*** **(L.) Pers.**	+	+	+	+	+	+	+	+	+	+	+	+	-	-	+	-	-	+
389.	***Sesbania javanica*** **Miq.** (=*Sesbania roxburghii* Merr.)	+	+	-	+	-	-	-	-	-	+	+	-	-	-	-	-	-	-
390.	***Sesbania procumbens*** **(Roxb.) Wt. & Arn.** (=*Aeschynomene procumbens* Roxb.)	+	+	-	+	-	-	-	-	-	-	-	-	-	-	-	-	-	-
391.	***Sesbania sesban*** **(L.) Merr.** (=*Sesbania aegyptiaca* Poir., *Sesbania pubescens sensu auct.)*	+	+	+	+	-	+	-	-	-	+	+	+	-	-	-	-	-	+
392.	***Sesbania speciosa*** **Taub.**	+	-	+	-	-	+	-	-	-	-	+	-	-	-	+	-	-	-
393.	***Shuteria involucrata*** **(Wall.) Wt. & Arn.** (=*Shuteria vestita* Wt. & Arn.)	-	-	+	-	-	-	-	-	-	-	-	+	-	-	-	-	-	-
394.	***Sindora sumatrana*** **Miq.**	-	+	-	-	-	-	-	-	-	-	-	-	-	-	-	-	-	-
395.	***Smithia bigemina*** **Dalz.**	-	-	+	-	-	-	-	-	-	-	-	+	-	-	-	-	-	-
396.	***Smithia blanda*** **Wall.**																		
	(=*Smithia racemosa* Wt. & Arn.)	+	-	-	-	-	-	-	-	-	-	-	-	-	-	-	-	-	-
397.	***Smithia conferta*** **Sm.**	+	+	-	-	-	-	-	-	-	-	-	-	-	-	-	-	-	-
398.	***Smithia hirsuta*** **Dalz.**	+	-	-	-	-	-	-	-	-	-	-	-	-	-	-	-	-	-
399.	***Smithia purpurea*** **Hook.**	+	-	-	-	-	-	-	-	-	-	-	-	-	-	-	-	-	-
400.	***Smithia sensitiva*** **Ait.**	+	+	-	+	-	-	-	-	-	-	-	-	-	-	-	-	-	-
401.	***Smithia setulosa*** **Dalz.**	+	-	-	+	-	-	-	-	-	-	-	-	-	-	-	-	-	-
402.	***Sophora alopecuroides*** **L.**	-	+	-	+	-	-	-	-	-	-	-	-	-	-	-	-	-	-
403.	***Sophora flavescens*** **Ait.**	-	+	-	-	-	-	-	-	-	-	-	-	-	-	-	-	-	-

Sl. No.	*Botanical Name*	1	2	3	4	5	6	7	8	9	10	11	12	13	14	15	16	17	18
404.	***Sophora interrupta* Bedd.** (=*Edwardsia maderaspatana* Wt.)	-	+	-	-	-	-	-	-	-	-	-	-	-	-	-	-	-	-
405.	***Sophora mollis* (Royle) Baker** (=*Sophora griffithii* Stocks.)	-	+	-	+	-	-	-	-	-	-	-	-	-	-	-	-	+	+
406.	***Sophora moorcroftiana* (Benth.) Baker**	-	+	-	-	-	-	-	-	-	-	-	-	-	-	-	-	-	-
407.	***Sophora secundiflora* (Ortega) DC.**	+	+	-	-	-	-	-	-	-	-	+	-	-	-	-	-	+	+
408.	***Sophora tomentosa* L.**	-	+	-	-	+	-	-	-	-	-	+	-	-	-	-	-	-	-
409.	***Sophora viciaefolia* Hance** (=*Sophora viccifolia* Hance)	+	-	-	-	-	-	-	-	-	-	+	-	-	-	-	-	-	-
410.	***Sophora velutina* Lindl.**	-	-	-	-	-	+	-	-	-	-	-	-	-	-	-	-	-	-
411.	***Sophora wightii* Baker**	-	+	-	-	-	-	-	-	-	-	-	-	-	-	-	-	-	-
412.	***Spartium junceum* L.**	-	+	-	-	+	+	-	-	-	-	+	-	-	-	+	-	-	-
413.	***Spartium littoralis* Hassk.**	-	+	-	-	-	-	-	-	-	-	-	-	-	-	-	-	-	-
414.	***Spatholobus parviflorus* (DC.) O. Kuntze** (=*Spatholobus roxburghii* Benth.)	+	+	-	-	+	+	+	+	-	+	-	-	-	-	-	-	+	+
415.	***Sphenostylis bracteata* (Baker) Gillett**	+	+	+	-	-	-	-	-	-	-	-	-	-	-	-	-	-	-
416.	***Strongylodon macrobotrys* A. Gray**	+	+	-	+	+	-	-	-	+	+	+	-	-	-	-	-	+	+
417.	***Stylosanthes fruticosa* (Retz.) Alston**	-	+	-	+	-	-	-	-	-	-	-	-	-	-	-	-	-	-
418.	***Stylosanthes guianensis* (Aubl.) Swartz** (=*Trifolium guianense* Aubl.)	-	-	-	+	-	-	-	-	-	-	-	+	-	-	-	-	-	-
419.	***Stylosanthes hamata* (L.) Taub.**	-	-	-	+	-	-	-	-	-	-	-	-	-	-	-	-	-	-
420.	***Stylosanthes humilis* Kunth** (=*Stylosanthes sundaica* Taub.)	-	-	-	+	-	-	-	-	-	-	-	+	-	-	-	-	-	-
421.	***Styphnolobium japonicum* (L.) Schott** (=*Sophora japonica* L.)	+	+	-	+	+	-	-	-	+	+	+	-	-	-	-	-	+	+
422.	***Sutherlandia frutescens* (L.) R.Br.**	-	+	-	+	-	-	-	-	-	-	-	-	-	-	-	-	-	-
423.	***Swainsona formosa* (G. Don) Thomps.** (=*Clianthus formosus* (G. Don) Ford & Vickery)	-	-	-	-	-	-	-	-	-	-	+	-	-	-	-	-	-	-
424.	***Tadehagi triquetrum* (L.) H. Ohashi** (=*Desmodium triquetrum* (L.) DC.)	+	+	+	-	-	-	-	+	-	-	-	-	+	-	-	-	-	-
425.	***Taverniera nummularifolia* DC.**	-	+	-	-	-	-	-	-	-	-	-	-	-	-	-	-	-	-
426.	***Tephrosia apollinea* (Delile) DC.**	-	+	-	-	-	-	-	-	+	-	-	-	-	-	-	-	-	-
427.	***Tephrosia bracteolata* Guill. & Perr.**	-	-	-	+	-	-	-	-	-	-	-	-	-	-	-	-	+	-
428.	***Tephrosia candida* (Roxb.) DC.**	-	+	+	-	-	-	-	-	-	-	+	+	-	-	-	-	+	+
429.	***Tephrosia grandiflora* (Ait.) Pers.**	-	+	-	-	-	-	-	-	-	-	+	-	-	-	-	-	+	+
430.	***Tephrosia maxima* (L.) Pers.** (=Tephrosia purpurea (L.) Pers. var. maxima (L.) Baker)	-	+	+	-	-	-	-	-	+	-	+	-	-	-	-	-	-	-
431.	***Tephrosia noctiflora* Baker** (=*Tephrosia hirta* Bojer., *Tephrosia hookeriana sensu auct.*)	-	+	+	-	+	-	-	-	-	-	-	+	-	-	-	-	+	+

Sl. No.	Botanical Name	1	2	3	4	5	6	7	8	9	10	11	12	13	14	15	16	17	18
432.	***Tephrosia pumila* (Lam.) Pers.** (=*Tephrosia purpurea* (L.) Pers. var., *pumila* (Lam.) Baker, *Tephrosia procumbens sensu acut.*)	+	+	-	-	-	-	-	-	-	-	-	-	-	-	-	-	+	-
433.	***Tephrosia purpurea* (L.) Pers.**	+	+	+	+	-	-	-	-	+	-	-	+	-	-	-	-	+	+
434.	***Tephrosia sinapou* (Buc'hoz) A. Chev.**	-	+	+	-	-	-	-	-	+	-	-	+	-	-	-	-	+	+
435.	***Tephrosia spinosa* (L. f.) Pers.**	-	+	-	-	-	-	-	-	-	-	-	-	-	-	-	-	-	-
436.	***Tephrosia strigosa* (Dalz.) Sant. & Mahesw.**	-	+	-	-	-	-	-	-	-	-	-	-	-	-	-	-	-	-
437.	***Tephrosia subtriflora* Baker**	-	-	-	-	-	-	-	-	-	-	-	-	-	-	-	-	-	+
438.	***Tephrosia tinctoria* Pers.**	-	+	+	+	-	-	-	-	+	-	-	+	-	-	-	-	+	-
439.	***Tephrosia uniflora* Pers.**	-	+	-	-	-	-	-	-	-	-	-	-	-	-	-	-	+	+
440.	***Tephrosia villosa* (L.) Pers.**	-	+	+	-	-	-	-	-	-	-	-	-	-	-	-	-	-	+
441.	***Tephrosia vogelii* Hook. f.**	+	+	+	+	-	-	-	-	-	-	+	+	-	-	-	-	+	+
442.	***Teramnus labialis* (L.f.) Spreng.**	+	+	+	+	-	-	-	-	-	-	-	+	-	-	-	-	-	+
443.	***Teramnus mollis* Benth.**	-	+	-	-	-	-	-	-	-	-	-	-	-	-	-	-	-	-
444.	***Teramnus volubilis* Swartz**	-	-	-	+	-	-	-	-	-	-	-	-	-	-	-	-	-	-
445.	***Thermopsis barbata* Benth.**	+	+	-	-	-	-	-	-	-	-	-	-	-	-	-	-	-	-
446.	***Tipuana tipu* (Benth.) Kuntze**	-	-	-	-	-	-	-	-	-	+	+	-	-	-	-	-	-	-
447.	***Trifolium alexandrium* L.**	+	+	+	+	+	-	-	-	-	-	-	+	-	-	-	-	-	-
448.	***Trifolium dubium* Sibth**	-	-	-	+	-	-	-	-	-	-	-	-	-	-	-	-	-	-
449.	***Trifolium fragiferum* L.**	+	-	+	+	-	-	-	-	-	-	+	-	-	-	-	-	-	-
450.	***Trifolium hybridum* L.**	-	-	-	+	-	-	-	-	-	-	-	-	-	-	-	-	-	-
451.	***Trifolium incarnatum* L.**	-	+	+	+	-	-	-	-	-	-	-	+	-	-	-	-	-	-
452.	***Trifolium lappaceum* L.**	-	-	-	+	-	-	-	-	-	-	-	-	-	-	-	-	-	-
453.	***Trifolium pratense* L.**	+	+	+	+	-	-	-	-	-	-	+	-	+	-	-	-	-	-
454.	***Trifolium repens* L.**	+	+	+	+	-	-	-	-	-	-	+	+	-	-	-	-	-	-
455.	***Trifolium resupinatum* L.**	-	-	+	+	-	-	-	-	-	-	+	-	-	-	-	-	-	-
456.	***Trifolium subterraneum* L.**	-	-	-	+	-	-	-	-	-	-	+	+	-	-	-	-	-	-
457.	***Trigonella balansae* Boiss. & Reuter** (=*Trigonella corniculata* (L.) L.)	+	+	-	-	-	-	-	-	-	-	-	-	-	-	-	-	-	-
458.	***Trigonella emodi* Benth.**	+	-	-	-	+	-	-	-	-	-	-	-	-	-	-	-	+	-
459.	***Trigonella foenum-graecum* L.**	+	+	+	+	+	-	-	-	+	-	-	-	+	-	-	-	+	-
460.	***Trigonella glabra* Thunb. ssp., *uncata* (Boiss. et Noe.) Lassen** (=*Trigonella uncata* Boiss. et Noe.)	-	+	-	-	-	-	-	-	-	-	-	-	-	-	-	-	-	-
461.	***Trigonella gracilis* Benth.**	-	-	-	+	-	-	-	-	-	-	-	-	-	-	-	-	-	-
462.	***Trigonella occulta* Ser.**	+	+	-	-	-	-	-	-	-	-	-	-	-	-	-	-	-	-

Sl. No.	Botanical Name	1	2	3	4	5	6	7	8	9	10	11	12	13	14	15	16	17	18
463.	***Ulex europaeus* L.**	-	-	-	+	+	-	-	-	+	-	+	-	+	-	-	-	-	-
464.	***Uraria crinita* (L.) DC.** (=*Hedysarum crinitum* L.)	-	+	+	-	-	-	-	-	-	-	-	+	-	-	-	-	-	+
465.	***Uraria lagopodioides* (L.) DC.** (=*Uraria alopecuroides* (Roxb.) Sweet)	-	+	+	-	-	-	-	-	-	-	-	-	-	-	-	-	-	-
466.	***Uraria lagopus* DC.**	-	+	-	-	-	-	-	-	-	-	-	-	-	-	-	-	-	-
467.	***Uraria picta* (Jacq.) DC.** (=*Hedysarum pictum* Jacq.)	-	+	-	-	-	-	-	-	-	-	-	-	-	-	-	-	-	-
468.	***Uraria prunellaefolia* Baker**	-	+	-	-	-	-	-	-	-	-	-	-	-	-	-	-	-	-
469.	***Uraria rufescens* (DC.) Schindl.** (=*Desmodium rufescens* DC., *Uraria hamosa* (Roxb.) Wt. & Arn.)	-	+	-	-	-	-	-	-	-	-	-	-	-	-	-	-	-	-
470.	***Vataireopsis araroba* (Aguiar) Ducke.** (=*Andira araroba* Aguiar)	-	+	-	-	-	-	-	-	-	-	-	-	-	-	-	-	-	-
471.	***Vicia bakeri* Ali**	+	-	-	+	-	-	-	-	-	-	-	-	-	-	-	-	-	-
472.	***Vicia benghalensis* L.**	+	-	+	+	-	-	-	-	-	-	-	+	-	-	-	-	-	-
473.	***Vicia biennis* L.**	-	-	-	+	-	-	-	-	-	-	-	-	-	-	-	-	-	-
474.	***Vicia cassubica* L.**	-	-	-	+	-	-	-	-	-	-	-	-	-	-	-	-	-	-
475.	***Vicia faba* L.**	+	+	+	+	+	+	-	-	-	-	+	+	-	-	-	-	-	+
476.	***Vicia hirsuta* (L.) S.F. Gray** (=*Ervum hirsutum* L.)	+	+	+	+	-	-	-	-	-	-	-	-	-	-	-	-	-	-
477.	***Vicia michauxii* Spreng.**	-	-	-	+	-	-	-	-	-	-	-	-	-	-	-	-	-	-
478.	***Vicia monantha* Retz.**	+	-	+	+	-	-	-	-	-	-	-	+	-	-	-	-	-	-
479.	***Vicia narbonensis* L.**	+	-	+	+	-	-	-	-	-	-	-	+	-	-	-	-	-	-
480.	***Vicia sativa* L.**	+	+	+	+	-	-	-	-	-	-	+	+	-	-	-	-	-	-
481.	***Vicia sativa* L. ssp., *nigra* Ehrhl.** (=*Vicia angustifolia* L.)	+	+	+	+	-	-	-	-	-	-	-	+	-	-	-	+	-	-
482.	***Vicia sepium* L.**	+	-	-	-	-	-	-	-	-	-	-	-	-	-	-	-	-	-
483.	***Vicia tenuifolia* Roth**	+	+	-	-	-	-	-	-	-	-	+	-	-	-	-	-	-	-
484.	***Vicia villosa* Roth**	+	-	+	+	-	-	-	-	-	-	+	+	-	-	-	-	-	-
485.	***Vigna aconitifolia* (Jacq.) Marechal** (=*Phaseolus aconitifolius* (Jacq.) Maréchal)	+	+	+	+	+	-	-	-	-	-	-	+	-	-	-	-	-	-
486.	***Vigna adenantha* (G.F.Meyer) Marechal & *al.*** (=*Phaseolus adenanthus* G.F. Meyer)	+	+	-	-	-	-	-	-	-	-	+	-	-	-	-	-	-	-
487.	***Vigna angularis* (Willd.) Ohwi & H. Ohashi**	+	+	-	+	-	-	-	-	-	-	-	-	+	-	-	-	-	+

Sl. No.	*Botanical Name*	1	2	3	4	5	6	7	8	9	10	11	12	13	14	15	16	17	18
488.	***Vigna dalzelliana* (Kuntze) Verdc.** (=*Phaseolus dalzellianus* Kuntz.)	+	+	+	+	-	-	-	-	-	-	-	-	-	-	-	-	-	-
489.	***Vigna glabrescens* Marechal & *al.***	+	-	-	-	-	-	-	-	-	-	-	-	-	-	-	-	-	-
490.	***Vigna grahamiana* (Wt. & Arn.) Verdc.**	-	+	-	-	-	-	-	-	-	-	-	-	-	-	-	-	-	-
491.	***Vigna hosei* (Craib.) Baker**	-	-	+	-	-	-	-	-	-	-	-	+	-	-	-	-	-	-
492.	***Vigna khandalensis* (Sant.) Sundararagh. & Wadhwa** (=*Phaseolus khandalensis* Sant.)	+	-	-	-	-	-	-	-	-	-	-	-	-	-	-	-	-	-
493.	***Vigna marina* (Burm.) Merr.** (=*Phaseolus marinus* Burm.)	+	+	+	-	-	-	-	-	-	-	-	+	-	-	-	-	-	-
494.	***Vigna mungo* (L.) Hepper** (=*Phaseolus mungo* L., *Phaseolus radiatus* Roxb. non L.)	+	+	+	+	-	-	-	-	-	-	-	+	-	-	-	-	-	-
495.	***Vigna radiata* (L.) R. Wilczek** (=*Phaseolus radiatus* L., *Phaseolus aureus* Roxb.)	+	+	+	+	-	-	-	-	-	-	-	-	+	-	-	+	-	-
496.	***Vigna sublobata* (Roxb.) Babu & Sharma** (=*Phaseolus sublobatus* Roxb. var., *setulosus* (Dalzell)Prain)	+	+	-	-	-	-	-	-	-	-	-	-	-	-	-	-	-	-
497.	***Vigna subterranea* (L.) Verdc.**	+	-	+	+	-	-	-	-	-	-	-	+	+	-	-	-	-	-
498.	***Vigna trilobata* (L.) Verdc.** (=*Phaseolus trilobatus* (L.) Schreber)	+	+	-	+	-	-	-	-	-	-	-	+	-	-	-	-	-	-
499.	***Vigna umbellata* (Thunb.) Ohwi & Ohashi** (=*Dolichos umbellatus* Thunb.)	+	+	+	+	+	-	-	-	-	-	-	-	+	-	-	-	-	-
500.	***Vigna unguiculata* (L.) Walp.** (=*Vigna catjang* Savi., *Vigna sinensis* (L.) Hassk.)	+	+	+	+	-	+	-	-	+	-	-	+	+	-	-	-	-	+
501.	***Vigna vexillata* (L.) A. Rich**	+	+	+	+	-	-	-	-	-	-	-	+	-	-	-	-	-	-
502.	***Virgilia oroboides* (Berg.) Salter**	-	-	-	-	-	-	+	-	-	+	-	-	-	-	-	-	-	-
503.	***Wisteria sinensis*** (Sims) Sweet (=*Wisteria chinensis* DC.)	+	+	-	-	+	+	-	-	-	-	+	-	+	-	-	-	-	-
504.	***Zornia gibbosa* Span.** (=*Zornia diphylla sensu auct.*, *Zornia graminea* Span.)	-	+	+	+	-	-	-	-	-	-	-	-	-	-	-	-	-	-

Table 3.3

Economic Potential of Legumes in India: Mimosaceae

Eighteen known uses, identified as important for Legumes, are indicated for each species by '+'; a '-' indicates that particular use is not known for that species. The correct binomials are used and the important synonyms, if any, are given parenthesis preceded by '=', for each species. A wider synonymy is given in the information profiles for those species for which profiles were compiled. The genera and species are in alphabetical order.

Sl. No.	*Botanical Name*	Edible 1	Medicinal 2	Manure 3	Fodder/Forage 4	Oil 5	Fibre 6	Gum/Resin 7	Tannin 8	Dye 9	Timber 10	Ornamental/Avenue/Hedge/ Shade/Support 11	Cover/Soil Binder/Windbreak 12	Beverage 13	Lac insect host 14	Paper industry 15	Soap substitute 16	Insecticide/Pesticide 17	Molluscicide/piscicide 18
1.	**Acacia auriculiformis** ***Benth.***	-	+	-	-	+	-	+	+	-	+	+	+	-	-	+	-	-	+
2.	***Acacia baileyana*** **F. Muell.**	-	-	-	-	-	-	-	-	-	-	-	+	-	-	-	-	-	-
3.	***Acacia caesia*** **(L.) Willd.** (=*Acacia intsia sensu auct.*, *Mimosa caesia* (L.) Willd.)	+	+	-	-	-	+	-	-	+	-	-	-	-	-	-	+	-	+
4.	***Acacia catechu*** **(L.f.) Willd.** (=*Mimosa catechu* L.f.)	+	+	-	+	-	+	+	+	+	+	-	+	-	+	-	-	-	-
5.	***Acacia chundra*** **(Rottler) Willd.** (=*Acacia sundra* (Roxb.) DC., *Mimosa chundra* Rottler., *Acacia catechu* (L.f.) Willd., *var. sundra* (Roxb.) Prain*)*	-	+	-	-	-	-	+	-	-	+	-	-	-	-	-	-	-	-
6.	***Acacia concinna*** **(Willd.) DC.** (=*Acacia rugata* (Lam.) Merr.)	-	+	-	-	-	-	-	-	-	-	-	-	-	-	-	-	+	+
7.	***Acacia dealbata*** **Link.** (=*Acacia decurrens* Willd. var., *dealbata* (Link.) Maiden)	-	+	-	-	+	-	+	+	+	+	+	-	-	-	+	-	-	-
8.	***Acacia decora*** **Rchb. f.**	+	-	-	-	-	-	+	-	-	-	-	-	-	-	-	-	-	-
9.	***Acacia decurrens*** **Willd.** (=*Mimosa decurrens* Wend.)	-	+	+	-	+	-	-	+	-	+	-	+	-	-	+	-	-	-
10.	***Acacia eburnea*** **(L.f.) Willd.** (=*Acacia roxburghii* Wt. & Arn., *Mimosa eburnea* L.f.)	-	+	-	-	-	-	-	-	-	+	-	-	-	-	-	-	-	-
11.	***Acacia farnesiana*** **(L.) Willd.** (=*Mimosa farnesiana.* L.)	+	+	-	+	+	-	+	+	+	+	+	+	-	+	-	-	-	+
12.	***Acacia ferruginea*** **DC.** (=*Mimosa ferruginea* Roxb.)	+	+	-	+	-	-	+	-	-	+	-	-	+	-	-	-	-	-
13.	***Acacia horrida*** **(L.) Willd.**	-	+	-	+	-	+	+	-	-	+	-	+	-	-	-	-	-	-
14.	***Acacia jacquemontii*** **Benth.**	+	+	-	+	+	-	+	+	-	+	+	+	+	-	-	-	-	-
15.	***Acacia lenticularis*** **Benth.**	-	-	-	-	-	-	+	-	-	+	+	-	-	-	-	-	-	-

Sl. No.	*Botanical Name*	1	2	3	4	5	6	7	8	9	10	11	12	13	14	15	16	17	18
16.	***Acacia leucophloea* (Roxb.) Willd.** (=*Mimosa leucophloea* Roxb.)	+	+	-	+	-	+	+	+	+	+	-	-	+	-	-	+	-	+
17.	***Acacia longifolia* (Andr.) Willd.**	+	-	-	-	-	-	-	-	-	-	+	+	-	-	-	-	-	-
18.	***Acacia macrantha* Willd.**	-	+	-	+	-	-	-	-	-	+	-	-	-	-	-	-	-	-
19.	***Acacia mearnsii* de Wild.**	+	+	+	-	+	+	+	+	-	+	+	-	-	-	+	-	-	-
20.	***Acacia melanoxylon* R.Br.** (=*Mimosa melanoxylon* R.Br.)	-	+	-	-	-	-	-	+	-	+	+	-	-	-	-	-	-	-
21.	***Acacia modesta* Wall.** (=*Mimosa obovata* Roxb.)	-	+	-	+	-	+	+	-	-	+	-	-	-	-	-	-	-	-
22.	***Acacia nilotica* (L.) Delile** (=*Acacia arabica* (Lam.) Willd.).	+	+	-	+	+	+	+	+	+	+	+	-	+	+	+	+	-	-
23.	***Acacia nilotica* (L.) Delile ssp., *tomentosa* (Benth.) Brenan**	-	+	-	-	-	-	+	+	+	+	-	-	-	-	-	-	-	-
24.	***Acacia pennata* (L.) Willd.** (=*Mimosa pennata* L.)	+	+	-	+	-	+	+	+	-	-	-	-	-	+	-	+	+	+
25.	***Acacia planifrons* Wt. & Arn.** (=*Acacia planiformis* Wt. & Arn.)	-	+	-	+	-	+	-	+	-	+	-	-	-	+	-	-	-	-
26.	***Acacia podalyriifolia* G.Don**	-	-	-	+	-	-	-	+	-	-	+	-	-	-	-	-	-	-
27.	***Acacia polyacantha* Willd.** (=*Acacia suma.* (Roxb.) J. Voigt, *Mimosa suma* Roxb.)	-	+	-	-	-	-	+	+	-	+	+	-	-	-	-	-	-	-
28.	***Acacia pycnantha* Benth.**	-	+	-	-	+	-	+	+	-	-	+	+	-	-	-	-	-	-
29.	***Acacia quadrimarginea* F. Muell.**	+	-	-	+	-	-	-	-	-	-	-	-	-	-	-	-	-	-
30.	***Acacia salicina* Lindl.**	+	-	-	-	-	-	-	-	-	-	-	-	-	-	-	-	-	-
31.	***Acacia saligna* (Labill.) Wendl.** (=*Acacia cyanophylla* Lindl.)	+	-	-	-	-	-	+	-	-	-	+	+	-	-	-	-	-	-
32.	***Acacia senegal* (L.) Willd.**	+	+	+	+	-	+	+	-	+	+	+	+	-	-	+	-	-	-
33.	***Acacia seyal* Del.**	+	+	-	+	-	-	+	+	-	+	-	-	-	-	-	-	-	-
34.	***Acacia sieberiana* DC.**	-	-	-	-	-	-	+	-	-	-	-	+	-	-	-	-	-	-
35.	***Acacia sinuata* (Lour.) Merr.** (=*Acacia concinna* non (Willd.) DC., *Acacia rugata* Benth., *Mimosa sinnuata* Lour.)	+	+	+	+	-	-	-	+	+	-	-	-	-	+	-	+	+	+
36.	***Acacia tenuifolia* (L.) Willd.** (=*Acacia paniculata* Willd.)	-	-	-	-	-	-	-	+	-	-	-	-	-	-	-	-	-	-
37.	***Acacia tomentosa* Willd.**	-	-	-	-	-	-	-	-	-	+	-	-	-	-	-	-	-	-
38.	***Acacia torta* (Roxb.) Craib.** (=*Mimosa torta* Roxb.)	-	+	+	-	-	-	-	+	+	-	-	-	-	-	-	+	-	-
39.	***Acacia tortilis* (Forssk.) Hayne**	-	+	-	+	-	-	+	+	-	+	-	-	-	-	-	-	-	-
40.	***Adenanthera pavonina* L.**	+	+	-	-	+	-	+	-	+	+	+	-	-	-	-	+	-	+
41.	***Albizia amara* (Roxb.) Boiv.**	-	+	+	+	+	-	+	-	-	+	+	-	+	+	-	+	-	+
42.	***Albizia chinensis* (Osb.) Merr.** (=*Mimosa chinensis* Osb.)	-	+	+	+	-	-	+	+	-	+	+	-	-	+	+	-	+	+

Sl. No.	*Botanical Name*	1	2	3	4	5	6	7	8	9	10	11	12	13	14	15	16	17	18
43.	***Albizia julibrissin* Duraz.** (=*Acacia julibrissin* (Duraz.) Willd.)	+	+	-	-	-	-	-	+	-	+	+	-	-	-	-	-	-	+
44.	***Albizia lebbeck* (L.) Benth.**	+	+	+	+	+	-	+	+	+	+	+	+	-	-	+	+	+	+
45.	***Albizia lucidior* (Steud.) Nielsen** (=*Inga lucidior* Steud.)	+	+	-	-	-	-	-	-	-	+	+	-	-	+	-	-	-	+
46.	***Albizia myriophylla* Benth.**	-	-	-	-	-	-	-	-	-	-	+	-	-	+	-	-	-	+
47.	***Albizia niopioides* (Benth.) Burk.** (=*Albizia richardiana* King & Prain)	-	-	-	-	-	-	-	-	-	-	+	-	-	+	-	-	-	-
48.	***Albizia odoratissima.* (L.f.) Benth.**	+	+	-	+	-	-	+	+	+	+	+	-	-	+	-	-	-	+
49.	***Albizia procera* (Roxb.) Benth.**	+	+	-	+	-	-	+	+	+	+	+	-	-	+	+	-	+	+
50.	***Albizia thompsonii* Brandis**	-	-	-	-	-	-	-	-	-	+	-	-	-	-	-	-	-	-
51.	***Archidendron clypearia* (Jack.) Nielsen** (=*Inga clypearia* Jack.)	-	+	-	-	-	-	-	+	+	+	+	-	-	-	-	+	-	+
52.	***Archidendron globosum* (Bl.) Nielsen** (=*Inga globosa* Bl.)	+	+	-	-	-	-	-	-	-	+	-	-	-	-	-	-	-	-
53.	***Archidendron monadelphum* (Roxb.) Nielsen** (=*Mimosa monadelpha* Roxb.)	+	+	-	-	-	-	-	-	-	+	-	-	-	+	-	-	-	+
54.	***Calliandra brevipes* Benth.**	-	-	-	-	-	-	-	-	-	-	+	-	-	-	-	-	-	-
55.	***Calliandra haematocephala* Hassk.**	-	-	-	-	-	-	-	-	-	-	+	-	-	-	-	-	-	-
56.	***Calliandra houstoniana* (Mill.) Standley**	-	-	-	-	-	-	-	-	-	-	+	-	-	-	-	-	-	-
57.	***Calliandra tergemina* (L.) Benth.**	-	-	-	-	-	-	-	-	-	-	+	-	-	-	-	-	-	-
58.	***Desmanthus virgatus* (L.) Willd.** (=*Mimosa virgata* L.)	+	-	-	+	-	-	-	-	-	-	-	-	-	-	-	-	-	-
59.	***Dichrostachys cinerea* (L.) Wt. & Arn.** (=*Mimosa cinerea* (L.))	-	+	-	+	-	+	+	+	+	-	-	-	-	+	-	-	-	-
60.	***Entada gigas* (L.) Faw. & Rendle** (= *Entada scandens* (L.) Benth., *Mimosa scandens* L.)	+	+	-	+	+	+	-	+	-	-	-	-	-	-	+	+	+	+
61.	***Entada rheedei* Spreng.** (=*Entada pursaetha* DC., *Entada scandens sensu auct.*,	+	+	-	-	-	-	-	-	-	-	-	-	+	-	-	-	-	+
62.	***Faidherbia albida* (Delile) A. Chev.** (=*Acacia albida* Delile)	+	+	-	+	-	-	+	+	-	+	+	-	-	-	-	-	-	+
63.	***Falcataria moluccana* (Miq.) Barn. & Grim.** (=*Albizia falcatoria* (L.) Fosberg)	-	+	+	+	-	-	-	-	-	+	+	+	-	-	+	-	-	-
64.	***Indopiptadenia oudhensis* (Brandis) Brenan**	-	-	-	+	-	-	-	-	-	+	-	-	-	-	-	-	-	-
65.	***Leucaena leucocephala* (Lam.) de Wit.** (=*Leucaena glauca* (Willd.) Benth.)	+	-	-	-	-	-	-	+	-	+	+	-	-	-	-	-	-	-
66.	***Lysiloma latisiliquum* (L.) Benth.** (=*Leucaena latisiliqua* (L.) Gillis & Stearn., *Mimosa latisiliqua* L.)	+	+	+	+	+	-	-	+	+	+	+	+	+	-	+	+	-	+
67.	***Mimosa bimucronata* (DC.) Kuntze** (=*Mimosa sepiaria* Benth.)	-	-	-	-	-	-	-	-	-	+	+	-	-	-	-	-	-	-

Sl. No.	Botanical Name	1	2	3	4	5	6	7	8	9	10	11	12	13	14	15	16	17	18
68.	***Mimosa hamata* Willd.**	-	+	+	+	-	-	-	-	-	-	+	-	-	-	-	-	-	-
69.	***Mimosa pudica* L.**	+	+	+	+	+	-	-	+	-	-	+	-	-	-	-	-	-	-
70.	***Mimosa rubicaulis* Lam.**	+	-	-	-	-	-	+	-	-	-	+	-	-	-	-	-	-	-
71.	***Mimosa rubicaulis* Lam. ssp., *himalayana* (Gamble) H.Ohashi** (=*Mimosa himalayana* Gamble., *Mimosa rubicaulis auct.*, non Lam.)	-	+	-	-	-	-	-	-	-	-	+	-	-	-	-	-	-	-
72.	***Neptunia oleracea* Lour.**	+	+	-	-	-	-	-	-	-	-	-	-	-	-	-	-	-	-
73.	***Neptunia triquetra* (Vahl) Benth.**	-	+	-	-	-	-	-	-	-	-	-	-	-	-	-	-	-	-
74.	***Parapiptadenia rigida* (Benth.) Brenan** (=*Piptadenia rigida* Benth.)	-	+	-	-	-	-	+	+	+	+	-	-	-	-	+	-	-	-
75.	***Parkia biglandulosa* Wt. & Arn.**	+	+	-	-	+	-	-	+	-	+	+	-	+	-	-	-	-	-
76.	***Parkia biglobosa* (Jacq.) G. Don** (=*Parkia clappertoniana* Keay.)	+	+	-	-	-	-	+	+	+	+	+	-	+	-	-	-	-	-
77.	***Parkia sumatrana* Miq.** (=*Parkia insignis* Kurz)	-	-	-	-	-	-	+	-	-	+	-	-	-	-	-	-	-	-
78.	***Parkia javanica* (Lam.) Merr.**	+	-	-	-	-	-	-	-	-	-	-	-	-	-	-	-	-	-
79.	***Parkia timoriana* (DC.) Merr.** (=*Parkia roxburghii* G. Don *Inga timoriana* DC.)	+	+	-	-	+	-	-	+	-	+	-	-	-	-	+	-	-	-
80.	***Pithecellobium dulce* (Roxb.) Benth.** (=*Mimosa dulcis* Roxb.)	+	+	-	+	+	-	+	+	-	+	+	+	-	+	-	-	-	+
81.	***Pithecellobium major* L.**	-	+	-	-	-	-	-	-	-	-	-	-	-	-	-	-	-	-
82.	***Pithecellobium unguis-cati* (L.) Benth.** (=*Mimosa unguis-cati* L.)	+	+	-	-	-	-	-	-	-	-	-	-	-	-	-	-	-	-
83.	***Prosopis cineraria* (L.) Druce** (=*Mimosa cineraria L.*, *Prosopis spicigera* L.)	+	+	+	+	-	+	+	+	-	+	+	-	-	-	-	-	-	-
84.	***Prosopis farcta* (Banks & Soland.) Macbr.**	-	+	-	+	+	-	-	+	-	-	-	-	-	-	-	-	-	-
85.	***Prosopis glandulosa* Torrey**	-	+	-	+	-	-	-	-	-	-	+	-	-	-	-	-	-	-
86.	***Prosopis juliflora* (Swartz) DC.**	+	+	+	+	-	+	+	+	+	+	+	+	+	-	+	-	-	+
87.	***Prosopis pallida* (Willd.) Kunth** (=*Prosopis limensis* Benth.)	+	-	-	+	-	-	-	-	-	-	-	-	-	-	-	-	-	-
88.	***Prosopis pubescens* Benth.**	+	-	-	-	-	-	-	-	-	-	-	-	+	-	-	-	-	+
89.	***Samanea saman* (Jacq.) Merr.** (=*Enterolobium saman* (Jacq.) Prain., *Mimosa saman* Jacq.)	-	+	+	+	-	-	+	-	-	+	+	-	+	-	-	-	-	+
90.	***Tetrapleura tetraptera* (Schum. & Thonn.) Taub.**	-	+	-	-	-	-	-	-	-	-	-	+	-	-	-	-	-	-
91.	***Xylia xylocarpa* (Roxb.) Taub.** (=*Mimosa xylocarpa* Roxb., *Xylia dolabriformis* Benth.)	+	+	+	+	+	-	+	+	-	+	-	-	-	+	+	-	-	-
92.	***Zapoteca portoricensis* (Jacq.) Hern.** (=*Calliandra portoricensis* (Jacq.) Benth.)	-	+	-	-	-	-	-	-	-	-	-	-	-	-	-	-	-	-

Table 3.4
Nomenclatural Equivalents of Legume Species

For scientific reasons, the botanical names of many legume species have been changed over a period of time. As per the rules of Botanical Nomenclature, the earlier names have now become synonyms that cannot be used. Since the now rejected names have been in popular use in literature for a long time, it would be difficult to locate the species by the scientifically correct names used in this book. In order to facilitate an easier location of common legume species by their familiar old names, and to provide links to names widely used in literature, the synonyms (before 'see') and the current equivalents (after 'see') are given here. The information profiles of species contain a wider synonymy.

CAESALPINIACEAE

1. *Bauhinia anguina* Roxb., see *Bauhinia scandens* L.
2. *Bauhuinia lawii* Baker see *Bauhinia faveolata* Dalz.
3. *Bauhinia macrostachya* Benth., see *Bauhinia ungulata* L.
4. *Bauhinia pubescens* DC., see *Bauhinia tomentosa* L.
5. *Bauhinia racemosa* Vahl., see *Bauhinia vahlii* Wt. & Arn.
6. *Bauhinia retusa* Roxb., see *Bauhinia semla* Wunder.
7. *Bauhinia triandra* Roxb., see *Bauhinia purpurea* L.
8. *Bauhinia variegata* L., var., *chinensis* DC., see *Bauhinia variegata* L.
9. *Caesalpinia bonducella* (L.) Flem., see *Caesalpinia bonduc* (L.) Roxb.
10. *Caesalpinia chinensis* Roxb., see *Caesalpinia crista* L.
11. *Caesalpinia crista sensu auct.,* see *Caesalpinia bonduc* (L.) Roxb.
12. *Caesalpinia major* (Medik.) Dandy & Excell., see *Caesalpinia globulorum* Bakh.f. & Royen
13. *Caesalpinia paniculata* (Lam.) Roxb., see *Caesalpinia crista* L.
14. *Cassia absus* L., see *Chamaecrista absus* (L.) H.S. Irwin & Barnb.
15. *Cassia alata* L., see *Senna alata* (L.) Roxb.
16. *Cassia angustifolia* M. Vahl., see *Senna alexandrina* Miller
17. *Cassia auriculata* L., see *Senna auriculata* (L.) Roxb.
18. *Cassia australis* Sims see *Senna odorata* (R. Morris) Randall
19. *Cassia bicapsularis* L., see *Senna bicapsularis* (L.) Roxb.
20. *Cassia corymbosa* Lam., see *Senna corymbosa* (Lam.) H.S. Irwin & Barnb.
21. *Cassia didymobotrya* Fresen., see *Senna didymobotrya* (Fresen.) H.S. Irwin & Barnb.
22. *Cassia divaricata* Nees & Blume see *Senna divaricata* (Nees & Blume) Lock.
23. *Cassia excelsa* Schrad., see *Senna spectabilis* (DC.) H.S. Irwin & Barnb.
24. *Cassia floribunda* Cav., see *Senna floribunda* (Cav.) H.S. Irwin & Barnb.
25. *Cassia hirsuta* L., see *Senna hirsuta* (L.) H.S. Irwin & Barnb.
26. *Cassia italica* (Mill.) Spreng., see *Senna italica* Miller
27. *Cassia leptocarpa* Benth., see *Senna hirsuta* (L.) H.S. Irwin & Barnb., var., *leptocarpa*
28. *Cassia marilandica* L., see *Senna marilandica* (L.) Link.
29. *Cassia mimosoides* L., see *Chamaecrista mimosoides* (L.) E. Greene
30. *Cassia montana* Navis & Violar see *Senna timoriensis* (DC.) H.S. Irwin & Barnb.
31. *Cassia multijuga* Rich., see *Senna multijuga* (Rich.) H.S. Irwin & Barnb.

32. *Cassia nigricans* Vahl see *Senna nigricans* (Vahl) Greene
33. *Cassia nodosa* Roxb., see *Cassia javanica* L., ssp., *nodosa* (Roxb.) K. Larsen & S. S. Larsen
34. *Cassia obovata* Colla., see *Senna italica* Miller
35. *Cassia obtusa* Roxb., see *Senna italica* Miller
36. *Cassia obtusifolia* L., see *Senna obtusifolia* (L.) H.S. Irwin & Barnb.
37. *Cassia occidentalis* L., see *Senna occidentalis* (L.) Link
38. *Cassia parahyba* Vell., see *Schizolobium parahyba* (Vell.) Blake
39. *Cassia phyllodinea* R.Br., see *Senna podocarpa* (Guill. & Perott.) Lock
40. *Cassia podocarpa* Guill. & Perrot., see *Senna podocarpa* (Guill. & Perrot.) Lock
41. *Cassia pumila* Lam., see *Chamaecrista pumila* (Lam.) K. Larsen
42. *Cassia senna* L., *Senna alexandrina* Miller
43. *Cassia sericea* Sw., see *Senna uniflora* (Mill.) H.S. Irwin & Barnb.
44. *Cassia siamea* Lam., see *Senna siamea* (Lam.) H.S. Irwin & Barnb.
45. *Cassia sophera* L., see *Senna sophera* (L.) Roxb.
46. *Cassia spectabilis* DC., see *Senna spectabilis* (DC.) H.S. Irwin & Barnb.
47. *Cassia surattensis* Burm.f., see *Senna surattensis* (Burm.f.) H.S. Irwin & Barnb.
48. *Cassia timoriensis* DC., see *Senna timoriensis* (DC.) H.S. Irwin & Barnb.
49. *Cassia tomentosa* L.f., see *Senna multiglandulosa* (Jacq.) H.S. Irwin & Barnb.
50. *Cassia tora* L, see *Senna tora* (L.) Roxb.
51. *Cassia uniflora* Mill., see *Senna uniflora* (Mill.) H.S. Irwin & Barnb.
52. *Cynometra polyandra* Roxb., see *Maniltoa polyandra* (Roxb.) Harms.
53. *Gleditsia inermis* Mill., See *Gleditsia aquatica* Marshall
54. *Gleditsia triacantha* Gaertner see *Gleditsia aquatica* Marshall
55. *Gleditsia triacanthos* (L.) var., *aquatica* (Marshall) Castigl., see *Gleditsia aquatica* Marshall
56. *Haematoxylon campechianum* L., see *Haematoxulym campechianum* L.
57. *Hardwickia pinnata* Roxb., see *Kingiodendron pinnatum* (DC.) Harms.
58. *Hymenaea candolleana* Kunth see *Hymenaea courbaril* L.
59. *Mezoneuron cucullatum* (Roxb.) Wt. & Arn., see *Caesalpinia cucullata* Roxb.
60. *Peltophorum ferrugineum* (Decne.) Benth., see *Peltophorum pterocarpum* (DC.) K. Heyne
61. *Phanera purpurea* (L.) Benth. see *Bauhinia purpurea* L.
62. *Phanera vahlii* (Wt. & Arn.) Benth., see *Bauhinia vahlii* Wt. & Arn.
63. *Phanera variegata* (L.) Benth., see *Bauhinia variegata* L.
64. *Piliostigma faveolatum* (Dalz.) Thoth., see *Bauhinia faveolata* Dalz.
65. *Piliostigma racemosa* (Lam.) Benth., see *Bauhinia racemosa* Lam.
66. *Poinciana elata* L., see *Delonix elata* Gamble
67. *Poinciana pulcherrima* L., see *Caesalpinia pulcherrima* (L.) Sw.
68. *Poinciana regia* Hook., see *Delonix regia* (Hook.) Raf.
69. *Senna angustifolia* (Vahl.) Batka see *Senna alexandrina* Miller
70. *Saraca indica sensu* Bedd., see *Saraca asoca* (Roxb.) Willd.
71. *Senna obtusa* Roxb., see *Senna italica* Miller
72. *Senna prostrata* Roxb., see *Chamaecrista pumila* (Lam.) K. Larsen
73. *Wagatea spicata* (Dalz.) Wight., see *Moullava spicata* (Dalz.) Nicolson

FABACEAE

1. *Abrus fruticulosus non* Wt. & Arn., see *Abrus pulchellus* Thwaites
2. *Alhagi pseudalhagi* Desv., see *Alhagi maurorum* Medik. & Boiss.
3. *Astragalus gummifer* Laibill., see *Astracantha gummifera* (Laibill.) Podl.
4. *Butea frondosa* Roxb., see *Butea monosperma* (Lam.) Taub.
5. *Cajanus indicus* Sprengel see *Cajanus cajan* (L.) Millsp.
6. *Canavalia ensiformis sensu auct.*, see *Canavalia gladiata* (Jacq.) DC.
7. *Canavalia gladiata* (Jacq.) DC., var., *leucosperma* Voigt see *Canavalia ensiformis* (L.) DC.
8. *Crotalaria mucronata* Desv., see *Crotalaria pallida* Aiton
9. *Crotalaria striata* DC., see *Crotalaria pallida* Aiton
10. *Cullen corylifolia* (L.) Medik., see *Cullen corylifolium* (L.) Medik.
11. *Dalbergia emarginata* Roxb., *see Dalbergia latifolia* Roxb.
12. *Dalbergia paniculata* Roxb., see *Dalbergia lanceolaria* L.f., ssp., *paniculata* (Roxb.)Thoth.
13. *Dalbergia robusta* DC., see *Derris robusta* (DC.) Benth.
14. *Dalbergia scandens* Roxb., see *Derris scandens* (Roxb.) Benth.
15. *Derris indica* (Lam.) Bennett see *Pongamia pinnata* (L.) Pierre
16. *Desmodium capitatum* (Burm.f.) DC., see *Desmodium styracifolium* (Osbeck) Merr.
17. *Desmodium gyroides* (Link) DC., see *Codariocalyx gyroides* (Link.) Hassk.
18. *Desmodium gyrans* (L.) DC., see *Codariocalyx motorius* (Houtt.) H. Ohashi
19. *Desmodium diffusum* DC., see *Desmodium laxiflorum* DC.
20. *Desmodium recurvatum* (Roxb.) Graham ex Wt. & Arn., see *Desmodium laxiflorum* DC.
21. *Desmodium retroflexum* var., *minus* Wt. & Arn., see *Desmodium styracifolium* (Osbeck) Merr.
22. *Desmodium triangulare* (Retz.) Merr., see *Dendrolobium triangulare* (Retz.) Schindler
23. *Desmodium triflorum* (L.) DC., var., *majus* Wt. & Arn., see *Desmodium heterophyllum* (Willd.) DC.
24. *Desmodium triflorum* (L.) DC., var., *minus* Wt. & Arn., see *Desmodium triflorum* (L.) DC.
25. *Desmodium triquetrum* (L.) DC., see *Tadehagi triquetrum* (L.) H. Ohashi
26. *Desmodium umbellatum* (L.) DC., see *Dendrolobium umbellatum* (L.) Benth.
27. *Dicerma pulchellum* (L.) DC., see *Desmodium pulchellum* (L.) Benth.
28. *Dolichos biflorus sensu auct.*, see *Macrotyloma uniflorum* (Lam.) Verdc.
29. *Dolichos lablab* L., see *Lablab purpureus* (L.) Sweet
30. *Dolichos umbellatus* Thunb., see *Vigna umbellata* (Thunb.) Ohwi. & H.Ohashi
31. *Erythrina corallodendron* Lour., see *Erythrina variegata* L.
32. *Erythrina mysorensis* Gamble see *Erythrina variegata* L.
33. *Erythrina orientalis* (L.) Merr., see *Erythrina variegata* L.
34. *Erythrina sublobata* Roxb., see *Erythrina suberosa* Roxb.
35. *Flemingia semialata* Aiton., see *Flemingia macrophylla* (Willd.) Merr.
36. *Gliricidia maculata* Kunth(Walp)., see *Gliricidia sepium* (Jacq.) Walp.
37. *Glycine soja sensu auct.*, see *Glycine max* (L.) Merr.
38. *Hedysarum crinitum* L., see *Uraria crinita* (L.) DC.
39. *Indigofera aspalathifolia* Roxb., see *Indigofera aspalathoides* DC.
40. *Indigofera echinata* Willd., see *Indigofera nummularifolia* (L.) Alston
41. *Indigofera endecaphylla* Jacq., *sensu auct.,* see *Indigofera hendecaphylla* Jacq.

42. *Indigofera pulchella* Roxb., see *Indigofera cassioides* DC.
43. *Medicago denticulata* Will., see *Medicago polymorpha* L.
44. *Melilotus alba* Medik., see *Melilotus albus* Medik.
45. *Melilotus altissimus sensu auct.,* see *Melilotus suaveolens* Ledeb.
46. *Melilotus suaveolens sensu auct.*, see *Melilotus suaveolens* Ledeb.
47. *Moghania bracetata* (Roxb.) Li, see *Flemingia strobilifera* (L.) Aiton
48. *Mucuna utilis* Wt., see *Mucuna pruriens* (L.) DC., var., *utilis* (Wt.) Burck.
49. *Ougeinia oojeinensis* (Roxb) Hochr., see *Desmodium oojeinense* (Roxb.) H. Ohashi
50. *Phaseolus aconitifolius* (Jacq.) Marech., see *Vigna aconitifolia* (Jacq.) Marech.
51. *Phaseolus adenanthus* G.F. Meyer see *Vigna adenantha* (G.F. Meyer) Marech. & *al.*
52. *Phaseolus aureus* Roxb., see *Vigna radiata* (L.) Wilcz.
53. *Phaseolus dalzellianus* Kuntz., see *Vigna dalzelliana* (O. Kuntz) Verdc.
54. *Phaseolus mungo* L., see *Vigna mungo* (L.) Hepper
55. *Phaseolus radiatus* L., see *Vigna radiata* (L.) Wilcz.
56. *Phaseolus trilobatus* (L.) Schreber., see *Vigna trilobata* (L.) Verdc.
57. *Pongamia glabra* Vent*.,* see *Pongamia pinnata* (L.) Pierre
58. *Pongamia elliptica* Wall., see *Derris elliptica* (Wall.) Benth.
59. *Psoralea corylifolia* L., see *Cullen corylifolium* (L.) Medik.
60. *Pteroloma triquetrum* (L.) DC., see *Tadehagi triquetrum* (L.) H. Ohashi
61. *Rhynchosia scarabaeoides* (L.) DC., see *Cajanus scarabaeoides* (L.) Thou.
62. *Sarothamnus scoparius* (L.) Koch., see *Cytisus scoparius* (L.) Link.
63. *Sesbania aculeata* (Willd.) Pers., see *Sesbania bispinosa* (Jacq.) W. Wight
64. *Sesbania aegyptiaca sensu* Bojer see *Sesbania bispinosa* (Jacq.) W. Wight
65. *Sesbania aegyptiaca* Poir., see *Sesbania sesban* (L.) Merr.
66. *Sophora japonica* L., see *Styphnolobium japonicum* (L.) Schott.
67. *Stylosanthes gracilis* Kunth., see *Stylosanthes guianensis* (Aubl.) Sw.
68. *Tephrosia hirta* Bojer*., see Tephrosia noctiflora* Baker
69. *Tephrosia purpurea* (L.) Pers., var., *maxima* (L.) Baker see *Tephrosia maxima* (L.) Pers.
70. *Tephrosia procumbens sensu auct.*, see *Tephrosia pumila* (Lam.) Pers.
71. *Trigonella corniculata* (L.) L., see *Trigonella balansae* Boiss & Reuter
72. *Trigonella incisa* Benth., see *Medicago monantha* (C. Meyer) Trautv.
73. *Uraria alopecuroides* (Roxb.) Sweet., see *Uraria lagopodioides* DC.
74. *Uraria hamosa* (Roxb.) Wt. & Arn., see *Uraria rufescens* (DC.) Schindler
75. *Uraria macrostachya* Wall., see *Uraria crinita* (L.) DC.
76. *Vigna catjang* Savi., see *Vigna unguiculata* (L.) Walp.
77. *Vigna pilosa* (Willd.) Baker see *Dysolobium pilosum* (Willd.) Marechal.
78. *Vigna sinensis* Hassk., see *Vigna unguiculata* (L.) Walp.
79. *Wisteria chinensis* DC., see *Wisteria sinensis* (Sims)Sweet
80. *Zornia gibbosa* Span see *Zornia diphylla sensu auct.*

MIMOSACEAE

1. *Abarema bigemina* (L.) Kosterm., see *Archidendron monadelphum* (Roxb.) Nielsen
2. *Acacia albida* Delile see *Faidherbia albida* (Delile) A. Chev.
3. *Acacia aneura* F. Muell., see *Acacia quadrimarginea* F. Muell.
4. *Acacia arabica auct.,* see *Acacia nilotica* (L.) Del.
5. *Acacia arabica* (Lam.) Willd., see *Acacia nilotica* (L.) Del.
6. *Acacia auriculaeformis* Benth., see *Acacia auriculiformis* Benth.
7. *Acacia catechu* (L.f.) Willd. var., *sundra* (Roxb.) Prain see *Acacia chundra* (Rottler) Willd.
8. *Acacia concinna non* (Willd.) DC., see *Acacia sinuata* (Lour.) Merr.
9. *Acacia concinna* (Willd.) DC., var., *rugata* (Benth.) Baker see *Acacia concinna* (Willd.) DC.
10. *Acacia cyanophylla* Lindl., see *Acacia saligna* (Laibill.) Wendl.
11. *Acacia decurrens* Willd., var., *angulata* (Desv.) Benth., see *Acacia decurrens* Willd.
12. *Acacia intsia sensu auct.,* see *Acacia caesia* (L.) Willd.
13. *Acacia farnesiana* (L.) Willd. var., *lenticellata* (F. Muell.) Bailey see *Acacia farnesiana* (L.) Willd.
14. *Acacia fistula* Schweinf., see *Acacia seyal* Del.
15. *Acacia julibrissin* (Duraz.) Willd., see *Albizia julibrissin* Duraz.
16. *Acacia leiophylla* Benth., see *Acacia pycnantha* Benth.
17. *Acacia leucocephala* (Lam.) Link see *Leucaena leucocephala* (Lam.) de Wit.
18. *Acacia melanoxylon* R.Br., var., *obtusifolia* Ser., see *Acacia melanoxylon* R.Br.
19. *Acacia odoratissima* (L.f.) Willd., see *Albizia odoratissima* (L.f.) Benth.
20. *Acacia paniculata* Willd., see *Acacia tenuifolia* (L.) Willd.
21. *Acacia planiformis* Wt. & Arn., see *Acacia planifrons* Wt. & Arn.
22. *Acacia pycnantha* Benth., var., *petiolaris* H. Vilm., see *Acacia pycnantha* Benth.
23. *Acacia raddiana* Savi see *Acacia tortilis* (Forssk.) Hayne
24. *Acacia roxburghii* Wt. & Arn., see *Acacia eburnea* (L.f.) Willd.
25. *Acacia rugata* (Lam.) Merr., see *Acacia concinna* (Willd.) DC.
26. *Acacia rugata* (Lam.) Voigt see *Acacia sinuata* (Lour.) Merr.
27. *Acacia suma* (Roxb.) J. Voigt see *Acacia polyacantha* Willd.
28. *Acacia sundra* (Roxb.) DC., see *Acacia catechu* (L.f.) Willd.
29. *Acacia sundra* (Roxb.) DC., see *Acacia chundra* (Rottler) Willd.
30. *Acacia tora* (Roxb.) Craib., see *Acacia caesia* (L.) Willd.
31. *Albizia falcatoria* (L.) Fosberg see *Falcatoria moluccana* (Miq.) Bern. & Grim.
32. *Albizia marginata* (Lam.) Merr., see *Albizia chinensis* (Osbeck.) Merr.
33. *Albizia odoratissima* (L.f.) var., *mollis* Baker see *Albizia odoratissima* (L.f.) Benth.
34. *Albizia richardiana* King & Prain see *Albizia niopoides* (Benth.) Burk.
35. *Calliandra emarginata* (Willd.) Benth., see *Calliandra tergemina* (L.) Benth., var., *emarginata* (Willd.) Barnb.
36. *Calliandra grandiflora* (L'Hert.) Benth., see *Calliandra houstoniana* (Mill.) Standley var., *anomala* (Kunth) Barnb.
37. *Calliandra portoricensis* (Jacq.) Benth., see *Zapoteca portoricensis* (Jacq.) Hern.
38. *Entada parrana* Spreng., see *Entada gigas* (L.) Fawcett & Rendle
39. *Entada phaseoloides* (L.) Merr., see *Entada gigas* (L.) Fawcett & Rendle
40. *Entada phaseoloides sensu auct.*, see *Entada rheedei* Sprengel

41. *Entada pursaetha* DC., see *Entada rheedei* Sprengel
42. *Entada pursaetha* DC., ssp., *sinohimalensis* Grierson & Long see *Entada rheedei* Sprengel ssp., *sinohimalensis* (Grierson & Long) Panigr.
43. *Entada scandens* (L.) Benth., see *Entada gigas* (L.) Fawcett & Rendle
44. *Entada scandens sensu auct.,* see *Entada rheedei* Sprengel
45. *Enterolobium saman* (Jacq.) Prain see *Samanea saman* (Jacq.) Merr.
46. *Inga clypearia* Jack., see *Archidendron clypearia* (Jack.) Nielsen
47. *Inga globosa* Bl., see *Archidendron globosum* (Bl.) Nielsen
48. *Inga lucidior* Steud., see *Albizia lucidior* (Steud.) Nielsen
49. *Inga timoriana* DC., see *Parkia timoriana* (DC.) Merr.
50. *Leucaena glauca* (Willd.) Benth., see *Leucaena leucocephala* (Lam.) de Wit.
51. *Leucaena latisiliqua* (L.) Gillis & Stearn see *Lysiloma latisiliquum* (L.) Benth.
52. *Mimosa caesia* (L.) Willd., see *Acacia caesia* (L.) Willd.
53. *Mimosa catechu* L.f., see *Acacia catechu* (L.f.) Willd.
54. *Mimosa chinensis* Osbeck see *Albizia chinensis* (Osbeck) Merr.
55. *Mimosa cineraria* L., see *Prosopis cineraria* (L.) Druce
56. *Mimosa cinerea* (L.), see *Dichrostachys cinerea* (L.) Wt. & Arn.
57. *Mimosa concinna* Willd., see *Acacia concinna* (Willd.) DC.
58. *Mimosa decurrens* Wend., see *Acacia decurrens* Willd.
59. *Mimosa dulcis* Roxb., see *Pithecellobium dulce* (Roxb.) Benth.
60. *Mimosa eburnea* L.f., see *Acacia eburnea* (L.f) Willd.
61. *Mimosa farnesiana* L., see *Acacia farnesiana* (L.) Willd.
62. *Mimosa ferruginea* Roxb., see *Acacia ferruginea* DC.
63. *Mimosa glauca* L., see *Lysiloma latisiliquum* (L.) Benth.
64. *Mimosa himalayana* Gamble see *Mimosa rubicaulis* Lam., ssp., *himalayana* (Gamble) H. Ohashi
65. *Mimosa leucocephala* Lam., see *Leucana leucocephala* (Lam.) de Wit.
66. *Mimosa leucophloea* Roxb., see *Acacia leucophloea* (Roxb.) Willd.
67. *Mimosa monadelpha* Roxb., see *Archidendron monadelphum* (Roxb.) Nielsen
68. *Mimosa nilotica* L., see *Acacia nilotica* (L.) Del.
69. *Mimosa pedunculata* Roxb., see *Parkia biglandulosa* Wt. & Arn.
70. *Mimosa pennata* L., see *Acacia pennata* (L.) Willd.
71. *Mimosa rubicaulis auct. non* Lam., see *Mimosa rubicaulis* Lam., ssp., *himalayana* (Gamble) H. Ohashi
72. *Mimosa saman* Jacq., see *Samanea saman* (Jacq.) Merr.
73. *Mimosa sepiaria* Benth., see *Mimosa bimucronata* (DC.) Kuntze
74. *Mimosa sinuata* Lour., see *Acacia sinuata* (Lour.) Merr.
75. *Mimosa suma* Roxb., see *Acacia polyacantha* Willd.
76. *Mimosa sundra* Roxb. ex Rottler see *Acacia chundra* (Rottler) Willd.
77. *Mimosa torta* Roxb., see *Acacia torta* (Roxb.) Craib.
78. *Mimosa unguis-cati* L., see *Pithecellobium unguis-cati* (L.) Benth.
79. *Mimosa virgata* L., see *Desmanthus virgatus* (L.) Willd.
80. *Mimosa xylocarpa* Roxb., see *Xylia xylocarpa* (Roxb.) Taub.

81. *Parkia clappertoniana* Keay., see *Parkia biglobosa* (Jacq.) G.Don
82. *Parkia insignis* Kurz see *Parkia sumatrana* Miq.
83. *Parkia roxburghii* G. Don see *Parkia timoriana* (DC.) Merr.
84. *Piptadenia oudhensis* (Brandis) Brenan see *Indopiptadenia oudhensis* (Brandis) Brenan
85. *Piptadenia rigida* Benth., see *Parapiptadenia rigida* (Benth.) Brenan
86. *Prosopis juliflora* (Sw.) DC., var., *constricta* Sarg., see *Prosopis glandulosa* Torrey
87. *Prosopis juliflora* (Sw.) DC., var., *glandulosa* (Torrey) Cockerell., see *Prosopis glandulosa* Torrey
88. *Prosopis limensis* Benth., see *Prosopis pallida* (Willd.) Kunth
89. *Prosopis spicigera* L., see *Prosopis cineraria* (L.) Druce
90. *Schrankia brachycarpa* Benth., see *Mimosa invisa* Colla
91. *Tetrapleura thonningii* Benth., see *Tetrapleura tetraptera* (Schum. & Thonn.) Taub.
92. *Xylia dolabriformis* Benth., see *Xylia xylocarpa* (Roxb.) Taub.

Table 3.5

The Relative Economic Importance of the Legume Families

The number of species and the percentage are given for the three families, and for all legumes, under each of the 18 categories of use.

SL. No.	CATEGORY OF USE	CAESALPINIACEAE (15 Species)		FABACEAE (505 Species)		MIMOSACEAE (92 Species)		TOTAL (722 Species)	
		Number	%	Number	%	Number	%	Number	%
1	**Edible**	42	33.6	200	39.5	38	41.7	280	38.9
2	**Medicinal**	87	69.6	363	71.7	64	70.3	514	71.5
3	**Manure**	15	12.0	146	28.9	16	17.6	177	24.6
4	**Fodder/Forage**	19	15.2	224	44.3	38	41.8	281	39.0
5	**Oil**	25	20.2	69	13.6	19	20.9	113	15.7
6	**Fibre**	17	13.6	46	9.0	15	16.5	78	10.8
7	**Gum/Resin**	15	12.0	32	6.3	38	41.7	85	11.8
8	**Tannin**	24	19.4	20	3.9	43	47.2	87	12.1
9	**Dye**	19	15.3	47	9.3	21	23.0	87	12.1
10	**Timber**	44	35.5	52	10.3	53	58.2	149	20.7
11	**Ornamental/Avenue/Hedge/ Shade/Support Plant**	68	54.4	108	21.3	44	48.4	220	30.6
12	**Cover/ Soilbinder/ Windbreak**	05	4.0	102	20.2	18	19.8	125	17.4
13	**Beverage**	14	11.2	23	4.5	12	13.1	49	6.8
14	**Host for lac insect**	04	3.2	04	0.8	17	18.7	25	3.4
15	**Paper Industry**	04	3.2	22	4.3	16	17.6	42	5.8
16	**Substitute for soap**	04	3.2	09	1.8	12	13.1	25	3.4
17	**Insecticide/Pesticide**	04	3.2	50	9.9	05	5.5	59	8.2
18	**Molluscicide/Piscicide**	11	8.8	82	16.2	24	26.4	117	16.2

Table 3.6
Edible Legumes in India

The legume species that are consumed in India, in different forms and contexts, are listed, indicating the edible parts. Other than the usual pulses and vegetables, the other species are used only in restricted local situations, some of them in times of food scarcity. Additional information is given in the respective species profiles.

Sl. No.	Species	Edible part
I	**CAESALPINIACEAE**	
1.	*Afzelia africana*	Seed
2.	*Bauhinia accuminata*	Flower
3.	*Bauhinia malabarica*	Young shoot, leaf, flower, seed
4.	*Bauhinia purpurea*	Young shoot, leaf, flower bud, pod
5.	*Bauhinia racemosa*	Leaf, flower, pod, seed
6.	*Bauhinia semla*	Gum
7.	*Bauhinia tomentosa*	Leaf, seed
8.	*Bauhinia vahlii*	Pod, seed
9.	*Bauhinia variegata*	Leaf, flower bud, pod
10.	*Caesalpinia pulcherrima*	Seed
11.	*Cassia fistula*	Flower, pod
12.	*Cassia grandis*	Pod pulp
13.	*Cassia javanica*	Leaf
14.	*Ceratonia siliqua*	Pod, gum, seed
15.	*Chamaecrista absus*	Seed
16.	*Chamaecrista mimmosoides*	Leaf
17.	*Cynometra cauliflora*	Pod
18.	*Delonix regia*	Gum, pod
19.	*Dialium guianense*	Pod
20.	*Dialium indicum*	Pod pulp, seed
21.	*Gleditsia aquatica*	Pod pulp, seed
22.	*Gymnocladus dioica*	Seed
23.	*Hymenaea courbaril*	Pod pulp
24.	*Lysidice rhodostegia*	Seed
25.	*Parkinsonia aculeata*	Pod, seed
26.	*Saraca declinata*	Flower, seed
27.	*Schotia brachypetala*	Seed
28.	*Senna auriculata*	Bark, leaf, flower
29.	*Senna obtusifolia*	Pod
30.	*Senna occidentalis*	Leaf, flower,pod, seed
31.	*Senna siamea*	Flower
32.	*Senna sophera*	Leaf, flower
33.	*Senna tora*	Leaf, flower, pod, seed
34.	*Tamarindus indica*	Leaf, flower, pod pulp, seed

Sl. No.	Species	Edible part
II	**FABACEAE**	
1.	*Abrus precatorius*	Leaf
2.	*Aeschynomene aspera*	Leaf
3.	*Amphicarpaea bracteata*	Pod, seed
4.	*Apios americana*	Tuber
5.	*Arachis hypogaea*	Young pod, seed, seed oil
6.	*Astracantha gummifera*	Gum
7.	*Astragalus christianus*	Root
8.	*Astragalus fatimensis*	Gum
9.	*Astragalus strobiliferus*	Gum
10.	*Baphia nitida*	Young twigs as chewing sticks
11.	*Butea monosperma*	Flower, pod
12.	*Cajanus cajan*	Leaf, flower, pod, seed
13.	*Cajanus goensis*	Leaf
14.	*Cajanus lineatus*	Pod
15.	*Canavalia gladiata*	Leaf, flower, young pod, seed
16.	*Caragana ambigua*	Flower
17.	*Caragana arborescens*	Seed
18.	*Caragana sinica*	Flower
19.	*Castanospermum australe*	Pod, seed
20.	*Cicer arietinum*	Stem, leaf, pod, seed
21.	*Cicer microphyllum.*	Shoot, leaf, seed
22.	*Clitoria ternatea*	Flower, pod
23.	*Crotalaria juncea*	Flower
24.	*Cullen corylifolium*	Seed
25.	*Cyamopsis tetragonoloba*	Leaf, pod, seed
26.	*Cytisus scoparius*	Flower bud
27.	*Dalbergia lanceolaria* ssp., *paniculata*	Leaf
28.	*Dalbergia pinnata*	Bark, root
29.	*Dalbergia stipulacea*	Seed
30.	*Dendrolobium triangulare*	Pod
31.	*Dendrolobium umbellatum*	Pod
32.	*Desmodium microphyllum*	Leaf
33.	*Desmodium repandum*	Young plant
34.	*Desmodium oojeinense*	Flower
35.	*Dysolobium pilosum*	Seed
36.	*Eriosema chinense*	Tuber
37.	*Eriosema himalaicum*	Tuber
38.	*Erythrina stricta*	Young leaf
39.	*Erythrina variegata*	Young leaf, pod, seed
40.	*Flemingia chappar*	Tuber
41.	*Flemingia macrophylla*	Pod
42.	*Flemingia procumbens*	Tuber, pod

Sl. No.	Species	Edible part
43.	*Flemingia tuberosa*	Tuber
44.	*Genista tinctoria*	Bud, seed
45.	*Gliricidia sepium*	Flower
46.	*Glycine max*	Seedling, leaf, pod, seed, seed oil
47.	*Glycyrrhiza glabra*	Root
48.	*Indigofera cassioides*	Leaf, bud, flower, pod
49.	*Indigofera coerulea*	Seed
50.	*Indigofera cordifolia*	Seed
51.	*Indigofera dosua*	Flower
52.	*Indigofera glabra*	Leaf
53.	*Indigofera glandulosa*	Seed
54.	*Indigofera heterantha*	Flower
55.	*Indigofera linnaei*	Leaf
56.	*Indigofera nummularifolia*	Young plant
57.	*Lablab purpureus*	Pod, seed
58.	*Laburnum anagyroides*	Leaf
59.	*Lathyrus aphaca*	Young seed
60.	*Lathyus odoratus*	Pod, seed
61.	*Lathyrus sativus*	Leaf, pod, seed
62.	*Lens culinaris*	Pod, seed
63.	*Lotus tetragonolobus*	Pod
64.	*Lupinus albus*	Seed
65.	*Lupinus luteus*	Seed, seed oil
66.	*Macrotyloma geocarpum*	Seed
67.	*Macrotyloma uniflorum*	Pod, seed
68.	*Medicago falcata*	Flower
69.	*Medicago lupulina*	Seed
70.	*Medicago monantha*	Plant, leaf, pod
71.	*Medicago polymorpha*	Plant, leaf
72.	*Medicago sativa*	Flower
73.	*Melilotus albus*	Young shoot, leaf
74.	*Melilotus indicus*	Plant, seed
75.	*Melilotus officinalis*	Herb, flower, seed
76.	*Melilotus suaveolens*	Leaf, seed
77.	*Mucuna bracteata*	Seed
78.	*Mucuna gigantea*	Seed
79.	*Mucuna monosperma*	Young pod, seed
80.	*Mucuna pachylobia*	Pod
81.	*Mucuna pruriens* var., *hirsuta*	Pod, seed
82.	*Mucuna pruriens* var., *pruriens*	Pod, seed
83.	*Mucuna pruriens* var., *utilis*	Young leaf, pod, seed
84.	*Mucuna urens*	Pod
85.	*Myroxylon balsamum*	Gum-resin

Sl. No.	Species	Edible part
86.	*Myroxylon peruiferum*	Gum-resin, seed
87.	*Neonotonia wightii*	Leaf
88.	*Ononis spinosa* ssp., *hircina*	Tender twig
89.	*Pachyrrhizus erosus*	Tuber, pod
90.	*Pachyrrhizus tuberosus*	Tuber
91.	*Phaseolus coccineus*	Pod, seed
92.	*Phaseolus lunatus*	Pod, seed
93.	*Phaseolus vulgaris*	Leaf, pod, seed
94.	*Pisum sativum*	Pod, seed
95.	*Pongamia pinnata*	Flower
96.	*Psophocarpus scandens*	Rhizome, pod
97.	*Psophocarpus tetragonolobus*	Leaf, flower, pod, seed, sprouts
98.	*Pterocarpus indicus*	Leaf, flower
99.	*Pterocarpus marsupium*	Flower, seed
100.	*Pterocarpus santalinus*	Dye from wood
101.	*Pueraria montana* var., *chinensis*	Tuber
102.	*Pueraria montana* var., *lobata*	Root, shoot, leaf
103.	*Pueraria peduncularis*	Flower
104.	*Pueraria phaseoloides*	Root
105.	*Pueraria tuberosa*	Tuber, leaf
106.	*Rhynchosia bracteata*	Seed
107.	*Rhynchosia volubilis*	Seed
108.	*Robinia pseudoacacia*	Leaf, flower, seed
109.	*Rothia indica*	Leaf, pod
110.	*Sesbania bispinosa*	Leaf, seed
111.	*Sesbania cannabina*	Seed
112.	*Sesbania grandiflora*	Young leaf, flower, pod
113.	*Sesbania javanica*	Leaf
114.	*Sesbania procumbens*	Seed
115.	*Sesbania sesban*	Leaf, flower, seed
116.	*Smithia blanda*	Leaf
117.	*Smitha conferta*	Leaf
118.	*Smithia hirsuta*	Leaf
119.	*Smithia purpurea*	Leaf
120.	*Smithia sensitiva*	Leaf
121.	*Smithia setulosa*	Leaf
122.	*Spatholobus parviflorus*	Seed, seed oil
123.	*Tadehagi triquetrum*	Root
124.	*Tephrosia purpurea*	Root, seed
125.	*Tephrosia vogelii*	Herb
126.	*Teramnus labialis*	Young leaf
127.	*Thermopsis barbata*	Root
128.	*Trigonella balansae*	Whole plant, seed

Sl. No.	Species	Edible part
129.	*Trigonella emodi*	Leaf
130.	*Trigonella foenum-graecum*	Shoot, leaf, seed
131.	*Trigonella occulta*	Leaf, seed
132.	*Vicia bakeri*	Seed
133.	*Vicia benghalensis*	Seed
134.	*Vicia faba*	Pod, seed
135.	*Vicia hirsuta*	Leaf, seed
136.	*Vicia monantha*	Seed
137.	*Vicia narbonensis*	Seed
138.	*Vicia sativa*	Seed
139.	*Vicia sepium*	Seed
140.	*Vicia tenuifolia*	Seed
141.	*Vicia villosa*	Plant
142.	*Vigna aconitifolia*	Pod, seed
143.	*Vigna adenantha*	Root, seed
144.	*Vigna angularis*	Pod, seed
145.	*Vigna dalzelliana*	Pod, seed
146.	*Vigna glabrescens*	Seed
147.	*Vigna khandalensis*	Seed
148.	*Vigna marina*	Leaf
149.	*Vigna mungo*	Pod, seed
150.	*Vigna radiata*	Pod, seed
151.	*Vigna sublobata*	Pod, seed
152.	*Vigna subterranea*	Seed
153.	*Vigna trilobata*	Pod, seed
154.	*Vigna umbellata*	Leaf, pod, seed
155.	*Vigna unguiculata*	Leaf, pod, seed
156.	*Vigna vexillata*	Root tuber, pod, seed
157.	*Wisteria sinensis*	Flower
III	**MIMOSACEAE**	
1.	*Acacia catechu*	Heart wood extract
2.	*Acacia decurrens*	Gum
3.	*Acacia farnesiana*	Gum, leaf
4.	*Acacia ferruginea*	Leaf
5.	*Acacia jacquemontii*	Gum, seed
6.	*Acacia leucophloea*	Pod, seed
7.	*Acacia longifolia*	Pod, seed
8.	*Acacia mearnsii*	Seed oil
9.	*Acacia nilotica*	Gum, pod, seed
10.	*Acacia quadrimarginea*	Seed
11.	*Acacia pennata*	Leaf
12.	*Acacia salicina*	Gum, pod, seed
13.	*Acacia senegal*	Gum, seed

Sl. No.	Species	Edible part
14.	*Acacia seyal*	Gum
15.	*Acacia sinuata*	Seed
16.	*Adenanthera pavonina*	Seed
17.	*Albizia julibrissin*	Leaf
18.	*Albizia lebbeck*	Pod
19.	*Albizia lucidior*	Seed
20.	*Albizia myriophylla*	Bark
21.	*Albizia odoratissima*	Leaf
22.	*Albizia procera*	Pod, seed
23.	*Archidendron globosum*	Pod
24.	*Archidendron monadelphum*	Seed
25.	*Entada rheedei*	Pod, seed
26.	*Leucaena leucocephala*	Shoot, pod, seed
27.	*Lysiloma latisiliquum*	Shoot, pod, seed
28.	*Mimosa pudica*	Seed
29.	*Neptunia oleracea*	Plant, pod, seed
30.	*Parkia biglandulosa*	Pod pulp, seed, seedling
31.	*Parkia biglobosa*	Pod, pulp, seed
32.	*Parkia timoriana*	Pod, seed
33.	*Pithecellobium dulce*	Seed aril, oil
34.	*Prosopis cineraria*	Gum, leaf, pod, seed
35.	*Prosopis juliflora*	Pod
36.	*Prosopis pallida*	Tuber
37.	*Prosopis pubescens*	Pod
38.	*Samanea saman*	Pod
39.	*Xylia xylocarpa*	Pod, seed

Table 3.7
Frequency Index of Economic Uses of Legumes in India

The percentage index of the frequency of uses of legume species, with seven or more of the 18 identified uses, are listed. More the uses for a species, the higher are the percentage index and so is the economic importance of the species.

No.of uses	Percentage index	Species
CAESALPINIACEAE		
13	72.2	*Tamarindus indica*
12	66.6	*Ceratonia siliqua, Senna siamea*
11	61.1	*Bauhinia purpurea, Hardwickia binata*
10	55.5	*Bauhinia variegata, Senna auriculata, Senna tora*
9	50.0	*Bauhinia racemosa*
8	44.4	*Bauhinia vahlii, Gleditsia aquatica, Parkinsonia aculeata, Peltophorum pterocarpum*
7	38.8	*Bauhinia racemosa, Bauhinia tomentosa, Cassia fistula*

No.of uses	Percentage index	Species
FABACEAE		
14	77.7	*Butea monosperma, Sesbania grandiflora*
13	72.2	*Erythrina variegata, Pongamia pinnata*
12	66.6	*Robinia pseudoacacia*
11	61.1	*Pterocarpus marsupium*
10	55.5	*Gliricidia sepium, Lupinus albus*
9	50	*Cytisus scoparius, Desmodium oojeinense, Medicago sativa, Pachyrrhizus erosus, Sesbania sesban, Spatholobus parviflorus, Styphnolobium japonicum, Vicia faba, Vigna unguiculata*
8	44.4	*Butea superba, Cajanus cajan, Clitoria ternatea, Crotalaria pallida, Cyamopsis tetragonoloba, Dalbergia latifolia, Glycine max, Lespedeza juncea, Mucuna pruriens* var. *pruriens, Myroxylon balsamum,Pueraria montana* var., *lobata, Sesbania bispinosa, Sesbania cannabina, Tephrosia purpurea, Tephrosia vogelii, Trigonella foenum-graecum*
7	38.8	*Arachis hypogaea, Crotalaria juncea, Crotalaria spectabilis, Dalbergia sissoo, Desmodium gangeticum, Kummerovia stipulacea, Lupinus angustifolius, Psophocarpus tetragonolobus,Pterocarpus indicus*
MIMOSACEAE		
15	83.3	*Albizia lebbeck*
14	77.7	*Acacia nilotica, Lysiloma latisiliquum, Prosopis juliflora*
13	72.2	(none)
12	66.6	*Acacia farnesiana, Albizia amara, Albizia procera*
11	61.1	*Acacia leucophloea, Acacia senegal, Albizia chinensis*
10	55.5	*Acacia catechu, Acacia jacquemontii, Aacia mearnsii, Albizia odoratissima, Pithecellobium dulce, Xylia xylocarpa*
9	50	*Acacia auriculiformis, Acacia pennata, Acacia sinuata, Adenanthera pavonina, Entada gigas*
8	44.4	*Acacia dealbata, Faidherbia albida, Parkia biglobosa, Prosopis cineraria*
7	38.8	*Acacia decurrens, Archidendron clypearia, Dicrostachys cinerea, Falcatoria moluccana, Parkia biglandulosa, Samanea saman*

CHAPTER-4

Cultivated Legumes

4.1 INTRODUCTION

Of about 719 species of economically important legumes discussed in this book, about 280 species (38.9 per cent) are edible. However, only a few are cultivated as crops for food, feed, fodder and green manure. Several of these species have multiple uses and appear in different categories of use (see Tables 3.1, 3.2 and 3.3 in chapter 3). The dried mature seeds of some of the species are used as gram/pulse while the tender fruits and seeds of the same species are used as vegetable. Many important edible legumes, such as tamarind, are obtained from wild and self-sown populations, and are rarely intentionally grown.

Most of the legume crops have been in cultivation for centuries. In the face of attraction of high commercial returns from crops like cotton, tobacco and sugarcane, the cultivation of some of the dry land legume crops such as the pulses and oil seeds has fallen significantly, causing an upset of rural economy and severe food grain and oil seed shortage. However, soybean (*Glycine max*) has gained enormous importance as a source of food, feed and oil the world over, during the past five or six decades. The extension of use and cultivation of soybean is the most significant event of the last century among not only the cultivated legumes but all agricultural crops.

There are several publications providing basic information on legume crops, about the varieties in cultivation in India, practices of cultivation, pests and diseases, among which the 'Handbook of Agriculture' (Anonymous, 2002) is very comprehensive, but unfortunately this is not up-to-date.

4.2 LEGUME CROPS

The following are the commonly cultivated legume crops:

1. Edible Oil Crops

1. *Arachis hypogaea* (Groundnut)
2. *Glycine max* (Soybean)

2. Fodder and Forage Crops

1. *Clitoria ternatea* (Butterfly pea)
2. *Crotalaria juncea* (Sunn hemp)
3. *Cyamopsis tetragonoloba* (Cluster bean)
4. *Lablab purpureus* (Field bean)
5. *Lathyrus sativus* (Chickling pea)
6. *Medicago sativa* (Leucerne)
7. *Melilotus parviflora* (Indian sweet clover)

8. *Mucuna pruriens* var. *pruriens* (Velvet bean)
9. *Phaseolus acutifolius* (Moth bean)
10. *Trifolium alexandrinum* (Egyptian clover)
11. *Trigonella foenum-graecum* (Fenugreek)
12. *Vigna mungo* (Black gram)
13. *Vigna radiata* (Green gram)
14. *Vigna unguiculata* (Cow pea)

3. Fibre Crop

1. *Crotalaria juncea* (Sunn hemp)

4. Gram and Pulse Crops

1. *Arachis hypogaea* (Groundnut)
2. *Cajanus cajan* (Red gram)
3. *Canavalia gladiata* (Sword bean)
4. *Cicer arietinum* (Chickpea)
5. *Glycine max* (Soybean)
6. *Lablab purpureus* (Lablab bean)
7. *Lens culinaris* (Lentil)
8. *Macrotyloma uniflorum* (Horse gram)
9. *Phaseolus acutifolius* (Moth bean)
10. *Pisum sativum* (Pea)
11. *Vigna mungo* (Black gram)
12. *Vigna radiata* (Green gram)
13. *Vigna unguiculata* (Cow pea)

5. Green Manure Crops

1. *Gliricidia sepium* (Gliricidia)
2. *Sesbania sesban* (Sesban)

and most of the crops in other groups above

6. Pasture Crops

1. *Clitoria ternatea* (Butterfly pea)
2. *Lablab purpureus* (Lablab bean)

and most of the crops in the above groups

7. Vegetable Crops

1. *Canavalia gladiata* (Sword bean)
2. *Cyamopsis tetragonoloba* (Cluster bean)
3. *Glycine max* (Soybean)
4. *Lablab purpureus* (Lablab bean)
5. *Phaseolus lunatus* (Lima bean)
6. *Phaseolus vulgaris* (Common bean)
7. *Pisum sativum* (Pea)
8. *Psophocarpus tetragonolobus* (Winged bean)
9. *Trigonella foenum-graecum* (Fenugreek)
10. *Vicia faba* (Broad bean)
11. *Vigna unguiculata* (Cow pea)

In addition to these, some others such as *Indigofera tinctoria* (dye) and *Cullen corylifolium* (medicinal seed oil) are cultivated on a small scale in certain parts of India.

4.3 CULTIVARS AND LAND RACES

There has been intensive and extensive research to improve legume crops in India over the decades. Consequently, each one of the cultivated legume species now includes a large number of cultivated varieties (Cultivars) and locally developed and preferred races (Land Races), bred to suit different agro-climatic conditions in the country.

The Indian Agricultural Research Institute, New Delhi, the Universities of Agricultural Sciences and the research stations of the Department of Agriculture of different States, have contributed immensely to improvement of legume crops. In addition, the International Crops Research Institute for Semi-Arid Tropics (ICRISAT), Hyderabad, is concerned with research and development of several legume crops, particularly for the drier areas.

Gram and pulse crop varieties are bred for different qualities such as high yield, nutritional enhancement, reduction of anti-nutritional factors and toxins, pest and disease resistance, tolerance of drought and salinity, visible features like size and acceptable colour, etc. There has been a continuous addition of new improved varieties, while some old favourites went out of cultivation. It is difficult to draw an up-to-date list of all current cultivars and the land races of legumes cultivated in different parts of India and a survey of this diversity is outside the scope of this book.

However, photographs of mature seeds of varieties of the following cultivated legumes are presented here, in order to provide a glimpse into the enormous genetic diversity in some edible legume crops in India: Red gram *(Cajanus cajan)*, sword bean *(Canavalia gladiata)*, chick pea *(Cicer arietinum)*, cluster bean *(Cyamopsis tetragonoloba)*, soybean *(Glycine max)*, field bean *(Lablab purpureus)*, horse gram *(Macrotyloma uniflorum)*, Lima bean *(Phaseolus lunatus)*, common bean *(Phaseolus vulgaris)*, pea *(Pisum sativum)*, fenugreek *(Trigonella foenum-graecum)*, broad bean *(Vicia faba)*, black gram *(Vigna mungo)*, green gram *(Vigna radiata)*, and cow pea *(Vigna unguiculata)* (Figs. 4.1 to 4.11). Photographs of legume vegetables are presented in Figs. 4.12 to 4.17

4.4 LEGUME IN GENETIC ENGINEERING

The recombinant DNA technology, has made it possible to select genes from any organism and to insert them into the genome of another organism, irrespective of the degree of genetic relationship. The resultant varieties are called transgenic varieties.

Transgenic varieties of soybean, with genes for pesticidal proteins from the ubiquitous soil bacterium *Bacillus thuringiensis* (Bt), are now in cultivation in several countries. Varieties of soybean resistant to specific herbicides have been developed to protect the crop from herbicides sprayed to control weeds. Genes for protein from the Brazil nuts and bacteria have been incorporated into soybean, to provide for the essential amino acids such as methionine, in which the normal plant foods are usually deficient.

ICRISAT has developed a transgenic groundnut variety resistant to Indian peanut clump virus and a Bt red gram variety that is resistant to pod borer. Both the diseases cause severe damage to the respective crops and traditional breeding techniques did not yield resistant varieties.

The iron content of normal rice grain is negligible. Genes for ferritin, isolated from soybean and common bean, were inserted into rice to produce iron-enriched rice. Such transgenic rice is expected to correct iron deficiency in our diet, which causes anemia.

While Bt soybean and herbicide resistant soybean are widely in cultivation in different parts of the world, it would take some more time for the release of transgenic protein rich soybean, virus resistant groundnut, pod borer resistant red gram and the iron rich rice, for commercial cultivation.

4.5 LEGUMES IN GARDENS

A number of species of legumes are cultivated in gardens as ornamental plants for their showy flowers and/or leaves. Some are planted as avenue trees and as hedges, while some are used as shade plants in plantations and for support of weak-stemmed cultivated non-legume species, as detailed in Chapter 3 (section 3.2.11, Tables 3.1, 3.2, 3.3). Species of the Caesalpiniaceae predominate the other two families in such uses. In some countries such as Thailand, the tamarind tree is trimmed and trained.

The following species are more commonly found in gardens:

1. Caesalpiniaceae

1. *Amherstia nobilis*
2. *Bauhinia acuminata*
3. *Bauhinia monandra*
4. *Bauhinia purpurea*
5. *Bauhinia racemosa*
6. *Bauhinia tomentosa*
7. *Bauhinia variegata*
8. *Brownea coccinea*
9. *Caesalpinia bonduc*
10. *Caesalpinia pulcherrima*
11. *Caesalpinia sappan*
12. *Cassia fistula*
13. *Cassia grandis*
14. *Cassia javanica*
15. *Cassia renigera*
16. *Cassia roxburghii*
17. *Ceratonia siliqua*
18. *Colvillea racemosa*
19. *Delonix elata*
20. *Delonix regia*
21. *Haematoxylum campechianum*
22. *Hardwickia binata*
23. *Parkinsonia aculeata*
24. *Peltophorum pterocarpum*
25. *Saraca asoca*
26. *Saraca declinata*
27. *Senna floribunda*
28. *Senna alata*
29. *Senna alexandrina*
30. *Senna bicapsularis*
31. *Senna siamea*
32. *Senna sophera*
33. *Senna spectabilis*
34. *Senna surattensis*
35. *Tamarindus indica*

2. Fabaceae

1. *Baphia nitida*
2. *Butea monosperma*
3. *Butea superba*
4. *Canavalia gladiata*
5. *Castanospermum australe*
6. *Clitoria ternatea*
7. *Crotalaria alata*
8. *Crotalaria spectabilis*
9. *Cytisus scoparius*
10. *Dalbergia latifolia*
11. *Dalbergia sissoo*
12. *Erythrina corallodendron*
13. *Erythrina crista-galli*
14. *Erythrina variegata*
15. *Gliricidia sepium*
16. *Medicago sativa*
17. *Melilotus albus*
18. *Millettia peguensis*
19. *Mucuna pruriens*
20. *Myroxylon balsamum*
21. *Pongamia pinnata*
22. *Pterocarpus marsupium*
23. *Pterocarpus santalinus*
24. *Robinia pesudoacacia*
25. *Sesbania grandiflora*
26. *Styphnolobium japonicum*
27. *Tephrosia candida*
28. *Trifolium pratense*
29. *Trifolium repens*
30. *Ulex europaeus*
31. *Wisteria sinensis*

3. Mimosaceae

1. *Acacia auriculiformis*
2. *Acacia cyanophylla*
3. *Acacia dealbata*
4. *Acacia decurrens*
5. *Acacia ferruginea*
6. *Acacia melanoxylon*
7. *Acacia nilotica*
8. *Acacia podalyriifolia*
9. *Adenanthera pavonina*
10. *Albizia amara*
11. *Albizia lebbeck*
12. *Dicrostachys cinerea*
13. *Leucaena glauca*
14. *Lysiloma latisiliquum*
15. *Mimosa hamata*
16. *Parkia biglandulosa*
17. *Parkia biglobosa*
18. *Pithecellobium dulce*
19. *Prosopis cineraria*
20. *Prosopis juliflora*
21. *Samanea saman*
22. *Xylia xylocarpa*

Fig. 4.1 Mature seeds of cultivars of *Cajanus cajan*

Var. GCP-103

Var. Annigeri 1996

Var. Local GKVK

Var. BG-373

Var. BG-390

Var. Kabul

Var. Local IAS

Fig. 4.2 Mature seeds of cultivars of *Cicer arietinum*

Cyamopsis tetragonoloba
Var. Pusa naubhar

Cyamopsis tetragonoloba
Var. Gujarath-U

Cyamopsis tetragonoloba
Var. F_1 hybrid

Glycine max
Var. KB-79

Glycine max
Var. Hardee

Glycine max
Var. KHSB-2

Lablab purpureus
Var. Local

Lablab purpureus
Var. Arka vijay

Lablab purpureus
Var. Arka ajay

Lablab purpureus
Var. Arka jay

Fig. 4.3 Mature seeds of cultivars of *Cyamopsis tetragonoloba, Glycine max* and *Lablab purpureus*

Fig. 4.4 Mature seeds of cultivars of *Macrotyloma uniflorum*

Canavalia gladiata
Var. Local Red

Canavalia gladiata
Var. Local White

Phaseolus vulgaris
Var. IIHR 220

Phaseolus vulgaris
Var. Sel 2.

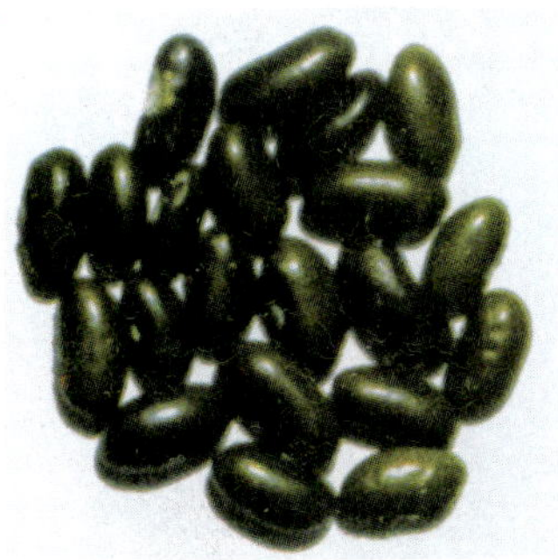

Phaseolus vulgaris
Var. Burpee stringles

Phaseolus vulgaris
Var. Open pollinated

Phaseolus vulgaris
Var. Sel. 909

Phaseolus *vulgaris*
Var. Arka komal

Phaseolus vulgaris
Var. Selection 9

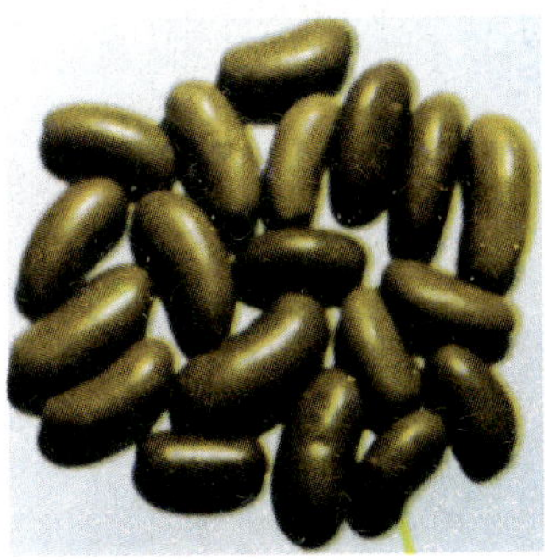

Phaseolus vulgaris
Var. Dwarf

Fig. 4.5 Mature seeds of cultivars of *Cananalia gladiata* and *Phaseolus vulgaris*

Var. Bonneville

var. Azad P_1

Var. Arka N-Delhi

Var. Arka Ajit

Fig. 4.6 Mature seeds of cultivars of *Pisum sativum*

Phaseolus lunatus
Var. IIHR-Sel-4

Phaseolus lunatus
Var. Local

Trigonella foenum-graecum
Var. Local

Vigna mungo
Var. T-9

Vigna mungo
Var. VB-4

Fig. 4.7 Mature seeds of cultivars of *Phaseolus lunatus, Trigonella foenum-graecum* and *Vigna mungo*

Fig. 4.8 Mature seeds of cultivars of *Vicia faba*

Fig. 4.9 Mature seeds of cultivars of *Vicia faba*

Var. MGG-341

Var. GM-84-4

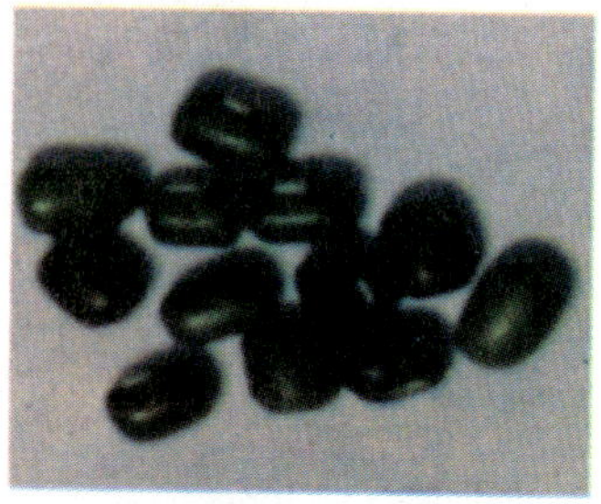

Var. LGG-460

Var. PBM

Var. SLM-100

Var. Pusa baisaki

Var. 9105

Var. Kopargoan

Var. PDM-86-199

Fig. 4.10 Mature seeds of cultivars of *Vigna radiata*

Fig. 4.11 Mature seeds of cultivars of *Vigna unguiculata*

Fig. 4.12 Pods of *Cyamopsis tetragonoloba*

Fig. 4.13 Pods of *Lablab purpureus*

Fig. 4.14 Pods of *Phaseolus lunatus*

Fig. 4.15 Pods of *Phaseolus vulgaris*

Fig. 4.16 Pods of *Pisum sativum*

Fig. 4.17 Pods of *Vigna unguiculata*

CHAPTER-5

Chemistry of the Legumes

5.1 INTRODUCTION

Organisms should be viewed as composed of systematically ordained packets of chemical compounds, in a complex, interrelated and interdependent milieu. How the organisms acquire their physical form and how they react and respond to beneficial and antagonistic internal and external factors are the consequence of chemical interactions. The physiological functions, such as nutrition, growth, reproduction and survival, are all manifestations of chemical interactions. So are the physiological disorders, diseases caused by pests and pathogens, as well as the cures and recovery. All these responses and interactions depend upon the genotype (genetic constitution) of the organisms, the environment in which they live and the interaction of the genotype and the environment.

An understanding of the chemistry of organisms is required in order to appreciate the potential of harm we suffer and the benefits we derive, from organisms.

5.2 CHEMICAL COMPOUNDS IN ORGANISMS

Organisms contain several thousands of organic and inorganic compounds belonging to almost every chemical group, now generally referred as biomolecules. Both the wild and cultivated plants contain certain compounds that are beneficial, anti-nutritional or even toxic, depending upon several factors, such as concentration, duration of intake and our own individual susceptibilities. With time, man has learnt to put many of these compounds to several uses in nutrition, medicine and industry. In spite of a two to three centuries of such effort, we have not yet realized the full economic potential of plant chemicals. We need to study and understand the detailed chemistry of our plant resources in order to derive the full benefit from them.

The species we use as food are the sources of the major nutrients such as carbohydrates, proteins and fats. In addition, we need several hundreds of other compounds present in plants in smaller quantities, and these chemicals play very crucial roles in our metabolism and health. During the process of domestication of plants and animals, our ancestors have painstakingly and cleverly chosen species and their varieties that contained more of beneficial chemical compounds, and less of hazardous compounds, in an effort spanning over 10,000 years of the history of agriculture. Our understanding of the chemistry of edible plant varieties will help us in choosing the right food and feed more judiciously, avoiding the hazardous chemicals. We can use the toxic chemicals to develop drugs to cure our disorders and diseases. And we need raw material from plants for our industries.

5.3 BIOINORGANIC COMPOUNDS

In addition to thousands of organic compounds that basically contain carbon and hydrogen, organisms also contain several elements and a large number of inorganic compounds. This area of biological chemistry is now rapidly developing into 'bioinorganic chemistry'. Plants absorb and accumulate a number of elements and inorganic compounds, from water and soil.

Elements and inorganic compounds may also influence biological processes either in a beneficial or a hazardous manner (bioactivity). Mineral deficiencies cause disorders in the plants themselves and also in our body system. Arsenic and mercury are cumulative poisons, used as medicine in small doses. Selenium absorbed by plants from the soil is a toxicant. Cyanogenic glycosides are frequently found in many plants including legumes, making such plants toxic to grazing animals.

5.4 PRIMARY AND SECONDARY METABOLITES

Metabolism is the sum total of all chemical interactions in an organism. The products of metabolism are metabolites, which in turn may be the players in other metabolic processes. Certain compounds with high molecular weight, such as the deoxyribonucleic acid (DNA), ribonucleic acid (RNA) and proteins (that include thousands of enzymes that are essential to every other metabolic function), are considered as the primary metabolites. Other compounds, most of which are of low molecular weights, are secondary metabolites. Such a categorization is not entirely appropriate, since certain carbohydrates that are high molecular weight compounds, and some low molecular weight compounds like the amino acids, are also involved in primary metabolism.

The secondary metabolites do not usually re-enter the metabolic processes. These are called the Natural Products, which either accumulate in certain tissues/organs or excreted (in animals). Over a million of natural products are known, mostly from plants. Organisms are virtually highly systematically organized chemical factories precisely turning out an enormous number of chemical compounds in varying quantities, in a short time and at ambient temperatures.

The majority of compounds occurring in the biological systems are endlessly in a metabolic flux, due to continuous interactions, fresh synthesis, utilization or inter-conversion of compounds or some of them going into the storage sites. The nucleotides, amino acids, carbohydrates, carotenoids and organic acids occur in pools with continuous additions and depletions. Against this dynamic state, it is an intellectual and technical feat to extract, isolate, characterize and identify such a large number of chemical compounds from organisms.

While it is not always possible to explain the exact individual role of each of the large number of secondary metabolites in plants, many of them have been shown to be the chemical defense against pests, pathogens and grazers, the common sources of biotic stress. Some compounds accumulate in response to abiotic stress, such as high temperature, drought or salinity.

5.5 BIOSYNTHESIS, SEMISYNTHESIS, STRUCTURAL MODIFICATION AND ARTIFICIAL SYNTHESIS OF BIOMOLECULES

The natural synthesis of different chemical compounds in biological systems utilizing the elements and/or other simpler chemical compounds in the environment of the cell is biosynthesis, also called *in vivo* (in the living organism) synthesis. The complicated routes of biosynthesis of different compounds are called biosynthetic pathways, which virtually constitute a maze. Semi-synthesis is synthesizing a biological compound in the laboratory, using naturally occurring precursor compounds. In certain situations biochemists modify a part of the structure of a naturally occurring compound in order to produce a more effective or more useful or less expensive product. This is structural modification. When compounds are synthesized totally artificially in the laboratory, starting from elements and/or simple artificially synthesized compounds, it is synthesis or artificial synthesis or *in vitro* (literally, in the test tube) synthesis. With the availability of extensive computer databases on the physico-chemical characteristics of chemical compounds and their functional properties, it is now possible to design a compound with desired functions, using procedures of bioinformatics, before one attempts to actually synthesize it in the laboratory. This bypasses the conventional route that involves a lot of time and money with the results being uncertain. This computer based procedure, described as *in silico* (because computer functions are based on silicon chips), is the modern tool for the synthesis of new compounds, particularly drugs, usually based on natural models, in a very short time and is far less expensive. The *in vitro* synthesis follows the *in silico* procedure. In actual practice, all these procedures come in handy, in one or the other situation, since there is no single universally convenient or economical method of designing and/or synthesizing biomolecules.

5.6 BIOACTIVITY OF METABOLITES

Many of the primary metabolites like proteins and a very large number of secondary metabolites, either produced within an organism (endogenous) or introduced from outside sources (exogenous), induce responses in biological systems, called bioactivity. Bioactivity can be either beneficial or hazardous to organisms. We take advantage of the bioactivity of such compounds and use them in various ways.

The beneficial compounds function as precursors of other metabolically significant compounds, micronutrients, growth promoters, antioxidants, pharmaceuticals, nutriceuticals, agriceuticals and/or participate in biosynthetic processes. In recent times, antioxidants have gained a lot of importance as they scavenge free radicals in our system. Free radicals form routinely during metabolic processes and are implicated in several diseases such as heart attack and cancer.

The hazardous compounds may function as anti-nutritional factors limiting the benefits we derive from food. Some others, called toxicants or toxins, interfere with metabolic processes by inhibiting enzymes or act in some other way to disrupt metabolic processes, causing minor to major disruption of physiological processes, or organic damage, leading to several disorders, or even death.

A very large number of bioactive compounds are beneficial at lower doses or concentrations, and toxic at higher levels. This factor is in fact crucial in the development of drugs. Many of the toxicants are useful as drugs, because usually the pathogens and the diseased cells/tissues are more sensitive to them than the healthy. Depending upon their effects, toxicants are also used to control pathogens, deranging their metabolism in some way, without affecting the host's metabolic processes.

5.7 CLASSIFICATION OF CHEMICAL COMPOUNDS IN ORGANISMS

In order to clearly understand different chemical compounds and their relationships, it is necessary to classify them systematically. A classification facilitates an orderly handling, understanding and conveying information. Classification is particularly helpful in dealing with the enormous number of secondary metabolites.

The classification of the multitude of biomolecules is a hard task. There are several different ways of classifying biomolecular compounds, but none is totally satisfactory or useful in all contexts. The complexity of the synthesis of thousands of chemical compounds via

Box 5.1

Outline of the Classification of Secondary Metabolites in Legumes Adopted in the Subsequent Chapters

The following outline of classification of chemical compounds excludes the nucleic acids, proteins, carbohydrates and lipids. It mainly deals with the secondary metabolites. The phytates, inhibitors of digestive enzymes, anti-vitamin factors and inorganic compounds are heterogenous assemblages and not cogent and related chemical groups.

A. NITROGENOUS COMPOUNDS

- a. **Amino acids** (protein and non-protein amino acids and their derivatives)
- b. **Amines**
- c. **Alkaloids**
  - i. Indole alkaloids
  - ii. Indolizidine alkaloids
  - iii. Piperidine alkaloids
  - iv. Pyrrolizidine alkaloids
  - v. Quinolizidine alkaloids
  - vi. Isoquinolizidine alkaloids
  - vii. Macrocylic alkaloids
- d. **Cyanogenic glycosides**
- e. **Pyrimidine glycosides**

B. PHENOLIC COMPOUNDS:

- a. **Phenolic glycosides**
- b. **Flavonoids**
- c. **Tannins**
- d. **Quinones**
- e. **Chromones**
- f. **Phenyl propanoids**
- g. **Anthrones**

C. TERPENOIDS

D. ORGANIC ACIDS

E. PHYTATES

F. INHIBITORS OF DIGESTIVE ENZYMES

G. VITAMIN INHIBITORS OR ANTI-VITAMINFACTORS

H. ELEMENTS AND INORGANIC COMPOUNDS

inter-dependent and inter-connected routes and the overlap of their physico-chemical properties make it difficult to arrive at a universally acceptable classification. With the result, the classification of biological chemicals is need based and depends upon the context.

The nucleic acids are polymers of nucleotides. Similarly, the proteins are polymers of amino acids, carbohydrates are polymers of monosaccharides (simple sugars) and lipids are polymers of fatty acids. These classes of compounds offer fewer problems though the classification within each group is not an easy matter. It is the secondary metabolites that are difficult to handle. In this book, a general-purpose classification of secondary metabolites used by plant chemists, is adopted as outlined in Box 5.1. The last four groups (E,F,G,H) in Box 5.1 do not strictly constitute chemical classes, but are put together basing on the similarity of their activity.

5.8 CHEMICAL COMPOUNDS IN LEGUMES

The Legumes, constituting one of the largest of the natural plant groups, contain an enormous number of chemical compounds belonging to almost every chemical group.

Harborne *et al.*, (1971) have provided the first comprehensive survey of the chemistry of the legumes, where different classes of compounds were reviewed by specialists under the groups flavonoids, alkaloids, non-protein amino acids, carbohydrates, lipids, terpenoids, seed proteins, lectins and enzymes. This was capped up by Gibbs (1974), who provided an exhaustive survey of chemical compounds in flowering plants, legumes forming a significant part of the coverage. Several articles on the chemistry of legumes, revised and updated after Harborne *et al.*, (1971), are included in a two-volume publication edited by Polhill and Raven (1981). Harborne and Baxter (1996) discussed toxic compounds in plants, of which many are from the legumes. Harborne (1973, 1984) also provided the techniques required for the analysis of phytochemical compounds. The legume database ILDIS contains information on the distribution of various chemical compounds in different legume taxa. NAPRALERT is a source of very useful chemical information on bioactive natural products, of which legumes form a significant part. Duke's phytochemical database provides information on the phytochemistry and bioactivity of legumes, along with other economically important plant species. There are several other sources on the Internet for obtaining need-based information of biomolecular compounds.

A comprehensive review of the chemistry of the legumes requires several volumes and is outside the scope of this book. Several chemical compounds of different chemical groups found in legume species and their potential uses are discussed throughout this book, more particularly in Chapters 3, 6, 7, 8, 9, 10, 11, 12, 13, 14, 15, 22 and 23 and in the species profiles.

CHAPTER-6

Nutritional Significance of Edible Legumes

6.1 FOOD AND NUTRITION

'Food' is any substance that organisms feed on, and which nourishes the body promoting growth and functioning. Staple food is the main commodity of food, such as rice, wheat, millets or tubers. The term 'feed' generally refers to the food of domesticated animals, the livestock and poultry, often including human non-vegetarian foods.

'Nutrition' is the process of nourishing the body, encouraging body growth and functions and a nutrient is a nourishing substance in food. 'Diet' is the planned selection of food under normal health conditions, or prescribed in ill health, designed to provide our body with what all it needs to be supplied from the outside and which the body cannot itself synthesize. For these reasons, our diet should include diverse food items. Indian culture has designed balanced diets as seen in different parts of the country.

The process of 'digestion' breaks down complex substances in the food into their components, such as starch (from plants) or glycogen (from animals) into glucose, proteins into amino acids or fats into fatty acids, so that they can be absorbed across the brush border of the small intestine. This process requires several hundred enzymes. Digestion also releases several other chemical compounds for the use of the body. Organisms utilize these exogenous chemical compounds to synthesize a number of endogenous chemical compounds, that are essentially required, but do not come through normal diet. The dietary requirements and the digestion process of both humans and mammals are largely similar.

In addition to the three major nutrients, the carbohydrates, proteins and fats that constitute the bulk of our food, our body needs several other nutrients such as vitamins and minerals, in varying quantities, for proper functioning. These are the micronutrients that are needed in smaller quantities. The human body uses these nutrients as precursors for, or promoters of, the synthesis of other compounds. Plants do not contain vitamin A, but only the three carotenoids that are precursors of the vitamin. The mammalian system uses β-carotene or α-carotene or β-cryptoxanthin as a precursor to synthesize vitamin A, whose deficiency leads to several diseases, more importantly blindness. Iron is required for the synthesis of haemoglobin, the critical component of blood. Iron deficiency leads to anaemia. Calcium is basic for healthy bones and teeth and also has other roles to perform. Compounds of phosphorous are the major players in energetics. Zinc is important for the functioning of nearly 200 enzymes. Our body is riddled with free radicals that form during normal metabolic reactions in the body. Free radicals are implicated in heart disease and cancer. Free radicals are removed from the system by antioxidants, such as phenolic compounds and carotenoids from plants. All these substances should come from food. Food deficient in nutrients leads to malnutrition and results in disease, or even death.

6.2 NUTRITIONAL FACTORS, BIOAVAILABILITY AND BIOCONVERSION

The nutritive potential of a particular food item is basically the compounded function of a) the relative concentrations of the nutritionally significant chemical compounds in the food,

b) the digestibility of the food consumed, c) the availability of these nutrients to the body system, and d) their convertibility by the body system into compounds that become metabolically involved in the body. All these are the normal functions of a healthy body system. Any factor that affects any of these processes is an anti-nutritional factor (ANF), which may reside in the food or the body system.

Certain compounds, such as proteinase inhibitors, present in plant foods, may prevent the digestive enzymes from functioning and thus affect digestion of foods, which will limit the release of nutrient compounds from the food that was consumed. In addition to such anti-nutritional factors, plant foods also contain hundreds of other toxic compounds some of which are harmless in low concentrations but are toxic in larger quantities. The anti-nutritional factors and toxins in edible legumes are discussed in Chapters 7 and 8.

Even if a particular nutrient is abundantly available in a particular food item, and the digestive process releases all or most of it, all that is released may not be available to the body system, due to the presence of certain other compounds in the food. This is 'bioavailability'. For example, phytic acid limits the bioavailability of iron. Phytase is the enzyme that is required to metabolize phytate formed by the complexing of phytic acid with iron, to promote the availability of iron.

'Bioconversion' is the process by which a compound that is available to the body system is converted into another compound needed for metabolism. For example, vitamin A is formed in the mammalian system through bioconversion of β-carotene, by the enzymes β-carotene-15, 15-deoxygenase and retinal reductase. If these enzymes are not available in adequate quantities, even if the precursor compound is bioavailable, the body does not benefit from it.

The stability in the body system, transport and storage of certain compounds is dependent upon the presence of some other compounds. For example, retinal (a form of vitamin A) requires an endogenous, retinal binding protein, and for its stability it requires palmitic acid to form an ester, the retinyl palmitate, the storage form of vitamin A. Some compounds such as palmitic acid are available through food, but endogenous substances must be synthesized by the body and cannot be supplied through food alone.

Thus, digestion, bioavailability and bioconversion are very complex phenomena that essentially determine our nutritional status. The factors that adversely affect digestion, bioavailability and bioconversion are all anti-nutritional factors, and much depends upon the state of health of an individual.

6.3 NUTRITIVE POTENTIAL OF PLANTS

All edible plants contain a certain quantity of carbohydrates, proteins, fats and other nutrients. The cereals (rice, wheat, corn), the millets (sorghum, pearl millet), and the tubers (potato, tapioca, taro), provide us most of the carbohydrates needed in the form of starch. Though pulses and the non-vegetarian foods are the chief sources of proteins, they also contain large quantities of carbohydrates. Different kinds and quantities of fats come from both vegetarian and non-vegetarian diets. Several items of plant foods such as vegetables, greens and fruits provide us with all the other nutrients we need, in some degree or the other. The grazing animals get all the nutrients they need from wild vegetation. We can use as dietary supplements, the nutrients extracted from wild or cultivated plants. Some of the nutrients can also be totally chemically synthesized. Semi-synthetic or synthetic nutrients that are marketed as dietary supplements are called nutriceuticals, like the drugs that are components of medicine are pharmaceuticals, the components of cosmetics are cosmeceuticals and those used in agriculture are agriceuticals. Both edible and non-edible plants constitute a rich source of nutriceuticals, pharmaceuticals, cosmeceuticals and agriceuticals.

Large sections of urban and semi-urban populations spend millions of rupees on nutriceuticals and cosmeceuticals, which can conveniently and inexpensively be provided by edible or by forage and fodder plants. Using these sources requires an understanding of the chemical constitution of the species.

Box 6.1

Nutritional Benefits from Legume Seeds

a) Legume seeds are unparalleled sources of plant proteins, which also help to conserve calcium, unlike animal proteins.
b) Legume proteins provide a wide range of amino acids, some of which (lysine, methionine) are essential to our health.
c) They provide a wide range of vitamins and/or their precursors.
d) Rich in nutritionally significant minerals such as calcium, iron, magnesium, potassium, and zinc.
e) Very low in cholesterol, and both total and saturated oils.
f) Rich in carotenoids which function as antioxidants to scavenge free radicals and reduce risk of such diseases as heart attack and cancer.
g) Rich in dietary fibres that lowers levels of blood cholesterol and blood glucose and reduces the risk of colon cancer.

6.4 NUTRITIVE POTENTIAL OF LEGUMES

The nutritional significance of legumes has been recognized for long (Aykrayd and Doughty, 1964). Nutritionally legumes are second only to cereals in their importance. They are a cheaper source of proteins when compared to animal proteins, particularly in the developing countries. Legumes are also an important source of carbohydrates, fats, dietary amino acids, vitamins, antioxidants, minerals and dietary fibre (Box 6.1; Table 6.1 to 6.3).

The nutrient composition of edible legumes is greatly influenced by genetic and environmental factors. The cultivated legumes are influenced by fertilizer application and agronomic practices.

Nutritional data on legumes: The nutrient component of edible legumes is given in Table 6.1. The bulk of the data provided in this table, with several gaps in information, came from general surveys conducted several decades ago, when the analytical techniques were not as sophisticated as they are today. In addition, the chemical composition of foods is controlled by genetic and environmental factors, resulting in variation from sample-to-sample and from time-to-time. Consequently, the information in Table 6.1 is only an approximate guide to the nutrient component of the respective items. Nutritionists have been analyzing specific foods or their combinations for their nutritional or toxic potential. Such specialized information is available in technical publications on the subject.

Proteins: Edible legume seeds generally contain 18 to 43 per cent protein. Grams and pulses cater to most of our protein needs, and no other plant food provides such large quantities of proteins.

Legume proteins have low digestibility, low in sulphur containing amino acids and contain antinutritional factors. Nevertheless, combining legume seeds with cereals compensates these deficiencies.

Amino acids: Legume peptides and proteins contain higher quantities of the amino acids such as cysteine, cystine, histidine, lysine and threonine, compared to other foods, but are generally low in tryptophan and methionine.

Carbohydrates: The current wave of non-carbohydrate diets for weight control emphasizes eating large quantities of grams and pulses (and vegetables), to the almost total exclusion of cereals and millets, on the presumption that the non-cereal and non-millet diet does not contain starch. Contrary to this general belief, legume seeds contain 12.6 to 66 per cent of carbohydrates in the form of starch (Table 6.1). In fact, the legume vegetables also provide large quantities of starch, though cannot be compared with the cereals and millets.

Fats: The fat content of legume seeds is 0.7 to 40 per cent. Other than groundnut and soybean, which are commercial sources of edible oils, the fat content of several legume seeds such as chickpea and green gram is considerable.

Vitamins: Pulses are a good source of vitamins. They usually contain 0.30 to 9.90 mg/100g thiamine and 0.13 to 0.39 mg/100g riboflavin, as well as considerable quantities of vitamin B_{12}, ascorbic acid, folic acid, tocopherol, pantothenic acid, biotin, and nicotinic acid. The terminology of vitamins has undergone radical changes leaving a lot of confusion

Box 6.2
Terminology of Vitamins

Common names	Scientific names
Vitamin A	Retinyl acetate
Vitamin A_1	Retinol
Vitamin B_1	Thiamine hydrochloride
Vitamin B_2	Riboflavin (also called vitamin G)
Vitamin B_3	Nicotinic acid, Nicotinamide, Niacinamide
Vitamin B_4	Adenine
Vitamin B_5	Calcium pantothenate, Pantothenic acid
Vitamin B_6	Pyridoxin
Vitamin B_7	Biotin (also called vitamin H)
Vitamin B_{12}	Cyanocobalamine
Vitamin B_x	*p*-aminobenzoic acid
Vitamin C	Ascorbic acid
Vitamin D_2	Ergocalciferol
Vitamin D_3	Cholecalciferol
Vitamin E	Tocopherol nicotinate
Vitamin G	Riboflavin (also called vitamin B_2)
Vitamin H	Biotin (also called vitamin B_7)
Vitamin H_1	*p*-aminobenzoic acid
Vitamin K_1	Phylloquinone
Vitamin K_2	Menatetrenone, Menaquinone
Vitamin K_3	Menadione
Vitamin M	Folic acid
Vitamin P	Rutin

Box 6.3
Edible Legume Seeds Rich in Dietary Fibres

Castanospermum australe (Black bean)
Cicer arietinum (Chickpea)
Glycine max (Soybean)
Lablab purpureus (Field bean)
Lens culinaris (Lentil)
Phaseolus lunatus (Lima bean)
Phaseolus vulgaris (Kidney bean)
Pisum sativum (Pea)
Vicia faba (Broad bean)
Vigna radiata (Green gram)
Vigna unguiculata (Black eyed pea)

between the old and the current terms both of which are given in Box 6.2, to help while looking into the current literature.

Interestingly, validating the traditional practice of eating soaked/ germinated gram, germinated gram seeds contain increased amounts of tocopherol, ascorbic acid, pantothenic acid, biotin, nicotinic acid, riboflavin, choline, thiamine and cyanocobalamine (vitamin B_{12}) (Chattopadhyaya *et al.*, 1950; Nandi and Banerjee, 1950a, b; Chattopadhyaya and Banerjee, 1951b, 1952; Banerjee *et al.*, 1954; Rohtagi *et al.*, 1955; Chen *et al.*, 1975; Fordham *et al.*, 1975; Kylen and McCready, 1975; Venugopal and Rama Rao, 1978), compared to the dry seeds.

Minerals: Pulses are also good source of minerals (Salunkhe, 1982). The total mineral content of legumes varies from 3.0 to 4.5 per cent. They generally contain 90 to 240 mg/100g calcium, 2.8 to 11.5 mg/100g iron and varying quantities of zinc, copper, manganese, magnesium, phosphorous, and potassium. Mineral composition of *Cicer arietinum, Vigna mungo, Vigna radiata, Pisum sativum, Cajanus cajan, Lens culinaris, Vigna aconitifolia* and *Lathyrus sativus* showed that *Vigna aconitifolia* and *Vigna mungo* are higher in calcium and magnesium than any other pulses. While *Vigna aconitifolia* was the poorest in its phytin phosphorous, containing hardly seven per cent of total phosphorus as compared to 40 to 50 per cent in other pulses. (Balasubramanian *et al.*, 1962).

Antioxidants: Legume seeds are rich in carotenoids, the antioxidants that are recommended to prevent heart attack and cancer. The caroteonoid content of germinated seeds is far greater than that in the dry seeds (Chattopadhyaya and Banerjee, 1951a). In fact, the green leaves and pods of both edible and fodder legumes are very high in carotenoids, calcium and iron (Table 6.1), but these parts are not usually consumed, except in a few cases like *Trigonella foenum-graecum.* They also contain significant amounts of phenolic compounds, which are also antioxidants.

Dietary fibres: In modern life styles, soft foods, which are low in dietary fibre, predominate. Dietary fibres are water soluble (pectin, hemicellulose and gum) or insoluble (cellulose and lignin) polysaccharides from the edible parts of plants.

The soluble fibres bind to bile salts resulting in an increased excretion of cholesterol. This reduces the quantity of cholesterol available for reabsorption into the body system, thus reducing the chances of coronary heart disease. The dietary fibre content of legume is 0.7 to 6.4 per cent. The legume seeds rich in dietary fibres are given in Box 6.3.

Dietary fibres also delay the passage of food through the alimentary canal and aid in the release of the gastric inhibitory polypeptide (GIP). GIP is a hormone that potentiates insulin response to glucose load in the stomach. This facilitates absorption of glucose more in the distal part of the small intestine, which helps in lowering glucose and lipid levels in the blood. Increase in the intake of dietary fibres helps in lowering body weight and blood pressure as they function as non-calorific fillers reducing food consumption. Dietary fibres also prevent constipation and offer protection against certain types of cancer, more particularly the colon cancer.

6.5 FOOD RIGHT FOR HUMAN BLOOD GROUPS

A comparison of primitive and modern diets started in the late 1930s (Price, 1939), developed into the concept of 'paleolithic nutrition' (Eaton and Konner, 1985). The subject attracted wide attention and now there is extensive literature, and several websites, the more useful among them is at http://www.panix.com/~paleodiet/.

Significance of the right food for each human blood group is traced to the evolution of mankind as each blood type contains a different genetic information inherited from our ancestors, influenced by their diet and way of living. Each blood type is programmed to accept the kinds of foods that were the major part of the diet of our ancestors, thousands of years ago, which continues to affect many of our traits, particularly the immune system.

Human immune system recognizes invading proteins and produces antibodies to counter them. Antibodies are also produced against some antigens in foods and this is related to the four blood types. Foreign antigens in food play a major role in the onset and progression of multiple sclerosis when genetic and environmental factors combine. A number of studies link tolerance to different kinds of foods to the immune system.

The concept of paleolithic nutrition holds that foods introduced into the human diet by the agricultural revolution over 8,000 years ago can potentially cause biochemical failures leading to chronic illness in genetically susceptible individuals who have had only an insufficient time to adapt to new foods. Thus, in a given population, a certain percentage of people are genetically incompatible with one or more of the newly' introduced foods and their immune system reacts to these foods. Plant proteins are one of these 'new foods', along with sugars, saturated fats and yeast. Most plant foods, particularly the legumes, consumed by us contain lectins, a special class of antigenic proteins, discussed in detail in Chapters 9 and 23.

Whether the immune system of an individual accepts a particular food or not largely depends upon the proteins, more particularly the lectins found in that food. Lectins incompatible with a particular blood group trigger the production of antibodies against them and may even agglutinate cells in the body and/or damage them. Prolonged intake of a certain initially incompatible food, may in course of time be tolerated by the body system, but not in all individuals.

People of human blood group A, constituting 42 per cent of the world's population, are considered as the 'cultivators'. They can flourish on legumes but should avoid meat and wheat. They have a high rate of diabetes and heart disease. Blood group A individuals by long time exposure to dietary lectins, acquired the ability to secrete an enhanced protective mucosal barrier. But these secretors may have more difficulty in dealing with tumour activity on account of fully sialated carbohydrate chain in the mucosal barrier (Timoshenko *et al.*, 1995).

The proteins in milk are not readily compatible with blood type A individuals, and are not digested by them, particularly the infants. Milk proteins may trigger the immune system into an allergenic response, causing stomach inflammation or damage the intestinal brush border, leading to an unrestricted macromolecular permeability or the proteins may be absorbed into blood stream resulting in cell agglutination causing major problems such as irritable bowel syndrome, liver cirrhosis or kidney malfunction.

Type B blood group occurs in seven per cent of the human population, who are the 'nomads'. Fish and dairy products best suit them. They should avoid chicken, corn and peanuts. They have a high incidence of kidney and bladder infection.

Type AB blood group, considered an 'enigma', occurs only in four per cent of the human population. They can tolerate a wide range of foods, such as vegetable proteins, dairy and seafood but should avoid red meat, corn and legumes. They are prone to heart diseases, cancer and anaemia.

People with type O blood group, the 'hunters', constitute about 46 per cent of human population. Type O blood group is considered more ancient than other blood groups. O blood group individuals are associated with high levels of stomach acid and cholesterol splitting intestinal enzymes and can metabolize animal proteins. They should avoid wheat and corn. They have greater incidence of blood clotting disorders and gastric ulcer (Adamo, 1998). Blood group O individuals do not have the enhanced mucous protective barrier. These individuals secrete higher amount of hydrochloric acid, which is probably the reason for the greater incidence of gastric ulcers in them than blood group A individuals which may also explain the susceptibility of the blood group O individuals to develop irritable bowel and celiac diseases (Boldt and Banwell, 1985; Freed, 1991).

The key to health appears to be the identification of foods that do not agree with one's own body system and avoiding their consumption and eat right for one's own blood group.

Table 6.1
Nutritive Potential of Food, Feed and Forage Legumes in India

Abbreviations :

1: Edible portion; 2: Moisture; 3: Protein; 4: Fat; 5: Minerals; 6: Fibres; 7: Carbohydrates; 8: Energy; 9: Calcium; 10: Phosphorus; 11: Iron; 12: Carotene; 13: Thiamine; 14: Riboflavin; 15: Niacin; 16: Vitamin C.

Sources:

*Anonymous, 1979; 1 : FAO, 1968 in Tindall, 1983; 2 : FAO, 1972 in Tindall, 1983; 3 : Klen and McReady, 1975 in Yamaguchii, 1983; 4 : Kay, 1979 in Tindall, 1983; 5 : Platt, 1962 in Tindall, 1983; 6 : Knott and Deanon, 1967 in Tindall, 1983; 7 : Watson, 1971 in Tindall, 1983; 8 : Duke, 1981; Unmarked : Gopalan *et al*., 1985.

(values given are per 100 g of edible portion)

Sl. No.	TAXON	PART Used	1 %	2 g	3 g	4 g	5 g	6 g	7 g	8 k.cal	9 mg	10 mg	11 mg	12 µg	13 mg	14 mg	15 mg	16 mg
I	**Caesapiniaceae**																	
1	*Bauhinia malabarica*	Leaf	-	82.7	6.1	0.7	1.5	-	9.0	67	112	122	12.1	1764	0.10	0.27	1.70	10
2	*Bauhinia purpurea*	Leaf	-	78.1	3.6	1.0	2.1	5.5	9.7	62	312	92	-	-	-	-	-	-
3	*Bauhinia vahlii*	Seed kernel	-	9.5	27.3	29.9	3.6	1.1	28.6	493	302	718	6.8	-	0.17	0.17	1.1	-
4	*Ceratonia siliqua*	Fruit* Aril* Seed*	- - -	- - -	05.3 05.1 19.9	03.8 01.6 02.8	- - -	10.7 08.7 08.6	42.6 91.2 66.8	- 203 -	- - -	- - -	- - -	- - -	- - -	- - -	- - -	- - -
5	*Delonix regia*	Seed[8]	-	-	35.4	5.2	-	-	51.7	378	673	416	14.7	4122	0.96	0.42	5.79	633
6	*Dialium guianense*	Pulp[8]	-	-	4.1	0.4	-	2.7	94.9	256	237	428	4.3	-	-	-	-	-
7	*Gleditsia aquatica*	Pod*	-	-	14.3	01.9	-	18.4	60.8	-	-	-	-	-	-	-	-	-
8	*Hymenaea courbaril*	Fruit[8]	-	-	6.9	2.5	-	15.7	88.1	361	32	167	3.7	1.0	0.27	0.20	4.80	13
9	*Senna alexandrina*	Dried seed[8]	100	10.3	18.0	1.3	-	6.5	66.9	346	205	184	-	-	1.04	-	-	-
10	*Senna siamea*	Leaf[8]	-	-	19.1	4.3	-	11.3	66.2	330	290	450	15.4	19256	0.11	2.06	3.77	162

Sl. No.	TAXON	PART Used	1 %	2 g	3 g	4 g	5 g	6 g	7 g	8 k.cal	9 mg	10 mg	11 mg	12 µg	13 mg	14 mg	15 mg	16 mg
11	*Senna tora*	Fresh leaf	-	84.9	5.0	0.8	1.7	2.1	5.5	49	520	39	12.4	10152	0.08	0.19	0.8	82
		Dried leaf	-	9.7	20.7	3.9	11.8	10.4	43.5	292	3200	292	-	-	-	-	-	-
12	*Tamarindus indica*	Tender leaf	100	70.5	5.8	2.1	1.5	1.9	18.2	115	101	140	5.2	250	0.24	0.17	4.1	3
		Dried leaf	-	8.9	8.6	3.0	8.5	10.1	60.9	305	1485	124	-	-	-	-	-	-
		Leaf[8]	100	78	4	12	-	3	16	75	250	40	2	2500	0.1	0.1	1.5	6
		Flower*	-	-	12.5	09.0	-	06.0	75.0	375	-	-	-	-	-	-	-	-
		Dried fruit[8]	100	21.3	5.0	0.6	-	18.3	70.7	270	166	190	2.2	60	0.18	0.09	0.6	9
		Fruit[8]	-	-	10.1	1.0	-	42.4	83.8	343	303	490	-	-	-	-	-	-
		Pulp	-	2.9	3.1	0.1	2.9	5.6	674	283	170	110	10.9	60	-	0.07	0.7	3
		Pulp[8]	100	38.7	2.3	0.2	-	1.9	56.7	214	8.1	86	1.3	10	0.22	0.08	1.10	3.0
		Seed	-	9.9	16.1	7.3	166	1.0	64.1	387	121	237	-	-	-	-	-	-
II	**Fabaceae**																	
1	*Abrus precatorius*	Leaf[8]	-	-	26.7	-	-	17.6	65.2	295	127	309	-	-	-	-	-	-
2	*Apios americana*	Root*	-	-	175	-	-	-	57.1	-	-	-	-	-	-	-	-	-
3	*Arachis hypogaea*	Leaf*	-	-	20.5	02.8	-	21.4	69.3	320	-	-	-	-	-	-	-	-
		Leaf[8]	100	78.5	4.4	0.6	-	4.6	14.9	69	262	82	4.2	7735	0.23	0.58	1.6	98
		Nut	73	3.0	25.3	40.1	2.4	3.1	26.1	567	90	350	2.8	37	0.90	0.13	19.9	0
		Fried nut	69	1.7	26.2	39.8	2.5	3.1	26.7	570	77	370	3.1	0	0.39	0.13	22.1	0
		Cake	-	7.2	40.9	7.4	2.5	3.2	38.8	386	213	548	-	-	-	-	-	-
4	*Butea monosperma*	Tuber Root	-	454	2.1	0.3	0.6	1.6	50.0	211	25	21	-	-	-	-	-	-
5	*Cajanus cajan*	Plant[8]	-	11.2	14.8	1.7	-	28.9	-	-	-	-	-	-	-	-	-	-
		Forage[8]	-	70.4	7.1	1.6	-	10.7	-	-	-	-	-	-	-	-	-	-
		Tender Seed	42	65.1	9.8	1.0	1.0	6.2	16.9	116	57	164	1.1	469	0.32	0.33	3.0	25
		Seed[1]	100	10	19.5	1.3	-	7.3	65	345	161	285	15.0	55	0.72	0.14	2.9	trace
		Dhal	100	13.4	22.3	1.7	3.5	1.5	57.6	335	73	304	5.8	132	0.45	0.19	2.9	0
6	*Canavalia ensiformis*	Green pod*	-	-	21.0	02.6	-	15.8	770	324	-	-	-	-	-	-	-	-
		Pod[1]	-	88	2.4	0.3	-	1.8	8	37	-	-	-	-	-	-	-	-
		Seed[1]	-	10	29.5	2.6	-	7.4	59	347	158	290	7.0	-	0.77	0.15	1.8	-
		Seed*	-	-	23.7	03.6	-	08.6	68.9	393	-	-	-	-	-	-	-	-
7	*Canavalia gladiata*	Fresh Pod	98	87.2	2.7	0.2	0.6	1.5	7.8	44	60	40	2.0	24	0.08	0.08	0.5	12
		Green seed*	-	-	23.7	01.8	-	13.2	56.1	-	-	-	-	-	-	-	-	-
		Seed	100	15	27.1	0.6	-	11.6	5.3	318	-	-	-	-	-	-	-	-
8	*Cicer arietinum*	Leaf	92	73.4	7.0	1.4	2.1	2.0	14.1	97	340	120	23.8	978	0.09	0.10	0.6	61
		Seed	100	9.8	17.1	5.3	3.0	3.9	60.9	360	202	312	10.2	189	030	0.15	2.9	3
		Dhal	100	9.9	20.8	5.6	2.7	1.2	59.8	372	56	331	9.1	129	0.48	0.18	2.4	1
		Fried Dhal	100	10.7	22.5	5.2	2.5	1.0	58.1	369	58	340	9.5	113	0.20	-	1.3	0
9	*Clitoria ternatea*	Pod	100	80.0	3.8	0.4	-	4.8	15.0	66	40	24	0.4	-	0.04	0.18	1.4	248
		Seed[8]																
		Fruit[8]	-	5.0	19.0	2.0	-	24.0	75.0	330	206	207	7.2	-	-	-	-	-
10	*Crotalaria juncea*	Stalk[8]	-	14.4	-	-	6.4	27.4	35.8	-	-	-	-	-	-	-	-	-
		Flower	-	78.9	4.8	0.6	1.4	3.9	10.4	66	200	100	-	-	-	-	-	-
		Seed[8]	-	8.6	34.6	4.3	-	8.1	-	-	-	-	-	-	-	-	-	-

Sl. No.	TAXON	PART Used	1 %	2 g	3 g	4 g	5 g	6 g	7 g	8 k.cal	9 mg	10 mg	11 mg	12 µg	13 mg	14 mg	15 mg	16 mg
11	*Cyamopsis tetragonoloba*	Forage[8]	-	80.8	3.1	-	-	4.4	40.7	-	0.61	0.07	-	-	-	-	-	-
		Bean	-	81.0	3.2	0.4	1.4	2.1	10.8	16	130	57	4.5	198	0.09	0.03	0.6	49
12	*Desmodium dicolor*	Leaf*	-	-	20.9	03.0	-	27.3	38.1	-	-	-	-	-	-	-	-	-
13	*Erythrina fusca*	Leaf[8]	-	-	24.9	4.3	-	22.2	63.3	325	308	222	5.2	-	0.91	0.52	6.54	78
14	*Flemingia procumbens*	Root*	-	-	09.3	02.3	-	01.1	83.9	-	-	-	-	-	-	-	-	-
15	*Gliricidia sepium*	Leaf[8]	-	-	15.7	3.2	-	8.5	77.8	340	111	264	12.8	150	0.79	0.57	4.28	314
		Flower[8]	-	-	17.1	1.4	-	11.4	77.1	328	152	426	11.9	1130	1.26	0.53	4.17	85
16	*Glycine max*	Leaf	-	79.5	6.0	0.5	3.2	-	10.8	72	180	190	8.0	-	-	0.16	-	-
		Seed[2]	-	68	13.0	5.7	-	1.9	11	139	78	158	3.8	360	0.40	0.17	1.5	27
		Green seed*	-	-	14.3	0.9	-	18.4	60.8	436	-	-	-	-	-	-	-	-
		Seed	-	8.1	43.2	9.5	4.6	3.7	20.9	432	240	690	11.5	426	0.73	0.39	3.2	-
		Sprout[3]	100	73.2	12.0	2.6	-	2.3	-	105	75.0	-	0.4	-	0.32	0.16	1.1	12
		Cooked Sprout[3]	100	67.2	13.1	-	-	2.5	-	-	81.7	-	0.4	-	0.42	0.19	1.1	12
17	*Hedysarum coronarium*	Forage[8]	-	-	2.26	0.38	-	5.16	-	-	-	-	-	-	-	-	-	-
		Hay[8]	-	-	13.5	2.4	-	30.6	-	-	-	-	-	-	-	-	-	-
18	*Indigofera glandulosa*	Seed[8]	-	-	34.8	2.4	-	8.5	-	-	-	-	-	-	-	-	-	-
19	*Indigofera hirsuta*	Hay[8]	-	10.7	13.7	1.4	-	21.0	-	-	-	-	-	-	-	-	-	-
20	*Indigofera spicata*	Leaf, Stem[8]	-	80.5	4.1	0.6	-	4.7	7.9	-	-	-	-	-	-	-	-	-
		Hay[8]	-	-	21.0	3.1	-	24.1	-	-	-	-	-	-	-	-	-	-
21	*Lablab purpureus*	Leaf*	-	-	22.0	03.7	-	61.4	55.9	284	-	-	-	-	-	-	-	-
		Fruit*	-	-	25.0	02.7	-	16.1	65.2	312	-	-	-	-	-	-	-	-
		Tender bean	93	86.1	3.8	0.7	0.9	1.8	6.7	48	210	68	1.7	187	0.10	0.06	0.7	9
		Dry bean	-	96	24.9	0.8	3.2	1.4	60.1	347	60	433	2.7	0	0.52	0.16	1.8	0
		Fresh pod	-	82	4.5	0.1	-	2.0	10	-	50	-	10	-	-	-	-	1
		Seed	-	10	22.8	1.0	-	8.6	62	340	90	328	9.0	-	0.54	0.14	2.3	trace
22	*Lathyrus sativus*	Leaf[8]	-	84.2	6.1	1.0	1.1	2.1	5.5	55	160	1000	7.3	3000	0.01	0.03	-	41
		Dhal[8]	100	10.0	28.2	0.6	2.3	2/3	56.6	345	90	317	6.3	120	0.39	0.17	2.9	0
23	*Lens culinaris*	Seed	100	12.4	25.1	0.7	2.1	0.7	59.0	343	69	293	4.8	270	0.45	0.20	2.6	0
		Raw Sprout[3]	-	72.7	8.4	0.3	-	1.1	-	104	12.0	-	3.0	-	0.21	0.09	1.1	24
		Cooked Sprout[3]	-	60.0	8.8	-	-	1.1	-	-	13.7	-	3.0	-	0.22	0.05	1.1	24
24	*Lespedeza juncea*	Hay[8]	-	-	18.0	3.8	-	227	-	-	1.5	0.2	0.03	-	-	-	-	-
25	*Kummerovia stipulacea*	Green forage[8]	-	-	16.5	2.7	-	30.1	-	-	1.1	0.34	-	-	-	-	-	-
26	*Lespedeza striata*	Hay[8]	-	-	13.7	1.7	-	32.0	-	-	1.0	0.20	0.03	-	-	-	-	-
		Forage[8]	-	-	2.1	2.1	-	36.1	-	-	1.1	0.27	0.03	-	-	-	-	-
27	*Lotus corniculatus*	Hay[8]	-	8.8	14.2	2.1	6.0	41.9	-	-	1.6	0.24	0.01	143.9 ppm	6.8 ppm	16.1 ppm	-	-
28	*Lotus tetragonolobus*	Fruit	-	90.6	1.9	0.4	0.5	1.8	4.8	30	30	20	-	-	-	-	-	-

Sl. No.	TAXON	PART Used	1 %	2 g	3 g	4 g	5 g	6 g	7 g	8 k.cal	9 mg	10 mg	11 mg	12 µg	13 mg	14 mg	15 mg	16 mg
29	*Lupinus albus*	Seed[8]	-	11.0	32.8	8.9	-	-	6.3	-	0.33	0.25	0.01	-	-	-	-	-
		Seed	-	-	36.7	10.0	-	-	-	-	336	258	10.1	-	-	-	-	-
30	*Lupinus angustifolius*	Seed[8]	-	13.8	31.0	6.4	-	11.5	-	-	-	-	-	-	-	-	-	-
		Seed	-	-	36.0	7.4	-	13.3	-	-	-	-	-	-	-	-	-	-
		Hay[8]	-	-	32.5	5.2	-	20.1	-	-	0.59	0.46	-	-	-	-	-	-
31	*Lupinus luteus*	Seed[8]	-	-	39.8	5.0	4.5	14.0	-	-	0.23	0.39	-	-	-	-	-	-
		Hay[8]	-	-	35.6	5.2	-	19.6	-	-	0.64	0.82	-	-	-	-	-	-
32	*Lupinus mutabilis*	Green seed	-	-	32.2	32.6	-	07.1	32.2	513	-	-	-	-	-	-	-	-
		Green seed[8]	-	-	48.0	17.8	-	-	-	441	-	-	-	-	-	-	-	-
		Seed[8]	-	-	47.8	17.8	-	-	30.4	440	-	-	-	-	-	-	-	-
		Seed	-	-	48.0	17.8	-	02.1	31.5	441	-	-	-	-	-	-	-	-
33	*Macrotyloma geocarpum*	Seed[1]	-	9	19.4	1.1	-	5.5	67	248	103	392	15.0	-	0.76	0.19	2.3	-
34	*Macrotyloma uniflorum*	Hay[8]	-	-	10.6	1.8	-	16.2	-	-	-	-	-	-	-	-	-	-
		Seed	100	11.8	22.0	0.5	3.2	5.3	57.2	321	287	311	8.4	71	0.42	0.20	1.5	1
35	*Medicago lupulina*	Forage[8]	-	-	23.3	3.3	-	24.7	-	-	1.0	0.34	-	-	-	-	-	-
36	*Medicago polymorpha*	Leaf[8]	-	-	35.8	2.2	-	6.7	50.0	284	992	462	380	18015	1.34	2.09	5.2	731
		Hay[8]	-	-	19.0	2.8	-	25.1	-	-	-	-	-	-	-	-	-	-
		Forage[8]	-	-	24.1	4.9	-	20.4	-	-	-	-	-	-	-	-	-	-
37	*Medicago sativa*	Shoot[8]	100	82.7	6	0.4	-	3.1	9.5	52	12	51	5.4	3.41	0.13	0.14	0.5	162
		Shoot	-	-	34.7	2.3	-	17.9	54.9	301	69	295	31.2	11825	0.75	0.81	2.89	936
		Forage[8]	100	80.0	5.2	0.9	-	3.5	-	-	-	-	-	-	-	-	-	-
		Leaf	-	-	34.7	02.3	-	17.9	54.9	301	-	-	-	-	-	-	-	-
		Seed[3]	-	7.4	35.1	12.6	-	7.9	-	389	136	-	12.9	-	1.08	0.58	1.8	-
		Sprout[3]	-	88.3	5.1	0.6	-	1.7	-	41	28.0	-	1.4	-	0.14	0.21	1.6	-
		Cooked sprout[3]	-	87.5	5.1	-	-	1.7	-	-	28.3	-	1.4	-	0.12	0.21	0.8	-
38	*Melilotus albus*	G. feed[8]	-	79.2	4.1	-	-	4.9	-	-	-	-	-	-	-	-	-	-
		Silage[8]	-	72.0	4.5	-	-	9.6	-	-	-	-	-	-	-	-	-	-
		Seed[8]	-	7.8	37.4	-	-	11.3	-	-	-	-	-	-	-	-	-	-
39	*Melilotus indicus*	Green feed[8]	-	78.4	3.3	-	-	6.4	-	-	-	-	-	-	-	-	-	-
		Silage[8]	-	70.2	2.1	-	-	14.1	-	-	-	-	-	-	-	-	-	-
		Hay[8]	-	10.0	18.6	-	-	23.8	-	-	-	-	-	-	-	-	-	-
		Pods[8]	-	0	25.3	-	-	14.8	-	-	-	-	-	-	-	-	-	-
40	*Mucuna deeringiana*	Seed[8]	-	-	24.0	5.0	-	5.2	67.2	398	149	361	-	-	0.14	0.11	3.39	2
41	*Mucuna pruriens*	Seed[8]	-	-	33.0	3.6	-	8.7	59.2	385	-	-	-	-	0.57	1.27	1.90	-
42	*Mucuna pruriens* var. *utilis*	Seed[8]	-	-	28.3	3.5	-	-	64.9	39.2	146	236	2.4	45	0.35	0.20	2.83	0
43	*Pachyrhizus erosus*	Tuber[5]	-	89	1.0	0.1	-	0.5	9	41	14	15	0.5	trace	0.03	0.02	0.02	20
		Tuber[2]	-	87	1.6	0.2	-	1.3	10	46	18	17	0.8	0	0.08	0.06	0.6	15
		Pod[8]	-	-	19.1	2.2	-	2.2	73.5	-	889	287	9.6	2536	0.81	0.66	5.88	-
		Fresh pod[6]	-	89	2.6	0.3	-	2.9	10	45	121	39	1.3	345	0.11	0.09	0.8	-
44	*Phaseolus acutifolius*	Hay[8]	-	6.6	9.9	1.9	-	29.3	-	-	-	-	-	-	-	-	-	-
		Dried beans[8]	-	9.5	24.5	1.5	-	3.7	65.5	-	112	310	-	-	0.33	0.12	2.8	-
		Pod[8]	-	8.0	4.1	0.5	-	37.0	-	-	-	-	-	-	-	-	-	-
		Seed[8]	-	-	20.7	1.3	-	5.1	72.5	378	120	332	-	-	0.35	0.13	2.99	0
		Seed[*]	-	-	24.5	01.5	-	03,7	65.5	378	-	-	-	-	-	-	-	-

Sl. No.	TAXON	PART Used	1 %	2 g	3 g	4 g	5 g	6 g	7 g	8 k.cal	9 mg	10 mg	11 mg	12 µg	13 mg	14 mg	15 mg	16 mg
45	*Phaseolus coccineus*	Fruit	59	58.3	7.4	1.0	1.6	1.9	29.8	158	50	160	2.6	34	0.34	0.19	0	27
		Fruit[8]	-	-	17.8	2.4	-	2.4	71.5	-	120	384	6.2	82	0.81	0.45	-	65
		Seed[*]	-	-	23.1	02.1	-	05.5	70.7	385	-	-	-	-	-	-	-	-
		Seed[8]	-	-	23.1	2.1	-	5.5	70.7	385	130	404	10.3	-	0.57	0.21	2.62	2
		Seed[4]	-	13	20.9	0.9	-	4.8	60	327	200	390	10.9	-	0.49	0.21	2.4	-
46	*Phaseolus lunatus*	Leaf[*]	-	-	21.4	60.0	-	-	60.7	286	-	-	-	-	-	-	-	-
		Seed[1]	-	12	21.4	1.4	-	4.3	61	335	116	387	4.9	-	0.33	0.16	2.1	1
		Seed[2]	-	10	19.8	1.3	-	-	65	346	90	24.0	5.6	trace	0.46	0.21	1.4	0
		Sprout[*]	-	-	36.1	02.2	-	01.7	55.6	306	-	-	-	-	-	-	-	-
47	*Phaseolus vulgaris*	Leaf	-	87	3.8	0.4	-	-	6.6	36	274	75	9.2	-	-	-	1.3	110
		Bean	94	91.4	1.7	0.1	0.5	1.8	4.5	26	50	28	1.7	132	0.08	0.06	0.3	24
		Bean[8]	-	-	21.1	1.7	-	15.9	69.9	318	381	425	12.4	6638	0.71	1.06	4.42	239
		Seed	-	12.0	22.9	1.3	3.2	-	60.6	346	260	410	5.8	-	-	-	-	-
48	*Pisum sativum*	Seed	53	72.1	7.2	0.1	0.8	4.0	15.9	93	20	139	1.5	83	0.25	0.01	0.8	9
		Dry seed	100	16.0	19.7	1.1	2.2	4.5	56.5	315	75	298	5.0	39	0.47	0.19	3.4	0
		Fried seed	100	10.1	22.9	1.4	2.4	4.4	58.8	340	81	345	6.4	18	0.47	0.21	3.5	0
49	*Psophocarpus tetragonolobus*	Root[1]	-	95	2.8	0.6	-	1.5	20	91	-	-	-	-	-	-	-	-
		Root[8]	-	-	11.4	2.4	-	6.1	82.6	370	-	-	-	-	-	-	-	-
		Leaf[8]	-	-	33.4	03.3	-	-	56.7	313	894	540	41.4	20977	1.87	-	-	-
		Leaf	100	85.0	5.0	0.5	-	-	8.5	47	134	81	6.2	3.1	0.28	-	-	28
		Pod[8]	100	89.5	1.9	0.1	-	1.6	7.9	34	53	48	0.2	340	0.19	0.08	1.0	21
		Pod[1]	-	92	2.1	0.3	-	1.7	4	25	-	-	-	-	-	-	-	-
		Seed[8]	-	-	36.4	18.8	-	-	4.5	450	88	222	2.2	-	-	-	-	-
		Seed[7]	-	9	31.2	17.0	-	6.6	33	420	210	410	15.0	-	0.08	-	-	trace
50	*Pueraria lobata*	Root[8]	-	-	6.7	0.3	-	02.2	88.4	359	47	57	1.9	-	-	-	-	-
		Root[8]	100	68.6	2.1	0.1	-	0.7	27.8	113	15	18	0.6	-	-	-	-	-
		Leaf	-	-	3.6	0.9	-	70.0	88.2	327	-	-	-	-	-	-	-	-
		Cooked Leaf[8]	100	89	0.4	0.1	-	7.7	9.7	36	34	20	4.9	-	0.03	0.91	0.8	-
51	*Pueraria phaseoloides*	Leaf , Stem[8]	-	80.9	3.8	0.4	-	5.5	7.9	-	0.41	0.03	-	-	-	-	-	-
52	*Sesbania grandiflora*	Leaf	-	73.1	8.4	1.4	3.1	2.2	11.8	93	1130	80	3.9	5400	0.21	0.09	1.2	169
		Leaf[8]	-	-	36.3	07.5	-	9.2	46.1	321	-	-	-	-	-	-	-	-
		Flower[8]	-	-	14.5	03.6	-	10.9	77.3	345	-	-	-	-	-	-	-	-
53	*Sesbania sesban*	Flower	-	92.9	1.0	0.5	0.4	0.8	4.4	26	9	5	-	-	-	-	-	-
54	*Styphnolobium japonicum*	Seed[8]	-	-	17.2	12.2	-	10.0	67.0	435	266	273	-	1069	-	-	-	-
55	*Trifolium alexandrium*	Hay[8]	-	-	15	2.9	-	25	-	-	2.5	0.3	-	-	-	-	-	-
56	*Trifolium hybridum*	Hay[8]	-	-	15.1	35	-	30.3	-	-	1.3	0.29	0.05	-	10 ppm	2	-	-
57	*Trifolium incarnatum*	Hay[8]	-	-	16.7	-	-	28.3	-	-	1.6	0.35	0.70	-	-	-	-	-
58	*Trifolium pratense*	Green forage[8]	-	81.0	4.0	0.7	-	2.6	-	-	1.76	0.29	0.03	-	-	-	-	-
		Hay[8]	-	12.0	11.8	2.6	-	27.2	-	-	1.61	0.22	0.01	-	-	-	-	-
59	*Trifolium repens*	Hay[8]	-	-	22.0	2.6	-	23.3	-	-	1.2	0.4	0.11	-	-	-	-	-
		Green Roughage[8]	-	-	27.8	3.2	-	15.6	-	-	1.5	0.5	0.03	-	-	-	-	-

Sl. No.	TAXON	PART Used	1 %	2 g	3 g	4 g	5 g	6 g	7 g	8 k.cal	9 mg	10 mg	11 mg	12 µg	13 mg	14 mg	15 mg	16 mg
60	*Trifolium resupinatum*	Leaf[8]	-	-	23.1	2.2	-	19.4	-	-	-	-	-	4000	-	-	-	-
		Hay[8]	-	10	16.3	1.8	-	30.8	-	-	1.66	0.23	-	-	-	-	-	-
61	*Trigonella foenum-graecum*	Leaf	59	86.1	4.4	0.9	1.5	1.1	6.0	49	395	51	16.5	2340	0.04	0.31	0.8	52
		Seed	-	13.7	26.2	5.8	3.0	7.2	44.1	333	160	370	14.1	96	0.34	0.29	1.1	0
62	*Vicia faba*	Leaf	-	77.6	5.6	1.3	1.3	3.7	11.5	71	111	149	-	-	-	-	-	-
		Dried seed[1]	-	10	26.2	1.3	-	6.8	59	344	104	301	6.7	130	0.38	0.24	2.1	1.6
		Seed[8]	-	-	29.0	1.4	-	5.9	66.0	380	121	461	4.9	975	0.52	0.22	27.8	0
		Seed	88	85.4	4.5	0.1	0.8	2.0	7.2	48	50	64	1.4	9	0.08	-	0.8	12
63	*Vicia sativa*	Hay[8]	-	11.1	13.3	1.1	-	25.2	-	-	-	-	-	-	-	-	-	-
		Seed[8]	100	9.5	29.5	0.8	-	5.7	-	343	137	240	9.3	-	0.15	0.10	2.7	-
64	*Vicia villosa*	Green Foliage[8]	-	-	23.1	2.7	-	27.5	-	-	-	-	-	-	-	-	-	-
		Hay[8]	-	-	22.6	3.2	-	28.7	-	-	1.1	0.33	-	-	-	-	-	-
65	*Vigna aconitifolia*	Hay[8]	-	10.0	16.2	2.4	-	16.0	-	-	-	-	-	-	-	-	-	-
		Seed	100	10.8	23.6	1.1	3.5	4.5	56.5	330	202	230	9.5	9	0.45	0.09	1.5	2
		Seed*	-	-	26.5	1.2	-	05.0	63.2	-	-	-	-	-	-	-	-	-
66	*Vigna angularis*	Seed[8]	100	10.8	19.9	0.6	-	7.8	64.4	336	136	260	9.8	-	0.06	0.09	2.0	2
		Boiled & Sweetened Seed[8]	-	45.3	3.0	0.1	-	0.6	50.7	218	26	66	3.9	-	0.05	0.01	0.1	-
67	*Vigna mungo*	Seed[3]	100	10.1	22.9	1.4	-	4.9	-	4	83	-	11.6	-	0.70	0.47	1.8	0
		Seed[8]	-	-	23.5	1.8	-	4.9	71.0	385	123	390	9.4	22	0.65	0.22	2.58	2
		Dhal	100	10.9	24.0	1.4	3.2	0.9	59.6	347	154	385	9.1	38	0.42	0.20	2.0	0
		Sprout[3]	100	85.9	4.3	0.2	-	0.6	-	53	13.0	-	1.9	-	0.14	0.18	1.1	20
		Cooked sprout[3]	100	84.3	4.3	-	-	0.7	-	-	13.1	-	1.9	-	0.14	0.8	1.2	16
68	*Vigna radiata*	Seed	100	10.4	24.0	1.3	3.5	4.1	56.7	334	124	326	7.3	94	0.47	0.27	2.1	0
		Seed[8]	-	-	25.6	1.3	-	4.9	69.2	381	118	370	7.9	62	0.59	0.29	2.80	4
		Dhal	100	10.1	24.5	1.2	3.5	0.8	59.9	348	75	405	8.5	49	0.47	0.21	2.4	0
		Sprout[8]	-	-	42.4	2.0	-	9.1	50.5	303	152	717	12.1	202	1.11	1.01	8.08	182
69	*Vigna subterranea*	Fresh seed[8]	100	57	7.8	3.1	-	3.0	30	152	14	258	1.2	-	-	-	-	-
		Dried seed[8]	100	10	18.8	6.2	-	4.8	61	367	62	276	12.2	10	0.47	0.14	1.8	trace
		Seed*	-	-	17.8	6.7	-	-	72.2	412	-	-	-	-	-	-	-	-
70	*Vigna trilobata*	Seed[8]	-	-	27.7	0.7	-	-	8.7	64.6	365	-	-	-	-	-	-	-
71	*Vigna umbellata*	Seed	-	9.6	21.5	0.3	3.5	4.2	60.9	332	302	297	-	-	-	-	-	-
		Seed[8]	-	-	21.5	1.2	-	01.2	71.6	390	93	444	5.8	0	0.35	0.24	0.28	10
		Seed[4]	-	13.0	20.9	0.9	-	4.8	6.0	327	200	390	10.9	-	0.49	0.21	2.4	-
72	*Vigna unguiculata*	Shoot[8]	100	89	4.8	0.3	-	-	4.4	30	73	106	2.2	-	0.35	0.18	1.1	36
		Shoot	-	-	43.6	2.7	-	-	40.0	273	664	964	20.0	-	3.18	1.64	10.0	327
		Leaf	-	89.0	3.4	0.7	1.6	1.2	4.1	38	290	58	20.1	6072	0.05	0.18	0.6	4
		Hay[8]	100	9.6	18.6	2.6	-	23.3	-	-	-	-	-	-	-	-	-	-
		Pod	-	85.3	3.5	0.2	0.9	2.0	8.1	48	72	59	2.5	564	0.07	0.09	0.9	14
		Immature pod[8]	100	85.3	3.6	0.3	-	1.8	10.0	47	45	52	1.2	170	0.13	0.10	1.0	22
		Seed[8]	100	11.4	22.5	1.4	-	5.4	61	338	104	416	-	-	0.08	0.09	4.0	2
		Seed	97	13.4	24.1	1.0	3.2	3.8	54.5	323	77	414	5.9	12	0.51	0.20	1.3	0
III	**Mimosaceae**																	
1	*Acacia aneura*	Leaf*	-	-	13.2	02.2	-	27.9	51.5	-	-	-	-	-	-	-	-	-
2	*Acacia farnesiana*	Leaf[8]	100	81.4	8.0	0.6	-	5.7	9.0	57	93	84	3.7	12255	0.20	0.17	8.5	49
		Seed[8]	-	7	12.6	4.6	-	9.5	72.4	377	-	-	-	-	-	-	-	-

Sl. No.	TAXON	PART Used	1 %	2 g	3 g	4 g	5 g	6 g	7 g	8 k.cal	9 mg	10 mg	11 mg	12 µg	13 mg	14 mg	15 mg	16 mg
3	*Acacia senegal*	Pod[*]	-	-	22.0	01.0	-	39.0	30.9	-	-	-	-	-	-	-	-	-
4	*Acacia tortillis*	Pod[*]	-	-	17.8	1.7	-	175	54.6	-	-	-	-	-	-	-	-	-
5	*Desmanthus virgatus*	Leaf[*]	-	-	22.4	-	-	-	-	-	-	-	-	-	-	-	-	-
6	*Enterolobium cyclocarpum*	Seed[8]	-	-	31.6	3.9	-	2.0	60.8	389	30	506	13.9	418	1.2	1.2	7.66	234
7	*Faidherbia albida*	Pod[*]	-	-	11.5	01.4	-	23.4	55.1	-	-	-	-	-	-	-	-	-
8	*Lysiloma latisiliquum*	Leaf[*]	-	-	14.2	03.9	-	8.8	74.7	332	-	-	-	-	-	-	-	-
		Tender tops & pods[8]	100	80.7	8.4	0.9	-	3.8	8.8	59	137	11	9.2	4730	-	0.09	5.4	8
		Green fruit[8]	-	-	43.5	04.7	-	19.7	45.6	306	710	57	47.7	24501	-	0.47	28	4.1
9	*Mimosa pudica*	Leaf[8]	-	-	20.0	0	-	-	72.9	314	500	757	20.0	93028	1.71	0.71	-	257
10	*Neptunia oleracea*	Leaf[8]	-	-	43.3	1.7	-	15.0	45.0	28.3	466	300	32.5	29613	1.0	1.17	9.16	175
11	*Pithecellobium dulce*	Leaf[8]	-	-	29.0	4.4	-	17.5	-	-	999	350	-	-	-	-	-	-
		Fruit	60	79.2	2.7	0.4	0.7	1.0	16.0	78	14	49	1.0	0	0.22	0.06	1.6	108
		Aril[8]	-	-	0.32	2.7	-	5.5	89.9	-	59	244	6.3	-	0.99	0.24	0.90	120
		Seed[8]	-	-	20.4	19.8	-	9.0	-	-	-	-	-	-	-	-	-	-
12	*Prosopis pallida*	Fruit[*]	-	-	10.3	00.7	-	26.7	58.6	-	-	-	-	-	-	-	-	-

CHAPTER-7

Antinutritional Factors in Edible Legumes

7.1 ANTINUTRITIONAL FACTORS

The concepts of bioavailability, bioconversion and the antinutritional factors (ANFs) were introduced in Chapter 6 and different groups of chemical compounds in the legumes were outlined in Chapter 5.

ANFs are naturally occurring substances that bind with enzymes and nutrients and inhibit their absorption and/or function. Various procedures of food preparation and cooking largely inactivate ANFs. However, even residual quantities of ANFs may result in some reduction in nutrient absorption and growth rate. The ANFs are not limited to seeds but are also found in other edible and inedible parts like root, stems, leaves and flowers. In fact, green leafy vegetables are at the lowest rung of the ladder of bioavailability of nutrients, due to a high concentration of ANFs in them.

Goitrogens, phytosterols, phytates, tannins, inhibitors of digestive enzymes, antivitamin factors, certain carbohydrates and some proteins including lectins, are identified as the major ANFs in plant foods (Cabrejas *et al*., 1995). The ANF content of a plant food depends upon the genetic factors inherent in crop variety and cultivar, as well as environmental factors that influence the expression of the genes (Muzquiz *et al.,* 1999).

Foods also contain another group of compounds, the toxins or toxicants discussed in Chapter 8, which are not exclusively and directly related to nutritional inhibition. The ANFs affect digestion, bioavailability and bioconversion, and restrict the realization of the full nutritive potential of the food we consume. The toxicants act in a more direct manner resulting in serious impairment of metabolic processes, organ deformities and consequent disorders and even death. Both the ANFs and toxicants are generated normally in routine metabolic processes of both edible and inedible plants alike. While some compounds can clearly be identified as ANFs, in practice the distinction between ANFs and toxicants is not an easy matter. The same compound may function as a nutrient, ANF or toxicant depending upon the dosage, duration of consumption, concentration with reference to body weight, and the mode of food processing. For these reasons, this Chapter should be read in conjunction with Chapter 8, which deals with toxicants in legumes.

7.2 ANTINUTRITIONAL FACTORS IN EDIBLE LEGUMES

Among several hundreds of species of legumes, only the black gram (*Vigna mungo*), chickpea (*Cicer arietinum*), common bean (*Phaseolus vulgaris*), cowpea (*Vigna unguiculata*), green gram (*Vigna radiata*), ground nut (*Arachis hypogaea*), horse gram (*Macrotyloma uniflorum*), lentil (*Lens culinaris*), red gram (*Cajanus cajan*), and soybean (*Glycine max*) are in common use as food, while some like the broad bean (*Vicia faba*), cluster bean (*Cyamopsis tetragonoloba*), lima bean (*Phaseolus lunatus*), winged bean (*Psophocarpus tetragonolobus*), and a few others are used regionally or less commonly. The reason for so few species being in wider use as food is that it was possible to select cultivated varieties of these species, with lower content of ANFs and toxicants, a process that took several centuries. An unmanageably higher content of ANFs and toxicants or other practical considerations, have deterred the use of the other legumes as food. Hippocrates (460-377 B.C.) noted that peas were toxic to human beings, but now peas are widely consumed without any ill effects, because over two millennia, man could select safer varieties of peas. A vast and long period of cultural experience and careful selection has given us the safer legumes and other foods. But improper methods of processing and cooking food or an individual's physiological deficiencies may result in an exposure to ANFs and toxicants that still remained in plant foods. Individual susceptibility also is an important factor, since many other people consume the same foods without any untoward effects. In times of food scarcity or due to an inability to afford safer foods, people are forced to consume species known to be unhealthy or even toxic, such as *Lathyrus sativus,* which causes lathyrism.

7.2.1 Goitrogens

Goitrogens induce goiter, a disorder caused when iodine levels are low in water or food sources or when iodine absorption was interfered for some physiological cause, leading to its deficiency. Goitre is histologically characterized by an increase in follicular epithelial cell height and foamy cytoplasm and externally by thyroid hypertrophy (van Etten, 1969), and is prevalent in several parts in India, more particularly the deep inland or high mountain States. Realizing the seriousness of the situation, the Government of India has made it mandatory to market only the salt that is fortified with iodine.

Naturally occurring goitrogens have been under study for a long time (Greer, 1962; Langer and Greer, 1977; van Etten, 1983; Heaney and Fenwick, 1987). Most goitrogens are phenolic glycosides. Glucosinolates or thiocyanates, derived directly from dietary indole glucosinolates or indirectly via the metabolism of isothiocyanates and nitrites, are goitrogens. Some non-protein amino acids like mimosine are also goitrogenic. These compounds induce a deficiency of iodine by blocking iodine uptake by the thyroid. Goitrogens occur in *Arachis hypogaea, Glycine max, Phaseolus vulgaris, Pisum sativum* and *Vigna mungo,* among several other edible legumes. The risk is more from raw foods.

7.2.2 Phytosterols

The triterpenoid alcohols in plants are called phytosterols. Like many plant foods, several edible legumes also contain sterols. Plant sterols are often added to processed foods to reduce cholesterol absorption. Milk enriched with plant sterols reduced cholesterol absorption by 60 per cent. But plant sterols also reduce the bioavailability of carotenoids, more particularly β-carotene, the precursor of vitamin A, by about 50 per cent and that of vitamin E (tocopherol) by 20 per cent. The effect was far greater with sterol esters than with free sterols. However, it is not yet understood how plant sterols affect the bioavailability of carotenoids and vitamin E. The following species are the important edible legumes that contain sterols: *Arachis hypogaea* (β-sitosterol, daucosterol), *Cajanus cajan* (β-sitosterol, phytosterol, stigmasterol, campesterol, cholesterol), *Cicer arietinum* (β-sitosterol, coumesterol), *Glycine max* (β-sitosterol, stigmasterol, campesterol) and *Phaseolus vulgaris* (stigmasterol and β-sitosterol). Phytosterols in legumes are listed in Table 13.2 in chapter 13.

7.2.3 Phytates

Phytic acid is myoinositol 1, 2, 3, 4, 5, 6-hexakis dihydrogen phosphate, which chelates (bind to and remove) mineral elements and reduces their bioavailability. Phytates are formed when phytic acid interacts with minerals. Phosphorous is commonly stored as a phytate, which is bound to inositol in the fibre of raw legumes and nuts. Hemicellulose and phytin form conjugates with certain proteins, affecting the absorption of calcium.

Phytates are not exactly toxic but are antinutritional, as they chelate mineral nutrients in foods, such as calcium, copper, iron, magnesium, manganese, phosphorous, potassium and zinc, rendering them unavailable for absorption. Raw foods particularly those with high fibre content such as green leaves, green fruit walls, seed husks and bran are rich in phytates. Phytates in *Glycine max* and *Pisum sativum* decreased the availability of zinc, manganese, copper and iron (Kienholz *et al.*, 1962; Liener, 1975). Prolonged deficiency in microelements leads to several disorders including heart disease (Reddy and Pierson, 1994).

However, phytates were shown to have the following beneficial effects (Peary and Peavy, 2000):

a) Small amounts of phytates in food slow down the absorption of sugars and regulate insulin levels, which is beneficial to diabetics and those suffering from hyperlipidemia (high blood fats).
b) As phytates are found in fibre rich foods, they are involved in offering protection against colon and breast cancer.
c) Phytates increase the activity of natural killer cells, which destroy cancer cells and have a role in controlling cell growth, thus becoming protective against wide range of cancers.
d) Phytates that bind to minerals such as calcium, iron and zinc in foods, function as antioxidants as they keep these minerals at safe levels in the body, reducing the formation of free radicals.
e) Antioxidant effects of phytates also inhibit the onset of cancer due to free radicals, by indirectly helping the immune system.
f) Excessive calcium in the body leads to stones in the kidneys and the gall bladder. Excess iron in the body leads to heart disease. Chelation of calcium and iron by phytates up to a particular level is beneficial.

7.2.4 Tannins

Tannins are water soluble phenolic compounds. Tannins bind to proteins forming hydrophobic complex. Ingestion of dietary tannins impairs feed efficiency due to a decreased digestibility of proteins and the availability of amino acids from protein digestion. Tannins also reduce intestinal absorption of galactose and leucine (Santidrian and Marzo, 1989).

Tannins are often considered to be nutritionally undesirable. Tannins form complexes with protein and starch. They also complex with digestive enzymes like trypsin, α-amylase and lipase and inhibit them (Longstaff and McNab, 1991; Jansman *et al.*, 1994). This causes a severe reduction in nutritional benefits from foods and also reduce urea-nitrogen concentration in serum and liver (Nishioka, 1983). Tannin-protein complexes, which are extremely hydrophobic (Hagerman and Butler, 1980; Mitaru *et al.*, 1984), are partly responsible for low protein digestibility and low amino acid availability and thus increased faecal nitrogen excretion in rats and decrease in its biological availability to the animals (Hlodversson, 1987; Moseley and Griffiths, 1980; Glick and Joslyn, 1970 a,b) and chicken (Ortiz *et al.*, 1993). Addition of tannic acid to a heated soybean diet of chickens also reduced the rate of *in vivo* intestinal absorption of galactose and leucine (Santidrian and Marzo, 1989). Supplementation of both tannase and normal digestive enzymes, is likely to improve the protein value of tannin containing grain legumes.

The condensed tannins in certain legumes inhibit plant protein degradation in the rumen and decrease the availability of sulphur, which in turn depresses the digestibility of plant cell walls. These tannins may inhibit microbial enzymes in the rumen and decrease the availability of peptides derived from plant proteins for digestion in the intestine.

Tannins have been reported to be responsible for decrease in feed intake, growth rate, and net metabolizable energy in experimental animals. Therefore, foods rich in tannin are considered to be of low nutritional value. Several studies have shown that tannins severely affect bioconversion (Chung *et al.*, 1998).

There was reduced digestibility of proteins when the experimental animals were fed either with *Vicia faba* or dark flowered cultivar of *Pisum sativum* (Moseley and Grifiths, 1980; Hlodversson, 1987; Santidrian and Marzo, 1989; Longstaff and McNab, 1991; Alzueta *et al.*, 1992; Jansman *et al.*, 1993). The greater negative effect of tannin with maximum reports is from *Vicia faba* (Griffiths, 1981). Tannins extracted from faba bean hull formed complexes with components of intestinal brush border and impaired sugar transport in the intestine.

Tannins are known to affect the utilization of vitamins and minerals. It was also reported that faba bean tannins exhibited high iron binding capacities. Tannins form insoluble complexes with divalent iron and render it less absorbable.

In studies on the biochemical mechanism of the antinutritional effects of tannins, it was demonstrated that the major effect of dietary condensed tannins was not related to inhibition of food consumption or digestion. Instead, the major effect was on the efficiency with which the digested and absorbed nutrients were converted into new body substances. This would be due to their direct inhibitory effect on the key metabolic pathway or the indirect effect related to diversion of metabolism into detoxification of polyphenols or their degradation products.

The dosage and the kind of tannin are very important in producing antinutritional or toxic effects (Chung *et al.*, 1998).

7.2.5 Proteinase Inhibitors

Plant proteinase inhibitors are proteins of wide distribution in plant kindgom which inhibit the activities of proteinases.

As per the current practice, the term 'protease (peptidase)' includes both 'endopeptidases' and 'exopeptidases' while the term 'proteinase' is used to describe only 'endopeptidases' (Barnett and McDonald, 1986). Ectopeptidases act towards the ends of C and N terminus, while endopeptidases act within the polypeptide chain, away from the C and N terminus.

Pepsin is secreted through the mucous membrane of the stomach, the principle site of its action. Trypsin, chymotrypsin and elastase, secreted by the pancreas are responsible for the digestion of proteins in the gut of most of the higher animal species (Olmedo *et al.*, 1987). Renin hydrolyzes casein, the milk protein. Carbohydrates are hydrolyzed by oral and pancreatic α-amylase. Several inhibitors of these enzymes occur naturally in our foods.

Now the approach to the classification of proteinase inhibitors has changed. Four mechanistic classes of plant proteinases have been proposed by Hartley 1989. They are serine proteinases, sulphyhydryl proteinases, acid proteinases and metallo proteinases which generally associate strongly with the enzymes they inhibit.

The serine proteinase inhibitors from plants are grouped into four families: The soybean trypsin inhibitor (Kunitz)/soybean proteinase inhibitor (Bowman-Birk); potato I and potato II inhibitors; squash inhibitors and α-amylase / trypsin and chymotrypsin inhibitors.

Proteinase inhibitors are present in the seeds of majority of members of leguminosae. They are also detected in the stems and flowers of *Vigna radiata, Phaseolus lunatus, Lablab purpureus* and *Psophocarpus tetragonolobus* and in the products of secretion like gum arabic, catechu gum and tragacanth gum (Fiho and Campos, 1984). Bowman-Birk proteinase inhibitor is typical of leguminosae. The trypsin and chymotrypsin inhibitors are found in tribes such as Phaseoleae, Glycineae and Viciae of Fabaceae. Mostly found in the seeds, it is also found in the leaves of alfa-alfa. Potato inhibitor I is found in broad beans, sulphhydryl proteinase inhibitors are found in the seeds of *Vigna radiata* and *Vigna unguiculata* (Filho and Campos, 1989).

Proteinase inhibitors were first isolated from *Glycine max* and later found to be abundant in legume seeds. Proteinase inhibitors were studied in the seeds of 250 samples of over 100 species of edible legumes, more particularly the species of *Cicer, Glycine, Phaseolus, Pisum, Psophocarpus, Vicia* and *Vigna* (Weder, 1981). Species containing inhibitors of digestive enzymes are given in Table 7.1.

Table 7.1

Inhibitors of Trypsin and Chymotrypsin in the Seeds of Indian Legumes

Species with edible parts are marked by an asterisk (*)

I Inhibitors of both trypsin and chymotrypsin

1. Caesalpiniaceae

1. *Bauhnia variegata*
2. *Caesalpnia bonduc*
3. *Caesalpinia sappan*
4. *Ceratonia siliqua**
5. *Delonix regia**
6. *Gleditsia aquatica*
7. *Parkinsonia aculeata**
8. *Senna tora*
9. *Tamarindus indica**

2. Fabaceae

1. *Abrus precatorius*
2. *Arachis hypogaea**
3. *Cajanus cajan**
4. *Canavalia ensiformis**
5. *Canavalia gladiata**
6. *Cicer arietinum**
7. *Dalbergia sissoo*
8. *Glycine max**
9. *Lablab purpureus**
10. *Lathyrus latifolius*
11. *Lathyrus sativus**
12. *Lens culinaris**
13. *Kummarovia stipulacea*
14. *Macrotyloma uniflorum**
15. *Melilotus albus*
16. *Mucuna deeringiana**
17. *Mucuna pruriens* var., *utilis**
18. *Phaseolus coccineus**
19. *Phaseolus lunatus**
20. *Phaseolus vulgaris**
21. *Physostigma venenosum*
22. *Pisum sativum**
23. *Psophocarpus tetragonolobus**
24. *Robinia pseudoacacia**
25. *Sesbania sesban**
26. *Trifolium pratense*
27. *Trigonella foenum-graecum**
28. *Vicia faba**
29. *Vicia tenuifolia**
30. *Vigna aconitifolia**
31. *Vigna angularis**
32. *Vigna mungo**
33. *Vigna radiata**
34. *Vigna subterranea**
35. *Vigna umbellata**
36. *Vigna unguiculata**

3. Mimosaceae

1. *Acacia nilotica* ssp., *indica**
2. *Acacia polyacantha*
3. *Albizia lebbeck*
4. *Albizia lophantha*
5. *Dichrostachys cinerea*
6. *Entada gigas*
7. *Entada rheedei*
8. *Faidherbia albida*
9. *Mimosa pudica**

II Inhibitors of Trypsin

1. Caesalpiniaceae

1. *Senna alexandrina*

2. Fabaceae

1. *Aeschynomene sensitiva*
2. *Caragana arborescens*
3. *Crotalaria juncea*
4. *Cyamopsis tetragonoloba**
5. *Lotus corniculatus*
6. *Lotus uliginosus*
7. *Lupinus albus**
8. *Medicago lupulina**
9. *Medicago sativa**
10. *Trifolium dubium*
11. *Trifolium hybridum*

III Inhibitors of Chymotrypsin

1. Caesalpiniaceae

1. *Senna occidentalis*

Proteinase inhibitors have a detrimental effect on the utilization of proteins. The inhibitors bind strongly to the active sites of proteolytic enzymes, thereby reducing digestibility that results in a high proportion of essential amino acids not being available to the body (Kakade *et al.*, 1969). This also triggers a hyper-secretion of proteinases by the pancreas leading to enlargement of the pancreas and impairment of growth. An increased pancreatic secretion depletes body tissues of methionine and cysteine that are already in short supply in legume seeds (Kakade *et al.*, 1969).

Hyperactive pancreas is one of the primary physiological effects produced by a prolonged or excessive consumption of raw seeds of *Arachis hypogaea, Cicer arietinum, Glycine max, Phaseolus coccineus, Phaseolus lunatus, Phaseolus vulgaris, Pisum sativum, Psophocarpus tetragonolobus, Vicia faba, Vicia sativa* and *Vigna unguiculata,* which in the long run damages the pancreas (Booth *et al.*, 1960).

Ingestion of raw soybean meal or heated soybean products containing trypsin inhibitors besides reducing proteolysis of dietary proteins cause hypersecretion of pancreatic enzymes into the intestines that are neither degraded nor reabsorbed. It passes into the caecum and large intestine resulting in diminished proteolysis of both exogenous and endogenous proteins which impair growth. Trypsin inhibitors also cause hypertrophic and hyperplasic pancreas. It is believed that trypsin and chymotrypsin inhibitors in the upper intestine suppresses the release of cholecystokinin-pancreozymin (CCK-PZ) resulting in increased flow of pancreatic juice.

There are also inhibitors of oral and pancreatic α-amylase. *Phaseolus vulgaris* contains such inhibitors, affecting the hydrolysis and absorption of carbohydrates.

Plants are considered to have developed inhibitors of proteolytic enzymes to prevent proteolysis by enzymes secreted into the plant tissues by fungal and bacterial pathogens. They are also a defense against insect pests, which on consuming the host plant tissue suffer from indigestion and nutritional impairment leading to death. The role of proteinases and their inhibitors, in the context of the control of pests and diseases in crop plants, is discussed in Chapter 15.

Inhibitors of digestive enzymes are mostly destroyed during food processing. They are not classic toxicants, but they reduce the nutritional value of foods leading to other serious problems of health. Inhibitors of digestive enzymes are put to good use in nutritional research, to study the diet-enzyme-nutrition relationships.

Since legume seeds are our chief source of vegetable proteins and their amino acids, the presence of inhibitors of digestive enzymes, which affects digestion of food and restricts the availability of amino acids, is a matter for concern. The importance for animals and humans of ingesting food containing proteinase inhibitors cannot be under emphasized. The kinds and levels of inhibitors, species specificity has to be kept in mind as these factors can influence the effects of inhibitors.

7.2.6 Antivitamin Factors

Some naturally occurring compounds function as antagonists to vitamins A, D, E and K. They belong to different chemical groups, but considered together here on account of their common functional feature, the antivitamin activity. The exact causes for the antivitamin effects are not evident, but possibly several factors are involved.

Vitamin A is essential for the functions of the eye as it is associated with regeneration of the protein concerned with vision. It is also required for the immune system and healthy epithelial and mucous membranes. Vitamin A deficiency impairs mobilization of iron from liver leading to anaemia. B group vitamins have more functions in our metabolism than most other enzymes. Vitamin C is implicated in several day-to-day ailments like colds and is a key to our health. Deficiency of Vitamin D causes rickets, osteomalacia and fat malabsorption. Vitamin E plays an important role in cellular respiration and is a naturally occurring antioxidant. Its deficiency exposes fatty acids to peroxidation, leading to metabolic problems. Vitamin K lowers the blood clotting time and its deficiency leads to prolonged haemorrhages.

Many vitamins in food are affected by the consumption of uncooked legumes resulting in their deficiency. The antivitamin A and D factors of *Glycine max* results in low levels of plasma tocopherol (vitamin E), and in turn lead to liver necrosis and muscular dystrophy (Liener, 1975; Yamaguchi, 1983). So is the antivitamin E effect of the seeds of *Medicago sativa, Phaseolus vulgaris* and *Pisum sativum.* Vitamin K is destroyed by an intake of the seeds of *Melilotus albus* (Yamaguchi, 1983).

7.2.7 Carbohydrates

A large number of carbohydrates are an essential part of the structure and physiology of plants. They are products of photosynthesis, and provide the required respiratory energy. Carbohydrates occupy a central position in plant metabolism. They conjugate with proteins to form glycoproteins. With lipids they form glycolipids and with several secondary metabolites they form glycosides. Such glycoconjugates, without the sugar molecule, are called aglycones. Glycoconjugates play major roles in plant and animal metabolism.

Sugars such as stachyose and raffinose, present in soybeans and several other legume cause flatulence (Rackis and Gumbman, 1981), impairing digestion and absorption of nutrients, even when the seeds were defatted and roasted (Padgetle *et al.*, 1996).

Non-starch carbohydrates such as cellulose, hemicellulose, pectin, gums and mucilage present in many edible legume seeds, but more particularly in *Acacia catechu* and *Trigonella foenum-graecum,* may impair digestion of proteins, lipids and carbohydrates and affect absorption of products of digestion in the gut. Hemicellulose forms conjugates with proteins that affect the absorption of calcium. The gumminess imparted by the non-starch carbohydrates reduces the rate of hydrolysis, transport and uptake of nutrients in the gut.

7.2.8 Proteins

Improperly cooked legume seeds cause stomach upsets and consequent impairment of nutrient absorption, principally on account of indigestible proteins. Some of the ANFs that influence the nutritional quality of legume seeds are directly related to globulins, the major storage proteins in legume seeds (Chapter 2, Section 2.7.1)

7.2.9 Lectins

Lectins are proteins or glycoproteins that bind to specific cell surface carbohydrates. Lectins in some of the edible legumes were identified as ANFs and this activity is due to a) growth inhibitory effect and impairment of absorption efficiency during digestion as was observed with *Glycine max* and *Phaseolus vulgaris,* and b) nondegradation of lectins in the gastrointestinal tract resulting in the enlargement of the small intestine by cellular hypertrophy and hyperplasia (Pusztai *et al.*, 1987). ANF activity of legume lectins is pronounced in the improperly cooked seeds.

The implications of legume lectins in health and disease are discussed in Chapters 9 and 23.

7.3 DENATURING ANTINUTRITIONAL FACTORS IN EDIBLE LEGUMES

The ANFs in the legume seeds can be inactivated to circumvent their effects on human health. Many of them can be denatured by simple food preparative procedures such as soaking, leaching, sprouting, fermentation and cooking (Sharma and Sehgal, 1992; Qui *et al.*, 1996). The effect of soaking and cooking at high temperature reduced the level of ANFs in *Mucuna pruriens* and *Glycine max.* (Vijayakumar *et al.*, 1996) Soaking and sprouting improves starch digestibility. Cooking improves protein digestibility and decreases the levels and effectiveness of trypsin inhibitors (Rodriguez *et al.*, 1998). Fermentation improves protein quality and the bioavailability of essential amino acids.

CHAPTER-8

Toxicants in Edible Legumes

8.1 INTRODUCTION

The focus of this Chapter is toxic chemical compounds in edible legumes, and their effects on humans and domesticated animals, when consumed. Toxicants are naturally occurring biomolecules that act in a more direct manner resulting in serious impairment of metabolic processes, organ deformities and consequent disorders, and even death. During the process of selection and domestication of plant species for use as food and feed, varieties were developed not only for a higher nutritional potential but also for their low levels of toxicants.

> **Box 8.1**
> **Substances Hazardous to Living Systems**
>
> **Poison** is a substance hazardous to living systems. Strictly speaking, the term poison refers to a hazardous inorganic substance, such as arsenic and mercury. In recent times heavy metals like lead, mercury and cadmium have gained significance as hazardous environmental pollutants. Carbon monoxide, an atmospheric pollutant is a poison.
>
> **Toxicant** (or toxin) is a substance of plant origin, as for example, strychnine and hemlock (alkaloids), cholera toxin, podophyllotoxin, abrin or ricin (proteins) and lathyrine (an amino acid).
>
> Some scientists restrict the term toxin to those hazardous organic substances that can elicit antibodies in mammalian systems. Toxoid is a toxin altered to remove its toxic potential but retaining its physical configuration, so as to use it as an antigen to produce antibodies (for example, tetanus toxoid).
>
> **Venom** is a substance of animal origin, such as cobra venom, scorpion venom and bee venom.
>
> However, such a precise usage of these terms is uncommon, and the terms poison and toxin are used generally and interchangeably, to all substances harmful to living systems.
>
> Mammalian systems produce antibodies against most venoms and against many toxins, which are proteins or non-protein compounds bound to proteins, facilitating immunological defense. Poisons do not elicit an antibody response.

8.2 POISONS, TOXICANTS AND VENOMS

The naturally occurring chemical compounds hazardous to biological systems go by the names poisons, toxicants (or toxins) and venoms (see Box 8.1). When even food, of an inappropriate quality and/or quantity, can be toxic, it often requires extensive research to determine whether a particular compound is toxic, in what way and to what extent. The most modern approach to understand the toxic effects of substances in terms of their influence on our genes is toxicogenomics (Box 8.2).

8.3 TRADITIONAL TOXICOLOGY IN INDIA

Toxicology is the study of substances hazardous to life and their antidotes.

The faculty of recognition of toxic species of plants is not confined to man. Herbivorous primates, and even non-primate mammals are known to avoid toxic plants.

Toxic and psychoactive plants were among the first of non-food plants man had learnt to use. Specific use of toxic plants was made in all cultures, as reflected in such uses as arrow poisons. The traditional medicine-man (or woman) has used them effectively, in keeping a hold on the population, occasionally curing a patient or two.

Poisonous plants were among the very important secret weapons in settling personal and political scores, in several parts of the world, all through history. Coupled with pathogens, these plants form the arsenal of biowarfare and bioterrorism.

Like in every other culture, in India too, toxic plants have occupied a prominent place on the minds of medical men and their literature. Toxicology was dealt in detail by both Charaka and Sushrutha and is one of the eight major aspects of *Ayurveda* (*Damstra chikitsa*).

Box 8.2

Toxicogenomics

The genome is one set of all genes of an organism. Genomics is the area of study of the structure and function of genes.

Different toxicants affect different vital processes in biological systems, such as DNA replication, transcription, translation, and membrane transport, leading to a variety of metabolic disruptions including blocked receptor sites and inhibited enzyme action.

Toxicogenomics is the study of a) the influence of toxic substances on gene expression, b) the sites and mechanism of their action in an organism, and c) the consequences of such effects. Such information will help in understanding the cause and effect factor of toxin-induced disorders and diseases. It also helps in finding antidotes to the toxins and/or to use them as drugs.

Box 8.3

Legumes Used in Traditional Toxicology in India

Caesalpiniaceae: *Cassia fistula* and *Senna auriculata.*

Fabaceae: *Butea monosperma, Canavalia ensiformis, Clitoria ternatea, Desmodium gangeticum, Glycyrrhiza glabra, Indigofera tinctoria, Lathyrus sativus, Pterolobium indicum, Pterocarpus santalinus* and *Uraria picta.*

Mimosaceae: *Acacia suma, Albizia lebbeck* and *Prosopis spicigera.*

Chanakya, also known as Kautilya, the master of chicanery and political intrigue, and the master of ancient Indian economics, is believed to have had an extensive knowledge of Indian poisonous plants. Koutilya is credited with a compilation of detailed information on over 180 species of poisonous plants, their effects, uses, dosage, mode of administration, camouflaged administration, slow poisoning and the antidotes for them.

In a sense, most medicinal plants are toxic, and the knowledge of medicinal and poisonous plants grew concurrently. In *Ayurveda,* poisonous plants and the disorders they cause and their therapeutic uses were recognized. For example, *Bhavaprakasha,* an Ayurvedic text, refers to *Triputa* (*Lathyrus sativus,* chickling pea) as crippling a man (making him lame) and irritating the nerves. Modern science has identified the amino acid lathyrine as the chief cause of this ailment, which affects the nerves and the bones and consumption of the seeds of *Lathyrus sativus* was long prohibited. This shows how carefully our ancients have selected plant species for use as food.

Traditional Indian toxicology has been discussed in a considerable detail (Raut, 1993; Sathyanarayana Bhat and Kameswara Rao, 1993a; Kameswara Rao, 2000). About 160 species of plants are in traditional toxicology in India (Kameswara Rao, 2000). The legumes used in traditional toxicology are given in Box 8.3. Significantly no important edible legume is in this list.

The deep scientific interest in toxic plants is to develop drugs. Plants that are effective against a number of ailments are toxic at certain doses. The hope is that cytotoxic and mitostatic effects of plant constituents would be of use in treating cancer. The importance of toxic species in drug development is indicated by the series of volumes published by CSIR on the poisonous plants of India.

8.4 TOXICANTS IN LEGUMES

Like species of all other plant groups, legumes also contain a large number of toxicants (Harborne *et al.*, 1971; Gibbs, 1974; Pollhill and Raven, 1981; Cheeke, 1989; Harborne and Baxter, 1996). Several edible legumes exert toxic effects when consumed in excess or for long periods of time or when an inappropriately processed food is taken. Animals are also affected from toxicants in feed plants, and in turn we may be affected by consuming meat, milk and eggs from such animals.

The occurrence of toxicants in plants is a function of the genotype (cultivar) and the environment. Geographical location, climatic conditions, soil type, water availability and several other ecological factors also have a determining role in this regard. The degree of toxicity of a plant chemical is relative and is dependent on a) the dose taken, b) concentration in the body with reference to the body weight, c) age, the state of health and susceptibility of the individual, d) the mechanism and rate of absorption and e) the mechanism and the rate of excretion. Some hazardous compounds are removed from our system through the liver and some through the kidneys. In the long run, either the liver or the kidneys, sometimes both, are put under additional stress, damaging them.

The various toxicants in the edible Indian legume species, and their effects are discussed here. The classification of these compounds was outlined in Chapter 5. Both the beneficial uses and toxic potential of some of the compounds are discussed in other appropriate places. Toxins in individual species are mentioned in the species profiles.

Tables 8.1 to 8.5 present the toxic compounds, their known sources and effects.

8.4.1 Nitrogenous Compounds

Nitrogenous compounds are nitrogen containing organic substances forming salts with mineral acids. The nitrogenous compounds include amino acids, amines, alkaloids, cyanogenic glycosides, purines and pyrimidines and their glycosides.

8.4.1.1 Amino Acids: Amino acids are colourless ionic compounds. There are over 20 amino acids that constitute the proteins and these are called the 'protein amino acids'. The concentration of protein amino acids in plant tissues varies from 20 to 200 mg/g fresh weight. Glutamine and aspargine, the acidic amides of glutamic acid and aspartic acid, occur in larger quantities as they are the storage form of nitrogen whereas histidine, tryptophan, cysteine and methionine are low in plant tissues.

In addition to the protein amino acids, there are over 400 amino acids that do not go into the structure of proteins, called the 'non-protein amino acids', among which only γ-aminobutyric acid is regularly found in plants. The remaining non-protein amino acids occur only in specific plant groups. For example canavanine occurs only in certain subgroups of the Fabaceae and not outside the Fabaceae.

A few protein amino acids or their derivatives and most of the non-protein amino acids in legumes are toxicants, usually present in their seeds (Table 8.1). Either these amino acids or their breakdown products, inhibit variety of essential cellular process, and cause liver lesions, bone and nervous disorders.

Lathyrogens, agents that cause lathyrism, are toxic amino acids found in the seeds of several legumes (Table 8.1). Lathyrism, that mostly affects the young between 20 and 30 years of age, is a disorder associated with the consumption of the seeds of certain species of *Lathyrus,* particularly of *Lathyrus sativus*, which are used as pulse in poverty ridden areas or in times of drought, in spite of a strong advice against such use. Lathyrism is more prevalent in some States like Madhya Pradesh in India. It is manifested in the affliction of central nervous system and paralysis of leg muscles (neurolathyrism) and/or of skeletal deformities (osteolathyrism) (Liener, 1975a). The causative amino acida of lathyrism is characterized as β-N-Oxalyl L-α, β-diaminopropionic acid (ODAP) and β-N-oxalylamino-L-alanine (BOAA) which produces severe neurotoxic effects (Adiga *et al.*, 1962). A water soluble, aliphatic amino acid glycoside with a nitrile group designated as N-β-D-glucopyranosyl-N-L-arabinosyl-α, β-diamino propionitrile that acts synergistically with ODAP was also isolated. ODAP is also known to cause ammonia toxicity in the brain causing convulsions (Cheema *et al.*, 1969). It drastically reduced glucose, ATP, seratonin and acetylcholine levels in the brain. Another symptom, angiolathyrism, is manifested in disorders of collagen and elastin metabolism of the heart, is caused by an amino acid β-aminopropionitrile (BAPN) occurring in the seeds of species of *Lathyrus*.

Canavanine, a non-protein amino acid found in several members of the Fabaceae is a toxin abundant in the seeds of *Canavalia ensiformis, Medicago sativa*, and several other species of the Fabaceae, causes lupus like anaemia due to alteration in red blood cells.

Although *Leucaena leucocephala* is primarily a forage crop, the Indonesians consume its leaves, pods and seeds. Excessive consumption results in hair fall. This effect was attributed to the sulphur containing amino acid mimosine that occurs also in *Mimosa pudica* (Liener, 1975b). Mimosine is also a goitrogen. Seeds of *Pithecellobium lobatum* cause kidney failure and formation of djenkolic acid as needle like clusters in the urine.

8.4.1.2 Amines: Plant amines are the products of decarboxylation of amino acids. They may also form from ammonia. Amines are divided into aliphatic and aromatic amines. The aliphatic amines are divided into mono, di and polyamines. Polyamines are less volatile than the other two. The aromatic amines are phenolic in nature. Histamine, seratonin and noradrenalin are hallucinogenic plant amines, which may also occur in animal brains.

Legumes are abundant in aliphatic and aromatic amino acids. Cadaverine, an aliphatic amine in the seeds of *Glycine max, Lathyrus sativus* and *Pisum sativum,* emits the odour of a human corpse and is a skin irritant and sensitiser. The leaves and seeds of *Mucuna pruriens* contain an aromatic amine, 3, N, N-dimethyl tryptamine (dopamine), which causes hypertension, papillary dilation, anxiety, hallucination and perceptual disorders. Another aromatic amine, N-methyl mescaline, in the leaves and seeds of *Alhagi maurorum* is a nervous system depressant and a psychomimetic.

8.4.1.3 Alkaloids: Alkaloids are the largest single class of very widely distributed secondary metabolites in plants, which can be identified by the presence of nitrogen in the heterocyclic ring. Over 10,000 plant alkaloids are known and they are one of the most interesting groups of compounds used in medicine due to their broad biological activities. The most common precursors of alkaloids are amino acids. The pathways of their formation, physico-chemical characteristics and bioactivity are very diverse and complex. Alkaloids interfere with DNA replication, transcription, translation, and membrane transport. They block receptor sites and inhibit enzyme action.

Various other alkaloids identified in species of *Erythrina* (Soto-Hernandez and Jackson, 1994) and *Spartium junceum* (Lurz *et al.*, 1990) are known to impair nutrition. The different alkaloids present in legume species, their adverse effects are given in Table 8.2.

Among several alkaloid subgroups, the indole alkaloids are not considered as the true alkaloids, as they do not contain heterocyclic nitrogen other than the indole ring. These are amines formed from similar precursor of true alkaloids and may interfere with DNA replication, transcription, translation, membrane transport and block receptor sites and so are generally toxic. Hypaphorine, an indole alkaloid from species of *Pterocarpus* and *Erythrina,* causes convulsions.

The other subgroups of toxic alkaloids are, Pyrrolizidine, Indolizidine, Quinolizidine, Piperidine, Isoquinolizidine and Macrocylic alkaloids.

Among these, the Pyrrolizidine alkaloids are the most important toxins and carcinogens. These are hepatotoxins, which are metabolized in liver to pyrroles, and aldehydes, which are responsible for toxicity. They cause veno-occulusive disease in children resulting from the consusmption of species of *Crotalaria.* The pyrrolizidine alkaloids found in legumes possess cytotoxic effects, seen mainly on the liver cells resulting in haemorrhagic necrosis, formation of giant liver cells leading to liver tumour. Apart from this, they are also pneumotoxic and cardiotoxic. Pyrrolizidine alkaloids interfere with the uptake of copper, iron, zinc and vitamin A. A potential route of ingestion of pyrrolizidine alkaloids is indirect through milk and meat of animals that graze on species of *Crotalaria.* Monocrotaline causes pulmonary arterial hypertension and right ventricular hypertrophy.

Indolizidine alkaloids cause gastrointestinal irritation. Swainsonine, present in species of *Astragalus* and *Swainsona canescens,* causes swainsona poisining characterized by neuro-visceral cytoplasmic vacuolation and neuroxonal dystrophy.

The quinolizidine alkaloids are compounds with bicyclic quinolizidine ring with a functional carbonyl group and are synthesized from lysine. Lupanine and albine from species of *Lupinus* are highly toxic causing nausea, ataxia, diaphoresis and paralysis. They affect the biosynthesis and metabolism of glycine and cause respiratory arrest and visual disturbances.

Piperidine alkaloids are derived from glycine by deamination and decarboxylation by pyrridoxyl phosphate dependent enzymes. Species of *Cassia, Senna, Prosopis* and *Genista* contain these alkaloids which are teratogenic.

The isoquinolizidine alkaloids cause neuromuscular disorders.

8.4.1.4 Cyanogenic Glycosides: Glycosides are compounds conjugated with carbohydrates. Cyanogenic glycosides occur widely in plants and virtually occur in every part of the plant. They originate from ammonia or structurally related amino acids and contain cyanide molecules as sugar precursors. The toxicity of cyanogenic glycosides is due to their ability to release hydrogen cyanide (HCN), on exposure to air following mechanical injury such as cutting and grazing. HCN inhibits cytochrome oxidase, an enzyme of the respiratory electron transport chain and thus affects respiration.

Legumes predominate among the other plants in terms of their cyanide producing potential. Cyanogenic glycosides such as linamarin, lotaustralin, phaseolunatin, vicianin, and several others, occur in legumes, particularly, plants belonging to Fabaceae and Mimosaceae.The ability of cyanogenic glycosides to release cyanide implicates them in number of disorders in populations that depend on food sources containing these glycosides. Excessive consumption of *Phaseolus lunatus* and *Vicia sativa* may result in cyanide poisoning causing inhibition of respiratory enzymes and paralysis (Liener, 1975). Cyanide released from the glycosides produce electro-encephalographic and electro-cardiographic conditions. It causes extensive alteration in cardiac physiology with an effect on contractile components of myocardium and abrupt loss of electrical activity. Cyanogenic glycosides were used as potent military chemical weapon during World War I and II.

Production of cyanogenic glycosides by plants is considered as a defense against herbivores, as grazing cattle are affected more commonly.

The cyanogenic glycosides in Indian legume species, the toxicity symptoms of these compounds are given in Table 8.3.

8.4.1.5 Pyrimidine Glycosides: These are closely related in structure to nucleic acid bases. Their low molecular weight forms occur in a bound state in plants.

The pyrimidine glycosides vicine, convicine and their respective aglycons (glycosides without the sugar molecule) divicine and isouramil (IU) pyrimidines trigger favism (Mudulie *et al.*, 1982; Mager *et al.*, 1980), a disease characterized by acute haemolytic anaemia triggered by ingestion of raw or cooked *Vicia faba* beans. The blood cells of affected individuals show diminished level of glutathione and glucose-6-phosphate dehydrogenase activity, resulting in a disturbance of cell membrane integrity (Liener, 1975). The levels of glucose, ATP, ceratonine and acetylcholine in the brain are also reduced.

8.4.2 Phenolic Compounds

Phenolic compounds are a class of aromatic (in the chemical sense) organic compounds, technically alcohols, in which one or more hydroxy (–OH) groups are attached directly to the benzene ring, as for example, phenol itself. Phenolic compounds are derived from this basic structure. Shikimate/arogenate pathway through phenylalanine leads to the majority of the plant phenolics.

Free phenols are relatively rare in plants. Phenolic compounds may accumulate as a result of stress and influence allelopathy, a phenomenon involving leached compounds inhibiting other plant species. In foods they are toxic to different degrees.

Phenolic compounds include phenolic glycosides, flavonoids, tannins, quinones, chromones, phenyl propanoids and anthrones.

8.4.2.1 Phenolic Glycosides: Phenolic glycosides such as 1-5, vinyl-2-thioxazolidone block the uptake of iodine by the thyroid and thus cause goiter, the condition discussed in Chapter 7. *Arachis hypogaea, Glycine max, Phaseolus vulgaris, Pisum sativum* and *Vigna radiata* contain phenolic glycosides in the seeds.

8.4.2.2 Flavonoids: Flavonoids are a group of aromatic, oxygen containing heterocyclic pigments that are derived from flavones, generally present in plants as glycosides, called the flavone glycosides. Auriculoside, a flavon glycoside from the aerial parts of *Acacia auriculiformis*, causes depression of the central nervous system.

Estrogens are steroidal compounds, functioning as sex hormones in our body system. They are related to cholesterol and circulate in blood, usually as protein complexes. Phytoestrogens have estrogen-like activity, and occur in both edible and inedible parts of several plant species. They are more widely distributed in the legumes, occurring more abundantly in the leaves than in the seeds.

Chemically phytoestrogens belong to one of several flavonoid groups such as lignans, isoflavonones, flavonones, chalcones, diterpenoids, triterpenoids, coumastans, or acyclics. Among these groups, isoflavonones have emerged in recent times as important phytoestrogens, as they are more related to estrogens structurally, than the others. The isoflavonones-genistein, biochanin A, diadzein, formononetin, equol and glycitein function as phytoestrogens. These compounds occur as glycosides in plants. Genistein, diadzein and equol from soybean, were shown to control menopausal symptoms, osteoporosis, hypertension and atherosclerosis as well as breast and prostate cancer (Goodman, 2004), and are now being commercialized (See also Chapter 22).

Many isoflavonones, a subgroup of flavonoids, may induce estrogenic and hypocholesterolemic effects, resulting in hormonal disturbances. Isoflavonone phytoestrogens have been reported from the following legume species, which occur in India: *Abrus precatorius, Arachis hypogaea, Cajanus cajan, Canavalia ensiformis, Cicer arietinum, Crotalaria pallida, Cyamopsis tetragonoloba, Entada scandens, Glycine max, Lablab purpureus, Lathyrus sativus, Lens culinaris, Macrotyloma uniflorum, Medicago sativa, Mucuna pruriens, Pachyrrhizus tuberosus, Phaseolus vulgaris, Pisum sativum, Senna occidentalis, Tamarindus indica, Trifolium pratense, Trigonella foenum-graecum, Vicia faba, Vigna mungo, Vigna radiata,* and *Vigna unguiculata* (Altern, 1997). *Glycine max* and *Trifolium pratense* have been identified as the richest source of phytoestrogens among all these species. Smaller quantities of lignans, coumastans and flavones occur along with much larger quantities of isoflavonones in these legumes.

A larger list of legume species with phytoestsrogens is given in Table 8.4

8.4.2.3 Tannins: Tannins are water-soluble phenolic compounds with a high potential to bind to proteins. Tannin-protein complexes are hydrophobic, cause low protein digestibility, low amino acid availability and reduce intestinal absorption of galactose and lucine (Mitaru *et al.*, 1984; Santidrian and Marzo, 1989).

Tannins in edible legumes decreased food intake, growth rate, protein digestibility and net metabolizable energy in experimental animals. Hence, food rich in tannin is considered to be of low nutritional value. Recent studies have shown that tannins greatly affect bioconversion also (Chung *et al.*, 1998). Ingestion of large amounts tannins may cause gastric irritation, nausea, and liver damage.

Among the edible legumes, *Arachis hypogaea, Cajanus cajan, Cicer arietinum, Cyamopsis tetragonoloba, Glycine max, Glycyrrhiza glabra, Lens culinaris, Macrotyloma uniflorum, Phaseolus lunatus, Phaseolus vulgaris, Pisum sativum, Sesbania grandiflora, Trigonella foenum-graecum, Vicia faba* and *Vigna unguiculata* contain tannins, particularly in the seed coats.

Proanthocyanidins (non-hydrolyzable tannins) in the seeds of *Vicia faba* and *Phaseolus vulgaris* decrease metabolizable energy and starch digestibility.

Tannins being of a wider interest, are discussed in Chapters 7 and 11.

8.4.2.4 Quinones: Quinones are phenolic compounds with two carbonyl rings. They are divided into benzoquinone, naphthaquinone, anthraquinone, plastoquinone and isoprenoid quinone groups. They are allergens, and cause bronchial asthma (Table 8.5). Species of the Caesalpiniaceae are rich in anthraquinones.

8.4.2.5 Chromones: Chromones are phenolic compounds with depressing effects on the central nervous system (Table 8.5).

8.4.2.6 Phenyl Propanoids: These are phenolic compounds with an aromatic ring to which a three-carbon side chain is attached. These propanoids are derived biosynthetically from aromatic protein amino acid phenylalanine. Present in species of *Melilotus,* the pasture legumes, and in *Cullen corylifolium*, they cause haemorrhagic disorders, respiratory tract irritation, allergy, and liver damage (Table 8.5).

8.4.2.7 Anthrones: Anthrones are allergens when bound to a protein irritate the respiratory tract (Table 8.5). Chrysarobin is found in species of *Cassia* and *Senna*

8.4.3 Terpenoids

Biosynthetically terpenoids and steroid compounds are closely related. Alkaloids are also related to terpenoids. Triterpenoids are compounds with a carbon skeleton based on six isoprene units and are derived biosynthetically from squalene. There are triterpenoidal and steroidal saponins, and sapogenins, which arise via mevalonate pathway.

Various terpenoid compounds are discussed in later Chapter 13.

8.4.3.1 Sterols: Sterols, a subgroup of steroidal compounds, are solid alcohols. They may occur either in the free state or as glycosides. Cholesterol, bile salts, adrenal cortical hormones, sex hormones and vitamin D are also steroidal compounds. α-spinasterol (isomer of stigmasterol), β-sitosterol (sitosterol), stigmasterol, campesterol and fucosterol are some examples of phytosterols. Cholesterol is of common occurrence in organisms as it is a component of biological membranes and is involved in several metabolic processes.

Steroidal compounds like lupeol, occurring in the seeds of *Lupinus luteus* may be hypotensive and anti-hyperglycaemic.

The anti-nutritonal effects of sterols were discussed in Chapter 7. Steroid compounds are discussed in Chapter 13, Table 13.2.

8.4.4 Organic Acids

The unique feature of plant metabolism is the ability of plants to accumulate organic acids in the cell vacuoles. The organic acids are water-soluble colourless liquids, formed as intermediate products of plant respiration through the citric acid cycle (Krebs' cycle). The organic acids occur universally, participate in several metabolic processes and are harmless. However, malonic acid in the seeds of *Phaseolus coccineus* is a strong irritant that damages the mucous membrane and the skin. Monofluro acetic acid in the fruits of species of *Oxylobium* can arrest human respiration.

8.4.5 Inorganic Compounds

Plants accumulate elements and inorganic compounds absorbed from water and soil, in all parts. These compounds are not actually synthesized by plants but form when the elements accumulate and form salts. When such plant foods are consumed excessively or over long periods of time, they cause various bodily disturbances. Such a situation may develop with aluminum, copper, zinc, iron, selenium and several other elements, when edible species are grown in soils rich in them. The symptoms of excessive levels of inorganic compounds in the human body include anorexia, indigestion, malnutrition, nausea, vomiting, diarrhoea, subcutaneous oedema, skin eruptions, skin lesions, discoloration of the skin and teeth, hair loss, and chronic arthritis. Several of these symptoms may manifest simultaneously, often confusing diagnosis.

Species of *Astragalus* accumulate selenium in the tissues as selenocystathionine, semethylselenocysteine and selenocysteine, which cause selenosis in the human body, characterized by some of the symptoms mentioned.

Geologists use species of plants that accumulate quantities of inorganic elements in their tissues as indicators of richness of the soils in the respective elements. This helps the mining industry in detecting soils rich in the respective elements.

8.4.6 Allergens: Allergy is an inflammatory disorder mediated by IgE type antibodies, which are elicited by certain proteins or non-protein compounds. Allergy is not a universal phenomenon. Only certain food items affect a few individuals and that too for a period of their lifetime. Some non-vegetarian foods such as fish, eggs and milk and several plant food items may cause allergy in some hypersensitive individuals. The symptoms vary from skin rashes to severe gastro-intestinal cramps and dysfunction. Allergy caused by peanuts is well known. Some varieties of soybean contain several allergenic protein fractions, at least some of which are lectins. Allergenic proteins of soybean cause adverse gastrointestinal reactions in hypersensitive individuals.

8.5 DENATURING TOXICANTS IN FOOD

Usually toxicants in plant foods are in such low quantities that they are harmless. Their concentrations may increase if the plants were grown under stress conditions like drought or excessive pest pressure. A repeated exposure to the toxins reduces susceptibility. But some individuals are more sensitive to toxins than others.

Like with the anti-nutritional factors, the small quantities of toxicants that occur in plant foods can be denatured by the usual food processing such as soaking, leaching, fermenting and cooking. Water-soluble compounds go as leachates and many others would be denatured by the heat of the cooking process. Sun drying the food grains also helps considerably.

8.6 MEDICINAL USES OF TOXICANTS

a) BAPN, a toxic amino acid is used as a tool in connective tissue research. Aminoacetonitrile (AAN) is used in research on old age diseases; β-cyano-L-alanine is a tool in the study of degenerative, neurological diseases.The demonstration of skeletal and aortic lesions by the administration of osteolathyrogens aroused great interest in lathyrism as a prototype for a number of acquired or congenital lesions. The fundamental changes in the the collagen or elastin of the lathyritic tissue in experimental animals prompted investigators to model this disorder to study normal connective tissue chemistry and function.

b) Indole alkaloids such as physostigmine is used to contract the pupil of the eye; When given intravenously it reverses the effect of sedatives.

c) Indolizidine alkaloids such as slaframine is used clinically as parasymptomimetic and cholinergic agonists; it increases salivation and pancreatic flow; used in treating cystic fibrosis; Swainsonin is medically used as an immunomodulator and in the treatment of allergy, tu mours, lysosomal storage diseases and steriod dependant diseases.

d) Cyanogenic glycosides in small doses over a long period of time is used to control pulmonary diseases, asthma, cough, catarrh and as sedative; they are also employed in cancer research.

e) Pyrimidine glycosides convicine and vicine are used to induce immunity against malaria

f) Phenolic glycosides are used in dietary and thyroid research.

g) Isoflavonones are used as chemopreventive agents to protect against cancer of the stomach, liver, colon and prostate.

Table 8.1

Toxic Effects of Amino acids in Legumes

Sl. No.	Toxin	Source species	Part	Symptoms of toxicity in humans
1.	L-Aspartic acid	*Glycyrrhiza glabra*	Root	Dietary amino acid,which in large doses produce neuro-excilatory symptoms
2.	L-Djenkolic acid	Species of *Acacia* and *Mimosa*	Pod, seed	Acute kidney malfunction and blocking of urine flow, painful swelling of genitalia in young children
3.	L-Homoarginine	*Lathyrus sativus*	Seed	Antimetabolite
4.	L-Mimosine	*Mimosa pudica* *Leucaena leucocephala*	Leaf, seed	Has goitrogenic effect when used as human food; results in poor growth, cataract and reproductive problems and loss of body hairs; causes oesophageal ulceration
5.	L-Indospicine	*Indigofera linnaei* *Indigofera spicata*	Leaf, seed	Teratogenic, abortifacient; causes cleft palate, dwarfism and liver damage
6.	β-(γ-glutamyl) aminopropionitrile		Leaf, seed	
7.	Aminoacetonitrile (AAN)	Amino acids 6 to 28 are present in species of *Lathyrus* and *Vicia*	Leaf, seed	Amino acids 6 to 26 cause osteolathyrism; Symptoms include abnormalities of bone, skeleton and vertebral column, disruption of connective tissue, disorders of growth of cartilage and bone, disorders of collagen and elastic metabolism, malformed bones, chest deformities, loose intervertebral disks, excessive flexure of carpal joints and contraction of tendons
8.	Methylene aminoacetonitrile (MAAN)			
9.	β, β'-Iminodipropionitrile (IDPN)			
10.	β-Hydrazinopropionitrile			
11.	Semicarbazide (SCA)			
12.	Acetone semicarbazide			
13.	Isonicotinic acid hydrazide			
14.	Nicotinic acid hydrazide			
15.	Cyanoacetic acid hydrazide			
16.	Benzhydrazide			
17.	γ-L-Glutamylhydrazide			
18.	Glycine hydrazide			
19.	Carbohydrazide			
20.	Thiocarbazide			
21.	Glutamic acid hydrazide			
22.	p-Nitrobenzine acid hydrazide			
23	Hydrazine hydrate			
24.	Unsymmetrical dimethyl hydrazine			

Sl. No.	Toxin	Source species	Part	Symptoms of toxicity in humans
25.	Symmetrical dimethyl hydrazine		Leaf, seed	
26.	Phenyl hydrazine			
27.	β-cyano-L-alanine	Amino acids 27 to 44 are present in species of *Lathyrus* and *Vicia*	Seed	Amino acids 27 to 44 cause neurolathyrism. The symptoms include liver dysfunction, tremors and convulsions in rats, neurological disorders, secondary brain damage and ammonia toxicity
28.	β-N-Oxalyl amino-L-alanine (BOAA)			
29.	γ-glutamyl-β-cyano-L-alanine			
30.	L-2,4-Diaminobutyric acid			
31.	Oxalyl diamino propionic acid			
32.	N-Oxalylglycine			
33.	N-oxalyl-β-alanine			
34.	L-α-Amino-γ-Oxalylamino butyric acid			
35.	L-α-Amino-β-Oxalylamino propionic acid			
36.	L-α,γ-diaminobutyric acid (DABA)			
37.	Monohydroxy pipecolic acid			
38.	Dihydroxy pipecolic acid			
39.	Trihydroxy pipecolic acid			
40.	5-Hydroxy pipecolic acid			
41.	β-N-oxalyl amino-α, β-di-aminopropionic acid (ODAP)			
42.	L-3-oxalylamino-2-amino-propionic acid (OAP)			
43.	N-oxalyl-lysine		Seed	
44.	β-nitropropionic acid			
45.	β-aminopropionitrile (BAPN)	Species of *Lathyrus*	Leaf, seed	Causes angiolathyrism. The symptoms include aneurysm, disrupted elastin formation in the arterial wall, abnormal proliferation of fibro blast and irregularly arranged collagen fibres
46.	2-Cyanopropyl amine (2-CPA)			
47.	L-Canavanine	*Canavalia ensiformis* *Medicago sativa*	Seed	Causes lupus like anaemia, inhibits placental alkaline phosphatase activity and blocks arginine uptake in humans
48	L-pipecolic acid	Species of *Lathyrus* *Vicia, Phaseolus vulgaris* and *Calliandra haematocephala*	Leaf, seed	Neurotoxic effect on foetal development causing neuronal damage, causes Zellweger's syndrome (cerebrohepato–renal syndrome) and infantile Refsum's disease

Table 8.2

Toxic effects of Alkaloids in Legumes

Sl. No.	Toxin	Source species	Part	Symptoms of toxicity in humans
1.	**Indole alkaloids**			
a.	Isophysostigmine	*Physostigma venenosum*	Seed	Highly toxic, used as "ordeal poison"; large doses fatal, counter acts the effects of anticholinergics such as atropine; interferes with DNA replication, transcription, membrane transport and blocks receptor sites
b.	Calabacine			
c.	Calabatine			
d.	Eseramine			
e.	Physostigmine			
f.	Physovenine			
2.	**Indolizidine alkaloids**			
a.	Castanospermine	*Castanospermum australe*	Seed	Causes gastrointestinal irritation in humans
b.	Swainsonine	*Astragalus oxytropis*	Plant	Causes lysosomal storage diseases, Swainsona poisoning
c.	Slarfamine	*Glycine max* *Trifolium pratense* *Vigna sinensis*	Seed	Causes Slobber's syndrome
3.	**Isoquinoline alkaloids**			
a.	Erysonine	*Erythrina suberosa*	Flower	Neuromuscular blocking agent
b.	β-Erythrodine	*Erythrina* species	Flower	Neuromuscular blocking agent, hypnotic, respiratory depressant, hypotensive
c.	Erythratidine	*Erythrina* species	Flower	
4.	**Macrocyclic alkaloids**			
a.	Pithecelobine	*Samanea saman*	Plant	Toxic
5.	**Piperidine alkaloids**			
a.	Ammodendrine *Sophora* species *Thermopsis lupinoides*	*Lupinus formosus*	Plant Seed Seed	Toxic
6.	**Pyrrolizidine alkaloids**			
a.	Anacrotine	*Crotalaria anagyroides* *Crotalaria incana* *Crotalaria laburnifolia*	Vegetative parts, seed	Hepatotoxic, hepatocarcinogen; causes liver lesions, cancer, fibrosis, biliary hyperplasia, jaundice, hyperbilirubinemia, hepatomegaly, hypoalbumineuria, veno-occlusive diseases, pulmonary arterial hypertension, right ventricular hypertrophy; cardiotoxic.
b.	Crotanecine	*Crotalaria* species		
c.	Crotaburmine	*Crotalaria spectabilis*		

Sl. No.	Toxin	Source species	Part	Symptoms of toxicity in humans
d.	Fulvine	*Crotalaria fulva* *Crotalaria paniculata* *Crotalaria madurensis*	Vegetative parts, seed	Alkaloids **d** to **n** are hepatotoxic and hepatocarcinogens; cause liver lesions, cancer and fibrosis, biliary hyperplasia, jaundice, hyperbilirubinemia, hepatomegaly, hypoalbuminemia, veno-occlusive disease, pulmonary arterial hypertension, right ventricular hypertrophy, oedema, renal retention of sodium, portal cirrhosis, cardiotoxicity and pneumotoxicity.
e.	Junceine	*Crotalaria juncea* *Crotalaria spectabilis*	Vegetative parts, seed	
f.	Monocrotaline	*Crotalaria laburnoides* *Crotalaria pallida* *Crotalaria quinquefolia*	Vegetative parts, seed	
g.	Retronecine	*Crotalaria* species	Vegetative parts, seed	
h.	Retrorsine	*Crotalaria trichotoma*	Vegetative parts, seed	
i.	Retusine	*Crotalaria retusa* *Crotalaria spectabilis*	Vegetative parts, seed	
j.	Seneciphylline	*Crotalaria juncea*	Vegetative parts, seed	
k.	Senkirkine	*Crotalaria laburnifolia*	Vegetative parts, seed	
l.	Spectabiline	*Crotalaria nana* *Crotalaria spectabilis*	Vegetative parts, seed	
m.	Trichodesmine	*Crotalaria juncea*	Vegetative parts, seed	
n.	Usaramine	*Crotalaria brevifolia* *Crotalaria incana*	Vegetative parts, seed	
7.	**Quinolizidine alkaloids**			
a.	Albine	*Lupinus albus* *Lupinus angustifolius*	Vegetative parts, seed	Causes nausea, ataxia, respiratory arrest, visual disturbances, coma, diaphoresis
b.	Allylcystisine	*Lupinus albus* *Lupinus angutifolius*	Vegetative parts, seed	Causes nausea, ataxia, respiratory arrest, visual disturbances, coma, diaphoresis
c.	Aloperine	*Sophora alopecuroides*	Vegetative parts, seed	Toxic, causing paralysis of central nervous system and respiratory organs
d.	Ammothamnine	*Sophora flavescens*	Vegetative parts, seed	Toxic
e.	Anagyrine	*Genista* species *Lupinus* species *Sophora* species *Ulex europaeus*	Vegetative parts, seed	Toxic, teratogenic, induces tachycardia
f.	Angustifolin	*Lupinus albus* *Lupinus angustifolius*	Vegetative parts, seed	Causes nausea, ataxia, respiratory arrest, visual disturbances, coma, diaphoresis
g.	Cystisine	*Anagyris* species *Baptisia* species *Genista* species *Laburnum anagyroides* *Sophora* species *Thermopsis* species	Vegetative parts, seed	Toxic, teratogenic

Sl. No.	Toxin	Source species	Part	Symptoms of toxicity in humans
h.	Cytisine	*Lupinus albus* *Lupinus angustifolius*	Vegetative parts, seed	Causes nausea, ataxia, respiratory arrest, visual disturbances, coma, diaphoresis
i.	Epilupinine	*Lupinus albus* *Lupinus angustifolius*	Vegetative parts, seed	Causes nausea, ataxia, respiratory arrest, visual disturbances, coma, diaphoresis
j.	Lupanine	*Cytisus scoparius* *Lipinus* species	Vegetative parts, seed	Toxic, anti-arrhythmic, hypotensive, hypoglycaemic
k.	Lupinine	*Lupinus albus* *Lupinus angustifolius*	Vegetative parts, seed	Causes nausea, ataxia, respiratory arrest, visual disturbances, coma, diaphoresis
l.	Multiflorine	*Lupinus albus*	Vegetative parts, seed	Central nervous system depressant
m.	Sparteine	*Lupinus albus* *Lupinus angustifolus*	Vegatative parts, seed	Causes nausea, ataxia, respiratory arrest, visual disturbances
n.	Thermopsin	*Lupinus albus* *Lupinus angustifolius*	Vegetative parts, seed	Causes nausea, ataxia, respiratory arrest, visual disturbances

Source: Cheeke, 1989.

Table 8.3

Toxic Effects of Cyanogenic Glycosides of Legumes

Sl. No.	Toxin	Source species	Part	Symptoms of toxicity in humans
1. 2.	Acacipetalin Heterodendrin	*Acacia* species	Pod	Bound toxin, releases cyanide on hydrolysis and causes tissue damage
3.	Hyrcanoside	*Securigera securidaca*	Leaf	Toxin
4.	Linamarin	*Lotus corniculatus* *Phaseolus lunatus* *Trifolium repens*	Seed	Causes neurological and endocrinological disorders; tropical ataxia, neuropathy (TAN), ambylopia, alteration in cardiac physiology, inhibition of cytochrome oxidase and respiratory enzymes, paralysis, tissue damage; long term effects due to formation of thiocyanate in the body; aromatic aldehydes formed due to hydrolysis are also toxic
5.	Lotaustralin	*Lotus corniculatus* *Phaseolus lunatus* *Trifolium repens*	Seed	Bitter tasting toxin causing acute poisoning due to thiocyanate formed in the body.
6.	Prunasin	*Acacia glaucescens*	Pod	Bound toxin, releasing poisonous cyanide on hydrolysis and causes tissue damage
7.	Vicianin	*Vicia angustifolia* *Vicia sativa*	Seed	Bound toxin, releasing poisonous cyanide on hydrolysis and causes tissue damage

Table 8.4

Phytoestrogens in Legumes

Species with edible parts in common use are marked by an asterisk (*)

Sl. No.	Family and Botanical Names
I	**Caesalpiniaceae**
1.	*Ceratonia siliqua*
2.	*Senna floribunda*
3.	*Senna occidentalis*
4.	*Tamarindus indica**
II	**Fabaceae**
1.	*Abrus precatorius*
2.	*Arachis hypogaea**
3.	*Cajanus cajan**
4.	*Canavalia ensiformis**
5.	*Canavalia gladiata**
6.	*Cicer arietinum**
7.	*Crotalaria laburnifolia*
8.	*Crotalaria pallida*
9.	*Cyamopsis tetragonoloba**
10.	*Glycine max**
11.	*Lablab purpureus**
12.	*Lathyrus sativus**
13.	*Lens culinaris**
14.	*Lupinus albus*
15.	*Lupinus luteus*
16.	*Macrotyloma geocarpum*
17.	*Macrotyloma uniflorum**
18.	*Medicago sativa*
19.	*Mucuna pruriens*
20.	*Pachyrrhizus erosus*
21.	*Phaseolus coccineus*
22.	*Phaseolus vulgaris**
23.	*Phaseolus lunatus**
24.	*Pisum sativum**
25.	*Psophocarpus scandens*
26.	*Psophocarpus tetragonolobus**
27.	*Sesbania bispinosa*
28.	*Trigonella foenum-graecum**
29.	*Vicia faba**

Sl. No.	Family and Botanical Names
30.	*Vicia monantha*
31.	*Vicia sativa**
32.	*Vigna aconitifolia*
33.	*Vigna adenantha*
34.	*Vigna angularis*
35.	*Vigna mungo**
36.	*Vigna radiata**
37.	*Vigna subterranea*
38.	*Vigna trilobata*
39.	*Vigna umbellata*
40.	*Vigna unguiculata**
III	**Mimosaceae**
1.	*Acacia longifolia*
2.	*Entada rheedei*
3.	*Lysiloma latisiliquum*
4.	*Parkia clappertoniana*

Table 8.5

Toxic Phenolic Compunds: Toxic Effects of Anthrones, Chromones, Phenyl Propanoids and Quinones of Legumes

Sl. No.	Toxin	Source species	Part	Symptoms of toxicity in humans
1.	**Anthrone**			
	Chrysarobin	*Andira inermis* *Cassia* species *Senna* species	Seed	Allergen and respiratory tract irritant
2.	**Chromone**			
	Rubrofusarin	*Senna tora*	Root	Depressant of central nervous system
3.	**Phenyl propanoids**			
a.	Coumarin	Pasture legumes	Leaf, seed	
b.	Dicoumarol	*Melilotus* species	Leaf, seed	Causes haemorrhagic disorders and liver damage
c.	Furanocoumarin	*Cullen corylifolium*	Seed	
d.	Psoralen	*Cullen corylifolium*	Seed	Phototoxin, photo-carcinogen
4.	**Quinones**			
a.	Acamelin (Benzoquinone)	*Acacia melanoxylon*	Heart wood	Allergen, causes dermatitis and bronchial asthma in workers handling the wood of this tree.
b.	Aloe–emodin (Anthraquinone)	*Senna alexandrina*	Leaf	Toxic to humans

CHAPTER-9

Lectins in Legumes

9.1 INTRODUCTION

In 1888 Stillmark observed agglutination of animal erythrocytes by proteins in aqueous extracts of seeds of *Ricinus communis*, the castor oil seed plant. This lead to the characterization and recognition of a distinct class of proteins, first in plants and then in almost every group of organisms. Basing on their most distinctive property of cell agglutination, these proteins were called agglutinins, haemagglutinins (blood cell agglutinins), or phytohaemagglutinins (as they come from plants). Subsequently, they were named lectins, from the Latin root '*legere*' to mean 'to pick or choose', as recognition by lectins is selective and specific to a cell type. The original term, phytohaemagglutinins, once used generally for lectins, is now restricted to the lectin of *Phaseolus vulgaris,* the red kidney bean.

Lectins are sugar binding proteins (or glycoproteins) of non-immune origin with the ability to agglutinate cells and/or precipitate glycoconjugates. They are constitutive and not induced like immunological proteins. Lectins have an affinity to specific sugars on cell surfaces resulting in agglutination of cells. Lectin induced agglutination or glycoconjugate precipitation is inhibited by a specific sugar and this distinguishes lectins from mammalian antigens, which also may agglutinate cells or precipitate glycoconjugates, but this activity is not inhibited by carbohydrates.

9.2 LECTIN DISTRIBUTION

Lectins are present in almost every group of organisms. Lectins are also present in mycoplasma, bacteria, fungi and algae. Among the animals, lectins are known from slugs, snails, eels, crabs, and cobras.

Lectins are readily detectable in different parts, such as root, bark, stem, leaf, fruit and seed, of a large number of species of plants, seeds being the richest source (Kameswara Rao, 2000). Lectins were found even in pollen and stigmas (Kameswara Rao and Sangeetaa, 1996). An intervarietal difference in the distribution of lectins was evident in several crop plant species more particularly rice and tomato (Sharu Raj, 1990).

Legumes are the most important source of lectins with diverse properties. As the number of legume lectins now known is very large and the area of applications of lectins is very vast, only the salient aspects relevant to the legumes are discussed here.

9.3 PHYSICO-CHEMICAL PROPERTIES OF LECTINS

A number of legume lectins have been chemically characterized, some of which are given in Table 9.1. Names of lectins are acronyms, based on the botanical or the common name of the species.

> **Box 9.1**
> **Conjugated Lectins**
>
> Taking advantage of the lectin-sugar specificity, lectins or the related carbohydrates are detected in cells and tissues. Like proteins, lectins conjugate with a variety of substances. Researchers design and produce lectin conjugates to serve as tools in their work Conjugated lectins do not suffer from any loss of activity.
>
> The following are some of the commonly used substances to conjugate with lectins:
>
> a) **Fluorochromes** (fluorescent chemicals detected by fluorimetry): fluoresceine-5(6)-isothiocyanate (FITC) and Tetramethylrhodamine-5-isothiocyanate (TRITC).
> b) **Electron-dense substances** (detected in electron microscopy): ferritin and colloidal gold.
> c) **Enzymes** (for lectin assay): peroxidase and alcohol dehydrogenase.
> d) **Insoluble matrices** (used in affinity studies): biotin, agarose and acrylic beads

Of the enormous range of physico-chemical properties of lectins, the important aspects of legume lectins are summarized here.

9.3.1 Cell Agglutination

Owing to their specificity to cell surface carbohydrates, different lectins can recognize and bind to specific cells. Lectins bind to cell membrane carbohydrates in intact cells. There are legume lectins that can recognize and bind to different species of bacteria and fungi, human sperm, erythrocytes, leucocytes, healthy and diseased tissues of different human organs, etc.

9.3.2 Precipitation of Glycoconjugates

Conjugation is the molecular association of two different classes of compounds, which is a very common natural phenomenon. Carbohydrates conjugate with proteins (glycoproteins), lipids (glycolipids), alkaloids (glycoalkaloids) or other compounds (glycosides). Depending upon the sugar specificity of the lectin and the glyco- part of the conjugate, lectins can precipitate glycoconjugates in cells or in solution.

The property of conjugation with other compounds is used to produce lectin conjugates for use in research (Box 9.1). Representative examples of over thirty conjugated lectins that are commercially available are given in Table 9.2.

9.3.3 Glycoprotein Lectins

Most lectins, such as those of *Canavalia ensiformis* and *Pisum sativum* are proteins without any constituent carbohydrate content. The lectins of some species contain four to six per cent carbohydrates as constituents and these lectins are glycoproteins. *Glycine max* and *Phaseolus lunatus* contain mannose and N-acetylglucosamine, while *Styphnolobium japonicum* and *Wisteria floridbunda* contain xylose. However, the carbohydrate moieties are not required for the biological activity of the lectins and the sugar specific for their activity is different from the constituent carbohydrate.

9.3.4 Subunits

Most lectins contain covalently bound two or four subunits (Table 9.1). Subunits are identical in most lectins, but in some, particularly those belonging to the tribe Vicieae (*Glycine max, Macrotyloma uniflorum, Lens culinaris, Pisum sativum* and *Phaseolus vulgaris*), contain unidentical subunits called α (A) and β (B) chains.

In some cases (*Abrus precatorius*) one subunit is responsible for haemagglutinating activity and the other for toxicity or mitogenicity (*Maackia amurensis*).

9.3.5 Molecular Weight

The molecular weights (MW) of legume lectins vary widely. The MW of the whole lectin is often a multiple of the MW of the subunits, if the subunits are identical. Subunit MW is often 11,000 Daltons. It is not so if the subunits are not identical as for example, in *Macrotyloma uniflorum.* The range of MWs of different legume lectins is given in Table 9.1.

9.3.6 Sugar Binding Sites

The number of sugar binding sites is variable, but there must be at least one binding site per subunit, as in *Arachis hypogaea, Glycine max* and *Phaseolus lunatus*. Two to four binding sites may be present in some cases (*Abrus prcatorius, Vigna mungo*). The sugars specific to these binding sites may be identical in a particular lectin or may be different as in *Macrotyloma uniflorum* and *Phaseolus vulgaris.*

9.3.7 Carbohydrate Specificity

Cell agglutination or glycoconjugate precipitation is the fundamental activity of lectins and this is based on the specificity of lectins to particular carbohydrates (sugars). The sugar specificity of lectins is as precise as the relationship between an enzyme and its substrate or an antigen and its antibody. Since cell surfaces contain membrane bound sugars, the lectins specific to these sugars recognize and bind to them. Examples of legume lectins and their specific sugars are given in Table 9.3. If the sugar specific to a lectin is present in the medium, cell agglutination is inhibited and this is a confirmatory test to identify a lectin. The subunits may have the same or different sugar specificities as in *Griffonia simplicifolia* or *Ulex europaeus*.

9.3.8 Metal Ions

Akin to the prosthetic groups of enzymes, many legume lectins such as those in *Lens culinaris* and *Pisum sativum* require metal ions for their biological activity. Lectin of *Macrotyloma uniflorum* requires Ca^{+}. However, some lectins as *Wisteria floribunda,* do not require metal ions for their biological activity. The two subunits of lectin in *Canavalia ensiformis* bind differentially to Mg^{2+} and Ca^{2+}.

9.3.9 Temperature and pH Sensitivity

Most lectin activity occurs at about pH 7.0 (So and Goldstein, 1967) and a temperature below 60°C (Hunt *et al.*, 1975; Sathyananda, 1989).

9.3.10 Multiple Lectins

Certain plant species contain distinctly different lectins, sometimes in the same part. Each lectin is different from the other in physico-chemical properties and activity, more so in molecular weight, isoelectric point and sugar specificity. These are s multiple lectins. While each lectin has its own inhibitory sugar, when crude extracts from such species are studied for inhibition of lectin activity, more than one sugar will inhibit the activity, though none completely. Some examples of species with multiple lectins are given in Table 9.4.

9.3.11 Isolectins

Isolectins are multiple molecular forms of the same lectin, like an enzyme and its isoenzymes. Isolectins have the same general physico-chemical properties and sugar specificity as the parent lectin. Isolectins may slightly differ from each other in molecular weight, but they cannot be distinguished in electrophoretic analysis. Since their isoelectric points are very distinct, isolectins are readily resolved in isoelectric focussing, a technique similar to electrophoresis. Isolectins, like isoenzymes, reflect genetic diversity. Some well established examples of legume species with isolectins are given in Table 9.5.

9.3.12 Mitogenicity

Mitosis, the type of cell division that occurs in the body cells, is essential for the growth of an organism. Mitosis is also important in research, as at this stage it is ideal to study the number and structural characteristics of chromosomes, an area called karyotypic analysis.

Fully differentiated cells do not normally undergo mitosis. Mitogenic stimulation or mitogenecity, is the ability of a factor to induce mitotic division in a mature quiescent cell that does not normally divide.

Human erythrocytes (red blood corpuscles, RBC) have no nuclei (enucleate). The differentiated leucocytes (white blood corpuscles, WBC) that occur in the circulatory system are nucleate, but do not divide. The absence of a lab procedure *(in vitro)* to stimulate the leucocytes into division has hindered karyotypic studies on mammals, till Nowell (1960) demonstrated *in vitro*, the dramatic induction of mitogenic activity in the cells of the lymphatic system (lymphocytes), by the seed lectin of the red kidney bean (*Phaseolus vulgaris,* PHA). This has revolutionized the study of mammalian karyotypes, richly contributing to both biology and medical diagnostics.

Mitogenic lectins such as PHA stimulate the transformation of lymphocytes from small resting cells into large blast like cells that may ultimately undergo mitotic division. Mitogenic lectins also induce cellular proliferation, in the small intestine (Pusztai *et al.*, 1988). Prolonged mitogenic activity results in cell proliferation as in cancer.

Now a number of plant sources of lectins with mitogenic activity are known, and most of them are legumes (Table 9.6). Nevertheless, PHA and Con A (seed lectin of *Canavalia ensiformis)* are the best-known and most widely used mitogens. All mitogenic lectins agglutinate erythrocytes of all the four human blood groups.

All legume lectins are not mitogenic. In the case of lectins of *Lens culinaris, Maackia amurensis* and *Phaseolus vulgaris*, only one of the fractions is responsible for mitogenecity (Tables 9.1, 9.6). Some lectin fractions such as succinyl-concanavalin do not agglutinate erythrocytes, but are mitogenic, while Con A is both agglutinating and mitogenic. Pre-treatment with an enzyme like neuraminidase may be needed to induce mitogenesis, as in the case of soybean lectin (SBA) on mouse lymphocytes and peanut lectin (PNA) on rat and human lymphocytes. Neuraminidase removes sialic acid residues from cell surfaces, which prevent lectin-cell binding. Calcium ions have been shown to be essential for the mitogenic activity of lectins (Sharon, 1976).

All mitogenic lectins stimulate the T cells (thymus dependent lymphocytes), except that of *Phytolacca americana* (Phytolaccaceae), which stimulates both T and B cells. Lectins that exclusively stimulate B cells are not known.

Like the other activities of lectins, mitogenic stimulation also can be influenced by a sugar specific to the lectin, as the presence of a lectin specific sugar inhibits mitogenecity.

Two signaling steps have been identified in mitogenesis. The 'first signal' constitutes the binding of the lectin to lymphocyte surface sugar receptors. This binding alters the positions of cell surface receptors and clusters them together (cap cell stage). This first signal can be inhibited by the sugar specific to the mitogenic lectin. The first signal triggers the 'second signal' transmitted into the cell, which initiates a series of biochemical events leading to cell growth and proliferation. The biochemical events that follow are: a) increased membrane fluidity, b) a striking increase in c-GMP concentration (Waton, 1975), c) uptake of K^+ ions and d) increased phosphatidylinositol turn over followed by an increased uptake of glucose and uridine (Cooper, 1972; Barnet *et al.*,1974). However, the phenomenon of mitogenic activity is still inadequately understood.

9.3.12.1 Applications of Mitogenic Lectins The potential of mitogenic lectins in inducing cell division has been put to the following uses:

1. To detect and understand congenital disorders rooted in numerical and structural variation and defects in the chromosomes. Using banding techniques, chromosomes are stained to unravel structural differences between healthy and diseased individuals, particularly in such disorders as Klinefelter's, Turner's, Down's, Patau's and Edward's syndromes.
2. Human karyotypes studied with the help of mitogenic lectins aid in the determination of the sex of the unborn.
3. As polyclonal reagents in immunology to investigate the molecular basis of lymphocyte activation and proliferation, irrespective of antigen specificity (van Damme, 1998).

4. To study the alterations in the number, distribution and mobility of cell surface receptors, which control cell proliferation, and as tools for studying the biochemical events accompanying cellular growth, differentiation and division.
5. As immunosuppressive agents *in vivo*. *Phaseolus vulgaris* lectin (PHA), *Lens culinaris* lectin (LCA) and *Canavalia ensiformis* lectin (Con A) could prolong the allograft survival time over weak histocompatibility barriers (Hildert *et al.* 1987).
6. As mimics of super antigens (usually obtained from bacterial or viral polyclonal activators), to facilitate antigen binding and immune activation.
7. Mitogenic stimulation by T- or B-cell specific lectins is useful in assessing immunocompetence of patient's sufferring from many diseases.
8. To enhance the rate of accumulation of biomass, due to the enhanced rate of cell division (Pineau *et al.*, 1990).

9.3.13 Antigenicity

Lectins are antigenic, i.e., they elicit antibodies specific to and against themselves in the mammalian systems. While most of them elicit IgG antibodies, certain lectins such as ricin from the castor seed elicit IgE antibodies and may cause allergy in human beings.

Several applications have been found for lectin antibodies. Many researchers raise antibodies in rabbits, for the lectins of their interest, for histochemical localisation of lectins, or to verify if the same lectin is present in different parts of the same organism.

With a wide use of lectin antibodies in research labs, antibodies to some legume lectins such as those of *Arachis hypogaea, Canavalia ensiformis, Glycine max, Phaseolus vulgaris,* and *Ulex europaeus,* are commercially available.

9.3.14 Enzyme Activity

Although some definitions of lectins exclude enzyme activity, instances of proteins with both agglutinating and enzymic activity are known. Activity of α-galactosidase was demonstrated with seed lectins of mung bean, soybean and the broad bean. But the fractions with enzyme activity do not agglutinate.

9.4 SPOROPHYTIC LECTINS

Etzler (1985b) reported the difference between lectins in the vegetative (sporophytic) tissues and the seeds (gametophytic), which perform different functions. Although very closely related to seed lectin, leaf lectins are not immunologically identical with them and differ in their molecular weight, carbohydrate content and pH sensitivity (Poretz *et al.*, 1974). Lectins from roots of *Pisum sativum* and *Glycine max,* roots, stems and leaves of *Macrotyloma uniflorum,* roots and hypocotyls of *Arachis hypogaea* are different from seed lectins in these species (Talbot and Etzler, 1978a, b, Etzler, 1985b). For these reasons, the possibility of sporophytic and gametohytic lectins being different was mooted (Kameswara Rao and Sangeetaa, 1996), but the issue is far from settled. Examples of sprophytic lectins are given in Table 9.7.

9.5 BIOSYNTHESIS OF LECTINS

Lectins are synthesized like any other proteins. Precursor components of many legume lectins are synthesized in the leaves and fully formed lectins are constituted at the sites of storage and/or function.

9.6 BIOLOGICAL ROLES OF LECTINS

The common occurrence of lectins in various parts of a large number of species has lead to ascribing a variety of roles to lectins in the organisms in which they occur. These include many physiological functions such as in embryo development and as protective agents against pests and diseases (van Damme *et al.*, 1998). Lectins are believed to have a role in the attachment of symbiotic nitrogen fixing bacteria to the legume host root hairs. Lectins are also implicated in biorecognition of pollen appropriate for a specific stigma (Sangeetaa, 1994).

9.7 TOXICITY OF LECTINS

The concept that food causes diseases as well as prevents the disease is not new in clinical medicine. It was in 1980, D'Adamo who introduced this concept in clinical medicine, emphasizing the role of food in the development of a disease. This was based largely on the effect of dietary lectins (Kruzel, 1998).

Until recently most of the work on the medical significance of lectins was focused on their role as immune system modulators or as markers of specific disease states (Kaufman, 1958; Powell *et al.*, 1972; Perlman and Epstein, 1990; Coppo *et al.*, 1995; Kaspers *et al.*, 1996).

Lectins can be toxic in human systems but only under a few specific conditions. Toxicity of lectins to humans is sometimes exaggerated, even though there are no concerted studies involving humans.

9.7.1 Toxicity of Dietary Lectins

Most plant species, more particularly the legume seeds, consumed as food or medicine, contain lectins. The human system is accustomed to the lectins in plants routinely consumed as food.

Among the vegetarian foods pulses and nuts are the richest source of proteins forming the chief substance in building of muscles and vital fluids like blood and are referred as "Proximate principles" that supply energy. The edible legumes constitute an important source of dietary proteins of the world's population contain lectins.

When raw legumes rich in lectins are consumed they may pass relatively intact into the intestinal tract.The ability of the body to digest lectins is due to the concentration of disulphide cross linkages in the lectins. Greater the number, the more resistant it is to denaturation. Additionally many lectins are enzyme inhibitors, which inhibit intestinal enzymes from denaturing the lectins. During the digestive process, the health of the digestive tract plays an important role to digest the lectins. Due to their affinity, the lectins irreversibly bind to many sites at once resulting in agglutination, which can cause gastrointestinal discomfort. The degree of discomfort depends on the susceptibility of the individual, digestive ability, the type of lectin and intestinal microflora.

Lectins may influence systemic metabolism by two different and simultaneous mechanisms. They can indirectly influence the endocrine system of the body by binding to neuroendocrine cells of gut and stimulating the secretion of gut peptide hormones into systemic circulation. Alternatively lectins can be transmitted through the gut wall into blood circulation and thus may directly influence peripheral tissues and body metabolism by mimicking the effect of endocrine hormones.

The organs which are most affected by this are pancreas, skeletal muscles, heart, liver, kidney and thymus. A high intake of antinutritional legume lectins of kidney beans is known to cause atrophy of thymus and spleen, reduction in weight of heart and liver, increase in size of kidneys and loss of skeletal muscle (de Oliveira *et al.*, 1988). The lectins affect on the heart and liver weight is due to loss in lipid and glycogen (Palmer *et al.*, 1987).

Most dietary lectins produce tropic changes in the pancreas, mainly mediated through cholecystokinin (CCK), which is a gut hormone, released from the duodenal enteroendocrinal cells by the action of lectins. The mechanism of lectin induced pancreatic growth appears to involve a route of hormonal mediation (Pustzai *et al.*, 1992).

A diet of *Phaseolus vulgaris* agglutinin has long been known to mimic the biological effects of insulin. It binds to insulin receptor of muscle cells and unlike insulin, it will not stimulate protein synthesis in the tissue instead it causes muscle catabolism leading to loss of muscle protein resulting in 30% reduction in skeletal muscles. It is because, the receptors sites for insulin are blocked by PHA (Palmer *et al.*, 1987; Knott *et al.*, 1992).

Some studies have indicated the possibility of toxic effects of lectins on mammalian (not necessarily human) systems. When beans were added to animal diet the following effects were observed (Pusztai *et al.*, 1979; Greer *et al.*, 1985): a) the lectins reacted with intestinal cells causing disruption of brush borders of duodenal and jejunal enterocytes; b) high nitrogen excretion through increased tissue catabolism; c) increased intestinal permeability to serum proteins resulting in its accumulation in the lumen and walls of the small intestine indicated by increased vascular permeability; d) loss of serum proteins; e) hypertrophy of pancreas; f) atrophy of thymus; g) doubling in weight of the small intestine; h) mucosal hyperplasia indicated by increased DNA content; i) changes in the tissue composition of the small intestine; and j) interference with insulin secretion and thus change in the hormone balance in the body (van Damme *et al.*, 1998).

Seed lectins of *Psophocarpus tetragonolobus, Glycine max, Phaseolus vulgaris* and *Canavalia ensiformis* bind to the intestinal mucosa and may cause dysfunction (Shet *et al.*, 1989). As a result of damage to intestinal mucosa by lectins, the absorption of nutrients across the intestinal wall may be impaired. Lectins may also interact with brush border hydrolase, which plays a role in the digestion of proteins and carbohydrates (Sharon, 1994).

In response to assault by lectins, the gastro-intestinal mucosa increases the secretion of sialic acid rich mucous protective layer. During an infectious disease, the enzyme neurominidase secreted by pathogenic bacteria as a survival mechanism removes the sialic acid from the mucosal barrier. Thus rendering it ineffective against gut microorganism and lectin exposure (Freed, 1991). A disruption of gut protective layer by the lectin is thought to be responsible for proliferation of abnormal bacteria resulting in inflammation. These bacteria establish linkage with epithelial cells and adhere causing peptic ulcers.

As the protective barrier becomes less effective in scavenging excess of lectin, the exposure of lectin to cell membrane becomes greater. When exposed to the cell membrane, mitogenic lectins exhibit a phenomenon of "positive cooperativity" (Freed, 1991). In this scenario the first few lectins binding to the cell membrane rearrange it in such a way that further lectin binding is enhanced (Freed, 1991; Miettinen *et al.*, 1980). As the lectin content of the cell surface increases, the ability of the cell to conform to different surroundings becomes impaired due to its inelasticity and stickiness. In other words, the cell becomes more rigid and less adaptable. Further changes take place at the cellular level resulting in decreased ability to absorb nutrients, increased gut permeability, a reduction in cellular enzyme systems and changes in the DNA structure and content (Freed, 1991; Nachbar and Oppenheim, 1980; Powers, 1991; Murata *et al.*, 1992).

Lectins can also mimic hormone function by attaching to and activating specific sites on the cell surface (Shechter, 1983). These factors contribute to apoptosis and possibly malignant transformation as the cell attempts to conform to the new environment (Kim *et al.*, 1993).

Vicia faba lectin stimulated proliferation of colon cancer cells. *Arachis hypogaea, Phaseolus vulgaris* and *Canavalia ensiformis* lectin is known to cause mitogenic stimulation of human colon cells (Ryder *et al.*, 1992); Hyperplasia of small intestine was noticed in rats by a diet of *Phaseolus vulgaris* beans. A diet of *Glycine max* induces hyperplasia, hypertrophy of pancrease and crypt cell proliferation. Consumption of *Arachis hypogaea* resulted in increased rectal proliferation due to mucosal expression of *Arachis hypogaea* lectin receptor indicating an association between diet and colorectal cancer.

Hypersensitivity reactions in the gut for the lectin occur due to the stability of lectins which persistantly bind to brush border cells. This binding increases vascular permeability due to degranulation of sub mucosal mass cells by the lectin and also cross linking of membrane glycans by the lectin resulting in increased allergic response in the body (Greer and Pusztai, 1985).

Legume lectin such as PHA when orally administered can induce overgrowth of *Lactobacillus* and *Proteus* bacteria, which are coliforms in the gut and the small intestine, which causes an imbalance in bacterial flora in the system. This results in diarrhoea and other digestive and absorptive problems (Pusztai *et al.*, 1993; Pusztai *et al.*, 1995).

Once bound to cellular surface, the lectin induces cell-mediated responses. When the foreign material attempts to pass through intestinal brush border epithelial cell, IgA effectively binds to it forming immune complex. In cases where lectin is bound to cellular surfaces an alternate pathway of the complement system is activated with the loss of immune complex and causing an inflammatory diseases (Brinck *et al.*, 1995).

Along with alternate complement activation, seratonin and histamine are activated resulting in an increase in capillary permeability and migration of blood cells into extravascular space and crossing of WBC, over the brush border and their entry into circulation. This results in potentiating the inflammatory reaction at other sites in the body, because non-secretors do not have sialic acid in their mucous secretion, they are more prone to prolonged inflammation. They may become more prone to autoimmune diseases.

However, these observations on mammalian systems do not completely reflect the situation in humans. Most ingested lectins are digested like any other protein, so long as the digestive potential of the consuming individual is normal. If a lectin is digested before reaching the intestine, the lectin is harmless. Digestion of ingested lectin can be impaired when unusually large quantities of lectin containing food are consumed or the process of cooking was incomplete or faulty.

Cooking procedures help in rendering the legumes safe for human consumption all though in practice complete destruction of lectins might not be achieved. However, not all lectins are toxic or even growth inhibitory as human food. Processing of food by boiling, decreased the acitivity of lectins in *Vicia faba, Phaseolus vulgaris* and *Pisum sativum* (Mollina *et al.*, 1976).

9.7.2 Injected Lectins

The main effect of the entry of lectins into the circulatory system is intracapillary erythrocyte agglutination (Sangeetaa, 1994), if the lectin is specific to the blood type. This can block the blood vessels and affect circulation.

When injected, protein synthesis may be inhibited with such lectins as abrin, that have a ribosome inhibiting component. In some cases, immune hypersensitivity and tissue damage may occur (Pusztai *et al.*, 1981a). Injected dietary lectins in legumes may induce histamine release from basophils (Haas *et al.*, 1999). However, these effects are not universal.

In animal model studies, when *Phaseolus vulgaris* lectin was injected, changes such as a) occurrence of pathological lesions, b) degeneration of various tissues, c) edema, d) necrosis of liver, e) haemorrhage in the stomach, intestinal wall and other organs, f) distension of capillaries in the kidney, and g) weight gain of kidney, heart, pancreas and liver were noticed.

9.7.3 Inhaled Lectins

Lectins may be inhaled through pollen or dried dusty lectin containing plant parts. Some of them such as the seed of the castor oil plant are known to cause allergy in sensitive individuals and lectins from some other species may also be allergenic (Sangeetaa, 1994). If the inhaled lectin is specific to the blood group of an individual, it may cause intracapillary agglutination in the nasal mucosa and simulate symptoms of allergy.

9.8 BENEFICIAL EFFECTS OF LECTINS

Oral administration of low doses of lectins can have many beneficial effects on the digestive/absorptive efficiency of the gut, its immune system, bacterial ecology, endocrine system with beneficial consequence for general metabolism (Pusztai, 1993). It has also been demonstrated that orally administered lectins, which are generally regarded as toxic, are not deleterious per se but most of adverse effects are due to their interaction with gut bacteria (Jayne-Williams and Burgess, 1974; Rattray *et al.*, 1974; Wilson *et al.*, 1980; Ceri *et al.*, 1988; Banwell *et al.*, 1988; Pusztai *et al.*, 1993).

Legume lectins can directly interact with and agglutinate some of the resident bacteria of digestive tract and selectively remove them from the lumen. Therefore legume diet containing abundant lectins can change bacterial ecology of the gut.

Phaseolus vulgaris lectin as dietary adjunct or therapeutic agent can be effectively used to reduce body protein. With a diet of high lectin content of *Phaseolus* a safe and effective dose range can be established to ameliorate obesity in human subjects (Pusztai *et al.*, 1998).

9.9 APPLICATIONS OF LEGUME LECTINS

A large number of applications have been found for lectins in diverse areas of biological and biomedical research and diagnostics. Applications of legume lectins in medicine and in cancer diagnostics are discussed in Chapter 23. Several other applications are given here.

9.9.1 Legume Lectins in Histochemistry

Histochemical studies using legume lectins have numerous research and diagnostic applications. Such studies have contributed towards a better understanding of important cellular processes in disease development.

Instead of antibodies, conjugated lectins are used in immunohistochemistry to detect immunohistochemical changes, to investigate the binding of lectin to cell surface. Lectin bound cells can readily be detected under the electron microscope and fluorescent activated cell sorter (FACS) (Walker, 1988).

The morphological features of cells and tissues, and the chemistry of the cell often form the basis of diagnosis (Rosai, 1989). In immunohistochemistry, differential expression and quantitative changes in various proteins and glycoconjugates in healthy and diseased tissues are utilized for diagnosis (Rademacher *et al.*, 1998).

The following examples illustrate the diversity of uses of legume lectins in histochemistry:

Soybean agglutinin (SBA) is used to distinghish B and T splenocytes (Maekawa and Nishimune, 1985).

Ulex europaeus agglutinin (UEA) is a good histological marker for the endothelium of small blood vessels in the human tissues (Holthofer *et al.*, 1982).

Lectins of *Canavalia ensiformis* agglultinin (Con A), *Peanut* agglutinin (PNA), *Ulex europaeus* agglutinin (UEA), *Dolichos biflorus* (*Macrotyloma uniflorum*) agglutinin (DBA), Soybean agglutinin (SBA), *Bandeiraea simplicifolia* agglutinin (BSA,), *Pisum sativum* agglutinin (PSA) and *Lotus tetragonolobus* agglutinin (LTA) (Walker, 1985; Dangui *et al.*, 1994) are useful in examining the binding pattern of lectins to lymphocytes by quantitative autoradiography.

Radioactively or flourescently labeled *Phaseolus vulgaris* agglutinin (PHA) is widely suited as tracer for mapping neuronal connections (van Damme, 1998), which is an interesting application in neurophysiology.

Canavalia ensiformis agglutinin (Con A), *Lens culinaris* agglutinin (LCA), *Phaseolus vulgaris* agglutinin (PHA) and are useful in the analysis of mucins of the respiratory airways of humans (Mazzuca *et al.*, 1982).

Arachis hypogaea agglutinin (PNA), *Canavalia ensiformis* agglutinin (Con A) and *Ulex europaeus* agglutinin (UEA) are useful to record changes in glycoproteins of endometrium during menstrual cycle (Bychkov and Toto, 1988).

Psoriatic skin shows subtle differences in lectin binding compared to normal skin with *Arachis hypogaea* agglutinin (PNA), *Canavalia ensiformis* agglutinin (Con A), *Glycine max* agglutinin (SBA), and *Ulex europaeus* agglutinin (UEA) which are considered as markers for squamous differentiation (Suter *et al.*, 1991; Neogrady *et al.*, 1994). *Ulex europaeus* agglutinin (UEA) has an altered binding in psoriasis indicating a defect in epidermal maturation (Kariniemi *et al.*, 1983).

Differential staining pattern using *Arachis hypogaea* agglutinin (PNA), *Bandieiraea simplicifolia* agglutinin (BSA), *Canavalia ensiformis* agglutinin (Con A), *Dolichos biflorus* agglutinin (DBA), *Glycine max* agglutinin (SBA), *Lens culinaris* agglutinin (LCA), *Lotus tetragonolobus* agglutinin (LTA), *Pisum sativum* agglutinin (PSA), *Ulex europaeus* agglutinin (UEA) and *Vicia villosa* agglutinin (VVA) gives an indication about the changes in the brain tissue of patients suffering from Alzheimer's disease (Bonshek *et al.*, 1988).

During wound healing there is cellular spreading followed by cell division. The pattern of lectin binding showed that saccharides with terminal α-manosyl residues appear specifically at the spreading margin of the cells (Stoddart *et al.*, 1988). *Glycine max* agglutinin (SBA) and *Macrotyloma uniflorum* agglutinin (DBA) bind to goblet cells in inflammation of the appendix (Yoshioka *et al.*, 1989; Brink *et al.*, 1995). Several examples of lectin based histochemistry in the management of cancer are given in Chapter 23.

9.9.2 Blood Typing

One of the earliest and widely used applications of plant lectins is their capacity to distinguish erythrocytes of different blood groups. Batteries of legume seed lectins of over 1,500 species were tested for blood group specificity of lectins by several workers (Kameswara Rao, 2000).

9.9.2.1 Identification of Human and Animal Blood Groups

Clinically human blood groups are determined on the basis of reactivity against sera specifically developed for this purpose. In view of the differences in the specificity of lectins to cell surface carbohydrates, and the differences between the erythrocytes of different groups of human and animal blood with respect to the surface carbohydrates, lectins can be used in human and animal blood group determination. This will eliminate the technically involved process of raising antisera, and the practical difficulties associated with using them. Besides, with little training semi skilled workers can undertake the determination of blood groups, using lectins.

9.9.2.2 Human Blood Group Lectins

Human blood groups A, B, AB and O, come under the ABO series. The accurate determination of these four blood types is very critical for blood transfusion. The four human blood groups are clinically identified on the basis of interaction with two antisera.

There are two subgroups, A1 and A2, in the human blood group A but there are no antisera to distinguish them. The seed lectin of horse gram (*Macrotyloma uniflorum*) is specific to the A1 subgroup, while the seed lectin of *Phaseolus lunatus* is specific to A2.

The Bombay Blood group is rare and is confined to parts of Gujarath and Maharashtra. In blood banks it is identified with difficulty when antisera are used but seed lectins of *Bauhinia tomentosa* and *Ulex europaeus* are useful in this regard.

A protocol for determination of human blood groups using legume seed is given in Table 9.8.

9.9.2.3 Identification of Non-ABO Factors in Human Blood

Human blood factors Rh, N and T are outside the ABO characteristics and are not significant in blood transfusion. Seed lectins of, *Bauhinia purpurea, Bauhinia variegata,* and *Vicia graminea* are anti-N, while seed lectins of *Arachis hypogaea* and *Tephrosia hookeriana* are anti-T (Toms and Western, 1971). No lectins to distinguish the two variants of the Rh factor are known.

9.9.2.4 Identification of Blood of Animals

Identification of the blood of different animals species is of great forensic utility. There are factors in the blood of animals, which make the blood species specific. Cross species blood transfusion among animals would be as disastrous as transfusion between the human blood groups. In many animals like the horse, sub-groups also occur within the species. Animals too have blood antigens but these are not critical for transfusion within the same species. Consequently, no antisera have been developed, to identify blood of even different animal species, which has been quite a disadvantage to forensic scientists. Legume seed lectins come very useful in determining the blood of certain animals. A protocol for the determination of blood of sheep, pig, cow, rabbit, dog and chicken is given in Table 9.9.

9.9.3 Detection, Identification and Inhibition of Human Pathogens

Lectins can induce bacterial cell agglutination and inhibit their growth. This characteristic may be useful as an effective tool in the identification of pathogens and control infections caused by them. In addition, lectins can pre-empt the binding sites on bacterial cell surfaces preventing the binding of the pathogen to human skin, gut and other surfaces and membranes, which is an important event in pathogenesis.

The lectin from *Glycine max* can be used to detect certain pathogens. For example, *Bacillus anthracis* can be identified using *Glycine max* agglutinin. Sharon (1994) identified certain plant lectins used in control of gastrointestinal disorders as several lectins inhibit the growth of intestinal pathogens. Three roles have been assigned to lectins in dental care, as a) haemostatic agents, b) microbial agglutinins and c) bacteriostatic and/or bactericidal agents.

Lectins of *Robinia pseudoacacia* and *Macrotyloma uniflorum* prevent *streptococci* from adhereing to cell surface, but higher doses may result in disturbance of microflora of the gut. *Pterocarpus angiolensis* and *Abrus precatorius* lectins are found to be lethal to *Schistosoma haematobium* which causes schistosomiasis, by obstructing its binding to the cell surface.

Con A prevents the attachment of herpes simplex virus to the cell surface by attaching itself to the viral capsid or it may cause a transformation in the virus itself (Freed, 1991). Lectin of *Lens culinaris* and *Phaseolus vulgaris* affords cellular protection by blocking the ability of HIV to bind to cell surfaces (Robinson *et al.*, 1987).

9.9.4 Determination of Secretor Status

Individuals who secrete glycoprotein antigens in the body fluids like sweat and saliva are called secretors and the others are non-secretors. The activity of a particular anti-H lectin is inhibited by a secretor's saliva (Toms and Western, 1971). Several anti-H lectins such as *Lotus tetragonolobus* and *Bauhinia tomentosa* are used in determining secretor status, which is a gene controlled trait (Race and Sanger, 1975; Sathyananda, 1989). This method is used in genetic analysis of human populations.

9.9.5 Sperm Agglutinating Lectins

Sperm agglutinating lectins are of interest as they bind the sperms and immobilize them. This makes sperm agglutinating lectins useful as topical contraceptives. As the surface carbohydrates of the head and tail of human sperm are different, some lectins (*Abrus precatorius*) bind the sperms by the head, while some others (*Bauhinia racemosa*) bind them both by the head and the tail.

9.10 Lectins as Vectors in Targetted Oral Administration of Drugs

Drugs are most effective as injections but their oral administration is an easier and convenient means ensuring maximum compliance. However, the passage of the drugs through the harsh gastric environment before they are absorbed in the intestines was a serious handicap and many drugs were denatured. Many orally administered drugs are encapsulated in gelatine tubes or coated with sugar or starch bound by vegetable gums. The drugs so administered are released in the intestine but are not membrane receptor targetted and may not overcome the mucous barrier. A very promising means of safe, effective and receptor targetted oral administration of drugs is by coupling them to ligands such as plant lectins, which recognize and bind specifically to the carbohydrate moiety of receptor sites on the epithelial cell surface, providing for a mucous independent bioadhesion (Junginger, 1990). As lectins are generally stable in the gastrointestinal tract (Pusztai *et al.,* 1995), the possibility of exploiting them for drug delivery and also to induce transcellular transport of the drugs administered orally, is being promoted. As the binding of different lectins is specific to different types of cells in the gastrointestinal tract, lectins can be used as targeting agents for different regions of the alimentary canal (Junginger, 1990).

Of the various possible strategies for improved delivery of macromolecular drugs into the systemic blood circulation, the receptor-mediated endo-transcytosis also is promising. Endocytosis is known to occur with many physiologically active macromolecules such as lipoproteins, vitamin B_{12}, tranferrin, immunoglobutins, etc. However, under the difficult conditions of the gastro-intestinal tract, lectins appear to be a better option. Lectin specific for galactose and N-acetyl galactosamine from *Glycine max* lectin (SBA), *Canavalia ensiformis* lectin (Con A) and many more, have an extensive spreading capacity (Pusztai, 1988).

The major drawback of the use of lectins in oral drug delivery is that some plant lectins may exercise undesirable effects. At high concentration *Phaseolus vulgaris* lectin (PHA) disrupts the luminal surface of the small intestine and induces morphological and metabolic changes in the enterocytes but the damage is unnoticeable at minimal concentrations that can be used in drug delivery (Bardocz *et al.*, 1995). *Phaseolus vulgaris* lectin (PHA) is a model lectin with high adhesiveness for drug delivery as it extensively interacts and binds to the brush border epithelium of the gut and also due to its ease in coupling to drug microparticles. Microparticles coated with lectins administered along with food were incorporated in the digestive system (van Damme, 1998).

However, great deal of work needs to be done before lectins can be used in the human system. Coating concentration of lectins on drug microparticles, pharmacological testing of drugs administered via lectins, the response of the gut-associated immune system to a repeated administration of the lectin and other questions have to be answered.

9.11 LECTINS AS THERAPEUTIC TOOL

Certain researchers and clinicians now theorize that the same properties of lectins, which have the ability to cause a disease process, should also be useful as a therapeutic agent if used in proper dosage and under particular conditions. This is because of the apparent specificity of lectins for a variety of mucopolysaccharides and their ability to alter the surface proteins of a variety of organisms thus affecting cellular function.

The therapeutic effects of lectins seem to take place at lower concentration rather than higher. The concentration of lectin needed to afford a therapeutic effect is in the order of microgram doses (Hammar *et al.*, 1994; Hammar, 1995; Ortega-Barria *et al.*, 1994). Lectins have a beneficial affect upon cells by its ability to modify the cell surface and alter the function of macrophages. Lectins can activate macrophage surface to determine which functions they will perform. Galactose and N-acetyl galactosamine specific lectins are involved in recognition and lysis of tumour cells (Koyko *et al.*, 1994).

Lectins have demonstrated antimicrobial activity. Many legume lectins are antiviral and antibacterial. Con A lectin prevents the attachment of the Herpes Simplex virus to the cell surface by attaching itself to the viral capsid (Freed, 1991) but sometimes the lectin causes a transformation in the virus itself resulting in diseases (Freed, 1991). Whether the lectin does this by a transformation in the virus by alteration in the cell surface is unknown. Here too the concentration of the lectin is an important factor.

Lectins with specificity for the terminal α-1-3 and α-1-6-mannose on HIV will cause a neutralization of the viral capsid *in vitro*. Lectins from *Lens culinaris* and *Phaseolus vulgaris* afford cellular protection by blocking the HIV's ability to bind to cell surfaces. Additionally, it may protect cellular surfaces by blocking HIV binding sites before the virus can attack (Robinson *et al.*, 1987).

Lectins from *Robinia pseudoacacia* and *Macrotyloma uniflorum* have antibacterial activity. They prevent *Streptococci* from adhering to human cell surface (Kellens *et al.*, 1994) but higher dose of these lectins result in the disturbance of microflora of the gut.

Chronic pancreatitis where there is severe reduction of digestive enzymes, administration of low doses of lectins can stimulate pancreatic growth, which may find use in clinical practice.

On account of the apparent specificity of lectins for a variety of mucopolysaccharides and their ability to alter the cell surface proteins in a variety of organisms affecting cellular functions, it is now believed that the same properties of lectins which cause a disease process should also be useful as therapeutic agents at proper dosage and under particular conditions.

9.12 LECTINS IN LEGUME SYSTEMATICS

Legume lectins can be used as chemical markers for evolutinary and taxonomical relationship (Toms and Western, 1971; Toms, 1981). The presence of specific lectin indicates an advanced position of plant from evolutionary viewpoint within a family, besides the taxonomic relationship between various genera is suggested by the lectin content.

It is suggested that the general evolutionary trend in the legumes is towards lectin production. Among the subfamilies, specific lectins are confined to Caesalpiniaceae and Fabaceae and not yet detected in Mimosaceae. Blood group A or B specific lectins are supposed to be found in relatively more evolved groups like the legumes. The H blood group specific lectin is considered to be found in primitive families. The high incidence of anti-H lectin in Fabaceae provides further evidence for the evolution of anti-A or anti-B from anti-H.

When a large number of cultivars are involved, identification of seeds becomes difficult.To distinguish the seeds of *Phaseolus vulgaris,* and *Phaseolus lunatus* conventional characters break down as the morphology of the seeds overlaps. *Phaseolus vulgaris* and *Vicia faba* seed lectins are nonspecific antihuman erythrocyte agglutinin while that of *Phaseolus lunatus* is Anti-A_2 and A_2B. This differentiation in agglutination of human erythrocytes enables the identification of even powdered seeds.

In general lectins from plants within particular taxonomical groups have distinctive properties that distinguish them from lectins of less closely related plants.

Evidence is accumulating that several types of plant lectins exist with more or less pronounced specific characteristics. Usually, related lectins occur in taxonomically related plant species and are supposed to represent evolutionarily closely related proteins. Extensive studies on lectins from a number of legume species revealed that amino acid homologies and immunological relationships correspond well with taxonomic relationship between these species (Sathyananda, 1989).

Table 9.1

Chemically Characterized Legume Seed Lectins

The seed lectins of the legume species in this table are fully characterized. The acronyms used for the lectins are based on abbreviated names of the species. In some cases, such as chickpea, lima bean, mung bean, pea, peanut and soybean, the name of the lectin is based on the common name of the species. In the acronyms, abbreviated terms lectin or agglutinin or haemaglutinin, are used. The term phytohaemagglutinin, which was originally used for lectins in general, is now restricted to the lectin of *Phaseolus vulgaris*. The source species, the acronym used for the lectin and the molecular weight (in Daltons) of the whole lectin or its fractions are given. Some of these lectins, marked by an asterisk (*), are extensively used as research tools and are commercially available.

---

1. **Abrus precatorius* ABRIN A, toxic lectin fraction, MW 62,500.
2. **Abrus precatorius* ABRIN C, toxic lectin fraction, MW 65,600.
3. **Abrus precatorius* agglutinin APA, MW 1, 34,000.
4. **Arachis hypogaea* agglutinin PNA, MW 1,10,000.
5. **Bandeiraea simplicifolia* BS-I, BSI-B4, BSI-AB3, BSI-A2B2, BSI-A3B, BSI-A4, BS-II, BS-IV; lectin fractions of BS-I, MW 1,14,000, BS-II 1,13,000, BS-IV 56,000; the correct name of this species is *Griffonia simplicifolia.*
6. **Bauhinia purpurea* agglutinin BPA, MW 1,95,000.
7. *Bauhinia tomentosa* lectin BTL , MW 66,000.
8. **Canavalia ensiformis* CON A , CONCANAVALIN A; one of the most known and most widely used lectins and the first to be crystallised by Sumner (1919); MW 55,000, 1,10,000.
9. **Cicer arietinum* agglutinin CAA, CPA, MW 44,000.
10. **Cytisus scoparius* lectins CSI, CSII, CSL Lectins of CSI, CSIA, MW 70,000, CSIB, CSII 75,000.
11. **Dolichos biflorus* agglutinin DBA. The correct name of this species is *Macrotyloma uniflorum;* fraction A, MW 1,13,000, fraction B, MW 1,09,000.

12. *Erythrina corallodendron* agglutinin EcorA, MW 60,000.

13. **Erythrina crista-galli* agglutinin ECA, MW 56,800.

14. **Glycine max* agglutinin SBA, MW 1, 20,000.

15. **Lathyrus odoratus* agglutinin LOA MW one fraction 37,000, another 42,800.

16. **Lens culinaris* (=*Lens esculenta*) agglutin, haemagglutin LcA, LcH, LcH-A, LcH-B, MW LcH-A 60,000, LcH-B 42,000.

17. **Maackia amurensis* agglutinin MAA, MW 1, 30,000.

18. **Maackia amurensis* haemagglutinin MAH, strongly haemagglutinating fraction MW 33,000.

19. **Maackia amurensis* mitogen MAL, MAM, strongly mitogenic fraction MW 75,000.

20. **Phaseolus coccineus* agglutinin PCA, MW 1,12,000.

21. **Phaseolus limensis* agglutinin, lectin LBA, LBL, MW 1,24,000.

22. **Phaseolus vulgaris* phytohaemagglutinin PHA; lectin fractions as PHA-E, PHA-L, PHA-P, PHA-M, etc.; MW PHA-E 1,43,000, PHA-L 1,14,000.

23. **Pisum sativum* agglutinin PSA, MW 53,000.

24. **Psophocarpus tetragonolobus* lectin, PTL, PT-I, PT-II.

25. **Robinia pseudoacacia* agglutinin RPA, MW 110,000.

26. **Styphnolobium japonicum* agglutinin SJA, MW 1,33,000.

27. **Ulex europaeus* agglutinin UEA, UEA I, UEA II, UEA I MW 1,70,000, UEA II MW 23,000.

28. **Vicia faba* agglutinin VFA, MW 50,000.

29. *Vicia graminea* agglutinin VGA, MW 1,05,000.

30. **Vicia sativa* agglutinin, VSA, VSL, MW 70,000, 40,000.

31. *Vicia villosa* agglutinin VVA, MW 1,20,000.

32. **Vigna radiata* (=*Phaseolus radiatus*) agglutinin VRA, MW 1,60,000.

33. **Wisteria floribunda* agglutinin WFA, MW 68,000.

Table 9.2
Conjugated Legume Seed Lectins

Sl.No.	Lectin source	Conjugates
1.	*Abrus precatorius*	FITC, TRITC, biotin, agarose
2.	*Arachis hypogaea*	FITC, TRITC, ferritin, colloidal gold, peroxidase, biotin, agarose
3.	*Canavalia ensiformis*	FITC, TRITC, ferritin, colloidal gold, alcohol dehydrogenase, biotin, peroxidase, acrylic beads, agarose.
4.	*Glycine max*	FITC, TRITC, colloidal gold, peroxidase, biotin, agarose
5.	*Lens culinaris*	FITC, TRITC, colloidal gold, biotin, agarose
6.	*Macrotyloma uniflorum*	FITC, TRITC, colloidal gold, peroxidase, biotin, agarose
7.	*Phaseolus vulgaris*	FITC, TRITC, biotin, agarose
8.	*Pisum sativum*	FITC, TRITC, biotin, agarose
9.	*Psophocarpus tetragonolobus*	FITC, TRITC, peroxidase, biotin
10.	*Styphnolobium japonicum*	FITC, biotin
11.	*Ulex europaeus*	FITC, TRITC, ferritin, colloidal gold, peroxidase, biotin, agarose
12.	*Wisteria floribunda*	FITC, biotin.

Table 9.3
Legume Lectin Specific Sugars

Lectin-sugar specificity is complex. Representive list of legume species and the sugars that bind to their seed lectins is given below. The lectins in some species are complexes of several different fractions (multiple lectins), each with its own physico-chemical properties and a different binding sugar. Some sugars bind with the lectins of more than one species, indicating similar physico-chemical proterties of those lectins. While D-sugars are the most common in nature, some lectins bind to L-sugars, which are rare.

Sl.No.	Species	Sugars
1.	*Abrus precatorius*	α-D-galactose
2.	*Bauhinia variegata*	L-rhamnose
3.	*Canavalia ensiformis*	α-D-glucose, α-D-mannose
4.	*Cicer arietinum*	N-acetylneuraminic acid (sialic acid)
5.	*Crotalaria juncea*	D-xylose, α-D-mannose, D-mannitol, lactose, raffinose
6.	*Cytisus scoparius*	α-D-galactose, D-galactosamine
7.	*Glycine max*	Maltose
8.	*Griffonia simplicifolia*	D-galactosamine
9.	*Lablab purpureus*	α-methyl-α-D-mannopyranoside, lactose, maltose

Sl.No.	Species	Sugars
10.	*Lathyrus odoratus*	α-D-mannose
11.	*Lens culunaris*	α-D-mannose, L-sorbitose
12.	*Macrotyloma uniflorum*	N-acetyl-α-D-galactosamine
13.	*Pisum sativum*	D-fructose, α-D-mannose, sucrose
14.	*Psophocarpus tetragonolobus*	α-D-galactose
15.	*Styphnolobium japonicum*	L-arabinose, α-D-fucose, N-acetyl-β-D-galactosamine
16.	*Ulex europaeus* UEA I	α-L-fucose, D-galactosamine, digioxose
17.	*Ulex europaeus* UEA II	N-acetyl-α-D-glucosamine, N,N'-diacetylchitobiose
18.	*Vicia faba*	α-D-mannose, sucrose
19.	*Vigna radiata*	α-D-galactose
20.	*Vigna unguiculata*	D-lactose
21.	*Wisteria chinensis*	D-raffinose

Table 9.4

Legumes with Multiple Lectins

Examples of legume species with multiple seed lectins.

Sl.No.	Species	Multiple lectins
1.	*Griffonia simplicifolia* (=*Bandeiraea simplicifolia*)	BS I (with five isolectins) is a tetramer with identical subunits and four binding sites, recognises human ABO groups and specific to α-D-galactose and its derivatives. BS II is mostly identical with BS I, but recognises the human Tk and T series and B group and is specific to α-D-N-acetylglucosamine. BS IV is a dimer with two unidentical subunits, two binding sites, recognizes the human Le system and is specific to L-fucose.
2.	*Lens culinaris*	LCH-A 60,000 Da and LCH-B 42,000 Da.
3.	*Phaseolus vulgaris*	PHA-E 1,43,000 Da and PHA-L 1,14,000 Da.
4.	*Ulex europaeus*	UEA I α-L-fucose, D-galactosamine, digioxose; UEA II specific to N-acetyl-α-D-glucosamine, N,N'-diacetylchitobiose
5.	*Vicia cracca*	One 1,25,000 Da, specific to α-D-N-acetylglactosamine, the other 44,000 Da, specific to α-D-mannose/α-D-glucose.

Table 9.5

Some Legume Species with Isolectins

Species	Number of Isolectins
Bauhinia tomentosa	3
Griffonia simplicifolia	5
Phaseolus limensis	3
Phaseolus vulgaris	5
Pisum sativum	2
Tetragonolobus purpureus	3
Ulex europaeus	2
Vicia villosa	3

Table 9.6

Mitogenic Legume Lectins

Legume species, well established for mitogenic activity, are given here. The known mitogenic fraction of some sources is given in parenthesis. No mitogenic lectins are known from the Mimosaceae.

Caesalpiniaceae

1. *Bauhinia purpurea*

Fabaceae

1. *Abrus precatorius*
2. *Arachis hypogaea*
3. *Canavalia ensiformis*
4. *Cytisus scoparius*
5. *Erythrina crista-galli*
6. *Erythrina variegata*
7. *Glycine max*
8. *Lathyrus odoratus*
9. *Lens culinaris* (LCH-A)
10. *Maackia amurensis* (MAM)
11. *Phaseolus coccineus*
12. *Phaseolus limensis*
13. *Phaseolus vulgaris* (PHA-L)
14. *Pisum sativum*
15. *Robinia pseudoacacia*
16. *Styphnolobium japonicum*
17. *Vicia faba*
18. *Vicia sativa*
19. *Wisteria floribunda*

Table 9.7
Sporophytic Lectins in Legumes

Lectins, with different physico-chemical characteristics are known from the seeds of all the species in this table. Lectins in the vegetative parts are considered as sporophytic lectins.

Roots

1. *Arachis hypogaea*
2. *Glycine max*
3. *Lens culinaris*
4. *Lotononis bainesii*
5. *Lupinus luteus*
6. *Macrotyloma uniflorum*
7. *Medicago sativa*
8. *Ononis spinosa*;
9. *Phaseolus coccineus*
10. *Phaseolus vulgaris*
11. *Pisum sativum*
12. *Robinia pseudoacacia*

Petioles and stems

1. *Arachis hypogaea*
2. *Glycine max*
3. *Macrotyloma uniflorum*
4. *Styphnolobium japonicum*

Leaves

1. *Arachis hypogaea*
2. *Bauhinia monandra*
3. *Glycine max*
4. *Griffonia simplicifolia*
5. *Macrotyloma uniflorum*
6. *Phaseolus vulgaris*
7. *Styphnolobium japonicum*

Table 9.8

Legume Lectins in Human Blood Typing

The seed lectins of the legumes listed here agglutinate the erythrocytes of specific human blood groups.

Human blood groups	Species
A1	*Dolichos biflorus*
A2	*Psoralea corylifolia* and NOT *Cassia biflora* or *Senna spectabilis*
B	*Senna spectabilis* and NOT *Cassia biflora*
AB	*Crotalaria pallida* and NOT *Senna alata*
O	*Cassia biflora* and NOT *Crotalaria pallida* or *Senna alata*
H	NOT *Bauhinia tomentosa* (Bombay group)

Source : Kameswara Rao, 2000

Table 9.9

Legume Lectins in Animal Blood Typing

There are no antisera to identify the blood of animals. The seed lectins of the legumes listed here specifically agglutinate the erythrocytes of the animals indicated.

Chicken	*Cassia fistula, Cassia grandis, Cassia renigera, Senna hirsuta*
Rabbit	*Acacia catechu, Parkia biglandulosa, Bauhinia galpinii, Alysicarpus monilifer, Crotalaria retusa*
Sheep	*Trifolium repens*
Pig	*Cassia javanica, Sophera tomentosa*
Cow	*Cassia carnavel*
Dog	*Tephrosia purpurea*

Source : Kameswara Rao, 2000

CHAPTER-10

Lipids in Legumes

10.1 INTRODUCTION

Oils, fats and waxes are collectively referred to as Lipids (Box 1.1), along with some other compounds that are, strictly speaking, not lipids (Section 10.7). In view of their chemical similarity and the ambiguity of physical form (Box 10.1), the terms oil, fat and lipid are used interchangeably in literature. In the following account, the term 'oils' is used to mean oils and fats, distinguished from 'waxes', and the term 'lipids' for lipoid substances in general. Fatty acids are the acidic components of the true lipids.

Plants use lipids as energy reserves. Lipids accumulate in liposomes, the oil globules in the cytoplasm of a cell. By contrast, the desert plants store lipids as wax esters. Waxes are usually found on the surfaces of plant organs.

> **Box 10.1**
> **Lipids, fats, oils and waxes**
>
> Lipid is a term applied to any one of a group of chemicals, found in the tissues of plants and animals, such as oils, fats and waxes, or their derivatives or conjugates such as triglycerides, phospholipids (lipid and phosphorous), glycolipids (lipids and carbohydrates) and lipoproteins (lipids and proteins).
>
> At the ambient temperature, oil is a liquid and fat is a solid. Ambient temperatures being very variable, ghee and coconut oil solidify in India, except in the summer, while groundnut oil and even gingelly oil, solidify in the temperate countries in the winter. Some consider those oils that are solid at our body temperature as fats. Consequently, the liquid or solid form is not an exact scientific criterion to differentiate between oils and fats.
>
> Waxes are solid, unless melted at high temperatures, with the exception of the endosperm of *Simmondsia californica* (*jojoba,* Euphorbiacaeae), which is a liquid wax. Beeswax is an animal product.
>
> Oils are found in all parts of plants but are particularly concentrated in the seeds. In addition to our daily use, plant lipids have several industrial applications, such as in the manufacture of soaps, detergents, nylon, cosmetics, and as fuel and lubricants.
>
> Indian legumes as sources of seed oils were discussed in general in Chapter 3 (3.2.5). Chemical and other contextual information on lipids in legumes is provided here.

10.2 LEGUMES AS SOURCE OF OILS

The seed oils of several legumes have been in use for centuries in indigenous medicine and now for decades in the industry as lubricants and in other ways (Chapter 3.2.5; Box 10.2). Seed oils of soybean (*Glycine max*) and groundnut (*Arachis hypogaea)* are amongst the three or four primary sources of edible oils. The seed oil of *Pongamia pinnata* is now being promoted as 'vegetable diesel'. There is sufficient incentive for detailed studies on seed oils in the Indian legumes, at least for industrial uses, if not as edible oils. However, not much work was undertaken on the Indian legume seed oils.

10.3 SEED OIL CONTENT OF THE LEGUMES

In terms of seed oil content, the legume family ranks 122nd among 137 families. Among 704 legume samples, the average lipid content was about 5.5 per cent (Jones and Earle, 1966). In many pulses and edible legume seeds, the lipid content was only 1 to 2 per cent (Aykrayd and Doughty, 1964). Due to such low lipid levels the legumes did not generally attract the level of scientific attention they deserve.

Among the legumes, species of the genus *Hedysarum* showed the highest oil content (47.5 per cent) while the lowest was found in the genus *Lathyrus* (0.8 per cent) (Wolff and Kwolek, 1971). The lipid content of the seeds of *Arachis hypogaea, Glycine max Pongamia pinnata* and *Cullen corylifolium* is above 20 per cent.

Wolff and Kwolek (1971) statistically treated the previous data and found that the tribes Dalbergieae (22 to 45 per cent lipid) and the

Box 10.2
Industrial Processes of Oils

Saponification: Saponification, the process that converts oils into soap, is hydrolysis of esters into an acid and an alkali. Saponification is an attribute of oils.

Hydrogenation: Hydrogenation is a chemical process of combining fatty acids with hydrogen. During hydrogenation, hydrogen atoms attach to the double bonds that link the adjacent carbons in the fatty acids. Hydrogenation produces trans-fatty acids, which are different from the parent type. Hydrogenation converts an olein type fatty acid with 18 carbon atoms and one double bond into a stearin type fatty acid, with 18 carbon atoms and no double bonds, in the presence of a catalyst such as nickel or palladium. Hydrogenation solidifies the oil and prolongs its shelf life, as oils high in unsaturated fatty acids (explained later on) tend to become rancid in a short time. A mixture of vegetable oils, including cotton-seed oil, is used for hydrogenation to produce the branded products like Vanaspati or Dalda. Hydrogenated oils are rich in saturated fatty acids like palmitic and stearic acids, which are not good for our health.

Galegeae (up to 39 per cent) are high in seed oil content among the legumes. Species of *Acacia, Aeschynomene, Albizia, Butea, Dichrostachys, Erythrina, Gliricidia, Mundulea, Myroxylon, Psophocarpus,* and *Spartium* are often cited as rich in seed oil (Kochaar, 1981; Hill, 1989).

In general, a tropical distribution and a tree habit positively correlated with high seed oil content in the legumes. An earlier suggestion of an inverse correlation between oil content on one hand and high starch, galactomannan gum, protein or tannin content on the other, did not always prove to be true.

10.4 CHEMISTRY OF SEED LIPIDS

10.4.1 Lipids

Lipids are unbranched polymers of fatty acids. Oils, fats and waxes are esters of fatty acids. Ester is a compound formed by the condensation of an alcohol and an acid, with the elimination of water. Such oils are called fixed oils, as they are non-volatile.

10.4.2 Fatty Acids

Fatty acids are carboxylic acids, with long hydrocarbon tails that terminate in a methyl group. There are several hundreds of fatty acids, but only a few constitute the common oils. Though the number of carbon atoms varies from 8 (caprylic acid) to 24 (lignoceric acid), most fatty acids in vegetable oils have 16, 18 or 20 carbon atom structures. One to three carbon atoms in some of the fatty acids have double bonds. The number of carbon atoms and the number of double bonds in a fatty acid determine its physico-chemical properties. Cardiac and skeletal muscles depend heavily up on fatty acids as a major energy source.

The following species contain appreciable quantities of fatty acids such as linolenic, linoleic and oleic acids in the seeds:

Caesalpiniaceae: *Bauhinia racemosa, Bauhinia variegata, Caesalpinia sappan, Ceratonia siliqua, Saraca asoca, Senna alata, Senna auriculata, Senna obtusifolia, Senna occidentalis, Senna siamea, Senna tora, Tamarindus indica.*

Fabaceae: *Cajanus cajan, Calapogonium mucunoides, Cicer arietinum, Clitoria ternatea, Crotalaria juncea, Crotalaria retusa, Cyamopsis tetragonoloba, Erythrina crista-galli, Glycine max, Lablab purpureus, Lens culinaris, Medicago sativa, Mucuna pruriens, Phaseolus lunatus, Pisum sativum, Pongamia pinnata, Psophocarpus tetragonolobus, Pueraria montana, Robinia pseudoacacia, Sesbania cannabina, Sesbania grandiflora, Spartium junceum, Styphnolobium japonicum, Tephrosia purpurea, Trifolium alexandrium, Trigonella foenum-graecum, Vicia faba, Vigna mungo, Vigna radiata, Vigna umbellata.*

Mimosaceae: *Acacia farnesiana, Acacia nilotica, Albizia procera, Entada rheedei, Leucaena leucocephala, Mimosa pudica, Prosopis cineraria, Pithecellobium dulce.*

10.4.3 Saturated and Unsaturated Fatty Acids

Those fatty acids without any double bonds in the carbon chain are called saturated fatty acids. Unsaturated fatty acids contain one or more double bonds. For example, palmitic acid has 16 carbon atoms and stearic acid has 18, both without any double bonds and so both are saturated fatty acids. Oleic acid has 18 carbon atoms with one double bond, linoleic acid has 18 carbons with two double bonds and α-linolenic acid has 18 carbon atoms with three double bonds. These are the unsaturated fatty acids.

Humans can synthesize all the unsaturated fatty acids needed from palmitate (an ester of palmitic acid), except the essential polyunsaturated fatty acids.

> **Box 10.3**
> **Lipids in our Health**
>
> Phospholipids are important integral components of the double-layered cell membranes, in which proteins are embedded, a structure that facilitates the passage of small molecules through cell membranes.
>
> Vitamins such as vitamin A that are insoluble in water require oils for absorption by our body. Vitamin A forms an ester with palmitic acid, the retinyl palmitate, that ensures the stability and storage of vitamin A in the mammalian systems.
>
> Epidemiologists have found positive correlation between high oil intake, on the one hand, and certain cancers, especially of the colon and prostate, and with coronary heart disease, on the other.
>
> Optimal levels of triglycerides in our blood are essential for health. Higher levels of triglycerides are indicative of hyperlipidaemias, which may be due to genetic causes or secondarily, due to nephrosis, diabetes and endocrine disturbance. Elevated levels of triglycerides are also a risk factor in atherosclerosis. Decreased levels triglycerides indicate protein malnutrition, hyperthyroidism, coronary artery disease and liver disease.
>
> Oils that pose a health risk come mostly from animal sources and hence the recommendation to use more of vegetable oils. In higher plants triglycerides are the major constituents of neutral lipids.
>
> Imbalance in the metabolism of fatty acids and triacylglycerols in the human body results in obesity, diabetes, ketoacidosis, Refsum's disease and hypercholesterolemia. Lipids are essential for maintenance of lung alveolar integrity.

10.4.4 Saturation of Lipids

The degree of saturation of oil is dependent upon the relative number of saturated and unsaturated fatty acids in the oil. The degree of saturation of oils is estimated by the amount of iodine the oil absorbs (iodine value). Iodine values of saturated oils are in the range of 10 to 20 while those of unsaturated oils are 180 to 200. The iodine value of the whole oil indicates the relative percentage of saturated and unsaturated fatty acids in the oil. Legume oils have iodine values ranging from 70 to 160. Iodine values of groundnut oil average to about 93 and those of soybean oil to 133.

10.4.5 Omega-3 Fatty Acids

α-linolenic acid with three double bonds is an omega-3 fatty acid, but not all fatty acids with three double bonds are omega-3 fatty acids as this number refers to the double bond at the third carbon atom counted upwards from the terminal methyl carbon. Human system cannot make the omega-3 fatty acids, which have to be obtained through food. However, some omega-3 fatty acids needed by us are synthesized from α-linolenic acid. Omega acids are abundant in peanut oil, several nuts and fish oils.

10.4.6 Triglycerides

Triglycerides (triglycerols) are formed when three fatty acids are ester linked to glycerol, a trihydric alcohol (commonly known as glycerine, which contains three hydroxyl groups) and are the major constituents of neutral lipids. Triglycerides are a means of storing fatty acids in organisms as energy reserves.

Each triglyceride has three hydrocarbon tails. The last carbon atom belongs to the methyl group (CH_3). A triglyceride can be formed of some saturated and some unsaturated fatty acids.

Simple triglycerides contain the same fatty acids in all the three positions. If different fatty acids are present it is a mixed triglyceride. Triglycerides from which one or two fatty acids are removed, are respectively called mono- and di-glycerides. They are not strictly speaking fats and are the ingredients of fat-free margarine, prepared from vegetable oils.

Triglycerides have serious implications on our health (Boxes 10.3, 10.4). They also have several applications in medicine and industry. In view of the importance of triglycerides in different contexts, a list of Indian legumes rich in them is given in Table 10.1.

10.4.7 Non-glyceride Constituents of Seed oils

Tocopherol (vitamin E), saponins and sapogenins, sterols, phospholipids, glycolipids, hydrocarbons, long-chain alcohols and pigments are among the non-fatty acid

Box 10.4

Healthy Edible Oils

All oils contain both saturated and unsaturated fatty acids. Saturated fatty acids such as stearic and palmitic acids contribute to health problems while monounsaturated fatty acid (oleic acid) and polyunsaturated fatty acid (linolenic acid) are good for health. It is the high proportion of unsaturated fatty acids and more particularly the omega-3 fatty acids such as linolenic acid that are of importance to our health. Generally speaking, vegetables oils are rich in mono- or polyunsaturated fatty acids. Oleic and linolenic acids constitute about 65 per cent of vegetable oil.

The consumption of a substantial portion of such oils would be beneficial to health (Willet, 1994). An approximate guide is that, the higher the proportion of unsaturated fatty acids in oil, the lower is the temperature required for its solidification. It was also recommended that an intake of hydrogenated vegetable oils should be minimized because, during hydrogenation, a saturated trans-fatty acid of the stearin type substituted oleic acid, an unsaturated fatty acid.

The key to our health lies in the omega-3 fatty acids and polyunsaturated oils. Unfortunately, the unsaturated fatty acids and hence oils with high proportion of them are unstable and spoil rapidly, imparting disagreeable smell and/or taste.

Vegetable oils also contain tocopherol (vitamin E) that functions as an effective antioxidant and helps in the prevention of some types of cancer and coronary hearth disease.

Box 10.5

Cholesterol

Cholesterol is an important component of cell membranes and is a precursor of bile salts and steroidal hormones. It is the major sterol in humans that is abundant in the brain and the central nervous system. Cholesterol is implicated in heart disease. Low density lipoprotein (LDL) cholesterol is called bad cholesterol and high density lipoprotein (HDL) cholesterol is called good cholesterol, on account of the relative damage they cause to the human arteries. Both LDL and HDL are chemically the same, but differ in their complexing with proteins and other components (Develin, 1997).

constituents of oils. Edible legume seeds of *Cajanus cajan, Cicer arietinum, Macrotyloma uniflorum, Lablab purpureus, Lens culinaris* and *Pisum sativum* contain considerable amounts of tocopherols.

10.5 WAXES

Waxes are usually found on the surface of stems, leaves and fruits, providing some protection from invasion of pathogenic organisms and desiccation. The grey-bluish ashy appearance of leaves and fruits of certain plants such as apples (called bloom) is due to surface waxes. Waxes can be in the form a uniform coating or may be organized into scales of different sizes, shapes and densities.

Among the legumes in India, leaf waxes of *Cassia fistula, Caesalpinia bonduc, Medicago sativa, Phaseolus aureus* and the garden plant *Acacia podalyriifolia* were chemically analyzed (Wolff and Kwolek, 1971).

Waxes are different from oils both in physical form and chemistry. They are solids and esters of a monohydric alcohol (with only one hydroxyl group), fatty acids, ketones and alkanes. Waxes are of a higher molecular size as they contain from 26 to 33 carbon atoms in the hydrocarbon tail. The larger the hydrocarbon content of a lipid, the greater is its potential as biodiesel.

10.6 LIPIDS FROM THE WOOD OF LEGUMES

In addition to seeds lipids are found in the wood. Lipids from the wood of several legumes such as *Dalbergia sissoo* and *Pongamia pinnata* were investigated. Interestingly, basing on the differences in the lipid content of the wood and other parts, *Pongamia pinnata* and *Pongamia glabra* were sought to be considered as two different species, while such a distinction is not possible basing on morphological evidence (Wolff and Kwolek, 1971).

10.7 LIPIDS THAT ARE NOT LIPIDS

Over a period of time certain natural chemical compounds that are abundant in plants along with oils, fats and waxes, have been referred to as lipids, though these compounds do not constitute a natural chemical group with true lipids, as they are not formed of fatty acids. Legumes also abound in them.

Terpenoids (Chapter 13), including essential oils, and steroids like cholesterol and its derivatives are not polymers of fatty acids. Essential oils are volatile and highly fragrant. They are generally the derivatives of terpenes and benzenes, mostly found in fragrant plant parts such as flowers. Steroidal compounds such as cholesterol (see Box 10.5), vitamins like vitamin D and sex hormones like testosterone, are terpenoid derivatives that are neither volatile nor fragrant. These compounds cannot be saponified (except those containing esters) like oils and fats. Lipases are ineffective against terpenoids and steroids.

All such compounds are incorrectly considered treated together as lipids, along with oils and fats, because they are insoluble in water and soluble in organic solvents like alcohol.

Table 10.1

Triglycerides in the Seeds of some Indian Legumes

All species of plants contain a certain quantity of triglycerides in the seeds. The species in this Table contain significant levels of triglycerides. Species with edible seeds are marked by an astrisk (*). The triglyceride content of species marked with a double asterisk (**) is over 20 per cent.

Sl. No.	Species
I	**Caesalpiniaceae**
1.	*Bauhinia variegata*
2.	*Caesalpinia pulcherrima**
3.	*Cassia fistula*
4.	*Senna auriculata**
5.	*Senna occidentalis*
6.	*Senna sophera*
7.	*Senna tora**
8.	*Tamarindus indica**
II	**Fabaceae**
1.	*Abrus precatorius*
2.	*Arachis hypogaea***
3.	*Clitorea ternatea*
4.	*Cullen corylifolium***
5.	*Erythrina variegata**
6.	*Gliricidia sepium***
7.	*Glycine max***
8.	*Millettia peguensis***
9.	*Mundulea sericea***
10.	*Pongamia pinnata***
11.	*Spatholobus parviflorus**
12.	*Trigonella foenum-graecum**
III	**Mimosaceae**
1.	*Adenanthera pavonina**
2.	*Entada rheedei**
3.	*Samanea saman*

❑❑❑

CHAPTER-11

Tannins in Legumes

11.1 INTRODUCTION

Since the earliest times, man has used animal skins to protect the feet and the body, to provide shelter and make containers for liquids, floats and many other articles. Man has also learnt to convert the stinking and rapidly deteriorating hides into leather, a durable product, by 'tanning'.

Archaeological investigations in Germany indicate that tanning was in practice for over 12,000 years. In China, it was in practice before the birth of Christ. The ancient Egyptians, Hebrews, Romans, American Red Indians and Indians were skilled at tanning (Kochaar, 1981).

Tanning is enabled by a group of plant secondary metabolites, called the 'tannins', a name given in 1796 by Seguin to the constituents of oak gall that converted animal skin into leather, which is soft and pliable, but strong and resistant to air, moisture and temperature changes, as well as microbial attack.

The importance of tannins in other contexts is discussed in this book in Chapters 3, 7 and 8. Tables 3.1, 3.2, 3.3 and 3.5, and the species profiles indicate the legumes that contain tannins.

11.2 CLASSIFICATION OF TANNINS

Tannins are classified variously on different criteria. They are often placed in two groups, catechol and pyrogallol tannins, on the basis of their origin. But most common tannins extracted from plants are actually a mixture of catechol and pyrogallol tannins.

The most commonly accepted basis of classification of tannins is whether they are hydrolysed by acids and enzymes and their structure. Accordingly, tannins generally fall into two groups: a) hydrolyzable tannins, and b) non-hydrolyzable tannins, also called the condensed tannins or proanthocyanidin tannins.

The hydrolyzable tannins can be split by acids, bases and in some cases, hydrolytic enzymes (tannases) into sugars, related polyols and a phenolic acid. They contain a central core of polyhydric alcohol such as glycose, and hydroxyl groups, which are esterified either partially or wholly by gallic or ellagic acids, to give rise to gallotannins and ellagitannins, respectively. Hydrolyzable products of tannins can be absorbed from the intestinal walls and enter systemic circulation.

Condensed tannins posses the general structure of polymeric flavan-3-ol where the interflavan bonds are commonly C-4 (Harborne, 1984).

Tannins can also be classified into wood tans, bark tans, leaf tans, fruit tans and seed tans, based on the plant part used for tannin extraction.

11.3 CHEMISTRY OF TANNINS

Tannins are a large group of amorphous, dark reddish brown to black, polyphenolic compounds present in a very large number of plants in varying quantities. The detailed chemical structure of tannin varies depending upon the source.

Tannins are soluble in water, dilute alkalis, alcohols, glycerol and acetone, but are only slightly soluble in other organic solvents and almost insoluble in ether, chloroform and benzene. The true tannins have molecular weight between 500 to 5000 Daltons. The range of molecular weight of a tannin to be an effective tanning agent is between 500 and 3000 Daltons.

Most tannins are complex organic acids, formed within the plant tissues by the metabolic breakdown of sugars. Tannins contain a large number of hydroxy or other functional groups that form cross-linkages with other compounds such as alkaloids, gelatin and precipitate them. Tannins complexed with sugars occur as polyphenolic glycosides forming a colloidal suspension in water. Tannins also complex with proteins and this property is used to quantify both tannins and proteins.

Synthetic tannins are also available and these are generally the products of condensation of phenol, cresol and naphthalene with an aldehyde such as furfural.

11.4 DISTRIBUTION OF TANNINS IN PLANTS

Both hydrolyzable and non-hydrolyzable (condensed) groups of tannins are widely distributed throughout the plant kingdom, both in the temperate and tropical climates. Tannin concentrations are high during the dry seasons and dry areas.

The hydrolyzable tannins are limited to dicotyledonous plants and found in relatively few families. They occur in the bark of the root and stem, wood, leaves, fruits and seeds of the Anacardiaceae, Caesalpiniaceae, Combretaceae, Fabaceae and Mimosaceae. Tannins are also a response product in insect galls on the vegetative parts and fruits in these families. In the roots, leaves, flowers, fruit walls and seed coats, tannins occur only in small quantities (Deshpande *et al.*, 1982; Singh, 1993).Within the plant organs, tannins are contained in the cell sap of the vacuole and are found in greater quantities in dead or dying plant cells.

Non-hydrolyzable tannins (condensed tannins) occur in ferns and gymnosperms universally and are widespread among the angiosperms, especially the woody species. The condensed tannins are widely distributed in fruits, vegetables, forage plants, cocoa, red wine and certain food grains such as sorghum, finger millets and pulses (Harborne, 1984).

The chief commercial sources of tannins are the stem bark and wood, though they may also occur in considerable quantities in the root bark.

The quality and quantity of tannins depends upon the species, genotype, ecological conditions, age and stage of development of the plant, and the specific part of the same plant. The amount of tannin is negligible in the white flowered or the white seeded cultivars of legumes (Griffiths, 1981; Bos and Jetten, 1989). Mature parts contain higher concentrations of tannins.

11.5 TANNINS IN LEGUMES

Tannin containing species in the three legume families are indicated in Tables 3.1 to 3.3. Tannins are found in about 12.1 per cent of the legume species (Chapter 3 ,Table 3.5). In the Caesalpiniaceae, 19.4 per cent of the species and 3.9 per cent of the species in the Fabaceae, contain tannins. The Mimosaceae include the largest percentage (41.7) of tannin containing species. Many species of *Acacia, Leucaena, Lysiloma* and *Prosopis* are rich in tannin concentrated in the mature bark and pods. Mostly, it is the species of *Acacia* that are the source of industrial tannins, as they contain up to 35 per cent of tannin, by weight. Several tannin rich Australian species of *Acacia* were introduced into India, and are widely grown, to cater to the needs of the industry.

The edible parts of many legumes also contain tannins (Table 11.1).

11.6 DIETARY TANNINS

Tannins have been found in a variety of plants utilized as food and feed. This includes food grains such as sorghum, millets, barley, dry beans, faba beans, peas, pigeon peas, winged beans and other legumes. Fruits such as apples, bananas, blackberries, cranberries, dates, grapes, peaches, pears, persimmons, plums, raspberries, strawberries and tea also contain an appreciable quantity of tannin.

Jadhav *et al.*, (1989) and Shalini (2002) observed that among grain legumes, black gram, faba bean, green gram, horse gram, kidney bean, moth bean and pigeon pea have a very high content of tannin (Table 11.1).

11.7 BIOLOGICAL ROLES OF TANNINS IN PLANTS

The usefulness of tannins to plants is uncertain. They probably afford protection against parasitic bacteria and fungi and they might play a role in wound healing (Kochhar, 1981).

Tannins exert an inhibitory effect on many enzymes due to their ability to complex with proteins and hence contribute to protection against pathogens in bark and heartwood (Trease and Evans, 1978). Tannins in fruits serve as a natural defense against microbial infection (Chung *et al.*, 1998).

Tannins also afford protection against grazing animals on account of their disagreeable taste.

11.8 BIOACTIVITY OF TANNINS

Many of the true tannins are glycosides exerting inhibitory action on enzymes. The activity of tannins in the leather tanning process and in biological systems is directly related to their ability to complex with proteins and to precipitate them, which inactivates a very large number of enzymes (see Chapter 7, Section 7.2.4). Tannins have been shown to inhibit virtually every digestive enzyme and reduce the bioavailability of iron and vitamin B_{12}. However, when the molecular weight is very large (>5000 Daltons) the tannins become insoluble and loose their capacity to precipitate other compounds.

Sequential inhibiton of nuclear RNA synthesis, nucleolar segregation, inhibition of protein synthesis, all related to enzyme inhibition, were observed on administration of tannic acid.

It was reported that condensed tannins caused dryness and reduced taste of oral mucosa, by their astringent action. Tannins exert an astringent effect in the duodenum, due to precipitation of proteins, inhibiting their absorption. The kinetics of tannins can thus be considered as a local action observable only in the intestinal lumen.

11.8.1 Toxicity

Ruminants have a high tolerance for condensed tannins, but the ingestion of excess of hydrolyzable tannins can cause death. Hydrolysable tannins cause varied toxic manifestations due to hydrolysis in ruminants. They are also believed to be responsible for foetal poisoning of domestic animals fed on high tannin containing feeds (Salunkhe *et al.*, 1990).

Tannins are reported to be deleterious to the alimentary canals of animals. Ingestion of high levels of tannins could cause gastroenteritis and congestion of intestinal wall in rats and hemorrhagic gastroenteritis in rabbits.

Tannic acid is known as hepatotoxic agent resulting in liver necrosis, caused by tannins absorbed into the liver (Enomoto, 1987).

Human fatalities have occurred from preparations with high concentration of tannic acid in enemas or burn remedies. Other deleterious effects of tannins on humans include a) damage to mucosal lining of gastrointestinal tract, b) alteration of excretion of certain cations, and c) increased excretion of proteins and essential amino acids (Reddy *et al.*, 1985). All these effects lead to severe disruption of several metabolic processes.

11.8.2 Carcinogenesis

It has been reported that a prolonged consumption of tannin rich food may result in various types of cancer. Chronic ingestion of high content of tannin may increase the risk of some tumourigenic diseases.

An association between neoplasia and tannins in humans was observed among the shoe and wood workers. In both these industries the workers are constantly in contact with tannins. Tannins and tannic acid are listed as tentative carcinogens of category I by the Occupational Safety and Health Administration (OSHA).

Tumours were induced in mice by subcutaneous injection of tannin extracts prepared from various plant materials. Tannins also promoted skin cancer in mice. Tannin containing betel nut is responsible for cheek and oesophageal cancer. Herbal tea showed a positive correlation with the occurrence of stomach, lung, kidney and oesophageal cancer. Tannin and its related compounds such as rutin, kaempferol and catechin, also are suspected to cause cancers (Chung *et al.*, 1998). Quercetin, a mutagen, induced intestinal and bladder carcinomas.

However tannins were also shown to exercise protective action against cancer of the colon, breast, prostate, stomach and liver. Anti carcinogenic potential of tannins is related to their anti oxidative property in protecting cellular components from oxidative damage.

11.9 APPLICATION OF TANNINS IN MEDICINE AND INDUSTRY

11.9.1 Tannins as Medicine

In spite of several deleterious effects on our system, tannins have some important uses in medicine.

a) The antimicrobial activity of tannins is well documented. They inhibited the growth of many fungi, yeasts, bacteria and viruses. Tannins inhibit dental caries. Many of the chewing sticks and tooth brushes made out of plant twigs prevent dental caries and most of them are species of *Acacia* containing considerable quantities of tannin in the bark.

b) Tea polyphenols and tannin components were suggested to be anticarcinogenic. Tannins exhibited anticancer effects in lab animals. Many tannin molecules have been shown to reduce the activity of a number of mutagens. The anticarcinogenic and antimutagenic potentials of the tannins may be related to their antioxidative property, which is important in protecting cellular oxidative damage, including lipid peroxidation.

c) Tannins have been reported to exert other physiological effects, such as acceleration of blood clotting, lowering blood pressure, decreasing serum-lipid level, reducing blood sugar levels and modulation of immuno-responses.

d) Because of the capability of tannins to bind proteins and RBCs tannic acid is used as a reagent for passive haemagglutination test and to stain cell surface carbohydrates in normal and pre-malignant epithelium of cancerous tissue.

e) In ruminants, condensed dietary tannins (2 to 3 per cent) were shown to reduce wasteful protein degradation in the rumen, through the formation of a protein-tannin complex.

f) Drugs that contain tannin are used as astringents and styptics for application to sprains, bruises and superficial wounds, burns and as an ingredient of eye lotions.

g) Tannins are used to treat diarrhoea, as chemical antidotes in poisoning and as local astringents.

h) Tannin-fluoride mixture is used to prevent dental caries.

i) Tannic acid reduces allergen levels in house dust and is marketed as a commercial preparation. It is also used in reducing mite allergens in carpet dust, causing asthma, rhinitis and pneumonitis caused by birds.

11.9.2 Tannins in Industry

While the most important commercial application of tannins is in the tanning industry, tannins are employed in several other industries.

a) Tannins are used as flavouring agents in most categories of processed food, including alcoholic and nonalcoholic beverages, frozen dairy products, candy, baked foods, gelatins, puddings, meat and meat products.

b) The antimicrobial activity of tannic acid increases shelf life of processed foods, such as catfish fillets.

c) Tannins are used as clarifying agents in alcoholic beverages, especially beer and wines.

d) Basing upon the property of formation of insoluble precipitates of tannins with protein, gelatin, starch and heavy-metal ions, the textile industry uses tannins as fabric treating agents.

e) Tannins form dark-bluish or blackish compounds with iron salts, the basis of our inks.

f) Tannins are used as mordants for fixing dyes.

g) Tannins are used as disperssant in the petroleum industry to control viscosity of mud in oil-well drilling.

h) The mining industry uses small amounts of tannins for floatation of ores.

Table 11.1

Tannins in Edible Legumes

Tannins were demonstrated in the following species. Tannins may not be present in the fresh condition in some species/parts, but accumulate in the same parts when dry.

Caesalpiniaceae

Tamarindus indica (fresh leaf, flower, fruit pulp)

Fabaceae

Arachis hypogaea (dry seed); *Cajanus cajan* (dry seed); *Cicer arietinum* (dry seed); *Cyamopsis tetragonoloba* (dry seed); *Glycine max* (dry seed); *Lablab purpureus* var. *lignosus* and var. *typicus* (fruit, fresh and dry seed); *Lens culinaris* (dry seed); *Macrotyloma uniflorum* (dry seed); *Phaseolus lunatus* (fresh and dry seed); *Phaseolus vulgaris* (fresh and dry seed); *Pisum sativum* (dry seed); *Trigonella foenum-graecum* (fresh leaf and dry seed); *Vicia faba* (dry seed); *Vigna radiata* (dry seed); *Vigna unguiculata* (fresh fruit, fresh and dry seed).

Mimosaceae

Pithecellobium dulce (fresh fruit and aril).

Source : Shalini, 2002

❐❐❐

CHAPTER-12

Pigments and Dyes in Legumes

12.1 INTRODUCTION

Pigments are the chemical compounds that impart colour. Pigments that impart colour to organisms or their parts are called natural pigments.

Some colours are related to the state of development, changing as the organs mature and/or degenerate. Some colours are related to the season of the year, such as the foliage colours of plants that shed all leaves either in spring or winter (deciduous species), when a new crop of leaves develop. Several deciduous legumes exhibit very bright colours.

12.2 NATURAL PIGMENTS

Britton (1983), who provided the most comprehensive review on the chemistry of the natural pigments, distinguished the colour of organisms or their parts into structural and chemical colours. Structural colours are not specifically associated with any pigments, but are produced by interference and scattering of light, while chemical colours are pigment dependent.

Chemical colours mostly belong to two groups of natural products: the flavonoids, which are phenolic compounds, or the carotenoids that are terpenoid compounds. Phenolic and terpenoidal compounds present in edible legumes may exercise anti-nutritional effects (Chapter 7 Section 7.2.2) and some are toxic Chapter 8 (Section 8.4.2 and 8.4.3).

Box 12.1
Natural pigments in our lives

The patterns and combinations of colour of animals and of the leaves, flowers and fruits make our planet look beautiful. Natural pigments have a cultural and aesthetic role in our lives as they are used in various ways such as to colour food, dye textiles, paint our own bodies and our houses, etc. The colours of plant species play an important role in landscaping our gardens and avenues. The contribution of legumes in all these contexts is quite rich.

Flavonoids, derived from the parent substance flavone, are exclusively higher plant pigments concentrated in cell sap or vacuoles. There are over 600 flavonoids, bound to sugars, which occur as glycosides or as aglycones. Of all the flavonoids, anthocyanins contribute maximum to plant colours. Anthocyanins are a wide spread group of pigments in plants responsible for pink, scarlet, red, mauve, violet and blue colours of the leaf, petals and fruits. They are synthesized in cell organelles called anthocyanoplasts and stored in vacuoles. The flavonoid pigments such as chalcones and aurones give yellow and orange colour to flowers along with carotenoids but these compounds are occasionally present in other tissues.

Flavones and flavonols are mainly colourless and make a contribution to flower colours by co-pigmentation. They occur along with anthocyanins and form complexes. Flavans are not involved directly in plant colouration. Flavones or flavonols are very essential to "body" of white or cream colours of flowers. They may look white to human eye but will appear coloured to bees and butterflies. Proanthocyanins are mainly colourless and present in heartwood and leaves of woody plants. Flavanones and isoflavones are colourless and are present in leaf and fruit and root respectively. Isoflavones are common in the legumes. Flavonols and flavones lacking in the 5-hydroxyl group are characteristic of the legumes.

Some flavonoids such as the leucoanthocynidins are themselves colourless, but develop colour upon oxidation. Due to complex interactions among different pigments, it is not always possible to explain plant colours in precise chemical terms.

As can be seen, the most common plant colour is green due to chlorophyll in the leaves and younger vegetative or even floral parts. All parts of plants, including those of legumes, show a variety of colours, the most enchanting being the flowers that exhibit a wide range of colours. The fruits and seeds themselves show a range of colours, as also the wood.

12.3 NATURAL DYES

Various parts of plants such as the root, bark, wood, leaves, fruits and seeds contain pigments extractable in water or other solvents. The extracted colours, called dyes, can be imparted to other materials. Dyes are used in various contexts such as colouring cloth and in medicine. The chemistry of dyes is more complex due to the simultaneous presence of several other compounds that are not pigments but which may influence colours. The diversity of flower colours and the principle groups of compounds responsible for them are given in Table 12.1. Dye yielding legumes were discussed in Chapter 2 (Section 2.9). Different dyes from the legumes are given in Table 12.2.

12.4 CHEMICAL GROUPS OF NATURAL PIGMENTS AND DYES

The following is the summary of classification of the chemical groups of the major pigments and dyes in the legumes. These groups can be related to the chemical classes detailed in other contexts in this book and to the pigments in the non-legume plant groups.

12.4.1 Terpenoids

12.4.1.1 Tetrapyrroles are diterpenes, the most common of natural pigments. The group includes chlorophylls and other related photosynthetic pigments; cytochromes a, b, c, and d, vitamin B_{12}, phycobilins, phytochromes Pr and Pfr.

12.4.1.2 Carotenoids are tetraterpenes present in wood, bark, flower, pod, and seed of legumes. Yellow, red and orange flower colours of a number of legumes such as species of *Acacia, Cytisus, Delonix, Genista, Astragalus, Laburnum, Lotus and Ulex* contain carotenoids (Harborne, 1971; Goodwin, 1980). Carotenoids comprise the yellow-red fat-soluble hydrocarbon carotenes and their oxygenated derivatives, the xanthophylls. Taraxanthin is exclusive to the species of *Ulex* and rubixanthin is found only in the genus *Astragalus*.

Many carotenoids are antioxidants. The α-carotene, β-carotene and β-cryptoxanthin are precursors of vitamin A. Carotenoids are important components in plants as they are energy absorbing compounds that mask and protect chlorophyll from photo-bleaching. Sporopollenin, the hardest of plant components that withstands the ravages of fossilization, is a polymer of carotenoid units, and is the principal component of pollen walls called exine, which imparts an yellow colour to the pollen.

12.4.2 Phenolic compounds

Phenols are a class of aromatic (in the chemical sense) organic compounds, technically alcohols, in which one or more hydroxyl (–OH) groups are attached directly to the benzene ring, as for example, phenol itself. Phenolic compounds are derived from this basic structure.

Phenolic compounds are among the most abundant and diverse groups of chemical compounds in plants. Many phenolic compounds in plants particularly the catechols, oxidize when exposed to air. This is seen when cut vegetables such as brinjals, plantains, potatos and fruits like apples when exposed to air develop a brown colour on the cut surfaces.

12.4.2.1 Quinones are unsaturated cyclic diketones derived by oxidation of dihydroxy phenols. They are not very visible as they are absent from the visible parts but common in bark, heart wood, roots, etc., some like henna are well known; melanin (dark brown or black), more common in animals than in plants, is a polymer of indole-5, 6-quinone.

12.4.2.2 Flavonoids: are a group of aromatic, oxygen containing heterocyclic pigments, responsible for most of the yellow, red and blue pigments of plant parts. Such pigments of dietary origin like pterin also occur in the wings of butterflies and moths (Britton, 1983). Flavonoids include the following subgroups:

a) **Catechols** are colourless dihydric phenols, but darken on oxidation.

b) **Leucoanthocyanidins** initially colourless aglycons of anthocyanins, but produce colour on oxidation.

c) **Anthocyanins** are responsible for brilliant crimson, scarlet and purple colour of leaves, flowers and fruits. These are replaced by α-cyanin-3 in beetroot. Anthocyanins are obtained by conversion of flavanonols and are used in colouring wines, jams and jellies.

d) **Flavonones** such as 2, 3-dihydroxyflavone produce orange colour.

e) **Flavones** are formed by oxidation of flavonones and differ from flavonols in lacking 3-hydroxyl groups and appear as white and pale yellow to human but appear coloured to bees and butterflies. They also give body to white and cream flowers without which they would be transparent. Some flavones such as apigenin and luteolin produce orange colour in flowers of herbaceous legumes.

f) **Isoflavones** are characteristic of the Fabaceae and are rarely found elsewhere. They are formed from chalcones.

g) **Flavanones** are not involved directly in plant colours, but their polymers such as catechin and pro-anthocyanidin produce brown colour in heart woods. They are formed by isomerization of chalkones

h) **Flavanols:** 3-hydroxyflavone (not identical with flavonols), produces bright yellow colour.

i) **Flavonols:** occur in woody legumes Flavon-3-ol (not identical with flavanols) produces the ivory and light yellow colours. They contribute to flower colours by co-pigmentation. Blue flowers owe their colour to co-pigmentation complexes with an-thocyanins. Flavonols such as quercetin, myricetin, rutin and kaempferol usually accompany anthocyanidins such as pelargonidin and delphinidin. Flavonols are formed from chemical oxidation of flavanonol.

j) **Flavanonols:** Derived from flavanones by hydroxylation. They give rise to other groups of flavonoids with 3-hydroxy group like anthocyanidins.

Aurones and chalcones, designated as minor flavonoids, are responsible for orange and yellow colours.

Table 12.1
Chemistry of Colours of Legume Flowers

Sl. No.	Colour of the flower	Compound responsible for flower colour	Species
I	**Caesalpiniaceae**		
1.	White	Flavones, flavonols	*Bauhinia acuminata, Bauhinia malabarica, Bauhinia vahlii, Humboldtia vahliana*
2.	Snow-white with lower petal lemon yellow	Chalcones, carotenoids	*Baikiaea insignis*
3.	Pale yellow	Chalcones, carotenoids	*Caesalpinia coriaria*
4.	Yellowish white	Chalcones, carotenoids, flavones, flavonols	*Bauhinia diphylla, Pterolobium hexapetalum*
5.	Yellow	Chalcones, carotenoids	*Bauhinia tomentosa, Caesalpinia crista, Caesalpinia sappan, Cassia fistula, Delonix elata, Hardwickia binata, Haematoxylum campechianum, Moullava spicata, Parkinsonia aculeata, Peltophorum pterocarpum, Senna hirsuta, Senna sulfurea*
6.	Bright yellow	Chalcones, carotenoids	*Senna alata*
7.	Greenish yellow	Chalcones, carotenoids	*Caesalpinia coriaria*
8.	Reddish yellow	Chalcones, carotenoids, anthocyanins	*Senna absus*
9.	Yellow turning reddish orange	Chalcones, carotenoids, anthocyanins	*Saraca cauliflora*
10.	Yellow turning brick red	Chalcones, carotenoids, anthocyanins	*Cassia moschata*
11.	Yellow or orange marked with spots	Chalcones, carotenoids, anthocyanins	*Caesalpinia cacalco*
12.	Red and yellow	Chalcones, carotenoids, anthocyanins	*Caesalpinia pulcherrima*
13.	Red and yellow	Chalcones, carotenoids, anthocyanins	*Tamarindus indica*
14.	Rich reddish or rose purple almost crimson	Anthocyanins	*Bauhinia blakeana*
15.	Crimson	Anthocyanins	*Humboldtia brunonis*
16.	Brilliant red edged with yellow	Anthocyanins	*Amherstia nobilis*

Sl. No.	Colour of the flower	Compound responsible for flower colour	Species
17.	Terracotta	Anthocyanins	*Senna polyantha*
18.	Orange	Chalcones, carotenoids, anthocyanins	*Colvillea racemosa*
19.	Orange to scarlet	Anthocyanins	*Delonix regia*
20.	Orange red	Anthocyanins	*Saraca asoca*
21.	Light pink	Anthocyanins	*Cassia hybrida*
22.	Pink	Anthocyanins	*Cassia renigera*
23.	Bright pink	Anthocyanins	*Senna nodosa*
24.	Purple-pink	Anthocyanins	*Bauhinia variegata*
25.	Tones of purple, white, mauve	Anthocyanins	*Bauhinia purpurea*
II	**Fabaceae**		
1.	White	Flavones, flavonols	*Dalbergia horrida, Derris scandens, Derris trifoliata, Dendrolobium umbellatum, Melilotus alba*
2.	Creamy white	Flavones, flavonols	*Dalbergia latifolia, Dendrolobium triangulare*
3.	White or rose	Flavones, flavonols anthocyanins	*Desmodium oogeinensis*
4.	White or purple to reddish purple	Anthocyanins	*Glycine max*
5.	Whitish pink	Anthocyanins	*Sesbania grandiflora*
6.	Yellow	Chalcones, carotenoids	*Aeschynomene aspera, Cajanus cajan, Crotalaria* species, *Desmodium pulchellum, Melilotus indica, Paracalyx scariosus, Pterocarpus marsupium, Rhynchosia* species, *Sesbania bispinosa, Smithia* species, *Trifolium* species, *Uraria* species
7.	Golden yellow	Chalcones, carotenoids	*Pterocarpus indicus*
8.	Bright yellow	Chalcones, carotenoids	*Tipuana tipu*
9.	Yellowish to purplish	Chalcones, carotenoids, anthocyanins	*Alysicarpus pubescens*
10.	Yellow to scarlet	Chalcones, carotenoids, anthocyanins	*Butea superba*
11.	Yellow to orange and reddish	Chalcones, carotenoids, anthocyanins	*Castanospermum australe*

Sl. No.	Colour of the flower	Compound responsible for flower colour	Species
12.	Yellowish pink	Anthocyanins	*Aeschynomene indica*
13.	Yellowish white	Chalcones, carotenoids	*Dalbergia sissoo*
14.	Red	Anthocyanins	*Erythrina stricta, Erythrina suberosa, Erythrina subumbrans, Erythrina variegata*
15.	Deep red	Anthocyanins	*Erythrina blakei*
16.	Crimson to scarlet	Anthocyanins	*Erythrina crista-galli*
17.	Dark crimson	Anthocyanins	*Schotia latifolia*
18.	Flaming orange	Chalcones, carotenoids, anthocyanins	*Butea monosperma*
19.	Pink	Anthocyanins	*Abrus precatorius, Alysicarpus monilifer, Alysicarpus vaginalis, Canavalia ensiformis, Codariocalyx motorius, Desmodium dichotomum, Desmodium heterophyllum, Fleminigia grahamiana, Flemingia wallichii, Ormosia travancorica*
20.	Pinkish white	Anthocyanins	*Derris scandens, Spatholobus parviflorus*
21.	White to pink to purplish	Anthocyanins	*Cicer arietinum, Gliricidia sepium*
22.	Rose	Anthocyanins	*Abrus pulchellus*
23.	Rose or lilac	Anthocyanins	*Pongamia pinnata*
24.	Purple	Anthocyanins	*Andira inermis, Dalbergia volubilis, Desmodium styracifolium, Mucuna pruriens, Tephrosia purpurea*
25.	Purplish	Anthocyanins	*Parochetus communis*
26.	Purplish white	Anthocyanins	*Piscidia erythrina*
27.	Purplish yellow	Anthocyanins, chalcones	*Flemingia macrophylla*
28.	Purplish red	Anthocyanins	*Shuteria vestita*
29.	Purplish blue	Anthocyanins	*Desmodium triflorum*
30.	Purple or mauve	Anthocyanins	*Millettia peguensis*
31.	Pale mauve	Anthocyanins	*Dalbergia lanceolaria*

Sl. No.	Colour of the flower	Compound responsible for flower colour	Species
32.	Blue	Anthocyanins	*Clitoria ternatea, Desmodium heterocarpon, Dolichos trilobus*
33.	Bluish white	Anthocyanins	*Millettia racemosa, Desmodium laxiflorum*
34.	Bluish purple	Anthocyanins	*Tadehagi triquetrum*
35.	Blue-violet	Anthocyanins	*Wisteria sinensis*
36.	Violet to white	Anthocyanins	*Desmodium gangeticum*
37.	Violet	Anthocyanins	*Desmodium velutinum*
III	**Mimosaceae**		
1	White	Flavones, flavonols	*Acacia* species, *Albizia odoratissima, Albizia procera, Archidendron monaldelphum, Lysiloma latisiliquum, Pithecellobium dulce*
2.	Creamy white	Flavones, flavonols	*Albizia lebbeck*
3.	Greenish turning cream yellow with age	Chalcones, carotenoids	*Prosopis juliflora*
4.	Pale yellow	Chalcones, carotenoids	*Acacia catechu*
5.	Yellow	Chalcones, carotenoids	*Acacia auriculiformis, Acacia decurrens, Acacia leucophloea, Acacia nilotica* ssp. *indica, Adenanthera pavonina, Albizia amara, Neptunia oleracea, Samanea saman*
6.	Yellowish white	Chalcones, carotenoids	*Xylia xylocarpa*
7.	Upper yellow flowers, lower pink flowers	Chalcones, carotenoids, Anthocyanins	*Dichrostachys cinerea*
8.	Pink	Anthocyanins	*Albizia julibrissin, Mimosa* species
9.	Pinkish white	Anthocyanins	*Albizia chinenis*

Table 12.2
Dyes from Legumes

Sl. No.	Species	Part yielding the dye	Colour of the dye
I	**Caesalpiniaceae**		
1.	*Bauhinia purpurea*	Bark	Brown
2.	*Bauhinia tomentosa*	Leaf	Yellow
3.	*Bauhinia variegata*	Bark	Fast shades of brown
4.	*Caesalpinia coriaria*	Wood	Red
5.	*Caesalpinia pulcherrima*	Root, Flower	Red
6.	*Caesalpinia sappan*	Root Wood Bark Pod	Yellow Red Black shades of red and orange yellow Red and brown
7.	*Cassia fistula*	Bark	Light red
8.	*Cassia surattensis* ssp. *glauca*	Plant	Ochre red
9.	*Ceratonia siliqua*	Wood	Brown
10.	*Cynometra ramiflora*	Wood	Purple
11.	*Haematoxylum campechianum*	Heart wood	Blue
12.	*Intsia palembanica*	Wood	Yellow
13.	*Peltophorum pterocarpum*	Bark	Yellowish brown
14.	*Senna alexandrina*	Leaf	Black, dark shades
15.	*Senna obtusifolia*	Plant	Yellow
16.	*Senna tora*	Seed	Blue
17.	*Tamarindus indica*	Leaf Seed coat	Yellow Brown and olive brown shades
II	**Fabaceae**		
1.	*Baphia nitida*	Heart wood	Red
2.	*Butea monosperma*	Bark exudates Flower	Blue, Yellow
3.	*Butea superba*	Flower	Yellow
4.	*Cicer arietinum*	Leaf	Blue
5.	*Clitoria ternatea*	Flower	Blue
6.	*Crotalaria pallida*	Plant	Black
7.	*Cullen plicata*	Pod	Yellow
8.	*Cytisus scoparius*	Leaf	Green
9.	*Erythrina variegata*	Twig, flower	Red
10.	*Genista tinctoria*	Flower	Yellow green

Sl. No.	Species	Part yielding the dye	Colour of the dye
11.	*Indigofera angustifolia*	Plant	Blue
12.	*Indigofera argentea*	Plant	Blue
13.	*Indigofera arrecta*	Plant	Blue
14.	*Indigofera coerulea*	Plant	Blue
15.	*Indigofera cordifolia*	Plant	Blue
16.	*Indigofera dosua*	Plant	Blue
17.	*Indigofera hirsuta*	Plant	Blue
18.	*Indigofera longiracemosa*	Plant	Blue
19.	*Indigofera oblongifolia*	Plant	Blue
20.	*Indigofera parviflora*	Plant	Yellow
21.	*Indigofera spicata*	Plant	Blue
22.	*Indigofera tinctoria*	Plant, leaf	Blue
23.	*Lotus corniculatus*	Leaf, flower	Orange yellow
24.	*Physostigma venenosum*	Leaf	Black
25.	*Pterocarpus dalbergioides*	Heart wood	Reddish yellow
26.	*Pterocarpus indicus*	Wood	Red
27.	*Pterocarpus marsupium*	Bark, gum	Brownish red
28.	*Pterocarpus santalinus*	Shoot, wood	Red
29.	*Robinia pseudoacacia*	Wood	Brown orange shades
30.	*Styphnolobium japonicum*	Flower, pod	Yellow
31.	*Tephrosia maxima*	Plant	Blue
32.	*Tephrosia purpurea*	Leaf	Fast shades of blue
33.	*Tephrosia sinapou*	Plant	Blue
34.	*Tephrosia tinctoria*	Plant	Blue
35.	*Trigonella foenum-graecum*	Seed	Yellow
36.	*Ulex europaeus*	Plant	Yellow
37.	*Vigna unguiculata*	Leaf	Green
III	**Mimosaceae**		
1.	*Acacia catechu*	Bark, heart wood	Pale yellow,brown,black
2.	*Acacia farnesiana*	Bark, pod	Black
3.	*Acacia leucophloea*	Bark, gum, leaf	Black
4.	*Acacia nilotica*	Bark, gum, leaf, pod	Dark brown
5.	*Acacia sinuata*	Bark, leaf	Green
6.	*Adenanthera pavonina*	Wood	Red
7.	*Albizia odoratissima*	Bark, leaf	Brown
8.	*Archidendron clypearia*	Leaf	Black

CHAPTER-13

Terpenoids in Legumes

13.1 INTRODUCTION

Terpenoids (or terpenes) are one of the most diverse and biologically important classes of compounds responsible for both colours and fragrance of plants parts and products. The terpenoids responsible for colours are discussed in Chapter 12.

13.2 TERPENOIDS

Terpenoids are derived biosynthetically from precursors containing two or more isoprene units. They are products of acetate metabolism synthesized along the mevolonate pathway. Basing on the number of double isoprene units contained, the terpenoids are classified into mono- (C_{10}), sesqui- (C_{15}), di- (C_{20}), tri- (C_{30}) and tetra- (C_{40}) terpenes. All these groups of terpenoids occur in the legumes (Table 13.1).

13.2.1 Mono- and Sesquiterpenes

Mono- and sesquiterpenes are fragrant and volatile compounds referred to as 'essential oils'. Occurring in flowers, wood and resinous exudates, they are widely used in cosmetic and pharmaceutical industry. Legume resins were discussed in Chapter 3 (Section 3.2.7)

Monoterpenes are simple unsaturated hydrocarbons comprising the majority of essential oils. They occur as ocimene, myricene, limonene and linaloid in the flowers of *Medicago sativa* and *Acacia farnesiana*. A monoterpene phenol, bakuchiol, is reported from the seeds of *Cullen corylifolium*, which is used in treating skin ailments.

Sesquiterpenes are found in the leaves, resin and pods of species of *Hymenaea* and in leaf resin of species of *Copaifera*, imparting characteristic odour (Harborne, 1971b).

13.2.2 Diterpenes

These are derived from four isoprene units and are non-volatile. They are synthesized from geranyl pyrophosphate (GPP), which is produced from the condensation of two GPP units. They occur in exuded resins or woods of several tropical legume species. Most of them are produced in response to environmental stress.

13.2.3 Triterpenes

Triterpenes are very diverse and occur in either free state or esters or glycosides. They are synthesized from six isoprene units. Some triterpenes like phytoalexins are stress compounds. Triterpenes are widespread in the wood, bark, flower, pod and seed of species of the Fabaceae and Mimosaceae, particularly in the timber species and less common in the Caesalpiniaceae.

Triterpenes include sterols, saponins, sapogenins, sapogenols and cardiac glycosides. Several steroidal compounds occur in legumes. Seeds of many legumes are also rich in cholesterol (Table 13.2). Steroidal compounds in edible legumes that may exercise anti-nutrient effects were discussed in Chapter 7 (Section 7.2.2) and toxic steroidal compounds were discussed in chapter 8 (Section 8.4.3.1)

Saponins, sapogenols and sapogenins, have distinctive properties and a wide range of applications in medicine and industry as discussed in Chapter 14.

13.2.4 Tetraterpenes

Tetraterpenes derived from eight isoprene units are present in wood, bark, flower, pod and seed of legumes. These include carotenoids, which occur as pigments in flowers. Carotenoids protect chlorophyll from photo-bleaching.

Table 13.1

Terpenoid Compounds in Indian Legumes

Species with edible parts are marked by an asterisk (*)

Sl. No.	Species	Location
A	**MONOTERPENES**	
I	**Fabaceae**	
1.	*Cullen corylifolium*	Seed
2.	*Medicago sativa*	Flower
II	**Mimosaceae**	
1.	*Acacia farnesiana*	Flower
B	**SESQUITERPENES**	
I	**Caesalpiniaceae**	
1.	*Kingiodendron pinnatum*	Wood
2.	*Sindora wallichii*	Wood
II	**Fabaceae**	
1.	*Dalbergia sissoo*	Heart wood
2.	*Myroxylon balsamum*	Bark, wood oil
C	**DITERPENES**	
I	**Caesalpiniaceae**	
1.	*Caesalpinia bonduc*	Seed
2.	*Copaifera officinalis*	Heartwood
3.	*Kingiodendron pinnatum*	Resin
4.	*Hymenaea courbaril*	Wood
5.	*Hymenaea verrucosa*	Wood
II.	**Mimosaceae**	
1.	*Mimosa pudica*	Leaf
2.	*Xylia xylocarpa*	Wood

Sl. No.	Species	Location
D	**TRITERPENES**	
I	**Caesalpiniaceae**	
1.	*Brownea ariza**	Bark, flower
2.	*Caesalpinia bonduc*	Seed
3.	*Cassia fistula**	Bark, leaf, seed
4.	*Cassia grandis*	Leaf, seed
5.	*Cassia javanica*	Leaf, seed
6.	*Cassia javanica* ssp., *nodosa*	Leaf, flower, seed
7.	*Cassia renigera*	Leaf, seed
8.	*Cassia roxburghii*	Leaf, seed
9.	*Chamaecrista kleinii*	Seed
10.	*Chamaecrista mimosoides*	Seed
11.	*Senna bicapsularis*	Leaf, seed
12.	*Senna fruticosa*	Leaf, seed
13.	*Senna hirsuta*	Leaf, seed
14.	*Senna occidentalis*	Leaf, seed
15.	*Senna sophera*	Leaf, seed
16.	*Senna spectabilis*	Seed
17.	*Senna surattensis*	Pollen
18.	*Senna tora*	Leaf
19.	*Peltophorum pterocarpum*	Seed
20.	*Saraca cauliflora**	Bark
II	**Fabaceae**	
1.	*Abrus precatorius*	Root
2.	*Andira inermis**	Leaf
3.	*Arachis hypogaea**	Seed
4.	*Astragalus hamosus*	Root
5.	*Butea monosperma*	Seed
6.	*Castanospermum australe*	Wood
7.	*Cicer arietinum**	Seed, sprout
8.	*Clitoria ternatea**	Root
9.	*Crotalaria verrucosa*	Pollen
10.	*Cyamopsis tetragonoloba**	Seed
11.	*Dalbergia lanceolaria**	Pod
12.	*Desmodium gangeticum*	Root
13.	*Erythrina suberosa*	Bark
14.	*Erythrina subumbrans*	Wood
15.	*Erythrina variegata**	Bark, leaf, pollen
16.	*Glycine max**	Seed
17.	*Glycyrrhiza glabra**	Root

Sl. No.	Species	Location
18.	*Lens culinaris**	Seed, sprout
19.	*Lotus corniculatus*	Leaf
20.	*Lupinus luteus*	Shoot, seed coat
21.	*Lupinus mutabilis*	Shoot
22.	*Medicago sativa*	Root, stem, leaf, flower
23.	*Melilotus albus*	Shoot
24.	*Ononis spinosa*	Root
25.	*Pachyrrhizus erosus*	Seed
26.	*Phaseolus coccineus*	Seed
27.	*Phaseolus lunatus**	Seed
28.	*Phaseolus vulgaris**	Seed, sprout
29.	*Pisum sativum**	Seed
30.	*Pueraria lobata*	Root
31.	*Robinia pseudoacacia*	Root
32.	*Securigera coronilla*	Seed
33.	*Sesbania aculeata*	Pod, seed
34.	*Sesbania grandiflora**	Stem, leaf, seed
35.	*Sesbania sesban*	Leaf, pod, seed
36.	*Spartium junceum*	Petal
37.	*Styphnolobium japonicum*	Root
38.	*Trifolium fragiferum*	Plant
39.	*Trifolium pratens*	Shoot
40.	*Trifolium repens*	Leaf, shoot
41.	*Trigonella corniculata**	Seed
42.	*Trigonella foenum-graecum**	Leaf, seed
43.	*Vicia faba**	Seed
44.	*Vigna angularis*	Seed
45.	*Vigna mungo**	Seed
46.	*Vigna radiata**	Seed, sprout
III	**Mimosaceae**	
1.	*Acacia auriculiformis**	Seed
2.	*Acacia concinna**	Pod, pod rind, seed
3.	*Acacia confusa*	Seed
4.	*Acacia intsia*	Bark, seed
5.	*Albizia amara*	Seed
6.	*Albizia chinensis*	Seed
7.	*Albizia falcataria*	Seed
8.	*Albizia lebbeck**	Bark, shoot, flower, pod, seed
9.	*Albizia lucidior*	Seed
10.	*Albizia odoratissima*	Seed
11.	*Albizia procera*	Seed
12.	*Entada rheedei*	Seed

Sl. No.	Species	Location
13.	*Enterolobium cyclocarpum**	Leaf
14.	*Lysiloma latisiliquum*	Plant
15.	*Mimosa pudica*	Seed
16.	*Pithecellobium dulce**	Fruit, seed
17.	*Prosopis pubescens**	Bark, leaf
18.	*Samanea saman**	Stem bark, pod, seed
E	**TETRATERPENES**	
I	**Caesalpiniaceae**	
1.	*Delonix regia*	Flower
II	**Fabaceae**	
1.	*Cytisus scoparius*	Flower
2.	*Labrunum anagyroides*	Flower
3.	*Lotus corniculatus*	Flower
4.	*Medicago sativa*	Plant
5.	*Phaseolus vulgaris*	Leaf
6.	*Ulex europaeus*	Flower
III	**Mimosaceae**	
1.	*Acacia acuminata*	Wood
2.	*Acacia decurrens*	Flower
3.	*Acacia discolor*	Flower
4.	*Acacia dealbata*	Pollen

Table 13.2

Phytosterols in Indian Legumes

The specific sterol and the part in which it was detected are indicated. Edible parts are marked by an asterisk (*)

Sl. No.	Family and Species	Location	Sterol
I	**Caesalpiniaceae**		
1.	*Bauhinia acuminata*	Stem bark	β-sitosterol
2.	*Bauhinia racemosa*	Stem bark	β-sitosterol
3.	*Bauhinia vahlii*	Leaf	Campesterol, stigmasterol, β-sitosterol
4.	*Bauhinia variegata*	Stem	β-sitosterol, stigmasterol
5.	*Brownea ariza*	Flower	β-sitosterol
6.	*Caesalpinia bonduc*	Seed	Phytosterol
7.	*Caesalpinia pulcherrima*	Plant, root, flower, bark and stem	β-sitosterol β-sterol
8.	*Caesalpinia sappan*	Heart wood	β-sitosterol
9.	*Cassia fistula*	Stem bark, flower*, pod*	β-sitosterol
10.	*Cassia javanica*	Wood, leaf, flower	β-sitosterol
11.	*Cassia javanica* ssp. *nodosa*	Wood, leaf, flower	β-sitosterol
12.	*Cassia roxburghii*	Seed	β-sitosterol

Sl. No.	Family and Species	Location	Sterol
13.	*Ceratonia siliqua*	Seed	β-sitosterol, stigmasterol, campesterol
14.	*Chamaecrista absus*	Seed*, seed oil	β-sitosterol,
15.	*Chamaecrista pumila*	Leaf	β-sitosterol
16.	*Delonix regia*	Wood	β-sitosterol
17.	*Hardwickia binata*	Heart wood	β-sitosterol
18.	*Kingiodendron pinnatum*	Oleoresin	β-sitosterol
19.	*Parkinsonia aculeata*	Plant	β-sitosterol
20.	*Peltophorum pterocarpum*	Flower	β-sitosterol
21.	*Saraca asoca*	Stem bark, flower	β-sitosterol
22.	*Senna alata*	Root Stem bark Leaf Seed Seed oil	β-sitosterol Phytosterol Sitosterol β-sitosterol, campesterol Stigmasterol
23.	*Senna alexandrina*	Plant, leaf, seed, seed oil	β-sitosterol
24.	*Senna auriculata*	Plant, flower, seed	β-sitosterol
25.	*Senna obtusifolia*	Root Leaf Seed	β-sitosterol β-sitosterol, stigmasterol β-sitosterol
26.	*Senna occidentalis*	Plant Root Seed Seed oil	β-sitosterol β-Sterol α-sitosterol, campesterol β-sitosterol, stigmasterol, campesterol
27.	*Senna siamea*	Bark, leaf, flower*, seed	β-sitosterol γ-sitosterol
28.	*Senna sophera*	Root Wood Flower	β-sitosterol β-sitosterol β-Sterol
29.	*Senna spectabilis*	Plant	β-sitosterol, stigmasterol
30.	*Senna tora*	Plant, Stem, leaf*, seed*	β-sitosterol β-sitosterol
II	**Fabaceae**		
1.	*Abrus precatorius*	Root*, stem, leaf Seed	β-Sterol Sitosterol, campesterol, cholesterol, stigmasterol
2.	*Aeschynomene indica*	Seed	β-Sterol
3.	*Alhagi maurorum*	Plant Root*, stem	Cholesterol Phytosterol

Sl. No.	Family and Species	Location	Sterol
4.	*Alysicarpus longifolius*	Leaf, pod	β-sitosterol
5.	*Andira inermis*	Bark, heart wood, leaf	β-sitosterol
6.	*Arachis hypogaea*	Plant, pod shell Seed*	β-sitosterol, daucosterol, stigmasterol β-sitosterol
7.	*Butea monosperma*	Bark, seed, seed oil	β-sitosterol
8.	*Cajanus cajan*	Root bark, plant Leaf* Seed	Sitosterol Phytosterol, stigmasterol Campesterol, cholesterol
9.	*Cajanus scarabaeoides*	Leaf	β-sitosterol
10.	*Calopogonium mucunoides*	Stem	β-sitosterol, stigmasterol
11.	*Cicer arietinum*	Root Seed*	Coumesterol β-sitosterol
12.	*Clitoria ternatea*	Leaf, seed	γ-sitosterol, β-sitosterol
13.	*Crotalaria laburnifolia*	Seed*	β-sitosterol
14.	*Crotalaria medicaginea*	Plant, seed	β-sitosterol
15.	*Crotalaria spectabilis*	Seed	Stigmasterol
16.	*Crotalaria verrucosa*	Seed	β-sitosterol
17.	*Cullen corylifolium*	Root Seed	Coumesterol, sitosterol β-sitosterol, coumesterol, stigmasterol
18.	*Cytisus scoparius*	Seed oil	β-sitosterol, cholesterol, campesterol, stigmasterol
19.	*Dalbergia latifolia*	Bark	β-sitosterol, γ-sitosterol
20.	*Dalbergia sericea*	Stem bark	β-sitosterol
21.	*Dalbergia sissoides*	Wood	β-sitosterol
22.	*Dalbergia volubilis*	Twig Leaf	Sitosterol β-sitosterol
23.	*Derris robusta*	Seed	β-sitosterol
24.	*Derris scandens*	Plant	β-sitosterol
25.	*Derris trifoliata*	Plant	Stigmasterol
26.	*Desmodium elegans*	Leaf	Sitoseerol
27.	*Desmodium pulchellum*	Root	β-sitosterol
28.	*Desmodium triflorum*	Plant	Fucosterol
29.	*Erythrina arborescens*	Bark	Stigmasterol, sitosterol, campesterol
30.	*Erythrina blakei*	Plant	Sitosterol
31.	*Erythrina stricta*	Bark	Sitosterol, stigmasterol
32.	*Erythrina suberosa*	Bark Seed coat	β-sitosterol, cholesterol, β-sitosterol, campesterol, stigmasterol

Sl. No.	Family and Species	Location	Sterol
33.	*Erythrina variegata*	Bark	Campesterol, stigmasterol, β-sitosterol, γ-sitosterol
34.	*Flemingia chappar*	Plant, root	β-sitosterol
35.	*Flemingia macrophylla*	Root	β-sitosterol
36.	*Flemingia strobilifera*	Root	β-sitosterol
37.	*Glycine max*	Seed*, seed oil*	Campesterol, fucosterol, stigmasterol, β-sitosterol
38.	*Gueldenstaedtia multiflora*	Root	β-sitosterol
39.	*Indigofera heterantha*	Plant	β-sitosterol
40.	*Indigofera suffruticosa*	Plant	β-sitosterol
41.	*Indigofera trifoliata*	Plant	β-sitosterol
42.	*Lespeadeza juncea*	Plant	β-sitosterol
43.	*Lonchocarpus sericeus*	Leaf	β-sitosterol
44.	*Macrotyloma uniflorum*	Stem, leaf Seed*	Coumesterol β-sitosterol
45.	*Medicago sativa*	Root Leaf Seed oil	Coumesterol Coumesterol, stigmasterol, spinasterol β-sitosterol, coumesterol, campesterol, β-spinasterol
46.	*Melilotus indicus*	Leaf	β-sitosterol
47.	*Millettia leucantha*	Bark, pod	Stigmasterol, β-sitosterol
48.	*Mucuna pruriens*	Seed	Sitosterol
49.	*Phaseolus lunatus*	Leaf, seed	Coumesterol
50.	*Phaseolus vulgaris*	Pod*	β-sitosterol, stigmasterol
51.	*Physostigma venenosum*	Seed	β-sitosterol, stigmasterol
52.	*Pongamia pinnata*	Bark, heart wood Flower	β-sitosterol λ-sitosterol
53.	*Pterocarpus santalinus*	Bark Leaf	Sitosterol Stigmasterol, sitosterol
54.	*Pueraria tuberosa*	Root*	β-sitosterol, stigmasterol
55.	*Robinia pseudoacacia*	Root Flower	β-sitosterol, stigmasterol β-sitosterol
56.	*Sesbania sesban*	Leaf*, pod	Cholesterol, campesterol, β-sitosterol
57.	*Spartium junceum*	Flower	β-sitosterol
58.	*Spatholobus parviflorus*	Seed	β-sitosterol
59.	*Styphnolobium japonicum*	Plant	β-sitosterol
60.	*Tephrosia candida*	Seed	Sitosterol
61.	*Tephrosia purpurea*	Plant, seed*	β-sitosterol, spinasterol
62.	*Trifolium repens*	Plant	Coumesterol
63.	*Trigonella balanse*	Seed*	β-sitosterol, cholesterol

Sl. No.	Family and Species	Location	Sterol
64.	*Trigonella corniculata*	Seed*	Cholesterol
65.	*Trigonella foenum-graecum*	Plant*, leaf*, seed*	β-sitosterol, stigmasterol, spinasterol, cholesterol
66.	*Vicia angustifolia*	Seed	β-sitosterol, stigmasterol
67.	*Vicia faba*	Leaf Seed	Cholesterol, β-sitosterol, desmasterol, fucosterol β-sitosterol, campesterol, campesserol, cholesterol
68.	*Vicia sativa*	Leaf Seed	Sitosterol Cholesterol
69.	*Vigna radiata*	Leaf, seed*	Stigmasterol, campesterol, β-sitosterol
70.	*Vigna trilobata*	Seed*	β-sitosterol, stigmasterol
71.	*Wisteria sinensis*	Stem bark	β-sitosterol, protosterol
III	**Mimosaceae**		
1.	*Acacia auriculiformis*	Aerial parts	α-spinasterol
2.	*Acacia farnesiana*	Plant Leaf* Seed	α-spinasterol Cholesterol, campesterol, stigmasterol, sitosterol α-spinasterol
3.	*Acacia caesia*	Root	β-sitosterol, stigmasterol
4.	*Acacia lenticularis*	Plant	β-sitosterol
5.	*Acacia leucophloea*	Stem bark*, wood, flower, pod*	β-sitosterol
6.	*Acacia nilotica* ssp. *indica*	Root Heart wood, plant	β-sitosterol α-sitosterol, β-sitosterol
7.	*Acacia pennata*	Bark Stem	α-spinasterol β-sitosterol
8.	*Acacia senegal*	Root, wood, stem bark, flower, pod	β-sitosterol
9.	*Acacia sinuata*	Plant, stem bark, pod*, seed*	α-spinasterol, β-sitosterol
10.	*Acacia tortilis*	Heart wood	β-sitosterol
11.	*Adenanthera pavonina*	Root, bark, seed* Leaf	Stigmasterol Stigmasterol, β-sitosterol
12.	*Albizia amara*	Stem bark, wood, leaf, seed	β-sitosterol
13.	*Albizia chinensis*	Stem bark Heart wood Leaf	β-sitosterol α-spinasterol β-sitosterol
14.	*Albizia julibrissin*	Stem bark	α-spinasterol

Sl. No.	Family and Species	Location	Sterol
15.	*Albizia lebbeck*	Bark Root, leaf, flower, seed, pod, seed	β-sitosterol λ-sitosterol, β-sitosterol
16.	*Albizia lucidior*	Bark, seed*	β-sitosterol
17.	*Albizia procera*	Root, leaf Seed	α-spinasterol β-sitosterol
18.	*Dichrostachys cinerea*	Root Bark, heart wood	β-sitosterol Sitosterol
19.	*Entada gigas*	Plant, stem bark	α-spinasterol
		Seed	β-sitosterol
20.	*Entada rheedei*	Plant	α-spinasterol
		Seed*	β-sitosterol
21.	*Leucaena leucocephala*	Plant	β-sitosterol
22.	*Lysiloma latisiliquum*	Plant*	β-sitosterol
23.	*Mimosa pudica*	Plant	β-sitosterol
		Leaf	Stigmasterol
24.	*Mimosa rubicaulis*	Plant	β-sitosterol
25.	*Pithecellobium dulce*	Leaf, flower	α-spinasterol
26.	*Prosopis cineraria*	Plant, Stem Bark* Heart wood	β-sitosterol β-Sterol β-sitosterol, campesterol, stigmasterol
27.	*Prosopis glandulosa*	Plant	β-sitosterol
28.	*Samanea saman*	Stem bark	Phytosterol, α-spinasterol
29.	*Xylia xylocarpa*	Leaf	β-sitosterol

❑❑❑

CHAPTER-14

Saponins in Legumes

14.1 INTRODUCTION

Saponins are natural products with the properties of soaps and detergents. The saponin rich fruits of *Acacia sinuata* (the species which is more commonly, but erroneously, known as *Acacia concinna; shikakai;* Mimosaceae) and *Sapindus emarginatus* (*reetha),* have been in use as the traditional natural soap for centuries. Saponin containing plants, particularly the legumes, are also used as fish poison and as a soap substitute (see Chapter 3, Table 3.1). Saponins have now emerged as an important group of natural products with a wide range of applications in research, medicine and industry. The legumes are among the important saponin containing plants.

Detailed accounts on saponins can be obtained from Birk (1969), Gibbs (1974), Oakenfull and Sidhu (1989), Hostettmann *et al*., (1991), Kameswara Rao and Sangeetaa, (1993, 1997); Sangeetaa and Kameswara Rao (1993), Sangeetaa (1994) and Kameswara Rao (2000), among numerous other publications.

14.2 DISTRIBUTION OF SAPONINS

Saponins are predominantly plant products, although they also occur in a few animal species such as snakes. Saponins occur in over a 100 plant families and in several common species including many food and medicinal plants, and distributed in various tissues of plants such as roots, bulbs, stems, bark, leaves, blossoms, fruits and seeds (Kameswara Rao, 2000). They are also present in pollen and stigmas (Sangeetaa and Kameswara Rao, 1993; Sangeetaa, 1994).

The distribution and quantity of saponins in different tissues of different species of plants are dependent upon genetic and developmental factors. There is a decrease in saponin content during flowering and fruit bearing seasons. Environmental factors such as temperature and season also influence saponin distribution and content.

14.3 CHARACTERISTICS OF SAPONINS

Saponins are readily soluble in water, methanol and aqueous ethanol and are generally identified by their bitter taste, ability to foam in aqueous solutions and to lyse erythrocytes. However, saponins from all sources may not exhibit all these features. While some saponins do not foam, a few others do not lyse erythrocytes. Glycyrrhizin (liquorice), the saponin from the roots of *Glycyrrhiza glabra* and leaves of *Abrus precatorius* (both Fabaceae), is 60 times sweeter than cane sugar and is a non-calorie alternate sweetener and is used in medicine, particularly in cough remedies, for centuries.

14.4 CHEMISTRY OF SAPONINS

Some saponins such as glycyrrhizin are steroidal compounds (C_{27}) and those from most other sources are triterpenoidal compounds (C_{30}). More than one kind of saponin may occur in the same species. *Medicago sativa* contains at least five different saponins while soybean has four (Hostettmann *et al.*, 1991). Both triterpenoidal and steroidal saponins may occur in the same species.

In nature, saponins complex with simple sugars and occur as glycosides. On hydrolysis by acids, strong bases or enzymes, saponins yield the sapogenin (the aglycone, without the sugar molecule) and the carbohydrate moieties. Linkage of an aglycone to a sugar increases its solubility, but the sugar moiety is not essential for the biological activity of saponins. The diversity of saponins is based on both the aglycone and carbohydrate moieties.

14. 5 BIOSYNTHESIS OF SAPONINS

Saponins are synthesized by common metabolic pathway starting from acetyl coenzyme A. Mevalonic acid and squalene are the intermediary products for both the triterpenoidal and steroidal saponins. Up to the point of formation of squalene, the synthesis of steroidal compounds (including cholesterol) and saponins proceeds on the same lines. It is suggested that sapogenins are derived from medicagenic acid, the initial aglycone.

14.6 TOXICITY OF SAPONINS

Saponins are toxic to a variety of organisms, from bacteria to higher plants and animals. Species of the lower taxonomic groups and cold-blooded organisms, such as molluscs and fish, are more readily affected.

The toxic saponins are called sapotoxins. The most significant action of saponins is solubilization of biological membranes that disrupts the integrity of a cell. Populations of aquatic animals such as molluscs and fish, are controlled by using plants containing saponins (see Chapter 3, Section 3.2.18).

Saponins are also toxic to different degrees to mammals, particularly when introduced into the circulatory system, where they mainly act by solubilizing and thus destabilizing, the membrane systems of blood cells, blood vessels and cells of tissues.

Steroidal saponins have a natural affinity for cancer cell membranes and inhibit the growth of cancer at the cellular levels.

14.6.1 Effect of Saponins on Cell Membranes

The action of saponins on biological memberanes causes regular hexagonal holes bound by hexagonal arrays of saponins, cholesterol and phospholipid molecules (Hostettman *et al.*, 1991). The holes are formed by the selective removal of membrane components like lipids, proteins, cholesterol and water. Such an alteration of the membrane structure would increase cell permeability, allowing efflux of ions and influx of water, leading to cell lysis.

The characteristic haemolytic activity of saponins is due to their interaction with the constituents of the cell membrane. Membranes rich in cholesterol are much more susceptible to damage by saponins.

Saponins modify the permeability of the mucosal epithelial cells of the small intestine to allow entry of intact protein molecules and affect the absorption of drugs and other various compounds present in the digesta.

14.6.2 Trophic Effects

Long term, sustained heavy ingestion of saponins cause tropic effects on the physiology of the gastrointestinal tract. Several factors influence the trophic response of the mucosa, which include the physical nature of the digesta, enteric hormones, pancreatic and biliary secretions, mucosal blood flow, neural effects, gut microflora and permeability changes in the mucosal cell membrane.

14.7 SAPONINS IN LEGUMES

A large number of legume species contain saponins in different tissues in different concentrations (Table 14.1). Most legume seed saponins are chemically similar to soyasaponins (Table 14.2). Soyasapogenol B is more widely distributed among the legumes.

14.8 DIETARY SAPONINS

Many of the food plants contain considerable quantities of saponins and most of such dietary saponins are triterpenoids (Kameswara Rao, 2000). Legume seeds are among the major sources of dietary saponins (Box 14.1).

Box 14.1

Saponin content of some edible legumes

Cicer arietinum (chickpea)	60 g/kg
Glycine max (soybean)	56
Cyamopsis tetragonoloba (cluster bean)	50
Trigonella foenum-graecum (fenugreek)	28
Phaseolus vulgaris (beans)	21
Arachis hypogaea (peanuts)	16
Vigna radiata (mung bean)	06
Lens culinaris (lentil)	05

Saponins are also present in considerable quantities in other edible or medicinal legumes such as *Abrus precatorius*, *Glycyrrhiza glabra*, *Phaseolus coccineus*, *Phaseolus lunatus*, *Vicia faba*, *Vigna angularis*, and *Vigna mungo*.

Saponins are thermostable. Fermentation of saponins in cooked food may reduce their level. The best benefits from saponins are derived from fresh and raw or soaked or steam-cooked foods.

Vegetarian diet seems to have 10 per cent of saponins, most of which have positive health benefits. Dietary saponins are widely used as they are safe to the human system, when not used in excess. Many of the dietary saponins modulate the CAMP system in the plasma, brain, adipocytes and adrenal glands, which in turn affects the activity of hormonal messengers in the enterocrine cells. The hormonal messengers control absorption and secretion of water, electrolytes and nutrients, gastric emptying, control appetite, feed intake and body growth. These effects explain how dietary saponins influence the tissues and their long-term implications of their effects.

14.8.1 Fate of Ingested Saponins

The fate of ingested soybean saponins was studied in detail and the same seems to apply to other dietary saponins (Birk, 1969). A diet of soybean meal revealed the presence of saponins and not of sapogenins in the small intestine. In the caecum and colon sapogenins were found but not saponins. This shows that saponins are not hydrolyzed in the upper part of the digestive tract. They are hydrolyzed in the caecum and colon, not by our digestive enzymes but those secreted by micro-organisms (Birk, 1969). Neither saponins nor sapogenins were detected in the blood, which indicates that saponins pass intact through the gastro-intestinal tract and are not readily absorbed into the blood stream.

Massive doses of saponins possibly can cause intestinal lesions allowing saponins to enter the blood stream. The result is membrane destabilization, haemolysis, liver damage, respiratory failure, convulsion and coma (George, 1965; Martindale, 1972).

Improperly processed saponin containing foods, when taken in a large quantity, may cause stomach bloating, abdominal pain, digestive disturbances, vomiting and diarrhoea. Nevertheless, the toxicity of saponins to humans has rather been exaggerated. We regularly consume large quantities of saponins through food or herbal medicine, without any serious adverse effects.

Humans generally do not suffer severe poisoning from oral ingestion of saponins. Cholesterin in blood inactivates saponins considerably so that only outer mucous membranes are affected. On account of this property, saponins, particularly glycyrrhizin, are widely used in cough syrups and emetics to facilitate expectoration.

14.9 DUAL ROLE OF SAPONINS

Curiously, saponins have dual effects on organisms, that as being toxins and as eliminators of other toxins. A significant build of toxins in our bodies leads to various diseases. Consumption of food with saponins appears to inhibit the absorption of toxic tannin in other foods, by the gastro-intestinal tract. The saponins from oats and spinach increase and accelerate the body's ability to absorb calcium that has many physiological roles including reducing toxins in the system.

The so-called junk foods, combined with irregular food habits, affect the digestive processes resulting in a coating of different toxins in the digestive tract, hampering the absorption of nutrients. Saponins from chickpea and soybean and other vegetable sources have the ability to loosen and dissolve the hardened toxin from the colon wall and reduce toxicity. Consumption of saponins containing food by arthritic patients gave some relief from pain and swelling, particularly in those cases where arthritis was due to intestinal toxicity. The elimination of toxins from the body results in more vitality and energy and also strengthens the immune system.

14.9.1 Control of Serum Cholesterol by Dietary Legume Saponins

One of the potentially beneficial uses of saponins is their ability to lower the concentrations of cholesterol in blood plasma and organs like liver. This was observed over fifty years ago (Peterson, 1950) but somehow did not gain the recognition it deserves.

High serum cholesterol levels are associated with coronary heart disease, hypolipoproteinemia, hypothyroidism, nephrosis, diabetes, liver disease and gallstones. The general medical recommendation, the world over, is that serum cholestserol levels should be maintained at less than 200 mg/100 ml.

Bile acids are made from cholesterol in the liver and pass into the small intestine mixing with water to facilitate digestion. As this mixture enters the colon, the water is re-absorbed back into the body carrying with it the cholesterol and bile acids. Saponins bind to the cholesterol making it unavailable for re-absorption. Cholesterol is then passed on into the colon and eventually excreted. This forces the liver to produce more bile acid for which the liver must remove cholesterol from the blood stream, leaving less of it to build up in the arteries, thus reducing the risk of heart disease.

Cholesterol levels are clinically regulated by cholestyramine, a synthetic resin that binds with the bile acids, or by compactin (lovastatin) and similar synthetic compounds collectively called statins, which inhibit cholesterol synthesizing enzymes. A combination of the resin and the inhibitor can reduce cholesterol levels by 50 to 60 per cent but both have adverse effects. The human system is not accustomed to these substances that affect digestion and the absorption of many nutrients, and can cause abdominal discomfort and other side effects, none of which were reported with saponins, making them a very useful in the control of cholesterol.

Dietary tomatine (saponin of tomato) lowered plasma cholesterol and inhibited its absorption in rats by forming an insoluble complex in the gut. This increased faecal excretion of intestinal cholesterol occurring as bile acids thus reducing the plasma levels of cholesterol. Now there is conclusive evidence that many dietary saponins interact with cholesterol and interfere with its reabsorption.

A physical property of saponins that is of particular significance is their interaction with cholesterol to form mixed micelles with bile acids. Once micelles are formed, cholesterol cannot be absorbed by the intestine, but has to be eliminated through defaecation from the system. As cholesterol/bile acids are not freely available for absorption and recycling, serum cholesterol levels drop. When chicken feed included saponin-containing alfalfa, faecal excretion of cholestrol increased, its level in plasma decreased along with the reduction in amount of cholesterol deposition in eggs.

Saponins from chickpea, soybean and some other legumes lowered plasma cholesterol, plasma triglycerides and blood pressure, in human subjects. There are differences in the efficiency of different saponins. Some saponins like those in soybean influence more of faecal excretion of bile acids than cholesterol from the intestine. Saponins from the soap bark tree (*Quillaia saponaria*) had the opposite effect. Interestingly, in humans with normal cholesterol levels, saponins did increase faecal excretion but the serum cholesterol levels were not significantly lowered, probably because the levels were made up by endogenous synthesis of cholesterol. This indicates that saponin intake would not lower cholesterol to dangerous levels. (Hostettiman *et al.*, 1991)

14.9.2 Advantages of Dietary Saponins in the Control of Serum Cholesterol

The human system is generally adjusted to the effects of saponins in raw or cooked plant foods. For ages people have been eating salads of soaked or sprouted gram, raw tomato, beetroot, onion, cabbage, etc. Oat porridge and Asparagus soup are also commonly consumed. All these contain appreciable amounts of saponins. Hence, the use of dietary saponins to regulate cholesterol and bile acids is without the complications that synthetic chemical substances produce, to which the human system is not accustomed. Most important sources of dietary saponins are commonly available legumes (Box 14.1; Table 14.2), used daily and do not result in additional effort or expenditure.

Saponins are extractable in boiling water. So long as the water in which the vegetables, pulses, etc., were cooked is not thrown out, preparation of food by boiling will not affect their activity. Since saponins react with cooking oil, deep and dry frying saponin containing food articles may defeat the purpose of their use in the control of serum cholesterol.

Dietary saponins can be administered without pharmaceutical processing. Even if extraction and processing is required, the process is simple. Dosage control is easy. Saponins are active at different temperatures. They are not affected by the acidic pH of the stomach and so are stable till reaching the intestine where they complex with cholesterol. If a non-edible source of saponins turns out to be more effective in controlling serum cholesterol, the saponin can be administered in a capsule. Legume saponins offer a great promise on all these counts.

14.10 BIOACTIVITY OF SAPONINS

Saponins are extraordinarily physiologically active compounds. The physiological activity of saponins appears to arise from their ability to interact with the membranes of mucosal cells, causing profound changes in the membrane biochemistry and loss of activity of membrane bound enzymes. Saponins exert powerful physical interactions with other components of the digesta and cause various effects on the human system. The bioactive potential of some legume saponins is given in Table 14.3.

14.11 APPLICATIONS OF SAPONINS

14.11.1 Traditional Uses

One of the most ancient uses of saponins is as a cleansing agent. The common soapnut *Acacia sinuata* and *Sapindus emarginatus* are minor forest products abundant in saponins, which have traditionally been used to wash hair and delicate cloth.

Liquorice and fenugreek saponins are used as flavouring herbs or spices in foods. Some 'health foods' in Japan contain saponin as ingredients. Saponins from *Glycyrrhiza glabra* are used as food flavouring agents in Australia, Britain and US. They are used as emulsifiers in the preparation of *halvah* and confectionar, and also as antimicrobial agents in pickles.

14.11.2 Industrial Applications

Saponins are used in the industry in several manufacturing processes such as a) in cosmetics (Osaka and Kenkyusho, 1984; Tomohisa and Shintara, 1984); b) in cigarette filters as impregnating agents; c) as stabilizers in soft drinks, beer, and confectionary (Rocher, 1965); d) in soaps and shampoos as the main ingredients; e) as food flavours (George, 1965); f) as emulsifiers, g) as antimicrobial agents; h) as antioxidants in manufactured and canned food items; i) as surfactants; j) in Dizzo copying materials to improve spreading and stability of coating layers; k) in biological waste treatment plants to improve oxygen transfer efficiency; l) in fire extinguishers, photographic emulsions and preparation of light weight cement.

Saponins are used as foaming agents in fire fighting fluids and in the production of lightweight concrete (Banks and Mumby, 1963) and are used to form permanent suspensions with oil or powders (Rocher, 1965; Birk, 1969).

Steroidal saponins such as diosgenin, smilagenin and glycyrrhizin (from the species of *Dioscorea*, *Smilax* and *Glycyrrhiza* respectively) are used as precursors in the industrial synthesis of steroidal hormones like cortisone (Smith, 1976; Hostettmann *et al.*, 1991).

14.11.3 Saponins in Research

Saponins are useful as buffers (Arshad *et al.*, 1965) and they also increase cell hybridization sensitivity (Yamawaki *et al.*, 1993). Treatment of tissue with saponin allows good preservation of tissue morphology, as determined by differential interference and contrast enhanced video microscopy.

Saponins of *Acacia sinuata* are currently promoted as human-safe immunological adjuvants to elicit higher titres of antibodies (Wu *et al.*, 1992; Kensil *et al.*, 1995).

14.12 THERAPEUTIC USES OF SAPONINS

14.12.1 Traditional Medicine

Several species rich in saponins have been a part of many formulations in *Ayurveda* and other alternative systems of medicine. The active principles in some of the therapeutically important plants like ginseng, liquorice, yams, digitalis, sarasparilla, etc., are saponins. Glycyrrhizin from *Glycyrrhiza glabra* was used in Chinese medicine for treating throat inflammations, to increase physical strength and to cure wounds.

Saponins in the much valued Indian medicinal plant, brahmi (*Centella asiatica* Apiaceae) or neerbrahmi (*Bacopa monnieri,* Scrophulariaceae) are reputed to enhance memory. *Dodonaea viscosa* (Sapindaceae), a plant rich in saponins, is used by bone setters and is believed to have the effects of anabolic steroids.

14.12.2 Modern Medicine

There are number of applications for saponins in medicine, though not all of them are derived from legume sources. Since the effects of all triterpenoidal or of all steroidal saponins are generally similar within each group, the most easily available and economical sources are normally utilized.

The triterpenoidal saponins, which have a potential for direct pharmaceutical use. The steroidal saponins are more useful as starting material for chemical synthesis of steroid hormones for therapeutic use.

Saponins are used as antiinflammatory, antidiuretic, antipyretic and analgesic agents, as CNS depressants and in treatment of ulcers. These effects are related to stimulatory effect of saponins on the hypothalamus or pituitary glands and synthesis of prostaglandins in the intestinal mucosa.

Saponins from *Glycyrrhiza* are used as immunosuppressants and as expectorants to promote mucus secretion from the respiratory tract and as immunosuppressants directed towards β-cell function and in ointments to treat hemorrhoids and venous congestion.

Crude saponins from *Albizia lebbeck* has been shown to protect mast cells from antigenic shock. Soy saponins counteract the effects of endotoxins or thrombin by inhibiting the formation of fibrin thrombi. Soy saponins have been used to prevent liver damage from consumption of peroxidized corn oil.

The demonstrated beneficial effects of saponins are given in Table 14.3.

Table 14.1

Legume Sources of Saponins

The parts containing detectable quantities of saponins are indicated. Other parts might also contain saponins, but were not in detectable quantities. Information on qualitative and quantitative variation is not available.

Sl. No.	Species	Parts containing saponins
I	**CAESALPINIACEAE**	
1.	*Bauhinia purpurea*	Leaf, seed
2.	*Brownea ariza*	Bark, flower
3.	*Caesalpinia bonduc*	Leaf, seed
4.	*Caesalpinia crista*	Leaf, seed
5.	*Caesalpinia pulcherrima*	Leaf, seed
6.	*Cassia fistula*	Bark, leaf, seed
7.	*Cassia grandis*	Leaf, seed
8.	*Cassia javanica*	Leaf, flower, seed
9.	*Cassia renigera*	Leaf, seed
10.	*Cassia roxburghii*	Leaf, seed
11.	*Chamaecrista absus*	Leaf, seed
12.	*Chamaecrista mimosoides*	Seed
13.	*Gymnocladus dioica*	Leaf, seed
14.	*Peltophorum pterocarpum*	Seed
15.	*Senna alata*	Leaf, seed
16.	*Senna auriculata*	Leaf, seed, pollen
17.	*Senna bicapsularis*	Leaf, seed
18.	*Senna hirsuta*	Leaf, seed
19.	*Senna occidentalis*	Leaf, seed
20.	*Senna siamea*	Leaf, seed
21.	*Senna sophera*	Leaf, seed
22.	*Senna spectabilis*	Seed
23.	*Senna tora*	Leaf
24.	*Senna surattensis*	Pollen
II	**FABACEAE**	
1.	*Abrus precatorius*	Root
2.	*Abrus pulchellus*	Root
3.	*Andira inermis*	Leaf

Sl. No.	Species	Parts containing saponins
4.	*Aeschynomene indica*	Leaf, seed
5.	*Alysicarpus monilifer*	Leaf, seed
6.	*Arachis hypogaea*	Seed
7.	*Astragalus hamosus*	Root
8.	*Butea monosperma*	Bark, leaf
9.	*Canavalia ensiformis*	Leaf, seed
10.	*Castanospermum australe*	Wood
11.	*Cicer arietinum*	Seed, sprout
12.	*Clitoria ternatea*	Root
13.	*Crotalaria incana*	Leaf, seed
14.	*Crotalaria juncea*	Leaf, seed
15.	*Crotalaria pallida*	Leaf, seed
16.	*Crotalaria retusa*	Leaf, seed
17.	*Crotalaria spectabilis*	Leaf, seed
18.	*Crotalaria verrucosa*	Pollen
19.	*Cyamopsis tetragonoloba*	Seed
20.	*Cytisus scoparius*	Stem, leaf, pod, seed
21.	*Dalbergia lanceolaria*	Pod
22.	*Dalbergia stipulacea*	Bark, leaf
23.	*Derris cuneifolia*	Leaf, seed
24.	*Derris elegans*	Leaf, seed
25.	*Derris elliptica*	Leaf, seed
26.	*Derris ferruginea*	Leaf, seed
27.	*Derris malaccensis*	Leaf, seed
28.	*Derris scandens*	Leaf, seed
29.	*Derris trifoliata*	Leaf, seed
30.	*Desmodium gangeticum*	Leaf, seed
31.	*Desmodium oojeinense*	Leaf, seed
32.	*Endosamara racemosa*	Leaf
33.	*Erythrina indica*	Leaf
34.	*Erythrina suberosa*	Bark
35.	*Erythrina subumbrans*	Wood
36.	*Erythrina variegata*	Bark, leaf, pollen
37.	*Gliricidia sepium*	Bark, leaf, pod, seed
38.	*Glycine max*	Seed
39.	*Glycyrrhiza glabra*	Root
40.	*Indigofera hendecaphylla*	Seed
41.	*Indigofera galegoides*	Seed
42.	*Indigofera glandulosa*	Seed
43.	*Indigofera hirsuta*	Seed

Sl. No.	Species	Parts containing saponins
44.	*Indigofera linnei*	Seed
45.	*Indigofera zollingeriana*	Seed
46.	*Lathyrus odoratus*	Leaf, pod, seed
47.	*Lathyrus sativus*	Leaf, pod, seed
48.	*Lens culinaris*	Seed, sprout
49.	*Lonchocarpus sericeus*	Bark, leaf, pod, seed
50.	*Lotus corniculatus*	Leaf
51.	*Lupinus luteus*	Shoot, seed coat
52.	*Lupinus mutabilis*	Shoot
53.	*Medicago polymorpha*	Root, stem, leaf, flower
54.	*Medicago sativa*	Root, stem, leaf, flower
55.	*Melilotus albus*	Shoot
56.	*Melilotus officinalis*	Shoot
57.	*Millettia extensa*	Bark, leaf
58.	*Millettia platycarpa*	Bark, leaf
59.	*Millettia piscida*	Bark, leaf
60.	*Millettia thonningii*	Bark. leaf
61.	*Mucuna gigantea*	Leaf, pod, seed
62.	*Mucuna pruriens*	Leaf, pod, seed
63.	*Mundulea sericea*	Leaf, seed
64.	*Pachyrrhizus erosus*	Root, leaf
65.	*Pachyrrhizus tuberosus*	Root, leaf
66.	*Phaseolus coccineus*	Seed
67.	*Phaseolus lunatus*	Seed
68.	*Phaseolus vulgaris*	Seed, sprout
69.	*Physostigma venenosum*	Leaf
70.	*Piscidia piscipula*	Leaf, pod, seed
71.	*Pisum sativum*	Seed
72.	*Pongamia pinnata*	Bark, leaf, pod, seed
73.	*Pueraria montana*	Root
74.	*Rhynchosia minima*	Leaf
75.	*Robinia pseudoacacia*	Bark, leaf, seed
76.	*Securigera securidaca*	Leaf, seed
77.	*Sesbania bispinosa*	Pod, seed
78.	*Sesbania grandiflora*	Stem, leaf, seed
79.	*Sesbania sesban*	Seed
80.	*Sophora mollis*	Leaf
81.	*Sophora secundiflora*	Leaf, seed
82.	*Sophora tomentosa*	Leaf, pod, seed
83.	*Spartium junceum*	Petal

Sl. No.	Species	Parts containing saponins
84.	*Spatholobus parviflorus*	Leaf
85.	*Styphnolobium japonicum*	Leaf, seed
86.	*Tephrosia candida*	Leaf, pod, seed
87.	*Tephrosia grandiflora*	Leaf
88.	*Tephrosia noctiflora*	Leaf, seed
89.	*Tephrosia purpurea*	Root, leaf, seed
90.	*Tephrosia sinapou*	Leaf
91.	*Tephrosia subtriflora*	Leaf
92.	*Tephrosia uniflora*	Leaf, seed
93.	*Tephrosia villosa*	Leaf, seed
94.	*Tephrosia vogelii*	Leaf, seed
95.	*Teramnus labialis*	Leaf, seed
96.	*Trifolium fragiferum*	Leaf, seed
97.	*Trifolium pratense*	Shoot
98.	*Trifolium repens*	Leaf, shoot
99.	*Trigonella balansae*	Seed
100.	*Trigonella foenum-graecum*	Leaf, seed
101.	*Uraria crinita*	Leaf, seed
102.	*Vicia faba*	Seed
103.	*Vicia sativa*	Leaf, seed
104.	*Vigna angularis*	Seed
105.	*Vigna mungo*	Seed
106.	*Vigna radiata*	Seed, sprout
107.	*Vigna unguiculata*	Leaf, seed
108.	*Wisteria sinensis*	Leaf, seed
III	**MIMOSACEAE**	
1.	*Acacia auriculiformis*	Leaf, pod, seed
2.	*Acacia caesia*	Leaf, pod, seed
3.	*Acacia farnesiana*	Leaf, pod, seed
4.	*Acacia leucophloea*	Leaf, pod, seed
5.	*Acacia pennata*	Leaf, pod, seed
6.	*Acacia sinuata*	Leaf, pod, seed
7.	*Adenanthera pavonina*	Bark, leaf, seed
8.	*Albizia adianthifolia*	Bark, leaf, seed
9.	*Albizia amara*	Leaf, seed
10.	*Albizia chinensis*	Leaf, seed
11.	*Albizia falcataria*	Leaf, seed
12.	*Albizia gummifera*	Leaf, pod, seed
13.	*Albizia julibrissin*	Leaf, pod, seed
14.	*Albizia lebbeck*	Bark, shoot, flower, pod, seed
15.	*Albizia lucidior*	Leaf, seed
16.	*Albizia odoratissima*	Leaf, seed
17.	*Albizia procera*	Leaf, seed
18.	*Entada rheedei*	Leaf, seed

Sl. No.	Species	Parts containing saponins
19.	*Enterolobium cyclocarpum*	Leaf
20.	*Faidherbia albida*	Leaf, seed
21.	*Lysiloma latisiliquum*	Bark, leaf, pod, seed
22.	*Mimosa pudica*	Leaf, seed
23.	*Pithecellobium dulce*	Fruit, seed
24.	*Prosopis juliflora*	Bark, leaf
25.	*Prosopis pubescens*	Bark, leaf
26.	*Samanea saman*	Bark, leaf, pod, seed

Source : Kameswara Rao, 2000

Table 14.2
Saponins in the Seeds of Common Edible Legumes

Sl. No.	Species and common name	Saponins
1.	*Arachis hypogaea* (Peanut)	Soysaponin I (four aglycones were isolated, one of them is indentical to soysapogenol B)
2.	*Cicer arietinum* (Chickpea)	Soysaponin I (soysapogenol B is aglycone)
3.	*Glycine max* (Soybean)	Soysaponins A1 and A2 (Soysapogenol A as aglycone); Soysaponins I, II and III (soysapogenol B as aglycone); Soysaponin IV
4.	*Glycyrrhiza glabra* (Liquorice)	Glycyrrhetinic acid is the major sapogenin and many minor ones were also isolated
5.	*Lens culinaris* (Lentil)	Soysaponin I (soysapogenol B is aglycone)
6.	*Medicago sativa* (Alfalfa)	A complex mixture of 33 saponins, the actual number varying from cultivar to cultivar; hederagenin and medicagenic acid are aglycones; soysapogenols A-F are minor compounds
7.	*Phaseolus coccineus* (Scarlet runner bean)	Soysaponin I (soysapogenols B and C are sapogenols)
8.	*Phaseolus lunatus* (Lima bean)	Soysaponin I (soysapogenols B and C are sapogenols)
9.	*Phaseolus vulgaris* (Kidney bean)	Soysaponin I (soysapogenols B and C are sapogenols)
10.	*Pisum sativum* (Pea)	Soysaponin I (soysapogenol B is aglycone)
11.	*Trigonella foenum-graecum* (Fenugreek)	Five saponins; fenugrins A to E (major aglycone is diosgenin)
12.	*Vicia faba* (Faba bean)	Soysaponin I (soysapogenol B is aglycone)
13.	*Vigna angularis* (Azuki bean)	The saponin contains four sapogenols, soysapogenol B, gypsogenic acid, sophoradiol, and azukisapogenol
14.	*Vigna radiata* (Green gram)	Soysaponin I (soysapogenols B and C)

Table 14.3
Bioactivity of some Legume Saponins

Sl. No.	Species	Saponins
1.	*Acacia sinuata*	Saponins from *Acacia sinuata* were demonstrated to be efficient and human-safe immunological adjuvants to elicit higher titres of antibodies (Sangeetaa, 1994; Kameswara Rao and Sangeetaa, 1997).
2.	*Albizia lebbeck*	Saponins from *Albizia lebbeck* protects mast cells against antigenic shock (Johri *et al.*, 1985).
3.	*Glycine max*	Soya saponins have anti-oxidant property, retard aging and prevent liver injury caused by peroxidized fats (Ominami *et al.*, 1984; Takashi *et al.*, 1986). The water soluble soya saponins protected mouse fibroblast cell from hydrogen peroxide damage (Yoshikoshi *et al.*, 1996). Physiological functions such as hypocholesterolemic (Oakenful, 1981), anti-oxidant (Ohminami *et al.*, 1984), anti-tumour promoting (Konoshima *et al.*, 1991,1992) and HIV infection inhibitory properties (Nakashima *et al.*, 1989) of soya saponins were reported. Hiai (1985) showed that soya saponins can stimulate production and release of neuropeptide hormones from the hypothalamus or pituitary. Soya saponin II inhibits human cyto-megalo-virus, influenza virus and HIV type I and herpes simplex virus type I (Hayashi *et al.*, 1997). Soya saponins have anti-nephritic and hepatoprotective activity (Ikeda *et al.*, 1998); they are known to prevent liver injury and hyperlipemia induced by highly peroxidized fat diet (Ohminami *et al.*, 1984, Takashi *et al.*, 1986). Soya saponin counteract the effect of endotoxins by inhibiting the formation of fibrin thrombi and increases fibrin degradation (Kereselidze, *et al.*, 1975). Soya saponins influence more of faecal excretion of bile acids than cholesterol from the intestine. Soya saponins are to a certain extent responsible for the inhibition of proteinase and cholinesterases activity of RBCs (Ishaaya and Birk, 1968) but their activity is counteracted by proteins which are present on the gut. Plasma and constituents of red blood cells inhibit the haemolytic activity of soya saponins.
4.	*Glycyrrhiza glabra*	Glycyrrhizin is used to treat gastric ulcer, Addison's disease (Conn *et al.*, 1968; Shibata, 1972) and is antiviral (Pompei *et al.*, 1979). In Chinese medicine,

Sl. No.	Species	Saponins
		licorice is administered to treat throat inflammation, to increase physical strength, to strengthen muscles and bone and to cure wounds. In western medicine liquorice has been used since the ancient Grecian age as expectorant and anti-tussive agent. Glycyrrhizin is hypertensive (Essman, 1984), immunosuppressive (Hikokichi *et al.,* 1973), anti-inflammatory, anti-pyretic, and analgesic (Shibata, 1972; Lacaille-Dubois and Wagher, 1996); clinical evidence indicates that liquorice could be used to treat gastric ulcers (Otttenjann and Rosch 1970). The side effects of glycyrrhizin are oedema and hypertension (Revers, 1946). Saponins from *Glycyrrhiza glabra* potentiate the action of gluco-corticoids, which is related to inactivation of hormones by certain liver enzymes.
5.	***Medicago sativa*** (Alfalfa)	Alfalfa saponins inhibit α-chymotrypsin and proteases, which are the digestive enzymes. They are partly responsible for depressing effect on the growth, diet consumption and diet utilization (Heywang and Bird, 1954). But these effects can be partly reversed by cholesterol or β-sitosterol (Peterson, 1950). Alfalfa saponins inhibit respiration in the rat diaphragm by inhibiting succinate oxidase in the liver and also inhibit trypsin and chymotrypsin.
6.	***Phaseolus vulgaris***	The saponins in ethanolic crude extract of *Phaseolus vulgaris* exhibited antifertility activity (Sun *et al.*, 1983).
7.	***Pueraria montana*** **var. *lobata***	Saponins from the root have hepatoprotective action (Arao *et al.*, 1998).

CHAPTER-15

Legumes with Antimicrobial and Pesticidal Activity

15.1 INTRODUCTION

Viruses, bacteria and fungi are an integral component of our environment and our lives. A number of these microbes are essential for our survival, but many of them also cause various diseases and disorders in humans, animals and crop plants.

Infectious diseases cause 25 per cent of all human deaths. In the low income countries 45 per cent of all deaths are due to microbial infections. Worldwide 65 per cent of child deaths are due to microbes. Microbial infections also cause the loss of millions of man-hours every year.

All plants, especially the crops, are susceptible to several diseases caused by microbes. Insect and nematodal pests infest our crops. Our farm and feed animals, and pets also suffer from various infections.

15.2 PLANT DERIVED CHEMICALS IN PATHOGEN AND PEST CONTROL

Man has sought to control pathogens and pests by various means. The use of plant products being the most important of these strategies. Plants synthesize several compounds that are deleterious to viruses, bacteria, fungi or other pathogens and pests. In the advanced countries plants constitute over 25 per cent of pest and pathogen control strategies, while in the developing countries it is over 75 per cent. Legumes constitute an important component of plant based pest and pathogen control.

15.3 ANTIMICROBIAL POTENTIAL OF LEGUMES

The antimicrobial potential of legumes has been used in different systems of human and veterinary medicine for ages (Kameswara Rao, 2000). Databases of plants with a) antimicrobial activity (Sathynarayana Bhat, 1993), and b) plants used in gastro-intestinal disorders (Sharon, 1994) discussed a large number of plants with antimicrobial activity. The legume species with antiviral, antibacterial and antifungal acivity are given in Table 15.1. The profiles of legume species in this book cite the cases of antimicrobial chemical compounds in legumes. Legumes have a great potential in the biocontrol of microbial diseases in plants, animals and humans and deserve to be studied further.

15.4 CROP LOSSES DUE TO PESTS AND DISEASES

Annual worldwide crop losses due to pests were estimated to be about 70 per cent of crop production, amounting to US $ 400 billion. In spite of using pesticides, the worldwide pre-harvest losses due to insect pests are 15 per cent of the total production amounting to over US $ 100 billion. The annual cost of insect control itself amounts to US $ 8 billion. It is essential that we prevent such huge loss of our agricultural produce.

15.5 SYNTHETIC PESTICIDES

The term pesticide is now applied collectively to all substances and measures that are applied to control all pathogens and pests. However, the specific means and measures against each of them are necessarily different due to biological factors inherent in different organisms.

Commercial pesticides are synthetic chemicals. Though certain chemicals are recommended against certain specific pests, chemical pesticides are not really target oriented. They affect the non-target organisms altering the biodiversity component. Several beneficial organisms like bees and butterflies, which help in pollination of the crops, are at a serious risk. There are some predator insects that feed on crop pests and these beneficial insects are also affected. Birds that contribute to certain degree of insect pest control by feeding on them, are affected too. Use of synthetic pesticides disturbs the natural balance between pests and their predators and is usually in favour of the pests (Metcalf, 1986).

The exclusive and excessive use of chemical pesticides also results in a rapid build-up of resistance by the pests to such compounds and this phenomenon is called acquired resistance, which makes pesticide application almost useless after a period of time.

Farm workers who handle and apply pesticides suffer from chemical poisoning, resulting in several deaths each year. Excessive use of pesticides causes soil contamination by the chemicals, which in course of time makes the soils unfit for cultivation. Pesticides that are being sprayed are carried by air drift to neighbouring crop fields, water bodies or other areas, resulting in an unintended exposure. More importantly, chemical residues in the food from crop plants treated with synthetic pesticides get into the food chain and seriously affect the health of the consumers.

In spite of the inherent drawbacks, synthetic chemical pesticides have dominated agricultural practices, during the past several decades.

In view of all these problems, scientists have sought alternatives to the use of chemical pesticides. In this context, pesticides from biological sources have gained great importance. This area of science uses entire organisms or their products. Such pesticides are called biological pesticides or biopesticides and the approach is biological control or biocontrol.

15.6 INTEGRATED PEST MANAGEMENT

Pests can be controlled more effectively, without the inherent risks associated with chemical pesticides, by breeding pest resistant varieties. Other contributory control measures are crop rotation, mixed crop cultivation, crop field sanitation and more importantly pesticides from biological sources. A modern strategy that judiciously combines all these approaches, seeking to minimize the use of chemical pesticides, is called Integrated Pest Management (IPM). Organic Farming is heavily dependent upon IPM.

15.7 SECONDARY METABOLITES OF PLANTS AS PESTICIDES

Plants synthesize several secondary metabolites of the chemical groups of glycosides, tannins, terpenoids, saponins, lignans, lactones, alkaloids and even proteins like lectins, to protect them from pests and diseases. Such chemical compounds are reasonably pest specific and being biodegradable natural products their side effects are negligible. We also have a wide choice in using them to control a variety of crop pests. Species containing such compounds have been used as sources of natural pesticides for a very long time even before the advent of chemical pesticides. Among the most potent insecticides of plant origin are pyrethrum from the members of the Asteraceae and rotenone from members of the Fabaceae. In recent times, the demand for natural pesticides has grown enormously with the global enthusiasm for Organic Farming, which prohibits the use of chemical inputs as fertilizers and pesticides, except for a few classical practices like using Bordeaux mixture that contains copper sulphate.

15.8 PESTICIDAL LEGUMES

Many phytochemicals in legumes are used as pesticides. Rotenone, a commonly used potent insecticidal flavonoid, is usually extracted from the legumes *Derris elliptica* and *Tephrosia purpurea*. Species such as *Albizia chinensis, Amorpha fruticosa, Astragalus gummifer, Castanospermum australe* and species of *Derris, Delonix, Lonchocarpus, Sophora* and *Tephrosia* are the important legume sources of pesticides (Kochaar, 1981; Hill, 1989). Farmers use water extractions or decoctions of such plants and spray them on the crops to control a variety of pests.

Among the legumes in India, the Fabaceae contain the largest percentage of species with pesticidal effects (9.9 per cent), while 3.2 per cent of the Caesalpiniaceae and 5.5 per cent of the Mimosaceae are put to such use Chapter 3, (Table 3.5). The Indian legume species with pesticidal or insecticidal activity are given in Chapter 3, (Tables 3.1, 3.2 and 3.3). The more commonly used species are given in Table 15.2.

15.9 MODERN APPROACH TO PEST CONTROL

In recent times the major innovation in the area of biocontrol of pests is the production of transgenic varieties by genetic engineering. This involves identification of compounds with a very specific action against a targeted pest, isolation of the genes responsible for the synthesis of these compounds and the production of genetically engineered crop varieties that contain the chosen genes to produce the respective compounds affording an inbuilt and effective non-synthetic chemical control. Such transgenic varieties could be an important component of IPM and make a significant contribution to sustainable agriculture (Boulter, 1993). Three of the more promising modern approaches to pest control, involving legumes are discussed here.

15.9.1 Lectins

In view of their diverse applications, lectins are treated in a separate Chapter 9. Only issues relevant to pesticidal effects are briefly mentioned here.

The lectin from the seeds of *Vicia cracca* inhibit the growth of soil micro-organisms that metabolize seed coats of this species. Chitin, an important constituent of many fungal walls, is a polymer of N-acetylglucosamine. Lectins specific to this sugar bind to hyphal tips of fungi such as *Trichoderma viride* and inhibit their growth (Mirelman *et al.*, 1975). Lectins from various legumes showed specificity to saprophytic pseudomonad bacteria and controlled their proliferation (Jasalavich and Anderson, 1981). Several legume species synthesize lectins that control bruchlid insect pests. Over time, a number of legume lectin sources, with inhibitory effects on different groups of pathogens and pests have been identified. Genes for the synthesis of such lectins from legumes like *Glycine max* are being introduced into crop plants, so that such transgenic varieties will resist the corresponding pest, just as the genes from *Bacillus thuringiensis* impart pest resistance in Bt varieties of several crop plants.

15.9.2 Polygalacturonase-Inhibiting Proteins (PGIPs)

Mansfield (1983) observed that certain proteins in plant cell walls inhibited polygalacturonases produced by some pathogenic fungi. The proteins were not exactly toxic to the invading fungi but prevented them from establishing a nutritional relationship with the host tissues.

Polygalacturonase is an enzyme secreted by the fungal pathogens to degrade pectin in the host cell walls (Bergey *et al.*, 1999), in order to gain entry into the host cells. Some host species have developed proteins that inhibit polygalacturonase. PGIPs from *Glycine max* inhibit the fungal pathogens *Solanum sclerotium* and *Aspergillus niger* and also the nematode

Heterodora glycines. PGIPs from *Phaseolus vulgaris* inhibit the pathogenic fungi *Aspergillus niger*, *Fusarium moniliforme* and those from *Pisum sativum* inhibit *Ascochyta pisii*.

The genes for PGIPs have been isolated from the legume sources and are now being introduced into other crop plants that are susceptible to these fungi and nematodes, to produce transgenic varieties resistant to them (Gomathy and Gnanamanickam, 2004).

15.9.3 Inhibitors of Proteinase

Inhibitors of proteinase were discussed in Chapter 7 (Section 7.2.5), in connection with their effects on human nutrition. Since several aspects of the action of proteinase inhibitors on pests overlap with the issues of human nutrition, Section 7.2.4 needs to be read in conjunction with the following account. For more details, the comprehensive review on plant proteinase inhibitors in the control of phytophagous insects by Lawrence and Koundal (2002) should be consulted.

Proteinase inhibitors were known for a long time. Mickel and Standish (1947) studied the possible role of proteinase inhibitors in plant protection, when it was observed that the larvae of certain insects were unable to develop normally on soybean products. The trypsin inhibitors present in soybean were shown to be toxic to the larvae of flour beetle, *Tribolium confusum*. Subsequently, several non-homologous families of proteinase inhibitors are recognized among the animals, micro-organisms and plant kingdom.

There have been many examples of proteinase inhibitors active against certain insect species, both in *in vitro* assays against insect gut proteases (Pannetier *et al.,* 1997; Koiwa *et al.*, 1998) and in *in vivo* artificial diet bioassays (Urwin *et al.*, 1997; Vain *et al.*, 1997).

Majority of proteinase inhibitors studied in plant kingdom originate from the plant families namely the Leguminosae, Solanaceae and Poaceae (Richardson, 1991). Currently proteinase inhibitors from potato, tomato, sunflower, and several legumes are under serious investigation, for possible application in pest control.

Proteinase inhibitors from plants exhibit a very broad spectrum of bioactivity. They suppress pathogenic nematodes like *Globodera tabaccum*, *Globodera pallida*, and *Meloidogyne incognita* (Williamson and Hussey, 1996). Trypsin inhibitors from *Vigna unguiculata* controlled several insect pests (Hilder, *et al.*, 1987). Spore germination and mycelial growth of *Alternaria alternata* were inhibited by buckwheat trypsin or chymotrypsin inhibitors (Dunaevskii *et al.*, 1997). Cysteine proteinase inhibitors from pearl millet inhibited the growth of many pathogenic fungi including *Trichoderma reesei* (Joshi *et al.*, 1998).

Such a broad range of activity makes proteinase inhibitors an ideal choice to be used in developing transgenic crops resistant to insect pests. The first transgenic was a variety of tobacco, with a trypsin inhibitor gene from *Vigna unguiculata,* which conferred resistance to a wide range of insect pests including lepidopterans, such as *Heliothis* and *Spodoptera*, coleopterans such as *Diabrotica*, *Anthonomnous* and orthoptera such as locusts (Hilder, *et al.*, 1987).

Proteinase inhibitors have become attractive because of their wide range of target of pests, and the absence of evidence that they have serious toxic or deleterious effects on mammals. Many of these proteinase inhibitors are rich in cysteine and lysine, contributing to better and enhanced nutritional quality (Ryan, 1989). Further, transformation of plant genomes with proteinase inhibitor-encoding cDNA clones appears attractive not only for the control of plant pests and pathogens, but also as a means to produce proteinase inhibitors, useful in medicine and for the use of plants as factories for the production of heterologous proteins (Sardana *et al.*, 1998).

Table 15.1

Indian Legumes with Antimicrobial Activity

Sl. No.	Species	Anti-Bacterial	Antifungal	Antiviral
I	**CAESALPINIACEAE**			
1.	*Bauhinia accuminata*	+	+	+
2.	*Bauhinia malabarica*	+	-	+
3.	*Bauhinia purpurea*	+	-	+
4.	*Bauhinia racemosa*	+	+	+
5.	*Bauhinia tomentosa*	+	-	+
6.	*Bauhinia vahlii*	+	-	-
7.	*Bauhinia variegata*	+	+	+
8.	*Bauhinia wallichii*	+	+	-
9.	*Caesalpinia bonduc*	+	-	+
10.	*Caesalpinia crista*	+	-	+
11.	*Caesalpinia decapetala*	+	-	-
12.	*Caesalpinia pulcherrima*	+	-	+
13.	*Caesalpinia sappan*	+	+	-
14.	*Cassia fistula*	+	+	+
15.	*Cassia grandis*	+	-	+
16.	*Ceratonia siliqua*	+	-	+
17.	*Chamaecrista mimosoides*	+	-	-
18.	*Cynometra ramiflora*	+	+	-
19.	*Delonix regia*	+	-	+
20.	*Gleditsia aquatica*	+	-	+
21.	*Haematoxylum campechianum*	+	-	+
22.	*Humboldtia vahliana*	+	-	-
23.	*Hymenaea verrucosa*	+	-	+
24.	*Saraca asoca*	+	+	+
25.	*Senna alata*	+	+	+
26.	*Senna alexandrina*	+	+	+
27.	*Senna auriculata*	+	+	+
28.	*Senna occidentalis*	+	+	+
29.	*Senna sophera*	+	-	+
30.	*Senna tora*	+	+	+
31.	*Tamarindus indica*	+	+	+
II	**FABACEAE**			
1.	*Abrus precatorius*	+	-	+
2.	*Alhagi maurorum*	+	-	-
3.	*Andira inermis*	+	+	-
4.	*Arachis hypogaea*	+	+	+
5.	*Butea monosperma*	+	+	+

Sl. No.	Species	Anti-Bacterial	Antifungal	Antiviral
6.	*Cajanus cajan*	+	+	+
7.	*Cicer arietinum*	+	+	-
8.	*Clitoria ternatea*	+	-	-
9.	*Crotalaria juncea*	+	+	+
10.	*Crotalaria pallida*	+	+	-
11.	*Crotalaria spectabilis*	+	+	-
12.	*Crotalaria verrucosa*	+	-	+
13.	*Cullen corylifolium*	+	+	-
14.	*Dalbergia cultrata*	+	+	-
15.	*Dalbergia lanceolaria*	+	-	-
16.	*Dalbergia sissoo*	+	-	-
17.	*Dendrolobium triangulare*	+	-	-
18.	*Derris robusta*	+	+	-
19.	*Derris trifoliata*	+	-	+
20.	*Desmodium gangeticum*	+	+	+
21.	*Desmodium oojeinense*	+	-	+
22.	*Desmodium pulchellum*	+	-	+
23.	*Desmodium triflorum*	+	-	-
24.	*Erythrina crista-galli*	+	-	+
25.	*Erythrina fusca*	+	-	+
26.	*Erythrina suberosa*	+	-	-
27.	*Erythrina variegata*	+	-	-
28.	*Flemingia macrophylla*	+	-	-
29.	*Genista tinctoria*	+	+	-
30.	*Gliricidia sepium*	+	+	-
31.	*Glycine max*	+	+	+
32.	*Glycyrrhiza glabra*	-	-	+
33.	*Indigofera aspalathoides*	+	+	-
34.	*Indigofera tinctoria*	+	-	-
35.	*Lablab purpureus*	+	+	-
36.	*Laburnum anagyroides*	-	+	-
37.	*Lathyrus hirsutus*	-	+	-
38.	*Lens culinaris*	+	+	+
39.	*Lupinus albus*	+	+	+
40.	*Lupinus luteus*	+	-	+
41.	*Macroptilum atropurpureum*	+	-	-
42.	*Macrotyloma uniflorum*	+	-	+
43.	*Medicago sativa*	+	+	+
44.	*Melilotus albus*	+	-	+
45.	*Melilotus indicus*	+	-	+

Sl. No.	Species	Anti-Bacterial	Antifungal	Antiviral
46.	*Mucuna pruriens*	+	+	+
47.	*Mundulea sericea*	+	-	-
48.	*Myroxylon balsamum*	+	+	+
49.	*Myroxylon peruiferum*	+	+	+
50.	*Phaseolus vulgaris*	+	-	+
51.	*Physostigma venenosum*	+	+	+
52.	*Pisum sativum*	+	+	-
53.	*Pongamia pinnata*	+	+	+
54.	*Psophocarpus tetragonolobus*	+	-	+
55.	*Pterocarpus indicus*	+	-	+
56.	*Pterocarpus marsupium*	+	-	+
57.	*Pterocarpus santalinus*	+	+	+
58.	*Pueraria phaseoloides*	+	-	+
59.	*Pueraria tuberosa*	+	-	-
60.	*Rhynchosia rufescens*	+	-	+
61.	*Robinia pseudoacacia*	+	-	+
62.	*Sesbania grandiflora*	+	+	+
63.	*Spartium junceum*	+	-	+
64.	*Styphnolobium japonicum*	+	-	+
65.	*Tephrosia purpurea*	+	+	+
66.	*Tephrosia spinosa*	+	-	+
67.	*Tephrosia uniflora*	+	-	+
68.	*Tephrosia vogelii*	+	+	-
69.	*Teramnus labialis*	+	-	-
70.	*Trifolium alexandrium*	+	-	+
71.	*Trifolium pratense*	+	+	-
72.	*Trigonella foenum-graecum*	+	+	+
73.	*Ulex europaeus*	+	-	+
74.	*Uraria picta*	+	-	+
75.	*Vicia faba*	+	+	+
76.	*Vigna aconitifolia*	+	-	+
77.	*Vigna angularis*	+	-	+
78.	*Vigna mungo*	+	-	+
79.	*Vigna radiata*	+	-	+
80.	*Vigna unguiculata*	+	+	+
III	**MIMOSACEAE**			
1.	*Acacia catechu*	+	-	+
2.	*Acacia decurrens*	+	-	+
3.	*Acacia farnesiana*	+	-	-
4.	*Acacia ferruginea*	+	+	-

Sl. No.	Species	Anti-Bacterial	Antifungal	Antiviral
5.	*Acacia melanoxylon*	+	-	+
6.	*Acacia nilotica*	+	-	+
7.	*Acacia senegal*	+	-	-
8.	*Acacia seyal*	+	-	+
9.	*Acacia sinuata*	+	-	+
10.	*Acacia tortilis*	+	-	+
11.	*Adenanthera pavonina*	+	-	-
12.	*Albizia amara*	+	-	+
13.	*Albizia chinensis*	+	-	+
14.	*Albizia julibrissin*	+	-	+
15.	*Albizia lebbeck*	+	-	+
16.	*Albizia lucidior*	+	+	-
17.	*Albizia odoratissima*	+	-	-
18.	*Albizia procera*	+	-	+
19.	*Archidendron clypearia*	+	-	-
20.	*Archidendron monodelphum*	+	+	-
21.	*Entada rheedei*	+	-	+
22.	*Leucaena leucocephala*	+	-	-
23.	*Mimosa pudica*	+	-	+
24.	*Parkia timoriana*	+	-	-
25.	*Prosopis juliflora*	+	-	+
26.	*Xylia xylocarpa*	+	-	-

Note : Based on Sathyanarayana Bhat (1993), Sharon (1994), Kameswara Rao (2000).

Table 15.2
Indian Legumes with Pesticidal/Insecticidal Activity

Sl. No.	Species
I	**CAESALPINIACEAE**
1	*Bauhinia purpurea*
2	*Caesalpinia pulcherrima*
3	*Cassia grandis*
4	*Cassia javanica*
5	*Ceratonia siliqua*
6	*Senna alata*
7	*Senna occidentalis*
8	*Senna tora*
II	**FABACEAE**
1	*Amorpha fruticosa*
2	*Andira inermis*

Sl. No.	Species
3	*Astracantha gummifera*
4	*Canavalia gladiata*
5	*Crotalaria retusa*
6	*Derris cuneifolia*
7	*Derris elegans*
8	*Derris elliptica*
9	*Derris malaccensis*
10	*Derris robusta*
11	*Derris scandens*
12	*Derris trifoliata*
13	*Desmodium oojeinense*
14	*Erythrina crista-galli*
15	*Euchresta horsfieldii*
16	*Flemingia strobilifera*
17	*Gliricidia sepium*
18	*Glycyrrhiza glabra*
19	*Lathyrus odoratus*
20	*Lonchocarpus sericeus*
21	*Millettia extensa*
22	*Millettia pachycarpa*
23	*Millettia pisidia*
24	*Millettia thonningii*
25	*Mucuna pruriens*
26	*Mundulea sericea*
27	*Myroxylon balsamum*
28	*Pachyrrhizus erosus*
29	*Pachyrrhizus tuberosus*
30	*Physostigma venenosum*
31	*Pongamia pinnata*
32	*Pueraria pedicularis*
33	*Sesbania bispinosa*
34	*Sophora secundiflora*
35	*Sophora tomentosa*
36	*Styphnolobium japonicum*
37	*Tephrosia bracteolata*
38	*Tephrosia candida*
39	*Tephrosia grandiflora*
40	*Tephrosia noctiflora*
41	*Tephrosia pumila*

Sl. No.	Species
42	*Tephrosia purpurea*
43	*Tephrosia sinapou*
44	*Tephrosia tinctoria*
45	*Tephrosia uniflora*
46	*Tephrosia vogelii*
47	*Trigonella emodi*
III	**MIMOSACEAE**
1	*Acacia concinna*
2	*Acacia pennata*
3	*Acacia senegal*
4	*Acacia sinuata*
5	*Albizia chinensis*
6	*Albizia procera*
7	*Entada rheedei*
8	*Lysiloma latisiliquum*

CHAPTER-16

Systems of Medicine and Health Care

16.1 INTRODUCTION

There are several different systems of medicine and indigenous practices of medical care in all countries and India has its own. It is necessary to understand the principles and practices of these systems in order to use plants to advantage in medicine.

Allopathy is the predominant, basically urban, system of medicine that has immense attraction even to rural population. All the other systems of medicine are collectively known as Alternative and Complementary Systems of medicine. In India, the indigenous systems of medicine such as *Ayurveda, Siddha, Unani* and the German system Homoeopathy, are in wide use. In addition, there are several minor medical concepts that engage the fancy of different sections of people at different times.

16.2 ALLOPATHY

Allopathy advocates therapy with remedies that produce effects differing from those of the treated disease, the principle of *contraria contraiis curentur* (Canary, 1983). In contrast, Homoeopathy is based on the principle that the remedies produce the same symptoms as the disease, *similia similibus curentur.* Allopathy is also called 'modern', 'western', or 'scientific' medicine. The term 'biomedicine' defined as the 'application of the principles of natural science, especially biology and physiology to clinical medicine', is also in use.

The development of Allopathy was mainly due to the contributions of Aretaeus of Cappadocia, Hippocrates of Greece and Ibn Sina (Avicenna) of Persia. These were followed by a host of European practitioners and observers, aided by the establishment of great Universities in Italy, France and England. Developments in the past 50 years or so in anatomy, biochemistry, physiology, pharmacology, physics, biology, electronics, engineering, and other disciplines have made Allopathy radically different not only from the other systems of medicine but also from its own earlier versions. Vast changes have occurred in the philosophy, methods of diagnosis, drug development, treatment and prognosis. Concomitant with the progress in medicine, there were very important developments in surgery, which also has an important place in *Ayurveda.*

Allopathy is firmly rooted in the products of synthetic chemistry, as its drug arsenal. Nevertheless, in the developed countries, 25 per cent of all medical drugs are still plant based and in the poorer world this is closer to 75 per cent (Principe, 1991). Allopathy will continue to depend on plants for its drugs and this dependence is more likely to increase rather than decrease.

Although almost every country has its own pharmacopoeia (the literature on drugs, their preparation and usage), there are no standard international pharmacopoeiae and *Materia Medicae* (drugs, their properties and usage) for Allopathy (Canary, 1983). The information on the numerous aspects of Allopathy is unmanageably voluminous. By and large, to the general practitioner as well as the specialist, other than those engaged in research, the drug manufacturers provide the functional clinical literature. Research oriented doctors obtain their information from journals, treatises, etc. Professional organizations and conferences also provide a fillip. Understanding the patient, the disease and the multifarious effects of the medicine, before administering it, is the cardinal principle of all systems of medicine, but in the current times it is easier pronounced than practiced.

Allopathy, with very impressive advancements in its numerous areas of specialization, has performed wonders and achieved a phenomenal popularity throughout the world, much to the detriment of the indigenous/traditional systems of medicine. However, it has become increasingly inaccessible to the majority of the world population due to its dependence on expensive instrumentation and very high costs of drugs and services. It has also accumulated a long list of inadequacies and failures. These factors are responsible for leaving a lot of ground for the other systems of medicine to be of considerable importance.

16.3 ALTERNATIVE AND COMPLEMENTARY MEDICAL SYSTEMS

The terms 'Alternative Medicine' and 'Complementary Medicine' are often used interchangeably. However, some make a distinction. Alternative Systems of Medicine are those that offer independent therapies for the full range of all the diseases, as Allopathy does, as for example, *Ayurveda*, *Siddha*, *Unani*, Chinese medicine, and Homoeopathy. Systems of medicine such as Naturopathy, Aromatherapy, Hydrotherapy, etc., that are only supportive of other systems, are regarded as Complementary Systems of Medicine. Most of these systems are being taught in specialized institutions and even in courses of modern medicine for a long time (Daly, 1974).

There arises a problem when a particular system of medicine falls in between. Some consider Acupuncture as a totally potent system, while others regard it as only supportive. Herbal medicine, not in the sense of the established systems like *Ayurveda* or Chinese medicine, but in that of the unwritten traditional medical practices in South America and Africa, was cited as Alternative or Complementary, by different authors.

There are over 125 systems of Alternative and Complementary medicine (Kameswara Rao, 2000). Listing all of them is not an easy task as some single plant remedies are also projected as systems of therapy, for example, Ginseng, Honey, Cucumber, Eucalyptus, Garlic, Peppermint, Misle-toe, Rose-water therapies.

The range and depth of the Alternative therapies is both illuminating and interesting. The plurality and the overlapping nature of these therapies make it a very difficult task to classify them. Representative examples of Alternative and Complementary therapies are given in Table 16.1.

Some of the Alternative therapies may go into disuse with time and many more may be proposed, but Alternative therapies, in general, will continue to occupy a very important place in our health care efforts.

16.4 PLANTS IN MEDICINE

All through the history of medicine, plants have formed a very important source of drugs in all systems and practices of medicine, throughout the world. The role of plants as the major source of drugs has not diminished with time but has actually increased. The potential of plants as drug sources has never been fully exploited. Plants should be studied for a better understanding of their use in medicine, and for wider use. This is in the interests of global health security and to provide effective and economical means of maintenance of health and cure of disease, for the masses.

16.5 LEGUMES IN MEDICINE

It is necessary to appreciate the philosophy and principles of different systems of medicine, in order to understand the use of plants in them. For this reason and also because of their importance in India, the systems of *Ayurveda, Siddha, Unani* and Homoeopathy are discussed in detail (see Chapters 18, 19, 20 and 21). Outside these structured and organized systems of medicine, indigenous populations in every country have built up vast amounts of knowledge on the use of plants as medicine. This largely unwritten wealth of information is referred to as Ethnomedicine, which has a very considerable component of legume species (Chapter 17).

Legumes being an important component of vegetation everywhere, they have a conspicuous place as source of drugs in different medical systems. The legumes used in formulations in these systems are identified in Chapters 18 to 21. The therapeutic uses of legume species used in different systems of medicine are indicated in the respective species profiles.

Table 16.1

Alternative and Complementary Systems of Medicine

Representative examples are given under each category. The plant-based therapies are marked by an asterisk (*)

1.	**Special implement based therapies**	
	a) Acupuncture	Insertion of special needles at specific points in the body
	b) Electro-acupuncture	Acupuncture with small direct current
	c) Laser-acupuncture	Acupuncture points stimulated by a weak laser beam
	d) Homoeopuncture	Dipping acupuncture needles in homoeopathic medicine before acupuncture
	e) Moxibustion	Application of moxa (*Artemisia moxa*) smoke and heat to acupuncture points
	f) Acupressure	Finger tip or thumb pressure on acupuncture points
	g) Cupping	Vacuum treatment on acupuncture points using a suction cup
2.	**Manual therapies**	
	a) Osteopathy and Chiropractice	Bone setting
	b) Applied Kinesiology	Motion or movement of external or internal organs by external manipulation
	c) Reflexology	Zone therapy to treat the whole body through contact with feet
	d) *Shiatzu*	Chinese massage similar to acupressure
	e) Rolfing	Correcting postural problems by structural integration
	f) Massage	Practice of applying structured pressure to relieve pain
	g) Rosen method	Psychotherapy combined with a massage

3.	**Oral therapies**	
	a) Homoeopathy*	German system of medicine
	b) Herbal medicine	South American, African, etc., other than the well established and recorded systems*
	c) *Ayurveda**	Indian system of medicine
	d) *Siddha**	Dravidian system of medicine using minerals and metals
	e) Tibetan medicine*	Tibetan system of medicine
	f) *Unani**	Arab system of medicine
	g) Chinese medicine*	Chinese system of medicine
	h) Bach's flower remedies*	Healing using dilutions of flower essences
	i) Naturopathy*	Therapy and treatment using natural remedies like sunlight, air, water etc.
	j) Nutrition therapy*	Holistic health through proper diet
	k) Specific therapies	*Aloe vera**, vitamin E, garlic*, folic acid, green-lipped mussel, fasting, etc.
4.	**Postural therapies**	
	a) *Yoga*	Using disciplines of asceticism and meditation
	b) Alexander technique	Correction of bad body postures
	c) Feldenkrais technique	Alexander technique combined with *yoga*
5.	**Psychosomatic therapies**	
	a) Hypnotherapy	Healing through hypnosis
	b) Autogenic therapy	Deep relaxation exercises
	c) Meditation	State of concentrated attention on some object of thought or awareness
	d) Biofeedback	Sensory control of involuntary body responses like blood pressure
6.	**Sensory therapies**	
	a) Aromatherapy*	Use of essential oils from plants
	b) Sound therapy	Use of pleasant sounds to relieve stress
	c) Colour therapy	Use of spectrum colours to stimulate the body's healing process
	d) Dance therapy	Therapeutic use of movement to improve physical and mental well being
	e) Primal scream therapy	To relieve trauma by screaming
7.	**Energy therapies**	
	a) Radionics	Healing by energy directed from a distance: also called 'pranic healing'
	b) Universal energy (Radiance technique, Reiki):	Exchange of universal energy between the patient and the therapist to reintroduce universal energy into the patient who lost it
	c) Polarity	Balance of energy flow between different organs in the body

8.	**Diagnostic aids**	
	a) Iridology	Study of iris
	b) Kirlian photography	Photographic record of energy flow about and around the body
9.	**Other therapies**	
	a) Anthroposophical	Inner feelings medicine
	b) Bates' method of eye testing	Exercises of the eye muscles
	c) Schuessler's Biochemic tissue salts	An adjunct of Homoeopathy, using salts extracted from plants*
	d) Bowen technique	Gentle and subtle massage
	e) Colonic irrigation	Enema and purgation
	f) Conductive education	Improvement of motor skills
	g) Crystal and gem therapy	Use of crystals and gems for healing
	h) Enzyme potentiated desensitisation	Immunotherapy for allergies
	i) Eurhythmy	Gentle music and dance
	j) Facial diagnosis	Evaluation of health by face reading
	k) Flotation therapy	Lying in a tank of saline
	l) Healing	Faith and spiritual healing
	m) Hydrotherapy	Use of water for soothing pains and treating diseases
	n) Integration therapy	Adult adjustment to one's own neo-natal reflexes
	o) Iscador therapy	Use of mistletoe extract*
	p) *Jin shin jyutsu*	Japanese method of touch and gentle massage
	q) Magnetotherapy	Curing ailments by applying magnets
	r) Manual lymphatic drainage	Massage of lymph nodes and glands
	s) Neurolinguistic programming	Programming neurological processes by communication
	t) Pilates method muscles	exercises to strengthen the abdominal and related
	u) Probiotics	Use of friendly bacteria like *Lactobacillus**
	v) *T'ai chi ch'uan*	A Chinese exercise
	w) Tea tree oil antibacterial, antifungal*	Oil of *Melaleuca alternifolia* (Myrtaceae),
	x) Zero balancing	Osteopathy plus energy flow
	y) Zinc therapy reduce recovery time	Addition of zinc to standard antibiotic treatments to

Note : *Based on Kameswara Rao (2000).*

CHAPTER-17

Legumes in Ethnomedicine

17.1 INTRODUCTION

The term 'ethnic' pertains to human racial groups, native to a particular geographical region, in contrast to the populations that are either immigrant or populations influenced by the influx of civilization. There are about 400 different ethnic groups in India, which constitute about 7.5 per cent of the Indian population, constituting a rich base for ethnic studies.

The prefix 'ethno-' is used in various contexts related to the ethnic groups, covering all traditional aspects of their culture and lives such as gods, rituals and festivals, customs, language, dress, agriculture, food, medicine, dwellings, etc. Ethnology is cultural anthropology, ethnography is the scientific description of the races of the earth, while ethnolinguistics, ethnomusicology, ethnobiology, ethnobotany, ethnomedicine, etc., deal with the respective areas of study. Martin (1995) outlined the principles and practices of ethnobotany. Kameswara Rao (2000) discussed the issues related to ethnobotany, ethnoecology, ethnomedicine and ethnotherapeutics, while Trivedi (2002) discussed Indian ethnobotany.

The main distinction is between knowledge that is indigenous (local and traditional) *vs* knowledge that is partly or wholly exotic (modern). The major objective of study of all ethnic areas is to accurately document all the unrecorded information before it is lost to us, in the face of cross-cultures and onslaught of modern civilization, in order to benefit from traditional wisdom, in our efforts towards the betterment of mankind.

17.2 ETHNOBOTANY

The term 'ethnobotany' is often used to cover all plant related areas of ethnoecology, other than ethnomedicine and ethnopharmacology. Some like Schultes and Reis (1995), include ethnobotany, ethnomedicine and ethnopharmacology together under ethnobotany.

Many people, especially in the poorer countries, depend upon plants collected from the wild neighbourhood localities for food, construction materials, fuel wood, medicine and many other purposes. Today, there is often a decrease in the availability of wild plant resources, related to increased human populations and the effects of competition with other forms of land use such as agriculture. An ethnobotanical study of these aspects is basic to an efficient use of our plant resources.

Ethnobotanical studies can help local communities define their needs for plant resources more clearly, thus assisting them to state their case for continuing access to certain areas of land or for provision of alternatives to wild gathering (Martin, 1995). They also bring to light, established but little known uses of plants.

Traditionally, communities worldwide are extremely knowledgeable about local plants and other natural resources, on which they are so intimately dependent. Much of this wealth of knowledge is being rapidly lost as traditional cultures are being eroded, a serious concern that prompts ethnoecological studies.

Ethnobotanists can play a useful role in rescuing the disappearing knowledge and returning it to the local communities. This will help conserving at least a part of ethnobotanical heritage, as a living-cultural ecosystem, helping to maintain a sense of pride in local cultural knowledge and practices, and reinforcing links between communities and the environment, which is essential for biological conservation (Hamilton, 1995). Ethnobotanical studies can also help identify conservation issues, such as cases where rates of harvest from plants exceed rates of regrowth.

One important current concern of ethnobotanical research is the potential use of plants as medicine, a knowledge that is often exclusive to specific communities and linked to the local flora.

17.3 ETHNOMEDICINE AND ETHNOTHERAPEUTICS

Therapeutics is that branch of the medical science dealing with the treatment of disease and the term is used interchangeably with medicine, but distinguished from surgery or obstetrics.

The importance of plants in our lives, in addition to food, is their therapeutic potential. Every culture in the world developed its own practices of treating the diseased. The fund of traditional knowledge, developed over the millennia by thousands of ethnic groups, is largely unrecorded and faces the danger of becoming extinct. Our urgent concern is to preserve, refine and use this information for a more effective management of health.

While the basic emphasis in ethnic studies is on unrecorded information, often even well-documented, but traditional systems such as *Ayurveda*, are also incorrectly included under ethnomedicine.

By all counts, ancient cultures throughout the world nurtured till recent times, vast amounts of knowledge pertaining to the uses of their flora. The loss of this knowledge is especially detrimental to our efforts in seeking potentially valuable chemical constituents in the great number of species of unstudied plants which is a major goal of ethnobotany and particularly ethnopharmacology. Research on ethnomedical uses of plants may lead to the discovery of new drugs, as happened in the case of strychnine from *Strychnos nux-vomica* (Bisset, 1995; Kameswara Rao, 2000).

17.4 ETHNOPHARMACOLOGY

Pharmacology is the study of drugs and a drug is any substance with action on the physiology and psychology of the human system. Pharmacology is concerned with the physico-chemical nature, physiological action, target organs, therapeutic benefits, dosage and mode of administration, toxicity and side effects, and contra-indications of drugs.

To be internationally acceptable, drugs need to be subjected to a very rigorous pharmacological testing. The stringent drug testing procedures have human safety at heart and should scrupulously be adopted even though they impose heavy time and financial inputs. Through a battery of simultaneous procedures, one would be able to evaluate the antibacterial, antiviral, analgesic, anti-inflammatory effects, along with the cytotoxic, mitostatic, mutagenic, or carcinogenic potential of the drug.

Ethnopharmacology is a subdivision of ethnobotany (or ethnoecology). It is the study of aboriginal knowledge of the physiological properties of plants or their products. Ethnopharmacology may be defined as 'the interdisciplinary scientific exploration of biologically active agents traditionally employed or observed by man' (Bruhn and Helmsted, 1981).

The concept of ethnopharmacology is now firmly established, with high promise as a means of scientific validation of traditional therapeutics and of discovering new drugs. The origin, development and current status of different aspects of ethnopharmacology were reviewed extensively (Bruhn and Helmstedt, 1981; several authors in Schultes and Reis, 1995; Kameswara Rao, 2000).

Pharmacological testing of indigenous plant based drugs will help in proving or disproving the claims made for them and places their local uses on a firm scientific footing.

Ethnobotanical, phytochemical, biochemical, *in vivo* physiological and clinical studies of the traditional medicinal plants is an area of research that is essential to promote our efforts in search of medicines. The end result of an ethnopharmacological study is either an authentication of a traditional drug or the discovery of a new drug. Vincristine, a very effective anti-leukaemic drug was discovered in *Catharanthus roseus* (*Vinca rosea*), while the species was being studied for its traditional use in Madagascar, as an antidiabetic drug.

The traditional aspect of ethnomedicine provides us with a reference and starting point for ethnopharmacological studies, without which we would be lost among the thousands of plant species.

Biomedicine and modern pharmacology deal with a single purified compound at a time, whether natural or synthetic. On the other hand, ethnomedicine and indigenous systems operate in terms of compound formulations simultaneously containing several plant species, though there are some single species drugs. A crude drug from even a single plant species used in traditional medicine, simultaneously contains a large number of chemical compounds, each with a different therapeutic effect. For example, neem and ginger have many different chemical compounds which exert over a dozen different therapeutic effects (Sharon, 1994; Shubharani, 1995). Many medicines in the indigenous systems are formulations containing several species of plants and other materials. For example, *Dashamoolaristha* has 68 species, *Mahaavishagarbhadi taila* 64 species, *Mahaanaarayana taila* 55 species and *Karpuraadiarka* 52 species (Anonymous, 1978). The number of chemical compounds in the formulations is enormous and it is impossible to analyze them chemically and subject them to a pharmacological study. Unless therapeutic effects are directly related to specific chemical compounds, modern medicine does not accept traditional drugs. A way out would be to identify the main ingredients in a formulation and subject these species of plants to pharmacological testing for further use as the situation indicates. In cases like reserpine or vincristine or other situations involving a single active chemical compound from a single species, there is no conflict in approach.

17.5 LEGUMES IN ETHNOMEDICINE

Legumes being a conspicuous and important component of the local vegetation everywhere in the world, they naturally constitute an important component of the medical arsenal of local populations and India is no exception. Often, the medicinal uses of plants in ethnomedicine and other systems are identical.

Over the past three or four decades, vast literature has been built up on ethnobotany in India, with emphasis on medicinal uses. The ethnomedical uses of Indian legumes are given in Table 17.1. These species should be subjected to ethnopharmacological investigation, in order to put them to a better use. Culled from a large number of diverse publications on ethnobotany and ethnomedicine, the list in Table 17.1 is only representative and not exhaustive. In the literature on ethnomedicine there are many problems of identification and nomenclature of the species of plants, which require a serious consideration by plant taxonomists.

Table 17.1

Legumes in Ethnomedicine

Sl.No.	Uses	Species
1.	**Abortifacient**	*Abrus precatorius, Acacia catechu, Butea monosperma, Caesalpinia pulcherrima, Clitoria ternatea, Crotalaria juncea, Desmodium microphyllum, Leucaena leucophloea, Rhynchosia minima, Tephrosia purpurea, Uraria lagopodioides*
2.	**Abscesses, apthae and sores**	*Acacia catechu, Acacia chundra, Acacia farnesiana, Albizia lebbeck, Astragalus aegacanthoides, Atylosia platycarpa, Atylosia scarabaeoides, Bauhinia purpurea, Bauhinia retusa, Bauhinia variegata, Butea monosperma, Caesalpinia mimosoides, Cajanus cajan, Cassia fistula, Crotalaria calycina, Desmodium triflorum, Entada rheedei, Eriosema himalaicum, Erythrina suberosa, Erythrina variegata, Flemingia chappar, Indigofera tinctoria, Leucaena leucophloea, Macrotyloma uniflorum, Mimosa rubicaulis* ssp., *himalayana, Mimosa pudica, Mucuna monosperma, Mucuna pruriens, Pongamia pinnata, Pterocarpus marsupium, Pueraria tuberosa, Senna tora, Sesbania sesban, Tamarindus indica, Tephrosia candida, Tephrosia purpurea, Trigonella foenum-graecum, Vigna vexillata*
3.	**Anthelmintic**	*Abrus precatorius, Crotalaria spectabilis, Erythrina variegata, Mucuna pruriens, Senna tora, Tamarindus indica.*
4.	**Antidote**	*Bauhinia acuminata, Bauhinia purpurea, Desmodium gangeticum, Tephrosia purpurea*
5.	**Antifertility**	*Abrus pulchellus, Acacia farnesiana, Aeschynomene indica, Alysicarpus vaginalis, Atylosia scarabaeoides, Bauhinia vahlii, Butea monosperma, Butea superba, Cassia fistula, Crotalaria juncea, Indigofera linnaei, Lablab purpureus, Mimosa pudica, Pisum sativum, Sesbania sesban, Smithia conferta.*
6.	**Antiseptic**	*Abrus precatorius, Cassia fistula, Senna tora*
7.	**Aphrodisiac**	*Abrus precatorius, Bauhinia vahlii, Butea monosperma, Mucuna pruriens.*
8.	**Appetizer**	*Mucuna pruriens*
9.	**Asthma**	*Acacia catechu, Albizia myriophylla, Alhagi maurorum, Cassia fistula, Desmodium gangeticum, Desmodium oojeinense, Mucuna monosperma, Pterocarpus marsupium, Pueraria tuberosa, Rhynchosia minima, Senna auriculata, Tephrosia purpurea.*

Sl.No.	Uses	Species
10.	**Astringent**	*Acacia catechu, Acacia farnesiana, Acacia ferruginea, Bauhinia purpurea, Butea monosperma, Leucaena leucophloea.*
11.	**Bites**	*Bauhinia purpurea, Butea monosperma, Cassia fistula, Pithecellobium dulce.*
12.	**Blood purifier**	*Abrus precatorius, Astragalus candolleanus, Cassia fistula, Crotalaria juncea, Mimosa hamata, Tephrosia purpurea.*
13.	**Body ache**	*Desmodium gangeticum.*
14.	**Bone fractures**	*Abrus precatorius, Alysicarpus vaginalis, Bauhinia purpurea, Butea monosperma, Indigofera tinctoria, Mucuna pruriens, Senna occidentalis, Senna tora, Spatholobus parviflorus.*
15.	**Bronchitis**	*Abrus precatorius, Acacia caesia, Acacia catechu, Acacia nilotica* ssp., *indica, Albizia myriophylla, Desmodium gangeticum, Desmodium oojeinense, Indigofera tinctoria, Mimosa rubicaulis* ssp., *himalayana, Pongamia pinnata, Tephrosia purpurea.*
16.	**Burns**	*Acacia nilotica* spp. *indica, Astragalus aegacanthoides, Atylosia scarabaeoides, Cajanus cajan, Cassia fistula, Crotalaria burhia, Crotalaria prostrata, Lens culinaris, Macrotyloma uniflorum, Pongamia pinnata, Pterocarpus marsupium.*
17.	**Burning sensation inside body**	*Desmodium pulchellum.*
18.	**Carbuncles**	*Acacia farnesiana, Crotalaria calycina.*
19.	**Cardiac complaints**	*Indigofera tinctoria, Saraca asoca.*
20.	**Carminative**	*Trigonella foenum-graecum.*
21.	**Child birth**	*Abrus precatorius, Acacia catechu, Atylosia scarabaeoides, Crotalaria alata, Crotalaria prostrata, Indigofera cassioides, Macrotyloma uniflorum, Mimosa pudica, Pterocarpus marsupium, Senna tora, Tephrosia purpurea, Uraria lagopodioides*
22.	**Cold**	*Abrus precatorius, Crotalaria juncea, Erythrina variegata, Flemingia macrophylla, Mucuna monosperma, Pongamia pinnata, Senna tora, Tadehagi triquetrum, Tamarindus indica, Tephrosia villosa.*
23.	**Conception**	*Codariocalyx gyroides, Indigofera cassioides.*

Sl.No.	Uses	Species
24.	**Convulsions, cramps**	*Abrus precatorius, Acacia farnesiana, Bauhinia purpurea, Caesalpinia cucullata, Crotalaria calycina, Erythrina variegata.*
25.	**Delirium**	*Acacia farnesiana, Butea monosperma.*
26.	**Dental problems**	*Abrus precatorius, Acacia catechu, Acacia chundra, Albizia lebbeck, Caesalpinia pulcherrima, Cassia fistula, Dalbergia lanceolaria, Desmodium pulchellum, Desmodium triflorum, Indigofera astragalina, Indigofera trita, Macrotyloma uniflorum, Mimosa pudica, Millettia extensa, Pterocarpus marsupium, Pueraria phaseoloides, Senna auriculata, Spatholobus parviflorus, Tamarindus indica, Tephrosia purpurea.*
27.	**Diabetes**	*Acacia nilotica* ssp., *indica, Butea monosperma, Mucuna pruriens, Pisum sativum, Pongamia pinnata, Pterocarpus marsupium.*
28.	**Diarrhoea**	*Abrus precatorius, Bauhinia racemosa, Bauhinia vahlii, Butea monosperma, Caesalpinia bonduc, Cajanus cajan, Cajanus scarabaeoides, Cassia fistula, Desmodium gangeticum, Desmodium oojeinense, Mimosa hamata, Pterocarpus marsupium, Pueraria tuberosa, Tadehagi triquetrum, Trigonella foenum-graecum, Vigna radiata.*
29.	**Digestive disorders**	*Acacia pennata, Acacia sinuata, Astragalus himalayana, Atylosia scarabaeoides, Bauhinia vahlii, Bauhinia variegata, Butea monosperma, Cassia fistula, Caesalpinia bonduc, Clitoria ternatea, Dalbergia latifolia, Dalbergia paniculata, Desmodium dichotomum, Desmodium triangulare, Desmodium triflorum, Endosamara racemosa, Erythrina stricta, Flemingia macrophylla, Flemingia procumbens, Flemingia strobilifera, Indigofera linnaei, Mimosa rubicaulis* ssp., *himalayana, Paracalyx scariosus, Parochetus communis, Pterocarpus marsupium, Pueraria tuberosa, Senna occidentalis, Senna tora, Sesbania sesban, Spatholobus parviflorus, Tadehagi triquetrum, Tamarindus indica, Tephrosia villosa, Tephrosia purpurea, Uraria picta.*
30.	**Diuretic**	*Butea monosperma, Mimosa pudica, Oxytropis tatarica, Senna occidentalis.*
31.	**Dropsy**	*Atylosia mollis, Atylosia scarabaeoides, Bauhinia purpurea, Cassia fistula, Entada rheedei, Indigofera tinctoria, Mucuna pruriens, Senna occidentalis, Tamarindus indica.*

Sl.No.	Uses	Species
32.	**Dysfunctional**	*Chamaecrista pumila, Dichrostachys cinerea, Pueraria tuberosa, Senna occidentalis.*
33.	**Dyspepsia**	*Cassia fistula, Cicer arietinum, Senna tora.*
34.	**Ear complaints**	*Acacia nilotica* ssp., *indica, Pongamia pinnata.*
35.	**Epilepsy**	*Acacia farnesiana, Cassia fistula, Entada rheedei, Fleminigia chappar, Flemingia strobilifera, Indigofera cassioides, Indigofera tinctoria, Mimosa pudica, Mimosa rubicaulis* ssp., *himalayana, Senna tora.*
36.	**Expectorant**	*Alysicarpus vaginalis.*
37.	**Eye diseases**	*Abrus precatorius, Acacia farnesiana, Acacia nilotica* spp. *indica, Albizia lebbeck, Atylosia platycarpa, Butea monosperma, Cassia fistula, Chamaecrista absus, Dendrolobium triangulare, Desmodium gangeticum, Entada rheedei, Erythrina variegata, Flemingia chappar, Flemingia macrophylla, Indigofera tinctoria, Mucuna pruriens, Senna auriculata, Senna tora, Tamarindus indica.*
38.	**Fever**	*Abrus precatorius, Acacia pennata, Aeschynomene indica, Alhagi maurorum, Alysicarpus bupleurifolius, Atylosia scarabaeoides, Butea monosperma, Butea superba, Caesalpinia bonduc, Dalbergia lanceolaria, Dendrolobium umbellatum, Desmodium dichotomum, Desmodium gangeticum, Desmodium heterocarpon, Desmodium laxiflorum, Erythrina stricta, Erythrina variegata, Flemingia macrophylla, Flemingia strobilifera, Indigofera tinctoria, Mimosa rubicaulis* ssp., *himalayana, Mimosa pudica, Mucuna monosperma, Pongamia pinnata, Pterocarpus marsupium, Pueraria tuberosa, Rothia indica, Senna occidentalis, Senna tora, Tadehagi triquetrum, Tamarindus indica, Tephrosia purpurea, Uraria picta.*
39.	**Fistula**	*Flemingia macrophylla, Mimosa pudica, Pongamia pinnata.*
40.	**Goiter**	*Clitoria ternatea.*
41.	**Gynaecological problems**	*Atylosia mollis, Prosopis cineraria, Sophora wightii, Uraria lagopodioides.*
42.	**Haemostatic**	*Alhagi maurorum, Butea monosperma, Mucuna bracteata.*
43.	**Haemoptysis**	*Acacia catechu.*

Sl.No.	Uses	Species
44.	**Hair care, dandruff, hair fall**	*Acacia nilotica* ssp., *indica, Acacia pennata, Atylosia scarabaeoides, Indigofera tinctoria, Millettia extensa, Pithecellobium dulce, Trifolium repens, Trigonella foenum-graecum.*
45.	**Hydrocele**	*Mimosa pudica.*
46.	**Hyperacidity**	*Albizia amara.*
47.	**Hysteria**	*Flemingia strobilifera, Mimosa rubicaulis* ssp., *himalayana, Mimosa pudica.*
48.	**Impotency**	*Aeschynomene indica, Mucuna bracteata, Mucuna pruriens, Senna hirsuta, Tephrosia purpurea.*
49.	**Inflammation**	*Abrus precatorius, Albizia lebbeck, Butea monosperma, Caesalpinia bonduc, Cassia fistula, Clitoria ternatea, Desmodium triflorum, Erythrina stricta, Indigofera cassioides, Mimosa hamata, Mimosa pudica, Pueraria tuberosa, Tamarindus indica.*
50.	**Injuries**	*Abrus precatorius, Atylosia scarabaeoides, Bauhinia purpurea, Butea monosperma, Butea superba, Caesalpinia bonduc, Chamaecrista absus, Codariocalyx motorius, Dalhousiea bracteata, Desmodium triflorum, Entada rheedei, Indigofera tinctoria, Mimosa pudica, Mucuna bracteata, Pongamia pinnata, Oxytropis mollis, Senna occidentalis, Senna tora, Uraria alopecuroides.*
51.	**Insecticidal**	*Alysicarpus monilifer, Desmodium pulchellum, Flemingia strobilifera, Senna hirsuta, Trigonella emodi.*
52.	**Insomnia**	*Butea monosperma.*
53.	**Intenstinal infections**	*Abrus precatorius, Acacia catechu, Acacia farnesiana, Acacia nilotica* ssp., *indica, Ablizia lebbeck, Atylosia scarabaeoides, Bauhinia racemosa, Bauhinia semla, Bauhinia vahlii, Bauhinia variegata, Bauhinia malabarica, Butea monosperma, Caesalpinia bonduc, Cassia fistula, Crotalaria calycina, Crotalaria saltiana, Dalbergia latifolia, Dalbergia sissoo, Dendrolobium triangulare, Desmodium gangeticum, Desmodium oojeinense, Desmodium triflorum, Erythrina suberosa, Flemingia macrophylla, Flemingia procumbens, Indigofera linnaei, Macrotyloma uniflorum, Mimosa rubicaulis* ssp., *himalayana, Mimosa pudica, Mucuna pruriens, Neptunia triquetra, Physostigma venenosum, Pterocarpus marsupium, Pueraria tuberosa, Senna auriculata, Senna occidentalis, Senna tora, Spatholobus parviflorus, Vigna vexillata.*

Sl.No.	Uses	Species
54.	**Joint disorders**	*Abrus precatorius, Acacia catechu, Alysicarpus vaginalis, Atylosia scarabaeoides, Bauhinia purpurea, Butea monosperma, Caesalpinia bonduc, Cassia fistula, Codariocalyx motorius, Crotalaria pallida, Crotalaria spectabilis, Cyamopsis tetragonoloba, Dendrolobium triangulare, Erythrina variegata, Indigofera cassioides, Indigofera prostrata, Pongamia pinnata, Pueraria tuberosa, Senna auriculata, Senna occidentalis, Senna tora, Spatholobus parviflorus, Trigonella foenum-graecum.*
55.	**Kidney diseases**	*Indigofera tinctoria, Macrotyloma uniflorum, Medicago sativa, Mimosa pudica.*
56.	**Leprosy**	*Acacia nilotica* ssp., *indica, Astragalus himalayanus, Bauhinia variegata, Cassia fistula, Cullen corylifolium, Dalbergia sissoo, Macrotyloma uniflorum, Pithecellobium dulce, Pongamia pinnata, Senna obtusifolia.*
57.	**Leucoderma**	*Abrus precatorius, Acacia chundra, Cullen corylifolium, Senna italica.*
58.	**Leucorrhoea**	*Abrus precatorius, Acacia catechu, Dichrostachys cinerea, Flemingia strobilifera, Mimosa pudica.*
59.	**Liver complaints**	*Acacia catechu, Acacia nilotica* ssp., *indica, Bauhinia purpurea, Caesalpinia decapetala, Cassia fistula, Cicer arietinum, Entada rheedei, Indigofera tinctoria, Senna tora, Tephrosia purpurea.*
60.	**Lung diseases**	*Abrus precatorius, Acacia caesia, Acacia nilotica* ssp., *indica, Albizia myriophylla, Alhagi maurorum, Astragalus candolleanus, Caesalpinia bonduc, Cassia fistula, Chamaecrista absus, Crotalaria juncea, Desmodium gangeticum, Desmodium heterocarpon, Erythrina variegata, Indigofera cassioides, Lablab purpureus, Macrotyloma uniflorum, Mimosa hamata, Mimosa rubicaulis* ssp., *himalayana, Mucuna monosperma, Mucuna pruriens, Pongamia pinnata, Senna auriculata, Senna tora, Sesbania sesban, Tadehagi triquetrum, Tamarindus indica, Tephrosia villosa.*
61.	**Malaria**	*Acacia farnesiana, Bauhinia variegata, Caesalpinia bonduc, Pongamia pinnata.*
62.	**Measles**	*Acacia caesia, Macrotyloma uniflorum, Mimosa rubicaulis* ssp., *himalayana.*

Sl.No.	Uses	Species
63.	**Menstrual complaints**	*Abrus precatorius, Acacia catechu, Butea monosperma, Caesalpinia bonduc, Desmodium triflorum, Dichrostachys cinerea, Erythrina stricta, Erythrina variegata, Indigofera cassioides, Indigofera linifolia, Macrotyloma uniflorum, Pterocarpus marsupium, Pueraria tuberosa, Saraca asoca.*
64.	**Mumps**	*Atylosia volubilis, Entada rheedei, Sesbania grandiflora.*
65.	**Nail diseases**	*Desmodium triflorum.*
66.	**Nasal problems**	*Acacia catechu, Vigna grahamiana.*
67.	**Nervous disorders**	*Abrus precatorius, Acacia farnesiana, Indigofera tinctoria, Mucuna pruriens, Spatholobus parviflorus, Teramnus labialis.*
68.	**Oedema**	*Atylosia mollis, Atylosia scarabaeoides, Bauhinia purpurea, Entada rheedei, Mucuna pruriens, Tamarindus indica, Trigonella foenum-graecum.*
69.	**Pain**	*Abrus precatorius, Acacia nilotica* ssp., *indica, Acacia pennata, Albizia procera, Alysicarpus vaginalis, Atylosia mollis, Atylosia scarabaeoides, Bauhinia purpurea, Caesalpinia pulcherrima, Caesalpinia spicata, Crotalaria calycina, Dalbergia latifolia, Dalbergia paniculata, Dalbergia sissoo, Desmodium pulchellum, Desmodium triflorum, Flemingia macrophylla, Flemingia strobilifera, Mimosa pudica, Pongamia pinnata, Pterocarpus marsupium, Pueraria tuberosa, Sesbania grandiflora, Smithia sensitiva, Tamarindus indica, Uraria picta, Uraria lagopodioides.*
70.	**Paralysis**	*Abrus precatorius, Entada rheedei, Erythrina variegata.*
71.	**Piles**	*Albizia lebbeck, Bauhinia variegata, Butea monosperma, Mimosa pudica, Pongamia pinnata, Rhynchosia minima, Senna sophera, Tephrosia purpurea, Trigonella foenum-graecum.*
72.	**Purgative**	*Abrus precatorius, Bauhinia variegata, Caesalpinia bonduc, Cassia fistula, Clitoria ternatea, Crotalaria albida, Indigofera oblongifolia, Senna occidentalis.*
73.	**Rabies**	*Acacia farnesiana, Pithecellobium dulce.*
74.	**Refrigerant**	*Abrus precatorius, Butea monosperma, Cassia fistula, Chamaecrista absus, Crotalaria burhia, Pterocarpus marsupium.*

Sl.No.	Uses	Species
75.	**Revival of taste**	*Cajanus cajan.*
76.	**Rheumatoid arthritis**	*Abrus precatorius, Delonix regia, Mimosa pudica, Prosopis cineraria, Senna auriculata.*
77.	**Sciatica**	*Abrus precatorius.*
78.	**Scorpion bite**	*Abrus precatorius, Bauhinia purpurea, Cassia fistula, Mimosa pudica, Mucuna pruriens, Senna tora, Tamarindus indica.*
79.	**Sedative**	*Flemingia strobilifera.*
80.	**Sexual disorders**	*Dalbergia volubilis, Senna hirsuta.*
81.	**Skin diseases**	*Abrus precatorius, Acacia ferruginea, Acacia nilotica* ssp., *indica, Acacia sinuata, Astragalus aegacanthoides, Astragalus candolleanus, Bauhinia variegata, Butea monosperma, Caesalpinia bonduc, Caesalpinia pulcherrima, Cajanus cajan, , Cassia fistula, Chamaecrista absus, Crotalaria juncea, Crotalaria spectabilis, Cullen corylifolium, Dalbergia latifolia, Dalbergia paniculata, Dalbergia sissoo, Dolichos trilobus, Erythrina arborescens, Erythrina variegata, Flemingia macrophylla, Lablab purpureus, Mucuna pruriens, Pongamia pinnata, Pterocarpus marsupium, Pueraria tuberosa, Senna alata, Senna auriculata, Senna obtusifolia, Senna occidentalis, Senna sophera, Senna tora, Sesbania sesban.*
82.	**Small pox**	*Acacia caesia, Atylosia scarabaeoides, Bauhinia purpurea, Desmodium laxiflorum, Flemingia macrophylla, Lens culinaris, Macrotyloma uniflorum, Mimosa rubicaulis* ssp., *himalayana, Vigna unguiculata.*
83.	**Snake bite**	*Acacia farnesiana, Acacia pennata, Alysicarpus monilifer, Atylosia scarabaeoides, Bauhinia purpurea, Bauhinia semla, Bauhinia variegata, Cassia fistula, Clitoria ternatea, Crotalaria alata, Entada rheedei, Erythrina stricta, Erythrina variegata, Mimosa rubicaulis* ssp., *himalayana, Mimosa pudica, Millettia extensa, Mucuna pruriens, Paracalyx scariosus, Senna occidentalis, Tadehagi triquetrum, Tamarindus indica, Tephrosia purpurea, Uraria picta, Uraria lagopodioides.*
84.	**Spermatorrhoea**	*Atylosia scarabaeoides.*
85.	**Spleen complaints**	*Desmodium triflorum, Flemingia macrophylla, Indigofera tinctoria, Millettia extensa, Tephrosia purpurea.*

Sl.No.	Uses	Species
86.	**Throat complaints**	*Abrus precatorius, Acacia nilotica* ssp., *indica, Cajanus cajan, Mimosa rubicaulis* ssp., *himalayana, Mucuna nigricans, Senna occidentalis.*
87.	**Tonic**	*Abrus precatorius, Acacia catechu, Acacia nilotica* ssp., *indica, Bauhinia vahlii, Butea monosperma, Caesalpinia bonduc, Caesalpinia spicata, Cassia fistula, Cicer arietinum, Crotalaria bialata, Dalbergia lanceolaria, Desmodium gangeticum, Desmodium heterocarpon, Erythrina stricta, Pongamia pinnata, Pueraria tuberosa, Saraca asoca, Senna occidentalis, Tephrosia purpurea, Trigonella foenum-graecum.*
88.	**Tuberculosis**	*Acacia caesia, Acacia catechu, Albizia myriophylla, Astragalus candolleanus, Bauhinia variegata, Butea monosperma, Caesalpinia bonduc.*
89.	**Tumours and ulcers**	*Acacia catechu, Acacia chundra, Acacia farnesiana, Albizia lebbeck, Astragalus aegacanthoides, Atylosia platycarpa, Atylosia scarabaeoides, Bauhinia purpurea, Bauhinia retusa, Bauhinia variegata, Butea monosperma, Caesalpinia mimosoides, Cajanus cajan, Cassia fistula, Crotalaria calycina, Desmodium triflorum, Entada rheedei, Eriosema himalaicum, Erythrina suberosa, Erythrina variegata, Flemingia chappar, Indigofera tinctoria, Leucaena leucophloea, Macrotyloma uniflorum, Mimosa rubicaulis* ssp., *himalayana, Mimosa pudica, Mucuna monosperma, Mucuna pruriens, Pongamia pinnata, Pterocarpus marsupium, Pueraria tuberosa, Senna tora, Sesbania sesban, Tamarindus indica, Tephrosia candida, Tephrosia purpurea, Trigonella foenum-graecum, Vigna vexillata.*
90.	**Tympanitis**	*Caesalpinia bonduc, Pueraria tuberosa.*
91.	**Urinary complaints**	*Abrus precatorius, Acacia nilotica* ssp., *indica, Butea monosperma, Butea superba, Crotalaria bialata, Crotalaria calycina, Indigofera cassioides, Indigofera heterantha, Macrotyloma uniflorum, Mimosa rubicaulis* ssp., *himalayana, Mucuna pruriens, Pterocarpus marsupium, Saraca asoca, Senna tora, Sesbania sesban, Spatholobus parviflorus, Tamarindus indica.*
92.	**Urinary diseases**	*Acacia catechu, Acacia farnesiana, Acacia nilotica* ssp., *indica, Albizia lebbeck, Atylosia scarabaeoides, Bauhinia variegata, Cassia fistula, Crotalaria calycina, Dalbergia sissoo, Eriosema himalaicum, Mimosa rubicaulis* ssp., *himalayana, Mucuna pruriens, Paracalyx scariosus, Pterocarpus marsupium, Pueraria tuberosa, Spatholobus parviflorus, Tamarindus indica, Tephrosia purpurea, Trigonella foenum-graecum.*

Sl.No.	Uses	Species
93.	**Vermifuge**	*Bauhinia vahlii, Bauhinia variegata, Butea monosperma, Canavalia gladiata, Cassia fistula, Crotalaria pallida, Erythrina stricta, Flemingia strobilifera, Lablab purpureus, Meizotropis buteiformis, Mucuna bracteata, Mucuna pruriens, Oxytropis tatarica, Senna tora.*
94.	**Veterinary**	*Abrus precatorius, Atylosia scarabaeoides, Butea superba, Caesalpinia decapetala,Cassia fistula, Cicer microphyllum, Clitoria ternatea, Crotalaria linifolia, Crotalaria spectabilis, Cullen corylifolium, Cyamopsis tetragonoloba, Dichrostachys cinerea, Dalbergia latifolia, Entada rheedei, Erythrina arborescens, Erythrina resupinata, Flemingia strobilifera, Glycine max, Indigofera linifolia, Indigofera linnaei, Millettia extensa, Mimosa pudica, Mucuna pruriens, Pterocarpus marsupium, Pueraria tuberosa, Senna occidentalis, Senna tora, Tephrosia tinctoria, Tephrosia uniflora.*
95.	**Vomiting**	*Crotalaria prostrata, Dalbergia sissoo, Desmodium gangeticum, Desmodium laxiflorum, Pueraria tuberosa, Uraria picta.*
96.	**Whooping cough**	*Desmodium triflorum, Indigofera tinctoria, Pongamia pinnata, Senna occidentalis.*

CHAPTER-18

Legumes in *Ayurveda*

18.1 THE INDIAN SCIENCE OF LIFE

Ayurveda, 'the science of life', is an Indian system of medicine, based on solid foundations that are fully sustained by long experimentation and philosophical propositions, dating back to about 1,000 BCE. Among innumerable publications, Sathyanarayana Bhat and Kameswara Rao (1993b) and Kameswara Rao (2000) presented a detailed account of different aspects of *Ayurveda*.

Ayurveda encompasses all aspects of life and disease, such as anatomy, physiology, pathology, symptomatology and medical prescription, to maintain health and cure disease.

Knowledge of Ayurvedic medicine has unfortunately been confined to India and the West is largely ignorant of it. Even in India, this traditional medical practice has lost a lot of its importance in the urban situations. One of the main reasons for this is that much of the early and core medical literature on *Ayurveda* is in *samskrith*, the ancient language which ceased to be a day-to-day language, except in extremely small groups of the vast Indian population. Even today a considerable bulk of Ayurvedic knowledge is in the form of ancient palm leaf manuscripts hidden in remote libraries and private collections, and as treasured personal knowledge of a few individuals. The net result is that *Ayurveda* has been away from limelight and does not enjoy the importance and popularity it deserves.

Ayurveda is a very rational system of medicine, conceptually differing from a) the folk or home medicine, which is largely a matter of localized community experience and individualized medical practice, and b) from *bhoothavidya* that is heavily loaded with occult practices.

The Ayurvedic system grew vigorously till about 1300 CE. The beginning of that century, however, marked the end of a glorious era of growth and standardization. In the medieval times, the *rasaayana* (alchemy) practices reappeared, and the drugs and remedies evolved in these systems got incorporated into *Ayurveda*.

Ayurveda had a glorious past even during the Buddhist times. It was taught in Universities of Nalanda and Takshashila. Subsequent cultural revolutions, spiritual renaissance and various political interventions have caused the decline of *Ayurveda*. Jainism and Buddhism, both preaching total nonviolence, have also affected *Ayurveda*, as many animal products were used as medicine. Surgery was totally banned in the cause of nonviolence. Human cadaver dissections were forbidden. Muslim rulers invaded India resulting in a serious cultural change. Moghal concepts of medicine were encouraged and *Ayurveda* faced a tough time. Further on, the European colonization, and import and domination of Allopathy have caused an immense damage to the practice of *Ayurveda*.

Even after India achieved independence, the situation did not improve immediately. The Indian systems of medicine received governmental recognition only after the National Health Policy was adopted in 1983. The status of Ayurvedic education has improved slowly, though never to the extent desirable. The Central Council of Indian Medicine now supervises this area. The issue is now focussed better after the setting up of the Department of Indian Systems of Medicine and Homoeopathy, in 1995, in the Ministry of Health and Family Welfare. The State Governments have similar departments. There are many colleges that teach *Ayurveda* at the degree and postgraduate level and conduct research. However, there is a great and urgent need to establish several centres of higher learning and research on *Ayurveda* in the country, as there are only very few such institutions now, such as the Gujarath *Ayurveda* University, Jamnagar, the Faculty of *Ayurveda* of the Banaras Hindu University, and the National Institute of *Ayurveda*, Jaipur, in India.

18.2 THE DIMENSIONS OF AYURVEDA

Ayurveda aims at all men being healthy, fit in body and keen in mind. Curing of diseased condition and the maintenance of health have not been the only aims of *Ayurveda*. It is concerned with harmonizing secular conduct and spiritual pursuit through a realization of the true relationships between and among the complex of body, mind and soul and the eternal Universe. The ultimate end is not merely mundane happiness but spiritual elevation. To achieve its objectives, *Ayurveda* developed into a comprehensive encyclopedia of knowledge system in medical subjects like genetics, gynaecology, obstetrics, aetiology, diagnosis, therapeutics, surgery, physiology, biology, dietetics, ethics, personal hygiene, preventive treatment and social medicine. The allied subjects like animal biology, botany, herbal cultivation, pharmacognosy, compounding and chemistry and also some subjects that are not usually considered as medical, like cosmology, climatology, psychology, parapsychology, philosophy and religion, have an important place in the Ayurvedic system. Mastery of *Ayurveda* presupposes knowledge in all these subjects.

The basic Ayurvedic texts (*samhitha*s) are distinguished by careful and penetrating observations, exhaustive and classified information, and an able presentation of the available knowledge in medicine and allied subjects that contribute to medicine.

There is a rigorous standard for the training of physicians and a meticulous code of personal ethics and social conduct for the medical profession. With deep insight, farsightedness and depth, Ayurvedic medical code stands a favourable comparison with the Hippocratic code.

18.3 PHILOSOPHICAL AND THEORETICAL BASIS OF AYURVEDA

The philosophical and theoretical basis of *Ayurveda* is rooted in the following:

a) the *Panchamahaabhootha*s: the five elements of the physical world; *prithvi* (the earth), *jala* (water/liquid), *tejas* (energy/light/heat), *vaayu* (wind), and *aakasha* (space);
b) the *Thridosha*s: three humours; *vaatha* (*vaayu* plus *agni,* the heat), pitta (*tejas*), and *kapha* (*prithvi* plus *jala*); and
c) the *Saptadhaathus*: seven elements of the body tissues; *rasa* (plasma), *raktha* (blood), *maamsa* (muscle), *maeda* (fat), *asthi* (bone), *majja* (marrow) and *shukra* (reproductive fluids).

The treatment and the drugs vary depending upon the physical element involved, the *dosha* that is deranged and the body tissue that is affected.

Ayurveda promoted the concept that food is medicine and that prevention of a disease is better than its cure.

18.4 STRATEGIES OF AYURVEDIC TREATMENT

Ayurveda deals with the disease and health management through the following strategies:

a) *Chikitsa dhatu satmyartha* (to bring about balance in tissues);
b) *Swabhaavoparma* (to augment the body's efforts to ward of disease, through fasting, warm water, rest);

c) *Nidana parivarjana* (avoidance of causative factors);
d) *Chikitsa rukpratikriya* (treatment that is antagonistic to disease);
e) *Saamanya* and *Vishesha Siddhantha* (treatment by the principle of similarity and dissimilarity, of the body, food, medicine, etc.);
f) *Satvavajaya, Aatmanigraha* (by the control of body and mind);
g) *Chayaeva jayedosha:* (to treat at the initial stage of the disorder); and
h) *Samprapti vighatana* (to disrupt pathogenesis).

18.5 BRANCHES OF AYURVEDA

Ayurveda has eight specialized branches, called *chikitsa*s (treatments):

a) *Kaya chiktsa* (diseases and symptamatology);
b) *Baala chikitsa* (paediatrics);
c) *Graha chikitsa* (*bhoothavidya*);
d) *Shalakya* or *Urdhwanga chikitsa* (Ottopharyngology and ophthalmology);
e) *Shalya chikitsa* (surgery);
f) *Visha chikitsa* (toxicology);
g) *Jara chikitsa* (geriatrics); and
h) *Vaajikara chikitsa* (invigouration and eugenics).

18.6 DIAGNOSIS IN AYURVEDA

The Ayurvedic physician examines (*pareeksha*) the patient (*rogi*) and the disease (*roga*), to obtain information regarding *dushya* (body tissues), *dasha* (humours), *dasha* (state of the body), *bala* (severity of the disease), *kaala* (time of the disease), *anala* (strength of digestion and metabolism), *prakriti* (constitution of the patient), *vaaya* (age of the patient), *satva* (psychological condition), *saathmya* (life style of the patient) and *aahaara* (dietary habits).

The examination of the patient constitutes the following aspects: *nadi* (pulse), *mootra* (urine), *pureesha* or *mala* (faeces), *jihwa* (tongue) and *netra* (eye) through *shabda* (auscultation, sound response), *sparsha* (tactile stimulation) and *aakriti* (the appearance and feel of different parts of the body).

18.7 TYPES OF THERAPIES IN AYURVEDA

Ayurveda administers palliative (*Shamana*), investigative (*Shodhana*) and rejuvenative (*Kaayakalpa*) therapies.

18.8 METHODS OF TREATMENT IN AYURVEDA

Ayurveda uses the following means in treatment:

a) Oral medication, single or compounded medicinal substances (formulations);
b) Eye drops, salves for eye disease and in loss of visual power, etc;
c) Gargles for disease of mouth, throat, teeth, gums;
d) Medicated cigars, smoking mixtures for diseases of head, nose, throat and bronchial tract;
e) Nasal medicinal powders, ointments, fumes and inhalation in fainting fits, epilepsy and insanity;
f) Liquid unguents, creams, lotions and medicinal oil for skin;
g) Suppositories, cotton swabs soaked in medicinal oil, for the ear passage, lower orifices, wounds;
h) Enemas of nutritive and healing fluids (*per* rectal, *per* vaginal, *per* urethral);
i) Sweating, sun bath, heated air, steam, contact with hot surfaces;
j) Douches to flush rectal/vaginal/urethral passages;

k) Bandages, splints, tourniquets and ligatures for poisonous bites;
l) Blood letting by incision, using leeches or other methods;
m) Eight types of surgical methods; and
n) Psychiatric methods.

18.9 MEDICAL ASTROLOGY

Ayurveda conceptualises the human being as a miniature replica of the Universe and with a close relationship with it. Astrology, therefore, runs parallel to the Ayurvedic doctrines. The balance or imbalance of the *dosha*s is influenced by the 12 zodiacs, 27 stars and nine planets. The position of the celestial bodies at the time of one's birth, is hence an important factor in determining the constitution of a person. The use of *mantra*s (incantation) to bring in an equilibrium in the disturbance caused by the astrological factors is recommended in *Ayurveda*. *Mantra*s are propitiatory hymns to Gods and celestial figures, now interpreted as packages of ordinated sound waves. They afford a psychological therapy combined with a form of sound therapy. Hence, a correct pronunciation and the proper style of chanting the *mantra*s is a crucial aspect, mastered only by a few.

18.10 THE PHYSICIAN-PATIENT RELATIONSHIP

By tradition, the *Vaidya*, the Ayurvedic physician, is well integrated with the society and knows the patient and family. He is the friend, guide and philosopher and is well respected by the community. This attitude is recommended in every system of medicine. Unfortunately, this ideal situation has undergone a vast change in recent times, with increased urbanisation, but by and large is still present in the rural places.

18.11 THE SAMHITHAS, THE ENCYCLOPAEDIC SOURCE TEXTS OF AYURVEDA

The treatises called *Samhitha*s prepared by Charaka, Sushrutha and Ashtangahridaya, are the major source texts of *Ayurveda*. There are scientific synoptic publications on the *samhithas* of Charaka and Sushrutha in English (Ray and Gupta, 1980; Ray *et al.*, 1980)

Charaka Samhitha deals mainly with anatomy, physiology, aetiology, prognosis, pathology, the influence of environmental factors, medicines, appliances, procedure and sequence of medication. These topics were discussed in 150 chapters, under the following eight sections: a) *Suthra sthana*, b) *Nidana sthana*, c) *Vimana sthana*, d) *Shareera sthana*, e) *Indriya sthana*, f) *Chikitsa sthana*, g) *Kalpa sthana* and h) *Siddhi sthana*.

Sushrutha Samhitha follows more or less the same pattern, but gives surgery, the place of honour. The Dhanvanthari school of medicine, to which Sushrutha belonged, believed surgery to be the most ancient and most efficacious of the eight branches of medical knowledge. *Sushrutha samhitha* contains six sections. It deals with fundamental postulates, pathology, embryology, anatomy, therapeutic and surgical treatment, and toxicology in five sections (*Sutra, Nidana, Shareera, Chikitsa,* and *Kalpa sthanas*). The final section is on subsequently gained and specialized knowledge of topics dealt with in earlier works. These are arranged in 184 chapters. *Sushruhta Samhitha* is very concise in language and is a repository of extensive factual information. It describes the necessity and *modus operandi* of dissections on human cadavers for gaining accurate anatomical knowledge. The elaborations of a large variety of medical prescriptions and their compounding, therapeutic methods including psychiatric and surgical procedures employing specific instruments for each type of operation are scientific, sound and exhaustive. Sushrutha described more than 300 different operations employing 42 different surgical processes and 121 different types of instruments.

Symptoms described in the *samhitha*s deal with pathological conditions and diseases, including diabetes mellitus, pulmonary tuberculosis, malignant growths, leprosy, gangrene, erysipelas, jaundice, diphtheria, tetanus, calculi, paralysis, insanity, epilepsy, epidemic diseases, bites of poisonous snakes and animals and hydrophobia. Rational and apparently efficacious surgery is described for many conditions including complicated fractures and dislocation, piles, fistulae and sinuses, tumours, carbuncles, malignant growths, cataract, complicated ophthalmic operations, strangulated hernia, urinary stones causing uremia, impacted gall stones, intestinal perforations and protrusions of the viscera, accidental injuries, amputations of the major limbs, tonsillitis, bone abscess and abscess of internal organs, serious head injuries with exudation of brain matter, and major obstetric operations including removal of the foetus. Interestingly even plastic surgery involving grafting of skin and muscular tissues from surrounding areas and/or other parts of the body was also discussed to help people whose noses and ears were removed for penal reasons, an apparently common practice of that period.

18.12 THE AYURVEDIC PHARMACOPOEIA

There are over 70 books that provide about 8,000 recipes of Ayurvedic medicines. The medicines are in the form of *arka* (distillates), *asava* and *arista* (fermented products), *avaleha* (linctus), *bhasma* (incinerated matter), *choorna* (powder), *ghritha* (in ghee), *taila*s (in oil), *vati* and *gutika* (tablets and pills) and *kwatha* and *kashaaya* (decoctions). In recent times, Ayurvedic medicines are also being dispensed as capsules or even injections, which naturally attracted a considerable criticism.

18.13 THE AYURVEDIC *MATERIA MEDICA*

Dravya is a substance, the drug, and *guna* is its attribute or property. *Virya* is the drug's potency, *vipaka* the taste, and *karma* its action. *Dravyaguna* is that aspect of *Ayurveda* that describes the sources and properties of substances used as food and medicine, and is equivalent to *Materia Medica*. The *Materia Medica* of *Ayurveda* is extensive and utilises substances of animal, vegetable and mineral origin. The Ayurvedic texts make it abundantly clear that the drugs included had each been selected after careful clinical experiments and long experience. The specific properties like tastes, assimilability, potency and physiological actions are described in almost all cases.

Such evaluation were naturally conducted mainly with materials available to the pioneers in their own environment, which explained the strong preference shown by classical *Ayurveda* for indigenous plant remedies. A second reason for this preference was that the plants being composed of all the five elements (the *panchamahaabhootha*s) had a natural flexibility to a wider range of application.

In *Charaka samhitha*, 341 drugs from plants, 177 drugs from animals and 64 from minerals and metals were discussed. The corresponding numbers, for *Sushrutha samhitha* are 395, 57 and 64, respectively. The animal products are found less and less in the later works, due to the prevalence of Buddhistic and Jainistic philosophies. Plants provided the natural and traditional drugs for most, particularly oral, medication. A clear understanding of their properties had led to a rational classification into different groups according to their therapeutic action.

18.14 LITERATURE ON AYURVEDA

There is extensive literature on diverse aspects of *Ayurveda,* in various languages, though the prime sources are in *samskrith.* A detailed list was compiled by Kameswara Rao (2000).

18.15 LEGUMES IN AYURVEDA

Among the vast number of plant species used in the Ayurvedic medicine, legumes have an important place. Many legumes are used as single species drugs or in numerous different formulations, each containing several different species. The legume species used in Ayurvedic formulations are given in Table 18.1. The therapeutic uses of individual legume species used in *Ayurveda* are given in the respective species profiles.

Table 18.1
Legumes in Ayurvedic Formulations

Sl. No.	Species	Formulations
1.	*Abrus precatorius*	*Bhallatakadi thaila, Goracandi gutika, Gunjabhadra rasa, Gunjadi thaila, Gunja pralepa, Ganjaphalagni choorana, Indralupta samana lepa, Mahalaxmi Narayana thaila, Mahavasanta kusumakara, Mritasanjeevani gutika, Nilibringadithaila, Svetagunjadi gutika*
2.	*Acacia catechu*	*Arimedadi thaila, Dasamoolarishta, Dasanasamskara choorana, Kadirarista, Khadira kalpa, Khadiradi gutika, Khadiradi thaila, Khadirashtaka choorana, Khadhiradi kwatha, Khadhiradi vati, Mahakhadiradya ghritha*
3.	*Acacia farnesiana*	*Irimedadya thaila, Khadiradi gutika, Khadiradi thaila, Khadiradivati, Ksaragada*
4.	*Acacia nilotica*	*Babbularista, Babbula choorana, Lavangadivathi, Malatyadyama ghritha, Mritasanjeevanisura, Oozhimathirai*
5.	*Albizia lebbeck*	*Devadarvadyarishta, Pancasirisa agada*
6.	*Alhagi camelorum*	*Agastyavaleha, Bhadrastiyadigana, Vasadya ghritha*

Sl. No.	Species	Formulations
7.	*Alhagi maurorum*	*Dasamoolaristam, Duralabharistam, Dvipanca muladya ghritha, Jivakadya sneha, Rasanachi kasaayam*
8.	*Bauhinia malabarica*	*Candanasava, Kanchanara guggulu, Kanchanaradi kwatha, Vidangarishta, Usheerasavam*
9.	*Bauhinia purpurea*	*Candanasava, Kanchanara guggulu, Kanchanaradi kwatha, Vidangarishta, Usheerasavam*
10.	*Bauhinia racemosa*	*Candanasava, Kanchanara guggulu, Kanchanaradi kwatha, Vidangarishta, Usheerasavam*
11.	*Bauhinia variegata*	*Candanasava, Kanchanara guggulu, Kanchanaradi kwatha, Vidangarishta, Usheerasavam*
12.	*Butea monosperma*	*Ayaskriti, Balathailam, Ceriyaaimaida thaila, Janamguti, Krimikuthara rasa, Mahanaarayana thaila, Palasabhijadi choorana, Palasakshara ghritha, Palasakshara thaila*
13.	*Caesalpinia bonduc*	*Misrakasneha, Pramehamihira thaila, Shakravati, Vayasyagudali lepa yoga, Vishamajwaraghni vati*

Sl. No.	Species	Formulations
14.	*Caesalpinia crista*	*Cangeryadi gutika, Ceriyaanthrakutaram gutika, Kazharchi chooranam, Kazharchi thaila, Kazharchi vittu chooranam, Meganatha thaila Paphanadi thaila, Ponkaradi gutika, Ullivettadukadi kwatha.*
15.	*Caesalpinia sappan*	*Chandanadya thaila, Karpuradyarka, Khadiradi gutika, Kumkumadi thaila, Patrangasava*
16.	*Cajanus cajan*	*Thuvarai verkuzti thaila*
17.	*Cajanus goensis*	*Bala thailam, Cyavanaprasa leham, Dhanvantaram kasaaya, Dhanvantararishtam, Vidaryadyasavam*
18.	*Cassia fistula*	*Aragvada shatapaka ghritha, Aragvada kvatha choorana, Aragvadhadi leha, Aragvadhadi thaila, Aragundharishta, Avittaladi bhasma, Brahmi ghritha, Kastharaksha thaila, Mahamarichyadi thaila, Mahamanjishtadyarishta, Manibhadaram Rasanadikvatha yoga, Valiya kasnadi kasaayam, Vasista rasaayanam*
19.	*Chamaecrista absus*	*Caksusyakalpa*
20.	*Cicer arietinum*	*Chanaka kshara, Gandhaka rasaayanam, Kravyada rasa*
21.	*Clitoria ternatea*	*Agastyahareethaki rasaayanam, Brahmi ghritha, Garbhapala rasa, Manasamitravatakam, Manjisthadi kasaayam, Misrakasneham,*

Sl. No.	Species	Formulations
		Somaraji thaila, Suvarnamakara dhwaja, Triphala ghritha, Visha thaila, Vishnu thaila
22.	*Crotalaria juncea*	*Dasamoola ghritha, Muktadichoorana, Nirgundyadi ghritha, Nirgundyadi gutika*
23.	*Crotalaria pallida*	*Dasamoola ghritha, Muktadichurna, Nirgundyadi ghritha, Nirgundyadi gutika*
24.	*Crotalaria retusa*	*Dasamoola ghritha, Nirgundyadi ghritha, Nirgundyadi gutika*
25.	*Crotalaria verrucosa*	*Dasamoola ghritha, Nirgundyadi ghritha, Nirgundyadi gutika*
26.	*Cullen corylifolium*	*Avalgujabijadichoorana, Avalgujadi lepam, Gopicandanadi gutika, Karappu thaila, Karpogi thaila, Khadirarishtam, Manjisthadi kasaayam, Rasagandhi moluku*
27.	*Dalbergia sissoo*	*Ayaskriti, Narasimha ghritha, Shalasaradileha*
28.	*Dendrolobium umbellatum*	*Dasamoola kwatha*
29.	*Desmodium gangeticum*	*Amritapasa ghritha, Amrtarista, Bramha rasaayana, Bandhika ghritha, Chitrakaharitaki, Cyavanaprasa, Dasamoola katutraya kwatha choorana, Dasamoola kwatha choorana, Dasamoola latpalaka ghritha,*

Sl. No.	Species	Formulations
		Dasamoola rasaayana, Dasamoola ghritha, Dasamularista, Dhanvantara ghritha, Dhanvantaram kulambu, Duntyadyarishta, Indukanta ghritha, Kalyanaka ghritha, Mahakalyanaka ghritha, Mahanarayana thaila, Mooshikadi thaila, Mritasanjivani sura, Shalaparnyadi kwatha, Siva gutika, Sukumara ghritha, Surendra thaila, Vastyamayantaka ghritha, Vyaghri thaila
30.	*Desmodium laxiflorum*	*Dasamoolarista, Dasamoola rasaayana, Dhanvantaram kulambu*
31.	*Desmodium triflorum*	*Manasamitra vatakam, Madhu yastadi thaila, Vidaryadi ghritha, Vidaryasavam*
32.	*Desmodium velutinum*	*Dasamoolarishta, Dasamoola rasaayana*
33.	*Dichrostachys cineraria*	*Veerataradi gana*
34.	*Erythrina stricta*	*Ksharagada, Maharaja prasarini thaila, Paribhadra valeha, Shikhari ghritha, Sudhakara thaila, Vidangamooladi kasaayam*
35.	*Erythrina variegata*	*Maharaja prasarini thailam, Paribhadravaleha, Vidangamooladi kasaayam*
36.	*Glycyrrhiza glabra*	*Ashwagandharishta, Asoka ghritha, Candana balalaksadithaila, Kalyani lehya, Kanakasava,*

Sl. No.	Species	Formulations
		Kutajarishta, Madhuka thaila, Phala ghritha, Pinda thaila, Prasarani thaila, Rhala ghritha, Sabdindu thaila, Satavari guda, Yastimadhu thaila, Yastiyadichoorana
37.	*Indigofera aspalathoides*	*Mahavallathy ivakam, Meganatha thailam, Sivanarvembukuzhi thailam*
38.	*Indigofera tinctoria*	*Neelabhringadi thailam, Nilinyadichoorana, Nilinyadighritha, Nilinyadigutika*
39.	*Lathyrus sativus*	*Kalayacurnagutika*
40.	*Lens culinaris*	*Masura ghritha, Propaundari kadi lepa, Rohinyadya ghritha, Sanatisaranashaakayavagu, Triphala yaga*
41.	*Macrotyloma uniflorum*	*Kulatthadya ghritha, Vanga bhasma*
42.	*Mimosa pudica*	*Pradarari lauha, Pusyanuga choorana, Rajatabhasma, Yuvatyadi thailam*
43.	*Mucuna pruriens*	*Amritaprasaghritham, Asvagandhadi leham, Atmagupta chooranam, Gandhaka rasaayanam, Stanyajananarasaayanam, Vanari gutika, Vidaryadi kasaayam*
44.	*Phaseolus mungo*	*Mahamasa thaila*
45.	*Pongamia pinnata*	*Aragvadhadi kwatha choorana, Jatyadi thaila, Karanjadya ghritha,*

Sl. No.	Species	Formulations
		Karanjadya thaila, Nagaradi thailam, Varanadi ghritham, Varanadi kasaayam, Vishnu thaila
46.	*Prosopis cineraria*	*Ksaragada*
47.	*Pseudarthria viscida*	*Agastya rasaayanam, Anuthailam, Dasamularishtam, Valiyapancagavya ghritham*
48.	*Pterocarpus marsupium*	*Asana bilwadi thaila, Ayaskriti, Bijakshara arishta, Khadirasaradighritham, Nyagrodhodichoorana, Varadi kasaayam, Vasista hareetaki*
49.	*Pterocarpus santalinus*	*Amritadi kwatha, Arakku thailam, Bhadrastiriyadi gana, Candanasava, Guducyadi modaka, Kiratadyarista, Laksadi ghritha, Lavangadichoorana, Maricadi thaila, Priyangavadiniruhabasti, Rhala ghritha, Sahacaradi thaila, Sarivadyasava, Thirathi chatty chooranam, Vallarainey*
50.	*Pueraria tuberosa*	*Cyavanaprasam, Dasamularistam, Dhatriyadi ghritha, Dvipanca moolodya ghritha, Jivakadya sneha, Satavari guda, Vidaryadi ghritha, Vidaryadi kasaaya, Vidaryadi asava*

Sl. No.	Species	Formulations
51.	*Saraca asoca*	*Asokaghritham, Asokarishta, Devadaryarishta, Kasisadi thaila, Madhukadyavalehya, Mahamarichyadi thaila, Pradarari rasa*
52.	*Senna alexandrina*	*Aragvadhadi kvatha, Aragvadhadi thaila, Aragvadhadi lehya, Aragvadharishta, Malak kudara neluku, Mahamanjishthadyarishta, Nilaavarai chooranam, Rasanadi kvatha yoga*
53.	*Senna auriculata*	*Meganatha thailam*
54.	*Senna occidentalis*	*Kasamardadi ghritha, Punarnavadi ghritha*
55.	*Senna tora*	*Cakramarda thaila, Dadrughnavati, Maharaja prasarini thailam, Saindhavadi lepa, Surasadi thailam, Yastimadhukadi thailam*
56.	*Sesbania grandiflora*	*Cempu parpam, Grahanikapata rasa, Pittakasantaka rasa, Rajata bhasma, Ratnagiri rasa, Sirungi parpam, Thanga parpam*
57.	*Sesbania sesban*	*Karunsembai thailam*
58.	*Tamarindus indica*	*Amlikasatva, Amlikasatwadi choorana, Chinchaksharadishankha vati, Cincadi leham, Hinguvacadi chooranam, Kottam chukkadi thaila, Mahajwarankusha rasa, Mahashankavati, Pancamla thailam, Shankavati, Shankarasa gutika, Tintidyadi kasaaya*

Sl. No.	Species	Formulations
59.	*Tephrosia purpurea*	*Argvadhadi kasaayam, Grihanyantaka ghritham, Viraatavadi kasaayam*
60.	*Teramnus labialis*	*Balathailam, Bramha rasaayana, Cyavanaprasa leham, Dhanvantaram kasaayam, Dhanvantararishtam, Dvipancamoolyadi ghritha, Jivakadya ghritha, Jivakadya thaila, Sastikadi gutika, Vidaryadyasavam*
61.	*Teramnus mollis*	*Vidaryadi kwatha*
62.	*Trigonella foenum-graecum*	*Bhakttatriya choorana, Brihat sowbhagyashunti, Chaturbeeja choorana, Mastakarishta, Methi modaka, Mritasanjeevani sura, Shiva modaka, Sundaivatral chooranam, Vendayum ilakam*
63.	*Uraria lagopodioides*	*Agastya rasaayanam, Anuthailam, Dasamularistam, Vatiyapancagavya ghritham*
64.	*Uraria picta*	*Agastya rasaayanam, Anuthailam, Dasamularistam, Dasamoola rasaayana, Sahacaradi thaila, Vatiyapancagavya ghritham*
65.	*Vigna adenantha*	*Amritaprasa, Mahakalyanaka ghritham, Matiamasa thailum, Vidaryadi leham*
66.	*Vigna angularis*	*Amritaprasa, Mahakalyanakaghritham, Matiamasathailum, Vidaryadi leham*

Sl. No.	Species	Formulations
67.	*Vigna marina*	*Amritaprasa, Mahakalyanakaghritham, Matiamasathailum, Vidaryadi leham*
68.	*Vigna mungo*	*Balathailum, Cyavanaprasa leham, Dhanvantaram kasaayam, Dhanvantararishtam, Mahamasa thaila, Pirandai vatakam, Vidaryadyasavam,*
69.	*Vigna pilosa*	*Amritaprasa, Mahakalyanakaghritham, Matiamasathailum, Vidaryadi leham*
70.	*Vigna radiata*	*Balathailam, Cyavanaprasa leham, Dhanvantaram kasaayam, Dhanvantararishtam, Vidaryadyasavam*
71.	*Vigna trilobata*	*Amritaprasa, Mahakalyanakaghritham, Matiamasathailum, Sastikadigutika, Vidaryadi leham*
72.	*Vigna umbellata*	*Balathailam, Cyavanaprasa leham, Dhanvantaram kasaayam, Dhanvantararishtam, Mahamasa thaila, Pirandai vatakam, Ulundu thailam, Vidaryadyasavam*

CHAPTER-19

Legumes in the *Siddha* System of Medicine

19.1 INTRODUCTION

The name *Siddha oushadha* (*siddha* medicine) relates to the earlier esoteric medicinal postulates concerning longevity, even immortality, and the later iatrochemical formulations as conceived and practiced by 18 *Siddhar*s, located in what is today, the Tamil Nadu. It is for this reason the entire *Siddha* medical literature is in Tamil. The *Siddha* system has about 800 texts of which 180 are in print. Kurup (1983), Sathyanarayana Bhat and Kameswara Rao (1993c), and Kameswara Rao (2000) have discussed the *Siddha* system of medicine in detail.

The *Siddha* system of medicine is rooted in the Dravidian culture, of the pre-*Vedic* period. The *Vedic Aryans* owed allegience to the cult of *Shiva*, which was later absorbed by the vedic culture (Kurup, 1983).

The *Siddha* tradition has incorporated minerals and metals, many of which are very toxic (mercury, sulphur, arsenic, etc.), and vegetable poisons. Independent of the *Siddha* system, India has developed its own tradition of alchemy and iatrochemistry, called *Rasashastra* or *Rasa vaidya*, that employed minerals, metals and some plants (Sathyanarayana Bhat and Kameswara Rao, 1993a).

The *Siddha* system of medicine is most predominant in Tamil Nadu. It is also popular in Andhra Pradesh, Karnataka and Kerala. Even outside India, the *Siddha* system is common in Sri Lanka, Malaysia and Singapore, where the Dravidian civilization had taken roots. In spite of the emphasis on iatrochemistry, there are many similarities between *Siddha* and *Ayurveda*.

19.2 ORIGIN AND DEVELOPMENT OF THE *SIDDHA* SYSTEM

The origin of alchemy in India has been traced to the seed ideas of the Chinese alchemy, established around the third and fourth centuries CE, giving a more important position to mercury and sulphur among other metals and minerals, not only in respect of aurification but also in the preparation of metallic/mineral elixirs and medical compositions. The *Siddha* alchemy came to the fore in India between 500 and 600 CE, and the *Siddha* medicine between 900 and 1000 CE.

The *Siddha* system received the patronage of Tamil Kings and Chieftans as well as the general public all through. *Siddha* practitioners were active also during the British colonial times. Since the Indian independence, the Government of India and that of the State of Tamil Nadu have encouraged the system very considerably. In fact, such close attention to an indigenous system is rare to find outside Tamil Nadu.

The Indian Medicine Practitioners' Co-operative Society, set up in 1945 in Chennai, has been engaged in the preparation and marketing of over 200 *Siddha* formulations. It had also published a Formulary of *Siddha* Medicines (Anonymous, 1989). The Central Government's

Siddha Research unit in Chennai has been conducting clinical trials of *Siddha* compositions. Single drugs were also standardized (Anonymous, 1992b, 1992c).

There are over 11,000 registered *Siddha* medical practitioners and nearly 500 licensed pharmacies. Two medical colleges in Tamil Nadu are exclusively devoted to teaching the *Siddha* system. There is a large number of hospitals and dispensaries, and three units of drug standardization. In addition, there are tribal health care centres dispensing *Siddha* medicine. There are clinical research centres, mobile clinics and farms to cultivate rare herbs needed by the drug manufacturers and practitioners.

19.3 PRINCIPLES OF THE *SIDDHA* SYSTEM

Siddha practitioners have adopted the Ayurvedic concept of *Tridosha* and give importance to the examination of the pulse. An examination of the urine too is considered important in the *Siddha* system. Astrology and incantation are also a part of the *Siddha* system.

Minerals, metals, salts, toxic substances and even herbs are classified in the *Siddha* system as male and female. These substances are also recognized as friendly or inimical, probably based on their mutual compatibility or otherwise. The *Siddha* system identifies 120 *uparasas*, 64 toxic substances, nine metals and nine gems, each requiring an elaborate processing, most often by high physical heat before they are considered fit for use in a medical composition. In many ways, this processing has similarities with that found in *Rasashastra* texts. Although there were attempts to categorise and interpret the ingredients of medicinal formulations in relation to the *Panchabhoothas*, they were not as detailed as in *Ayurveda* in which the *Panchabhootha* concept is a fundamental principle. The formation of the embryo, physiological processes, six *rasas*, *tridhathu, tridosha*, curative practices and the like, find rational relative theories in *Siddha* as in *Ayurveda*.

In the *Siddha* system, *Kaayakalpa* has a very conspicuous place. Inherent in *Kaayakalpa* is the belief of the *Siddha* physicians that the human body consists of 72,000 veins and nerves, six vital centres, 10 vital airs, and 10 vital pulses. These appear to relate to tantrik and yogic concepts. One of the notable characteristics of *Kaayakalpa* is the intake of *uuppu*, the three salts, besides the administration of meticulously processed minerals and other rejuvenating compositions, use of potent herbal extracts, breathing regulation, conservation of sperm and others. *Muppu* is believed to enhance the efficacy of any *Siddha* medicine, but its preparation, and even the composition are a closely guarded secret. In fact a very considerable part of the *Siddha* system appears to be shrouded in a cloak of secrecy; the texts being only in Tamil and the formulations being referred mostly by numbers, promotes security.

The *Siddha* system differs from pure alchemical medication in its insistence that the medicine be prepared through very elaborate processes incorporating a number of herbs in the formulations. In this respect the *Siddha* system is similar to the *Rasashastra* of *Ayurveda*, though the *Materia Medica* of the *Siddha* system is not extensive. A *Siddha* medical practitioner uses the same formulation for different ailments but varies its adjuvant, called *anupanam*, such as milk, ghee, honey, herbal extracts, ginger juice, betel leaf juice, cold water, warm water, etc. It is believed that, if proper, the adjuvant itself would modify the therapeutic properties and potency of the drug leading to the desired effect.

19.4 LEGUMES IN *SIDDHA* MEDICINE

A standardised formulary for 242 drug compositions that also involve about 100 species of plants was prepared over a quarter century ago. A number of legume species find important place in *Siddha* medicines. The legumes used in *Siddha* formulations are given in Table 19.1. Some species are used in several formulations. The therapeutic uses of different species of legumes in *Siddha* are given in the respective species profiles.

Table 19.1
Legumes in *Siddha* Formulations

Sl. No.	Species	Formulations
1.	*Abrus precatorius*	*Mahavasantha kusumaakaram, Moosambara patru*
2.	*Acacia catechu*	*Karisalai ilakam*
3.	*Acacia nilotica*	*Amukkara chooranam, Inbooral maathirai, Kabaada maathirai, Karuppu vittunuch chakkaram, Pancha deepaakkini chooranam, Parangi pattai chooranam, Thaaleesadi chooranam, Thiraatchaadi chooranam, Vallarai, Yelaathi chooranam*
4.	*Acacia pennata*	*Maanda ennai* No.1
5.	*Bauhinia tomentosa*	*Venpoosuananey, Laghu chandanadi thailam, Padiga linga chendooram*
6.	*Butea monosperma*	*Kshaya kulanthaga chendooram, Meghanatha tailam, Murukkan vidai maathirai, Vaata surak kudineer*
7.	*Caesalpinia bonduc*	*Kabaada maathirai, Kazharthi thailam, Meghanaatha thailam, Venkaara maathirai*
8.	*Caesalpinia crista*	*Kazharthi chooranam Kazharthi thailam, Kazharthi vittu chooranam, Meganatha thailam, Vaathasura kudineer, Venkaara maathirai*
9.	*Cajanus cajan*	*Thangaparpam, Thuvarai verruzti thailam*
10.	*Cassia auriculata*	*Avrai kudineer, Meghanaatha thailam*
11.	*Cassia fistula*	*Karunai legiyam, Nerinchil kudineer, Thiraatchai kudineer*

Sl. No.	Species	Formulations
12.	*Cicer arietinum*	*Kanthaka rasaayanam*
13.	*Clitoria ternatea*	*Kaakkana maathirai*
14.	*Cullen corylifolium*	*Erasa kenthi mezhugu, Karappam thailam, Karpogi pasai, Karpogi thailam, Maantha Ennai* No.2, *Maha vallaathi ilakum, Mahaavallaathi legiyam, Mahaavallaathi kulikai, Megaviranak kalimbu, Rasagandhi meluku*
15.	*Desmodium gangeticum*	*Sarapunga vilvaathi, Thiraatchaadi chooranam*
16.	*Erythrina variegata*	*Karuppu vittunuch chakkaram*
17.	*Glycyrrhiza glabra*	*Amirthaathi kulikai, Arakku thailam, Karisaalai legiyam, Mahayelaathi kulikai , Malakkudaara mezhugu , Naari roga naasa thailam , Peenasi thailam, Pitta sura kudineer, Thaaleesaadi chooranam, Thanneer vittan ney, Ulundu thailam, Vasantha kusumaakaram, Venpoosani legiyam*
18.	*Indigofera aspalathoides*	*Mahavallathi ilakam, Meganaatha thailam, Shivanaar vembu kuzhi thailam*
19.	*Indigofera enneaphylla*	*Nandi mezhugu, Pancha sootha mezhugu*
20.	*Indigofera linnei*	*Cheppu nerungil, Pancha sootha mezhugu*
21.	*Indigofera tinctoria*	*Karisaalai legiyam, Nandhi mezhugu*
22.	*Macrotyloma uniflorum*	*Erasa kenthi mezhugu*
23.	*Mimosa pudica*	*Yuvathyadi thailam*
24.	*Mucuna pruriens*	*Amritaprasaghritham, Atmagupta chooranam, Gandhaka rasaayanam, Ulundu thailam*

Sl. No.	Species	Formulations
25.	*Pongamia pinnata*	*Kankasa maathirai, Meghanatha thailam, Pungathy maathirai, Pungathailam*
26.	*Pterocarpus santalinus*	*Arakku thailam, Chukku thailam, Gorochana maathirai, Laghu chandarathi thailam, Nasiroganasa thailam, Nilavembu kudineer, Senthil ney, Thiraatchaadi chooranam, Vallaraney*
27.	*Saraca asoca*	*Ashoka pattai*
28.	*Senna alexandrina*	*Malakkudara mezhugu, Malakkudara neluku, Nilaavaarai chooranam, Pitta sura kudineer, Thiraatchai kudineer*
29.	*Senna angustifolia*	*Nilaavaarai chooranam, Pitta sura kudineer*
30.	*Senna auriculata*	*Avarai kudineer, Meghanaatha thailam*
31.	*Sesbania grandiflora*	*Cempu parpam, Sirungi parpam, Thanga parpam*
32.	*Sesbania sesban*	*Karunsembai thailam, Mandoora chendooram, Sirungi parpam*
33.	*Tamarindus indica*	*Hinguvacadi chooranam, Karuvanga parpam, Pancamla thailam, Thaalaka parpam*
34.	*Tephrosia purpurea*	*Sarapunga vilvaathi legiyam,*
35.	*Trigonella foenum-graecum*	*Chundai vatral chooranam, Kabaada maaththirai, Kanath thailam, Kummatti mezhugu, Vendayum ilakam*
36.	*Uraria picta*	*Sarapunga vilvaathi legiyam*
37.	*Vigna mungo*	*Ulundu thailam*

CHAPTER-20

Legumes in the *Unani* System of Medicine

20.1 INTRODUCTION

The *Unani* system of medicine is traced to the system of Greek medicine developed during the Arab civilization, is also called the Greco-Arab system. The name *Unani* (*Ionian*) is considered indicative of the Greek origin of the system, though the European historians call it Arab medicine. *Unani* system is now prevalent in India, Pakistan and Bangladesh. It had gone also to places where the Asian Muslims took it along. Detailed information on various aspects of the *Unani* system of medicine can be obtained from Baquai (1977), Said (1983), Sathnarayana Bhat and Kameswara Rao (1993d) and Kameswara Rao (2000). A very useful glossary of plants used in *Unani* medicine was compiled by Fathima (1994), which includes sources of therapeutic information.

20.2 ORIGIN AND DEVELOPMENT

The theoretical framework of *Unani* is based on the teachings of Hippocrates, the Greek philosopher-physician (460 to 377 BCE). The Greek scholar Galen (131–210 CE) has established the fundamentals of *Unani* on which the Arab physicians like Rhazes (850–925 CE) and Avicinna (980–1037 CE) constructed an imposing edifice.

The *Unani* system owes its development to the Arab and Iranian physicians and to a very considerable extent to the Indian *Hakims*. It has absorbed what was best in the contemporary medicine in Egypt, Syria, Iraq, Persia, India, China and other middle and far eastern countries.

The Arabs have introduced the *Unani* system into India and soon it took to deep roots. When Mongols ravaged the Persian and Central Asian cities like Siraz, Tabrez and Geelan, scholars and *Unani* physicians fled to India. They received State patronage from the Delhi Sultans. Some of the *Unani* scholars and physicians have even been court physicians. The *Unani* system had its heyday during the period between the 13th and the 17th centuries in India.

The scholars and physicians of the *Unani* system who settled in India were not content with the known drugs but they subjected the Indian drugs to trials and as a result of their experimentation added numerous native drugs to their own system, thus enriching it. The system found immediate favour with the masses and soon spread all over the country and continued to hold an unchallenged sway for a long period even after the down fall of the Mughal Empire.

During the British Rule, the *Unani* system sufferred a setback and its development hampered due to the withdrawal of governmental patronage. But since the system enjoyed the faith of the masses, it continued to be practiced. It was mainly due to the efforts of the

Sharifi family in Delhi, the Azizi family in Lucknow and the Nizam of Hyderabad, the *Unani* system survived in India during the British period. An outstanding physician and scholar of the *Unani* medicine, Hakim Ajmal Khan (1868-1927), championed the cause of the system in India. The *Hindusthani Dawakhana* and the *Ayurvedic* and *Unani Tibbia* College in Delhi are the two living examples of the *Hakim*'s immense contribution to the multi-pronged development of the two systems of medicine.

Unani as well as the other Indian systems of medicine gained momentum after India became independent. In order to streamline education and regulate the practice of the Indian systems of medicine, the Government of India has established by an Act of the Parliament, The Indian Medicine Central Council Act, 1970, the Central Council of Indian Medicine. In 1995, the Government of India has also set up a full-fledged Department of Indian Systems of Medicine and Homoeopathy, in the Ministry of Health and Family Welfare, to accelerate the pace of development of these systems.

20.3 PRINCIPLES OF *UNANI* SYSTEM

The *Unani* medicine, as said earlier, was based on the principles of Hippocrates, who first established that disease is a natural process, that its symptoms were the reactions of the body. He advocated that the chief function of a physician is to aid the natural forces of the body in combating the disease. Hippocrates was the first physician, and the only one on record from antiquity to introduce the practice of recording medical histories of patients.

The chief contribution of Hippocrates to medical realm is the humoural theory, which passed on to *Unani*. The humoural theory in *Unani* presupposes the presence of four humours in the body: *Dam* (blood), *Balgham* (phlegm), *Safra* (yellow bile) and *Sauda* (black bile). The humours were assigned temperaments. Blood is hot and moist, phlegm is cold and moist, yellow bile is hot and dry and black bile is cold and dry. Every person is supposed to have the unique humoural constitution that represents his/her healthy state. To maintain the correct humoural balance, there is a power of self-preservation or adjustment called *Quwwate-Mudabbira* (*vis medicatrix naturae*) in the body. If this power weakens, an imbalance of the humoural composition is bound to occur, causing the disease. The medicines used in fact help the body to retain this power to an optimum level and thereby restore the humoural balance, thus restoring health. In addition, a correct diet and proper digestion are considered to be important in the maintenance of the humoural balance.

In the *Unani* system, temperament (*mizaj*) has an important place, and forms the basis of pathology, diagnosis and treatment. The Galenic concept of temperament being sanguine, phlegmatic, choleric or melancholic, finds expression in the *Unani* system, which considers each individual as unique. The modern psycho-neuro-endocrinical concept is often taken in support of the idea that temperament is unique to an individual and that a shift in the temperament brings about a change in the person's state of health (Said, 1983). Thus, disease is the consequence of humoural imbalance in the body and of the failure of one or more parts of the body to get rid of pathogenic waste from the body.

As the humours, the drugs are also assigned temperaments. A drug considered hot, produces a temperament that is hot. Hence drugs are principally used to correct the abnormal or pathological temperament of the body or of any system or organ.

The basic philosophy of the *Unani* system is that the body, composed of matter and spirit, is taken as a whole since a harmonious life is possible only when there is a proper balance between the physical and spiritual functions. It aims not only to correct the present disturbance but also to make the individual emerge after recovery with a greater power of resistance to future disturbances.

Another distinctive feature of the *Unani* system is its emphasis on the diagnostic importance of *Nabz*, the pulse, the rhythmic expansion of the arteries felt by the fingers of the physician. Other methods of diagnosis include examination of *Baul* (urine) and *Boraz* (stool).

The *Unani* system recognizes the influence of one's surroundings and the ecological conditions on the state of health of human beings. The system aims at restoring the equilibrium of various elements and faculties of the human body. It has laid down six essential pre-requisites for the prevention of disease and places great emphasis, on the maintenance of proper ecological balance and on keeping water, food and air free from pollution. The air, food and drinks, bodily movements, psychic movements, repose, sleep, wakefulness, excretion and retention are known as *'Asab-e-sistta Zarooriya'*,

In the *Unani* system, various types of treatment are employed such as *ilaz bit tad bur* (regimental therapy), *ilaz bil Ghiza* (diet therapy), *ilaz bid Dawa* (pharmaco-therapy) and *jarhat* (surgery). The regimental therapy includes, vivisection, cupping, diaphoresis, diuresis, Turkish bath, massage, cautery, purging, emesis, exercise, leaching, etc. Dietotherapy aims at treating certain ailments by the administration of specific diets or by regulating the quality and quantity of food. The pharmaco-therapy deals with the use of naturally occurring drugs of predominantly mineral or animal origin. Surgery has been in use in *Unani* practice for long, and employs procedures for which certain instruments and techniques have been designed.

20.4 *UNANI* MEDICINE

In *Unani* medicine, single drugs or combinations in the raw form are preferred, over compound formulations. The *Unani Materia Medica* is very vast, but most of the ingredients are locally available and easy to obtain. The naturally occurring drugs used in the system are symbolic of life and are generally free from side effects. Drugs are processed from crude toxic material, and are purified in many ways before use.

The medicines are polypharmaceuticals in the form of decoctions, infusions, tablets, powders, confections, syrups, and *aqua*s.

The *tabib*, the *Unani* medical practitioner, is not merely a doctor; he is the guide, on even moral and social values. There are *tabiba*s, the lady doctors, to treat women.

The *Unani* pharmacopoeia (Said, 1969) has a rich armamentarium of natural drugs, consisting of mostly herbs but also material of animal, mineral and marine origin. There are over 2,000 species of plants in the *Unani Materia Medica* (Daljithsimha, 1974), of which many species of plants occurring in India found a place (Fathima, 1994).

Research was conducted on the efficacy of several plant and animal drugs. Nevertheless, the *Unani* pharmacopoeia lacks in a detailed experimental, physicochemical and bio-mathematical data (Said, 1983), but the medication is considered nearly always safe.

20.5 LEGUMES IN *UNANI* MEDICINE

The *Unani* system of medicine employs several legumes. Some legume species are used in several different formulations (Table 20.1). The therapeutic uses of legume species, most of them used as single species drugs in *Unani,* are given in the respective species profiles.

Table 20.1

Legumes in *Unani* Formulations

Sl. No.	Species	Formulations
1.	*Acacia nilotica*	*Habbe Sil* *Linctus cordial* *Sharbat banajshak murakkab*
2.	*Alhagi maurorum*	*Joshanda* *Khaisanda* *Qurs mulaiyan* *Qurs-e-tabasheer*
3.	*Astracantha gummifera*	*Dayagoza* *Habb-e-baqla* *Habb-e-shibiyar* *Habb-e-surfa* *Lauq-e-nafusu-ud-dum* *Lauq-e-shamoon* *Linctus cassia* *Majoon-e-teewaj* *Qurs-e-kahruba* *Qurs-e-kaknaj* *Qurs-e-tabashir* *Qurs mulaiyan* *Sufool-e-istehada*
4.	*Butea monosperma*	*Habb-e-didaan* *Majoon supari pak* *Sufoof-e-sailan-ur-rehm*
5.	*Cassia acutifolia*	*Majoom-e-suranjan* *Majoon-e-anjir* *Majoon-e-makki* *Majoon-e-shahtera* *Majoon-e-shibbiyar* *Majoon-e-ushbe* *Sufoof-e-mushil*
6.	*Cassia fistula*	*Hunool varam-e-rehm* *Lauq khiyar-shanbar* *Linctus cassia* *Linctus cordia*
7.	*Cullen corylifolium*	*Linctus cassia*

Sl. No.	Species	Formulations
8.	*Glycyrrhiza glabra*	*Dayaquozah* *Habb-e-baaqla* *Habb-e-surfa* *Jawarish-e-Asli-soos* *Linctus Cassia* *Lauq-e-sapistan* *Matbhook-e-sulaal* *Namak-e-Suleimani* *Sharbat-e-Eijay*
9.	*Macrotyloma uniflorum*	*Sufoof-e-hajar-ul-yahood*
10.	*Senna tora*	*Narde qoba*
11.	*Tephrosia purpurea*	*Araq-e-gulnar* *Araq-e-murakkab musaffi* *Araq-e-musaffi-e-khoon* *Matbhook-e-musaffi*
12.	*Trigonella uncata*	*Raughan-e-nakhuna* *Zimad-e-jiger* *Zimad-e-kabeer* *Zimad-e-muhallil* *Zimad-e-varm unsaiya*

CHAPTER-21

Legumes in Homoeopathy

21.1 INTRODUCTION

Homoeopathy originated in 1796 in Germany, with the historical publication, *Versuch uber ein neues Prinzip zur Auffindung der Heilkräfte der Arzneisubstanzen* (Essay on a New Curative Principle) by the German Doctor, Christian Samuel Freidrich Hahnemann, based on the Principle of Similarity. The Principle of Similarity was not new and was traced back to Hippocrates, who said that 'some diseases are best treated by similars and some by contraries' see Chapter 16 (Section 16.2). Nevertheless, Hahnemann was the first to systematize it and also to introduce the concept of dilutions of remedies (homoeopathic potencies). Hahnemann introduced Homoeopathy to the western world in 1810 through the book *Organon der Heilkunst* (Organan of the Art of Healing), which went into six editions. The principles of the homoeopathic system generated long lasting controversies that are still continuing today. However, Hahnemann was credited with 'disturbing and discrediting indefensible modes of practice' in medicine of his time. The System of Homoeopathy and the controversies surrounding it, were reviewed in great detail by Kameswara Rao (2000).

The primary emphasis of Homoeopathy is on therapeutics, in the treatment of both acute and chronic diseases. One of the major contributions of Homoeopathy is the successful treatment of chronic illnesses that were difficult to manage by orthodox systems. It was once a low cost system but currently it is becoming more and more expensive. It was emphasized that Homoeopathy employs non-toxic drugs, but not really so, as several plant and animal toxins have entered its *Materia Medica* a long time ago.

Homoeopathy takes a holistic approach and treats a patient on physical, emotional and mental levels at once, to bring back the body's equilibrium and to strengthen the defense mechanism such as the autoimmune system, the reticulo-endothelial system, the hormonal system, the sympathetic-parasympathetic system and the physiological mechanism that responds to stress.

There is extensive literature on various aspects of Homoeopathy in the form of *Materia Medicae*, Repertories, treatises on various ailments, and such others.

21.2 HOMOEOPATHIC MEDICAL PRACTICE IN THE WORLD

Homoeopathy is in use in every country of the world, except in China, Taiwan and Maldives. Israel is the only country that officially banned Homoeopathy. Homoeopathy has low popularity in the Muslim countries like Pakistan and Bangladesh where the use of ethyl alcohol in the preparation of medicine is a religious taboo. The degree of popularity of Homoeopathy varies from country to country in the western world but it is more prevalent in the developing countries, particularly because it is relatively inexpensive and the practitioners are far more informal and accessible. Homoeopathy is very much Indian now, not only by its strong presence for over a century but also from the fact that it is increasingly been Indianized by including Indian plants in its *Materia Medica*.

21.3 *MATERIA MEDICA* OF HOMOEOPATHY

The principle base of homoeopathic *Materia Medica* is the flowering plants. Even the inorganic salts of the 12 tissue remedies, are expected to be extracted from plants. The 38 flower remedies proposed in 1930 by Edward Bach, to treat 38 pathological states of the mind, form an important part of the medicine cabinet of a homoeopath. One interesting aspect is that Homoeopathy sources its medicines also from non-flowering plants, which have been largely ignored by most other systems of medicine.

Homoeopathy uses material from diverse sources, in addition to plants, for its over 2,000 remedies, as indicated by the following examples: a) Elements (sulphur, iodine), b) Minerals (silica), c) Inorganic compounds (alum, ammonium bromide, ammonium carbonate), d) Inorganic acids (hydrofluoric, sulphuric, hydrocyanic, nitric), e) Metals/ oxides (aluminium, arsenic, gold, iron, mercury, vanadium, titanium), f) Organic compounds (naphthalene, urea, glacial acetic acid, benzoic acid, butyric acid), g) Animal products (ambegris, lecithin, cod liver oil, bee venom, star fish, cockroach, snake venom, toad venom, plant lice, spiders), and h) Hormones (adrenalin).

21.4 MOTHER TINCTURES AND HOMOEOPATHIC POTENCIES

Homoeopathic medicines are extracted from the source material in ethyl alcohol. The weight/volume proportions of the raw material and the solvent are generally standardized for most of them. This extraction gives the 'mother tincture', which is serially diluted by 1:9 (solvent) proportion at each step, to obtain potencies that contain progressively smaller and smaller quantities of the active principle. This process is Potentisation. The smaller the quantity of the active principle in the medicine, the higher is the potency and stronger is the effect of medication.

The mother tincture (1x) is infrequently administered. So are potencies higher than 200x, while potencies of the order of 500x, 1000x or 2000x are used extremely rarely. The medicine is absorbed into sucrose or lactose (or sometimes starch) pills or small tablets, the size of which also determines the dosage.

21.5 THE LAWS OF HOMOEOPATHY

Hahnemann extensively researched the toxicological literature of his day and experimented, on healthy volunteers comprised of a group of doctors and himself, substances from minerals, animals and plants. He aimed his system to be a complete scientific method based on demonstrable Laws and Principles, the most important of them being:

a) **The Law of Similars:** One of the cardinal Principles of Homoeopathy is '*Similia similibus curentur*'. As per this Principle, a particular substance that produces a specific symptom in a healthy individual, cures the same symptom in a sick individual, although the causes that generate the symptoms are not the same. The body is made to reinforce its defense against a symptom caused by a medicine.

b) **The Law of Direction of Cure:** There is an initial aggravation of the symptoms, which move from the most vital functional centers of the body to the least vital centers, like the skin. Both the aggravation and movement of symptoms in a particular direction are an indication that the medicine is effective.

c) **The Law of Single Remedy:** The Physician should find one and the only remedy that has produced in its proving, the greatest similitude to the symptom complex existing in the patient.

d) **The Law of Minimum Dose:** The Physician has to prescribe a very minute dose of the medicine, in order to prevent an enormous aggravation of the patient's symptomatology. Potentisation, an important principle of Homoeopathy, is aimed at controlling the dosage.

e) **Individualization:** Every case, even of the same illness and set of symptoms, is treated as peculiarly individual. It may mean that a different medicine for the same illness, or same medicine for different illnesses, in different individuals.

Homoeopathy has been criticized on all these counts. Homoeopathic Principles were put to scientific validation often with strongly negative results. Nevertheless, it is undeniable that Homoeopathy has done enormous service to the masses in the developing world for over a century.

21.6 LEGUMES IN HOMOEOPATHY

Homoeopathy uses nearly a thousand species of plants, including a large number of legumes. The following legume species that occur in India are used in Homoeopathy: *Abrus precatorius, Caesalpinia bonduc, Cicer arietinum, Cytisus scoparius, Desmodium gangeticum, Genista tinctoria, Haematoxylum campechianum, Indigofera tinctoria, Lathyrus sativus, Medicago sativa, Melilotus albus, Melilotus indica, Melilotus officinalis, Mimosa pudica, Mucuna pruriens, Myroxylon balsamum, Phaseolus vulgaris, Physostigma venenosum, Piscidia piscipula, Robinia pseudoacacia, Saraca asoca, Senna acutifolia, Senna alexandrina, Senna obovata, Trifolium pratense, Trifolium repens* and *Ulex europaeus.*

The uses of Indian legumes in Homoeopathy are indicated in the respective profiles of species.

CHAPTER-22

Legume Flavonoids in Medicine

22.1 INTRODUCTION

The first clinical use of flavonoids dates back to Rusznyak and Szent-Gyorgy (1936), who studied the role of flavonoids in increasing vascular permeability and vitamin C sparing effect. Subsequently many flavonoids have been used in clinical research and in treatment of various disorders. Flavonoid containing plant parts have also been in use in traditional medicine for ages.

In this book, flavonoids are discussed in several contexts. In addition, flavonoid containing legumes are indicated in the species profiles.

22.2 FLAVONOIDS

Flavonoids (flavanoids), a group of aromatic, oxygen containing heterocyclic pigments, derived from the parent substance flavone, are exclusive to the higher plant groups. There are over 600 flavonoids, bound to sugars, which occur as glycosides or as aglycones (base molecules without the sugar). Flavonoids are classified into the subclasses Catechols, Leucoanthocyanidins, Flavonones, Flavanols, Flavones, Anthocyanins and Flavonols, often simultaneously occurring in diverse groups of higher plants.

22.3 FLAVONOIDS IN LEGUMES

Like the species of other groups of plants, legumes also contain the whole range of flavonoids. Legume flavonoids were reviewed by Harborne (1971), Gibbs (1974) and Polhill and Raven (1981). Flavonoid containing Indian legume species are listed in Table 22.1. With several flavonoids of different subclasses being widely distributed in the legumes, it is often the chemistry and structure of the flavonoid compound that becomes more important than the plant source. Like many other secondary metabolites, flavonoids accumulate in the root and stem bark, leaves, pod walls and seed coats. A very large number of flavonoid containing parts of legume species are used in indigenous medicine and several are commonly consumed as food in different forms.

22.4 THERAPEUTIC EFFECTS OF FLAVONOIDS

Flavonoid compounds may be antinutritional or even toxic at higher concentrations but they also exert a variety of beneficial therapeutic effects. The following are some representative examples:

a) Flavonoids act as scavengers of free radicals and regulate immune response (Pignol *et al.*, 1988).

b) Flavonoids modulate proliferation of T lymphocytes.

c) Flavonoids inhibit human platelet aggregation (Bourdillat *et al.*, 1988) and reduce the risk of heart attack.

d) Flavonoids are effective in the treatment of chronic liver disorders (Daniel *et al.*, 1988).

e) Flavonoids control haemolytic anaemia and fever during radiotherapy of malignancies (Berg and Daniel, 1988).

f) Flavanols inhibit metastasis of malignant tumours (Bracke *et al.*, 1988) and act as antitumour drugs (Bibby *et al.*, 1988).

g) Several isoflavonones were shown to exercise protective action against cancer of the stomach, liver, colon and prostate (Messina and Barnes, 1991).

h) Flavonoids act as antiviral agents by interfering with conformational changes during the uncoating process of the viral capsid in the host cells (Vlietinck *et al.*, 1988).

i) Flavonoids have a protective effect against lipid peroxidation (Salvayre *et al.*, 1988), which is involved in various physiological and pathological disorders such as ageing, inflammation, oxygen toxicity and ischaemia.

j) Flavonoids affect the transport, metabolism and action of thyroid hormones (Kohrle *et al.*, 1988), which regulate growth in animals and humans, and also control metabolic rate and body temperature.

k) Flavonoids are used as antigonadotrophic agents (Gumbinger *et al.*, 1988).

l) Substances that are not normally found in the body (xenobiotics) are metabolized in the human body by different enzyme systems which may be either beneficial leading to detoxification or detrimental such as carcinogen activation. Flavonoids are known to modulate the activities of xenobiotics (Siess *et al.*, 1988). Many of the dietary flavonoids are known to inhibit the metabolizing enzymes derived from the xenobiotic gut flora.

m) Flavonoids have a hypocholesterolemic effect and reduce serum cholesterol (Gabor and Gyorgi, 1988).

n) Flavonoids are used in treatment of thrombosis and atherosclerosis (Beretz and Cazenave, 1988).

o) Flavonoids have antifertility activity (Srivastava *et al.*, 1986) and antiosteoporotic effect (Gabor and Gyorgi, 1988).

p) Flavonoids are used as anti-inflammatory and antihepatotoxic agents. as gastric cytoprotectors (Robert *et al.*, 1979; Alcaraz and Jimenez, 1988) and as antiulcer drugs.

q) Quercetin, chalcone, catechin and rutin suppress the oxidants released by human neutrophils (Busse *et al.*, 1984). They enhance lymphocyte proliferation (Daniel *et al.*, 1986), stabilize cell membrane by altering membrane permeability and fluidity, which affects immune response, and have an inhibitory action on cytotoxic T lymphocytes (CTL) (Schwartz and Middleton, 1984).

r) Isoflavonone phytoestrogens like genestein, daidzein, glycetein and biochanin, from edible legumes, particularly soybean, were shown to control menopausal symptoms, osteoporosis, hypertension and atherosclerosis as well as breast and prostate cancer (Goodman, 2004), and are now being commercialized. They are also considered as antioxidants and anticarcinogenic. Legumes containing phytoesterogens are listed in Chapter 8 (Table 8.4).

s) Bark extract of *Piscipula erythrina* has spasmolytic properties and is used in many traditional drugs for gynaecological disorders (Loggia *et al.*, 1988).

t) Rutin derivatives of *Styphnolobium japonicum* (*Sophora japonica*) are used in circulatory diseases as vasoprotective agents (Bidet *et al.*, 1987).

u) Flavonoid compounds from *Lespedeza* species are used in hepatobiliary diseases and as diuretic agent.

v) *Butea monosperma* flowers are used in hepatic disorders and as antifertility agent.

w) Wood and bark of *Pterocarpus* species are used in diabetics and diarrhoea.

x) Stem bark of *Acacia tortilis* is used for asthma (Anton, 1988a).

y) Auriculoside, a flavan glycoside from *Acacia auriculiformis*, which depresses the central nervous system, is used as sedative (Anton 1988b).

Table 22.1
Flavonoids in Indian Legumes

Most of the species in this table are used in indigenous medicine. Species with edible parts in common use are marked by an asterisk (*)

Sl. No.	Species
I	**Caesalpiniaceae**
1.	*Bauhinia malabarica*
2.	*Bauhinia purpurea*
3.	*Bauhinia semla*
4.	*Bauhinia tomentosa*
5.	*Bauhinia vahlii*
6.	*Caesalpinia pulcherrima*
7.	*Caesalpinia sappan*
8.	*Cassia fistula*
9.	*Cassia javanica*
10.	*Ceratonia siliqua*
11.	*Delonix elata*
12.	*Delonix regia*
13.	*Gleditsia sinensis*
14.	*Haematoxylum campechianum*
15.	*Peltophorum pterocarpum*
16.	*Senna absus*
17.	*Senna alata*
18.	*Senna floribunda*
19.	*Senna occidentalis*
20.	*Senna sophera*
21.	*Senna tora*
22.	*Tamarindus indica**
II	**Fabaceae**
1.	*Abrus precatorius*
2.	*Aeschynomene indica*
3.	*Alhagi maurorum*
4.	*Andira inermis*
5.	*Arachis hypogaea**
6.	*Cajanus cajan**
7.	*Canavalia ensiformis**

Sl. No.	Species
8.	*Canavalia gladiata**
9.	*Castanospermum australe*
10.	*Centrosema pubescens*
11.	*Cicer arietinum**
12.	*Crotalaria laburnifolia*
13.	*Crotalaria pallida*
14.	*Crotalaria retusa*
15.	*Crotalaria verrucosa*
16.	*Cullen corylifolium*
17.	*Cyamopsis tetragonoloba**
18.	*Dalbergia lanceolaria*
19.	*Dalbergia latifolia*
20.	*Dalbergia sissoides*
21.	*Dalbergia sissoo*
22.	*Dalbergia volubilis*
23.	*Derris elegans*
24.	*Derris elliptica*
25.	*Derris malaccensis*
26.	*Derris scandens*
27.	*Derris trifoliata*
28.	*Glycine max**
29.	*Indigofera arrecta*
30.	*Lablab purpureus**
31.	*Lathyrus odoratus**
32.	*Lathyrus sativus**
33.	*Lens culinaris**
34.	*Lotus corniculatus*
35.	*Lupinus albus*
36.	*Lupinus luteus*
37.	*Macrotyloma uniflorum**
38.	*Medicago sativa*
39.	*Mucuna pruriens*
40.	*Pachyrrhizus erosus**
41.	*Pachyrrhizus tuberosus**
42.	*Phaseolus coccineus**
43.	*Phaseolus lunatus**
44.	*Phaseolus trilobatus*
45.	*Phaseolus vulgaris**
46.	*Pisum sativum**
47.	*Psophocarpus scandens*
48.	*Psophocarpus tetragonolobus**

Sl. No.	Species
49.	*Robinia pseudoacacia*
50.	*Sesbania bispinosa*
51	*Styphnolobium japonicum*
52.	*Trifolium pratense*
53.	*Trigonella foenum-graecum**
54.	*Trigonella incisa*
55.	*Ulex europaeus*
56.	*Uraria lagopodioides*
57.	*Vicia faba**
58.	*Vicia hirsuta*
59.	*Vicia monantha*
60.	*Vicia sativa**
61.	*Vigna aconitifolia*
62.	*Vigna adenantha*
63.	*Vigna angularis*
64.	*Vigna mungo**
65.	*Vigna radiata**
66.	*Vigna subterranea*
67.	*Vigna trilobata*
68.	*Vigna umbellata*
69.	*Vigna unguiculata**
70.	*Wisteria sinensis*
III	**Mimosaceae**
1.	*Acacia auriculiformis*
2.	*Acacia decurrens*
3.	*Acacia farnesiana*
4.	*Acacia julibrissin*
5.	*Acacia leucocephala*
6.	*Acacia longifolia*
7.	*Acacia nilotica* ssp. *indica**
8.	*Acacia senegal*
9.	*Adenanthera pavonina*
10.	*Albizia amara*
11.	*Dichrostachys cineraria*
12.	*Entada rheedei*
13.	*Entada scandens*
14.	*Lysiloma latisiliquum*
15.	*Parkia biglobosa*
16.	*Parkia clappertoniana*
17.	*Pithecellobium dulce**
18.	*Prosopis cineraria*
19.	*Prosopis juliflora*
20.	*Tetrapleura tetraptera*

CHAPTER-23

Legumes in the Management of Cancer

23.1 INTRODUCTION

In the current times, cancer is the most dreaded and prevalent disease, without yet a sure shot cure or even satisfactory management protocols in sight, in spite of the enormous amounts of time, money and energy spent on cancer research in the past three or four decades. This is one disease that engaged the attention of a very large number of experts in every system of medicine, molecular biology, physiology and genetics. A popular overview of the dimensions of cancer can be obtained from several review articles in the special issue of Scientific American of September 1996.

A number of plant species have been investigated for use in different contexts of cancer management and these include many legumes, as indicated in the respective contexts here.

23.2 LEGUMES WITH ANTINEOPLASTIC ACTIVITY IN ANCIENT MEDICAL LITERATURE

Cancer must have been at least of sporadic occurrence in ancient times, if not prevalent. Since cancer was not understood in the same manner as today, we cannot expect its direct mention in classical medical literature. Nevertheless, pathologists found that some Egyptian mummies of over 2,000 years old showed evidence of cancer. The Egyptian Ebers Papyrus (1,500 BCE) recorded the use of yeast in treating cancer-like diseases. A number of folk therapies of ancient Greece, Egypt and Rome against such diseases, have scientific approval (Kreig, 1965). Hippocrates (450 CE) suggested *Allium* extract as an effective medicine against uterine tumours. American Red Indians have employed an extract of *Podophyllum peltatum* against warts and skin cancer at the beginning of the last century (Hartwell, 1967). Interestingly, podophyllotoxin is now an across the counter anticancer drug.

Certain of the symptoms described in classical Indian medical literature, such as chronic and recalcitrant internal or external abscesses, ulcers, and tumours, are indicative of cancer. Several species of plants were cited, for use against these disorders, in *Charaka Samhitha* and *Sushrutha Samhitha*. On the basis of interpretation of symptoms, these species can be considered as antineoplastic. There are about 75 species of plants with such activity, which are also used in other therapeutic contexts (Kameswara Rao, 2000). Among these, there are such legumes as *Sesbania grandiflora* (internal abscesses), *Butea monosperma, Caesalpinia sappan, Cajanus cajan, Cassia fistula, Indigofera tinctoria, Mucuna pruriens, Pongamia pinnata* (internal tumours) and *Indigofera tinctoria* (malignant ulcers) (Ray *et al.*, 1993; Kameswara Rao, 2000). The legumes that can be considered as antineoplatstic in *Ayurveda*, *Siddha*, *Unani*, Chinese medicine and Homoeopathy, are given in Table 23.1. The potential of these species in the treatment of cancer is hardly recognized.

Antineoplastic activity of drugs is based on toxicity to the human system. Kameswara Rao (2000) cited 158 species of plants used in Indian traditional toxicology. A fairly large number of these species were not tested for anticancer activity. The legume species from this list are given in Table 23.2.

23.3 CRITERIA FOR AN ANTINEOPLASTIC DRUG

An antineoplastic (anticancer) drug should be:

a) **Cytotoxic:** to inhibit the cancer cell's metabolism particularly synthesis of proteins and nucleic acids, in order to prevent cell division, growth, differentiation, vascularization of the new growth, etc;

b) **Mitostatic:** to disrupt the process of cell division to prevent the uncontrolled number of cycles of cell division and growth, to retard the proliferation of the cancerous tissue;

c) **Selectively toxic:** the drug should be toxic only to the cancerous tissue and nontoxic to the rest of the body of the patient; it should not cause any side effects such as neural, renal or hepatic dysfunction, neurotoxicity, hypersensitivity, etc; and

d) **Target oriented:** it should be site-selective targetting in its action to the cancerous region and not cause the same effects (or at least not cause them to the same degree as on cancer) on the other parts of the patient's body.

In addition, the drug should be effective in small and/or few doses, should not be expensive, should have longer shelf life, unrestricted availability and accessibility, etc.

These criteria for an anticancer drug are a tall order. Hardly there is any drug, synthetic or natural, that meets with all these qualifications and so the choice is dictated by the maximum compliance of the criteria and the philosophy of 'the lesser evil'. There is a large number of different types of cancer, each a kind of a syndrome. Consequently, no single drug can be expected to be effective against more than one type of cancer or at the best a few of related types of cancer.

The development of anti-cancer drugs, synthetic or natural, involves a lot of time and financial inputs and the drugs are very expensive. In the developed world there are very strict regulations governing drug development and clinical testing, enforced by the respective Governments. These regulations add to the time and financial loads on drug development but have public safety at heart. Thus, several factors have made cancer drugs a multimillion dollar and multinational business, hardly within the reach of the poor, even in the western world.

In the face of a largely depressing scenario of the chemotherapeutic approach in developing safe, effective and broad range synthetic or natural drugs, alternative therapies like hormones, bone marrow transplants, immunotherapy, chemotherapy, radiation therapy, etc., have come into practice. These approaches are also very expensive, without yielding uniform or equally satisfactory results, and are fraught with severe side effects. It has now become evident that a miracle cancer cure is a myth. Hence, the current focus is on prevention, early detection, and pre- and post therapy management of cancer.

23.4 LEGUMES IN THE TREATMENT OF CANCER

In the face of failure to find synthetic drugs against cancer, thousands of species of plants were screened for over three decades, for antineoplastic activity, in the hope of discovering effective natural products. Compounds were isolated from hundreds of species and their activity in suppressing tumours induced in laboratory animals was evaluated. Such work is still going on in several laboratories throughout the world.

Basing on *in vitro* data, a large number of 'promising' species were taken to clinical trials. However, products of hardly a handful of plant species, such as the vinca alkaloids, taxanes, camptothecin, podophyllotoxin, etc., have passed through the rigorous tests to be officially used against certain types of cancer and are now available in the market. Yet there are severe side effects and other problems associated with the use of even these largely 'successful' drugs, which are among the most expensive plant products. This reflects the complexity of the scenario of cancer drugs in general and plant based drugs in particular.

The Natural Product Drug Development Program of the U.S. National Cancer Institute has identified, about 3,000 species of plants and animals as useful in dealing with one or the other aspect of cancer management (Cassady and Douros, 1980). Hartwell (1982) cited a number of species with promise, including 29 species of Caesalpiniacaeae, 140 of Fabaceae and 24 of the Mimosaccae (Table 23.3).

Frequent disappointments have set surveys for anticancer plants run in cycles of mounting and dwindling interest. Nevertheless, plants (legumes included) still hold the promise of providing an effective and practicable cure, at least against certain types of cancer.

23.5 LEGUMES IN THE MANAGEMENT OF CANCER

In addition to treatment to eliminate cancer from the body, a host of synthetic or natural drugs are used in the biomedical system in the management of cancer, which is distinct from treatment aimed at a cure.

Several symptoms are inherent in cancer and several others are incidental to its treatment. These symptoms include pain, inflammation, fever, nausea, emesis, purgation, renal and hepatic dysfunction, infections, hair loss, skin eruptions, etc. Plants have an appreciable potential in treating these symptoms to provide relief and to keep the patient in a reasonable comfort. Kameswara Rao (2000) discussed long lists of such plants, which include many legumes.

The usefulness of plants in the management of cancer such as the recovery of the body to full and normal functioning, prevention of remission, management of symptoms of incurable cancers, to keep the patients in the maximum possible comfort, is now largely recognized.

23.6 LEGUMES IN THE PREVENTION OF CANCER

There is a prevailing hope that 'someday people should be able to avoid cancer or at least delay its onset by taking specially formulated pills or foods' (Greenwald, 1996).

Chemoprevention is the use of natural and synthetic chemical compounds to intervene in the early stages of cancer, before the invasive disease becomes otherwise perceptible, as prevention of cancer is an immensely better option than its uncertain cure. Several chemopreventive agents have been identified in many vegetables and pulses.

During the normal metabolic processes in our body, free radicals are routinely generated. Defective metabolic state or some modern life styles of an individual enhance the turnover of free radicals. Free radicals may combine with other chemical entities and form harmful compounds. Among the factors that promote cancer, free radicals are of a very serious concern. Free radicals are usually eliminated from the body through the excretory processes. However, when the turn over of free radicals exceeds the excreted quantities, they may trigger such conditions as heart disease, arthritis and cancer. The use of antioxidants to scavenge free radicals from the system is a viable option in maintaining health and preventing the onset of disease. Many of the chemopreventive compounds are efficient antioxidants.

Plant foods, that include many legumes, have been identified as one of the most promising sources of the following natural chemopreventive agents:

a) **Vitamins:** vitamin A (and its analogues), C and E, obtained only from plants.
b) **Carotenoids:** Carotenoids, which can be primarily obtained only from plants, are rated high as efficient antioxidants. α- and β-Carotene, β-cryptoxanthin and several other carotenoids are present in many plant foods, particularly green leafy vegetables and pulses. A list of carotenoid rich legume species is given in Table 23.4.

c) **Indoles, isothiocyanates, dithiolthiones and organosulphur compounds:** dithiolthiones and organosulphur compounds are abundant in broccoli, cauliflower, cabbage and legumes.

d) **Isoflavones**: legume seeds are rich in many isoflavones such as daidzein and genestein, known as phytoestrogens (see Chapter 8, Table 8.4); these are abundant in pulses, but more well known from soybean; current research is focussed on soya's role in the prevention of breast, prostate, colon and possibly stomach and urinary tract cancers.

e) **Phytosterols:** the indigestible phytosterols prevent the growth of cancer cells in the colon; the seeds of many legumes contain phytosterols (see Chapter 13, Table 13.2).

f) **Phytates:** products of phytic acid, present in many legume foods slow down absorption of nutrients, particularly iron, calcium and sugars; they regulate levels of insulin and hyperlipidema; reduces the output of free radicals. Binding ability of phytates reduces cellular damage and inhibits cancer growth.

g) **Proteinase inhibitors:** proteinase inhibitors are among most potent anticarcinogens present in many legume seeds; proteinase inhibitors that have suppressed several types of cancer cells in cultures; they are particularly protective against cancer of the colon, breast and prostate. Legume seeds are rich in such inhibitors (see Chapters 7 and 15).

h) **Tannins:** Tannins are known to be protective against onset of cancer of the colon, breast and prostate; many legume vegetables and seeds contain tannins (see Chapter 11).

It is not yet clear how the chemopreventive agents function. They are believed to prevent the mutations that can lead to cancer, halt the process of excessive proliferation of altered cells, prevent neovascularization, and hasten apoptosis (programmed death) of altered cells.

Several different preparations from plants rich in chemopreventives are patented and sold across the counter in many developed countries. As it is too expensive to obtain chemopreventives in the form of processed pills/tablets, plant foods are both the safest and the cheapest source. They also provide the dietary fibre that is believed to prevent colon cancer. For these reasons, people are now encouraged to consume foods rich in antioxidants and chemopreventives.

23.7 LEGUME LECTINS IN THE DIAGNOSIS AND PROGNOSIS OF CANCER

An early detection of cancer, when the chances of cure are high, has been one of the objectives of current research efforts. Such methods also help in following the response of the cancer to treatment.

Lectins (see Chapter 9) have an affinity to the surface carbohydrates of cells. Consequently, lectins agglutinate free cells or bind to the membrane components of the cells in tissues. Agglutination of mouse tumour cells by wheat germ lectin was first discovered by Aub *et al.*, (1963) and since then several plant lectins have been identified for use as diagnostic and prognostic agents in cancer.

Lectins are more prone to bind to diseased cells than healthy cells. Healthy cells of the same tissue often require 250 times more of lectin to agglutinate. Lectins can be conjugated with a number of substances like fluorochromes, enzymes, antibodies, etc., to facilitate their detection in the cell bound form.

The binding of the lectin to the glycoprotein cell surface receptors of a malignant cell differs from a normal cell in irregular staining of the receptor sites due to clustering of the sites in the malignant cells. The changes in the architecture of the tumour cell surface are brought out in a staining differentiation between normal cells, primary cancer cells and invasive cancer cells.

Lectins can be used not only to determine the extent and specificity of pathology of cancer but also the progression of stages during malignant transformation. The ability of lectins to preferentially agglutinate malignant and embryonic cells has led to the use of these glycorpoteins in elucidating a transition from normal cells to malignancy (Drachenberg and Papadimitrou, 1995).

Lectins can also be used in studying the sensitivity of cells to radiation as they reflect the structural integrity of cell membrane. Radiation induced changes in membrane structure of normal as well as malignant cells are readily detected as the radiosensitive cells have a reduced number of lectin binding sites (Vijayakumar and Bhattathri, 1995).

Legume lectins are of value in the early diagnosis and prognosis of different types of cancer due to a characteristic pattern of lectin binding to malignant cells. Legume lectins show specific reactivity to tumour associated glycoproteins and so are extensively used in characterizing cell and tissue glycoconjugates as markers of normal and neoplastic tissue as their binding is specific and cytotypic (Sehwechhimer *et al.*, 1984). They also are helpful in detecting the changes in glycoconjugate expression in benign and malignant cancers (Zang *et al.*, 1989).

The following are the advantages of using legume lectins in cancer therapy:

a) A range of lectins with different carbohydrate specificities are readily available and are inexpensive;

b) Lectin-carbohydrate specificity makes them precise markers;

c) Lectins can be purified and conjugated to any diagnostic markers as potential reagents in histopathological investigations;

d) Lectins can be used in understanding both tumour biology and clinical oncology;

e) Simple histochemical procedures to study lectin binding to cancerous tissue can be performed routinely, both prospectively and retrospectively, as cytological smears, cryostat sections and paraffin embedded sections;

f) Staining procedures can be applied to small fragments of tissue or a few isolated cells;

g) Lectin specificity to certain types of cancer can be used as a therapeutic tool without affecting the normal tissues.

The following legume lectins have a variety of applications in cancer diagnosis and prognosis:

a) **Peanut agglutinin (PNA) (*Arachis hypogaea*):** serves as prognostic marker in gastric cancer (Kakegi *et al.*, 1994); binds strongly with primary pulmonary adenocarcinoma (Kawai and Suzuki, 1994); discriminates benign, premalignant and malignant condition of cervical epithelium and endometrium (Yen *et al.*, 1986); by exhibiting differential binding pattern, it selectively eliminates breast cancer cells from human bone marrow without affecting the normal cells (Shoshana *et al.*, 1988; Drachenberg and Papadimitriou, 1995); agglutinates only the cancerous cells of the breast at the site of Thomson-Friedereich antigens; exhibits differential positive binding to malignant lesions of the thyroid (Vijayakumar *et al.*, 1992); is used in the detection of malignant lesions of the thyroid (Walker 1985a); differentiates invasive carcinoma of urinary bladder; detects tubular necrosis and atrophy with thyroidization of epithelial cells of kidney (Nadasdy *et al.*, 1995); binds to the sites in the epithelial cells of the kidney during transitional cell tumour (Limas and Lange, 1986); binds to anaplastic and papillary meningiomas (Marafioti *et al.*, 1994); binds to prostatic intra epithelial neoplasia; reacts with lymphoid leukaemia cells by binding to only the monocytes and very strongly to differentiated lymphoid cells; can be used to study the carbohydrate composition of leukaemic cell membranes which differ from normal lymphocytes (Remani *et al.*, 1998).

PNA coupled to magnetic beads can be used to purge human bone marrow of plasma cells (Reisner, 1983); a technique has been developed for removing bone marrow cells of b-lineage from the marrow of multiple myeloma patients by combining immobilized PNA and immobilized anti-CD19 monoclonal antibodies (Rhodes *et al.*, 1991); this procedure removes all detectable plasma cells, b-cells and allows 55 per cent recovery of normal haemopoietic progenitor cells; the purging of marrow of myeloma patients undergoing autologous bone marrow transplantation after chemotherapy decreases the relapse rate; used as marker of gastric cancer.

PNA is used as histochemical probe to detect malignant lesions of the oral mucosa (Mazumdar, 1993); in early prediction of malignancy (Kannan *et al.*, 1993); used as renal epithelial marker to identify kidney diseases (Silva *et al.*, 1993).

PNA reactive N-acetyl-galactosamine containing glycoproteins and LCA reactive mannose/N-acetylglycosamine (-fucose)/sialic acid containing glycoprotein were detected in osteosarcoma and malignant fibrous histocytoma (MFH) (Stastny and Gupta, 1994). Neither of these glycoproteins were detected in the normal tissues. The high specificity of these lectins for sarcoma was confirmed by studies using monoclonal antibodies.

b) ***Bauhinia purpurea* agglutinin (BPA):** exhibits specific reactivity with tumour associated glycoproteins of cancerous pancreatic cells and not with the glycoproteins of healthy cells (Kawa *et al.*, 1992; Shue *et al.*, 1993); used in detecting pulmonary neoplasia; as a diagnostic marker for endometrial carcinoma (Sarker *et al.*, 1994).

c) ***Canavalia ensiformis* agglutinin (ConA):** is employed in detecting malignant seminoma; used in discriminating premalignant and malignant condition of cervical epithelium and endometirum and transition from benign to malignant state (Bychkov and Toto, 1987); conjugated with HRP (Horse radish peroxidase) ConA detects cervical cancer; during malignancy of cervical epithelium, there is an increase in sialation of the galactose group during secretory phase of menstrual cycle followed by expression of fucose containing glycosubstances in the malignant tissue of endometrium which is detected by ConA; it identifies the type of monosaccharide and its shift in the cell membrane and cytoplasm during malignancy; when bound to benign and malignant lesions of the breast, ConA showed affinity towards neoplastic cells (Walker, 1984); exhibits binding to meningiomas, chordomas, papillary meningiomas and oligodendrogliomas (Cruz-Sanchez *et al.*, 1991).

d) **Soybean agglutinin (SBA) (*Glycine max*):** is employed in detecting malignant seminoma; in eliminating selectively the breast cancer cells from bone marrow without affecting normal cells (Shoshana *et al.*, 1988); SBA binding is higher in primary tissue of the cancer than the secondary which helps in predicting the metastatic potential of cancer cells; alteration of distribution of binding sites of SBA and its intensity helps in detecting prostatic intraepithelial neoplasia; it also reduces 5'-α-reductase activity which converts testosterone to dihydrotestosterone which is implicated in prostate cancer; is employed in clinical practice to distinguish T-cells from stem cells to reduce the risk of rejection in graft-versus-host disease; is also used in removal of mature T-cells from human bone marrow and isolation of haemopoietic stem cells suitable for transplantation into haploidentical recipients; is employed in treating patients with combined immuno deficiency (Lis and Sharon, 1986); used to separate and transplant autologous bone marrow stem cells used in cancer therapy.

e) ***Lens culinaris* agglutinin (LCA)**: used in detecting osteosarcoma and fibrous histocytoma as it binds to N-acetyl galactosamine and glucosamine containing glycoproteins; used in the detection of hepatic neoplasia (Shiraki *et al.*, 1995).

f) ***Lotus tetragonolobus*** **agglutinin (LTA)**: is used as marker in the detection of hepatocellular carcinoma (Tu *et al.*, 1992), metaplasia of bladder and transitional cell carcinoma (Aragona *et al.,* 1991); used to distinguish stem cells from tumour cells in the bone marrow of cancer patients (Girgert *et al.*, 1988).

g) ***Dolichos biflorus*** **(DBA) agglutinin (*Macrotyloma uniflorum*):** can be used in detecting transitional cell carcinoma (Aragona *et al.*, 1991); labeled DBA is used in detecting adenomas; serves as prognostic marker in gastric cancer (Kakegi *et al.*, 1994).

h) ***Phaseolus vulgaris*** **agglutinin (PHA)**: is used to preserve progenitors of haemopoietic cells which are utilized in stem cell transplantation and tumour cell purging before autologous transplantation and stem cell gene therapy (Colucci *et al.*, 1999); helps in detecting leukaemia by reduced binding of PHA to lymphatic cells of patients with chronic leukaemia than to normal lymphocytes (Astaldi *et al.*, 1966).

i) ***Pisum sativum*** **agglutinin (PSA)**: is used as marker in the early detection of hepatocellular carcinoma (Tu *et al.*, 1992).

j) ***Psophocarpus tetragonolobus*** **agglutinin (PTA)**: binds to cervix, lymph nodes, oral, breast cancerous cells and leukaemic cells (Vijayakumar, 1998).

k) ***Tephrosia purpurea*** **agglutinin (TPA)**: Binding pattern of TPA is used as renal epithelial marker in identifying kidney diseases (Silva *et al.*, 1993); and in detecting tubular necrosis and atrophy with thyroidisation of epithelial cells of kidney (Nadasdy *et al.,* 1995).

l) ***Ulex europaeus*** **agglutinin (UEA)**: identifies specific cell types in tumours (Ree, 1988); can detect malignant seminoma; indicates difference between primary adeno carcinoma of lungs and their metastatic lesions; diagnoses angiosarcoma (Hattori *et al.*, 1991), cardiac myxoma and colonic adenocarcinoma (Kellokumpu *et al.*, 1986); differentiates basal cell carcinoma from squamous cell carcinoma of skin (Heng *et al.*, 1992); diagnoses follicular carcinoma of the thyroid (Gonzale-Campora *et al.*, 1986); identifies the type of monosaccharide in the cell membrane and cytoplasm during malignant transformation (during malignancy a shift of glucose and fucose from cellular membrane to intracytoplasmic area is noticed) (Jin and Lin, 1989); identifies malignant haemangioendothelioma of the thyroid (Beer, 1992); binds to colonic adeno carcinoma and colorectal cancerous cells; distinguishes bone marrow stem cells from tumour cells in bone cancer and as marker for vascular tumours (Miettinen *et al.*, 1983).

m) ***Vicia villosa*** **agglutinin (VVA):** exhibits specific reactivity with tumour associated glycoproteins during malignant pancreatitis (Shue *et al.*, 1993) and differentiates the pancreatic ductular adenocarcinoma with 80 per cent binding, from chronic pancreatitis with only 18 per cent binding (Itz Kowitz *et al.*, 1991).

The legume lectins that are being routinely used in the diagnosis and prognosis of cancer are given in Table 23.5.

23.8 TARGET ORIENTED DRUG DELIVERY BY LEGUME LECTINS

Lectins can be conjugated to drugs. Since lectins with specificity to a particular human tissue would bind to the respective tissues, lectin conjugated drugs can be delivered to the site of cancer. The legume lectins that are useful in the diagnosis of cancer are also useful as ligands to drugs.

Ligand binding of lectin is the most important feature in designing lectin mediated drug delivery system. Lectins coupled to drugs act as carrier molecules competing for the receptor binding sites with other endogenous ligands. Subsequent internalization of the drug complex leads to accumulation of the drug in target cells and exclusion from non target cells lacking the specific receptors. The lectins selectively recognize their saccharide ligands when presented in multiple geometries. It is possible to use these glycoconjugates to target any type of drug molecule.

Bioadhesion, the binding of a drug and its carrier such as a lectin, makes it easy to deliver macromolecular drugs directly to specific target cells. This has implications for tissue engineering, gene delivery and nanotechnology, three important areas of medical research.

Many examples of lectin mediated target delivery of drugs have been reported (Depierreux *et al.*, 1992). Lectins have been used as *in vitro* targeting agents for binding to lesions in the eye and mouth (Banchonglikitkul *et al.*, 2002).

Another use of lectins is in the enhancement of drug effectiveness. Lectin bound methotrexate and chlorambucil, common synthetic cancer drugs, were found to be far more effective than the drug administered without the lectin. The lectin enhances target orientation and the choice depends upon the cancerous organ.

Lectins are also used as selective carriers of monoclonal antibodies against certain cancer antigens. Treatment of cancer with monoclonal antibodies conjugated to anticancer drugs is less toxic and more sensitive. Such a procedure was found to be effective in treating B-cell leukaemia, but it is very expensive as monoclonal antibodies have to be prepared against the cancer antigens of each patient.

23.9 CYTOTOXIC LEGUME LECTINS IN CANCER THERAPY

Cytotoxicity is one of the interesting bioactivities of the lectins in mammalian cells. Ricin from *Ricinus communis* (Euphorbiaceae) and abrin from *Abrus precatorius* (Fabaceae) were the earliest cytotoxic lectins discovered. Among the legumes, the toxicity of Con A, seed lectin of *Canavalia ensiformis,* was known from many years. The cytotoxicity of the lectins of *Phaseolus acutifolius, Phaseolus lunatus, Phaseolus vulgaris, Robinia pseudoacacia,* and *Vicia germanica* was also well studied.

Potentially toxic lectins such as abrin, consist of two subunits, joined by disulphide bonds, one subunit is responsible for attachment to the cell membrane receptors, facilitating the penetration of a second toxic subunit via cell membranes (Nicolson, 1974). The subunit with the carbohydrate binding site is heavier than the toxic subunit, which is a ribosome inhibiting protein that inhibits protein synthesis.

Cytotoxic lectins have a potential therapeutic value in the treatment of certain types of cancer. Cells affected by a disease such as cancer undergo transformation in the configuration of their cell membranes and so are transformed cells. Transformed cells are ten times more sensitive to toxic lectins than normal cells, although both types of cells bind to similar quantities of lectin (Nicolson, 1974). The cytotoxic lectins kill both dividing and resting cancer cells and are specially suited to retard growing tumours (Shoham *et al.*, 1970; Ozanne and Sambrook, 1971). Cytotoxic effects are more pronounced in embryonic and undifferentiated cells than the differentiated tissues (Aubery and Bourrillon, 1972). Cytotoxic T lymphocytes (CTC) can lyse target cells (TC) nonspecifically in the presence of certain plant lectins notably Con A and PHA (Bradly and Bonavide, 1980).

Besides Con A, PHA and abrin, the cytotoxicity of *Vicia germanica* lectin against mammary carcinoma (Codington *et al.*, 1975), *Phaseolus lunatus* lectin against sarcoma (Tunis, 1964) and *Robinia pseudoacacia* lectin against mammalian tumours (Brown and Hunt, 1978). was also demonstrated.

Several legume lectins can be used to cause programmed death of tumour cells (apoptosis). Galactose and N-acetyl galactosamine specific lectins were used to lyse tumour cells (Maldomado *et al.*, 1994). *Griffonia simplicifolia* agglutinin induced, programmed cell death in several tumour cell types (Kim *et al.*, 1993). *Canavalia ensiformis* agglutinin (Con A) can be bound to liposomes (lipid vesicles in cells) resulting in an increased phagocytosis, leading to cell death.

Table 23.1
Legumes in the Indigenous Systems of Medicine to Treat Cancer

I. ***AYURVEDA***

1. Caesalpiniaceae

Bauhinia racemosa
Caesalpinia bonduc
Ceasalpinia sappan
Cassia fistula
Cyanometra ramiflora
Senna alexandrina
Senna tora
Tamarindus indica

2. Fabaceae

Abrus precatorius
Butea monosperma
Cajanus cajan
Cullen corylifolium
Cyamopsis tetragonoloba
Dalbergia sissoo
Indigofera tinctoria
Macrotyloma uniflorum
Mucuna pruriens
Pongamia pinnata
Pseudarthria viscida
Pterocarpus marsupium
Sesbania grandiflora

3. Mimosaceae

Acacia catechu
Mimosa pudica

II. ***SIDDHA***

1. Caesalpiniaceae

Caesalpinia bonduc
Senna obtusifolia

2. **Fabaceae**
   Butea monosperma
   Cajanus cajan
   Codariocalyx motorius
   Cullen corylifolium
   Indigofera aspalathoides
   Pongamia pinnata

3. **Mimosaceae**
   Mimosa pudica

III. *UNANI*

1. **Caesalpiniaceae**
   Chamaecrista absus
   Cassia fistula
   Senna tora

2. **Fabaceae**
   Astragalus boeticus
   Cicer arietinum
   Lablab purpureus
   Lens culinaris
   Lupinus albus
   Macrotyloma uniflorum
   Melilotus officinalis
   Tephrosia purpurea
   Trigonella foenum-graecum

3. **Mimosaceae**
   Acacia nilotica

IV. CHINESE

1. **Caesalpiniaceae**
   Gleditsia sinensis

2. **Fabaceae**
   Indigofera tinctoria
   Trigonella foenum-graecum
   Vigna radiata

V. HOMOEOPATHY

1. **Fabaceae**
   Phaseolus vulgaris
   Trifolium pratense

Note: *Based on Kameswara Rao (2000).*

Table 23.2

Legumes in Traditional Toxicology

The legumes in this table are used in traditional toxicology. The toxic activity of these species has a promise in the development of an anticancer drug.

1. Caesalpiniaceae:

Cassia fistula
Senna auriculata

2. Fabaceae:

Butea monosperma
Canavalia ensiformis
Clitoria ternatea
Desmodium gangeticum
Glycyrrhiza glabra
Indigofera tinctoria
Pterolobium indicum
Pterocarpus santalinus
Uraria picta

3. Mimosaceae:

Acacia polyacantha
Albizia lebbeck
Prosopis cineraria

Table 23.3

Legumes with Potential for Use in the Treatment of Cancer

This list is based on Hartwell (1982). The legumes in this table are also indicated in the indigenous medical literature for use against the cancer-like symptoms and are a potential choice for the development of an anticancer drug.

Sl. No.	Species	Plant part used	Cancer/tumour
I	**Caesalpiniaceae**		
1.	*Bauhinia malabarica*	Leaf	Tumours
2.	*Bauhinia purpurea*	Bark, flower, pod	Stomach cancer, in flammatory tumour
3.	*Bauhinia racemosa*	Stem, bark	Epidermal carcinoma
4.	*Bauhinia tomentosa*	Root, bark, seed	Glandular tumours
5.	*Bauhinia variegata*	Stem	Cancer, abdominal tumours
6.	*Caesalpinia bonduc*	Root, bark, leaf, pod, seed	Sarcoma, phantom abdominal tumours
7.	*Caesalpinia crista*	Root, leaf, pod	Cancer, tumours
8.	*Caesalpinia pulcherrima*	Flower	Tumours
9.	*Caesalpinia sappan*	Leaf, seed	Carbuncles

Sl. No.	Species	Plant part used	Cancer/tumour
10.	*Cassia acutifolia*	Leaf juice	Cancer, tumours
11.	*Cassia fistula*	Pod	Uterus and breast cancer, throat, stomach, liver and abdominal tumours
12.	*Chamaecrista absus*	Leaf	Tumours
13.	*Cynometra ramiflora*	Root, seed	Tumours
14.	*Dialium guianense*	Leaf	Tumours
15.	*Gleditsia sinensis*	Seed	Breast cancer, abscesses
16.	*Haematoxylum campechianum*	All parts	Cancer
17.	*Saraca asoca*	Bark	Tumours
18.	*Senna alexandrina*	Leaf, seed	Cancer, tumours
19.	*Senna auriculata*	Leaf	Tumours
20.	*Senna obtusifolia*	Leaf, pod	Cancer
21.	*Senna occidentalis*	Leaf	Tumours
22.	*Senna sophera*	Leaf	Tumours
23.	*Senna tora*	All parts	Cancer, phantom tumours
24.	*Tamarindus indica*	Leaf	Abdominal, glandular and uterovaginal tumours
II	**Fabaceae**		
1.	*Abrus precatorius*	Leaf, pod, seed	Epithelioma of skin and mucosa, tumours
2.	*Alhagi maurorum*	All parts	Abdominal and glandular tumours, nose polyps
3.	*Alysicarpus glumaceus*	All parts	Foot tumours
4.	*Arachis hypogaea*	Seed oil	Plantar warts
5.	*Astragalus hamosus*	Seed	Tumours, swellings
6.	*Butea monosperma*	Root, leaf, flower	Carcinoma, tumours
7.	*Cajanus cajan*	Leaf, pod, seed	Tumours, swellings
8.	*Cicer arietinum*	Flower	Tumours, swellings
9.	*Clitoria ternatea*	Seed	Abdominal tumours
10.	*Codariocalyx motorius*	All parts	Cancerous affections
11.	*Crotalaria sessiflora*	All parts	Tumours
12.	*Cullen corylifolium*	Plant, Seed	Uterine cancer, tumours, glandular swellings
13.	*Cyamopsis tetragonoloba*	Seed	Cancer, tumours
14.	*Dalbergia sissoo*	Root, bark, leaf	Tumours
15.	*Desmodium gangeticum*	Root	Abdominal tumours, nasal polyps
16.	*Desmodium triflorum*	All parts	Tumours
17.	*Erythrina crista-galli*	Ail parts	Stomach cancer
18.	*Erythrina fusca*	Seed	Cancer

Sl. No.	Species	Plant part used	Cancer/tumour
19.	*Erythrina suberosa*	Leaf	Sarcoma, tumours
20.	*Erythrina variegata*	Bark, leaf, seed	Cancer, tumours
21.	*Euchresta japonica*	Root	Cancer
22.	*Flemingia chappar*	Root	Cancer
23.	*Genista tinctoria*	Leaf, seed	Cancer, tumours
24.	*Gleditsia sinensis*	Stem	Rectal cancer
25.	*Glycine max*	Pod, seed	Cancer
26.	*Glycyrrhiza glabra*	Root	Cervical and intestinal cancer, abdominal tumours, indurations of kidney, bladder, uterus, and nasal polyps
27.	*Glycyrrhiza uralensis*	Root	Tumours
28.	*Gueldenstaedtia multiflora*	Root	Tumours
29.	*Indigofera argentea*	Leaf	Tumours
30.	*Indigofera arrecta*	Stem, leaf, pod	Cancerous affections
31.	*Indigofera aspalathoides*	All parts	Cancerous affections, tumours
32.	*Indigofera mysorensis*	All parts	Tumours
33.	*Indigofera tinctoria*	Root, stem, leaf	Cancer, tumours, abdominal and ovarian tumours
34.	*Indigofera uniflora*	Leaf, flower, twig	Phlegmatous tumours, cancerous ulcers
35.	*Lablab purpureus*	Leaf	Tumours
36.	*Laburnum anagryoides*	Leaf	Tumours
37.	*Lathyrus sativus*	All parts	Intestinal, face and gum cancer, genital, abdomen, ear, rectum, feet, breast and parotid gland tumours
38.	*Lens culinaris*	Seed	Tumours
39.	*Lupinus albus*	Bark, seed	Liver tumours
40.	*Lupinus angustifolius*	Leaf, seed	Cancer, glandular and parotid tumours
41.	*Lupinus hirsutus*	Bark, seed	Hard inflammatory tumours
42.	*Lupinus termis*	Bark	Hard inflammatory tumours
43.	*Macrotyloma uniflorum*	Seed	Cancer, abdominal tumours
44.	*Medicago sativa*	Juice	Cancer
45.	*Melilotus albus*	Juice	Induratons, scirrhous tumours
46.	*Melilotus indica*	Leaf	Glandular and scirrhous tumours
47.	*Melilotus officinalis*	Oil	Eye, anus, reproductive organ, stomach, uterine and scirrhous tumours

Sl. No.	Species	Plant part used	Cancer/tumour
48.	*Mucuna pruriens*	Pod, seed	Internal tumours, malignant ulcers
49.	*Myroxylon peruiferum*	All parts	Indolent tumours
50.	*Myroxylon balsamum*	Leaf	Cancer, tumours
51.	*Ononis arvensis*	Root	Ulcers, hardness
52.	*Ononis hircina*	Root	Tumours, ulcers
53.	*Ononis spinosa*	All parts	Cancerous ulcers
54.	*Paracalyx scariosus*	Root	Tumours
55.	*Phaseolus lunatus*	Seed	Tumours
56.	*Phaseolus vulgaris*	Seed	Cancer, testis and breast tumours
57.	*Pisum sativum*	Seed	Testicular tumour, warts, nasal polyps
58.	*Pongamia pinnata*	Leaf, bark, pod, seed	Internal and phantom tumours, glandular swellings
59.	*Pseudarthia viscida*	Root	Tumours
60.	*Psophocarpus tetragonolobus*	Seed	Tumours
61.	*Pterocarpus indicus*	Bark	Tumours, abdominal tumour
62.	*Pterocarpus marsupium*	Root, leaf	Tumours
63.	*Pterocarpus santalinus*	All parts	Mouth cancer, inflammatory tumours
64.	*Pueraria lobata*	Flower	Malignant tumours
65	*Pueraria tuberosa*	Root	Joint tumours
66	*Sesbania grandiflora*	Root	Tumours, abscesses
67.	*Sesbania sesban*	Leaf	Infammatory tumours
68.	*Sophora flavescens*	Bark, leaf	Breast cancer
69.	*Sutherlandia frutescens*	Bark, leaf	Cancer, malignant tumours
70.	*Tephrosia candida*	All parts	Tumours
71.	*Tephrosia purpurea*	Root	Tumours, scirrhous tumours
72	*Tephrosia sinapou*	Root	Tumours, scirrhous tumours
73.	*Trifolium arvense*	Flower	Cancer, Brain tumours
74.	*Trifolium pratense*	Leaf	Stomach, bowel, breast and pancreatic cancer, breast tumours, hard swellings
75.	*Trifolium repens*	Flower, seed	Cancer, tumours, whitlows
76.	*Trigonella foenum-graecum*	Leaf, seed, juice, oil	Uterus, rectum and intestinal cancer, carcinomata, uterus, bladder, thoracic organ, uvula, fauces, abdomen,

Sl. No.	Species	Plant part used	Cancer/tumour
			parotid gland, genital and breast tumours, indurations of stomach, uterus, joints, mammary glands, parotid, liver and spleen, sclerosis of spleen and liver
77.	*Uraria lagopodioides*	All parts	Abdominal tumours, nasal polyps
78.	*Vicia faba*	Stem, pod, seed	Cancer of eye, breast, testicles, genitals, penis, liver, spleen and eye lid tumours, parotid glandular tumours, warts, hardness of liver, stomach and spleen felons, scirrhous spleen, indurations of heels and hard swellings
79	*Vicia sativa*	Leaf	Cancerous ulcers, bladder tumours, warts
80.	*Vigna aconitifolia*	Seed	Tumours
81.	*Vigna mungo*	Seed	Abdominal and lachrymal tumours
82.	*Vigna radiata*	Leaf, seed	Carcinomatous tumours, tumours, purulent swelling
83.	*Vigna unguiculata*	Seed	Tumours
84.	*Wisteria sinensis*	Seed	Oesophagus and gastric cancer
III	**Mimosaceae**		
1.	*Acacia catechu*	Bark, wood, gum	Tumours
2.	*Acacia farnesiana*	Root, pod	Stomach cancer, tumours
3.	*Acacia ferruginea*	Bark	Cancerous affections
4.	*Acacia nilotica*	Bark	Testicular cancer, eye, ear and liver tumours
5	*Acacia senegal*	Bark	Cancer
6.	*Acacia sinuata*	Bark, pod	Tumours
7.	*Adenanthera pavonina*	Wood	Tumours
8.	*Albizia lebbeck*	Root, stem, bark, leaf	Cancer, epithelial carcinoma, abdominal tumours
9.	*Albizia procera*	Root, leaf	Sarcoma
10	*Entada rheedei*	Bark, seed	Cancer
11.	*Mimosa pudica*	Root, leaf, twig	Glandular tumours
12.	*Neptunia oleracea*	All parts	Inflammatory tumours
13.	*Prosopis juliflora*	Plant juice	Superfluous flesh tumours
14.	*Samanea saman*	Root	Stomach cancer

Table 23.4
Legume Species Containing Carotenoids

The legume species in this table contain different carotenoids, including α- and β-carotene, but the relative quantities are unknown.

Sl. No.	Species	Common name	Parts used	Quantity μg/100g
1.	*Arachis hypogaea*	Ground nut	Nut	37
2.	*Bauhinia malabarica*	Jerusalem date	Leaves	1764
3.	*Cajanus cajan*	Red gram	Dhal	132
4.	*Cajanus cajan*	Red gram	Tender seed	469
5.	*Canavalia gladiata*	Sword beans	Bean	24
6.	*Cicer arietinum*	Bengal gram	Gram	189
7.	*Cicer arietinum*	Bengal gram	Dhal	129
8.	*Cicer arietinum*	Bengal gram	Leaves	978
9.	*Cyamopsis tetragonoloba*	Cluster beans	Bean	198
10.	*Glycine max*	Soybean	Bean	426
11.	*Lathyrus sativus*	Chickling pea	Dhal	120
12.	*Lathyrus sativus*	Chickling pea	Leaves	3000
13.	*Lens esculenta*	Lentil	Dhal	270
14.	*Lepidium sativum*	Garden cress	Seed	27
15.	*Macrotyloma uniflorum*	Horse gram	Whole seed	120
16.	*Macrotyloma uniflorum*	Horse gram	Tender bean	187
17.	*Phaseolus vulgaris*	French bean	Seed	132
18.	*Pisum sativum*	Pea	Seed	39
19.	*Pisum sativum*	Pea	Roasted seed	18
20.	*Sesbania grandiflora*	Swamp pea	Leaves	5400
21.	*Tamarindus indica*	Tamarind	Leaves	250
22.	*Trigonella foenum-graecum*	Fenugreek	Leaves	2340
23.	*Trigonella foenum-graecum*	Fenugreek	Seed	96
24.	*Vicia faba*	Broad bean	Seed	9
25.	*Vigna aconitifolius*	Moth bean	Whole bean	9
26.	*Vigna coccineus*	Scarlet bean	Seed	34
27.	*Vigna mungo*	Black gram	Dhal	38
28.	*Vigna radiata*	Green gram	Whole seed	94
29.	*Vigna radiata*	Green gram	Dhal	49
30.	*Vigna coccineus*	Scarlet bean	Seed	34
31.	*Vigna mungo*	Black gram	Dhal	38
32.	*Vigna unguiculata*	Cow pea	Dahl	12
33.	*Vigna unguiculata*	Cow pea	Gram	564
34.	*Vigna unguiculata*	Cow pea	Leaves	6072

Table 23.5

Legume Seed Lectins in the Diagnosis of Cancer

Legume seed lectins and the type of cancer specific to them are given here. These lectins are now routinely used in the diagnosis of respective cancer.

Type of cancer	**Species**
Neurological cancer	*Lens culinaris, Pisum sativum*
Cervical cancer	*Glycine max*
Lung cancer	*Bauhinia purpurea, Macrotyloma uniflorum*
Prostate cancer	*Ulex europaeus*
Colon cancer	*Arachis hypogaea*
Endothelial cancer	*Ulex europaeus*
Endometrial cancer	*Ulex europaeus*
Thyroid cancer	*Bauhinia purpurea*

From Brooks *et al.*, 1997

PART-II

Information Profiles of Legume Species in India

Part-II

Information Profiles of Legume Species in India

FORMAT OF INFORMATION PROFILE

INTRODUCTION

There is an enormous amount of diverse kinds of information on the legume species in India, dispersed in several thousand publications of various categories, and websites. It requires a lot of time and effort to cull any kind of information from this vast mass of literature. Some species that have been in long use and/or have diverse uses were studied extensively, while some other species received very little attention. Some species are used for similar purposes in different countries. Some species that occur outside India are put to uses that are not known in India. It would help to enhance the range of uses of the species in India if information from different sources is put together and compared. This has to be done in the form of a database, in order to assess the state of knowledge and utility potential of the species. The objective of the information profiles presented here is to provide a platform that aids in understanding the chemical constitution, therapeutic potential and other uses of each species, as well as to indicate the gaps in information, which may require further research to fill in. Such a database helps in recognizing potentially important legume bioresources in different contexts of use. It would also help to identify areas of information that require verification and evaluation.

In Chapter 3, information on the economic potential of 125 species of the Caesalpiniaceae, 505 of the Fabaceae and 92 of the Mimosaceae, was presented for 18 different uses. Of these species that occur in India, Profiles of Information on 45 species of the Caesalpiniaceae, 198 species of the Fabaceae and 46 species of the Mimosaceae, considered important are provided here. The choice of species for the profiles was based on the importance of the species as food and/or medicine, and the availability of phytochemical and therapeutic information. These profiles indicate the amount of information available (or its paucity) on different aspects of the species.

FORMAT OF THE PROFILES

The profiles were compiled in the following format, and arranged alphabetically under each family:

Botanical name

The current scientific name, that has to be applied to a particular species, under the Rules of the International Code of Botanical Nomenclature (ICBN), is used. The names of a large number of species of legumes have changed in course of time, on account of either a better scientific understanding of the species or the application of ICBN. The current names of species used here were adopted from the online database of the International Legume Database Information System (ILDIS). The nomenclature in ILDIS was determined through a collaborative and consensual effort of international legume taxonomists and so more acceptable than individual opinions. For species that are not included in ILDIS, names given in relevant taxonomic literature are used.

Synonyms

Synonyms are the names that were once used for a species but now discarded, as they cannot be used under the rules of ICBN, are given to provide leads to the earlier literature. For some species there is a large number of synonyms, but those synonyms that are relevant on scientific grounds or those that were used in utilitarian literature over a long period of time, are given in the Tables of economic potential (Chapter 3) and in the profiles here. In order to facilitate location of species by their synonyms, nomenclatural equivalents of synonyms and current names of species are provided for each of the three families (Chapter 3, Table 3.4).

Notes: Information on the recognition of subspecies and/or varieties of the species or other points of interest is provided here.

Common names

Common English names, so far as available are given.

VERNACULAR NAMES

Names in different Indian languages, so far as available, are given. Vernacular names provide leads to the uses of the respective species. Connecting scientific names of species with vernacular names is often a difficult job. The availability of a vernacular name indicates that the species, whether native or introduced, is known in the region of that language. Species with several uses often have many names in each vernacular language, some of them restricted to specific parts of the linguistic area. Unfortunately, there is a considerable amount of uncertainty and confusion about vernacular names, particularly in the case of less known species.

Geographical distribution

Information on geographical distribution of species was summarized from ILDIS. For species that are not treated in ILDIS, distributional information was taken from relevant botanical literature. So far as available, distribution within India is also given. Information on the origin and nativity of species is both interesting and important in certain contexts, but such information is often patchy and/or controversial.

Botanical description of species

Botanical descriptions of species are needed to provide an idea of the features of a species and to identify the species. Species with wider ranges of distribution were studied in great detail and there is extensive descriptive information in Floras, the botanical works that deal with various aspects of vegetation. It is difficult to obtain descriptive information on a large number of species, particularly those that were introduced or those that have a narrow range of distribution.

Phytochemistry

The chemistry of a species is the key to its utility in different contexts, particularly in medicine and as food. One of the main efforts of this book is to provide phytochemical information. There is extensive literature on phytochemistry of legumes, but is widely dispersed. There is enormous information on some species and very little or none at all, on others. Phytochemical information is provided for the plant, root/bark, stem/bark/wood, leaf, flower, pod and seed.

Edibility

One of the most important uses of legumes is as food. In addition to the species used commonly as pulses or vegetables, there is an enormous number of species that are consumed in different forms in different regions, particularly in the rural areas. There is a lot of scope to enhance the use of such legumes as food on a wider scale.

Medicinal uses

The general therapeutic potential of different legume species is given for the plant, root/bark, stem/bark/wood, leaf, flower, pod and seed. In some cases all the parts of the plant are used for the same therapeutic effect. But in several species each part has a different therapeutic effect, much of this is related to its chemistry.

Bioactivity

Bioactivity indicates the potential of the chemical constituents of a species to modify metabolic processes. For several species research was conducted to study bioactivity, in order to use the specific chemical compounds to medicinal use. However, information on this aspect is patchy and a lot more work needs to be done.

Toxicology

Extensive research was conducted to study the toxic effects of several legume species, with the objective of using them in medicine, particularly as anticancer drugs and also not to use them in excess.

Clinical studies

While bioactivity and toxicology are studied using animal models, clinical studies are conducted using human volunteers. Since the results of animal studies do not ensure exactly the same effects in human subjects, clinical studies are very important in understanding how exactly phytochemical compounds can be used in disease treatment. There are certain hurdles in India to conduct clinical studies using human volunteers. Till this situation is resolved, which is going to be soon, a lot of information on clinical studies comes from studies conducted outside India.

Uses in alternative systems of medicine

One of the most important and crucial applications of Indian plant wealth is in medicine. The three indigenous systems of medicine prevalent in India are Ayurveda, Siddha and Unani, which have an age-old tradition of using plants. A very broad based literature has been built up on the plants used in these systems not only in the classical texts but also in the form of numerous commentaries, dictionaries and other works. Legumes are an important part of the arsenal of the three indigenous systems. It is important to know how legumes are used in traditional medicine, in order to use the leads in modern drug development. For these reasons, the therapeutic uses of legumes in Ayurveda, Sidha and Unani are provided, as well as for the Chinese and Tibetan systems, which have a number of similarities with the three indigenous Indian systems.

ETHNOMEDICINE

Ethnomedicine, as already discussed, is an important component of folk medicine and home remedies and served this country well. This fund of knowledge has to be preserved and its use promoted, in order to provide an affordable first line defense against disease. Strictly speaking ethnomedical uses are those that have not been recorded, but practiced in rural situations and passed on from one generation to the other by the word of mouth. However, in the past two or three decades there were several efforts to record these uses, for the fear that they would be otherwise lost to posterity. In most cases, uses in indigenous systems and ethnomedicine are similar. Together with classical literature on indigenous systems, ethnomedical information forms an important basis for modern uses of medicinal plants.

Medicinal uses outside India

A number of species are common to several countries, particularly in the tropics, as either native species or as introduced and naturalized. Comparison of medicinal uses of the same species in different countries affords strength to the uses that are common among different countries. In some cases, the same species is put to specific uses in one country but unknown in another. Such information would widen the range of medicinal uses in India.

Non-medicinal uses

Plants have been put from time immemorial to different uses, other than as food and medicine, as detailed in Chapter 3. The importance of a species is also dependent upon such uses.

SOURCES OF INFORMATION

Information for these profiles was culled from a very large number of printed sources, the following being more important among them:

Printed volumes

Abbiw, 1990; Agarwal and Ghosh, 1985; Agarwal, 1970; Ahuja, 1965; Ali, 1994; Ambasta, 1986; Anonymous, 1948, 1950, 1952, 1956a,b, 1959, 1962, 1966, 1969, 1972a,b, 1973, 1975, 1976a,b,c,d,e, 1977, 1978, 1979a,b, 1983, 1988a,b, 1991, 1992a,b,c, 1995; Arnold and Krochmal, 1984; Arora and Pandel, 1996; Asolkar *et al.*, 1992; Atkinson, 1980; Badhwar *et al.*, 1964; Bailey, 1949; Bakhru, 1997; Baldev *et al.*, 1988; Basu, 1918; Beal and Reinhard, 1981; Bentley and Trimen, 1980; Bhattacharjee, 1988; Birkhill, 1935; Biswas and Chopra, 1982;

Boericke, 1991; Bogdan, 1977; Bose and Som, 1986; Caius, 1989; Chauhan, 1968; Chopra and Chopra, 1955; Chopra and Handa, 1961; Chopra *et al.*, 1965,1986a,b; Choudhury, 1990; Clarke, 1982; Clay and Hubbard, 1977; Conway, 1973; Cooke, 1967; Dastur 1962, 1964; Dey and Bahadur 1972; Dhungat, 1972; Drury, 1878; Duke, 1981; Dummer, 1994; Dymock *et al.*, 1890; Ernst and Lehner, 1973; Farooqi, 2001; Fathima 1994; Fitzpatrick, 1964; Fluck, 1988; Gamble, 1979; Gibbs, 1974; Gopalan *et al.*, 1985; Graves, 1990; Grieve, 1967; Handa and Kapoor, 1995; Harborne and Baxter, 1996; Hartwell, 1982; Hasan, 1989; Hedrick, 1972; Herklots, 1972; Hermann, 1973; Heywood, 1978; Hill, 1989; Irvine, 1961; Jackson, 1893; Jain, 1968, 1981,1991,1996; Jain *et al.*, 1991; Joshi, 2000; Kachroo, 1970; Kameswara Rao, 2000; Kaufman, 1999; Kent, 1975; Kays and Dias, 1995; Keshavamurthy and Yoganarasimhan, 1990; Keys, 1995; Kirthikar and Basu 1975; Kochhar, 1981; Kreig, 1965; Krishna Murthy, 1993; Krochman, 1969; Kurup, 1977; Latha, 1992; Lewis and Lewis, 1977; Lindley 1981; Livingstone, 1978; Mabey, 1988; Maheshwari and Singh, 1965; Maheshwari *et al.*, 1981, 1986; Mahindru, 1992; Mahishi, 1993; Maiti, 1997; Morton, 1977; Nadkarni, 1976; Narayanan and Dabadghao, 1972; Nayar *et al.*, 1989; Newdick, 1991; Ochse, 1977; Oliver-Bever, 1986; Painter and Power, 1982; Patnaik and Dhawan, 1986; Perry, 1980; Premnath, 1976; Purkayastha, 1990; Purseglove, 1991; Radha, 1992; Ramaswamy and Razi, 1973; Rastogi and Mehrotra, 1990; Ray and Gupta 1980; Ray *et al.*, 1993; Robbins and Ramaley, 1937; Ross, 1999, 2000; Saldanha, 1982; Saldana and Nicolson, 1976; Sambamurthy and Subrahmanyam, 1989, 1998; Sanjappa, 1992; Santapau and Henry, 1973; Sathyanarayana Bhat, 1993; Sharon, 1994; Shah and Qadry, 1985; Shanmugavelu, 1989; Sharma, 1993; Sharu Raj, 1990; Shubha Rani, 1995; Simpson and Conner-Ogorzaly, 1986; Singh and Arora, 1978; Singh *et al.*, 1965; Sinha, 1998; Sivaranjan and Indira Balachandra, 1994; Skerman, 1977; Smartt, 1976; Srivastava, 1989; Stuart, 1985, 1989; Swiader *et al.*, 1992; Thakur *et al.*, 1991; Thomson, 1976; Tindall, 1983; Troup, 1986; Uma, 1993; Uphof, 1968; Variers, 1994, 1995, 1996; von Reis Altschul, 1973; Watt, 1972; Watts and Watts, 1944; Willis, 1973; Yoganarasimhan, 1996, 2000; Yamaguchi, 1983; Yeshidanden, 1986; Zafal, 1994; Zeven and Zhuykovsky, 1975.

Journals

Numerous articles in journals, more particularly the journal Economic Botany, Planta Medica and Abstracts of Medicinal and Aromatic Plants, have been very useful.

Websites

While there are over 700,000 links on the Internet to legumes, only a very small fraction of them provides authentic and reliable information. Seed and plant suppliers in the horticulture sector, and dealers of herbal products are hosting most of these websites. Much of the information on these websites is unscientific and unreliable. The website of the International Legume Database & Information Service (ILDIS, website at http://www.ildis.org/) for taxonomic and distributional information and Dr. Duke's Phytochemical and Ethnobotanical Databases (http://www.ars-grin.gov/duke/) were the most useful sources, used in this book.

Index of Species Profiles

Caesalpiniaceae

Fabaceae

CHAPTER-24

Caesalpiniaceae

1. *Afzelia africana* Pers.

Synonym: *Pahudia africana* (Pers.) Prain.

Common name: African mahagony.

Distribution: Native of Africa; introduced into India; cultivated in gardens (Lalbagh, Bangalore).

Description: Large tree with a spreading crown and scaly grey-brown bark, with a pink or red slash; twigs glabrous, with lenticels; glabrous leaves up to 30 cm long with 7-17 pairs of elliptic or ovate leaflets with an obtuse or sub-acuminate tip; nerves protruding with 8-12 pairs of narrow, secondary, connecting nerves; flowers with one single white petal, with red striae, set in terminal panicles up to 20 cm long; pod flat, glabrous, black, woody, bursting open at maturity spreading the seeds; seeds poisonous, with a sweet edible aril; tree is associated with ectomycorrhizal fungi.

Edibility

Aril: Edible.

Medicinal Uses

Plant: Analgesic, antihaemorrhagic, aphrodisiac, emetic, emmenogogue, febrifuge and laxative.

Root and leaf: Decoction with *Tamarindus indica* is used as antidote to poison.

Toxicology: Wood workers are susceptible to respiratory disorders due to inhalation of saw dust.

Medicinal Uses Outside India

Ivory Coast: Hot water extract of *Tamarindus indica* leaf and root, with root and leaves of *Afzelia* and *Ficus* is used as a vapour bath for sleeping sickness.

New Guinea: Decoction of *Tamarindus indica* bark with the bark of *Afzelia* is given during pregnancy as a remedy for weakness.

Ethnomedicine

Plant: Is used as antiemetic and diuretic; used in edema, fever, gonorrhoea, hernia, lumbago, sleeping sickness and stomach disorders.

Non-Medicinal Uses

Wood: Is used in construction of wooden houses as it is termite proof; used as substitute for mahagony.

Leaf: Used as manure and provides good fodder.

Pod: Used in making soap.

Seed: Considered as highly poisonous.

2. *Bauhinia acuminata* L.

Synonyms: *Bauhinia candida sensu auct.*
Bauhinia grandiflora sensu auct.
Bauhinia linnaei Ali

Common Name: White bauhinia.

Vernacular Names: Hin: Kunchun, Safed kachanaar; **Kan:** Bile kanchuvaala, Bili mandaara; **Mal:** Mandaaram, Velluta mandaaram; **Sans:** Kaanchanaara, Karubudarah; **Tam:** Mundareh, Valli mandaaram maram; **Tel:** Kaanchana, Kaanshana.

Distribution: Origin indeterminate; cultivated in parts of Africa, South East Asia including India and Sri Lanka; in India it occurs in the North West.

Description: Small tree or shrub, young parts pubescent; leaves united about 2/3 way upwards ending in obtuse or sub-obtuse lobes, base cordate, glabrous above, pubescent below, with subulate stipules; leaflets acute at the apex; flowers white; calyx limb acuminate, spathaceous, 5 toothed at the apex; pod long, ridged on each side along the upper suture, beaked.

Edibility

Flower: Eaten raw.

Medicinal Uses

Root: Extract is used in treating cold; root decoction with oil is applied on burns.
Stem bark: Decoction of the bark is given for asthma, biliousness, bladder stone and leprosy; it improves digestion; used as antidote to poison.
Leaf: Leaf decoction is used in treating asthma, biliousness and bladder stone; it improves digestion; crushed leaves are used as cataplasm for ulcers in the nose.
Flower: Is cooling.

Uses in Alternative Systems of Medicine

Ayurveda:
Part used: Bark and flower.
Uses: Useful in treating cough, glandular swellings, goitre, gout, haemorrhage, leprosy, obesity, scrofula, ulcers, urinary disorders and wounds.

Siddha:

Part used: Root, bark and flower.
Uses: Useful in constipation, cough, haemorrhage, obesity, scrofula and ulcers.
Ethnomedicine:
Plant: Is used in cold, sores and ulcer.

Non-Medicinal Uses

Plant: Grown as hedge plant and as an ornamental tree.

3. *Bauhinia malabarica* Roxb.

Synonyms: *Bauhinia acida* Korth.
Bauhinia malabarica Roxb. var., *reniformis* Baker
Bauhinia platyphylla Miq.

Common names: Butterfly flower, Jerusalem date, Malabar Mountain ebony.

Vernacular Names: Ass: Kattra; **Ben:** Karmai; **Hin:** Amli, Amlosa, Imli, Khat papri; **Kan:** Basavanapaada, Chippuru, Holeippe, Huliaachalu, Kogele, Kogge pukka, Nimbeda, Nimpidi, Nimpuli, Pulikogelepukku, Sappara, Shadhu; **Mal:** Aaraampuli, Kokkavaali, Minpuli; **Mar:** Aamlapathra, Asmanthaka, Asmayoktha, Korala; **Tam:** Malaiyaatti, Puliyaatti, Pullaninta, Vattaatti; **Tel:** Aare, Peddaare, Pullaare, Pullagoddukura, Pullandanda, Pullaninta.

Distribution: Native of India and other South East Asian countries; occurs throughout India.

Description: A medium sized tree with brownish bark; leaves broader than long, with 2 obtuse lobes reaching 1/3 of the length, slightly cordate and glabrous; bracts triangular; flowers yellow on racemes; calyx clothed with brown tomentum, limb 5 clefted; petals spathulate; stamens 10, all fertile; pod stalked, acuminate, firm, glabrous, valves marked with diagonally placed parallel veins.

Phytochemistry

Plant: Isoquercitroside, kaempferol glycosides, quercetol glycosides, quercitroside, rutoside and taxifolin.

Edibility

Root: Used for flavouring meat and fish.

Shoot: Edible.

Leaf: Edible; used in flavouring fish and meat.

Flower and tender seeds: Edible.

Medicinal Uses

Bark: Useful in bloody dysentery and diarrhoea.

Flower: Infusion given for dysentery.

Uses in Alternative Systems of Medicine

Ayurveda:

Part used: Bark and flower.

Uses: Useful in cough, glandular swellings, goitre, haemorrhage, leprosy, menorrhagia, scrofula, urinary disorders and wounds.

Non-medicinal uses

Young shoots and leaves: Used as fodder.

Bark: Is used as tanning material.

Wood: Is used for making scantlings and rafters; yields firewood.

4. *Bauhinia purpurea* L.

Synonyms: *Bauhinia coramandeliana* DC.
Bauhinia triandra Roxb.
Phanera purpurea (L.) Benth.

Common Names: Butter fly tree, Camel foot tree, Pink butterfly tree, Purple bauhinia, Purple butterfly tree, Purple mountain ebony.

Vernacular Names: Ben: Deva kanchan; **Guj:** Kovidara; **Hin:** Gairal, Kaliar, Kandan, Kaniar, Khairwal, Koilari, Koinar, Sona; **Kan:** Kempu kanchivaala, Kempu mandaara, Kanchi vaala, Raktha mandaara, Sarul, Ulipe; **Mal:** Suvarna mandaaram; **Mar:** Atmatti, Deva kanchana, Ragthachandan, Raktha kaanchan; **Sans:** Rakthapushpakovidara, Vanaraaja; **Tam:** Areka, Kalavilaichi, Mandaarai, Neelatheruvatti, Periyavaatti, Segappu mandarai; **Tel:** Bodanta chettu, Deva kaanchanamu, Kaanchanamu, Peddaare.

Distribution: Native of Hainan, Puerto Rico and Asia including India; occurs throughout India, often cultivated.

Description: A medium sized tree with ash coloured or brownish bark; leaves longer than braod, slightly cordate at the base, subcoriaceous, glabrous; flowers deep pink, on terminal and axillary corymbose racemes; bracts deltoid, minute; calyx tube split into 2 segments, lower one emarginate, the upper 3 toothed; fertile stamens 3 or 4, elongate; stigma large and oblique; pod linear, flat, decurved, pointed, greenish tinged with purple; seeds 12-15.

Phytochemistry

Plant: Yields gum.

Bark: Tannin.

Flower: Anthocyanins, astragalin, essential oil, isoquercetin, pelargonidin-3-glycoside, pelargonidin-3-triglycoside and quercetin.

Seed: Apigenin, 4-O-β-L-arabinopyranosyl, galacto pyranoside, butein-4'-O-β-arabino-O-β-D-galactoside, chalcone glycoside, 3,4-dihydroxychalcone and quercetin.

Edibility

Leaf: Cooked and eaten.

Flower bud and flower: Used in pickles.

Pod: Fresh pods are consumed.

Medicinal Uses

Plant: Is astringent, anthelminitic, carminative and diuretic; useful in anasarca, convulsions, delirium, diarrhoea, dropsy, haemorrhages, pain, *Datura* poisoning, prolapsed rectum, rheumatism, septicaemia, swelling and ulcers; used as antidote for scorpion sting and snake bite.

Root: Carminative and tonic; root and flower paste in rice water is applied on abscesses and boils as a maturant and also on cuts and bruises.

Stem: Used in setting bone fracture.

Bark: Alterative, astringent and tonic; used in diarrhoea; poultice of leaf and bark is used as haemostatic; decoction is used to wash ulcers and taken internally for goiter and stomach cancer; bark paste is applied on small pox pustules.

Leaf: Is hypoglycaemic; leaf paste with latex of *Jatropha curcas* is administered in jaundice.

Flower bud: Is anthelmintic and laxative; used in treating cancer.

Uses in Alternative Systems of Medicine

Ayurveda:

Part used: Bark, flower and pod.

Uses: Bark and flower useful in cough, consumption, glandular swellings, goitre, gout, haemorrhage, leprosy, menorrhagia, piles, scrofula, urinary disorders and wounds; pod used in fever and as antidote to poisons.

Siddha

Part used: Root, bark and flower.

Uses: Cooling and laxative; useful in bleeding piles, cough, menorrhagia, scrofula, stomach ache and ulcer.

Ethnomedicine

Plant: Is used as anodyne; useful in anasarca, convulsions, diarrhoea, dropsy, epilepsy, haemorrhages, rheumatism, septicemia, swelling and tumour; as antidote to fish, snake and scorpion poison.

Non-Medicinal Uses

Plant: Grown as an ornamental plant; yields gum.

Stem bark: Source of brown dye and tannin; used in leather industry; yields fibre, which is used to make ropes and as thatching for houses.

Wood: Used for making agricultural implements, rafters, scantlings, match boxes and match sticks.

Leaf: Used as fodder and as insecticide.

5. *Bauhinia racemosa* Lam.

Synonyms: *Bauhinia parviflora* Vahl.
Piliostigma racemosa (Lam.) Benth.

Vernacular Names: Hin: Ashta, Asoda, Dhorara, Ghila, Gurial, Jhanjhora, Jhinjheri, Kachnal, Makkuna, Marvil, Maula, Mawal, Papri, Shirahuta, Thaur; **Kan:** Aapata, Aaralu, Aaralu mandaara, Araginda, Arise, Arisile, Arthi, Bannae, Chitaachi, Ishae, Iyuta, Kaadu mandaara, Kitthi rasa, Vana sampige, Zingije; **Mal:** Kotapuli, Mandaaram; Mar: Aapata, Aapta, Kanraja, Seyara, Shiara, Sonan, Vanaraja; **Sans:** Anupushpaka, Ashmanthaka, Kushali, Shlashnathvaka, Shvetakaanchana, Vanaraaja, Yamalapathraka, Yugmapathra; **Tam:** Aar, Aarai, Aatthithi, Aram, Archi, Arikka, Atcha, Atti, Kokkumantharai, Sallagi, Sittacha,Taadagi; **Tel:** Aare, Are chettu, Ada, Adivi yavise, Manchiyaare, Pachaare.

Distribution: Native of Asia including India; occurs in North East India, Punjab and Rajasthan.

Description: A small crooked tree with thick rough brownish bark and drooping branches; leaves deeply cleft, broader than long, slightly cordate at the base, 7-9 nerved, densely grey, pubescent beneath; flowers, small, pale yellow on terminal leaf opposed racemes; calyx limb spathaceous, 5 toothed, finely reflexed; stamens 10, perfect; pod stalked, falcate, firm, torulose and indehiscent.

Phytochemistry

Stem bark: β-Amyrin, β-sitosterol and octacosane.

Heartwood: Dibenzol, gum, pacharin, racemosol and tetracyclic phenol.

Seed: Linoleic acid, mucilage, myristic acid, oil, oleic acid, palmitic acid, pentosan and stearic acid.

Seed oil: Lauric acid, linoleic acid, myristic acid, oleic acid, palmitic acid and stearic acid.

Edibility

Leaf, flower, bud, pod and seed: Edible.

Medicinal Uses

Root bark: Cholagogue; used in dysentery, diarrhoea, inflammations, skin diseases and ulcers as an astringent; decoction used as gargle for aphthae; taken internally, the decoction reduces liver inflammation and expels worms; bark paste is applied on wounds and swellings.

Stem: Is anticancer, CNS active and hypothermic.

Stem bark: Alexipharmic, anticancer, antipyretic, hypothermic and refrigerant; cures anal fistula, biliousness, diseases of blood, headache, quartan fever, thirst and urinary discharges; as an astringent the bark decoction is administered in chronic dysentery, diarrhoea, glandular inflammation, goiter, skin diseases and ulcer.

Leaf: Alexipharmic, anticephalalgic, antimalarial, antipyretic and refrigerant; cures biliousness, fistula, headache, intestinal worms, quartan fever, thirst and urinary discharges; paste of young leaves is applied for headache during fever; leaf decoction is given in malaria; leaf juice with pepper and onion is given in dysentery and diarrhoea.

Flower: Anthelmintic; used in cough, diarrhoea, dysentery, haemorrhage and piles.

Uses in Alternative Systems of Medicine

Ayurveda:

Part used: Bark, leaf, flower and pod.

Uses: Flower and bark useful in glandular swellings, goitre, gout, leprosy, menorrhagia, scrofula, urinary disorders and worms; leaf and flower is useful in anal fistula, biliousness, diarrhoea, dysentery, fever, headache, skin diseases, thirst, throat trouble, tuberculous glands, tumours and urinary discharges; pod is astringent to bowels, cooling and removes phlegm.

Unani

Part used: Bark and leaf.

Uses: Bark is used as astringent in diarrhoea and dysentery; decoction of the leaves is used in headache and malaria.

Non-Medicinal uses

Plant: Grown as ornamental or hedge plant.
Bark: Yields fibre used for making ropes.
Wood: Is used for making ploughs, posts, carts, axles, yokes and turnery works.
Leaf: Useful as fodder; used in making bidi wrapper.

6. *Bauhinia semla* Wunderlin

Synonym: *Bauhinia retusa* Roxb.

Vernacular Names: Hin: Semla; **Pun:** Kural; **Tel:** Nirpa.

Distribution: In India, it grows in Bihar, Himachal Pradesh, Orissa, Punjab and Uttar Pradesh.

Description: A medium sized tree with dark brown bark and branches with lenticels; leaves broad, cordate, 9 nerved; leaflets connate; flowers small or white with purple streaks or dots; fertile stamens three, silky hairy; pod glabrous, flat with obtuse tip; seeds 6-8 orbicular, smooth.

Phytochemistry

Stem bark: Quercetin-3,O-β-D-glucoside and rutin.

Edibility

Flower bud: Edible.
Gum: Edible.

Medicinal Uses

Plant: Hypoglycaemic, hypocholesterolemic and CNS active; used in cholera and as antidote to snake poison.
Gum: Is diuretic and emmenogogue; is used as an external application to sores; useful in cholera and diarrhoea and as antidote to snake poison.
Seed: Hypocholesterolemic and hypoglycaemic.

Ethnomedicine

Plant: Is used in cholera and as antidote to snake bite.

Non-Medicinal Uses

Wood: Used for general construction work and as firewood; used in making scantlings, hammer handles, posts and rafters.

Gum: Tree yields gum used for sizing paper and cloth; used as binder and for water proof terraces.

7. *Bauhinia tomentosa* L.

Synonyms: *Bauhinia pubescens* DC.
Bauhinia volkensii Taubert
Bauhinia wituensis Harms.

Common Names: Bell bauhinia, Holy mountain ebony, Yellow bauhinia, Yellow butterfly tree.

Vernacular Names: Hin: Kachnar, Kanchana; **Kan:** Kaadathi, Kaadu mandaara, Karanaapu, Vana sampige; **Mal:** Kaanjana, Kanschenapou, Kaattatthi; **Mar:** Aptu, Chan, Pivalakonchan; **Sans:** Phalgu, Peethakaanchana, Ushmadughaa; **Tam:** Iruvaatchimaram, Iruvaji, Kaanchini, Kaattatthi, Kaattumandarai, Kanjani, Siruvaatthi, Tiruvaatthi; **Tel:** Adavi mandaaramu, Deva kaanchanamu, Kanchani, Kanjelapa, Kanjini, Mandaara, Paalepi.

Distribution: Considered to be a native of India and Angola in Africa; introduced into several other countries of Africa, Asia, Caribbean Islands, Pacific Ocean Islands and South America; occurs in North and North West India and Andaman islands; often planted.

Description: A shrub with ash coloured branches and zigzag downy branches; leaves with a cleft less than half way down forming 2 obtuse lobes, mucronate in the sinus, truncate at the base, shining and glabrous above, pubescent and whitish beneath; flowers large in leaf opposed pairs; bracteoles linear; calyx limb long, spathaceous, splitting down to funnel shaped tube, finely pubescent; petals obovate, spatulate, yellow, the upper half with a red blotch on the face; stamens 10, all fertile; pod dehiscent, finely pubescent, becoming glabrous when ripe; seeds 10-16, flat, dark brown and shining.

Phytochemistry

Plant: Aspartic acid, fatty oil, glutamic acid, octadecadienoic acid and quercetin.
Flower: Isoquercetin, rutin and quercetin.
Seed: Fatty oil and quercetin.

Edibility

Leaf: Leaves are sour in taste and used as a substitute for vinegar.

Seed: Edible.

Medicinal Uses

Plant: Antidysenteric and anthelminitic; used in liver complaints and as antidote for snake bite.
Root: Vermifuge.
Root bark: Decoction is administered in dysentery, liver inflammation and as vermifuge.
Stem bark: Astringent and vermifuge; bark infusion is used as gargle for sore throat; ground into a paste it is applied externally to scrofulous sores, tumours and wounds.
Leaf and flower: Given to check dysentery.
Pod: Diuretic.
Seed: Aphrodisiac and tonic; the seed paste with vinegar is applied to cuts, wounds and bites of snake and scorpion.

Uses in Alternative Systems of Medicine

Siddha:
Part used: Leaf, flower and pod.
Uses: Useful in cough, diarrhoea, digestive disorders and leucorrhoea.

Ethnomedicine

Plant: Is used in boils and tumours.

Non-medicinal Uses

Plant: Cultivated as an ornamental plant.
Stem bark: Yields fibre used for making cordage.
Wood: Used for making knife handles and handles of cutting tools.
Leaf: Yields an yellow dye.

8. *Bauhinia vahlii* Wight & Arn.

Synonyms: *Bauhinia racemosa* Vahl.
Phanera vahlii (Wight & Arn.) Benth.

Common name: Camel's foot climber.

Vernacular Names: Hin: Jallur, Maaljan, Maholan, Mahul, Mahwal, Malghan, Maljhan, Malo, Malu, Maulein, Maurain; **Kan:** Aane paada, Hepparige, Kambi hoovu; **Mal:** Mottanvalli; **Mar:** Chambal, Chambil, Chambuli, Chambura, Charbor, Maljan; **Sans:** Asmanthaka, Malanjhana, Moorva; **Tam:** Mandaarai, Mandaaraiyilai; **Tel:** Addaku, Aadathige, Madapu.

Distribution: Native of Bhutan, India, Nepal, Pakistan and Srilanka; introduced into other countries; occurs in South, Central and North India.

Description: Climber with thick cable like branches; branchlets ending in a pair of strong, revolute tendrils; leaves deeply cordate, 11-13 nerved, glabrescent above, tomentose beneath; stipules obtuse, falcate; flowers white on long peduncled corymbose racemes; calyx tube cylindrical, with limb splitting into 2 lobes; petals white, hairy outside; stamens 3; pod pendulous, broad, flat, woody with 8-12 flat brown seeds.

Phytochemistry

Plant: Gum, isoquercitroside, rutoside, quercitroside and tannin.
Leaf: β-Sitosterol, betulinic acid, campesterol, flavone, kaempferol, quercetin, quercetin-3-glucoside and stigmasterol.
Flower: Kaempferol.

Edibility

Pod: Young pods are eaten raw; mature pods are roasted and eaten.

Seed: Fried and eaten.

Medicinal Uses

Plant: Useful in dysentery and stomach ache.

Root: Decoction cures fever.

Stem: Antiarrhythmic, CNS active and spasmolytic.

Leaf: Demulcent; used in headache, malaria and skin disease.

Pod: Antifertility agent.

Seed: Aphrodisiac, tonic and vermifuge; useful in dysentery.

Ethnomedicine

Plant: Is used in cholera, dropsy, dysentery and stomach ache.

Non-Medicinal Uses

Plant: Grown as an ornamental tree; yields fibre used in making mats and baskets; yields tannin used in tanning industry.
Stem bark: Yields fibre used in making ropes and suspension bridges.
Leaf: Used as fodder; used in making plates, cups, birdis, rain hats, caps, basket, umbrellas and for thatching.

9. *Bauhinia variegata* L.

Synonyms: *Bauhinia chinensis* (DC.) Vogel
Bauhinia decora Uribe
Bauhinia variegata L. var., *chinensis* DC.
Phanera variegata (L.) Benth.

Notes: Two varieties *candida* and *variegata* are recognized.

Common Names: Buddist bauhinia, Camel's foot, Poor man's orchid, Variegated bauhinia, Variegated mountain ebony.

Vernacular Names: Ben: Bidul, Raktachanchan; **Guj:** Kachnaar; **Hin:** Barial, Gurial, Gwiar, Kachnaar, Kandan, Khairlid, Koilar, Kovidara, Lal kanchnaar, Sona; **Kan:** Arishina tega, Kempu kanchivaala, Kogile pukka, Kondaalka, Kovidaara; **Mal:** Suvanna mandaaram, Unna; **Mar:** Kanchan, Raktha kanchan; **Sans:** Asphota, Gandari, Kanakaprabha, Kovidara, Kundali, Mahapushpa, Raktapushpa, Shenapushpa, Tamarapushpa, Uddalaka, Yugapatraka; **Tam:** Mandaarai, Segappu mandaarai, Vellaimatharai, Segappu manchari; **Tel:** Bodanta, Deeva kanchanamu, Mandaara.

Distribution: Native of Asia including India; introduced and cultivated in the tropics; grows throughout India.

Description: Medium sized tree with ash or brown bark; leaves broader than long, deeply cordate, subcoriaceous, 9-11 nerved; flowers large, fragrant, creamish purple or reddish purple with variegated petals on lateral corymbs; bracts minute, deltoid; calyx limbs spathaceous, 5 toothed, hairy outside; corolla with 4 white petals, one upper petal red with blue or purple veins or all the petals red with upper petal deeply coloured and variegated with cream and purple; pod hard, flat, glabrous, decurved; seeds 10-15.

Phytochemistry

Plant: β-Sitosterol, 5,7-dihydroxyflavanone-4'-O-α-L-rhamnopyranosyl-β-D-glucopyranoside, rhamnoglucoside 5,7-dimethoxyflavonone-4'-O-α-L-rhamnopyronosyl-β-D-galactopyranoside, isoquercitroside, lupeol, naringenin, quercitroside, rutinoside.

Bark: Hentriacontane, octacosanol, stigmasterol and tannin.

Flower: α-Ketoglutaric acid, alanine, aspartic acid, cyanidin-β-O-D-glucoside, glutamic acid, glycine, kaempferol-3-glycoside, kaempferol-3-galactoside, kaempferol-3-rhamnoglucoside, malvidin-3-β-O-D-glucoside, oxaloacetic acid, peonidin-3-β-O-D-glycoside, pyruvic acid and serine.

Seed: Glycosides, linoleic acid, myricetin, myricetol, oleic acid, palmitic acid, quercetin and stearic acid.

Edibility

Leaf, flower and pod: Edible; flower buds are pickled.

Medicinal Uses

Plant: Alterative, anthelmintic, astringent, carminative and laxative; used in glandular diseases, diarrhoea, dysentery, leprosy, malaria, piles, scrofula, skin diseases and ulcer; used as an antidote for snake poison.

Root: Anthelmintic, anti-inflammatory, astringent, carminative, constipating, cooling, depurative, styptic and vulnerary; useful in cough, diarrhoea, dysentery, dyspepsia, glandular diseases, haemoptysis, haemorrhoids, intestinal worms, leprosy, proctoptosis, prolapsed rectum, scrofula, skin diseases, swelling, tumours, ulcers, vitiated conditions of "pitta" and "kapha" and wounds; ground root with rice water is used as a cataplasm to promote suppuration of boils and sores; decoction of the root is used in dyspepsia, flatulence, obesity, tumours and as antidote for snake bite.

Bark: Alterative, antibacterial, anti-inflammatory, antileprotic, antirheumatic, anthelmintic, astringent, cooling, depurative, emmenogogue, hypothermic, styptic, tonic and vulnerary; used in "pitta" and "kapha", bleeding, cancer, cough, diabetes, diarrhoea, dysentery, dyspepsia, glandular diseases, haematuria, haemoptysis, haemorrhoids, intestinal and abdominal tumours, leprosy, menorrhagia, proctoptosis, prolapsed rectum, scrofula, skin diseases, sore throat, syphilis, throat diseases, ulcer and wounds; the juice of the fresh bark with the juice of the flowers of

Strobilanthes citrata is given as an expectorant; the bark given with ginger is an internal remedy for scrofula; a decoction of the bark with *Acacia* pod and Pomegranate flowers is used as a remedy for excessive salivation and sore throat; bark with gum resin of *Boswellia serrata* and Myrobalan is given in scrofulous enlargement of cervical glands and of glands on the neck; poultice of ground bark with rice water hastens suppuration of boils and sores.

Flower bud: Alterative, antibacterial, astringent, cooling, emmenogogue, laxative, vermifuge and tonic; useful in bleeding piles, diarrhoea, dyspepsia, dysentery, haematuria, menorrhagia, scrofula and ulcers.

Flower: With sugar it acts as a laxative; flowers with rice water act as a cataplasm to promote suppuration of boils and sores; dried and powdered it is given in menorrhagia; decoction of flower is given for cough, haematuria, menorrhagia, piles and tumours.

Toxicology: Plant contains chymotrypsin inhibitors.

Uses in Alternative Systems of Medicine

Ayurveda:

Part used: Root, bark and flower

Uses: Bark and flower is useful in haemorrhage, leprosy, menorrhagia, scrofula, urinary disorders, wasting diseases, worms and wounds; root of plants of red coloured variety is given in snake bite; bark is cooling, laxative and astringent to bowels; cures "kapha", "vata", biliousness, diarrhoea, dysentery, leprosy, piles, tuberculous glands and ulcer; bark of plant with white flowers is anthelmintic, astringent to bowels and cooling; cures anal troubles, asthma, biliousness, blood disorders, burning sensation, cough, goiter, gout, leucoderma, strangury, thirst, tuberculous glands, ulcers and vaginal discharges; flower is astringent, cooling and galactogogue; cures biliousness, diseases of blood, bronchitis, consumption, head ache, haemoptysis and vaginal discharges; root is used internally as emetic.

Siddha:

Part used: Root, bark and flower.

Uses: Useful in constipation, cough, haemorrhages, obesity, scrofula and ulcer.

Unani:

Part used: Bark and flower.

Uses: Bark is aphrodisiac, astringent and blood purifier; useful in asthma, dysmenorrhoea, leprosy, leucoderma, menorrhagia, stomatitis and as tonic to the liver; cures biliousness, tuberculous glands, ulcers and wounds; flower bud is used in cough, eye diseases, liver complaints and piles; astringent to bowels; used as styptic in haematuria and menorrhagia; flower paste useful in skin diseases.

Medicinal Uses outside India

China: Flower, bark and root is pounded and made into a paste and used for scrofula.

Myanmar: Root is used as a remedy for dyspepsia and bark decoction is used as a tonic.

Philippines: Leaf paste is applied on the head for relief from headache.

Ethnomedicine

Plant: Useful in bone fracture.

Bark: Is astringent; useful in smallpox and tumour in the intestine and wounds; antidote to snake bite, scorpion sting and *Datura* and *Rheum* poisoning.

Leaf: Useful in jaundice.

Flower: Cures indigestion.

Non-Medicinal Uses

Plant: Cultivated in gardens as an ornamental tree; is a source of gum and fodder.

Bark: Yields a dye and fibre; dye is used in giving brown colour to leather.

Wood: Useful in making agricultural implements.

Leaf: Used as bidi wrapper.

10. *Caesalpinia bonduc* (L.) Roxb.

Synonyms: *Caesalpinia bonducella* (L.) Fleming
Caesalpinia crista sensu auct.
Caesalpinia sepiaria sensu auct.

Notes: Originally coastal but widely naturalized in India.

Common Names: Bonduc tree, Cockspur, Fever nut, Gray nickel, Indian nut, Nicker bean, Nicker nut, Sappan liane, Sea pearl, Sea pod, Yellow nicker.

Vernacular Names: Hin: Kanja, Kantekaranja, Karanjava, Karanju, Kattkaranj, Kathaleja, Kaatkaranga; **Kan:** Gajjege, Gajjugada balli, Gajjuga, Heggejuga, Kitta, Thapasigajjuga; **Mal:** Aavil, Kalanji, Kazhangi, Kalimaarakam; **Mar:** Gajaga, Kanchaki, Karbath, Rahedo, Saagaragota; **Sans:** Gruthakaranja, Kakachika, Kantakaranja, Kantaki karanji, Kantakini, Karanja, Krakachika, Kuberaakshi, Lathakaranja, Poothikaranji, Prakriya, Shraangeshta, Thinagachhika, Thirini, Valli, Varini; **Tam:** Aavil, Gajji, Kachukkai, Kalal, Kalarsi, Kalargodi, Kalichikkai, Mulal, Suriyindu; **Tel:** Gachha, Gachhaakaaya, Gutsakai, Sukajambuka, Thellagachha.

Distribution: Native of Africa, Asia including India, Australia, Caribbean Islands, Central America, Indian Ocean and Pacific Ocean Islands and South America; occurs throughout India.

Description: A scandent shrub with hard yellow prickles on branches; prickles on rachis and branches usually in pairs and recurved; stipules bifid; pinnae 6-8 pairs; leaflets opposite, 6-10 pairs, oblong, obtuse, mucronate, membranous; racemes axillary; bracts linear, recurved at the apex; petals yellow recurved at the apex; pod densely armed on the faces with sharp wiry prickles, dehiscent with globose or ovoid, lead coloured shining seeds.

Phytochemistry

Plant: Ononitol and methyl myoinositol.

Wood: Brazilin and brasilein.

Seed: α,β,γ and Σ Caesalpin, bonducin, fatty oil, ethylidene, glutamic acid, phytosterols, saponin, starch and sucrose.

Seed kernel: Albuminoids, bonducellin, bonducin, caesalpin-F, fatty oil, ononitol, phytosterols, saponin, starch and sucrose.

Seed oil: Bonducin, ethylidene and glutamic acid.

Medicinal Uses

Plant: Anti-inflammatory and tonic; useful in hydrocele.

Root: Anthelmintic, antipyretic, antiviral, diuretic, emmenogogue, expectorant, stomachic and tonic; used as an astringent in blenorrhagia, dysentery and expulsion of placenta after child birth, stone in the bladder and tumours; decoction of the root is given with honey for leucorrhoea.

Root bark: Anthelmintic, antiperiodic, antispasmodic, bitter, emmenogogue, expectorant, febrifuge, stomachic and tonic; useful in amenorrhoea, asthma, colic, cough, dysmenorrhoea, dyspepsia, elephantiasis, flatulence, hepatomegaly, intestinal worms, pharyngodynia and splenomegaly.

Stem bark: Anthelmintic, antiviral, emmenogogue, expectorant, febrifuge and stomachic; useful in amenorrhoea, asthma, colic, dysmenorrhoea, dyspepsia, flatulence, intestinal worms and tumour.

Leaf: Anthelmintic, deobstruent, emmenogogue and febrifuge; useful in amenorrhoea, dysmenorrhoea, elephantiasis, hepatomegaly, inflammation, intermittent fever, intestinal worms, malaria, pharyngodynia and splenomegaly; leaves boiled in castor oil and applied on painful and swollen testicles; water in which the leaves are boiled is used as a gargle for sore throat; fomentation by roasted leaves with castor oil cures inflammatory, hydrocele, orchitis and piles; oil from the leaves is used as a massage in convulsions, palsy, nervous complaints and filariasis; leaf decoction is used to bathe infants.

Pod: Antipyretic and tonic.

Seed: Acrid, anodyne, anthelmintic, antibiotic, anti-inflammatory, antiperiodic, antipyretic, antispasmodic, aphrodisiac, astringent, bitter, blood purifier, contraceptive, depurative, digestive, diuretic, expectorant, laxative, liver tonic, stomachic, styptic, thermogenic and vesicant; useful in arthralgia, asthma, colic, cough, diabetes, diarrhoea, dysentery, dyspepsia, haemorrhoids, hepatopathy, hiccup, hydrocele, inflammation, intermittent fever, intestinal worms, leprosy, leucoderma, malarial fever, rheumatism, splenopathy, vitiated conditions of tridoshas and vomiting; seed decoction or paste is applied on inflammatory swellings, leprosy, skin diseases and ulcer; promotes digestive power, relieves congestion of blood; burnt seeds are used as an ingredient in dentrifice; seed powder with black pepper is applied on external inflammations; an ointment made of burnt seed powder with castor oil is externally applied on glandular hydrocele, orchitis and swellings, it acts as a antidote for snake bite.

Seed kernel: Is used in anasarca, hydrocele, intestinal worms, liver and spleen diseases and malaria; the kernel and egg are fried in ghee and eaten in orchitis, ovaritis and scrophula.

Seed oil: Oil is used for convulsions, palsy and paralysis; used as emollient and applied on the face to remove freckles and in skin diseases.

Bioactivity: The alcoholic extract of the seeds yielded 4 fractions – A, B, C and D; fraction A and B were antibiotic, fraction C was antipyretic and diuretic and fraction D was anthelmintic; alcoholic extract of root and stem was found to be antiviral against *Vaccinia* virus; aqueous extract of root, stem and leaves showed inhibitory action on sarcoma 180; seed extract showed antiestrogenic and antifertility action in mice, rabbits and dogs.

Toxicology: Plant contains chymotrypsin and trypsin inhibitors; taken in excess harmful to lungs and intestines.

Clinical Studies: Patients having loose motions with mucous and blood were treated with seeds in doses of 2 gm, 3 times a day, for 15 days showed eradication of *Entamoeba histolytia* from the stool samples.

Uses in Alternative Systems of Medicine

Ayurveda:

Part used: Plant, root, leaf, pod and seed.

Uses: Plant is beneficial in abdominal lump, constipation, deranged "kapha", gynaecological disorders, intermittent fever, phlegm, piles, skin diseases and ulcers; root and bark are antipyretic and emmenogogue; seed promotes digestive power, heals ulcers, cures diabetes, hiccup, spleenic disorders, intermittent fever, leprosy, piles and vomiting; leaf is used in "vatha", deranged "kapha", intestinal worms, edema and piles; pod is astringent, aphrodisiac, anthelmintic and healing; is useful in deranged "kapha" and "vatha", intestinal worms, piles and skin diseases.

Siddha:

Part used: Root, leaf and pod.

Uses: Useful in anuresis, hydrocele, neuralgia, orchitis and urinary calculi.

Unani:

Part used: Root, Leaf and seed.

Uses: Root is antiperiodic, anti-inflammatory, styptic and tonic; useful in chronic fever, chronic pain and leprosy; is used as antidote to snake bite. Leaf is used in liver disorders and skin diseases; seeds used in asthma.

Medicinal Uses Outside India

Ceylon: Young leaves are used in tooth ache and taken internally for worms.

Indonesia: Root infusion is given in lithiasis.

La Reunion and Madagascar: The roots are considered as anthelmintic and febrifuge; used as astringent in leucorrhoea and blenorrhagia; seeds are given as anthelmintic, febrifuge and tonic; seed oil is used in paralysis.

New Guinea: Decoction of the root is used as antipyretic and leaf decoction is used as gargle for sore throat; ground seeds are used as vesicant.

Nigeria: Decoction of the root bark is used as rubefacient and used as a cure for dyspepsia; seed oil is given as a remedy for convulsions.

Philippines: Seed decoction is given as emetic and is also helpful in treating dysentery and stomach disorders.

Ethnomedicine

Plant: Is used as antiperiodic, depurative, emmenogogue, emollient, laxative, rubefacient, tonic, vermifuge and vesicant; useful in colic, congestion, convulsion, diabetes, dropsy, fever, gonorrhoea, hydrocele, leprosy, malaria, palsy, piles, pimples, skin diseases, swelling, spasms and tooth ache.

Non-Medicinal Uses

Plant: Cultivated as hedge plant.

Pod: Piscicidal.

Seed: Used in making jewellery such as bracelets and necklaces; used for decorative articles, curtains, rosaries and charms; used in place of marbles by children; tribals hang the seeds on fruiting trees to stop early fruit fall.

Seed oil: Used in cosmetics and as illuminant.

11. *Caesalpinia coriaria* (Jacq.) Willd.

Synonyms: *Caesalpinia thomaea* Spreng.
Libidibia coriaria (Jacq.) Schltdl.

Common Names: American sumac, Divi Divi.

Vernacular Names: Kan: Alde kaayi, Divi-divi, Seeme alde mara, Vilaayathi alde kaayi,; **Mal:** Vilaattikadukku; **Tam:** Dividivi, Inkimaram, Kodichittal, Kodivaelam, Konakkaay, Konapuli, Konvel, Mullillaadavel, Sumikki kaay, Tividivi, Tividur; **Tel:** Dividivi, Dividivithumma.

Distribution: Native of Caribbean Islands, Central America and South America; introduced into India; grows abundantly in South India, West Bengal and North West India.

Description: A small unarmed tree; bark thin with fine vertical fissures; leaves 7-8 pairs, glabrous, oblong, petiole hairy; leaflets 25-30 pairs, narrow; flowers small, pale yellow, on dense panicles; pod twisted, thin, oblong or S shaped and indehiscent.

Phytochemistry

Plant: Ellagitannin and gallotannin.

Leaf: Catechol, ellagic acid, gallic acid and tannin.

Pod: Rich in tannin.

Seed: Has high sugar content.

Medicinal Uses

Stem bark: Antiperiodic; used in chronic fevers.

Pod: Astringent, antiperiodic, semen coagulant and tonic; used in chronic and intermittent fever; and as decoction, used for washing bleeding piles.

Toxicology: Seeds contain chymotrypsin and trypsin inhibitors.

Uses in Alternative Systems of Medicine

Ayurveda:

Part used: Pod.

Uses: Pod decoction is used in bleeding piles.

Siddha:

Part used: Pod.

Uses: Useful in stomatitis and ulcers.

Ethnomedicine:

Plant: Antiseptic, astringent, sedative and tonic; given in colic, diarrhoea, dysentery, dyspepsia, fever, gangrene, heart affections, malaria and piles.

Non-Medicinal Uses

Plant: Yields tannin used in tanning industry; grown as an ornamental plant.

Wood: Yields a red dye used in dyeing; wood is useful for general carpentry.

Leaf: Used as manure and fuel.

Pod: Rich in tannin and is used in tanning industry; used as a mordant in textile industry and dyeing industry; used to obtain dark blue or black colour; used in making black or blue ink.

Seed: Used in making divi-divi liquor.

Seed oil: Used in body lotions, skin creams, bath oil and other cosmetics.

12. *Caesalpinia crista* L.

Synonyms: *Caesalpinia chinensis* Roxb.
Caesalpina paniculata (Lam.) Roxb.

Vernacular Names: Ben: Nata, Shingrilota; **Hin:** Karanju, Naktamala, Kat karanj; **Kan:** Kiri gejjuga, Sanna gejjuga; **Mal:** Kaaka mullu; **Mar:** Gajaga, Sagargota; **Sans:** Puti karanja, **Tam:** Kalichikai, Kakumullu; **Tel:** Mulluthige.

Distribution: Native of Asia including India, Australasia, Andaman Islands and Pacific Ocean Islands; grows throughout India.

Description: A large, scandent prickly shrub; pinnae 3-4 pairs; leaflets 2-3 pairs, ovate, acute; stipules foliaceous; bracts lanceolate, acuminate; flowers yellow, fragrant on racemes; pod obliquely ovoid, beaked, covered with wiry prickles; seed solitary.

Phytochemistry

Plant: Bonducellin and caesalpin.

Medicinal Uses

Plant: Is antiperiodic, depurative, emmenogogue, emollient, laxative, tonic, vermifuge and vesicant; useful in colic, congestion, convulsion, dropsy, fever, gonorrhoea, hydrocele, leprosy, malaria, palsy, pimples, skin diseases and swelling.

Root: Is anthelmintic, antiperiodic, diuretic, emmenogogue and febrifuge; used as an astringent in blenorrhagia and leucorrhoea; useful in gravel in bladder; decoction is used in fever.

Stem: Stem juice is used externally in eye diseases.

Stem bark: Anthelmintic, emmenogogue, febrifuge and tonic; decoction is given for easy expulsion of placenta.

Leaf: Antiperiodic, deobstruent, emmenogogue and tonic; useful in asthma, colic, ear diseases and intermittent fever; leaf paste applied externally for inflammatory swellings and tooth ache; dried and powdered leaves used as tonic after child birth; tender leaves are eaten for intermittent fevers, intestinal worms and liver disorders; leaf decoction is used as gargle for sore throat.

Pod: Antipyretic and tonic; roasted pods are used as fomentation in eye diseases.

Seed: Anthelmintic, antidiabetic, antimalarial, antiperiodic, blenorrhagic, diuretic and tonic; useful in asthma, colic, inflammatory swelling, intermittent fever and stones in urinary bladder; seed ointment is applied externally on hydrocele, paralysis and orchitis; seed powder with black pepper is used as febrifuge and tonic; seed powder with coconut milk is given to cure hydrocele; seed paste applied on the face removes black heads, freckles and moles; given as antidote for snake bite.

Seed oil: Used in cosmetics and dropped into the ear to check discharges and applied on foul ulcers.

Seed sprout: Given for tumours.

Uses in Alternative Systems of Medicine

Ayurveda:

Part used: Leaf, flower, pod and seed.

Uses: Leaf decoction is given in ascites, filariasis, piles and small pox; flower cures ascites, "vatha" and "kapha"; pod is anthelmintic, aphrodisiac and astringent; cures leucorrhoea, urinary discharges and wounds; seed is useful in dysentery and fever.

Siddha:

Part used: Seed kernel.

Uses: Useful as antipyretic, anthelmintic and tonic.

Unani:

Part used: Leaf and seed.

Uses: Leaf is used in wound healing; seed is anti-inflammatory, antiflatulant, antipyretic, anthelmintic and styptic; is used in colic, chronic fever, hydrocele, leprosy, malaria and skin diseases.

Non-Medicinal Uses

Plant: Is a source of fibre used in cordage.

Root, stem and pod pulp: Used as fish poison.

Seed: Used as marbles by children.

Seed oil: Used in cosmetics.

13. *Caesalpinia decapetala* (Roth.) Alston

Synonyms: *Biancaea scandens* Tod.
Biancaea sepiaria (Roxb.) Tod.
Caesalpinia crista auct. non L.
Caesalpinia decapetala (Roth) Alston var., *japonica* (Siebold and Zucc.) H.Ohashi
Caesalpinia japonica Siebold and Zucc.
Caesalpinia sepiaria Roxb. var., *japonica* (Siebold and Zucc.) Makino

Common Names: Bahama sappan, Black bonduc, Mysore thorn, Wait-a-bit tree.

Vernacular Names: Guj: Alal, Ari; **Hin:** Aila, Arlu, Kando, Karanji, Raelu, Uri, Urn; **Kan:** Gajalige, Hotasige, Hotsige, Hunnula, Kurudu gejjuga, Kurutu gajjika; **Mal:** Inna; **Mar:** Chillara, Chillari, Chillur; **Tam:** Indu, Kadiyindu, Pulitadukki, Pulittodakki, Sirukalachhi; **Tel:** Gaddakorinda.

Distribution: Native of Asia including India; cultivated and naturalized in tropical Africa and America; grows in Bihar, Himalaya, Himachal Pradesh, Jammu & Kashmir, Rajasthan, Uttar Pradesh and West Bengal in India.

Description: A bushy armed straggler with recurved, hooked prickles on the petioles; leaflets 8-10 pairs, oblong; flowers yellow on simple racemes at the ends of lateral branches; pod oblong, flat, turgid with ellipsoidal seeds; seeds 8.

Phytochemistry

Leaf: Betulinic acid.

Medicinal Uses

Plant: Anthelmintic, antibacterial, anticancer, anti-inflammatory, antileukemic, antimalarial, antimelanomic, antinociceptive, antitumour, antiviral, apoptotic, contraceptive and cytotoxic.

Root: Emmenogogue and purgative.

Stem bark: Astringent.
Leaf: Emmenogogue and laxative; leaf infusion is used as cathartic.

Bioactivity: Plant is inhibitor of phospholipase and prostaglandin synthesis.

Uses in Alternative Systems of Medicine

Unani:
Part used: Leaf.
Uses: Antipyretic, carminative, emmenogogue, laxative and tonic; cures biliousness.

Ethnomedicine:
Plant: Is used in jaundice.

Non-Medicinal Uses
Plant: Grown as hedge plant.
Bark: Useful in tanning.

14. *Caesalpinia digyna* Rottler

Synonyms: *Caesalpinia gracilis* Miq.
Caesalpinia oleosperma Roxb.

Vernacular Names: Ben: Umulkuchi; **Hin:** Baankaerimool, Vakeri mul; **Mar:** Vakerimula; **Ori:** Gilo; **Sans:** Ashmabhaedi, Ghrita karanja, Gruth karanj, Vakeri; **Tam:** Nuni gatacha; **Tel:** Mulanune gaccha, Nooni gaccha.

Distribution: Native of Asia including India; grows in Andhra Pradesh, Assam, Bihar, Eastern Himalaya, Orissa, Tripura, Tamil Nadu and West Bengal.

Description: Prickly climbing shrub with hooked prickles; leaflets 8-10 pairs, oblong, membranous, oblique at base; flowers yellow, upper petal streaked with red; pod torulose, thick, fleshy, oblong with a short recurved beak; seeds 2-4.

Phytochemistry

Root: Bergenin and vakerin.
Leaf: Caesalpinine A and B, celallocinnine.
Pod: Gallotannin and gallic acid.
Seed: Bonducin and saponin.

Medicinal Uses

Root: Used as astringent; paste is used as an external application to boils and sores; root powder with milk, ghee, cumin and sugar is given for diabetes, diarrhoea, emaciation, goitre, pthisis and scrofula; pounded root is taken as febrifuge.

Uses in Alternative Systems of Medicine

Ayurveda:
Part used: Plant.
Uses: Pacifies deranged *tridosha*, cures emaciation, goitre, polyuria and tuberculosis.

Ethnomedicine:
Plant: Astringent; given in diabetes, pthisis and scrofula.

Non-Medicinal Uses
Plant: Grown as ornamental plant.
Root and Pod: Used for tanning.

Seed: Used as cattle feed.

Seed oil: Used as illuminant.

15. *Caesalpinia pulcherrima* (L.) Sw.

Synonyms: *Poinciana bijuga* Lour.
Poinciana elata Lour.
Poinciana pulcherrima L.

Notes: Widely cultivated in the tropics; several colour forms occur.

Common Names: Bird of paradise, Paradise flower, Peacock flower, Petit flamboyant, Pride of Barbados, Spanish carnation.

Vernacular Names: Hin: Gulthura; **Kan:** Channakeshava gida, Eejimullu, Kenjige mara, Kenjuga, Komaari, Nalligaane, Rathnaganchi, Rathnagandhi, Rathnagantige, Shankeshvari, Shee mullu, Hota seege; **Mal:** Setti mandaram, Techi mandaram; **Mar:** Sankeshvara; **Sans:** Krishnachooda, Rathnagandhi, Sidhakhya, Sidhanasha, Sidheshwara; **Tam:** Konrai, Mayil konnay, Mayirkonrai, Mayooram, Naalal, Sirumayir; **Tel:** Pamidi thangedu, Paidithangedu, Rathnagandhi, Sinnathuraayi, Thuraayi.

Distribution: Native of Africa and Asia including India; cultivated in India.

Description: Glabrous shrub or a small tree with glossy branches; stipules subulate; leaves 6-9 pairs; leaflets 8-12 pairs, sessile, opposite, obovate oblong, base obliquely round, apex mucronate; flowers yellowish orange on terminal and axillary racemes; pod straight, beaked, flat, 8-10 seeded; seeds black, rectangular.

Phytochemistry

Plant: 2,6-Dimethoxy benzoquinone, 4'-methyl isoliquiritigenin, 6-β-hydroxy vouacapen-5-α-Ol, 6-methoxy pulcherrimin, 8,9,11,14-didehydrovouacapen-5-α-Ol, 8-methoxy bonducellin, bonducellin, x-caesalpin, ellagitannin, gallic acid, leucodelphinidin, myricitrin and tannin.

Root: 6-β-Cinnamoyl-7-α-hydroxy-vouacapen-5-α-Ol and vouacapen-5-α-Ol.

Stem: 2,6-Dimethoxy benzoquinone, 6-methoxypulcherrimin, 8-methoxybonducellin, bonducellin, caesalpin and pulcherralpin.

Stem bark: α-Caesalpin, 2,6-methoxybenzoquinone, 4'-methyl isoliquiritigenin, 6-methoxy pulcherrinin, 8-methoxybonducellin, ellagic acid, ellagitannin, leucodelphinidin, pulcherralpin, pulcherrinin, quercimeritrin and sebacic acid.

Wood: Tannin.

Leaf: Benzoic acid, gallic acid, myricetol, myricitrin-3-rhamnoside and tannin.

Flower: β-Sitosterol, gallic acid, lupeol, lupeol acetate, myricetin, quercetin, rutin and sucrose.

Seed: Galactomannan, galactomannose and mucilage.

Edibility

Seed: The green seeds are eaten raw and taste like pea.

Medicinal Uses

Plant: Is analgesic, androgenic, anaesthetic, anthelmintic, antiallergenic, antibacterial, antiestrogenic, antifibrinolytic, antihistaminic, anti-inflammatory, antinociceptive, antioxidant, antiperoxidant, antipyretic, antiseptic, antitumour, antiviral, apoptotic, astringent, bronchodilator, cardiotoxic, capillary protective, demulcent, diaphoretic, diuretic, expectorant, hepatoprotective, spermicide, styptic, teratogenic and vulnerary; useful in arthritis, asthma, bronchitis, cancer, catarrh, colitis, convulsions, cold, dementia, dermatitis, diabetes, dysentery, diarrhoea, edema, epilepsy, erysipelas, arrhythmia, fibrosarcoma, fever, flu, gastritis, gingivitis, haematuria, herpes, hyperglycaemia, leukaemia, lymphoma, malaria, myocarditis, nephritis, pharyngitis, purpura, rheumatism, thyroid disorders, tumour of the breast, cervix, colon, lung, ovary and skin and varicose veins.

Root: Astringent; root decoction is given in cholera and convulsions.

Stem bark: Emmenogogue, febrifuge, purgative, stimulant and tonic; used as abortifacient.

Leaf: Abortifacient, emmenogogue, purgative, stimulant and tonic; infusion of the leaves is used as antiperiodic and cathartic; leaf decoction is used to relieve fever and used as mouthwash for mouth ulcers.

Flower: Febrifuge and pectoral; infusion useful in asthma, bronchitis, fever and intestinal worms; decoction is given for cough and chronic catarrh.

Seed: Powdered seeds used in stomach ache; seed paste with salt is applied on ringworm; decoction of fresh seed is used for inflammation of the gums.

Bioactivity: Plant is inhibitor of 5-lipoxygenase, ACE, ATPase, aldose-rectuctase, COMT, cyclo-oxygenase, cytochrome-P450, deiodinase, FPTase, glucosyl-transferase, MAO-A, MMP-9, metalloproteinase, NADH-oxidase, NO-synthase, ornithine-decarboxylase, oxidase, PTK, phospholipase, prostaglandin synthesis, protein kinase-C, reverse-transcriptase, topoisomerase-I and II, tyrosinase, tyrosine-kinase, xanthine-oxidase and CAMP-phosphodiesterase; is also copper chelator and mast cell stabilizer.

Toxicology: Seeds contain trypsin inhibitors which reduce proteolysis of dietery proteins, cause hypersecretion of pancreatic enzymes into the intestine resulting in growth impairement and pancreatic enlargement.

Uses in Alternative Systems of Medicine

Ayurveda:

Part used: Whole plant, bark, leaf and flower.

Uses: Whole plant cures fever, tumours, ulcers and "vatha"; bark is used as abortifacient; leaves are given as emmenogogue, purgative and tonic; flower infusion is prescribed in asthma, bronchitis and malaria.

Ethnomedicine

Plant: Plant is used as abortifacient, astringent, bechic, cathartic, laxative, pectoral, sudorific, tonic, vermifuge and vulnerary; useful in biliousness, cold, convulsions, cough, diarrhoea, epilepsy, erysipelas, fever, gastromegaly, inflammation, liver troubles, measles, opthalmia, skin diseases, sores, small pox, stomatitis, swelling, throat infections, tooth problems, tumour, ulcers and venereal diseases.

Seed: Seeds are used in gum trouble and ringworm.

Flower: Used in body ache.

Non-Medicinal Uses

Plant: Grown as hedge plant and ornamental.

Root: Yields a red dye.

Wood: Indian ink is made from charred wood; the wood is used in boat making and other general constructions.

Leaf: Is used to stupefy fish.

Flower: Red dye is extracted which is used in dyeing.

Pod: Yields a red dye.

Seed: Used as a substitute for senna.

16. *Caesalpinia sappan* L.

Synonym: *Biancaea sappan* (L.) Tod.

Notes: Origin unknown; considered as a sacred plant in India.

Common Names: Brazil wood, Sappan wood tree.

Vernacular Names: Hin: Aal, Bakam, Kaat, Pathang, Tairi; **Kan:** Chappanga, Pathanga, Pathanga chakke, Pathangada mara, Pathranga, Sappanga; **Mal:** Chappannam, Pathimugham, Patraangam, Tsja panagam; **Mar:** Pathang; **Sans:** Bharyavriksha, Kashtha, Kuchandana, Lohitharanga, Pathanga, Pathraanga, Pattanga, Pattaranjaka, Pattranya, Pattura, Rakthaka, Rakthakashtha, Rakthasaar, Ranjana, Rogakashtha, Suranga, Surangada; **Tam:** Padangam, Patinga, Patrangi, Patungam, Sappangu.

Distribution: Native of Asia including India; in India, it grows in Himalayas, Kerala, Maharashtra and Tamil Nadu.

Description: A large tree with prickly stem and branches; leaves 8-12 pairs, with small prickles at the base; leaflets 10-18 pairs, oblong, rounded at the apex, inequilateral, glabrous above, pubescent below; flowers yellow, in the axil of upper leaves; petals orbicular, subequal, the standard petal with a red spot at the base; stamens densely wooly at the base; pod woody, subcompressed, shiny, indehiscent with hard recurved beak at the apex.

Phytochemistry

Plant: β-Sitosterol, 3'-O-methyl sappanol, brazilin I and II, caesalpin J and P, dibenzoxocin, octacosanol, 3-deoxy sappanone and taraxerol.

Root: β-Sitosterol and furano terpenoids.

Heart wood: α-Amyrin, β-amyrin, β-amyrin glucoside, β-sitosterol, 2-deoxyribose, 3'-O-methyl brazilin, 3'-O-methylepisappanol, 3'-O-methylsappanol, 3'-deoxysappanol, 3,7-dihydroxy-3-(4'-hydroxybenzyl) chroman-4-one, deoxysappanone-B, 3,4,7-trihydroxy-3-(4'-hydroxy benzyl) chroman, 3,4-dimethyl galactose, 4,4'-dihydroxy-2'-methoxy chroman, dihydroxy-methoxychalcone, methylgalactose, 8-methoxy-bonducellin, alanine, aspartic acid, brazilin I and II, caesalpin J and P, dibenzoxocin, embuin, episappanol, fatty acid, galactose, glucose, glycine, lactose, leucine, linoleic acid, monosaccharide, norvaline, octacosanol, oleic acid, palmitic acid, proline, protosaponin A, B, C, quercetin, sappanol, sappanone B, sorbose, stearic acid, taraxerol, threonine and valine

Leaf: Contains 0.16-0.25% pleasant smelling essential oil containing α–phelandrene as a chief constituent along with oscimene.

Seed: Amino acids, arachidic acid, brazilin, caprytic acid, lauric acid, linoleic acid, linolenic acid, myristic acid, oleic acid, palmitic acid and palmitoleic acid

Medicinal Uses

Plant: Anticancer, antihypocholesterolemic, astringent, sedative and semen coagulant; useful in rheumatism.

Stem bark: Used in wide range of dental problems such as dental carries, gum pain, infected gum and tooth, shaking teeth and tooth ache.

Wood: Acrid, antiseptic, astringent, bitter, depurative, emmenogogue, refrigerant, sedative and vulnerary; used in diabetes, diarrhoea, dysentery, epilepsy, erysipelas, goiter, haemoptysis, haemorrhages, vitiated conditions of "pitta", leucorrhoea, menorrhagia, odontopathy, skin diseases, stomatitis, ulcers, and wound; decoction of the wood is anti-inflammatory, astringent and emmenogogue; useful in diarrhoea, dysentery, menstrual disorders and skin diseases.

Toxicology : Seeds contain trypsin and chymotrypsin inhibitors; excess usage may cause vomiting.

Uses in Alternative Systems of Medicine

Ayurveda:

Part used: Plant and wood.

Uses: Plant is used in the treatment of boils, bronchitis, deranged "vatha", "pitta", mental disorders, erysipelas, fever, goiter and wounds; wood is used in biliousness, blood disorders, delirium, fever, ulcers and urinary diseases.

Chinese:

Part used: Wood.

Uses: Used as astringent and haemostatic in metrorrhagia; useful in dysentery and enterorrhagia; externally the paste is applied for bruises and orchitis; used as sedative.

Unani:

Part used: Root and pod.

Uses: Used as blood purifier, emetic and haemostatic; checks bleeding from wounds and ulcers; cures rheumatism, improves complexion and removes impurities in blood and phlegm.

Ethnomedicine

Plant: Is used as astringent, antiseptic, depurative, emmenogogue and haemostatic; useful in contusion, diarrhoea, dysentery, dysmenorrhoea, enterorrhagia, haemorrhages, hematemesis, mastalgia, metrorrhagia, orchitis, skin diseases, syphilis and wounds.

Non-Medicinal Uses

Plant: Cultivated as hedge plant and is the source of red wood.

Root: Yields a yellow dye.

Stem bark: Contains tannin used to produce black shades in dyeing; also yields red and orange dye.

Wood: Yields a red dye used for colouring cotton, silk and woolen fabric and in calico printing; used in making red ink; used along with garcine, a chocolate coloured dye is obtained; used in making red powder used as "gulal" during holy festival; pieces of wood is dropped into drinking water to make it sterile and bacteria free; the wood takes a fine polish, does not crack; it is used in making walking sticks, cabinets, fancy decorative articles and inlay work.

Pod: Contains tannin which in combination with iron yields red and black dye; tannin is also used for curing leather articles.

17. *Cassia fistula* L.

Synonyms: *Cassia excelsa* Kunth

Cassia fistuloides Collad.

Common Names: Budding stick, Golden shower¸ Indian laburnum, Pudding pipe, Purging cassia, Purging fistula, Shepard's pipe.

Vernacular Names: Ass: Sonaru; **Ben:** Sonali, Bandarlati; **Guj:** Garmala; **Hin:** Amalthaas, Bandarlauri, Girmalah, Kirala, Sinar, Siyarlaatee, Sanaay; **Kan:** Araghvadha, Aragina mara, Arevata mara, Kaadu konde mara, Kakkaayi, Kakke mara, Kakkikaayi, Konde mara, Kokke, Konne, Raajadanda, **Mal:** Kanikkonna, Konna, Krithamalam, Saturangulam, Svarnnakam, Svarnavira; **Mar:** Bawa, Baya, Bahara, Bhaav, Bhawabaya, Boya, Chimkani, Garmala, Girimala, Gurmala, Samaba; **Ori:** Sunari; **Pun:** Amattas; **Sans:** Aaragvadha, Amaha, Anthadru, Aarevatha Arogyashimbi, Aruja, Chakraparivyadha, Chathurangula, Deerghaphala, Himapushpa, Jatharanut, Jvaraanthaka, Kandughna, Karnaabharanaka, Kratamala, Krithamaala, Krithamaalaka, Kundali, Kushtaghna, Kushthasudana, Mahaakarnikara, Mahaaraajdruma, Manthana, Naktamala, Nripadruma, Pramcha, Raaja, Raajatharu, Raajavriksha, Rechena, Rochana, Sampaka, Samyaak, Saraphala, Shamyaka, Shephalika, Suvarnaka, Svarnabhooshana, Svarnadra, Svarnanga, Svarnasthali, Svarnavriksha, Vyaadigaatha, Vyathantaka; **Tam:** Konai, Irjviruttam, Kontarikkai, Sarakonnai; **Tel:** Kondrakayi, Melachettu, Relaguji.

Distribution: Native of India, Indonesia, Malaysia, Myanmar, Pakistan and Sri Lanka; introduced into China, Egypt and several tropical countries; grows throughout India; often cultivated.

Description: Moderate sized deciduous tree with greenish grey, smooth bark when young, rough when old; branches glabrous; leaves acute, eglandular, pinnately compound; leaflets 4-8 pairs, bright green, ovate, subacute at the apex, cuneate at the base, distinctly stalked; bracts deciduous; flowers large, bright yellow on pendulous racemes; stamens 10, the lowest 3 with long filaments and large anthers, 4-6 intermediate ones with short filaments and small diverging anthers, the remaining with very short filaments and indehiscent anthers lobes; pod indehiscent, pendulous, cylindrical, shortly stipitate, smooth and shiny, brownish black, transversely divided into numerous 1 seeded compartments; seeds broadly ovate, many; seed pulp orange brown.

Phytochemistry

Plant: (–) Epiafzelechin, chrysophanic acid, clitorin, emodin, epicatechin, kaempferol-β-glucoside, kaempferol-3-neohesperidoside, oxymethylanthraquinone, phlobaphene, procyanidin and sennosides.

Root bark: 2,7,4-Trihydroxy-6,8,3'-trimethoxyflavone-3-O-α-L-rhamnosyl (1→2)-β-D-glucopyranoside, 5,7,3,4-tetrahydroxy-6,8-dimethoxyflavone-3-O-α–L-arabinopyranoside, fistucacidin, flavonoids and phlobaphenes.

Stem bark: β-Sitosterol, fistucacidin, hexacosanol, leucopelargonidin and lupeol.

Heartwood: β-Sitosterol, barbaloin, fistucacidin, hexacosanol, leucoanthocyanidin-5,4'-dihydroxyflavan-3,4-diol, lupeol and rhein.

Sap wood: (–) Epiafzelechin, (+) catechin, 1,8-dihydroxy-3-methylanthraquinone, kaempferol, leucoanthocyanidin-5,4'-dihydroxyflavan-3,4-diol and proanthocyanidin.

Leaf: Biflavonoids, rhein, rheinglucoside, sennoside A & B and triflavonoids.

Flower: β-Sitosterol, ceryl alcohol, clitorin, fistulin, kaempferol, kaempferol-3-β-glucoside, kaempferol-3-β-neohesperidoside, leucopelargonidin, rhein and rhein dianthraquinone glucoside.

Pod: (–) Epiafzelechin, (+) catechin, β-sitosterol, 1-8-dihydroxy-3-methylanthraquinone, arginine, aspartic acid, cathartic acid, dihydrokaempferol, epicatechin, fistulic acid, fructose, glucose, glutamic acid, hexacosanol, kaempferol, leucine, lupeol, methionine, oxalic acid, phenylalanine, proanthocynanidins, rhein, sucrose, tartaric acid and tryptophan.

Pod pulp: Arginine, aspartic acid, carbohydrates, fructose, glucose, glutamic acid, hexacosanol, leucine, methionine, phenylalanine, proteins, sucrose and tryptophan.

Seed: Amino acids, anthraquinones, D-galactose, D-mannose, mannohexose, mannopentose and tocopherols.

Seed oil: Dinoic acid, malvalic acid, monoenoic acid, sterculic acid, trienoic acid and vernolic acid.

Edibility

Flower and pod pulp: Are edible.

Medicinal Uses

Plant: Astringent, aperient, laxative, purgative, sedative and vermifuge; used in burns, cancer, constipation, convulsions, delirium, diarrhoea, dysuria, epilepsy, flatulence, gravel, gripping, haematuria, intestinal ulcers, liver diseases, syphilis and pimples; decoction is used as gargle for sore throat; every part of the plant is useful as an antidote in snake bite, scorpion sting and dog bite.

Root: Anti-inflammatory, febrifuge, laxative and tonic; used in burning sensation, dyspepsia, leprosy, skin diseases, syphilis, and tuberculous glands; smoke from burnt root is inhaled for running nose, encouraging copious discharge and providing relief; powdered root is an antidote to snake poison.

Root bark: Purgative; used in black water fever.

Stem bark: Anticancer, antifungal, anthelmintic, antiviral, astringent, depurative, emetic, febrifuge, hypoglycaemic, laxative and stomachic; useful in anaemia, boils, cardiopathy, colic, diabetes, diarrhoea, dyspepsia, dysentery, indigestion, jaundice, leucorrhoea, facial paralysis, pustules, ringworm, rheumatism, skin diseases and stangury; antidote to snake bite, scorpion sting and insect bites.

Gum: Useful as astringent in dysentery.

Leaf: Antiperiodic, antiseptic, blood purifier, depurative and laxative; used in treatment of abdominal pain, anaemia, arthritis, boils, carbuncles, constipation, dropsy, indigestion, intermittent fever, inflammation, jaundice, leprosy, leucorrhoea, polyuria, rheumatism, ringworm, skin diseases, swellings, ulcers, urticaria, vitiated "vatha"; leaf juice is useful as external application in erysipelas, ringworm, skin diseases and irritation caused by marking nut juice; poultice of leaves relieves chilblain, facial paralysis and rheumatic swellings; tender leaf paste is used internally as laxative and for bacterial infections; leaf paste in oil is rubbed on pustules; antidote to snake bite.

Flower: Antibilious, astringent, bitter, cathartic, cooling, demulcent, emollient, febrifuge and laxative; useful in biliousness, burning sensation, bronchitis, dry cough, jaundice, leprosy, polyuria, pruritus, skin diseases, urticaria, vitiated "kapha" and "pitta"; flower paste is applied on ringworm; decoction given in stomach affections; flowers with honey and sugar is made into "gulkan" and given as cooling and febrifuge; flower bud improves taste.

Pod: Abortifacient, anodyne, anti-inflammatory, antipyretic, antiviral, astringent, cathartic, carminative, depurative, demulcent, digestive, diuretic, emollient, hypoglycaemic, laxative, ophthalmic and tonic; improves appetite and digestion; useful in abdominal pain, anorexia, biliousness, constipation, cough, colic, debility, diabetes, diarrhoea, dysentery, eczema, flatulence, gout, hapatomegaly, indigestion, intermittent fever, jaundice, liver complaints, leprosy, leucoderma, ophthalmopathy, pruritus, pimples, rheumatism, strangury, swelling in the throat, tooth ache and vitiated conditions of "pitta"; pod shell with saffron, sugar and rose water relieves difficult micturition.

Pod pulp: Aperient, astringent, cathartic, cooling, emetic, emollient, febrifuge, laxative and tonic; useful in biliousness, colic, constipation, diabetes, epistasis, fever, flatulence, gastritis, gout, inflammation, piles and ringworm; pod pulp with sugar, without seeds acts as purgative; pod pulp with almond oil relieves thoracic obstruction; it is a safe aperient for children and pregnant women.

Seed: Cathartic, cooling, emetic, laxative and vermicidal; used in jaundice; pulp with hot milk is used as mouth wash to remove distaste or loss of taste due to opium or cocaine abuse; the sweetish pulp around the seed is used as a mild laxative.

Bioactivity: Aqueous extract of pod pulp exhibited antibacterial and laxative activities.

Toxicology: Excessive use of pod pulp may cause flatulence, nausea, purgation, respiratory problems and vomiting.

Clinical Studies: Clinical studies on psoriatic patients where infusion of the plant was given internally and applied externally for 2-6 months showed complete relief; external application of root bark paste was very effective in epidermophytosis and dermatophytic infections.

Uses in Alternative Systems of Medicine

Ayurveda:

Part used: Root, leaf, flower bud and pod.

Uses: Root is used in adenopathy, burning sensation, fungal skin diseases, leprosy, syphilis and tubercular glands; leaf juice is used in amoebiasis, anaemia, ascites, bronchitis, chancre, cough, diabetes, dropsy, dysuria, erysipelas, fever, gout, herpes, leprosy, musculoskeletal disorders, obesity related disorders, purpura, rat bite, skin diseases, sprue, syphilitic ulcers, tumours and vitiligo; leaf paste is used as wick in fistula; it is antifungal and antiviral; flower bud is used for biliousness, constipation, fever, leprosy, skin diseases; pod is used for abdominal pain, constipation, fever, heart diseases and leprosy.

Chinese:

Part used: Root.

Uses: Purgative.

Siddha:

Part used: Root, stem bark and leaf.

Uses: Useful in anaemia, constipation, diabetes, indigestion, intestinal worms, jaundice, leprosy, leucorrhoea, pain, piles, skin diseases and venereal diseases.

Unani:

Part used: Leaf, flower, pod and seed.

Uses: Leaf is used for inflammation; juice of the leaves is applied in skin diseases; flowers used as purgative; pod is used as anti-inflammatory, antipyretic, abortifacient, cathartic, demulcent, purgative and refrigerant; used in eye ailments, heart and liver disorders and rheumatism; application of pod pulp cures eczema and when eaten hastens child birth; pod paste is applied in snake bite and rheumatism; seeds emetic.

Medicinal Uses Outside India

Dominican Republic: Pod is used as a laxative and vermifuge.

Gold Coast: Pod pulp is used as purgative in constipation and kidney stones.

Java: Pod pulp is used in treating carbuncles, dermatitis, herpes and wound; used as purgative.

Panama: Pod plup is used in diabetes.

Rhodesia: Pod pulp is used for anthrax, blood poisoning, black water fever, dysentery and malaria.

Venenzuela: Used as astringent and laxative.

West Indies: Pod pulp and leaves are used as poultice for inflammation; bark and leaves are used for skin diseases; flower is used for fever; root is used for fever, gout and rheumatism and as diuretic.

Ethnomedicine

Plant: Is astringent, aperient, laxative and vermifuge; useful in cancer, carbuncles, constipation, convulsions, delirium, dermatitis, diarrhoea, dysuria, epilepsy, flu, gravel, haematuria, pimples, syphilis and tumourous glands.

Root: Used as purgative; useful in leprosy and skin diseases; used as antidote to snake bite.

Bark: Cures blindness; used as stomachic; antidote to scorpion sting and snake bite.

Leaf: Antiseptic, blood purifier and purgative; useful in abdominal pain, dropsy, rheumatism, ringworm and skin diseases.

Pod: Cures abdominal pain, asthma, chest infections, constipation, cough, diarrhoea, dysentery, indigestion, liver complaints, pimples, swellings and tooth ache; used as purgative; antidote to scorpion and snake poison.

Seed: Is cooling, purgative and vermicidal; used in constipation and gastric ailments.

Non-Medicinal Uses

Plant: Grown as an ornamental and avenue plant; is a source of firewood.

Twigs: Used as cattle feed.

Bark: Yields a red dye used in tanning industry; used in furnitures, decorative panelling, carving, bridge post, carts, shafts, scantlings, beams, axles, naives, spokes, felloes, yokes, agricultural implements, bowls, boats, pounders, tool handles, crushers and turnery.

Wood: Is reddish, hard, heavy and durable; is suited for cabinet work, farm implements, inlay work, wheels, mortars; is used as fuel and yields charcoal; ash is used as mordant in dyeing.

Pod pulp: Is used to flavour tobacco.

18. *Cassia grandis* L.f.

Synonyms: *Cassia brasiliana* Lam. var. *tomentosa* Miq.
Cassia brasiliensis Buc'hoz.
Cassia pachycarpa de Wit

Common Names: Coral shower tree, Horse cassia, Pink shower, Stinking toe.

Distribution: Native of tropical South America; introduced into India.

Description: Tree of 30m hight; leaves 15-25 cm long, short petioled, eglandular with 8-20 pairs of leaflets, deciduous; leaflets elliptic-oblong, 2.5-6 cm long, rounded at the apex and base, glabrous with minute hairs; stipules falcate or subulate, caducous; inflorescence is lateral raceme; bracts and bracteoles ovate, acute, tomentose; sepals tomentose, obovate; petals pink or white, orbicular to obovate; stamens 10, unequal, lower 3 filaments 2 cm long, sigmoid, others shorter; pod indehiscent, terete, transversely rugose; one margin with prominent obtuse rib, the other with 2 ribs; 70-80 brown obovoid-ellipsoid seeds.

Phytochemistry

Plant: Aloe emodin and kaempferol-3-O-β-D-mannopyranosyl (1→4) O-β-D-glucopyranoside.

Leaf: Aloe emodin.

Pod: Cinnamic acid, kaempferol-3-O-β-D-mannopyranosyl (1→4)-O-β-D-glucopyranoside and sucrose.

Seed: Galactose.

Edibility

Pod pulp: Is edible.

Medicinal Uses

Plant: Analgesic, antibacterial, antiherpetic, antileukaemic, antitumour, antiviral, cathartic, cytotoxic, genotoxic, laxative, mutagenic and tonic; useful in anaemia, intestinal worms, respiratory and pulmonary affections, skin diseases and sores.

Leaf, pod and seed: Astringent, blood cleanser and purgative; useful in cold, cough, and skin diseases; pod juice enriches the blood.

Bioactivity: Plant is arylamine-N-acetyl transferase inhibitor, caclium antagonist, cytotoxic, genotoxic and topoisomerase-II inhibitor.

Ethnomedicine

Plant: Laxative; is used in cold and fever.

Non-Medicinal Uses

Plant: Is grown as hedge plant and as an ornamental plant in gardens; used as fungicide and pesticide.

19. *Cassia javanica* L.

Synonyms: *Cassia bacillus* Gaertner
Cassia megalantha Decne.

Common Names: Java cassia, Pink and white shower, Pink cassia.

Distribution: Native of South East Asia including India and Malaysia; usually cultivated in gardens.

Description: Tree of 7-15m high with grey fissured bark; leaf lets 10-16 pairs, ovate; flowers pink on terminal and axillary raceme; pod thick, cylindrical and black.

Phytochemistry

Stem bark: Javanin, 1,3,5,8-tetrahydroxy-6-methoxy-2-methyl anthraquinone and 1,2,-dihydro-1, 3-dihydroxy-6-8-dimethoxy-2-methyl anthraquinone.

Root bark: Anthraquinones, 5-hydroxy emodin, emodin, leucoanthocyanins, leucocyanidins, rhamnopyranoside and rhamnoside.

Wood: Dihydroxy-dimethoxy-methyl anthraquinone, tetrahydroxy-methoxy-methyl anthraquinone, ceryl alcohol, chrysophenol, epiafzelechin, hentriacontane, javanin, piceatannol and β-sitosterol.

Leaf: Arachidate, behenate, behenic acid, butyrospermone, chrysophanic acid, emodin, heptacosane, kaempferol-3-O-β-D-glucosyl-6-O-α-L-rhamnopyranoside, myricyl alcohol, nonacosane, palmitate, pentacosanoic acid, rhein, tetracosane, triacontane, α-amyrin and β-sitosterol.

Flower: Peonidin-3-O-α-L-rhamnopyranoside, ceryl alcohol, dihydrorhamnetin-3-O-β-D-glucopyranoside, hentriacontanol, leucocyanidin-4-O-methyl-3-O-β-D-galactopyranoside, octacosanol, peonidin-3-O-α-L-rhamnopyranoside, quercetin-3',4',7-O-trimethyl-3-O-α-L-rhamnopyranoside and α-sitosterol.

Seed: 1,5-dihydroxy-4,7-dimethoxy-2-methyl anthraquinone-3-O-α-L-rhamnopyranoside, 1,3,6,7,8-pentahydroxy-4-methoxy-2-methyl anthraquinone, crysophanol, galactose, mannose and physcion.

Medicinal Uses

Plant: CNS active and semen coagulant.

Non-Medicinal Uses

Plant: Is grown as ornamental plant; used as insecticide.

20. *Ceratonia siliqua* L.

Common Names: Bread bean, Carob tree, Locust bean, St. Johns bean.

Distribution: Native of Africa, Europe and Middle East; introduced into India; cultivated in Punjab, Tamil Nadu and West Bengal.

Description: Large tree with paripinnate leaves; leaflets 3-5 pairs, opposite, elliptic, broadly obovate, emarginate, glabrous above, minutely pubescent beneath; flowers unisexual, green, minute on spike like raceme; pod curved, compressed, dark red; seeds pale brown, shining, embedded in a pulp.

Phytochemistry

Pod: (+) Catechin, (–) epicatechin, (–) epicatechin-gallate, absicic acid, benzoic acid, butyric acid, capronic acid, cellulose, ceratose, concanavalin-A, D-galactose, D-mannose, ferulic acid, fructose, gallic acid, glucose, hemicellulose, hydroxyproline, isobutyric acid, leucoanthocyanins, leucodelphinidin, lignin, maltose, mucilage, myo-inositol, myricetin, neochaftoside, pectin, pentosane, pinitol, phloroglucinol, pyrogallol, quercetin, saponins, sucrose, starch and tannin.

Seed: α-Amino pimelic acid, alanine, arginine, aspartic acid, β-sitosterol, campesterol, ceratoniase, cellulose, D-galactose, hemicellulose, galactomannan, glutamic acid, glycine, gum, leucine, isoleucine, lignoceric acid, linoleic acid, lysine, methionine, oleic acid, palmitic acid, phenylalanine, salicylate, serine, stearic acid, stigmasterol, threonine, tocopherol, tyrosine, valine and tannin.

Edibility

Gum: Is an excellent stabilizer used in food industry; is an ingredient of sauces; used as flavouring agent in dog biscuits and chewing tobacco.

Pod: Is rich in sugar and is used in juices, sauces and jams; pod pulp is sold as a confection called St. John's bread.

Seed: Used as a part of breakfast food; used as substitute for coffee.

Medicinal Uses

Plant: Is abortifacient, analgesic, androgenic, antiallergenic, antianginal, antibacterial, anticoronary, antifibrinolytic, antifibrosarcomic, antihepatitic, antihistaminic, antihyperglycaemic, anti-ischaemic, antilymphomic, antimetastatic, antimutagenic, antinociceptive, antioxidant, antiperiodontic, antiperoxidant, antipruritic, antiseptic, antitussive, antiviral, aphrodisiac, apoptotic, astringent, bronchodilator, CNS active, cancer preventive, capillaryprotective, cardioprotective, cerebroprotective, cholagogue, cytotoxic, demulcent, diaphoretic, diuretic, emetic, emollient, expectorant, fungicide, haemolytic, hepatoprotective, hypercholesterolemic, hypoglycaemic, hypolipidemic, hypotensive, immunostimulant, laxative, memory enhancer, mutagenic, nephrotoxic, neruroprotective, phagocytotic, spermicide, styptic, uterosedative and vasoconstrictor; is used in treating acidity, alopecia, arteriosclerosis, arthritis, asthma, atherosclerosis, bronchitis, cancer of the breast, colon, liver, skin, stomach, cataract, dermatitis, diabetes, diarrhoea, dysentery, dysmenorrhoea, eczema, edema, epilepsy, gastritis, gingivitis, herpes, intestinal worms, leukaemia, malaria, menorrhagia, nausea, pharyngitis, psoriasis, tumour of the bladder, breast, cervix, colon, stomach, liver, lung, ovary, prostate, rectum, skin, ulcer and varicose veins.

Pod: Astringent, bechic, pectoral, demulcent and purgative; useful in asthma, catarrh, cough and diabetes.

Seed husk: Antacid, astringent, pectoral and purgative.

Gum: Used as demulcent.

Bioactivity: Plant contains inhibitors of 5-α-reductase, 5-lipoxygenase, ACE, ATPase, aldose-reductase, COMT, COMP, cyclo-oxygenase, cytochrome-P_{450}, deiodinase, glucosyl-transferase, lipoxygenase, MAO, metalloproteinase, NADH-oxidase, NEP, ornithine-decarboxylase, oxidase, PTK, phospholipase, prostaglandin synthesis, protein-kinase, reverse-transcriptase, topoisomerase, tyrosine-kinase and tyrosinase; is angiotensin receptor blocker, CNS active, glutathionigenic, gonadotrophic, goitrogenic, mast cell stabilizer, metal chelator, pituitary stimulant and quinone-reductase inducer.

Toxicology: Plant contains chymotrypsin and trypsin inhibitors.

Non-medicinal Uses

Plant: Grown as a shade tree; yields gum called carob gum.

Stem bark: Used in tanning industry.

Wood: Is hard and heavy; used to make furniture and cabinets; is a source of dye used for dyeing fabrics to brown colour; used as a mordant for coal tar dyes.

Gum: Is used as lubricant and suspending agent; used in sizing rayon and cotton; as a thickener for colours used in calico printing; as a hydrating agent to improve paper strength; used in rubber industry to facilitate creaming; used the manufacture of chemicals, paints, inks, polishes, ceramics, synthetic resins, adhesives, insecticides and fungicides; is used as sweetening agent to mask the bitterness of medicine.

Pod: Infusion is employed in making alcoholic beverages and in pharmaceuticals as a sweetening agent; as an alternative sweetener for diabetic patients; used as cattle feed.

Seed: Used as a substitute for coffee called carob coffee; as cattle feed; as weighing measure for jewels.

21. *Chamaecrista absus* (L.) H. S. Irwin & Barneby

Synonyms: *Cassia absus* L.
Cassia babylonica Schrank.
Cassia exigua Roxb.

Notes: Much of the information on this taxon treats it as *Cassia absus*; two varieties *absus* and *meonandra* are recognized.

Vernacular Names: Hin: Bankulthee, Baanar, Chaksi, Chakso, Chakut; **Kan:** Chaksu, Cheemaara, Elehuri, Haalu beeja, Kaadu huliga, Kaadu hurali, Kanbelaga, Kanna beeja, Kannu kutukina beeja, Kulathanjana; **Mal:** Karinkolla; **Tel:** Agnisikha, Kooshumba chettu, Kusuma chettu, Kusimbe chettu.

Distribution: Native of Africa, Asia including India, Australasia, Madagascar, Yemen, and USA; grows throughout India.

Description: An erect annual clothed with bristly viscous hairs; leaf rachis with a small linear gland between each pair of leaflets; leaflets 4, membranous, oblong, subacute, unequal at the base, hairy beneath; stipules small, linear, acute, persistent; flowers very small, reddish yellow on erect racemes; sepals linear, lanceolate, and bristly; stamens 5, all fertile; pod ligulate, oblique with grey bristly hairs; seeds 5, compressed, shining and dark brown.

Phytochemistry

Plant: 5-O-D-Glucopyranosyl gentisic acid, aloe-emodin, chrysophanol, apigenin, ethyl-L-D-galactopyranoside, hydnocarpin and luteolin.

Root: Chaksine, isochaksine, aloe-emodin and chrysophanol.

Leaf: Chaksine, isochaksine, quercetin and rutin.

Flower: Abrin, aloe-emodin and apigenin.

Pod: Rhein.

Seed: Apigenin, β-sitosterol, chaksine, chaksine iodide, chrysophanol, D-galactomannan, fatty acids, fixed oil, galactose, isochaksine, mannose, raffinose, rhein, and tuteolin.

Seed oil: Chaksine.

Edibility

Seed: Edible.

Medicinal Uses

Plant: Used for bilious headache, catarrh, dyspepsia, eye complaints and granular eyelids.

Root: Is ground with pepper and taken internally with palm wine to cure constipation.

Leaf: Alexipharmic, astringent, bechic, bitter, cholagogue, expectorant, haematinic and thermogenic; used in asthma, cough, hiccup, "kapha", piles and "vatha"; enriches the blood, heals ulcers; leaf paste or decoction is applied in leprosy, skin affections and tumours.

Seed: Acrid, alexiteric, alexipharmic, astringent, bitter, cathartic, constipating, cooling, haemostatic, ophthalmic and vulnerary; useful in vitiated condition of "kapha", "pitta", boils, bronchitis, cataract, diarrhoea, dysentery, dyspepsia, epistasis, flatulence, haematemesis, hiccup, nephrtis, piles, ringworm, ulcers and wounds; powdered seeds are cooked and applied on the eye lids in purulent ophthalmia; seed paste is applied on sores and wounds on the penis.

Bioactivity: Chaksine and isochaksine act as anaesthetic in Guinea pigs when injected intradermally; they also produced hypotensive and deperessant action on parasympathetic nerve terminals of the intestine, bladder and neuromuscular junctions; it also depresses heart and respiratory activity and causes a fall in blood pressure in mice and ileum contraction in Guinea pigs.

Toxicology: Plant extract when used indiscriminately for catarrhal ophthalmia and granular eye lids produces pain; chaksine in seeds cause capillary dilation and CNS depression; excess usage of seeds may cause lumbar pain.

Uses in Alternative Systems of Medicine

Ayurveda:

Part used: Leaf and seed.

Uses: Leaf enriches the blood, is astringent to bowels; cures "kapha", "vatha", asthma, cough, hiccup and tumours; seeds are alexipharmic, astringent and cooling to bowels; cures bronchitis, constipation, itching, piles, ulcers and wounds.

Siddha:

Part used: Leaf and seed.

Uses: Useful in constipation, eye diseases, hiccup, skin diseases and retention of urine.

Unani:

Parts used: Seed.

Uses: Used as astringent, blood purifier, emmenogogue, diuretic, resolvent, styptic and stimulant; useful in strengthening eye sight and to resolve tumours; cures asthma, conjunctivitis, cataract, eye sores and piles; seed decoction is used as eye wash for eye pain and as wash for leprosy, leucoderma, skin eruptions and ulcers.

Medicinal Uses Outside India

Gold coast: Root ground with pepper and palm wine is a remedy for chronic constipation.
Nigeria: The leaves are used as dressing for ulcers and swellings of venereal origin.

Ethnomedicine

Plant: Is used as astringent and cathartic; useful in cataract, cough, ophthalmia, ringworm, skin diseases and tumour.

Seed: Used in eye infection, as cooling agent in ophthalmia and skin diseases.

Leaf: Is used in cough and wound healing.

Non-medicinal Uses

Plant: Is used as fish poison.

22. *Chamaecrista mimosoides* (L.) E. Greene

Synonyms: *Cassia angustissima* Lam.
Cassia mimosoides L.
Cassia sensitiva Roxb.

Notes: Extremely variable and taxonomically difficult species.

Common Names: Fish bone cassia, Mimosa senna.

Vernacular Names: Hin: Patua gaas; **Kan:** Nelathangadi; **Tel:** Nelaponna.

Distribution: Native of Africa, Asia including India and Australasia; grows thoughout India.

Description: A diffuse, glabrous or pubescent perennial; leaves with a small sessile gland below the lowest pair of leaflets; stipules straight, subulate; leaf lets 40-60 pairs, linear, obliquely mucronate, glabrous; flowers yellow, solitary or 2-3 together in leaf axils; bracteolate; stamens 10, alternately long and short; pod linear, flat; seeds 20-25.

Phytochemistry

Plant: Chrysophanol and n-hentriacontanol.
Root: Emodic acid and physcion.
Stem: Chrsysophanol.
Leaf: Chrysophanol, emodin and luteolin-7-glucoside.
Seed: Emodin, emodic acid, luteolin, luteolin-7-glucoside and physcion.

Edibility

Leaf: Is eaten as vegetable.

Medicinal Uses

Plant: Used as a cure for colic.
Root: Given in colic, delirium and stomach spasms.
Leaf: Ground and mixed with corn is given to relieve colic.

Uses in Alternative Systems of Medicine

Chinese:
Part used: Pod.
Uses: Used in diseases of the eye both internally and externally; useful in herpes and sores.

Medicinal Uses Outside India

Indochina: Plant decoction is applied on swollen abscesses.
Japan, Korea and Taiwan: A decoction of the whole plant is given as diuretic and stomachic.
Philippines: Root decoction is used to treat diarrhoea.

Ethnomedicine

Plant: Used in eye troubles, herpes and mouth ulcers.

Non-medicinal Use

Branches: Decoction of young branches and leaves are used as herbal tea in Japan.

23. *Chamaecrista pumila* (Lam.) K. Larsen

Synonyms: *Cassia pumila* Lam.
Senna prostrata Roxb.

Common Names: Black solah, Dwarf senna.

Vernacular Names: Kan: Nelathagache, Nelathagasi, Saramala; **Mar:** Sarmal; **Tel:** Nallajilluga.

Distribution: Native of Asia including India and Australia; grows throughout India.

Description: A diffuse procumbent shrub with finely downy branches; leaves with a gland at the base of petiole; stipules semi lanceolate, rigid, persistent; leaflets 20-40, linear, acute, strongly veined, mucronate, hairy, midrib close to the upper border; flowers solitary, in the axil of leaves; corolla exerted; stamens 5, anthers unequal; pod straight, thin, flat, torulose, glabrous with 6-10 seeds.

Phytochemistry

Plant: Anthraquinones, chrysophanic acid, dihydroxanthyletin, emodin, hentriacontanol, physcion and tetratriacontanol.

Leaf: β-Sitosterol, emodin, hexacosanol, physcion and rutin.

Medicinal Uses

Plant: Galactagogue.

Seed: Purgative.

Non-medicinal Uses

Plant: Is grown as cover crop and for green manure.

24. *Cynometra ramiflora* L.

Synonyms: *Cynometra bijuga* Span.
Cynometra carolinensis Kaneh.
Cynometra hosinoi Kaneh.
Cynometra ramiflora L. var., *mimosoides* Wallich

Vernacular Names: Ben: Shingar, Shingri; **Hin:** Itap, Shinge; **Kan:** Kanaga; **Mal:** Irripa; **Ori:** Madhuka; **Sans:** Madhuka; **Tam:** Irudbu, Irubu, Irapa, Nayppuduken.

Distribution: Grows throughout South East Asia including India; it occurs in Andaman and Nicobar Islands in India.

Description: A small tree; leaf rachis grooved, glabrous; leaflets 2 pairs, obliquely obovate, emarginate, subsessile, glabrous, with obliquely cuneate base; flowers white on small axillary corymbose racemes; bracts broad, ovate; bracteoles small, lanceolate; pod fleshy, wrinkled, irregularly oblong or ovoid with fleshy pointed tip.

Medicinal Uses

Root: Used as cathartic and purgative.

Leaf: Leaves boiled in cow's milk and mixed with honey is an effective remedy for scabies, leprosy and scaly cutaneous diseases.

Seed oil: Useful in leprosy, scabies and scaly cutaneous diseases.

Medicinal Uses Outside India

Myanmar: Roots are used as purgative.

Philippines: Leaf paste is applied for herpes.

Ethnomedicine

Plant: Is used in leprosy, scabies, skin diseases and as purgative.

Non-medicinal Uses

Plant: Grown as an ornamental plant in gardens.

Wood: Used in building wooden houses, carts, cabinets and decorative articles; used as fuel; is a source of purple dye.

25. *Delonix elata* (L.) Gamble

Synonym: *Poinciana elata* L.

Common Name: White gold mohur.

Vernacular Names: Hin: Sankesar; **Kan:** Kempu kenjiga, Vatanarayana; **Mar:** Sankasura; **Ori:** Simamondaro; **Tam:** Vadanarayana; **Tel:** Sankesula,Vatanarayana.

Distribution: Grows wild in Ethiopia and Arabia; cultivated throughout India.

Description: Tree with ash coloured bark; leaves abruptly 2-pinnate, opposite, 10-20 pairs; leaflets 10-20 pairs, subsessile, linear, oblong, with apiculate apex and round base; flowers yellow on finely pubescent pedicels, forming raceme; pod attenuated at both ends, reticulately veined; seeds 4-8.

Phytochemistry

Root: β-Amyrin, heperitin and neohesperidin.

Stem bark: (-) L-Aspargine and aspartic acid.

Flower: Isoquercetin, quercetin-3-0-rhamnoglucoside and quercetin 3-O-galactoside.

Medicinal Uses

Plant: Is spasmogenic; used in flatulence and rheumatism.

Stem bark and leaf: Is antiperiodic; useful in flatulence and rheumatism; prevents hairs loss; decoction applied externally prevents tanning of the skin.

Uses in Alternative Systems of Medicine

Ayurveda:

Part used: Bark.

Uses: Useful in glandular swellings and sinusitis.

Siddha:

Part used: Leaf.

Uses: Useful in arthritis, constipation, hemiplegia and inflammation.

Ethnomedicine

Plant: Is used as carminative; cures fever, malaria and rheumatism.

Non-medicinal Uses

Plant: Grown as an ornamental and avenue tree.

Wood: Useful in making cabinet, churns and combs; used in cosmetic industry.

Leaf and twigs: Useful as manure.

26. *Delonix regia* (Hook.) Raf.

Synonyms:
Delonix regia Hook. var., *flavida* Stehle
Delonix regia Hook. var., *genuina* Stehle
Poinciana regia Hook

Notes: One of the most widely cultivated ornamental species in the tropics; endangered in its native Malagasy habitat.

Common Names: Flamboyant tree, Flame tree, Gold mohar, Gulmohar, Madagascar royal gold mohar, Royal gulmohar, Royal Poinciana.

Vernacular Names: Hin: Gulmohar; **Kan:** Doddarathnagandhi, Gulmohar; Katthikaayi, Kempu thorana, Kempu thuraayi, Kodu, Seeme sankeswari; **Mal:** Alasippu; **Mar:** Gul mohar; **Tam:** Mayirgonrai, Mayirkkonrai, Mayooram, Mayarum, Perumayirgonrai, Punchaadi, Vaadanaaraayaan; **Tel:** Agnipoolu, Errathuraayi, Peddaseribiseri, Peddathuraayi, Seemesunkesvaramu, Shima sankesula.

Distribution: Native of Madagascar; introduced into India and other Asian, African, American, Middle Eastern countries and Caribbean and Pacific Ocean Islands; cultivated throughout India.

Description: Tall tree of 10-15m height with grey bark; leaves 10-20 pairs; leaflets 5-25 pairs, narrowly oblong; flowers scarlet or reddish orange on terminal axillary raceme; calyx lobes 5, unequal, green coloured externally, red internally; petals shortly clawed; pod large, lignose, laterally compressed; seeds 20-40, mottled.

Phytochemistry

Stem bark: Leucocyanidin.

Wood: β-Sitosterol, alanine, azetidine-3-carboxylic acid, galactose, glucose, lupeol, lysine, proline, quercetin, rhamnose, tyrosine and valine.

Seed: Contains gum.

Edibility

Gum: Used in food industry.

Medicinal Uses

Stem bark: Antiviral, CNS depressant and emetic.

Toxicology: Leaf and seeds contain toxic amino acid, azetidine-3-carboxylic acid, which causes human skin fibroblast growth and inhibits collagen production; seeds contain trypsin and chymotrypsin inhibitors.

Non-medicinal Uses

Gum: Used in textile and food industries.

27. *Gleditsia aquatica* Marshall

Synonyms: *Gleditsia inermis* Mill
Gleditsia triacantha Gaertner
Gleditsia triacanthos L.var., *aquatica* (Marshall) Castigl.

Common Names: Honey locust, Sweet locust, Three thorned acacia.

Distribution: Native of Canada and United States; introduced into India; in India, it is cultivated in gardens of Himachal Pradesh, Punjab and Uttar Pradesh.

Description: A deciduous tree of 60-80m height; trunk and branches with branched thorns; unipinnate to bipinnate yellow green leaves; leaflets 10-15 pairs; flowers white to greenish yellow on racemes; pod long, flat, twisted, purple to brown; seeds many.

Phytochemistry

Plant: 6-Amino-7-(r,r-dimethyl alyl) purine, 5-(3,4-dihydroxycinnamoyl)-quinic acid, gleditschioside, cryptochlorogenic acid, neochlorogenic acid, dihydroxy cinnamoyl quinic acid and triacanthosides.

Wood: Fisetin and fustin.

Leaf: Acrammerin, gleditschine, hypoxysin, n-methyl-β-phenylethylamine, oleanolic acid, olmelin, saponins, stenocarpine, tannin, triacanthine and tyramine.

Pod: (–) Epicatechin-3-glucoside, (+) catechin, dihydroxy-4-methoxy isoflavone, glucose, leucocyanidin, polyphenols, quercetin-diglucoside and saponins.

Seed: Acetic acid, dihydroxystearic acid, epicatechol-3-D-glucoside, galactomannan, galactose, mannogalactan, mannose, phlobaphene, phosphorous, protein, sativic acid, tetrahydroxystearic acid, tetrahydroxyisoflavanol, tocopherol and isoflavanol.

Edibility

Pod: Used as a sweetener; a pleasant drink is made from pod.

Seed: Used as a substitute for coffee; eaten raw or cooked.

Medicinal Uses

Plant: Is abortifacient, antibacterial, anthelmintic, anti-inflammatory, antimutagenic, antiperoxidant, antiviral, cardioprotective, cytotoxic, diuretic, expectorant, haemostatic, hepatoprotective, immunostimulant, laxative, oxytocic, phagocytotic and vasodilator; useful in cystitis, depression, diabetes, diarrhoea, dysentery, edema, herpes, hypertension, insomnia, nephritis, osteoporosis, tumour and ulcer.

Stem bark: An infusion is used in treating dyspepsia, measles, small pox and whooping cough.

Pod: A tea prepared from pod is taken internally for catarrh, indigestion, measles and as antiseptic.

Bioactivity: Plant is inhibitor of ACE, 5-lipoxygenase, cyclo-oxygenase, glucosyl-transferace, iodothyronine-deiodinase, lipoxygenase, MAO, ornithine-decarboxylase, topoisomerase, xanthine-oxidase and cAMP-phosphodiesterase; is angiotensin receptor blocker, β-blocker, β-adrenenergic receptor blocker and calcium channel blocker.

Toxicology: Excessive use results in stupor and loss of reflex activity; contains trypsin inhibitor.

Medicinal Uses Outside India

China: The pod and seed used as diuretic and for loosening of mucous in the chest; the root and bark are anthelminitic and antifebrile; pod is used as bechic, depurative, emetic and purgative; used to treat chronic dysentery, flatulence and prolapsed rectum; as an antidote for metal poisoning; seeds are used in chronic consumption and rectal cancer; the thorns on the plant are dried and a drug is prepared for lumps, tumours, swellings and unusual growth on the breast, mouth abscesses, ulcers and wounds and taken internally for malignant boils; the ash of the thorns is taken with hot water for indigestion.

Ethnomedicine

Plant: Is used as anodyne, oxytocic and narcotic.

Non-medicinal Uses

Tree: Grown as a ornamental tree.

Wood: Yields tannin used in tanning industry; used for making fence posts, rails, wheel hubs, farm implements and construction work.

Pod: Is used as livestock feed.

28. *Haematoxylum campechianum* L.

Synonyms: *Cymbosepalum baroni* Baker

Haematoxylon campechianum L.

Common Names: Campeche wood, Logwood.

Vernacular Names: Ben: Bekkan; **Hin:** Patang; **Kan:** Patanga; **Mar:** Partanga; **Tam:** Artanga; **Tel:** Gabbi chettu.

Distribution: Native of Central America; introduced into India, Africa, China, Pakisthan, Sri Lanka, Taiwan, Caribbean Islands, Indian Ocean Islands and America; in India it is cultivated in the gardens of Karnataka and West Bengal.

Description: A small tree of 3m height covered with thorns; leaflets 2-5 pairs, obovate, oblong with obtuse apex; flowers yellow on crowded spikes; pod broad, membranous, narrow at both ends; seed solitary.

Phytochemistry

Plant: Ellagic acid, ethylgallate, haematoxylin, isovitexin, myricetin, oxalic acid, phlobaphene, quercetin and tannin.

Wood: Acetic acid, haematoxylol – A, resin, tannin and tetra-methyl hematoxylol.

Medicinal Uses

Plant: Analgesic, anthelmintic, antiallergenic, antibacterial, anticancer, antiestrogenic, antihepatotoxic, antihistaminic, antihypertensive, anti-inflammatory, antioxidant, antiperiodontic, antiseptic, antispasmodic, antitumour, antiviral, apoptotic, CNS paralytic, cytotoxic, diaphoretic, haemostat, hepatoprotective, immunomodulator and neuroprotective; useful in arthritis, catarrh, colitis, cystitis, dermatitis, diabetes, diarrhoea, dysentery, fibrosarcoma, gastritis, gingivitis, herpes, leukaemia, malaria, melanoma, myocarditis, nephritis and ulcer.

Wood: Antiseptic, antidysenteric, anti-inflammatory, astringent, sudorific and tonic; useful in cancer, chronic diarrhoea and leucorrhoea; wood paste is applied for hospital gangrene.

Bioactivity: Plant is inhibitor of 5-lipoxygenase, aldose-reductase, COMP, COMT, COX-2, cytochrome-oxidase, cytochrome-P_{450}-1A2, deiodinase, glucosyl-transferase, MAO, MMP-9, metalloproteinase, NADH-oxidase, NEP, NO, PTK, phospholipase, prostaglandin synthesis, protein-kinase, topoisomerase I and II, tyrosine- kinase, xanthine-oxidase and CAMP-phosphodiesterase; is mast cell stabilizer and P_{450} inducer.

Uses in Alternative Systems of Medicine

Homoeopathy:

Part used: Heartwood.

Uses: Useful in colic, diarrhoea, dysmenorrhoea, excessive perspiration, headache, indigestion, nausea, pain, sore throat and tymphinitis.

Medicinal Uses Outside India

China: Plant is used as anodyne and decoagulant; used in amenorrhoea, purpura, swelling and trauma.

Haiti: Plant is used in amenorrhoea, anaemia, diarrhoea, dysentery, leucorrhoea, sores, sprain, trauma and tooth ache.

Mexico: Plant is used as astringent and antidysenteric; cures anaemia and tuberculosis.

Turkey: Plant is used as astringent and haemostat.

Ethnomedicine:

Plant: Is used as anodyne, antiseptic, astringent, antibacterial, decoagulant, depurative and tonic; used in bed sore, cancer, diarrhoea, dysentery, dyspepsia, gangrene, leucorrhoea, purpura and swelling.

Non-medicinal Uses

Plant: Grown as ornamental and hedge plant.

Heart wood: Yields a dye used in dyeing leather, fur, jute, wool, silk, rayon and nylon; used in the manufacture of ink; used as a histological stain; as an indicator for alkaloid filtration; used in the manufacture of fancy articles, furniture and in calico printing.

Flower: Is a source of honey of high quality.

29. *Hardwickia binata* Roxb.

Common Name: Malabar mahagony.

Vernacular Names: Kan: Jenu yenne; **Mal:** Kiyavu, Kodapalla, Kolla, Kulayu, Shirali, Shurami; **Mar:** Anjana; **Tam:** Enney kolavu, Kodapalai, Madana chambrani.

Distribution: Endemic to India (Central India, Bihar, Rajasthan and Uttar Pradesh).

Description: Tree of 6-7m height; leaflets single pair, subsessile, obliquely ovate; flowers yellowish green, on axillary and terminal panicle; pod strap shaped, veined, narrow at both ends; seed solitary.

Phytochemistry

Heart wood: (+) Catechin, (+) epicatechin, (+) mopanol, β-sitosterol, eriodictyol and taxifolin.

Medicinal Uses

Plant: Yields balsam of Hardwickia used in gonorrhoea and other venereal sores.

Stem bark: Is used in biliousness, epilepsy, leprosy and ulcers.

Non-medicinal Uses

Plant: Grown as an ornamental plant and provides good fodder.

Stem: The stem bark yields fibre used for ropes and cordages; used in paper making.

Wood: Useful in making house and bridge posts, beams, flooring, furniture, bowls, billiard tables, cabinet, ornamental veneers, rafters, frames, carts, shafts, spokes, felloes, ploughs, crushers, bridges, tent pegs, vessels, tool handles, ornamental articles, oars and railway sleepers.

Resin: The balsam obtained from the plant is a good substitute for clove oil; the oil is used as preservative and in the manufacture of varnishes.

Leaf: Useful as fodder for cattle and horses and manure.

30. *Hymenaea courbaril* L.

Synonyms: *Hymenaea candolleana* Kunth.
Hymenaea retusa Hayne
Inga megacarpa M.E. Jones

Notes: Six varieties, *courbaril, altissima, longifolia, stilbocarpa, subsessilis and villosa* are recognized.

Common Names: Stinking tree, West Indian locust.

Distribution: Native of Costa Rica; introduced into India, Africa, China, Taiwan, Central and South America, Caribbean and Indian Ocean Islands; cultivated in the gardens of Karnataka, Kerala, Tripura and West Bengal.

Description: A tall tree of 20m height; leaves two foliolate, coriaceous, oblique; inflorescence terminal, corymbose to paniculate; flowers white to yellow with glandular hairs; stamens 10, free; pod oblong, indehiscent; seeds numerous, oblong and large.

Phytochemistry

Plant: Yields gum.

Edibility

Pod: Pod pulp is sweet and edible.

Medicinal Uses

Plant: Is useful in diarrhoea, fever, pain, and respiratory disorders.

Bark: Decoction is given as vermifuge.

Gum: Given in bloody vomiting, cough, pthisis, respiratory disorders, rheumatism and spasmodic asthma.

Ethnomedicine

Plant: Is used as antiseptic, expectorant, laxative, sedative, stomachic and vermifuge; used in arthritis, asthma, blenorrhagia, bronchitis, catarrh, cystitis, diarrhoea, dyspepsia, fracture, headache, indigestion, kidney stones, laryngitis, malaria, respiratory disorders, rheumatism, sores, spasms, ulcer and venereal diseases.

Non-medicial Uses

Pod pulp: An intoxicating drink is prepared from fermened pod pulp.

31. *Kingiodendron pinnatum* (DC.) Harms

Synonym: *Hardwickia pinnata* Roxb.

Notes: Endemic to India.

Vernacular Names: Hin: Samprani; **Kan:** Yenne mara; **Mal:** Kiyavu, Kodapalla, Shurali; **Tam:** Kodapalai, Kolavu, Madayan samprani.

Distribution: Native of India; introduced into other Asian countries; grows in the Western Ghats in South India.

Description: A very large evergreen tree of 100 feet height; bark dark brown or green, rough; leaves abruptly pinnate; leaflets 4-7, lower leaflets alternate; flowers very small, white, numerous in panicles of racemes; bracts and bracteoles small; calyx lobes with a disc; stamens 10, equal; pod coriaceous, nearly woody, obovate, 1-seeded; seed pendulous, compressed.

Phytochemistry

Plant: Caryophyllene, copaene, hardwickiic acid, humulene, humulene oxide, kolovenic acid, kolavelool, kolavenol, kolavenolic acid, kolovic acid, norditerpene and oleoresin.

Seed: Vernolic acid.

Non-medicinal Uses

Plant: Yields oleoresin which is a source of volatile oil which is used as a substitute for clove oil; resin after distillation is used in varnishes.

Wood: Is used for beams, rafters, batten, boards, flooring and furniture; used in making bowls, croquet balls, billiard tables, cabinets and ornamental veneers.

32. *Parkinsonia aculeata* L.

Synonym: *Parkinsonia thornberi* M.E. Jones

Notes: Originated in tropical Central America.

Common Name: Jerusalem thorn.

Vernacular Names: Ben: Balati kikar; **Guj:** Pardeshi baval, Ram baval; **Hin:** Vilayati kikar, Vilayati babul; **Mar:** Adanti; **Tel:** Sima tumma.

Distribution: Native of Bahamas, Costa Rica, Mexico, Panama, Argentina, Bolivia and South America; introduced into India; cultivated throughout arid belts of India.

Description: Armed shrub; leaves bipinnate, pinnae 1-3 pairs, 25-35 cm long; leaflets numerous, minute, oblanceolate; flowers 1-2cm, yellow, on lax, short, axillary racemes; pod up to 14cm, torrulose, moniliform, finely veined; seeds 1 to 6, oblong.

Phytochemistry

Plant: β-Amyrin, β-amyrin acetate, β-amyrone, β-sitosterol, isoorientin, isovitexin, orientin and vitexin.

Leaf: 5-O-Methyl luteolin-β-D-glucoside, epiorientin and parkinsonin A & B.

Seed: Erythritol and glycerol.

Edibility

Pod and Seed: Edible.

Medicinal Uses

Plant: Is antipyretic.

Leaf: Is abortifacient and diaphoretic.

Toxicology: Plant contains chymotrypsin and trypsin inhibitors.

Ethnomedicine

Plant: Is abortifacient, diaphoretic and sudorific; used in epilepsy, fever and stomach ache.

Non-medicinal Uses

Plant: Grown as hedge plant, wind break and for reafforestation; used as fodder.

Bark: Yields fibre used in paper making.

Wood: Used as fuel and yields good charcoal.

Flower: Yields essential oil used in perfumery.

33. *Peltophorum pterocarpum* (DC.) K. Heyne

Synonyms: *Peltophorum ferrugineum* (Decne.) Benth.
Peltophorum inerme (Roxb.) Naves.

Common Names: Copper pod tree, Golden flamboyant, Rusty shield bearer, Yellow flame tree, Yellow Poinciana.

Vernacular Names: Kan: Bettadahuruli, Bettada hunise, **Tam:** Avalvagai, Ivalvagai, Perungondrai; **Tel:** Kondachinta, Ragipapuram.

Distribution: Native of Asia and Australasia; introduced into India; cultivated throughout India.

Description: Tree of 15-30m height with black, fissured bark; leaves bipinnate; leaflets 8-20 pairs; flowers golden yellow on axillary, terminal racemes; pod ellipsoid, oblong, flat, dark reddish brown with 1-4 seeds.

Phytochemistry

Stem bark: (–) Epicatechin, (+) leucocyanidin, bergenin, leucocyanidin-3-O-α-D-galactopyranoside and quercetin.

Leaf: Epicatechin, leucocyanidin and quercetin.

Flower: β-Sitosterol, bergenin, lupeol and naringenin-7-glucoside.

Pod: (–) Epicatechin, (+) leucocyanidin, bergenin, hirsutidin, quercetin-3-diglucoside and rhamnetin-3-glucoside.

Medicinal Uses

Bark: Used as astringent and tonic; used in dysentery, intestinal disorders; and externally for bruises, sprains, swellings, muscular pains and sores; bark powder is one of the constituents of tooth powders, mouth wash and eye lotions.

Bioactivity: Bergenin isolated from pods inhibited carrageenan induced paw edema in rats; ethanolic extract of the seed showed inhibitory activity against *Alternaria alternata* and *Aspergillus niger*; it also showed hypotensive action in mice.

Ethnomedicine

Plant: Is used as collyrium and dentrifice; useful in dysentery, myalgia, opthalmia and sore throat.

Non-medicinal Uses

Plant: Grown as an ornamental tree and shade plant in coffee and cocoa plantations and as a windbreak; host plant for lac insect.

Bark: Contains tannin used in tanning industry; yields a dye used in colouring fabric to yellowish brown.

Wood: Used for making planks, coaches, furniture and cabinets.

Leaf: Useful as cattle feed; contains tannin used in tanning industry.

34. *Saraca asoca* (Roxb.) Willd.

Synonym: *Saraca indica sensu* Bedd.

Notes: *Saraca indica* L., is a little known species introduced into India; the name is incorrectly used as a synonym of *Saraca asoca,* cultivated and protected in temples of South India as a sacred tree; it is the tree of Ramayana where Seeta sat under the Ashoka tree in Ashoka Vatika; Ashoka means 'free from pain'.

Common Name: Ashoka tree.

Vernacular Names: Ben: Ashoka; **Hin:** Anganapriya; **Kan:** Kenkali mara, Ashokada mara; **Mal:** Ashoka, Asogam, Asokha, Hemapushpam, Vanjulum; **Mar:** Ashoka, Jasundi; **Ori:** Ashoka; **Pun:** Asohok; **Sans:** Anganapriya, Ashoka, Chaaya, Gandapushpa, Ghandhapushpa, Kaamabaana, Kankeli, Kankelli, Krimikaaraka, Nalaaki, Vanjula, Vichitra, Vithasoka; **Tam:** Asogam, Asoka, Kaagoli, Malaikkarunai, Sukubam; **Tel:** Asoka, Asokamu, Kankeli, Seethammaasokamu, Vanjulamu.

Distribution: Native of India; grows throughout India; introduced into Mauritius and Reunion.

Description: A 20 to 30 feet tall tree with glabrous branches; leaves abruptly pinnate; petioles short, stout, wrinkled; leaflets 4-6 pairs, oblong lanceolate, obtuse or acute, glabrous with round or cuneate base, slightly oblique; stipules intrapetiolar, completely united, scarious, ovate-oblong, obtuse, parallel nerved and deciduous; flowers numerous, fragrant, bright orange or red, on dense axillary corymb; pedicels long, red, glabrous; bracts ovate, subacute; bracteoles 2, appearing like calyx, spathulate, subacute, ciliolate, amplexicaul, coloured; calyx petaloid, yellow, red to orange in colour; calyx tube cylindrical, with 4 segments, ovate, oblong; petals, absent; stamens 7 or 8, exerted, filaments filiform, thrice as long as the calyx segments; anthers purple; pod black, linear, oblong, tapering at both ends, compressed, veined, glabrous; seeds 4-8, ellipsoidal, oblong and compressed.

Phytochemistry

Bark: (–) epicatechin, (–) epicatechol, (+) catechol, β-sitosterol, 11´-deoxyprocyanidin, alkanes, catechin, esters, leucocyanidin, leucopelargonidin, n-octacosanol and tannin.

Pod: (+) Catechol and leucocyanidin.

Flower: Anthocyanin, apigenin-7-O-β-D-glucoside, cyanidin-3, 5-diglucoside, gallic acid, kaempferol-3-O-β-D-glucoside, linoleic acid, palmitic acid, pelargonidin-3 5-diglucoside, quercetin, quercetin-3-O-β-D-glucoside, sitosterol and stearic acid.

Edibility

Leaf: Edible.

Medicinal Uses

Plant: Antibacterial, antiestrogenic, astringent and antitumour; used in the treatment of all gynaecological disorders such as bloody leucorrhoea, menorrhagia and menopausal disturbances; imparts strength to the uterus; used also as antidote for scorpion sting.

Bark: Is the most curative part of the plant; it is alexiteric, anthelmintic, antibilious, astringent, cardiotonic, constipating, demulcent, diuretic, emollient, febrifuge, hypothermic, oxytocic, refrigerant, spasmodic, stomachic, styptic, uterine tonic and uterine sedative; it is useful in burning sensation, colic, diabetes, dysentery, dyspepsia, fever, leucorrhoea, metropathy, piles, pimples, scorpion sting, ulcer and visceromegaly; used mainly for cysts in the cervix and uterus, bulky uterus, bleeding haemorrhoids, bloody leucorrhoea, menorrhagia due to fibroids, cervical and vaginal tumours and uterine cancer; stimulates endomitrium and ovarian tissues; strengthens and tones up cervical and uterine muscles, corrects menopausal disturbances, menstrual disorders and uterine haemorrhage.

Leaf: Alterative, blood purifier, depurative and colic; leaf juice with cumin seeds used in stomachalgia.

Flower: Antibilious, antidiabetic, antisyphilitic and uterine tonic; useful in burning sensation, cervical adenitis, dysentery, haemorrhages, haemorrhoids, hyperdipsia, inflammation, retention of urine, scabies and vitiated "pitta".

Pod: Antidote for snake bite and scorpion sting.

Seed: Antifungal; useful in fracture, strangury, urinary discharges and vesicular calculi.

Bioactivity: Crude bark extract exhibited spasmogenic activity on uterei of rat, dog and humans and showed consistent oxytocic activity; it was non toxic upto 250mg/kg.

Uses in Alternative Systems of Medicine

Ayurveda:

Part used: Bark, flower and seed.

Uses: Bark is used as alexipharmic, alexiteric, antimenorrhagic, antiobese, astringent, demulcent, emollient, refrigerant and styptic; useful in biliousness, burning sensation, colic, constipation, dysmenorrhoea, dyspepsia, emaciation, enlarged abdomen, cervical glands, fever, haemorrhage, leprosy, menorrhagia, piles, polyuria, thirst, tumours, ulcers, uterine disorders and uterine inflammations; adds shine/luster to the skin, bark powder is dusted on wounds for quick healing; flower used in the treatment of bleeding piles and scabies; seed is used in poisoning and urinary discharges; the decoction of the seed and flower bud is prescribed for haemothermia and haemorrhages; used as sedative and purgative.

Homoeopathy:

Part used: Bark.

Uses: Useful in amenorrhoea, haemorrhoids, headache, jaundice, leucorrhoea, menorrhagia, nasal catarrh and nausea.

Siddha:

Part used: Bark.

Uses: Used as abortifacient; cures bleeding piles, diarrhoea and menorrhagia.

Medicinal Uses Outside India

Myanmar: Bark is used in treating menorrhagia.

Nepal: Bark is used as astringent in excessive menstrual flow; used as antidote for scorpion sting.

Ethnomedicine

Plant: Is considered as androgenic, antibacterial, depurative, oxytocic, sedative, stimulant and uterine tonic; used in biliousness, diabetes, dysentery, dyspepsia, haemorrhages, leucorrhoea, menorrhagia, piles, pimples, sores, stomach ache, syphilis and tumour.

Bark: Used as tonic and given in menstrual disorders.

Leaf: Is used in heart diseases and menstrual disorders.

Non-medicinal Uses

Plant: Grown for its showy scarlet-orange flowers; plant is considered as sacred; it is a anthropogonic tree of the Vaishya community in South India and a branch is brought during marriages.

Wood: Useful in plough and shaft making and house building.

Flower: Used as offering to god in religious ceremonies by Hindus and Buddists.

35. *Senna alata* (L.) Roxb.

Synonyms: *Cassia alata* L.
Cassia alata L. var., *perennis* Pampan.
Cassia alata L. var., *rumphiana* DC.

Notes: More popularly known as *Cassia alata* L.

Common Names: Candle Bush, Emperors candle stick, King of the forest, Ringworm weed.

Vernacular Names: Ben: Dadmari; **Hin:** Dadmurdan, Dadmari; **Kan:** Agase gida, Sheemigida; **Mal:** Elakajam, Seemagati; **Mar:** Dadamardana; **San**: Chakramarda, Dadrughna, Dvipaagasti; **Tam:** Anjali, Seemaiyagathi, Vendulholli; **Tel**: Mettata mara, Seemaavisi.

Distribution: Native of Caribbean Islands, Central and South America; introduced into tropical countries; naturalized in Southern and Eastern India.

Description: Glabrous shrub, 3-5m high; young stems and leaves puberulous with minute spreading hairs, eglandular; leaves spirally arranged, pinnately compound; leaflets 6-12 pairs, broad, oblong, blunt at the tip, unequal at the base, terminal pair much larger, obliquely round or truncate at the base, retuse at apex; stipules lanceolate to obliquely triangular, greenish orange or yellowish; flowers golden yellow, showy on dense elongate spikes borne terminally; bracts petal like, ovate-orbicular, orange coloured; calyx yellow or straw colored; corolla with equal obovate clawed petals; stamens 10, the 3 uppermost reduced, the lower 3 with large anthers, 4 with small anthers; pod four winged, dark brown, transversely septate with a crenate margined wing running longitudinally along middle of each valve; seeds 50 or more, flattened at right angles to the axis of the pod, deltoid with a longitudinal ridge along each face.

Phytochemistry

Plant: Alatonol, aloe-emodin, anthraquinone, chrysophanic acid, chrysophenol, dalbergin, daucosterol, methyl anthraquinone, emodin, 5-hydroxy-2-methyl dalbergin, kaempferol, luteolin, rhein, santalin, tetrahydroflavone and trihydroxy flavone.

Root: Anthraquinone, β-sitosterol and trihydroxy-methyl-anthraquinone.

Stem bark: Phytosterols and tannins.

Leaf: Aloe-emodin, anthraquinones, chrysarobin, chrysophanol glycoside, chrysophanol, thiamine, emodin, isochrysophanol, kaempferol, phenolic compounds, physcion-7-glucoside, trihydroxy methyl anthraquinone, rhein, sesquiterpenes, sitosterol and sennoside.

Pod: Aloe-emodin, chrysophanic acid, chrysophanol, emodin, oxymethyl anthraquinones and rhein.

Seed: Aloe-emodin, arginine, aspartic acid, β-sitosterol, campesterol, chrysophanol, emodin, fat, galactomannan, glutamic acid and hydroxyanthraquinones.

Seed oil: β-sitosterol, campesterol, dihydrospinasterol, linoleic acid, oleic acid, palmitic acid, proteins, stigmasterol and xanthone.

Medicinal Uses

Plant: Plant is allergenic, anthelmintic, antiandrogenic, antibacterial, anticarcinogenic, antiestrogenic, antihepatotoxic, antihypertensive, anti-inflammatory, antioxidant, antipyretic, antiradicular, antispasmodic, antiviral, cathartic, cytotoxic, emetic, immunosuppressant, laxative, psychotropic, purgative, spermicide and styptic; useful in breast cancer, diarrhoea, dysentery, edema, herpes, leukaemia, lymphoma, nephritis, sarcoma and tumour.

Root: Root infusion is used as antidote for snake poison; useful in rheumatism, stomach ache, venereal diseases, constipation and skin diseases.

Bark: Anti-inflammatory, CNS depressant and diuretic; is used to treat asthma, bronchitis, eczema and skin diseases; infusion of the bark cures "dhobis itch".

Wood: Decoction is used as a mild purgative.

Leaf: Analgesic, anthelmintic, anticlastogenic, antifungal, antihistaminic, antihypoglycaemic, anti-inflammatory, antiparasitic, antispasmodic, antitumour, aperient, astringent, expectorant, purgative and tonic; used in asthma, bronchitis, dyspnoea, eczema, herpes, ringworm and wound healing; roasted leaves are laxative; pounded leaves with lime juice is applied on skin diseases and ringworm; leaf decoction is astringent; cures dyspnoea, herpes, poisonous insect bites and venereal affections; promotes expectoration and urine output; tincture of leaves is antifungal and purgative; fresh leaf juice is used as eye drop to cure eye disorders and is applied on leucoderma patches; an ointment of ground leaves and vaseline is a recognized remedy for ringworm and other parasitic skin diseases; leaf poultice with black pepper is applied to "dhobis itch"; it also hastens suppuration of foul ulcers; leaves are boiled and water is drunk by women to hasten child birth.

Flower: Antibacterial, antimutagenic, purgative and tonic; decoction of the flower is useful in eczema and ringworm; it acts as an astringent, and expectorant in asthma and bronchitis; relieves dyspnoea, promotes urine output.

Seed: Is used as vermifuge in treating ringworm and discoloration of skin.

Bioactivity: Plant is inhibitor of arylamine-N-acetyl transferase, cyclo-oxygenase, glucosyl-transferase, lipoxygenase, MAO, ornithine-decarboxylase, PTK, prostaglandin synthesis, proteinase, topoisomerase; is calcium antagonist.

Toxicology: High dose of hot water extract of plant or leaves may cause abortion.

Uses in Alternative Systems of Medicine

Ayurveda:

Part used: Leaf and flower.

Uses: Flower extract is used internally for bronchitis and cough, tea made of flowers is given for diabetes; leaf decoction with lemon juice is used in ringworm and eczema; as a wash it is antimicrobial and anti-inflammatory; leaf poultice is used for eczema.

Siddha:

Part used: Root, leaf and flower.

Uses: Used as purgative and in skin diseases; ground leaves with lemon juice applied on ringworm.

Medicinal Uses Outside India

Australia: Hot water extract of dried leaves is taken orally as cathartic.

Bangladesh: Fresh leaves are crushed and applied on ringworm.

Brazil: Decoction of the leaves is taken as abortifacient and emmenogogue; decoction of the root bark is taken orally for malaria.

Fiji: Leaf juice is applied on skin affections and ringworm; leaf infusion is taken orally for diarrhoea and as blood purifier.

Gautemala: Hot water extract of the dried bark is applied externally for ringworm.

Indonesia: Leaves, onion and fennel is given for costiveness.

Ivory Coast: Decoction of the leaves is applied externally on yeast infection caused by *Candida albicans* and dermatophytes and orally taken to treat bacterial infections caused by *Escherichia coli.*

Jamaica: Hot water extract of leaves is taken orally for diabetes.

Malaysia: Root decoction is taken orally to ease stomach ache; hot water extract is taken as laxative; leaf paste is applied against scabies and ringworm.

Mexico: Hot water extract of the plant is used externally as an astringent and also against inflammation and rashes; used as anthelmintic and febrifuge.

Nicaragua: Fresh leaves are used as a cure for athlete's foot and ringworm; taken orally for stomach ache.

Nigeria: Dried, powdered leaves with pepper is taken orally for indigestion; fresh leaf juice acts as purgative and externally applied for ringworm and skin affections; leaves with fruit pulp of *Cucurbita pepo* and *Termitomyces microcarpus* is taken orally to treat gonorrhoea; ground flowers are taken orally to treat constipation.

Papua-New Guinea: Fresh leaves are used as cure for ringworm and skin eruptions.

Philippines: Fresh leaves are used to treat fungal infections of the skin and typhinitis.

Sierra-Leone: Decoction of the leaves is taken as laxative.

Tanzania: Decoction of the leaves is taken orally as purgative.

Thailand: Decoction of the leaves is taken for asthma; hot water extract of the entire plant is taken as cathartic; hot water extract of the leaves is used as febrifuge; flowers are taken orally as anthelmintic.

West Africa: Hot water extract of dried leaves is taken orally as ecbolic and emmenogogue; hot water extract in large doses is taken to induce abortion; hot decoction of leaves is applied on parasitic skin diseases.

West Indies: Hot water extract of flower is antibacterial; seeds are consumed as vermifuge; tea made from leaves is also used for intestinal worms.

Ethnomedicine

Plant: Aborifacient, bacterioside, depurative, diuretic, purgative, soporific, sudorific, taenifuge, tonic, vermifuge and vulnerary; useful in ascites, asthma, bronchitis, constipation, dhobis itch, diarrhoea, dysentery, eczema, fever, gonorrhoea, herpes, itch, leprosy, measles, mycosis, parturition, prurigo, psoriasis, rheumatism, ringworm, scabies, scurvy, shingles, skin diseases, sore throat, stomach ache, syphilis, venereal diseases and vitiligo.

Leaf and seed: Is used in eczema and skin diseases.

Non-medicinal Uses

Plant: Is used as a insecticide, larvicide and piscicide; plant extract is used in tattooing.

Root: Juice of the root is used for tattooing in Africa.

Bark: Used in tanning industry.

Leaf: Used as insecticide and fungicide; roasted leaves are boiled to prepare tea like beverage.

36. *Senna alexandrina* Miller

Synonyms: *Cassia angustifolia* M. Vahl.
Cassia senna L.
Senna angustifolia (Vahl.) Batka.

Notes: Leaves and pods are a source of the drug Alexandrian senna, preferred over the East Indian senna.

Common Names: Alexandria senna, Lotus senna and True senna.

Vernacular Names: Hin: Bhukhakhasa, Bhuyikhaakhsa, Hindisana, Hindisana-ka-pat, Sanaay; **Kan:** Nelaavare, Nelaavarike, Sonaamukhi, Sunnaa mukhi; **Mal:** Chinnamukki, Nilavaaka; **Mar**: Bhitharvada, Bhuhtharvad, Mulkacha, Shonamakhi, Sonaamukhi; **San**: Bhumiari, Bhumihaari, Bhumivalli, Bhupadma, Hemapathri, Kemapathrika, Jalathika, Kalyaani, Maarkandika, Mahaushadhi, Malaharini, Mridurechani, Peethapushpi, Rechani, Sthoolotpala, Svarnamukhi, Svarnapathri, Svarnapathrika, Svarnini; **Tam:** Kaattunilavirai, Nilavaagai, Nilaavirai, Ponna avarai, Tinnevelly senna; **Tel**: Nelaponna, Nelathangedu.

Distribution: Native of Africa, Asia including India, Pakisthan and Saudi Arabia; introduced into Central America and Mexico; in India, it grows in Gujarath, Maharashtra, Rajasthan and Tamil Nadu.

Description: Small shrub with pale yellowish brown bark; leaflets 4-7 pairs, opposite, narrow, lanceolate; flowers yellow on long racemes; pod flat, compressed, strongly nerved, stipitate, with ovoid, membranous seeds; seeds 8.

Phytochemistry

Plant: Aloe-emodin, anthraquinones, β-sterol, cathartomannite, chrysophanic acid, isorhamnetin, kaempferin, kaempferol, rhein, sennacrol, sennapicrin, sennosides C and D.

Leaf: Aloe-emodin, anthraquinone, arabinose, emodin, emodin dianthrone diglucoside, essential oil, fructose, galactose, galacturonic acid, glucose, kaempferol, kaempferin, mannose, mucilage, myricyl alcohol, palmitic acid, phytosterol, pinitol, rhamnose, rhein, sennosides A, B, C and D, siamine, dianthrone, siaminine A, B and C, stearic acid and sucrose.

Pod: Aloe-emodin, anthraquinone, chrysophanic acid, chrysophanol, emodin, essential oil, isorhamnetin, kaempferin, methylchromone, oxymethyl anthraquinone, physcion, rhein, sennosides A, B, C and D.

Seed: β-Sitosterol, aloe-emodin, chrysophanol, emodin, galactose, mannose and rhein.

Seed oil: Aloe-emodin, β- sitosterol, chrysophanol and rhein.

Medicinal Uses

Plant: Yields senna, which is antiallergenic, antianginal, anticancer, antibacterial, antidepressant, antineuralgic, antimetastatic, antispasmodic, antiseptic, antioxidant, antiradicular, antiviral, astringent and febrifuge; used in anorexia, anaemia, arrhythmia, arteriosclerosis, arthritis, atherosclerosis, canker, cataract, dermatitis, diabetes, dysmenorrhoea, eczema, edema, encephalopathy, epilepsy, gingivitis, glossitis, glaucoma, glycosuria, hepatitis, herpes, haemorrhages, hypoglycameia, insomnia, keratitis, leukaemia, lupus, melanoma, menorrhagia, orchitis, osteoarthritis, osteoporosis, pneumonia, polio, psoriasis, pityriasis, pneumonia, stomatitis, tumour of the breast, lung and brain, ulcer, vaginitis and vertigo; antidote to aluminum, cadmium, lead, mercury poisoning; infusion of tender tops is taken as laxative to allay fever.

Leaf: Acrid, astringent, athelmintic, bitter, cathartic, cholagogue, depurative, expectorant, febrifuge, laxative, stimulant, thermogenic and tonic; useful in abdominal disorders, anaemia, biliousness, constipation, costiveness, cough, dyspepsia, gout, helminthiasis, hepatomegaly, jaundice, leprosy, leucoderma, loss of appetite, rheumatism, skin diseases, splenomegaly, tumours, typhoid, vitiated conditions of "pitta" and "vatha"; dried and powdered leaves mixed with vinegar is applied locally on burns, skin eruptions, skin diseases and wounds; leaves with rose petal and tamarind is taken as purgative; leaf infusion in the form of tea is mild laxative; dried and powdered leaves are sprinkled to heal burns and wounds.

Pod: Laxative, purgative, stimulant and vermifuge; useful in costiveness.

Bioactivity: Plant is inhibitor of ACE, aldose-reductase, β-glucuronidase, ornithine-decarboxylase, polyamine, protein-kinase, proteinase, topoisomerase I and II; is β-adrenergic receptor blocker, β-blocker, calcium antagonist and calcium channel blocker; Water soluble polysaccharide fraction has shown antitumour activity against sarcoma–180 in mice.

Clinical Studies: Used medically for chronic constipation.

Toxicology: Causes abdominal discomfort such as colic, cramps, diarrhoea along with excessive loss of potassium, finger clubbing, development of cachexia, reduced serum globulin concentration and development of atonic, non-functioning colon; it irritates the bowel wall, causes griping and dermatitis due to anthraquinones and saponins; cannot be prescribed in spastic constipation; strong dose causes uterine and bladder contraction and uterine haemorrhage in pregnant women; should not be given in inflammatory conditions of alimentary canal, fever and during menstruation; seeds contain trypsin inhibitors.

Uses in Alternative Systems of Medicine

Ayurveda:

Part used: Leaf and pod.

Uses: Useful in abdominal disorders, anaemia, bronchitis, constipation, dyspepsia, fever, foul breath, gout, jaundice, leprosy, liver complaints, loss of appetite, skin diseases, spleen enlargement, tumours and typhoid.

Chinese:

Part used: Leaf.

Uses: Used as cathartic, laxative and purgative.

Siddha:

Part used: Leaf and pod.

Uses: Useful as anthelmintic and purgative.

Unani:

Part used: Leaf and pod.

Uses: Anthelmintic, antispasmodic, cleanser, diuretic and emmenogogue; useful in asthma, biliary disorders, colic, fever, gout, lumbago, wound healing, migraine, pleurisy, rheumatism and sciatica.

Ethnomedicine

Plant: Is used in treating ascites, dyspepsia and as laxative.

Non-medicinal Uses

Leaf: Senna leaves with *Lawsonia inermis* leaves used in hair dyes.

37. *Senna auriculata* (L.) Roxb.

Synonym: *Cassia auriculata* L.

Notes: More popularly known as *Cassia auriculata* L.

Common Name: Tanner's cassia.

Vernacular Names: Ben: Tarwar; **Guj:** Awal; **Hin:** Tarwar, Tharoda; **Kan:** Chookusina gida, Honnavare, Honnarike, Thangadi, Tharavada gida; **Mal:** Aveeram, Ponnaveeram, Jamute; **Mar**: Chambhaagaavadi, Tharavada, Tharoda, Tharwad; **San**: Avarthaki, Charmaranga, Mayahari, Peethakilaka, Peetha pushpa, Thimirihaari; **Tam**: Avaarai, Avaram, Semmalai; **Tel**: Avaray, Merakathangedu, Thangedu, Thangera.

Distribution: Native of India, Malaysia, Myanmar, Pakisthan and Sri Lanka; introduced into Africa, Indian Ocean Islands and Middle East; grows throughout except North Eastern India.

Description: A tall branched shrub with smooth bark; bark cinnamon brown with numerous brownish spots; branches finely pubescent; leaves abruptly pinnate; leaf rachis densely pubescent with an erect linear gland between each pair of leaflets; leaflets 8-12 pairs, slightly overlapping, oblong, obovate, obtuse, elliptic or emarginate, mucronate at apex, glabrous, dull green; stipules foliaceous, reflexed, reniform, or cordate found at the base on the side next to petiole with a long subulate point and a round base; flowers bright yellow, large, terminal, on axillary corymbose raceme; 2-8 flowers on a raceme; bracts ovate, acuminate; calyx glabrous, segments leathery, concave, outer two smaller than the other; corolla with clawed petals; stamens 10, the lower 3 with large anthers and long filaments and the upper three with reduced anthers (staminodes); pod flat, thin, papery, oblong, obtuse, strap shaped, pale brown, mucronate, deeply depressed between the seed with a crumpled appearance, transversely veined, pubescent; seeds 10-12 with distinct areole on each face.

Phytochemistry

Plant: Alkaloids, auriculacacidin, β-sitosterol, leucoanthocyanin, pyrrolizidine and tannin.

Root: Flavone glucoside.

Leaf: β-Sitosterol, emodin, fatty ketoalcohol, kaempferol and mucilage.

Flower: β-Sitosterol and kaempferol.

Pod: Nonacosane.

Seed: β -Sitosterol, fatty oils, linoleic acid, oleic acid and palmitic acid.

Edibility

Bark: Liquor is prepared by adding bruised bark to a solution of molasses and allowing the mixture to ferment.

Leaf and flower: Infusion is used as substitute for tea or coffee.

Flower and flower bud: Cooked and eaten.

Pod: Edible.

Seed: Is used as substitute for coffee.

Medicinal Uses

Plant: Is antiviral, anthelmintic, astsringent, diuretic, hypoglycaemic and spasmolytic; used in conjunctivitis, dysentery, ophthalmia, skin diseases and urinary complaints; decoction of the plant with honey is an excellent remedy for chylous urine and for regulating menstrual cycle; infusion of fresh and dried plant is a cooling drink.

Root: Alexiteric, alterative, antiviral, astringent, cooling, depurative, spasmolytic and tonic; useful in asthma, leprosy, skin diseases, tumours and urethrorrhoea; root is boiled with *Maytenus emarginatus* root and the decoction is taken in rheumatic pain.

Stem Bark: Alterative, astringent and tonic; decoction is used as gargle in sore throat and taken orally for dysentery and internal bleeding.

Leaf: Anthelmintic, astringent, depurative and tonic; recommended for leprosy, skin diseases and ulcers; leaf infusion is a cooling drink; it relieves feverishness and restores unbalanced nutritional status in the body.

Flower: Flower bud anthelmintic, astringent and tonic; used in pharyngopathy, spermatorrhoea and urethrorrhoea; decoction is given in asthma and chylous urine; paste of the bud cooked with unboiled rice is rubbed on the body as cooling and astringent; buds are used as pessaries to check excessive menstrual flow and its decoction is a remedy for diabetes; crushed flowers with bark of the plant, cubbebs, poppy seeds, black pepper, white mustard seeds and flower buds of *Musa paradisiaca* and kept overnight and in the morning mixed with bark decoction of *Ficus glomerata*, cloves and nut meg and dried roots of *Hemidesmus indicus* is given to reduce menorrhagia; decoction prepared from the flowers given to diabetic patients; syrup made from flowers, jasmine and Indian sarsaparilla is given to check nocturnal emissions.

Pod: Anthelmintic.

Seed: Alexiteric, alterative, anthelmintic, antibilious, antiscorbutic, aphrodisiac, astringent, attenuant, constipating, depurative, diuretic, expectorant, refrigerant and stomachic; used in chronic conjunctivitis, chylous urine, diabetes, diarrhoea, dysentery, leprosy, ophthlmia, skin diseases and swelling.

Uses in Alternative Systems of Medicine

Ayurveda:

Part used: Bark, leaf, flower, pod and seed.

Uses: Bark, flower and seed useful in asthma, diarrhoea, dysentery, intestinal worms, leprosy, polyuria, skin diseases and tumours; leaf is anthelmintic; cures anaemia, leprosy, skin diseases and ulcers; leaf poultice is applied on bone fracture; flower is used in diabetes, nocturnal emissions, throat trouble and urinary discharges; pod anthelmintic; cures thirst, vomiting and urinary discharge; seed is alexipharmic and used in diabetes.

Siddha:

Part used: Root, bark, leaf, flower and seed.

Uses: Astringent and anthelmintic; useful in chancre, chylous urine, colic, diabetes, diseases of "pitta", excessive body heat, gastritis, haematuria, leprosy, leucorrhoea, ophthalmia, rheumatic affections, skin diseases, thirst and urinary disorders; decoction of seeds is used as eye wash in conjunctivitis and ophthalmia; seed powder mixed with coconut oil is applied to sore eyes; powdered seeds prescribed in electuary and giddiness due to heart problems; all parts of the plant mixed with honey are specific for diabetes.

Kernel: Used in conjunctivitis.

Unani:

Part used: Root, bark, leaf, pod and seed.

Uses: Root is used in skin diseases; leaf and pod anthelmintic; seed used in chylous urine, conjunctivitis, diabetes and ophthalmia; bark and root is astringent.

Medicinal Uses Outside India

Ceylon: Bark and root used as astringent and alterative.

Myanmar: Leaf is used as astringent; seed is used as refrigerant.

Ethnomedicine

Plant: Is antidysenteric, astringent and vermifuge.

Root: Is used in rheumatism.

Stem: Is used in tooth ache.

Seed: Seeds are used in asthma, cough, ophthalmia, polyuria and skin diseases.

Non-medicinal Uses

Plant: Is grown for rejuvenating barren tracts; used as green manure crop and fodder crop; also grown as ornamental plant.

Twigs: Twigs are used as chewing sticks and tooth brushes.

Root: Is used in tempering steel and iron.

Bark: Tannin from the bark is used in colouring heavy leather and to dye the leather to buff colour; also yields a fibre which is used in making ropes; bruised bark with molasses is allowed to ferment and a liquor is prepared in India.

Leaf: Is used as a substitute for "Matar tea" which is a cooling drink; used as silk worm feed.

38. *Senna floribunda* (Cav.) H.S. Irwin & Barneby

Synonyms: *Cassia floribunda* Cav.
Cassia laevigata Willd.

Notes: Probably a hybrid of *Senna septemliriolinis* X *Senna multiglandulosa.*

Distribution: Native of Mexico; introduced into India; grows and naturalized throughout India.

Description: Shrub with glabrous branchlets and tomentose to silky leaves; leaflets 3-4 pairs, oblong, sharply mucronate; flowers red; pod thick woody, torulose especially at the end with numerous seeds; seeds variably shaped, ivory white, arranged around a gummy foul smelling pulp.

Phytochemistry

Plant: Anthraquinone, chrysophanol, emodin, glycoside, quercetin, physcion, leucoanthocyanins, leucocyanidin and leucocyanidin dimethyl-rutinoside.

Leaf: Floribundone 1 and 2, dibenzoyl spermudine and physcion.

Flower: Ceryl alcohol, docosyl alcohol, octacosanol, quercetin, rhamnetin-3-galactosyl (1→4) galacto peridoside and rhamnetin-3-galactosyl (1→4) galactopyranoside.

Seed: Chrysophanol, physcion and quercetin-3,7-dirhamnopyranoside.

Medicinal Use

Plant: Hypotensive.

39. *Senna italica* Miller

Synonyms: *Cassia italica* (Miller) F.W. Andrews
Cassia ligustrina Miller
Cassia obtusa Roxb.
Cassia obovata Colla
Cassia porturegalis Bancr.
Senna obtusa Roxb.

Common Names: Country senna, Dog senna, Italian senna, Jungle senna, Nubian senna, Official senna, Senegal senna, Tropoli senna.

Vernacular Names: Hin: Chhotatharoda, **Kan:** Adavithangadi, Kaadu sonaamukhi, Neladavarike; **Mal:** Seruvanni, Sunnamaki, Vatthantakara; **Mar:** Bhomyatharavad, Mendial; **San:** Bhumyahuli, Bhutalapota, **Tam:** Kaattunilavirai, Kaattu nilavaraichedy, Nilavaakai, Nilavirai, Vellaiponnavirai; **Tel**: Nellaponna, Nallathangedu, Sunamukhi.

Distribution: Native of Africa, Asia including India and Middle East; in India, it grows in Gujarat, Karnataka, Maharashtra, Punjab, Rajasthan and Tamil Nadu.

Description: A diffuse subglabrous under shrub of 1-4 feet height with angular branches; leaf rachis glabrous without glands; stipules lanceolate, persistent; leaflets 4-6 pairs, ovate-oblong, mucronate, unequal at the base, membranous, glaucous beneath; flowers pale yellow; stamens unequal; pod short stalked, curved, rounded at the ends, flexible, glabrous, valves cartilaginous, with prominent crests over the seeds; seeds 6-12, obovate, cuneate, separated by thin partitions.

Phytochemistry

Plant: Anthraquinone, chrysophanic acid, sennosides A and B.

Leaf: Oxymethyl anthraquinone and sennoside A and B.

Pod: Oxymethyl anthraquinone and sennoside A and B.

Medicinal Uses

Plant: The plant is a substitute for true senna; is antibacterial, antifungal, anti-inflammatory, antiviral, antibiotic, purgative and stimulant; is a laxative along with tamarind; useful in constipation, gastroenteritis and stomach cramps.

Twigs: Twigs used as tooth brush to cure spongy and bleeding gums.

Root: Infusion of the root mixed with milk is given for influenza and phlegm in chest.

Leaf: Antibiotic, laxative, purgative and stimulant; useful in constipation, gastroenteritis and stomach cramps; leaf paste is applied as dressing to burns and ulcers; when applied on hair, promotes healthy hair growth and applied to skin for healthy and glowing skin and for eczema and psoriasis.

Pod: Infusion is given as purgative.

Medicinal Uses Outside India

Myanmar: Seeds are used as purgative.

Ethnomedicine

Root: Is used in leucoderma.

Non-medicinal Uses

Leaf: Leaf powder is used as an excellent conditioner for hair; it makes hair glossy, thick and black.

40. *Senna obtusifolia* (L.) H.S. Irwin & Barneby

Synonym: *Cassia obtusifolia* L.

Common Name: Sickle pod.

Vernacular Name: Kan: Dundu thangadi.

Distribution: Native of Madagacar; introduced into India; naturalised throughout India.

Description: Annual undershrub of 2m high; stipules linear, caducous; leaf rachis with a single conical gland between the lowest pair of leaflets; leaflets 3 pairs, obovate, oblong, broadly deltoid at the apex, obliquely rounded at the base, membranous, glabrous on both the surfaces; flowers yellow, in subsessile pairs in the leaf axils; calyx 5 lobed; stamens 7, the upper 3 reduced to staminodes; pod obliquely septate, transversely reticulate with 30-35 seeds.

Phytochemistry

Root: Aloe-emodin, betulinic acid, β-sitosterol, leucopelargonidin-3-O-β-L-rhamnopyranoside, linoleic acid, O-methyl chrysophanol, palmitic acid, stearic acid and 1,3,5-trihydroxy-6,7-dimethoxy-2-methyl anthraquinone.

Leaf: Astragalin, β-sitosterol, chrysophanic acid, emodin, friedelin, isoquercetin, juglanin, kaempferol, myoinositol, ononitol, palmitic acid, querecetin, stearic acid, stigmasterol, tartaric acid, triacontan-1-ol and uridine.

Seed: β-Sitosterol, 1,3-dihydroxy-8-methyl anthraquinone, 2,5-dimethoxybenzoquinone, alanine, alaternin-1-O-β-D-glucopyranoside, anthraquinone, arabinose, arachidic acid, aspartic acid, betulinic acid, caffeic acid, cassialactone, cassioside, chlorogenic acid, chrysophanol, cholesterol, chrysoobtusin-2-O-β-D-glucopyranoside, chrysoobtusin, chrysophanic acid, chrysophanic acid-8-methyl-ester, cysteine, cystine, emodin, gadoleic acid, galactose, glucose, glutamine, helminthosporin, histidine, isochlorogenic acid, isoleucine, isotoralactone, leucine, linoleic acid, linolenic acid, lysine, m-cresol, methionine, methyloleate, methyl palmitate, naphtha-L-pyrone-toralactone, neochlorogenic acid, obtusifolin, obtusin, oleic acid, palmitic acid, physcion, rubrofusarin, rubrofusarin-6-β-gentiobioside, selenium, serine, sodium, stachyose, stearic acid, stigmasterol, threonine, toralactone, torasachrysone, tryptophan, tyrosine, xylose and γ-hydroxyarginine.

Edibility

Pod: Eaten as vegetable.

Medicinal Uses

Plant: Is analgesic, antiadenomic, antiandrogenic, antianginal, antihistaminic, antibacterial, anticancer, anticoronary, antioestrogenic, antifertility, antifibrinolytic, antigranular, antihepatotoxic, anti-inflammatory, anthelmintic, antigranular, antimetastatic, antinociceptive, antiperiodontic, antiradicular, aphrodisiac, antiviral, aperient, astringent, bronchoconstrictor, cancer preventive, cathartic, cholagogue, cytotoxic, demulcent, detoxicant, diuretic, hypoglycaemic, laxative, memory enhancer, metastatic, nephrotoxic, neurostimulant, purgative, radioprotective, spasmogenic, styptic, vasodilator and vulnerary; cures allergy, alopecia, anaemia, anorexia, arrhythmia, arteriosclerosis, antidote to aluminum, cadmium, iodine and lead poison, arthritis, atherosclerosis, cancer of the breast, colon, liver and skin, cataract, chilblain, cirrhosis, cold, colic, convulsions, dementia, depression, diabetes, dysmenorrhoea, eczema, edema, encephalopathy, epilepsy, gastritis, glaucoma, hepatitis, herpes, hypercholesterolemia, hypertension, hyperthyroidism, hyperglycaemia, impotency, lymphoma, malaria, melanoma, menopausal disorders, menorrhagia, migraine, myalgia, osteoarthritis, osteoporosis, rheumatism, retinopathy, stomatitis, stroke, tumour of the breast, cervix, colon, liver, lung and skin, tachycardia, ulcer and vitiligo.

Root: Root paste with lime juice is remedy for ringworm; useful in joint pain, sciatica, skin disease and tuberculous glands.

Stem bark: Useful as an antidote for scorpion sting.

Leaf: Antipyretic, aperient, digestive, diuretic, laxative and refrigerant; poultice is applied on foul ulcers to hasten suppuration; is a remedy for asthma, biliousness, bronchitis, gout, headache, joint pains, piles and sciatica; used as lotion for itching, leprosy, psoriasis, ringworm, skin diseases and urinary discharge.

Pod: Anthelmintic, alexiteric, alterative, astringent and antidysenteric; useful in asthma, cough, hemicrania, leprosy, ophthalmia, scabies, skin diseases and tumours; astringent to bowels and cures burning sensation.

Seed: Seed paste is useful in boils, eczema, itching, leprosy, psoriasis, ringworm and scabies; taken internally for asthma, bowel complaints, cough, intestinal worms, liver complaints and tumours.

Bioactivity: Plant is a inhibitor of 5-α-reductase, ACE, aldose-reductase, α-reductase, lipoxygenase, ornithine-decarboxylase, COX-2, PTK, phospholipase-A, polyamine, prostaglandin synthesis, protein-kinase, topoisomerase and xanthine-oxidase; is β-blocker, calcium channel blocker, CNS depressant, copper antagonist, DNA-protective, monoamine precursor and pituitary stimulant.

Uses in Alternative Systems of Medicine

Ayurveda:

Part used: Root, leaf, pod and seed.

Uses: Root is anthelmintic, antipyretic, diuretic and laxative; useful in ringworm, skin diseases and tuberculous glands; leaf is refrigerant; cures biliousness, bronchitis, leprosy, piles, skin disease and urinary discharges; pod and seed are anthelmintic, alexiteric, alterative and astringent; cures asthma, cough, leprosy, scabies, skin disease and tumour.

Unani:

Part used: Leaf and seed.

Uses: Leaf is demulcent, laxative and maturant; used in indolent ulcers, leprosy and skin disease; seed is useful in itch, ringworm, skin diseases and ulcers.

Medicinal Uses Outside India

Indochina: Pod is used in dysentery and ophthalmia.

Malay Peninsula: The leaves are ground and applied on the head for insomnia and on skin for skin diseases; leaf decoction is given for stomach ache and vomiting.

Ethnomedicine

Plant: Is considered as aperient and tonic; used in constipation, dermatitis, leprosy, nausea, parturition, stomach ache and tumour.

Seed: Is used in eczema, itch, leprosy, ringworm and other skin disease.

Non-medicinal Uses

Plant: Yields a yellow dye used in dyeing industry; seed is roasted, powdered and made into a bewerage.

41. *Senna occidentalis* (L.) Link

Synonyms: *Cassia caroliniana* Walter
Cassia ciliata Raf.
Cassia falcata L.
Cassia foetida Pers.
Cassia laevigata sensu auct.
Cassia macradenia Colladon
Cassia obliquifolia Schrank.

Cassia occidentalis L.

Cassia occidentalis L. var., *aristata* Colladon

Cassia planisiliqua L.

Notes: More popularly known as *Cassia occidentalis*.

Common Names: Arsenic bean, Cafe campo, Coffee senna, Negro coffee, Rubbish cassia, Stinking weed, Styptic weed.

Vernacular Names: Hin: Barikasondi, Chakunda, Kasonda, Kasondi, Kasunda; **Kan:** Aanesogate, Doddathagase, Eleuri, Elluri, Kolthagache, Kolthagasi, Urisogate, Urisonte; **Mal:** Karintakara, Nattantakara, Peyaaveeram, Ponnaaveeram; **Mar:** Kaasvindaa, Kasoda, Rankasvindaa, Rantaankala; **San:** Arimarda, Deepana, Jarana, Kaasamardaka, Kaashamarda, Kaasamardah, Kala, Kalankata, Kanaka, Karkasha, Kaasari, Vimard; **Tam:** Naatthamtakarai, Natthandagarai, Paayavarai, Peyaavirai, Ponnaavirai, Thagarai; **Tel:** Eddhukommaku, Kasindha, Kasintha, Kasivinda.

Distribution: Native of China, Caribbean Islands, Madagascar and South America; introduced into India; naturalized throughout India.

Description: A diffuse subglabrous under shrub; petiole with a single gland at the base; leaf rachis channelled; leaflets 3-5 pairs, opposite, ovate-oblong, lanceolate, acuminate, rounded at the base, glabrous above, pubescent beneath, foetid; stipules, obliquely cordate, acuminate; flowers yellow on axillary corymbs on a terminal panicle; bracts ovate acuminate, white, tinged with pink; stamens 7, upper 3 reduced to staminodes, the 3 lower longer with large anthers; pod falcate, compressed, transversely septate, torulose with 20-30 pale brown seeds.

Phytochemistry

Plant: β-sitosterol, achrosine, aloe-emodin, anthrones, anthraquinones, apigenin, apigenin-C-glycoside, campesterol, cassiollin, chrysoeriol, chrysarobin, chrysoobtusin, chrysophanic acid, chrysophanol, emodin, funiculosin, helminthosporin, islandicin, kaempferol, lignoceric acid, linoleic acid, linolenic acid, mannitol, obtusifolin, obtusin, oleic acid, physcion, quercetin, rhein, rubrofusin, tannin, xanthone and xanthorin.

Root: Cassiollin, chrysophanol, 1,8 dihydroxyanthraquinone, 1,4,5-trihydroxyanthraquinone, emodin, emodol, flavonoid, germichrysone, glycosides, helminthosporin, islandicin, methylgermichrysone, occidentalol I and II, phytosterol, pinselin, quercetin, tocopherols and xanthorin.

Leaf: Chrysophanic acid, dianthronic heteroside, jaceidin-7-rhamnoside, kaempferol, matteucinol-7-rhamnoside and oxymethyl anthraquinone.

Flower: Physcion-β-D-glucopyranoside.

Pod: Anthraquinone, tetrahydroxy-7-methoxyflavone and trihydroxy trimethoxyflavone.

Seed: α and β-Sitosterol, alanine, aloe-emodin, arginine, aspartic acid, chrysarobin, chrysophanol, emodin, fatty oil, histidine, isoleucine, leucine, methionine, 3-methyl-6-methoxy-1,8 dihydroxyanthraquinone, mucilage, n-methylmorpholine, physcion, physcion-1-β-D-glucopyranoside, physciondianthrone, phytosterolin, phytotoxin, quercetin, rhein, tannic acid, toxalbumin and xanthorin.

Seed oil: β-Sitosterol, campesterol, dihydrospinasterol, linoleic acid, linolenic acid, oleic acid and stigmasterol.

Edibility:

Leaf, flower, pod and seed: Edible; seeds roasted and used as coffee substitute.

Medicinal Uses

Plant: Alexipharmic, analgesic, antibacterial, anticancer, antifungal, antihaemorrhagic, anti-inflammatory, antimutagenic, antioxidant, antispasmodic, antitumour, antiviral, astringent, cardiotonic, contraceptive, detoxifier, diaphoretic, digestive, diuretic, expectorant, febrifuge, haemostat, hepatoprotective, hepatotonic, hypotensive, immunostimulant, memory enhancer, muscle relaxant, neuroprotector, purgative, restorative, stimulant, sudorific, thermogenic, tonic, uterine stimulant, vasoconstrictor and vermifuge; useful in alopecia, anaemia, anorexia, arrhythmia, arteriosclerosis, arthritis, asthma, atherosclerosis, canker, carcinoma of the breast, cataract, colitis, convulsions, cystitis, depression, dermatitis, diabetes, diarrhoea, dysentery, edema, eczema, encephalopathy, epilepsy, fibrosarcoma, gastritis, gingivitis, glaucoma, haemorrhoids, herpes, hypertension, hypoglycaemia, leprosy, leukaemia, malaria, menorrhagia, myocarditis, nephritis, osteoporosis, otitis, pharyngitis, retinopathy, rheumatism, rhinitis, sarcoma, stomatitis, tonsillitis, tumour of bladder, breast, lung, ovary and skin, ulcer and varicose veins; cures cirrhosis, constipation, convulsions, dropsy, epilepsy, hysteria, jaundice and whooping cough; it lowers blood pressure, strengthens the liver, tones up digestion, cures skin ailments and prevents cellular damage.

Root: Acrid, alexipharmic, anti-inflammatory, antiperiodic, bitter, digestive, diuretic, purgative, stomachic, thermogenic and tonic; useful in colic, convulsions, diabetes, dropsy, dyspepsia, elephantiasis, epilepsy, fever, flatulence, gastric complaints, neuralgia, ringworm, skin diseases, strangury, vitiated conditions of "vatha" and whooping cough; decoction of the root is given to stimulate lymphatic system; is used as substitute for quinine and used in treating fever; decoction of root is highly prized in hysteria to relieve the spasm in nervous women; also useful in relieving flatulence and obstruction in stomach; used as an antidote to scorpion sting and snake bite and other poisons.

Leaf: Acrid, alexiteric, alexipharmic, anodyne, aphroidisiac, bitter, cathartic, depurative, diuretic, expectorant, febrifuge, purgative, stomachic, sweet, thermogenic and vulnerary; useful as local application in abscesses, blisters, bone fractures, erysipelas, itch, leprosy, pruritus, ringworm, skin diseases, sores, ulcers and wounds; given in asthma, bronchitis, cough, fever, hiccup, hydrophobia, pharyngodynia, throat infection and tooth ache; used as a substitute for quinine; leaf decoction is given to relieve spasms in hysteria, and to soothen excited nerves; leaf paste with candle grease is used as adhesive plaster for wounds; leaf juice is given in bronchitis, cough, gynaecological problems, helminthiasis and rhinitis; used as bowel binder and as digestive.

Flower: Is diuretic, febrifuge, puragative, restorative, sudorific and tonic; flower decoction is given in hysteria to relieve spasms.

Pod: Is diuretic, febrifuge, puragative, restorative and sudorific.

Seed: Is acrid, alexiteric, antibiotic, bitter, depurative, diuretic, expectorant, febrifuge, purgative, stomachic, tonic and vulnerary; useful in bronchitis, cataract, constipation, convulsions, cough, dropsy, dyspepsia, eczema, erysipelas, fever, flatulence, hiccup, leprosy, liver diseases, pityriasis, psoriasis, rheumatism, ringworm, skin diseases, strangury, ulcers and whooping cough; seed decoction is given for constipation as a mild laxative; pounded seeds with cows milk is given for convulsions in children; seed decoction is applied on ringworm.

Bioactivity: Plant is a inhibitor 5-α-reductase, 5-lipoxygenase, aldose-reductase, acetylamine-N-acetyl transferase, COMT, COX-2, cytochrome-P_{450}, deiodinase, metalloproteinase, lipoxygenase, MAO, MMP-9, NADH-oxidase, NEP, tyrosinase, tyrosine-kinase, NO, NO-synthase, ornithine-decarboxylase, phospholipase, prostaglandin synthesis, protein-kinase, proteinase, CAMP-phosphodiesterase, topoisomerase I and II and xanthine-oxidase; is mast cell stabilizer and quinone reductase inducer.

Leaf extract when injected into mice significantly increased immunostimulant activity and protected them from chemically induced immunosuppression; in *in vitro* studies, leaf extract has shown a significant anti-inflammatory, hypotensive, antispasmodic, vasoconstrictor, smooth muscle relaxant activity in laboratory animals; ethanolic extract of the leaf showed hepatoprotective activity against paracetamol and ethyl alcohol induced hepatotoxicity in rats; leaf extract exhibited uterine stimulant activity in rats.

Toxicology: Excessive use causes headache and myodegeneration; should not be taken during pregnancy and should not be used by those with low blood pressure; use of the plant may interfere with some drugs in the liver by reducing their half-life; contains chymotrypsin inhibitors.

Clinical Studies: Patients with hepatitis and liver toxicity were treated with leaf extract, which normalized the damaged liver and restored liver enzymes.

Uses in Alternative Systems of Medicine

Ayurveda:

Part used: Leaf, root, and seed.

Uses: Leaf is alexiteric, aphrodisiac, bowel binder, digestive, emetic and stomachic; cures asthma, biliousness, cough, fever, hiccup and sore throat; leaf juice is given in bronchitis, cough, gynaecological disorders, helminthiasis and rhinitis; root and seed useful in cough, dyspnoea, eczema, filariasis and polyuria.

Siddha:

Part used: Whole plant, root, leaf and pod.

Uses: Whole plant is used in constipation, diseases of "vatha", dropsy, fever, headache, joint pain, rheumatism, sciatica, tooth ache and uterine pain; root juice or paste is applied for black scars, eczema, pimples, ringworm, skin diseases and heals wounds; when taken internally clears voice; cures dry cough, fever, gonorrhea, indigestion and phlegm; leaf juice is applied on eye for conjunctivitis and sores; pod cures asthma.

Unani:

Part used: Whole plant, leaf, root, and pod.

Uses: Leaf is used in fever, gonorrhoea, indigestion and phlegm; leaf juice is given in dropsy; root paste is applied on eczema and scars on the skin; pod is fried and given in asthma and as antidote to scorpion sting; all parts of the plant are given for tooth ache and to strengthen the gums.

Medicinal Uses Outside India

Africa: Plant is used for abscesses, bile complaints, bronchitis, bruise, cataract, constipation, edema, dysentery, erysipelas, eye infections, fever, gonorrhoea, headache, haematuria, hemorrhages, hernia, jaundice, leprosy, malaria, menstrual disorders, sore throat, stomach ulcers, syphilis, tetanus and wounds.

Amazon: Flower infusion is used for bronchitis.

Brazil: Root is considered as febrifuge, diuretic and tonic; used in anaemia, liver complaints, menstrual problems, tuberculosis and as a tonic for general weakness; the leaves are used for constipation, edema, fever, gonorrhoea, menstrual problems and urinary tract infections.

Central America: Plant is used for athlete's foot, constipation, diarrhoea, fungal infections, headache, menstrual disorders, respiratory infections, ringworm, spasm, uterine pain, urinary tract infections and worms.

China: Seed decoction is given to treat diarrhoea and haemorrhoids.

Dutch Indies: Leaf is used as poultice for tooth ache.

Guana: Leaf paste is applied to wounds and ringworm.

Haiti: Plant is used in acne, asthma, burns, colic, constipation, dropsy, gonorrhoea, eye infections, headache, malaria, rheumatism and skin infections.

Japan: Leaves are applied to insect bites and headache; leaf decoction is given in leucorrhoea, stomach ache and as anthelmintic.

Malaya: Leaf poultice is used for headache.

Mexico: Plant is used in dyspepsia, ear ache, eczema, edema, fatigue, fever, headache, inflammation, leprosy, nausea, pain, rash, rheumatism, ringworm, skin diseases, sluggish liver, sores, stomach ache, tumours, ulcers, venereal diseases, worms and yellow fever.

Nicaragua: Fresh plant decoction is used for premenstrual pain, general pain in the body and constipation.

Panama: The tea made from brewing the leaves is used for colic; crushed leaves are used as poultice in inflammations; taken internally for intestinal worms and as analgesic; applied locally on abscesses, fungal skin infections and wounds.

Peru: Roots are considered as diuretic and decoction of the root is given in asthma and fever; plant is used in asthma, bronchitis, fever and liver disorders.

Philippines: Seeds are used as febrifuge.

Trinidad: Plant is used in cold, constipation, inflammation, liver disorders and palpitation; used as abortifacient and contraceptive.

Venezula: Plant is used in asthma, cold, fever, intestinal gas, malaria, menstrual disorders and skin infections.

Ethnomedicine

Plant: Is considered as abortifacient, anodyne, antibilious, antiseptic, carminative, diaphoretic, diuretic, emmenogogue, febrifuge, purgative, stomachic, sudorific and vermifuge; used in treating abscesses, acne, asthma, bone fracture, bronchitis, burns, cataract, chancre, chill, cold, colic, dermatitis, diarrhoea, dropsy, dysentery, dyspepsia, ear ache, eczema, erysipelas, fever, gonorrhoea, head ache, heart diseases, haematuria, inflammation, jaundice, kidney problems, nausea, ophthalmia, palpitation, rheumatism, ringworm, sore throat, sores, spasms, stomach ache, swellings, tetanus, tooth ache, tumour, typhoid, ulcer, venereal diseases, wounds and yellow fever.

Twigs: Are used as tooth brush.

Root: Is used in gastric complaints and whooping cough.

Leaf: Is used in eczema, ringworm, skin diseases, throat infections and wounds.

Seed: Is boiled and used as stimulative drink and in whooping cough.

Non-medicinal Uses

Plant: Is used as green manure; young branches are used as tooth brush.

Seed: Is roasted and used as coffee substitute; finely ground seeds is mixed with charcoal to blacken the writing board; also used as insecticide.

42. *Senna siamea* (Lam.) H.S Irwin & Barneby

Synonyms: *Cassia siamea* Lam.
Cassia gigantea DC.

Common Names: Siamese senna, Sumatra senna.

Vernacular Names: Kan: Avarike, Hirethangadi, Seemethangadi; **Mal:** Mannakkona; **Tam:** Karungonrai, Karunkonne maram, Manja konnai, Pirambukkonrai, Seemaiyaavirai; **Tel**: Kurumbi, Seemathangedu.

Distribution: Native of Malaysia and China; introduced into tropical Africa, Asia including India and Sri Lanka, America, and the Pacific Island regions; cultivated and naturalized through out India.

Description: A moderate sized tree with grey bark; leaves pinnate with glands on the leaf rachis; leaflets 8-14 pairs, oblong, emarginate, shortly mucronate; flowers yellow on corymbs forming a terminal panicle; pod broad, apiculate with thick sutures indented between the seeds.

Phytochemistry

Plant: Alkaloids, anhydrobarakol, apigenein, apigenin-7-O-galactoside, barakol, β-sorbitol, cassiachromone, cassiasin, chrysophanol, dianthrone, D-pinitol, lupenone, lupeol, oleonic acid, p-coumaric acid, physcion, rhein, siameanin and tetrahydroxy stilbene.

Rootbark: Cassianin, A, B, C, chrysophanol, lupeol and emodin.

Stembark: β-Sitosterol, cassianin, chrysophanol, chrysophanol anthrone, chrysorhein, lupeol, lupenone, oleanolic acid, physcion, siameadin, siameanin and sucrose.

Wood: Anhydrobarakol, apigenin-7-O-galactoside, cassiachromone A, cassianin, chrysophanol and tetrahydroxystilbene.

Leaf: Anhydrobarakol, apigenin-7-O-galactoside, apigenin, β-sitosterol, cassiachromone, cassianin A, chrysophanol, kaempferol, p-coumaric acid and rhein.

Flower: β-Sitosterol, 5-acetonyl-7-hydroxy-2-methylchromone, arabinose, betulin, cassiadinine, dihydroisocoumarin, D-pinitol, flavone glucoside, friedelin, fructose, glucose, glucuronic acid, lupeol, sucrose and γ-sitosterol.

Pod: Alkaloids.

Seed: γ-Sitosterol and tannins.

Seed oil: Linoleic acid, palmitic acid, oleic acid, stearic acid, stercuilc acid and malvic acid.

Edibility

Flower, flower bud and leaf: Are used as vegetable.

Medicinal Uses

Leaf: Useful as tonic.

Pod: useful as tonic.

Medicinal Uses Outside India

Myanmar: The leaves, flowers and pods are made into a soup and drunk as a tonic and to relieve stomach ache.

Philippines: A decoction of the young leaves cures malaria.

Non-medicinal Uses

Plant: Is cultivated as ornamental and avenue tree; used as fodder; host plant for lac insect.

Wood: Is useful in making house posts, beams, scantlings, planks, mallets, walking sticks, furniture and inlay work; wood pulp is used for making writing and printing paper.

Leaf: Is used as fodder and manure.

43. *Senna sophera* (L.) Roxb.

Synonyms: *Cassia canca* Cav.
Cassia esculenta Roxb.
Cassia frutescens Miller.
Cassia geminiflora Schrank.
Cassia linearis Michaux.
Cassia occidentalis L. var., *glabra* DC.
Cassia occidentalis L. var., *sophera* (L.) Kuntze.
Cassia sophera L.
Cassia sophera L. var., *ligustrinoides* Benth.
Cassia torosa Cav.

Common Name: Foetid senna.

Vernacular Names: Hin: Banar, Baskikasondi, Kaalakasundi, Kaasunda, Kaasundi; **Kan:** Alloori gida, Alvari, Ekahuri, Elabadige, Elavarige, Kaasamarda, Kulakaasina; **Mal:** Ponnam tagera, Ponna virem, Pounaantakara; **Mar:** Kaalaakaasa vindaa, Kasodi, Ranthankala; **San:** Kaasamarda, Kaasaripu, Susha, Suvarnamayahari, Talapota; **Tam:** Naalal, Peravirai, Periatakarai, Ponnaaverai, Sulaarai, **Tel:** Kaasamardakamu, Kondakashinda, Nittykashindakoora, Noothikashinda, Paidithangedu, Tagara.

Distribution: Native of India; introduced into Africa, Asia, Carribean Islands, Indian Ocean and Pacific Ocean Islands and America; grows throughout India.

Description: A diffuse subglabrous shrub with foetid leaves; rachis with a single gland at the base; leaflets 8-12 pairs, lanceolate, acuminate, cuneate at the base, glabrous; flowers yellow on axillary corymbs; bracts green, small, ovate, acute; calyx lobes, elliptic, ovate, pubescent outside; petals oblong obovate; fertile stamens 7, upper 3 reduced to staminodes, the lower 3 with long filaments and large anthers; pod turgid, straight or curved, transversely septate with 30-40 dark brown seeds.

Phytochemistry

Plant: 1,8-Dihydroxy-3,6-dimethoxy-2-methyl-7-vinylanthraquinone, chrysophanic acid and emodin.

Root bark: Anthraquinones, 1,3-dihydroxy-5,7,8-trimethoxy-2-methyl anthraquinones, chrysophanol, physcion and sitosterol.

Heart wood: Anthraquinones, β-sitosterol, chrysophanol, emodin, n-octadecyl alcohol, quercetin and sopheranin.

Leaf: L-rhamnopyranoside.

Flower: Chrysophanic acid, chrysophanol, emodin, rhamnetin, rhamnetin-3-O-β-D-glucoside and sterols.

Seed: Anthraquinones, chrysophanol, emodin, fatty oil, mucilage, physcion, sitosterol, tannic acid and toxalbumin.

Edibility

Leaf and flower: Are edible and used as vegetable.

Medicinal Uses

Plant: Is alterative, depurative, diuretic, expectorant, febrifuge, purgative and tonic; used in asthma, bronchitis, respiratory obstructions and ringworm; it improves digestion, clears the throat and voice, purifies blood.

Root: Diuretic, expectorant and purgative; root paste is applied on eczema, itch, ringworm and skin affections; root paste with honey is a cure for dhobhi's itch and ulcers; root bark with seeds is ground into a paste and mixed with lime juice or sandalwood and applied for dhobi's itch and ringworm; root paste with pepper when taken internally is a remedy for snake poison; root juice is useful in scorpion sting.

Stem bark: Is antiseptic, cathartic and expectorant; infusion of the bark with honey is useful in diabetes.

Leaf: Is anthelmintic, antiseptic, cathartic and expectorant; useful in curing bronchitis, piles, ringworm and skin diseases; leaf paste with sandal wood or neem oil relieves itching and skin eruptions; leaf decoction is useful in asthma, fever and gout; leaf infusion is applied externally for inflammation, itch, rheumatic pain, ringworm and syphilitic sores; is taken internally for cough, fever, gonorrhoea and intestinal worms; leaf infusion or leaf decoction with sulphur and honey is applied on ringworm and psoriasis; leaf infusion is also given as injection for gonorrhoea; leaves with sugar is a cure for jaundice.

Seed: Is cathartic; powdered seeds with honey is given for diabetes; seed paste with sulphur is a cure for itch, skin disease, ringworm and psoriasis.

Bioactivity: Alcoholic extract of the plant relaxed the bronchial and intestinal muscles in Guinea pigs.

Uses in Alternative Systems of Medicine

Ayurveda:

Parts used: Root, bark, leaf and seed.

Uses: Root paste is applied for eczema, elephantiasis, ringworm, skin disorders and as antidote for scorpion sting; bark, leaf and seed is useful in asthma, cough, dyspnoea, eczema, filariasis, indigestion, polyuria and respiratory troubles; leaf cures biliousness, fever, hiccup, sore throat and *tridosha*.

Homoeopathy:

Part used: Root, leaf and seed.

Uses: Used in treating abdominal distension, anxiety, back ache, blurred vision, boils, burning of the eyes, burning pain in the umbilicus and hypogastrium, chill, constipation, coryza, cough, depression, dyspnoea, ear pain, eruptions, excessive perspiration, salivation, thirst, hunger, fever, flatulence, forgetfulness, insomnia, itching, leg and hip pain, lethargy, menstrual pain, micturition, mood swings, mouth and lip ulcers, nausea, pain and redness of the eyes, pimples, pustules, rashes, ringworm, shooting pain in head, teeth, gum and throat, urticaria, vertigo, vesicular eruption and watery discharge.

Siddha:

Part used: Root, leaf, flower and seed.

Uses: Useful in constipation, diseases of "vatha" and as rejuvenator.

Ethnomedicine

Plant: Is useful in abdominal tumours, cough, fever, headache, herpes, malaria, piles, pneumonia, ringworm, scabies, skin affections and sores.

Non-medicinal Uses

Plant: Is grown as an ornamental.

Leaf: Used as fodder and manure.

44. *Senna tora* (L.) Roxb.

Synonyms: *Cassia tora* L.
Cassia tora L. var., *borneensis* (Miq.) Miq.

Notes: More popularly known as *Cassia tora*; origin uncertain.

Common Names: Foetid cassia, Sickle senna.

Vernacular Names: Hin: Chakavad, Paamar, Panevar; **Kan:** Chagache, Chakramarda, Gandha thagase, Gandu thagase, Holasangi, Pumbadiyar, Thakkarike, Thagarisyaana gida, Thagache, Thajanaka, Thakala, Thagase, Thagate, Tharutagi gida; **Mal:** Chakramandrakam, Thakara; **Mar:** Thaakala, Thankli, Tharanvata; **San:** Avudham, Chakragaja, Chakramard, Chakramardaka, Chakravaha, Chakri, Dadrughna, Dridhabeeja, Edagaja, Edahasthi, Gajakhya, Kharjughna, Kusuma, Meshakshi, Meshalochana, Meshavhaya, Padmata, Prabhunatha, Prapunnala, Prapunnatha, Prishnaparni, Punnata, Sakramardhaka, Shukanaashana, Taga, Tarkila, Tarvata, Uranakhyaka, Urnaksha, Vimardhaka, Vyavartaka, Yadagaja; **Tam:** Senaavu, Siruthagarai, Thagarai, Thagashai, Usitthagarai, Vanamaavaaram, Vindu; **Tel:** Thagirise, Thantemu, Thantipu chettu, Thantiyamu.

Distribution: Native of Bhutan, China, India, Pakisthan, Thailand and Taiwan; introduced into Africa, Australia, Indian Ocean Islands, Middle East, Pacific Ocean Islands and Papua New Guinea; grows throughout India.

Description: An herbaceous annual weed growing to 90 cm height; leaves pinnately compound with a subulate gland between lowest pair of leaflets; leaflets 3 pairs, obovate, oblong, obtuse, apiculate at the tip, base rounded, rarely oblique, foul smelling; flowers yellow in axillary pairs; bracts linear; stamens 6-7, staminodes 3; pod apiculate, subtetragonal, obliquely septate; 20-30 seeds per pod.

Phytochemistry

Plant: β-Sitosterol, anthraquinones, arginine, aspartic acid, calcium, chrysophanol, coumestran, D-mannitol, fibre, glucose, fixed oil, histidine, iridoid glycosides, iron, magnesium, mannitol, myricyl alcohol, naphthopyrone, nor-rubrofusarin, oil, phosphorus, physcion, rubrofusarin, potassium, sterol, rubrofusarin glycoside, sodium and sulphur.

Leaf: Amino acids, chrysophanol, D-mannitol, myricyl alcohol, torachrysone, fatty acids, hydrocarbons, emodin, rhein and β-sitosterol.

Seed: Anthronc, anthraquinones, aurantiobtusin, β-gentiobioside, β-sitosterol, cassioside, chrysoobtusin, chrysophanic acid, chrysophanol-β-gentiobioside, chrysophanol, copper, emodin, glucoobtusifolin, glucosides, lactones, lignoceric acid, linolenic acid, magnesium, manganese, myricyl alcohol, naphthapyrone glycosides, nor-rubrofusarin, obtusifolin, obtusin, oleic acid, palmitic acid, phenolics, physcion, physciondiglucoside, potassium, protein, rhein, quercetin, rubrofusarin, rubrofusarin-6-β-gentiobioside, sennosides, sitosterol, sodium, stearic acid, terpenoids, torachrysone, toralactone, triglucoside and zinc.

Seed oil: Chrysophanic acid, emodin, lignoceric acid, linoleic acid, oleic acid and palmitic acid.

Edibility

Young leaves, pod and seed: Are edible; used as vegetable; the young leaves are cooked and eaten by Hindu Brahmins on fourth Saturday of month of shravan; seeds used in the preparation of sweets; leaves and seeds are used as substitute for coffee; leaves and immature pods boiled and eaten with rice.

Flower: Edible.

Medicinal Uses

Plant: Is analgesic, androgenic, anthelmintic, antiacne, antiallergenic, antianginal, antiarrhythmic, antiarthritic, antiasthmatic, antiatherosclerotic, antibacterial, anticanker, anticarcinogenic, anticephalopathic, antidiabetic, antihepatotoxic, antiherpetic, anti-inflammatory, antilymphomic, antimutagenic, antineoplastic, antinephritic, antioxidant, antisarcoma, antiseptic, antispasmodic, antistomatitic, antitumour, antiviral, astringent, antidote to cadmium and lead poison, cancer preventive, cardioprotective, cathartic, diuretic, emetic, haemolytic, haemostatic, hypercholesterolemic, hyperglycaemic, hypertensive, hypolipidemic, hypotensive, laxative, memory enhancer, mutagenic, nephrotoxic, purgative, spasmolytic, styptic, uterorelaxant, vermifuge, vasodilator and vulnerary; used in curing alcoholism, anaemia, anorexia, anxiety, candidiasis, cataract, cold, colic, constipation, convulsions, cough, dandruff, depression, dermatitis, dysmenorrhoea, dyspnoea, eczema, edema, epilepsy, fever, fibrinolysis, gastrointestinal disorders, glaucoma, helminthiasis, impotency, insomnia, itch, leprosy, leukaemia, menorrhagia, migraine, obesity, retinopathy, rheumatism, scabies, skin diseases, sores, tuberculosis, ulcer, respiratory disorders and varicose veins; antidote to snake bite.

Root: Vermicide and tonic; root paste with lime juice is specific for buboes, fever and ringworm; taken internally for stomach ache and wounds; powdered root with water is an antidote to scorpion bite.

Leaf: Alterative, anodyne, anthelmintic, antiparasitic, antiseptic, aperient, cardiotonic, depurative, diuretic, expectorant, ferbifuge, laxative, liver tonic, maturant and thermogenic; useful in bone fracture, bronchitis, cheloid psoriasis, colic, constipation, dyspepsia, flatulence, gout, haemorrhoids, helminthiasis, hepatopathy, intermittent fever, itch, joint pain, leprosy, night blindness, rheumatism, scabies, sciatica, skin diseases and tumour; young leaves are eaten for rheumatic pain; leaves fried in castor oil is applied on foul ulcers; leaf juice used as emollient to inflammation caused by *Semecarpus anacardium* juice; leaf decoction is a mild laxative and useful for fever in children while teething; pounded leaves or leaf paste acts like tincture iodine when applied on cuts and wounds; it is also applied against eczema and to promote suppuration.

Seed: Acrid, anodyne, anthelmintic, antiperiodic, cardiotonic, depurative, expectorant, laxative, liver tonic and ophthalmic; useful in arthritis, boils, bronchitis, cardiac problems, colic, dyspepsia, constipation, cough, eczema, flatulence, gout, haemorrhoids, hemicrania, intermittent fever, itch, leprosy, liver complaints, scabies, sciatica, scorpion sting, stomach ache and ulcer; it is used against boils and liver complaints; seed infusion is taken for cough; seeds steeped in the juice of *Euphorbia nerifolia* and cow's urine is applied to cheloid tumours, itch, leprosy, psoriasis and skin eruptions; seed paste with lime juice is applied on to itch and ringworm.

Bioactivity: Plant is inhibitor of 5-β-reductase, ACE, arylamine-N-acetyltransferase, PTK, prostaglandin synthesis, proteinase and topoisomerase-II; is angiotensin receptor blocker, β-blocker, CNS depressant, calcium antagonist and copper antogonist.

Toxicology: Excessive consumption of leaves as pot herb causes purgation; acts as central nervous system depressant in animals; contains chymotrypsin and trypsin inhibitors.

Uses in Alternative Systems of Medicine

Ayurveda:

Part used: Leaf.

Uses: Leaf juice is applied on dhobis itch, ringworm and other skin diseases.

Siddha:

Part used: Root, leaf and seed.

Uses: Used in fever, glandular swelling, inflammation, skin disease, ulcers, urticaria and venereal disease.

Unani:

Part used: Leaf and seed.

Uses: Leaf is used to check excessive urination and to prevent infectious diseases; seed decoction taken internally for asthma and to remove blood impurities; seed paste cures freckles, itch, leprosy, leucoderma, piles and ringworm.

Medicinal Uses Outside India

China: Seed infusion is used in conjunctivitis, gonorrhea and for gonorrheal sores.

Indochina: Pods are used in dysentery and eye diseases; the ground leaves are applied on head for sleeplessness; cures itches and skin eruptions.

Nigeria: Leaves are used in the treatment of ringworm, ulcers and parasitic skin affections.

Philippines: Plant is used as anthelmintic and purgative.

Ethnomedicine

Plant: Is used as cathartic, laxative, oxytocic and purgative; used in cheloid tumours, cold, constipation, ear ache, epilepsy, fever, headache, hepatitis, herpes, itch, kidney troubles, leprosy, liver disorders, ophthalmia, photophobia, plague, psoriasis, pterygium, ringworm, scabies, skin eruptions, tumours, typhoid, ulcer and vertigo; used as antidote to snake bite.

Root: Is used to cure fever and ringworm; used as antidote for scorpion bite.

Leaf: Used as anthelmintic, antiseptic and vermifuge; useful in bone fracture, cuts, indigestion, cold, intestinal troubles, jaundice, night blindness, rheumatism, skin diseases and wounds.

Seed: Is used to cure cold, eczema, itch and skin diseases.

Non-medicinal Uses

Plant: Is used as fungicide, nematicide and pesticide; used as a substitute for soap, sweetener and deodorant.

Bark: Yields tannin used in tanning industry.

Leaf: Used as a substitute for tea; as green manure.

Seed: Gives a blue dye; used as livestock feed.

45. *Tamarindus indica* L.

Notes: Tamarind shows enormous diversity in fruit characters, such as size (length 5 to 20 cm), colour (greenish brown to dark brown), very sweet to very sour and pulp-fibrous to free from fibre.

Common Name: Tamarind.

Vernacular Names: Ass: Tetuli; **Ben:** Ambli, Kidamar, Tentuli; **Guj:** Ambli; **Hin:** Amli, Imli; **Kan:** Hunise; **Mal:** Amlam, Puli; **Mar:** Chinch; **Ori:** Konya, Tentuli; **Sans:** Ambia, Amlavriksha, Amlika, Chincha, Tintida, Tintili, Tintiri, Tintrani; **Tam:** Amilam, Puli, Puliyan, Puliyam, Puzham; **Tel:** Chinthapandu, Chinthachettu.

Distribution: Native of Africa; introduced into most of the tropical countries; cultivated throughout India as an avenue tree on roadside.

Description: Large ever green tree with brownish grey bark; leaves abruptly pinnate; leaflets 20-40 pairs, oblong, small, round on both ends; flowers on a few flowered raceme at the end of the branch lets; flowers yellow, pink lined; calyx with 4 sepals, ovate lanceolate, tip acute; petals wrinkled at the margins; stamens 3; pod straight or curved, brown, flat, constricted at intervals, with a brittle shell; pod pulp with fibre; seeds 4-6.

Phytochemistry

Plant: α-L-rhamnopyranoside, cardenolide, citric acid, furanone, furfural, galactose, glucose, hordenine, lactic acid, malic acid, methyl furfural, niacin, oxalic acid, Pipecolic acid, proline, phenyl acetaldehyde, riboflavin, succinic acid, tamarindinal, uzarigenin-3-O-β-D-xylopyranosyl, xylose.

Bark: Hordenine, proanthocyanidin and tannin.

Leaf: α-oxo-glutaric acid, ascorbic acid, β-Carotene, calcium, carbohydrates, copper, fibre, glyoxylic acid, homoorientin, hordenine, iron, isovitexin, magnesium, malic acid, niacin, orientin, oxalic acid, oxalosuccinic acid, oxaloacetic acid, phosphorous, polyose-glycosidase, potassium, proteins, riboflavin, saponaretin, sodium, sulphur, tartaric acid and vitexin.

Flower: α-oxo-glutaric acid, β-Carotene, calcium, carbohydrates, fat, fibre, glyoxylic acid, hordenine, iron, niacin, oxaloacetic acid, oxalosuccinic acid, phosphorous, potassium, riboflavin, sodium and thiamine.

Pod: α-copaene, α-lumulene, α-murolene, α-oxo-glutaric acid, α-pinene, β-alanine, β-carotene, β-caryophyllene, β-pinene, acetic acid, alanine, ascorbic acid, cardenolide, cellulose, citric acid, fibre, fructose, galactose, galacturonic acid, glucose, glyoxylic acid, gum, hordenine, iron, lactic acid, leveine, limonene, linoleic acid, lysine, magnesium, malic acid, methionine, myrcene, myristic acid, niacin, oleic acid, oxalic acid, oxaloacetic acid, oxalosuccinic acid, palmitic acid, pantothenic acid, phenylalanine, phosphorus, pipecolic acid, potassium, potassium oxide, proline, protein, pyridoxine, quinic acid, riboflavin, serine, sodium, sodium potassium tartarate, stearic acid, succinic acid, sugar, tartaric acid, tryptophan, uronic acid and xyloglucan along with α-L-rhamnopyranoside, aromadendrene, γ-cadenine, carvacrol, cellulose, cinnamaldehyde, ethyl cinnamate, methylsalicylate, myrcene, N-octadecane, p-cresol, pectins, phenols, piperitone, potassium bitartarate, safrole, tartarate, tamarindienol.

Seed: 3-4-Dihydroxy methyl ester, 3-4-dihydroxyphenyl acetate, albuminoids, arabinose, benzoic acid, calcium, carbohydrates, D-arabinose, D-galactose, D-gluose, D-xylose, epicatechin, fat, fibre, non-drying oil, oleic acid, pectin, proteins, tannin and trypsin.

Seed testa: Rich in tannin.

Seed oil: Arachidic acid, behenic acid, lauric acid, lignoceric acid, linoleic acid, myristic acid, oleic acid, palmitic acid and stearic acid.

Seed kernel: Rich in polysaccharides.

Edibility

Bark: Is chewed as masticatory.

Young seedling, fresh shoot and leaf: Edible; new shoots are used in curries; is an excellent source of Vitamin B.

Flower: Edible; used in salads; a dish called "chingar" is made from the flowers; a good source of honey with slight acidity peculiar to the flower.

Pod: Most important edible part of the plant; it was an important item of diet of sailors in olden times as its water infusion helped them to overcome nausea and vomiting; pod juice is consumed as beverage during summer as a cooling drink; the pod pulp is used as preservative for pickling fish; "Jugo" or "frescoed tamarindo" is a favourite beverage in many Latin American countries; in Africa the pods are added to dishes made out of *Dioscorea* to remove any toxic substance; unripe pod is a rich source of tartaric acid, lactic acid and citric acid and is used in food industry; unripe and ripe pod are used in most of the south Indian dishes; pod is a principal souring agent in sauces, syrups and chutneys.

Seed: Edible; roasted and eaten; is used as food during famine; raw seeds are eaten with betel nut as masticatory; polyose obtained from the seed is a good substitute for fruit pectin which is used in jams and jellies; the reddish brown seed coat is powdered and used as adulterant to coffee powder.

Seed kernel: Tribals eat the kernels mixed with flower of *Madhuca latifolia;* jellose from the kernel is used in making jujubes which is used as a stabilizer in ice creams, mayonnaise and marmalades; kernel is fried, powdered and mixed with rice and made into balls with ghee and eaten.

Medicinal Uses

Plant: Is analgesic, antiandrogenic, anorexic, anthelmintic, antiadenomic, antibacterial, antidiuretic, antifibrinolytic, antihaemorrhagic, antihepatotoxic, antihistaminic, antihypertensive, anti-inflammatory, antimastalgic, antinociceptive, antioxidant, antiperoxidant, antipellargic, antispasmodic, antiviral, bronchorelaxant, cancer preventive, cardioprotective, carminative, cytotoxic, demulcent, detoxicant, diuretic, emetic, expectorant, febrifuge, goitrogenic, haemolytic, hypnotic, immunostimulant, keratolytic, laxative, refrigerant, spermicide, tranquilizer, vasodilator, vasomotor stimulant and vulnerary; cures allergy, acne, anorexia, aphthae, arrhythmia, arteriosclerosis, arthritis, asthma, atherosclerosis, beri-beri, calculi, cancer of the breast, cervix and pancreas, cardiospasms, cataract, cephalagia, delirium, dementia, dermatitis, diabetes, diarrhoea, dysentery, dysmenorrhoea, eczema, edema, encephalopathy, epilepsy, gastritis, gingivitis, glaucoma, hepatitis, herpes, hydrocele, haemorrhoids, hypercholesterolemia, hypoglycemia, insomnia, keratitis, leukaemia, lupus, lymphoma, measles, melanoma, menorrhagia, myocarditis, nausea, neuralgia, neuropathy, orchitis, pityriasis, pneumonia, retinopathy, rheumatism, sinusitis, tumour of the breast and prostate.

Tiwgs: Used in dysentery, fever, leprosy, ophthalmia, oral infections, respiratory ailments and wounds.

Root: Aphrodisiac and laxative; useful in dysentery, glandular tumours, indurations of spleen, utero-vaginal tumours and whitlow.

Root bark: Astringent, emmenogogue and tonic; useful in asthma, diarrhoea, constipation, gingivitis and ulcers.

Stem bark: Astringent, antiparalytic, febrifuge and tonic; given in amenorrhoea, diarrhoea, ulcers; bark infusion is useful in aphthous sores and sore throat; decoction is useful in diarrhoea, fever, gingivitis and as tonic; used as eye wash; bark lotion or poultice is applied to boils and sores; bark ash is antiscorbutic and digestive; is kept in earthen pots with salt and given in colic and indigestion.

Leaf: Antibilious, antifungal, anti-inflammatory, anthelmintic, anodyne, astringent, aperient, diuretic, febrifuge, ophthalmic and thermogenic; useful in treating boils, conjunctivitis, fever, gastropathy, helminthiasis, jaundice, otalgia, scabies, scalding urine, ringworm, small pox, ulcers and wounds; it reduces inflammatory swellings; tender leaf paste is used as external application for burns, ringworms, ulcers and inflammed joints to relieve pain and swelling; it is effective against abdominal, glandular and uterovaginal tumours, indurations of spleen and whitlows; leaf juice as liniment is applied over fractured bones, fungal infections on skin and inflamed eyes; tender leaf juice is taken in bilious fever and scalding urine; boiled leaf pulp is used as dressing for healing open sores; leaf infusion is cooling and useful in bilious fever; leaf decoction is used as gargle in throat infections and as wash for indolent ulcers and taken internally as anthelmintic; the ash of burnt leaves is a remedy for acidic urine and gonorrhoea.

Flower: Antibilious, appetizer and cooling; it is applied as a paste or lotion on eyes for conjunctivitis; taken internally for cancer, excessive perspiration, glandular tumours, indurations of spleen, urinary discharges, uterovaginal tumours and whitlows; flower juice cures bleeding piles.

Pod: Anthelmintic, antiseptic, anti-inflammatory, antiscorbutic, astringent, digestive, diuretic, cardiotonic, carminative, laxative and refrigerant; useful in alcoholic intoxication, bilious vomiting, Datura posisoning, glandular tumours, haemorrhoids, indurations, jaudice, ophthalmopathy and whitlow; pod with pepper is made into a soup and given as a home remedy for cold and cough; pod with sugar acts as a gentle laxative; pod made hot on fire is applied on part bitten by scorpion; boiled pod with dates and milk is a good antiscorbutic; ripe pod is used in atony of liver, stomach and intestines and as carminative; it cures biliousness, constipation and gastric disorders; infusion of the pod pulp is used in loss of appetite, lack of taste and interest in food; pod pulp with salt is used as liniment for rheumatism; pulp paste is applied on head for apoplexy and sunstroke.

Pod shell: Is antiasthmatic, astringent and tonic; useful in amenorrhoea, diarrhoea, dysentery, enlargement of spleen, ulcers and urinary discharges; shell is burnt and its ash is used in gonorrhoea.

Seed: Aphrodisiac, astringent, constipating and cooling; useful in abscesses, burning sensation, colitis, cough, chronic ulcers, debility, diabetes, diarrhoea, dysentery, haematuria, hepatopathy, inflammation, intestinal disorder, liver complaints, ulcers, vaginopathy and vertigo; powdered seeds are mixed with honey to cure acute dysentery; seed paste applied on the head cures cough.

Bioactivity: Plant is inhibitor of acetyl-cholinesterase, 5-α-reductase, aldose-reductase, α-amylase, β-glucuronidase, COMP, COX-I, glucosyl-transferase, lipoxygenase, MAO, ornithine-decarboxyglase, prostaglandin synthesis, reverse transcriptase, thyroid-peroxidase, tyrosinase and xanthine oxidase; is β-adrenergic receptor blocker, β-blocker, CNS depressant, calcium antagonist, cytochrome-P_{450} and P_{480} inducer, DNA binder, interferon synergist and monoamine precursor.

Allergenic activity: Pod powder showed weak activity to patch test for dermatitis.

Antibacterial activity: Acetone extract of pod is active on *Salmonella typhimurium*; ethanol extract is active on *Bacillus cereus, Bacillus subtilis, Bacillus megaterium, Escherichia coli, Pseudomonas aeruginosa, Salmonella typhimurium, Staphylococcus albus, Staphylococcus aureus* and *Vibrio cholerae;* hot water extract of dried pod is active on *Escherichia coli;* ethanol and hot water extract of dried leaves is active on *Staphylococcus aureus.*

Antifungal activity: Hot water extract of dried pod pulp is active on *Microsporum canis*, *Epidermophyton floccosum;* aqueous extract of leaves was active on *Ustilago muda*; ethanol extract of pod is active on *Aspergillus fumigatus; Aspergillus niger, Penicillium digitatum, Rhizopus nigricans* and *Trichophyton* species.

Antiviral activity: Ethanol extract of dried pod is active on herpes virus and Ranikhet virus.

Antimalarial activity: Ether extract of leaves is active on *Plasmodium falciparum*.

Antioxidant activity: Methanol extract of seed produced antioxidant activity.

Molluscicidal activity: Methanol extract of seeds is active on *Bulnus glososus*; dried pod extract is active on *Bulnus trunctus*.

Radical scavenging effect: Aqueous extract of dried leaves showed radical scavenging effect.

Toxicology: Plant contains chymotrypsin and trypsin inhibitors.

Uses in Alternative Systems of Medicine

Ayurveda:

Part used: Bark, leaf, flower, pod and seed.

Uses: Bark is used as astringent, cooling and used topically for loss of sensation in paralysis; bark ash is given for gonorrhoea and urinary discharges; leaf paste is applied to reduce inflammations, ringworm and swelling; leaf juice as liniment is applied over fractured bones; useful in colic, constipation, cough, dyspepsia, ear ache, fever, flatulence, gastrointestinal diseases, ophthalmia, smallpox and urinary affections; leaf infusion is used as gargle for sore throat and for washing indolent ulcers; flowers are used as antibilious, appetizer and cooling; cures "vata", "kapha", bad body odour, bleeding piles, constipation, colic, cough, dyspepsia, fever, flatulence, gastrointestinal disorders and urinary discharges; flower infusion is useful in conjunctivitis; pod is anthelminitic, antiscorbutic, appetizer, carminative, laxative, stomachic and stimulant; used in colic, constipation, diarrhoea, dysentery, dyspepsia, fever, edema, flatulence; "vata", piles and urinary affections; seed is astringent, heals fractures and wounds; useful in vaginal discharges and ulcers; seed testa is an effective remedy for diarrhoea, dysentery and delirium due to alcoholic intoxication.

Siddha:

Part used: Bark, leaf, flower, pod and seed.

Uses: Astringent, carminative, cooling, diuretic and laxative; useful in anaemia, indigestion, rheumatic pain and swelling.

Unani:

Part used: Bark, flower and pod.

Uses: Bark is used as anti-inflammatory, aperient, aphrodisiac, astringent, cathartic, cardiotonic, demulcent, laxative and stomachic; flower paste is laxative and used in bleeding piles; pod is used in excessive vomiting and biliousness, liver complaints, sore throat, stomatitis, thrist, and vertigo; used externally on inflammations.

Medicinal Uses Outside India

Africa: Beverage of pod is a valued refreshing drink.

Brazil: Decoction of the pod is taken for fever and haemorrhoids; used as emollient and purgative.

Cambodia: Pod pulp used for conjunctivitis; bark is used as an astringent in diarrhoea and gingivitis; used as tonic; wood decoction is used as diuretic and purgative; leaves used in conjunctivitis and scabies.

Canary Islands: Dried pod is eaten as choleretic.

Columbia: Hot water extract of dried pod is abortifacient.

Dominican Republic: Water extract of leaves is taken for liver complaints.

Eritrea: Pod pulp is used in dysentery.

Fiji: Infusion of the bark, pod and leaves is taken for piles; leaf poultice with mustard oil is applied for sprain; leaves soaked in water overnight and the water is used as eye wash for eye troubles; pod pulp is eaten for diarrhoea and sore throat.

Gold Coast: Bark tincture is used in curing diarrhoea.

Guatemala: Hot water extract of the pod is febrifuge and sudorific; used in treating erysipelas, ringworm, skin infections and urinary tract infections.

Guinea: Bark extract is taken by women as tonic during pregnancy and after child birth; is used as astringent; boiled leaves are used in washing wounds, dressing ulcers, swellings and sprains; pulp is used in diarrhoea, dysentery, dyspepsia, fever and gastritis.

Indo China: Pod extract is diuretic and laxative; used to treat loss of appetite and constipation.

Indonesia: Water extract of pod is used to relieve itching of boils and wounds.

Ivory Coast: Hot water extract of leaf and root is taken orally to treat sleeping sickness.

Madagascar: Hot water extract of bark is taken orally for amenorrhoea; bark decoction is useful in asthma; leaf infusion is anthelmintic and stomachic; pulp is anthelmintic and taken for stomach disorders.

Malaysia: Hot water extract of stem bark is taken orally for amenorrhoea.

Mauritius: Cooked bark is used as antiasthmatic; pod pulp is used as liniment for rheumatism; seeds used in dysentery.

Nigeria: Fresh leaves ground with leaves of *Prosopis africana* and taken with water to cure malaria; crushed leaves with potash is soaked in water and the extract is used as laxative; hot water extract of bark, pod shell with leaves and bark of *Diospyros mespiliformis* is a remedy for leprosy.

Peru: Hot water extract of dried pod is taken as laxative.

Senegal: Hot water extract of the bark is used externally as cicatrizant.

Sri Lanka: Tender leaves are applied as poultice for boils; pod pulp is used in jaundice and ulcers; seeds used in diarrhoea.

Sudan: Dried pod pulp is taken as laxative; used in treating bacterial infections and malaria.

Tanzania: Root decoction is taken orally for painful distended abdomen and for dysentery; decoction of dried bark and root together with *Stereospermum kunthianum* is taken for leprosy; fresh leaf juice cures bloody diarrhoea.

Thailand: Hot water extract of dried leaves is cathartic; hot water extract of dried pod pulp is taken for cough as expectorant; hot water extract of seed is used as anthelmintic.

West Indies: Pod pulp is taken as laxative.

Ethnomedicine

Plant: Is used as antiseptic, aperient, carminative, depurative, digestive, diuretic, emetic, laxative, refrigerant, styptic, suppurative, tonic and vermifuge; used in head ache, apoplexy, asthma, biliousness, boil, carbuncle, cold, colitis, conjunctivitis, cough, diabetes, diarrhoea, dropsy, dysentery, fever, haemorrhages, hypertension, inflammation, jaundice, liver disorders, piles, rheumatism, scurvy, sore throat, sores, swellings and tumour.

Bark: Is used in stomatitis.

Leaf: Is used as anthelmintic and stomachic; useful in anasarca, eye complaints, swellings, toothache and urinary complaints.

Pod: Is used in cold, cough and digestive disorders; pod infusion is given to lessen the effects of sunstroke.

Seed: Useful in boils.

Non-medicinal Uses

Plant: Is a host plant for lac insect and of a silk worm of the genus *Anaphe*; grown as shade tree; used as fuel; yields a dirty black gum containing galactan.

Stem: Yields fibre.

Wood: Timber is resistant to insect attack and so is mainly used for making agricultural implements, wheels, rice pounders, mortar and pestle, ploughs, tool handles, cart shafts, axles, side planks of boats, printing blocks, tent pegs, furniture, decorative panelling and used as a substitute for teak wood for beams, rafters and posts; gives an excellent charcoal used in gun powder.

Wood ash: Useful for tanning and dehairing of goatskin.

Leaf: Yields a yellow dye used for colouring silk and woollen clothes; also used as auxiliary in dyeing; useful as fodder and green manure.

Flower and pod: Used as auxiliary in dyeing.

Pod pulp: Used in cleaning brass and copper vessels and articles; infusion of pulp is used for cleaning silver ware.

Seed: Finely ground and mixed with ordinary glue is one of the strongest wood cements; seed powder is used in sizing country made blankets, paper and jute; yields a gum used in food processing; ground seeds used as cattle feed; used in making ornaments.

Seed testa: Used in dyeing; it gives brown to olive brown shades to wool, silk and cotton; also used for making plywood adhesive.

Seed oil: Used in the preparation of paints and varnishes; used in painting idols, also used as illuminant.

Kernel: Kernel powder after decorticating the seeds is used as sizing material in textile industry; used also as creaming agent for rubber latex, stabilizer in bricks, adhesive in book binding, cardboard and plywood industry and in sizing leather; boiled in water and mixed with boric acid and phenol makes a good paper glue; it is recommended as printing paste in calico printing on cotton and artificial silk fabric; it can be used alone or with gum tragacanth as printing paste; as a substitute for corn steep liquor; jellose from the kernel is useful as a thermoplastic resin; used in pharmaceuticals and insecticides; used as manure.

Fig. 1. *Acrocarpus fraxinifolius*

Fig. 2. *Amherstia nobilis*

Fig. 3. *Amherstia nobilis*

Fig. 4. Pods of *Amherstia nobilis*

Fig. 6. *Bauhinia malabarica*

Fig. 5. *Bauhinia galpinii*

Fig. 7. *Bauhinia tomentosa*

Fig. 8. *Brownea coccinea*

Fig. 9. Pods of *Caesalpinia sappan*

Fig. 10. *Cassia fistula*

Fig. 11. *Cassia javanica*

Fig. 12. *Cassia sericea*

Fig. 14. *Moullava spicata*

Fig. 13. *Delonix elata*

Fig. 15. *Peltophorum pterocarpum*

Fig. 16. *Peltophorum pterocarpum*

Fig. 17. *Saraca asoca*

Fig. 18. *Saraca declinata*

Fig. 19. *Saraca taipingensis*

Fig. 20. *Senna alexandrina*

Fig. 21. *Senna tora*

Fig. 22. *Tamarindus indica*

Fig. 23. *Tamarindus indica*

CHAPTER-25

Fabaceae

1. *Abrus precatorius* L.

Notes: Two subspecies, ssp. *precatorius*, (Asia) and ssp., *africanus* (Africa) are recognized. Widespread in the old world; the African subspecies is widely cultivated and naturalized in the new world; Abrin-A is a potent toxic lectin, used as a molecular probe in biomedical research; due to its strong cytotoxic activity; abrin is used in cancer research; the leaves contain more glycyrrhizin than the roots of *Glycyrrhiza glabra* and is a potential source of glycyrrhizin.

Common Names: Crab's eye, Jequirity bean, Lucky bean, Paternoster bean, Precatory bean, Rosary bean.

Vernacular Names: Hin: Chirmiti, Gaungchi, Ghungachi, Gunja; **Kan:** Gulaganji, Gundumani, Madhukavalli; **Mal:** Irattimadhuram, Kaakani, Klitakkam, Kunni, Kunnikkuru, Madhukam, Madhumulam, Shekunni; **Mar:** Chanoti, Gunchi, Gunja; **Sans:** Angarvallari, Aruna, Bhilabhooshana, Chakrashalya, Chataki, Choodala, Choodamani, Dhvankshanaka, Durmogha, Gunja, Gunjika, Kaaka, Kaakachinchi, Kaakadani, Kaakajangha, Kaakananthika, Kaakashimbi, Kaakathundika, Kaakavallari, Kaakini, Kaaksha, Kaamboji, Kanchi, Krishna choodika, Krishnala, Mirintika, Rakhta, Saumya, Shangushtha, Shikandi, Shyamalachooda, Tamrika, Tulaabeejam, Ucchatta, Vakrashalya; **Tam:** Adisaamiyai, Adhimadhur, Kunjuram, Kunri, Kunirimani, Kuruvindam, Maduragam, Singli, Siridam, Sittilai; **Tel:** Athimadhuramu, Gulivinda, Gurinja, Guriginja, Gurivinda, Kukkutama, Nallagulivinda, Rakhthika, Sinnaguruginja.

Distribution: Native of Africa and Asia including India, Australasia, Caribbean islands, Central America and other tropical countries; grows throughout India.

Description: A deciduous wiry climbing shrub with slender branches; leaves paripinnate, alternate; rachis ending in a spine; leaflets opposite, 12-16 pairs, small, linear-oblong, apiculate, sparsely hairy; stipules 3 mm long, linear; inflorescence is a raceme on axillary peduncle and shorter than leaves; flower white to pink, crowded; calyx sparsely pubescent, lobes short; standard petal suborbicular, auricled, wings narrow, oblong-falcate; stamens 9+1; ovary subsessile, many ovuled; pod compressed, linear-oblong, thick, septate between the seeds, with a sharp deflexed beak; seeds round, hard, bright red with a black spot and a basal aril.

Phytochemistry

Plant: Abraline, abrin, abrisin, abrusgenic acid, abruslactone, abrussic acid, anthocyanins, arabinose, ash, calcium, campesterol, choline, cycloartenol, 5-β-cholanic acid, glucodelphinidin, N, n-di-methyl-tryptophan methyl ester, delphinidin, gallic acid, galactose, hypaphorine, precatorine, pectin, pentosans, phosphorus, polygalacturonic acid, precasine, precatorine, trigonelline, tryptophan, tryptophan-methyl ester and xylose.

Root: 5, β-Cholanic acid, abrasine, abrol, abruquinone A, B and C, choline, glycyrrhizin, hypaphorine, isoflavonoquinone, precasine, sterols and trigonelline.

Leaf: β-Cholanic acid, abrine, abrusoside A and D, arabinose, choline, flavonoids, galactose, glycyrrhizin, hypaphorine, pinitol, sterols, taxifolin-3-glucoside, trigonellin, 4,5,7-trihydroxy flavone and xylose.

Seed: Abrine A, B and C, 5, β-cholanic acid, abralin, abranin, abrectorin, abricin, abridin, abruslactone A, abriolin, abrisapagenol, abrusgenic acid, alanine, arabinose, arachidic acid, pelaragonidin-3-glucosides, campesterol, carbohydrates, cholesterol, choline, cyanidin-3-glucoside; delphinidin, cycloartenol, demethoxycentaureindin-7-O-rutenoside, fatty acids, gallic acid, glycyrhizine, hencicosane, hydrocarbons, hypaphorine, pelargonidin 3,5-diglucoside, methylabrusgenate, N-dimethyl tryptophan, precatorine, serine, squalene, trigonellin, sterols, stigmasterol, sugars, quinone I, II, III, valine, xylose, β-amyrin, β-sitosterol.

Seed oil: Arachidic acid, behenic acid, linoleic acid, linolenic acid, oleic acid, palmitic acid and stearic acid.

Seed coat: Abranin, anthocyanins, delphinidin-3,5-diglucoside, P-coumaryl-galloyl-glucosyl delphinidin, pelargonidin 3,5-diglucoside and xyloglucosyl delphinidin.

Edibility

Leaf: Leaves are used as sweetening agent; glycyrrhizin in the leaves is 60 times sweeter than cane sugar.

Medicinal Uses

Plant: Is abortifacient, aphrodisiac, antiallergenic, analgesic, antiandrogenic, anodyne, anorexic, antibacterial, anticarcinogenic, anticancer, antidepressant, antidysenteric, antiestrogenic, antifertility, antifibrinolytic, antihepatotoxic, antihistaminic, anti-inflammatory, antimutagenic, antinociceptive, antiobesity, antioxidant, antiperoxidant, antiseptic, antistomachic, antitumour, antitussive, antiviral, apoptotic, astringent, bronchodilator, cardioprotective, cerebrotonic, chemopreventive, demulcent, detoxicant, diuretic, emetic, emollient, estrogenic, expectorant, febrifuge, haemolytic, immunomodulator, oxytocic, pectoral, purgative, sedative, styptic, thermogenic, tonic and vermifuge; useful in anorexia, alopecia, arteriosclerosis, arthritis, atherosclerosis, biliousness, bladder complaints, blenorrhagia, boils, bronchitis, cancer of breast, cervix and liver, cataract, cirrhosis, chest complaints, cough, cold, colic, conjunctivitis, consumption, convulsions, diabetes, diarrhoea, dysuria, diphtheria, edema, enteritis, eczema, encephalopathy, epithelioma, fever, freckles, gingivitis, gonorrhoea, gravel, heart trouble, herpes, hyperglycaemia, hypercholesterolemia, hypertension, inflammation of gums, insomnia, jaundice, joint pain, kidney complaints, leukaemia, lymphoma, malaria, menorrhagia, migraine, myalgia, myopia, ophthalmia, polio, rheumatism, stomatitis, skin diseases, tumour of the breast, cervix, colon, liver and lung, throat problems, trachoma, urinogenital disorders, ulcer and venereal diseases; it enhances memory; removes mucous in the urine; protects the pancreas and the capillaries.

Root: Abortifacient, alexiteric, antiseptic, anthelmintic, astringent, blood purfier, diuretic, emetic and tonic; useful in bronchitis, chest pain, cold, colic, cough, diarrhoea, gonorrhoea, inflammations, jaundice, pectoralgia, pharyngodynia, rheumatism, sore throat and strangury; root paste with rhizome of *Curcuma longa* cures wounds when applied locally and causes abortion if taken internally; root paste made with pure ghee cures cough.

Leaf: Abortifacient, alexiteric, analgesic, anti-inflammatory, antiphlogistic, antispasmodic, anthelmintic, astringent, blood purifier, demulcent, diuretic, emetic, nervous system depressant, uterine stimulant and tonic; useful in cough, freckles, leucoderma, ophthalmia, paludism, pharyngodynia, pectoralgia, rheumatism, strangury, swellings, ulcer and vitiated "vata"; the leaf juice is taken internally for chest pain ; the leaves are chewed as a remedy for hoarseness of voice; fresh leaf juice with oil is applied externally for local pain and taken internally act as emetic after snake bite.

Seed: Abortifacient, analgesic, antifertility, aphrodisiac, astringent, emetic, purgative, trichogenous and tonic; useful in alopecia, asthma, cholera, corneal opacity, diarrhoea, dysentery, eye diseases, fever, granular lid, hyperdipsia, leucoderma, paralysis, palludism, vitiated "pitta", stomatitis, trachoma, tubercular glands, ulcers and wounds; seed paste is applied for paralysis, sciatica, stiffness of shoulder joints, skin diseases and ulcers; leaves and seeds are ground into a paste and applied on head for 3 to 4 days for graying hair and baldness; seed paste is taken internally for spermatorrhoea; seeds without the seed coat when eaten with banana for 3 days after the menstrual cycle acts as a contraceptive; mixed with sugar it is anthelmintic; seed paste with *Plumbago* root is applied as dressing for wounds; seed powder is taken as snuff in violent headache due to cold.

Bioactivity: Plant is inhibitor of 5-β-reductase, ACE, cyclo-oxygenase, MAO, prostaglandin synthesis, reverse-trascriptase, topoisomerase-I and xanthine-oxidase; is angiotensin receptor blocker, β-blocker, calcium channel blocker, metal chelator, monoamine precursor, NO-scavenger and ribosome inactivator.

Alcoholic extract of the seeds showed parasympathomimetic effect on smooth muscle of guinea pigs, rabbits and skeletal muscles of frog; abrin A at 0.8 ug/ml agglutinated human O-type erythrocytes where as abrin B showed no such activity; both A and B agglutinated sarcoma 180 and Ehrlich ascites tumour cells at 5 and 150 ug/ml respectively; abrin injected into mice at 0.7 mg/kg body weight is lethál; 2 ounce of seeds are fatal to horse and 2-4 ounce are required for goats, cows and dogs.

Toxicology: Seeds are highly poisonous; ingestion of seeds is fatal; in humans, the ingestion of abrin results in excessive salivation, colic, bloody diarrhoea, seizures, prolonged vomiting and shock leading to acidosis due to renal failure, lethargy, muscle weakness, drowsiness, disorientation, cyanosis, stupor, circulatory failure, haematuria, oligosuria, hallucinations, cardiac dysrrhythmia, muscle cramps, haemolysis of RBC; cellular action of abrin results in ribosome inactivation, inhibition of bilirubin and serum level of aspartic transferase, alanine transferase, lactic dehydrogenase and protein synthesis; abrin effects rapidly dividing cells, causes anorexia, violent vomiting, chills and disorientation in animals; no specific antidote is available; emesis with syrup of ipecacuanha is recommended to maintain normal blood circulation followed by correction of electrolyte disturbance.

Ornaments made from seeds may cause contact dermatitis; seeds contain chymotrypsin and trypsin inhibitors.

Uses in Alternative Systems of Medicine

Ayurveda:

Part used: Root, leaf and seed.
Uses: Used in alopecia, hemiplegia, leprosy and nervine troubles.

Chinese:

Part used: Seed.
Uses: Used as a remedy for dropsy, fever, malaria and worms.

Siddha:

Part used: Root, leaf and seed.
Uses: Used in arthritis, eye diseases, fainting spells, jaundice, leucoderma and pain.

Unani:

Part used: Seed and root.
Uses: Seed is aphrodisiac, emetic, purgative and tonic; useful in nervous disorders; used as suppository to bring about abortion; root is alterative and emetic.

Medicinal Uses Outside India

Ambonia: Leaf is used as a remedy for swollen tonsils and sprue.
Ceylon: The root is taken for rheumatism and sore throat.
China: Seed is given for fever, headache, dropsy, malaria and worms.
Indochina: Seed is used as a remedy for malaria; root is used in diarrhoea; bark is used for dysentery and leaf for conjunctivitis.
Malay Peninsula: Decoction of leaf with betel nut is used for colic, cough and cold.
Philippines: Decoction of stem and leaf is employed to bathe newborn children and to treat cough, cold and colic.
South America: Plant infusion is useful in ophthalmic inflammation.

Ethnomedicine

Root: Used as abortifacient, antiseptic, blood purifier and febrifuge; useful in bronchitis, cold, cough, leucoderma and nervous complaints; used as antidote to snake poison.
Leaf: Useful in body pain, pneumonia, scabies, skin diseases, sore throat, swelling and urinary complaints.
Pod: Useful in eye complaints.
Seed: Abortifacient, antiseptic and aphrodisiac; useful in eczema, pneumonia, sciatica, skin diseases and stomach complaints; used as a tonic.

Non-medicinal Uses

Stem: Yields fiber used in basket making.

Seed: Used for homicidal purpose due to the toxic chemical abrin in the seeds; used by gold smiths as weights; used as ornamental beads for ear rings, necklaces, pendants and rosaries; to decorate boxes and baskets; seed garland is worn by tribals to ward off evil spirits.

2. *Abrus pulchellus* Thwaites

Synonyms: *Abrus fruticulosus auct. non* Wight & Arn.
Abrus stictosperma Berhaut

Notes: Five subspecies are recognized; two of which are in Africa and three are in Indonesia.

Vernacular Names: Sans: Angaravallika; **Tam:** Venkundri, Vaidathari.

Distribution: Native of Africa and Asian countries; introduced into India; grows throughout India upto Himalaya, Nepal, Bhutan, Burma, Sikkim and Andaman Islands.

Description: Climbing woody shrub; leaflets paripinnate, 10-20 in number, opposite, ligulate, oblong, membranous, glabrous above, silky beneath; flowers pink or white on raceme; calyx silky; pod large, flat, incurved; seeds numerous, compressed, black or light brown.

Medicinal Uses

Root: Diuretic, emetic and tonic; its decoction is useful in cough, cold and colic.

Leaf: Decoction of the leaf is useful in cough and cold.

Seed: Powdered seeds are given in affections of the nervous system and seed paste is applied in paralysis, sciatica and stiffness of shoulder joints.

Toxicology: Seeds are poisonous when taken in excess.

Uses in Alternative Systems of Medicine

Ayurveda:

Part used: Root, leaf and seed.

Uses: Used in alopecia, diseases of nervous system, hemiplegia and leprosy.

Siddha:

Part used: Root, leaf and seed.

Uses: Useful in arthritis, eye diseases, jaundice, leucoderma and pain.

Non-medicinal Uses

Seed: Crushed seeds are used for poisoning and other homicidal purpose.

3. *Aeschynomene aspera* L.

Synonym: *Aeschynomene trachyloba* Miq.

Common Names: Hat plant, Singapoor sponge wood, Sola pith plant.

Vernacular Names: Ben: Sholar; **Hin:** Sola; **Kan:** Alagina gida, Bendu, Karebendu; **Mal:** Kadessum; **Sans:** Damana; **Tam:** Aathuneddi; **Tel:** Nir jiluga.

Distribution: Native of Cambodia, India, Laos, Sri Lanka, Thailand and Vietnam; in India it grows throughout and in Western Himalaya.

Description: A perennial aquatic shrub growing to a height of 3m and is unbranched; stem glabrous, soft with white and spongy pith; leaves sessile or subsessile; leaflets alternate or opposite, linear, oblong, obtuse, apiculate, glabrous, set obliquely on the rachis; stipules lanceolate, base auriculate, setaceous; inflorescence 2-8 flowered axillary raceme with a hairy peduncle; flowers yellow; bracteate; bracteolate; calyx hispid, bilabiate, upper lip round, the lower lip 3 lobed; corolla with keel petal longer than the standard and wing petal, hairy; pod jointed, indented and echinulate on both the surfaces; seeds subreniform, estrophiolate.

Edibility

Leaf: Leaves are eaten as pot herb.

Medicinal Uses

Leaf: Useful in pain and swellings.

Uses in Alternative Systems of Medicine

Siddha
Part used: Leaf.
Uses: Used as poultice in pain and swellings.

Non-medicinal Uses

Plant: Is a source of fodder, fuel and charcoal.

Stem: Stems are tied together and used as rafts.

Bark: Yields fiber.

Pith: Useful as insulator; used in making toys, ornamental beads, sun hats, artificial flowers, medals, bottle corks, swimming jackets, life belts and paper; it is a substitute for surgical lint; used for embedding plant materials to take free hand sections of plant parts.

4. *Aeschynomene indica* L.

Synonyms: *Aeschynomene indica* L. var., *punctata* Pers.
Aeschynomene indica L. var., *viscosa* Miq.
Aeschynomene pumila L.

Common Names: Hard sola plant, Sensitive vetch.

Vernacular Names: Ass: Kuhila; **Ben**: Bhat shoal, Kath shola; **Hin:** Laugauni, shola, Phulan; **Kan:** Bendukasa; **Mal:** Nellitthaali; **Tam:** Nellithalli; Kottiram; **Tel:** Thigejiluga, Jiluga kaada.

Distribution: Native of Africa, Asia including India, Australasia, Indian Ocean Islands and United States; grows throughout India up to Western Himalaya.

Description: Annual undershrub, with glabrous stem; branches terete; branches and leaf rachis furnished with glandular excretions looking like prickles; leaflets alternate, 30-70 in number, subsessile, overlapping, linear, obtuse, mucronate, glabrous; stipules linear, lanceolate, acuminate with auricled base; inflorescence is a 1-4 flowered axillary raceme, peduncle and pedicel glandular and viscid; bracteate; calyx bilabiate, glabrous, upper lip bifid, lower minutely 3-toothed; corolla yellow, vexillum veined with purple; pod slightly curved, 6-10 jointed, glabrous or slightly hairy, readily separating into 1 seeded parts; seeds 8-10.

Phytochemistry

Plant: Myricitrin, neynoutrin, robinin, saponin and vicenin-2.
Leaf: Flavone and neynoutrin.
Seed: Alkaloids, nujol, oleyl alcohol and sterols.

Medicinal Uses

Plant: Spermicidal.

Uses in Alternative Systems of Medicine

Siddha

Part used: Leaf.

Uses: Useful in leprosy.

Non-medicinal Uses

Plant: Yields pith of inferior quality; used as fodder but toxic in the fruiting stage.

Stem: Yields 'sola' wood which is inferior to wood of *Aeschynomene aspera*; used in making hats, floats for fishing nets, fishing baskets and rafts; wood is also, used as fuel in firing pottery and charcoal, and in making gun powder and fire works.

5. *Alhagi maurorum* Medikus & Boiss.

Synonyms: *Alhagi graecorum* Boiss.
Alhagi camelorum Fischer
Alhagi pseudalhagi (M.Bieb) Fisch

Notes: *Alhagi maurorum* Medikus & Boiss. and *Alhagi graecorum* Boiss., are two distinct species, however in literature there is a serious confusion between them and it is not possible to differentiate information on phytochemistry and medicinal uses for these species.

Common Names: Arabian manna plant, Camel thorn, Persian manna plant.

Vernacular Names: Hin: Hingua, Javasana, Javaasa, Junvasa, Juvasa, Yavaasa; **Kan:** Balli duraba, Balli thurachi, Dhurlava, Jayasi, Thore ingala; **Mal:** Kappattumpa, Kotittva; **Mar:** Javaas, Kantechumbaka, Kas, Yavas; **Sans:** Adikantaka, Anantha, Bahukantaka, Balapathra, Duramoola, Durlabha, Dusparsha, Dheerghamoola, Gandhari, Girikarnika, Kachhura, Kantakaluka, Kantaki, Kshudrengudi, Marudbhava, Rodanika, Samudraantha, Sookshmapathra, Theekshnakantaka, Thriparnika, Thuranjbeen, Vanadarbha, Vasantha, Vishaghna, Vivarnaka, Yaasa, Yavaasa, Yavaasaka; **Tam:** Thurinjibin; **Tel:** Girikarnika, Thella giniya chettu.

Distribution: Native of Afghanistan, China, Iran, Iraq, Pakistan, former USSR and Middle Eastern countries; introduced into India; in India it grows in Bihar, Goa, Gujarat, Haryana, Jammu and Kashmir, Madhya Pradesh, Maharashtra, Punjab, Rajasthan and Uttar Pradesh.

Description: Undershrub, glabrous or pubescent, armed with hard, sharp spines; leaf simple, obovate or elliptic-oblong, glabrous or pubescent, entire, apiculate, petiolate; stipules minute; inflorescence is a lateral axillary raceme ending in a spine; bracteoles 1-2, minute; calyx glabrous, teeth almost obsolete to triangular; corolla pink or reddish violet; pod glabrous, more or less constricted between the seeds; seeds 1-9, blackish, falcate, brown, smooth and shiny.

Phytochemistry

Plant: β-Phenethylamine, 22-dien-3β-Ol, 24(28)-dien-3β-Ol, 24-ethylcholesta-5-en-3β-Ol, 24-methylcholesta-5-en-3β-Ol, 3,4-dihydroxy-β-phenethyl trimethyl ammonium hydroxide, alhalgin, avenasterol, catechin, cholesterol, epigallocatechin, essential oils, gallocatechin, hordenine, leucodelphinidin, N-methyl-β-phenethylamine, N-methylmescalin, quercetin, rutin, salsolidine, sterols and tannin.

Root: 3,4-Dihydroxy-β-phenethyl trimethyl ammonium hydroxide, β-phenethylamine, colouring matter, hordenine, N-methyl tyramine, N-methyl-β-phenethylamine, phytosterols, resin, salsolidine, tannins, tetrahydroxy quinoline and wax.

Stem: β-Phenethylamine, 3,4-dihydroxy-β-phenethyl-trimethyl ammonium hydroxide, betaline, choline, hordenine, hydrozide, N-methyl-β-phenethylamine, N-methyl mescaline, N-methyl tyramine, phytosterols and salsolidine.

Pod: Tannin.

Medicinal Uses

Plant: Alexipharmic, antibilious, antibiotic, anticancer, antiprotozoal, antiarrhythmic, antiseptic, aperient, bitter, cooling, diuretic, expectorant, laxative and spasmolytic; used in abdominal and glandular tumour; twigs are used as poultice or burnt and the fumigation is given for piles; decoction of the twigs is a remedy for cough; the fresh juice is diuretic; it is also applied on the eye for corneal opacity; the plant infusion is diaphoretic; the dried plant is smoked with *Datura* and ajwain seeds for asthma.

Root: Decoction of the root is used to treat abscesses and swellings.

Manna: The sugary exudate from the plant is antiemetic, aperient, blood purifier, cholagogue, diuretic, expectorant, fattening and laxative; given with milk it is aphrodisiac and restorative.

Leaf: Is antibacterial, anti-inflammatory and used in diseases of mouth; used as poultice for rheumatism.

Flower: Dried and burnt flower is used as fumigation in piles.

Toxicology: The leaf and seed contains N-methyl mescaline, a toxic amine, which depresses the central nervous system and is psychomimetic; may cause nausea.

Uses in Alternative Systems of Medicine

Ayurveda:

Part used: Whole plant and root.

Uses: Plant juice is given in constipation, coryza, cough, dyspnoea, excess of thirst, fainting, giddiness, gout, haemorrhage, headache, piles, skin diseases and vomiting; root is useful as astringent and blood purifier; cures asthma, burning sensation in the chest, cold, diarrhoea, fever, skin diseases, vomiting, vitiated "kapha" and "pitta".

Siddha:

Part used: Whole plant.

Uses: Useful in chest pain, cough, pyrexia and as aphrodisiac.

Unani:

Part used: Whole plant, leaf and flower.

Uses: Whole plant is given in bleeding piles, bloody urine, chest pain, cough, corneal opacity, dysuria, fainting, hemicrania, nausea, small pox and vomiting; leaf is used in rheumatic pain; flower used in piles.

Ethnomedicine

Plant: Used in bilious affections, glandular tumour and nose polyps.

Non-medicinal Uses

Plant: Used as fodder; plant extract is used in tanning; twigs are used in making tatties (cooling mats) and screens.

Leaf : Useful as fodder.

6. *Alysicarpus longifolius* (Sprengel) Wight & Arn.

Synonyms: *Hedysarum longifolium* Sprengel.
Alysicarpus bupleurifolius. var., *erecta.* Haines

Vernacular Names: Kan: Naamada soppu; **Mar:** Jangli gailia; **Tam:** Naamappondu; **Tel:** Pedda kandikar aaku.

Distribution: Native of Australia, India and Pakistan; introduced into Africa; in India it grows through out plains.

Description: Stout glabrous herb; leaves linear, subcordate, mucronate, lanceolate, glabrous beneath; leaflets oblong or lanceolate, obtuse or subacute, base subcordate, glabrous above, appressly hairy on the nerves beneath; petiolules very short; flowers with orange standard petal and keel, wing petal crimson; calyx hairy, teeth longer than the calyx tube; stipules lanceolate and acuminate; flowers on spicate raceme, appressed to the rachis; bracts ovate, acuminate, cuspidate; pod moniliform, short stalked, apiculate with 3-4 joints, reticulately veined.

Phytochemistry

Leaf: β-Sitosterol, aliphatic ester, caffeic acid, chrysoeriol-7-0-xyloside, esculetin, ferulic acid, flavones, isoquercetrin, kaempferol-3-0-xylosyl-7-0-rhamnoside, ρ-coumaric acid, pinitol, quercetin-7-0-rhamnoside, rutin and soya saponins.

Seed: Arginine, lysine, phytoalexins, propionic acid and stilbene.

Pericarp: β-Sitosterol, apigenin, behenic acid, chrysoeriol, dancosterin, eriodictyol, lignoceric acid, luteolin, myristic acid, palmitic acid, pratensein, soyasaponin, stearic acid and vallic acid.

Medicinal Uses

Root: Is used as a substitute for liquorice and to stop uterine bleeding.

Non-medicinal Uses

Plant: Is useful as fodder.

7. *Alysicarpus vaginalis* (L.) DC.

Synonyms: *Alysicarpus nummularifolius* (L.) DC.
Alysicarpus nummularifolius (L.) DC. var., *angustatus* Ohwi
Alysicarpus vaginalis (L.) DC. var., *diversifolius* Chun.
Alysicarpus vaginalis (L.) DC. var., *nummularifolius* Miq.
Hedysarum cylindricum Poiret
Hedysarum vaginale L.

Common Names: Alyce clover, Buffalo clover, One leaf clover.

Vernacular Names: Guj: Gaadar samerwo; **Kan:** Namadasoppu; **Mar:** Dhaka dhampta; **Tam:** Kuranthiadakki; **Tel:** Namapandu.

Distribution: Native of Africa, Asia including India, Madagascar, Mauritius, Reunion and Rodrigues; it grows throughout India.

Description: Erect or spreading procumbent perennial herb; stem densely or sparsely pubescent; leaves unifoliolate; leaflets ovate, oblong to lanceolate, mucronate at the apex, cordate at the base, lower leaflets lanceolate, upper oblong or the lower oblong and upper ovate; stipules lanceolate, ovate, acute, parallel veined; inflorescence terminal and opposite a leaf, densely pubescent; bracts ovate, lanceolate; calyx teeth linear, longer than the tube; flowers 8-12, pink or yellow tinged with pink on a raceme; pod subterete, with 4-8 joints, reticulately veined, slightly thickened at each end, finely pubescent; seeds many.

Medicinal Uses

Plant: Antipyretic and expectorant; used in bone setting, cough, pain in joints and uterine bleeding.

Root: Root decoction is used for cough, irregular menstruation, leprosy, urinary trouble and uterine bleeding.

Leaf: Leaf infusion is anticancer and purgative.

Non-medicinal Uses

Plant: Is used as forage and pasture plant; is an excellent soil improver, green manure and cover crop in rubber plantations; though used as fodder, in excess it causes diarrhoea in animals.

Seed: Is a source of industrial gum.

8. *Andira inermis* (Wright) DC.

Synonyms: *Andira grandiflora* Guillemin & Perrott
Andira jamaicensis (Wright) Urban.
Geoffoea inermis (Wright) Wright.
Geoffoea jamaicensis Wright var., *inermis* Wright.

Notes: Three subspecies – *inermis, grandiflora* and *rooseveltii* are recognized; the tree yields araroba powder which is contained in the vessels of the wood; is yellow in colour and bitter in taste and turns brown or purple on exposure to light; has medicinal properties.

Common names: Angelin tree, Bastard mahagony, Cabbage bark tree, Dog almond, Jumbie bead, West Indian walnut, Worm back.

Distribution: Native of Africa, Caribbean Islands, Central America, North and South America; introduced into Asia, Mauritius, Fiji and India; in India it is cultivated in the Botanical gardens.

Description: A tall tree with branches towards the top in the form of a head; it has smooth grey fibrous bark; leaves imparipinnate, ovate, lanceolate; leaflets glabrous; flower purple.

Phytochemistry

Plant: Berberine, biochanin A, inermin and N-methyl tyrosine.

Bark: Andrine, berberine, inermin, N-methyl tyrosine and tyrosine.

Heart wood: β-Sitosterol, biochanin A, fatty acids, inermin and isoflavonoids

Araroba powder: Chrysarobin, chrysophenol anthrone, chrysophenoldianthrone, emodin, physciondianthrone and physicion.

Leaf: β-Amino butyric acid and β-sitosterol.

Medicinal Uses

Plant: Is a powerful anthelmintic.

Root: Anthelmintic; used in treating mental diseases.

Wood and Seed: Powerful anthelmintic, narcotic and purgative.

Araroba powder: Mixed with lime juice, is a remedy for ringworm and other skin diseases.

Seed: Anthelmintic, febrifuge and purgative; large doses cause vomiting and delirium.

Toxicology: Anthrones such as chrysoarabin in the seeds is allergen causing respiratory tract irritation and narcosis. Bark is used as arrow poison in Northern Nigeria; its infusion called as 'red water', is highly toxic causing diarrhoea, delirium, vomiting but when used with caution it is anthelmintic, febrifuge and purgative.

Non-medicinal Uses

Plant: Ornamental, wind break and shade plant in coffee plantation; the bark is used for Trial by Order as it is highly poisonous in large doses.

Root bark: Insecticidal.

Wood: Is hard, heavy and elastic, used in making furniture and boats.

9. *Arachis hypogaea* L.

Notes: A major crop in 80 tropical countries and it ranks 13th among the food crops; probable centers of origin and diversity are Northwest Argentina, Southern Brazil and Bolivia; there are over 13,000 accessions in the groundnut germplasm repository at ICRISAT; two subspecies are recognized; *hypogaea* contains two botanical varieties, *hypogaea* and *hirsuta,* and ssp., *fastigata* contains four varieties *fastigata, vulgaris, peruviana, aequatoriana.*

Common Names: Chinese almond, Earthnut, Groundnut, Monkeynut, Peanut.

Vernacular Names: Ben: Chinerbadam; **Hin:** Bhuising, Mungphali; **Kan:** Kadale kayi, Nelagadale; **Mal:** Nelakadala; **Mar:** Bhuichane, Bhuimuga; **Sans:** Bhoochanaka, Bhoomija, Bhooshimbika, Rakthabeeja, Snehabeejaka; **Tam:** Verkadalai, Manila kottai; **Tel:** Verasenagalu, Nelasenagalu.

Distribution: Native of Argentina, Bolivia and Brazil; cultivated throughout India.

Description: A branched hairy annual herb with erect branches; branches glabrous or hirsutous; the first two nodes (sometimes three) of every vegetative branch is subtended by cataphylls; leaves stipulate, pinnate, with two opposite pairs of leaflets, alternately arranged in 2/5 phyllotaxy on the axis; flowers yellow, enclosed between two bracts, one simple subtending the peduncle and the other bifid, subtending the pedicel; calyx lobes are two, in number – one awl like and opposite the keel and the other broad, four notched, opposite the standard petal; standard petal yellow with red veins; wing petal yellow at the base, brick red apically; keel petal pale yellow and clasps the staminal column; stamens 10, sterile anthers 2 in number; the tip of the ovary bearing the ovules grows out between the floral bracts bearing the dried petals, calyx, hypanthium and fruit creating a unique structure; the peg turns towards the soil and thrusts its tip into the soil and the tip turns horizontally and develops into a thick, moniliform pod; seed pink, red, purple, brown, or pink marked with purple streaks; 2-15 seeds per pod.

Phytochemistry

Plant: 11-Eicosenoic acid, 13-eicosenoic acid, 13-hydroperoxy-cis-9,trans-11-octadecadienoic acid, 15-docosenoic acid, 24-methylene-cholesterol, 5,7-dihydroxychromone, 7-hexadecenoic acid, 9,10-dihydroxystearic acid, α-amino-β-hydroxy butyric acid, α-resorcyclic acid, arachin-c, calcium, caprylic acid, carbohydrate, cerebronic acid, cholesterol, chrysoeriol, cis-10-octadecenoic acid, cis-11-octadecenoic acid, D,L-amino butyric acid, daucosterin, dihydroxystearic acid, docosanoic acid, esculetin, fat, γ-amino-α-methylene butyric acid, γ-aminobutyric acid, glyceric acid, heptadecanoic acid, hexacosaenoic acid, hexadecenoic acid, hydroxybehenic acid, isooleic acid, isoquercetrin, lauric acid, luteolin, magnesium, malvalic acid, nonahexadecanoic acid, nicotinic acid, O-coumaric acid, octadecadienoic acid, phloretic acid, phosphorus, potassium, pratensin, protein, quercetin, rutin, sterculic acid, stigmasterol, tetradecenoic acid, trans-10-octadecenoic acid and trihydroxystearic acid.
Root: Pterocarpinoids and soyasaponins
Leaf: Caffeic acid, esculetin, ferulic acid and p-coumaric acid.
Pod: Arginine, 2,3-diaminopropionic acid and lysine.
Pod shell: β-Sitosterol, daucosterol, dihydroxychromone, luteolin, quercetin and saponins.
Seed: Alanine, arachin, α-cephalin, α-ketoglutaric acid, α-linolenic acid, α-tocopherol, aluminium, arachidic acid, arachidoside, arachin, arginine, ascorbic acid, aspartic acid, behenic acid, β-carotene, β-sitosterol, biotin, boron, cadmium, caffeic acid, calcium, carbohydrates, cellulose, cis-p-coumaric acid, trans-p-coumaric acid, chlorogenic acid, choline, chromium, cobalt, conarachin, conglutin, copper, cystine, tocopherol, fat, ferulic acid, flavones, fibre, foliacin, fumarin, fumarase, gadoleic acid, γ-methylene-α-ketoglutaric acid, gentisic acid, glutamic acid, glutathione, glycine, histidine, p-hydroxy benzoic acid, iron, isochlorogenic acid, isoferulic acid, isoleucine, lead, lecithin, leucine, leucocyanidin, lignoceric acid, linoleic acid, lysine, magnesium, manganese, mercury, methionine, methylene proline, myristic acid, niacin, nitrogen, o-coumaric acid, oleic acid, p-coumaric acid, p-hydroxy benzoic acid, palmitic acid, palmitolic acid, pantothenic acid, phenylalanine, phloretic acid, protocatechuic acid, phosphoric acid, potassium, proline, protein, resorcyclic acid, riboflavin, rubidium, salicylic acid, sarkosin, selenium, saponin, serine, sinapic acid, sodium, stilbene, sphingoglycolipid, stearic acid, stizolamine, sulphur, syringic acid, thiamine, tannin, threonine, tryptophan, tyrosine, valine, vanillic acid and vitamin-B_6.
Pericarp: β-Sitosterol, apigenin, arachidoside, behenic acid, chrysoeriol, dancosterin, dihydroxychromone, eriodictyol, lignoceric acid, luteolin, myistic acid, palmitic acid, potassium, pratensein, soyasaponin I, stearic acid and vallic acid.
Seed cake: Aflatoxins may be found in cakes formed under faulty storage, which is highly toxic.
Seed skin: Chromogen of anthocyanin, estrogenic factor, pyrocatecholic acid and tannin.

Edibility

Pod: Young pods are edible.

Seed: Consumed raw, boiled or roasted; used in the preparation of chocolates, peanut butter, oil and peanut meal.

Seed oil: Is edible, used in cooking, frying and as a substitute for olive oil and hydrogenated fats.

Groundnut lecithin: Used in confectionary, bakery products, salad dressing and to give extra smoothness to ice cream.

Medicinal Uses

Plant: Is analgesic, antiandrogenic, anthelmintic, antianginal, antiapoplectic, antiarrhythmic, antibacterial, anticardiospermic, antiestrogenic, antifibrinolytic, antigastric, antihepatotoxic, antihyperthyroid, antihypoglycaemic, anti-inflammatory, anti-ischaemic, antikeratitic, antimastalgic, antimitotic, antinociceptive, antioxidant, antipellargic, antiperiodontal, antiperoxidant, antiprostaglandin, antipsychotic, antipyretic, antiradicular, antiscorbutic, antiseptic, antispasmodic, antitussive, antiviral, aphrodisiac, apoptotic, astringent, bacterioside, CNS-active, cardiodepressant, cardioprotective, catabolic, cerebrotonic, chemopreventive, cholinergic, cytotoxic, DNA protective, diaphoretic, diuretic, emetic, estrogenic, fungicide, haemolytic, haemostat, hepatoprotective, hepatotonic, hypercholesterolemic, hypnotic, hypolipidemic, hypothermic, immunostimulant, insulinogenic, laxative, memory enhancer, metastatic, nephrotoxic, neuroexcitant, phagocytotic, radioprotective, sedative, spermicide, teratogenic, tranquilizer, uterorelaxant, vasodilator and vulnerary; is used in treating acne, asthma, adenoma, alcoholism, allergy, anaemia, anorexia, anxiety, arthritis, atherosclerosis, backache, beri-beri, cancer of breast, colon, liver and skin, canker, capillary fragility, cardiospasms, cataract, cephalgia, chilblain, cold, colic, convulsions, cystitis, dandruff, delirium, dementia, depression, dermatitis, diabetes, dysentery, diarrhoea, eczema, edema, epilepsy, fibroids, fibrosarcoma, gastritis, gingivitis, glaucoma, glycosuria, heart burn, haematuria, haemorrhages, haemorrhoids, hepatitis, herpes, hydrophobia, hyperactivity, hypertension, insomnia, leprosy, lethargy, leukaemia, lupus, lymphoma, malaria, melanoma, myalgia, menorrhagia, migraine, nephritis, neuralgia, neuropathy, obesity, orchitis, osteoarthritis, osteoporosis, pharyngitis, photophobia, phthisis, pneumonia, polio, purpura, pleurisy, psoriasis, retinopathy, rheumatism, spasms, sprue, stomatitis, stress, sunburn, thrombosis, tumour of bladder, breast, stomach, cervix, colon, liver, lung, mouth, ovary, skin and brain, tymphinitis, ulcers, varicose veins, vitiligo and vertigo; useful in bradycardia, cancer, cold, dermatitis, fistula, haematuria, scabies, tremors, urinary complaints and uterine disorders; as antidote to aluminium, lead and mercury poisoning.

Pod: Astringent and emollient to bowels.

Shell: Used in folk medicine for hypertension.

Seed: Antibilious, aperient, aphrodisiac, emollient, galactagogue, haemostatic and tonic; useful in agalactia, diarrhoea and general debility; unripe nuts are used as lactagogue.

Seed oil: Aperient, astringent and emollient; used in nephropathy and as a massage oil for joint pains; used in catarrh of the bladder; as an ingredient of ointment and to promote blood clotting in external wounds; in coronary atherosclerosis and as anticholesterolemic; a drug for haemophilia is prepared from the oil.

Bioactivity: Plant is a inhibitor of β-glucuronidase, COMT, cyclo-oxygenase, cytochrome-P_{450}, deiodinase, glucosyl-transferase, histamine, insulinase, leukotriene, lipoxygenase, lyase, MAO, metallo-proteinase, NADH-oxidase, NEP, NO-synthase, phospholipase, polyamines, prostaglandin synthesis, protein-kinase, topoisomerase I and II, tyrosinase, tyrosine-kinase, xanthine-oxidase, CAMP-phosphodiesterase; it is CNS active, calcium antagonist, calcium channel blocker, interferon synergist, mast cell stabilizer, metal chelator, monoamine precursor, pituitary stimulant, quinone reductase inducer and neurotransmitter.

The goitrogenic principle was found to be insoluble in oil and thermostable in the roasted seeds and oil cake; when normal diet of rat is supplemented with 20% of oil cake, the weight of the thyroid gland increased from 9 to 24 mg; but addition of 1 ml of water per day re-established the thyroid metabolism; the estrogenic factor in refined peanut oil when introduced in the food of immature mice increased uterine weight from 9.5 to 15.9 mg; peanut oil also showed marked aldosterone antagonistic action compared to other vegetable oils; the antagonist is contained in the unsaponifiable portion of the oil.

Toxicology: Seed contains chymotrypsin and trypsin inhibitors.

Clinical Studies: The presence of haemostatic factor in groundnuts was first reported in 1960. Daily intake of 180g of peanut flour or its alcoholic extract produces the effect after 48 hours; the action is attributed to a chemical similar to 5-hydroxytryptamine which produces vasoconstriction of blood vessels and stimulates smooth muscles; certain forms of haemophilia such as thrombopenic haemorrhage, chronic myelosis, vascular purpura and metrorrhagia are benefitted by this treatment; the groundnut skin also contains lipoxidase and protease inhibitors.

Uses in Alternative Systems of Medicine

Ayurveda:

Part used: Pod and seed oil.

Uses: Used as astringent to the bowels.

Chinese:

Part used: Seed.

Uses: Used as demulcent, nutritive, peptic and tonic; used in chronic cough and pectoral diseases.

Siddha:

Part used: Seed and oil.

Uses: Useful as abortifacient and in treating cystitis.

Non-medicinal Uses

Plant: Highly nutritious as cattle feed; increases the amount of milk in cattle; used as green manure; used as fungicide.

Seed: Roasted seeds are used as a substitute for coffee; seeds are a rich source of protein and oil and seeds are consumed either raw or roasted or boiled; used in various foods and confectionary; they are ground and made into peanut butter.

Seed coat: Husks are pulverized and mixed with groundnut meal and molasses to increase the bulk of cattle feed.

Seed cake: Is used as cattle feed and manure; seed cake is used in preparing glue for bonding plywood.

Seed shell: Used as fuel and in the manufacture of coarse boards and corks, insulation blocks and bedding the stables; used as fertilizer and live stock feed.

Seed lecithin: Useful in printing inks and for treating leather; used as stabilizer for emulsions and suspensions.

Seed oil: Used in the manufacture of soaps, creams, paints and dyes; as burning oil for lamps, for lubricating watches and delicate machinery; used as an emulsion for insect control and as massage oil; it is one of the major cooking oils in the world; used as a supplement in the culture medium for *Streptomyces;* peanut oil extracted from the seeds is used predominantly for cooking; also used in hydrogenated fats (*Vanaspathi*).

10. *Astracantha gummifera* (Laibill.) Podl.

Synonyms: *Astragalus gummifer* Laibill.
Astragalus rayatensis Eig.

Notes: Tragacanth gum, from the plant, was first used by man as survival food; it is one of the oldest natural emulsifier known to man; the plant serves as 'indicator species' in geobotanical prospecting.

Common Name: Gum tragacanth.

Distribution: Native of Asia Minor, Syria, Iraq, Lebanon, Former USSR and Turkey; introduced into India.

Description: Umbraciform shrub, thorny and branching; leaf rachis spiny, sparsely pubescent or glabrous; leaflets elliptic, mucronate, glabrescent, 4-7 paired; stipules triangular and ovate; flowers 2-3 per leaf axil; bracts navicular, bilobed, glabrous; calyx lobes divided almost to the base; pod bilocular due to intrusive membrane; seeds reniform.

Phytochemistry

Plant: Yields a gum, which has tragacanthin and bassorin and is called as tragacanth gum; also contains canavanine, cellulose, glycyrrhizin, malonic acid, saponin, S-methylcysteine, trigonnelline and starch.

Gum: Acetic acid, aldibiuronic acid, arabino galactan, bassorin, cellulose, D-galactose, D-galacturonic acid, D-xylose, L-arabinose, L-fucose, L-rhamnose, starch, tragacanth, tragacanthic acid, tragacanthin and xenognosin.

Edibility

Tragacanth gum: Used in salad dressing, sauces, condiments, desert topping, pie filling; as stabilizer in ice cream, cheese, milk powder, sherbets and citrus oil emulsions.

Medicinal Uses

Plant: Is antibacterial, emollient, expectorant, fungicide, haemostat, immunostimulant, mitogenic and spermicide; useful in treating cough and ulcer; used as an antidote to iodine poising.

Gum: Astringent, antimicrobial and cooling.

Toxicology

All the parts of the plant contain seleno compounds, which cause selenium poisoning; seeds contain trypsin inhibitor.

Ethnomedicine

Plant: Is used as emollient.

Non-medicinal Uses

Plant: Used as fodder.

Tragacanth gum: Tragacanth gum is one of the most valuable of all gums; is used in cosmetic and pharmaceutical industry as emulsifier for steroid preparations, cardiac glycosides, barbiturates and fat soluble vitamins; as binding agent in pills, tablets, lozenges, tooth paste and diabetic syrups; as emulsifier in cod liver oil, mineral oil, paraffin oil, hand lotions, leather polish and insect repellents; in ceramics and calico printing; as stiffner in textiles; used in making of tragacanth films which are insoluble in water; in lubricating jellies; used to give body to liquors and used in plastics, matches, perfumes; used as an adhesive powder for fixing artificial teeth and in tooth paste; it is used for polishing sewing threads and twine threads; with gum arabic and terpineol it is used as plasticizer; glycerate of gum tragacanth is used for lubricating surgical instruments; used in confectionary.

11. *Astragalus hamosus* L.

Vernacular Names: Ass: Asabeaulmalik; **Hin:** Purtuk; **Mar:** Akhlilelmalik; **Pun:** Akhilulmalik.

Distribution: Native of Africa, Asia, Europe and Middle East; introduced into Australia and India; in India it grows in the plains of Punjab.

Description: Annual with erect, ascending or prostrate stem; leaf imparipinnately compound, petiole very short; petiole and rachis appressed and hairy; leaflets oblong to oblong-obovate, retuse, glabrous above, pubescent below; stipules connate and adnate to petiole, ciliate on the margins; inflorescence is a pedunculate dense raceme; bracteate; calyx pubescent; corolla white or pale yellow; pod sessile, linear, falcate to subfalcate, glabrescent or minutely appressed-pubescent, bilocular, many seeded.

Medicinal Uses

Plant: Antiperiodic, aphrodisiac, demulcent, diuretic, emmenogogue, emollient, galactagogue, laxative, maturant, pectoral and stomachic; useful in the catarrh, irritation of mucous membranes, nervous affections, thirst, wounds and ulcers; heals inflammation of the eye, leucoderma, reduces pain and enlargement of liver and spleen; made into paste with vinegar, it is applied externally for headache.

Gum: Demulcent and emollient; useful in the irritation of mucous membrane.

Leaf: Juice of the leaves cures ear ache.

Pod: Tonic; useful in bronchitis, inflammation, headache, intestinal trouble and paralysis.

Uses in Alternative Systems of Medicine

Ayurveda:

Part used: Plant, leaf, pod, oil.

Uses: Plant is used as antiperiodic, aphrodisiac, galactagogue, emmenogogue, maturant, pectoral, diuretic and stomachic; used in enlargement of liver, heals wounds and ulcers, inflammation of the eye and leucoderma; leaf juice is dropped into the ear for ear ache; pod is used for bronchitis, inflammation, intestinal troubles, headache, paralysis and oil is used in paralysis and cold in the head.

Unani:

Part used: Plant, leaf and pod.

Uses: Plant is used as anti-inflammatory, antipyretic, aphrodisiac, galactagogue, tonic and wound healing; leaf juice is good for ear ache; pod is used in bronchitis, headache, intestinal troubles and paralysis.

Ethnomedicine

Plant: Is astringent, carminative and emetic; used in tumour.

Non-medicinal Uses

Gum: Used for glazing in dyeing and calico printing.

12. *Astragalus multiceps* Benth.

Synonyms: *Astragalus jacquemonti* Bunge
Astragalus leptocentrus Bunge

Vernacular Names: Mar: Peruvidulkol; **Pun:** Kandiara, Sarmul; **Sans:** Mashaparni.

Distribution: India (Grows in temperate West Himalaya), Pakistan and Nepal.

Description: Densely congested, armed with very crowded ascending old leaf-rachises; stipules, lanceolate, adnate only at the very base; leaflets 21-31, mostly crowded, obovate-oblong, long, thick in texture, dull green, clothed with grey silky hairs; flowers 1-2, together in leaf-axils usually not peduncled; bracts linear, exceeding the pedicels; calyx, thinly silky, teeth linear-setaceous, half as long as the tube; corolla, twice as long as the calyx; standard exceeding wings and keel; pod sessile, oblong, turgid, bilocular, 12-14 seeded, clothed with fine grey silky hairs.

Medicinal Uses

Root: Is useful in biliousness, consumption, fever, rheumatism and swelling.

Seed: Demulcent and emollient; given in colic and leprosy.

Toxicology: Nitro compounds like karakin and miserotoxin in the plant are bound toxins causing methaemoglobinaemia; the seleno compounds cause selenium poisoning.

Uses in Alternative Systems of Medicine:

Unani

Part used: Plant.

Uses: Used as demulcent and emollient.

13. *Baphia nitida* Lodd.

Synonyms: *Baphia angolensis sensu* Lest-Garl
Baphia hematoxylon (Schum & Thonn.) Hook.

Common Names: African red wood, Barwood, Cam wood.

Distribution: Native of Africa; introduced into India; cultivated in the Botanical Gardens in India.

Description: Large erect shrub with terete glabrous branches; leaves simple, ovate or oblong, subcoriaceous, glabrous, round at the base; flowers 1-4, white, on glabrous pedicels; bracts round, united at the base; calyx membranous, glabrous, straight and coriaceous, pod narrow upto 12cm; seeds 4-10 per pod.

Phytochemistry

Plant: Contains dimethoxy trihydroxy isoflavone, gum and santarubin.

Gum: Pyrocatechin.

Edibility

Twigs: Used as chew sticks.

Medicinal Uses

Root: Pounded and mixed with palm oil, is applied externally for foot diseases.

Bark: Decoction is given as enema for constipation.

Wood: Wood paste with butter is applied to sprains and swollen joints.

Leaf: Hot leaf poultice is applied on psoriatic skin and sores; leaf infusion is given as an enema for constipation.

Ethnomedicine:

Plant: Is carminative and laxative; used for arthritis, jaundice, dysentery, enteritis, gastritis, jaundice, rheumatism, skin diseases, sprain, stiffness, swelling, wounds and venereal diseases.

Non-medicinal Uses

Plant: Planted as hedge or fence plant; used as forage.

Wood: Dark red heartwood is the source of dye, which is used for dyeing and in cosmetics; wood is durable and used for construction of rafters, posts, mortars, axe handles, agricultural implements and walking sticks.

14. *Butea monosperma* (Lam.) Taubert

Synonym: *Butea frondosa* Roxb.

Notes: There are several colour variants not worthy of taxonomic recognition; the tree has long been known to the Hindus under the Sanskrit name *Palasha*, meaning possessing valuable medicinal properties; the red flowers are sacred and are offered in temples at bloody sacrifices to Goddess Kali; the leaf is supposed to represent the Hindu trinity; the name *monosperma* refers to a single seed in the pod.

Common Names: Flame of the forest, Bastard tree, Bengal Kino.

Vernacular Names: Ben: Palas; **Guj:** Khakro; **Hin:** Dhaki, Palas; **Kan:** Muttugada Mara; **Mal:** Palas, Palasinsamatha, Brahmavriksham; **Mar:** Palas, Kakracha-jhada, Phalas-cha-jhada; **Ori:** Kimjuko; **Sans:** Palasha, **Tam:** Kattumurukku, Kinjugam, Kirumisatturu; **Tel:** Moduga, Tella modugu, Palasamu, Paladulu.

Distribution: Native of Asia including India and Pakistan; grows in dry deciduous forest of Central, South and Western India.

Description: Moderate sized deciduous tree of 40-50 feet height with cracked trunk and irregular branches; a red colored gum similar to kino exudes from plant; young shoots densely pubescent and swollen at the base; stipules small, linear-lanceolate, deciduous; leaves pinnately trifoliolate; leaflets unequal, terminal is the largest, rhomboidal and orbicular, lateral ones, ovate, dialated in the lower half, obtuse, glabrous above when mature, finely tomentose, with raised reticulation beneath; flowers red, large silky pubescent outside, 2-3 together from swollen nodes of rigid, stout racemes on velvety drooping pedicels; bracts small, deciduous; calyx finely velvety out side with silvery hairs; calyx segments acute, deltoid; corolla with boat shaped acute keel petal; bracts and bracteoles small; pod velvety, transversely veined, finely pubescent on pendulous woody stalk, obtuse, thickened at sutures; seed flat, broadly oval, smooth, reddish brown; only one seed per pod at the apex.

Phytochemistry

Plant: Aleuritic acid, butrin, cantharic acid, coumaranone, gum, isobutrin, isomonospermoside, L-β-phenylalanine, monospermosides, palasonin, palastrin, proanthocyanidins, sucrose, α-amyrin and β-sitosterol.

Root: Glucose, glycine, lactone, monospermin, N-henecosanoic acid, (–) palasonin, and proteolytic enzymes.

Stem bark: 5-Methoxygenistein, amino acids, butrin, isobutrin, leucocyanidin, lupenone, lupeol, medicarpin, riboflavin, sitosterol, sugars and thiamine.

Gum: Jalaric esters I and II, laccijalaric esters I, II, III and IV, leucocyanidin, leucocyanidrin, mucilage, gallic acid and tannin.

Flower: Aurones, butrin, chalcone, coreopsin, isocoreopsin, isobutrin, isospermoside, jalaric esters I and II, laccijalaric esters I and II, monospermoside and sulphurein.

Seed: Butrin, coreopsin, isobutrin, isocoreopsin, isomonospermoside, methyl allophonic acid, lactone, monospermine, monospermoside, n-heneicosanoic acid, palasonin, proteolytic and lipolytic enzymes, sucrose, sulfurein, α-amyrin, β-D-glucopyranoside, α-sitosterol and its glucoside.

Seed oil: Butrin, coreopsin, isobutrin, isocoreopsin, lignoceric acid, linoleic acid, oleic acid, palasonin, palmitic acid, stearic acid, sulfurein, α-amyrin and β-sitosterol.

Seed coat: Allophanic acid, methyl allophonic acid and monospermin.

Edibility

Root: Is baked and eaten by tribals.

Flower and Pod: Edible.

Medicinal Uses

Plant: Anthelmintic, aphrodisiac, astringent, depurative, febrifuge and rubefacient; useful in diarrhoea, dysentery, eye complaints, haemorrhages, herpes, menstrual disorders, piles and tumours; the plant juice is given as anthelmintic; ash of the tender branches cures colic and flatulence; juice of the plant cooked with ghee and mixed with honey is a cure for haemorrhage; burnt plant is used for cautering; taken internally cures calculi, internal tumours, jaundice, leprosy, obesity, persistent dysentery, piles, seminal weakness and ulcers; when used as poultice, promotes adhesion of fractured bones.

Root: Cardiotonic, hypotensive, estrogenic and rejuvenating; useful in dog bite, elephantiasis, eye defects and night blindness; the Santal tribe in Bihar use it for tuberculosis; juice of the palasha root mixed with yellow mustard is given for filariasis.

Root bark: Analgesic, anthelmintic and aphrodisiac; applied externally on piles, sprue, ulcers and tumours; is used in treating high blood pressure and dropsy; root bark with leaves is used in medicinal ghee in treating colic, constipation, dysuria and skin diseases.

Stem bark: Acrid, antibacterial, anthelmintic, astringent, alterative, aphrodisiac, appetizer, bitter, digestive, diuretic, emollient, haemostatic, thermogenic and tonic; useful in anorexia, biliousness, bleeding piles, bone fracture, diabetes, diarrhoea, dysentery, dysmenorrhoea, dyspepsia, gonorrhoea, hepatopathy, hydrocele, inflammations, rectal diseases, tumours and ulcers; decoction of the bark is given in catarrh, cold, cough, haemorrhages, haemorrhoids, bloody diarrhoea and menstrual disorders; bark pieces chewed with sugar relieves abnormal thirst; with ginger it is given as an antidote for snake bite.

Gum: Aphrodisiac, astringent, constipating, depurative, diuretic, febrifuge and haemostatic; useful in back ache, debility, diabetes, dyspepsia, fever, haemataemesis, haemorrhagic affections of the stomach and bladder, hyperacidity, leprosy, leucorrhoea, piles and pharyngodynia; fresh gum is applied on bruises, erysipelas, ringworm, septic sore throat, skin diseases and ulcers; gum mixed with rock salt is used as a remedy for corneal opacity and pterygium; wet gum is applied over abdomen in colic.

Leaf: Anodyne, anti-inflammatory, aphrodisiac, astringent, depurative, diuretic, vermifuge and tonic; useful in arthralgia, arthritis, boils, colic, delirium, diarrhoea, diabetes, dysmenorrhoea, flatulance, heart burn, haemorrhoids, pimples, skin diseases, swellings and tumours; juice of leaves with curds and turmeric is beneficial in heat eruptions in children and to reduces blood sugar; infusion or decoction is useful as rectal enema in diarrhoea and dysentery due to bacterial infections; it is used as vaginal douche in leucorrhoea; as a mouth wash or gargle for septic and congested throat; fresh juice taken internally for phthisis, fever and haemorrhagic affections; hot poultice of leaves is useful in dispersing boils, pimples, tumourous haemorrhoids, ulcers, buboes and swelling; in case of retention of urine the region of the bladder is fomented with hot leaves; it is used as poultice to reduce swellings and to promote diuresis and menstrual flow; tender leaves with lemon juice and sour gravel is ground into a paste is applied to the body for burning sensation.

Flower: Antifungal, aphrodisiac, astringent, cooling, constipating, depurative, diuretic, febrifuge, emmenogogue, haemostatic and tonic; useful in arthritis, boils, bone fractures, burning sensation, diarrhoea, fever, gout, haemorrhoids, hyperdypsia, haemoptysis, leprosy, leucorrhoea, menorrhagia, pimples, skin disease, strangury, stomach ache and swellings; flower juice induces sterility in women; decoction of flower is given to reduce colic, diarrhoea, flatulance, purpura, worms and piles; it acts as vermicide against tape worm and ringworm; a hot poultice of boiled flowers is used to reduce bruises, sprain and orchitis and taken internally promotes diuresis and menstrual flow; the lotion prepared from flower is used as eye wash for various eye diseases; flowers taken with *Hygrophila auriculata* leaves and root with milk cures leucorrhoea; flowers with honey is rubbed on the eye for conjunctivitis.

Pod: Anthelmintic and aperient; used in "kapha", "vata", piles, scorpian sting, skin diseases, tumours and urinary discharges.

Seed: Abortifacient, anthelmintic, antibacterial, aperient, aphrodisiac, depurative, ophthalmic, purgative, rubefacient, sedative, stimulant and tonic; useful in arthritis, constipation, diabetes, epilepsy, flatulence, herpes, ophthalmia, piles, ringworm, snake bite, skin diseases, sore throat and tumours; seed paste applied on skin acts as a rubefacient to cure "dhobi's itch" and ringworm; seed powder is sprinkled on wounds and ulcers to kill maggots; decoction of the seeds with warm milk is given for diarrhoea; decoction of seeds with butter milk or rice water is a cure for worms; seed decoction is taken for gravel complaints and used as contraceptive; seed paste with honey and ghee is inserted into the vagina as contraceptive and when taken internally acts as anthelmintic.

Seed oil: Anthelmintic, astringent and hypotensive; applied on boils, pimples, wounds and on the part bitten by snake; taken internally it acts as vermifuge.

Bioactivity: Alcoholic extract of seed when fed to rats at 300mg/kg body weight prevented pregnancy, but at higher dose it was toxic; hot alcoholic extract of seeds produced anti-implantation and anti-ovulatory action in animals; alcoholic extract of whole plant produced persistent vasodepression in cats; alcoholic extract of flowers is antiestrogenic in mice; aqueous extract of flowers showed anti-implantation activity in rats; alcoholic extract of bark and seeds was inhibitory to *Escherichia coli* and *Micrococcus pyogenes;* palasonin isolated and characterized as hexahydro-3-α-Me-4,7-epoxyisobenzofuran-1, 3-dione is more effective than piperazine against round worms in humans and shows stimulatory action on smooth muscles.

Uses in Alternative Systems of Medicine

Ayurveda:

Part used: Root, bark, gum, leaf, flower.

Uses: Root cures elephantiasis, night blindness and sight defects; bark is anthelminitc, appetizer, aphrodisiac and laxative; used in, anal diseases, bleeding piles, bone fracture, dysentery¸dysmenorrhoea, hydrocele, haemorrhages, tumour and ulcers; gum is astringent to bowels and is used in treating diarrhoea, dysentery and haemorrhages; given to cure corneal opacity, cough, stomatitis and excessive persipiration; used as an external application to boils, inflammation and ulcers; leaf is used in curing eye diseases; flower astringent to bowels; cures "kapha", burning sensation, gout, leprosy, skin diseases and thirst; flower juice is used as eye wash.

Siddha:

Part used: Bark, gum, leaf, flower and seed.

Uses: Used as anthelmintic, astringent and contraceptive; used in abscesses, carcinoma, cough¸ dropsy, diarrhoea, dysentery, dysuria, eczema, skin diseases and worm infestation.

Unani:

Part used: Bark, gum, leaf, flower, pod and seed.

Uses: Anthelmintic, anti-inflammatory, aphrodisiac, appetizer, astringent, carminative, diuretic, emmenogogue, expectorant, resolvent and tonic; useful in fractures and piles; bark is useful in biliousness, boils¸ dysmenorrhoea, gonorrhoea, hydrocele, liver disorders, lumbago and piles; gum is useful in diseases of lung and chest, leucorrhoea and syphilis; leaf is useful in lumbago; flower is useful in purifying blood and phlegm, reduces inflammation of the bladder and uterus; used in burning urine and gonorrhoea; pod useful in diseases of bladder, eye and piles; seed used in diarrhoea, intestinal worms, piles and used externally for orchitis.

Tibetan:

Part used: Bark,gum, leaf.

Uses: Given in fever, intestinal worms and to reduce phlegm.

Medicinal Uses Outside India

Indochina: Gum is used in diarrhoea, for dressing boils, sores and ulcers; seed infusion is vermifuge; bark boiled with ginger is given as antidote for snake bite and scorpion sting.

Indonesia: Leaf poultice is useful in scorpion sting.

Myanmar: Gum and leaves used as astringent.

Nepal: Gum is used as astringent and given in diarrhoea and dysentery; flower is used as aphrodisiac, astringent, diuretic and depurative; leaf is used as astringent and tonic; bark decoction is useful in snakebite.

Tibet: Plant is used as anthelmintic; cures fever and phlegm.

Ethnomedicine

Root: Is used in dog bite and tuberculosis.

Bark: Is used in bone fracture, diarrhoea and dysentery.

Gum: Is useful as astringent, diuretic and tonic; used in diarrhoea, dysentery and piles.

Leaf: Is used in boils, delirium, inflammation, skin diseases and as vermifuge.

Flower: Useful in diabetes, diarrhoea, dog bite, dysentery, eye complaints, piles, pimples, stomach ache, urinary troubles and wounds.

Seed: Is anthelmintic, abortifacient, cooling and vermifuge.

Non-medicinal Uses

Plant: Has ornamental value; yields charcoal; host plant for lac insect; yields gum called Butea gum or Bengal kino.

Young shoot: Yields fiber used in making ropes.

Twigs: Dry twigs used to feed the sacred fire in temples.

Root: Yields a fiber used in making cordage, paper and caulking boats.

Bark: Yields fiber used for making paper and cordage; yields a dye used in tanning industry; used as fish poison.

Wood: Used for packing cases, water scoops, boards, ply work, gunpowder; wood pulp is useful in newsprint manufacture and wrapping paper; used as a source of charcoal and fuel.

Gum: Red gum is used in tanning.

Leaf: Leaves are sewn into plates and cups.

Flower: Yields a yellow dye, used for colouring food stuff to yellow or orange.

15. *Butea superba* Roxb.

Vernacular Names: Kan: Muttugada balli; **Mal:** Vallimurukka; **Mar:** Beltivas, Palasavela, Yelparas; **San:** Latha palaasha, Palaasha niryaasam, Valli palaasha; **Tam:** Kodi purasu, Pilachi valli; **Tel:** Baaranki chettu, Modugaige, Theegemoduga, Thivvamobuga.

Distribution: Native of Asia including India; grows in India in Andhra Pradesh, Bihar, Karnataka, Madhya Pradesh, Maharashtra, Orissa, Tamil Nadu, Uttar Pradesh, West Bengal.

Description: Giant climbing shrub; leaflets rhomboid or broadly ovate, obtuse; flowers are bright orange or scarlet yellow on a raceme; pedicels thrice as long as calyx; calyx velvety, lowest tooth as long as lateral; pod stalked, but in all other respects it resembles *Butea monosperma*.

Phytochemistry

Plant: Gum Kino.

Root: Glucosides.

Flower: Butrin and isobutrin.

Seed: Fatty oil and allophanic acid.

Medicinal Uses

Plant: Useful in dysuria, gravel and night fever.

Root: Cardiotonic, estrogenic, hypotensive and tonic, prescribed for snake bite and bites of animals; root with root of *Nyctanthes, Woodfordia floribunda,* seeds of *Senna tora* and *Veronia anthelmintica,* juice of *Trichosanthes* is made into a paste and applied for snake bite.

Root bark: Aphrodisiac and rejuvenative; useful in abdominal and indolent tumours and carcinoma.

Stem: Decoction is emollient and sedative; used in piles.

Stem bark: Is prescribed for snake bite; used as tonic.

Gum: Tonic; used as poultice on abscesses and wounds.

Leaf: Leaf decoction emollient and sedative; leaf juice with curds given for heat eruptions on skin.

Flower and seed: Antihepatotoxic, anthelmintic, antifertility and sedative; recommended as antidote for scorpion sting.

Seed: Anthelmintic, antifertility, emollient and sedative; used topically for piles.
Seed oil: Anthelmintic and hypotensive.

Toxicology: High doses can cause nephrotoxicity, anaemia and congestion of liver, lungs and spleen.

Uses in Alternative Systems of Medicine

Ayurveda:

Part used: Root, bark, gum, leaf, flower and pod.

Uses: Root cures defects of sight and elephantiasis; bark is useful in amenorrhoea, bleeding piles, bone fracture, dysentery, dysmenorrhoea, haemorrhages, hydrocele and ulcer; gum is given as astringent in diarrhoea, hemorrhages and as an external application to inflammation, ulcers and sores; leaves are used in eye diseases; flower is useful in burning sensation, gout¸leprosy, painful micturition and skin disease; pod is useful in curing 'kapha' 'vatha', piles, tumours, skin diseases and as antidote for scorpion sting.

Siddha:

Part used: Bark, gum, leaf, flower and seed.

Uses: Useful in diarrhoea, dysentery, dysuria, haematemesis and skin diseases; is anthelmintic and contraceptive.

Tibetan:

Part used: Gum.

Uses: Used in fever, intestinal worms and phlegm.

Unani:

Part used: Bark, leaf, gum, flower, pod and seed.

Bark uses: Bark is used as anthelmintic, anti-inflammatory, aphrodisiac, astringent, carminative, diuretic, emmenogogue, expectorant, resolvent, vermifuge and tonic; useful in piles, fractures, biliousness, boils, dysmenorrhoea, fracture, gonorrhoea, hydrocele, liver disorders and lumbago; leaf useful in lumbago; gum is useful in leucorrhoea and syphilis; flower is useful in burning micturition, phlegm and swelling; pod and seed is useful in diarrhoea, intestinal worms, orchitis and piles.

Medicinal Uses Outside India

Cambodia: Gum is used to treat diarrhoea; as dressing for boils, sores and ulcers; infusion of the seed given as vermifuge; bark boiled with ginger is used to treat snakebite and scorpion sting.

Indonesia: Leaf poultice applied for scorpion sting; bark decoction is used to treat convulsions, haemorrhoids and used for bathing.

Myanmar: Gum and leaves used as astringent.

Non-medicinal Uses

Plant: Grown as ornamental and hedge plant, yields a gum called 'Kino' gum.
Root: Used in making ropes.
Leaf: Used as fodder; in making dining plates and cups.
Flower: Yields a red dye used for dyeing.

16. *Cajanus cajan* (L.) Millsp.

Synonym: *Cajanus indicus* Sprengel.

Notes: Three botanical varieties *bicolor, cajan* and *flavus* are recognized. Originated in India but now cultivated in all tropical countries; red gram urease can be used with greater advantage than BHD urease tablets for the estimation of blood urea.

Common Names: Congo pea, No eye pea, Pigeon pea, Red gram.

Vernacular Names: Ben: Arhar, Tawar; **Hin:** Arhar, Tuvar, Thoor; **Kan:** Thogari, Turuku thogari; **Mal:** Aadhaki, Kaakshi, Thuvara; **Mar:** Thoora, Thoovar; **Sans:** Aadhaki, Kakshi, Karvirabhuja, Mritana, Mrittalaka, Peethapushpa,

Shanapushpika, Surasthaja, Thuvarika, Vritha bheeja; **Tam:** Andagi, Karapu woolandu, Paruppu, Thovarai; **Tel:** Aadhaki, Errakandulu, Kandipappu, Kondakandi, Peddakandi, Polukandi, Potukandalu, Sinnakandi.

Distribution: Native of Bangladesh, Bhutan, India and Sri Lanka; cultivated throughout India.

Description: An erect shrub growing to 3m; branches with silky hairs; leaves alternate, pinnately trifoliolate; leaftlets elliptic, oblong acute apically and basally, resinous on lower surface; stipulate; leaflets lanceolate to elliptic; flowers red or orange or yellow in colour with purple streaks borne on a axillary raceme; pod glandular, compressed; seeds 2-9, yellow, red, lenticular to ovoid in shape.

Phytochemistry

Plant: β-Amyrin, α-dimethylether, β-sitosterol, 2-hydroxygenistein, cajaisoflavone, cajaflavonone, cajanin, cajanol, cajanone, calcium, copaene, dihydroxy isoflavone, dipinostrobin, ferreirin, fibre, genistein, hentriacontane, isoflavone, phosphorus, protein, salicylic acid, tongistylin A and C and vitexin.
Root: α-Amyrin, β-amyrin, 7-0-β-D-glucoside, cajaflavanone, cajaisoflavone, cajaflavonone, cajanone, cajaquinone, cajonene, dihydroxy isoflavone, genistein, lupeol, methyl cajanone, sitosterol, sitosterol-β-D-glucoside and trihydroxy isoflavone.
Leaf: β-Amyrin, β-eudesmol, α-eudesmol, β-selinene, β-sitosterol, 2'-methyl cajanone, 5,2'-dihydroxy isoflavone-7-0-β-D-glucoside, cajaflavanone, cajanol, cajanone, campesterol, cholesterol, copaene, genestein, hentriacontane, laccerol, longistylin A and C, lupeol, methyl cajanone, naringinin-7,4'-dimethyl ether, pinostrobin, salicylic acid, sesquiterpenes, sitosterol-β-D-glucoside, stigmasterol, stilbene, vitexin and γ-eudesmol.
Pod: Ascorbic acid, β-carotene, calcium, carbohydrates, fat, iron, niacin, phosphorus, potassium, riboflavin and thiamine.
Seed: α-Amyrin, β-amyrin, β-carotene, β-galactosidase, β-L-arabinosidase, 5-methyl-L-cysteine, alanine, alkanes, arginine, ascorbic acid, aspartic acid, cajaflavonone, cajaisoflavone, cajanin, cajanol, cajaquinone, calcium, campesterol, canavinine, carbohydrates, cholesterol, citrulline, concajanin, copper, cystine, daidzein, esters, formononetin, genistein, glutamic acid, glycine, histidine, iron, isoleucine, leucine, linoleic acid, linolenic acid, lupeol, lysine, manganese, methionine, methylcajanone, niacin, nicotinamide deamidase, oleic acid, oxalic acid, palmitic acid, pantothenic acid, phenylalanine, phosphorus, potassium, proline, proteins, pyridoxine, raffinose, riboflavin, serine, sodium, stearic acid, stigmasterol, stillbene-2-carboxylic acid, succinic acid, sucrose, tannin, thiamine, threonine, tryptophan, tyrosine, urease, valine, verbascose, vitamin B_6 and zinc.
Seed oil: Oleic acid, linoleic acid, linolenic acid.

Edibility

Plant: Used as vegetable.

Leaf: Tender leaves used as potherb.

Flower: Used in the preparation of sweets.

Pod: Used in curries.

Seed: Nutritious, wholesome, rich source of proteins forming major part of vegetarian diet; used in soups and curries; germinated and eaten; ground into flour and used in making sweets.

Medicinal Uses

Plant: Is antiadenomic, antiallergenic, antiandrogenic, antibacterial, anticancer, anticoronary, antifibrinolytic, antihepatotoxic, antihistaminic, antihypertensive, anti-inflammatory, antilymphocytic, antimutagenic, antioxidant, antiperiodontic, antipyretic, antiseptic, antispasmodic, antiviral, astringent, apoptotic, cardio-protective, choleretic, diuretic, haemolytic, hepatoprotective, immunostimulant, laxative, memory enhancer, narcotic, sedative, tonic, vasodilator and vulnerary; useful in acne, alopecia, anaemia, arthritis, asthma, atherosclerosis, back ache, beri-beri, breast cancer, cephalalgia, chilblain, cold, convulsion, dandruff, delirium, dementia, depression, dermatitis, diabetes, eczema, edema, epilepsy, flu, flatulence, gastritis, gingivitis, glaucoma, heart burn, hepatitis, herpes, hypoglycaemia, hypertension, impotency, inflammation, insomnia, leprosy, leukaemia, lupus, lymphoma, malaria, measles, melanoma, migraine, myocardia, nephritis, neuralgia, neuroblastoma, obesity, osteoarthritis, osteoporosis, photophobia, pneumonia, polio, pruritus, psoriasis, rheumatism, tumour of breast, cervix, lung, ovary and stomach, ulcer, vertigo and vitiligo; antidote to aluminum, cadmium and lead poisoning.

Root: Is alexiteric, anthelmintic, expectorant, sedative and vulnerary.

Leaf: Is anodyne, anti-inflammatory, astringent, cooling, diuretic and laxative; useful in gingivitis, odontalgia, oral ulcers and strangury; boiled leaves are applied on wounds and sores as vulnerant; juice is given in haemorrhage and jaundice; warmed leaf paste is applied over the breast to check secretion of milk; decoction of leaves is used to cleanse ulcers; tender leaves are eaten for spongy gums.

Flower: Is bechic; flower infusion is pectoral; flower paste is applied on oral ulcers and sores; flower buds are pectoral.

Pod: An infusion of green pod is used as pectoral.

Seed: Is alexiteric, anthelmintic, astringent, cooling, constipating, expectorant, febrifuge, pectoral and resolvent; useful in bronchitis, cough, fever, haemorrhoids, intestinal worms, oral ulcers, vitiated "pitta", tumours and vomiting; seed paste is warmed and applied over the mammae to check excessive secretion of milk; powdered seeds are sprinkled on small pox pustules to cleanse it; expressed juice with salt is given in jaundice.

Bioactivity: Plant is inhibitor of 5-α-reductase, ACE, aromatase, β-glucuronidase, methyl-transferase, DOPA-decarboxylase, FPTase, histidine-kinase, insulinase, MAO, NADH-ubiquinone-oxidoreductase, ornithine-decarboxylase, PTK, toposiomerase, tyrosine-kinase and tyrosine; is a β-adrenergic receptor blocker, β-blocker, calcium channel blocker, interferon synergist, monoamine precursor, neuro inhibitor and quinone-reductase inducer.

Administration of protein fraction at 5-10% to rats fed with high cholesterol diet resulted in significant lowering of the total and free cholesterol, phospholipids and triglycerides in the serum and liver; urease fraction when injected intravenously to dogs, hydrolyzed the blood urea within the body; formononetin from the seeds exhibited hypolipidemic activity in male albino rats; the ethanolic extract of seeds showed hypoglycaemic activity when injected intraperitonially to albino rats.

Toxicology: When raw seeds are consumed in excess, it has a soporific effect; they contain chymotrypsin and trypsin inhibitors.

Uses in Alternative Systems of Medicine

Ayurveda:

Part used: Seed and leaf.

Uses: Seed is nutritious; given in diarrhoea and gout; leaf juice with salt is given in jaundice.

Chinese:

Part used: Root.

Uses: Is anthelmintic, expectorant, sedative and vulnerary; an excellent antidote for all poisons.

Siddha:

Part used: Leaf and seed.

Uses: Useful in dental problems, gingivitis, swelling and stomatitis.

Unani:

Part used: Leaf and seed.

Uses: Leaf paste is applied as poultice for inflammations and piles; seed is alexiteric, anti-inflammatory and expectorant; boiled seeds are used for liver improvement; cures biliousness, constipation and diarrhoea.

Medicinal Uses Outside India

Argentina: Leaf juice is applied for genital and other skin irritations in females; flower decoction is used for bronchitis, cough and pneumonia.

Brazil: Used for cough, fever, inflammation, sore throat, swelling and ulcer.

China: Dried roots are used as alexiteric, anthelmintic, expectorant, sedative and vulnerary; leaves used in dysentery, sore gums and tooth ache; leaf juice is used as mouth wash; fried seed powder is added to coffee to relieve head ache and vertigo; immature pods are used in liver and kidney ailments.

Cuba: Flower is used in bronchitis and cold.

Dominican Republic: Plant is used as pectoral and vulnerary.

Haiti: Plant is used in jaundice, urticaria and as vulnerary.

Indochina: Leaf paste cures bladder stones and baldness; salted leaf juice is taken for jaundice; seeds used as astringent and bechic.

Indonesia: Crushed leaves are applied for herpes, itch and other skin diseases; seeds used as soporific.

Java: Young leaves as poultice is applied to sores.

Malaya: Leaf is used in cough, dermatosis, diarrhoea, ear ache, enteritis and sores.

Malay Peninsula: Leaf infusion is given for abdominal troubles; leaf paste with salt is remedy for cough.

Mexico: Plant is used as astringent, diuretic, laxative and vulnerary; useful in dysentery.

Trinidad: Leaf is used in flu and sunstroke.

Non-medicinal Uses

Plant: Grown as shade plant; used as manure and fodder; is a host plant for lac insect; used in contour planting to check erosion; cover crop for coffee plants.

Stem: Dried stems are used for roofing, in basket making and wickerwork; used as fuel.

Wood: Is used as fuel; wood pulp is used for making straw boards.

Leaf: Is used as manure and fodder and for rearing silk worms.

Seed: Is used as fodder and as detergent for washing.

Husk: Is used as fodder.

17. *Cajanus scarabaeoides* (L.) Thouars

Synonyms: *Atylosia scarabaeoides* (L.) Benth.
Rhynchosia scarabaeoides (L.) DC.

Distribution: Native of Asia including India and Australia.

Description: A perennial climber with stems covered by reddish pubescence; leaves trifoliolate; leaflets elliptic, obovate, lateral leaflets oblique, obtuse to subacute, velvety pubescent, gland dotted; inflorescence with 2-6 yellow flowers; calyx densely pubescent; corolla yellow with flushes of crimson, vexillum glabrous; pod broad, densely pubescent, glandular, furrowed between the seeds 3-6 in number.

Medicinal Uses

Plant: Abortifacient, anticancer, antiestrogenic, antifertility, anti-inflammatory, anti-ischaemic, antilymphomic, antimicrobial, antimitogenic, antioxidant, antispasmodic, antitumour, antiviral, apoptotic, cytotoxic, hypocholesterolemic and immunostimulant; useful in atherosclerosis, cancer, flatulence, neuroblastoma and ulcer.

Bioactivity: Plant is inhibitor of catechol-O-methyl-transferase, DOPA-decarboxylase, histidine-kinase, lipase, MAO, MDR, NADH-ubiquinone-oxidoreductase, ornithine-decarboxylase, PTK, peroxidase, topoisomerase I and II and tyrosine-kinase; is calcium antagonist, pituitary sensitizer and quinone-reductase inducer.

Toxicology: Seeds contain chymotrypsin and trypsin inhibitors.

18. *Calopogonium mucunoides* Desv.

Synonyms: *Calopogonium brachycarpum* Hemsley
Calopogonium orthocarpum Urban
Stenolobium brachycarpum Benth.

Common Name: Wild ground nut.

Distribution: Native of Central and South America; introduced into Africa, Asia including India, Indian ocean and Pacific ocean Islands; introduced and naturalized in Nilgiri Hills, Tamil Nadu in India.

Description: Ferrugenous, pubescent, twining annual; leaves trifoliolate; leaflets ovate, elliptic, subacute, pilose on both surfaces, apiculate at the apex, round at the base; lateral leaf lets oblique; flowers blue on axillary raceme with pilose peduncles; bracteate and bracteolate; pod pubescent, transversely depressed between the many seeds.

Phytochemistry

Stem: β-Sitosterol, dihydroxy isoflavone, pinitol and stigmasterol.
Leaf: Calopocarpin, medicarpin, methylenedioxyisoflavone, neodunol, sophorapterocarpan A and tuberosin.
Seed: Calopogonium isoflavone, methylenedioxyisoflavone, pyranoisoflavone, tetramethoxy isoflavone and trimethoxy isoflavone.
Seed oil: Dimethyl chromene, linoleic acid, palmitic acid and stearic acid.

Medicinal Uses

Leaf: Is used in filariasis.

Non-medicinal Uses

Plant: Is used as green manure and cover crop for coconut and rubber; grown as soil binder.
Leaf: Is used as forage.

19. *Canavalia ensiformis* (L.) DC.

Synonyms: *Canavalia ensiformis* (L.) DC. var., *albida* DC.
Canavalia ensiformis (L.) DC. var., *truncata* Ricker
Canavalia gladiata (Jacq.) DC. var., *leucosperma* Voight.
Dolichos acinaciformis Jacq.
Dolichos ensiformis L.

Notes: Prehistoric American Indian crop; seed is a source of urease, an enzyme used as specific reagent for urea in biochemistry.

Common Names: Horse bean, Jack bean, Overlook bean, Sword bean.

Vernacular Names: Hin: Gajaasema, Kadsambu, Khadsambai, Lalkudsumbai; Safedkudsumbai, Sema, Surasema; **Kan:** Kaadavare, Shambe avari, Shimbe balli, Shimbe, Thamate balli, Thumbe kanji; **Mal:** Baramarecha, Kanavaala, Kaattuvaalamarakkaya, Vaalamara, **Mar:** Abai, Gavara, Gouari, Khadgashimbi, Koshaphala, Mahaashimbi, Neelashimbika, Sthodashimbi, **Sans:** Asishimbi, Brihathshimbi, Khadgashimbi, Koshaphala, Mahaashimbi, Neelashimbika, Sthoolashimbi; **Tam:** Kaattuthambattan, Kodithambattan, Kolhiyavarai, Kottuhambattam, Peyavarai, Sivapputthambattai, Thamattankodi, Thambattai, Valavaraj, Vellaitthambattai; **Tel**: Adavithamma, Chambakaasya, Errathamma, Kaaranuputhige, Karuthamma, Shimbi.

Distribution: Possibly a native of America; introduced into Asia including India; cultivated through out tropics; grows in India, Indonesia, Taiwan, Tanzania, Kenya, Hawai and tropical East Africa; grows throughout India.

Description: Perennial climber with adpressed-pubescent climbing stem; leaves trifoliolate, ovate-elliptic to oblong, broad, more or less cuneate, sparsely pubescent or glabrous on both the surfaces; raceme axillary, pendulous, with rose or purple flowers; calyx long, pubescent with truncate upper lip; pod linear oblong, flat, sword shaped with 12-20 seeds; seeds white or ivory in color with a brown mark near the grey hilum.

Edibility

Flower and Leaf: Is steamed and eaten in Indonesia.
Pod and Seed: Used as vegetable; roasted beans are used as substitute or adulterant for coffee.

Medicinal Uses

Plant: Is used in biliousness, burning sensation, indigestion and ulcer; given as appetizer and refrigerant.

Toxicology: Seeds contain chymotrypsin and trypsin inhibitors.

Uses in Alternative Systems of Medicine

Ayurveda:

Part used: Pod.

Uses: Given as appetizer and tonic; useful in biliousness and ulcer.

Chinese:

Part used: Pod.

Uses: Given to promote digestion and to strengthen the kidney.

Tibetan:

Part used: Pod

Uses: Cures fever and kidney disorders.

Ethnomedicine

Plant: Anodyne, digestive and tonic; useful in kidney disorders.

Non-medicinal Uses

Plant: Provides a good manure and forage; grown as soil cover for erosion control.

20. *Canavalia gladiata* (Jacq.) DC.

Synonyms: *Canavalia ensiformis* (L.) DC. var., *alba* Makino
Canavalia ensiformis (L.) DC. var., *gladiata* (Jacq.) Kuntze
Canavalia ensiformis sensu auct.
Canavalia gladiolata Sauer.
Canavalia incurva (Thunb.) DC.
Canavalia loureirii G.Don
Canavalia machaeroides (DC.) Steudel
Canavalia maxima Thouars
Canavalia plagiosperma sensu auct.
Dolichas gladiata Jacq.

Common Name: Sword bean.

Vernacular Names: Ben: Makham shim; **Hin:** Bara sem, Lalkudusumpal, Kadsumbal; **Kan:** Shembi avare, Tumbekai; **Mal:** Baramareha, Valpayar, Valvara; **Mar:** Abai; **Sans:** Asisimbi, Mahasimbi, **Tam:** Segappu thampattai; **Tel:** Karochikkudu, Thammakaaya, Yerra tamma.

Distribution: Native of Cambodia, China, India, Myanmar, Thailand, Vietnam and North America; introduced into Africa, Caribbean Islands, Central America, Indian Ocean and Pacific Ocean Islands and South America; grows throughout India.

Description: Annual or perennial climber with woody stem; leaves compound; leaflets ovate, acuminate, round or cuneate at the base, membranous, sparsely pubescent on both the surfaces; raceme axillary with numerous flowers; flowers inverted with standard petal on bottom, not on top, lilac or white; bracteoles obtuse; calyx glabrous; pod linear-oblong, compressed, and sword shaped with ridges, an extra rib of 2 mm is present below the sutural rib; seeds 8-12, red, pink, black, reddish brown, oblong and ellipsoid.

Phytochemistry

Seed: Alanine, arabinose, ascorbic acid, aspartic acid, choline, canavanine, canavaline, concanavaline, cystine, glucose, galactose, glycine, histidine, lupeoside, norvaline, proline, raffinose, rhamnose, serine and trigonelline.

Edibility

Pod: Edible.

Seed: After roasting, eaten as delicacy; roasted seeds used as substitute for coffee.

Medicinal Uses

Pod and Seed: Astringent, appetizer, cooling, digestive and vulnerary; useful in anorexia, burning sensation, dyspepsia, hyperdipsia, ulcers and wounds.

Toxicology: Seeds contain chymotrypsin and trypsin inhibitors; imperfectly cooked seeds when eaten cause intestinal troubles.

Uses in Alternative Systems of Medicine

Tibetan

Part used: Whole plant.

Uses: Cures fever and kidney disorders.

Medicinal Uses Outside India

China: Pods and seeds are bechic, stomachic and tonic; used in treating lumbago and swellings.

Ethnomedicine

Plant: Antitussive; used in conjunctivitis.

Non-medicinal Uses

Plant: Grown for green manure, forage and as cover crop.

21. *Castanospermum australe* Cunn. & Fraser

Common Names: Australian bay chestnut, Beam tree, Black bean, Moraton bay chestnut.

Distribution: Native of Australia; introduced into India; it is cultivated as a avenue tree in Karnataka, Kerala, Tamil Nadu and West Bengal in India.

Description: An ever green tree of 40m height; bark dark brown; leaves imparipinnate; leaflets 11-17, ovate elliptic or broadly oblong; flowers yellow, orange to red on racemes; pod turgid with chest nut like black seeds.

Phytochemistry

Plant: Bayogenin and castanogenol.

Bark: Bayogenin and castanogenol.

Wood: Afromosin, bayin, bayogenin, castanogenin, formononetin, hederagic acid, koparin, trihydroxy-4'-methoxyisoflavone and triterpenoidal sapogenins.

Flower: Cyanidin-3-monoside.

Pod: Ochratoxin A and saponin

Seed: 6-Epicastanospermine, alexines, australine, castanospermine, fagomine.

Edibility

Pod and Seed: Edible after boiling or roasting; the roasted seeds have the flavour of Spanish chestnut.

Medicinal Uses

Pod: Is used as astringent; castanospermine is used to treat hyperglycaemia.

Bioactivity: (+) castanospermine from the seeds at 2.5 mM/l prevents the formation of glycoprotein coating around the AIDS virus and at 0.62 mM/l reduced the infection capacity of the virus; has potential for treating AIDS; it altered glycogen metabolism when fed to rats; it also induced dose dependent decrease in glucosidase activity in liver, spleen and kidney.

Toxicology: The raw pods and seeds are poisonous; indolizine alkaloid-castanospermine in the seed causes gastrointestinal irritation in humans; plant is an inhibitor of amyloglucosidase and β-galactosidase.

Non-medicinal Uses

Plant: Grown as a shade tree.

Wood: Used for panelling, furniture, cabinets and other decorative work; suitable for switch boards and electrical fittings requiring high insulation, flooring, carving, veneers, inlay work, jewel boxes, glove boxes, ink stands, rulers, fancy paper weights, umbrella, trays and doors.

22. *Cicer arietinum* L.

Notes: Probably originated in Southwest Asia; but its native area is obscured by extensive cultivation; an important pulse crop which is widely cultivated; two main cultivars, '*Desi*' and '*Kabuli*' types are recognized in India; highly nutritious with 14 to 30 per cent protein.

Common Names: Bengal gram, Chickpea,

Vernacular Names: Ben: Chola; **Guj:** Chana, Chania; **Hin:** Channa, Chola, Rahilaa; **Kan:** Kadale; **Mal:** Kadalai; **Mar:** Chano, Harbharaa; **Sans:** Balabhaishajya, Balabhojya, Chanaka, Harimantha, Jeevana, Kanchuki, Krishnachanchuka, Sakalapriya, Sungandha, Vaajimantha; **Tam:** Chundakadalai, Kadalai, Uppukkadalai; **Tel:** Chanagalu, Senaga, Senagalu, Harimandakamu.

Distribution: Native of Afghanistan, China, India, Iran, Iraq, Israel, Jordan, Lebanon, Nepal and Pakistan; widely cultivated pulse crop throughout tropics and European countries; cultivated throughout India.

Description: Annual herb; erect or spreading, branched, glandular, pubescent; leaves imparipinnate, glandular, pubescent with 3-8 pairs of leaflets; leaflets ovate, margin serrate; stipules 2-5 toothed; flowers solitary or 2-3, pink or purple or blue; pod rhomboid, ellipsoid; seed round to angular, yellowish or brown in colour.

Phytochemistry

Plant: (–) Maackianin, 7-hydroxy-4-methoxyflavone, acetic acid, benzoic acid, biochanin-7-glucoside, biochanin-A, coniferin, daidzein, formononetin, garbenzol, isorhamnetin, kaempferol, malic acid, medicarpin, oxalic acid, p-coumaric acid, p-hydroxy benzoic acid, phytins, pratensein, quercetin, saponins, vanillic acid, vitamin A, D and E.

Root: 9-0-Methylcoumesterol and medicagol.

Leaf: Astragalin, biochanin-A, butanoic acid, formononetin, garbenzol, isoquercitrin, isorhamnetin, kaempferol-3-glucoside, ononin, p-hydroxybenzoic acid, populnin, pratensein and vanillic acid.

Seed: β-Carotene, 3,4,7 trihydroxyflavone, acetic acid, alanine, arginine, ascorbic acid, biochanin B and C, biochanin-7-glucoside, calcium, carbohydrates, carotenoids, copper, cystine, daidzein, formononetin, garbenzol, genistein, glutamic acid, glycine, histidine, iron, isoleucine, isoliquiritigenin, isoliquiritigenin-4'-glucoside, lecithin, leucine, linoleic acid, linolenic acid, lysine, magnesium, malic acid, manganese, medicarpin, methionine, myristic acid, niacin, oleic acid, oxalic acid, palmitic acid, palmitoleic acid, pantothenic acid, p-coumaric acid, phenylalanine, phosphorus, phytin, phytosterol, potassium, pratensein¸ proline, riboflavin, salicylate, saponin, serine, soyasaponin, stearic acid, thiamine, tryptophan, tyrosine, uracil-5-β-D-fructofuranosyl-4'-monophosphate, uracil-5-β-D-ribofuranosyl-2'-3'-cyclic monophosphate, valine and zinc.

Seed sprout: Biochanin-A, biochanin-A-7-0-glucoside, daidzein, (–) maackianin, medicarpin, pratensein.

Edibility

Plant, Leaf, Green pod: Edible.

Seed: Eaten raw or cooked or added to various dishes; highly nutritious and widely consumed as protein adjunct to starchy diets; seed flour is mixed with sugar and ghee to make sweet meats; is valued for its nutritive content; sprouted seeds are also eaten; in Chile, chick pea milk is given to infants to control diarrhoea.

Medicinal Uses

Plant: Is analgesic, antiacid, antiandrogenic, antiangiogenic, antiarrhythmic, antiatherosclerotic, antibacterial, anticancer, anticoronary, antidepressant, antiestrogenic, antifibrinolytic, antihaemolytic, antihepatotoxic, antihistaminic, antihypercholesterolemic, antilithic, antimicrobial, antimitogenic, antinociceptive, antioxidant, antipyretic, antiradicular, antiscorbutic, antiseptic, antispasmodic, antiviral, aphrodisiac, apoptotic, astringent, diuretic, CNS active, cardioprotective, immunomodulator, neuroexcitant, phagocytotic, refrigerant, sedative, tranquilizer, uterotropic, vasodilator and vulnerary; it is useful in allergy, alopecia, anaemia, anorexia, anxiety, arthritis, asthma, back ache, beri-beri, bronchitis, cancer of breast, canker, cardiospasms, cataract, catarrh, cervical dysplasia, chilblain, cholera, cold, convulsions, costiveness, constipation, cutamenia, delirium, dementia, depression, dermatitis, diabetes, dyspepsia, diarrhoea, dysmenorrhoea, eczema, edema, encephalopathy, epilepsy, fatigue, fibroids, gastritis, gingivitis, glaucoma, heart burn, haemorrhages, hepatitis, herpes, hyperactivity, hypertension, hyperglycaemia, impotency, indigestion, insomnia, ischaemia, jaundice, keratitis, leprosy, leukaemia, lupus, lymphoma, mastalgia, measles, melanoma, menorrhagia, nephritis, neuralgia, neuroblastoma, neuropathy, obesity, orchitis, osteoarthritis, osteoporosis, photophobia, pneumonia, polio, pruritus, psoriasis, retinopathy, rheumatism, spasms, stomatitis, stress, sun stroke, tuberculosis, tumour of breast, lung, ovary, prostate, stomach, ulcer, vertigo, vitiligo and warts; antidote to aluminium, cadmium and lead poison.

Glandular secretion of the plant was collected in the early days by covering the plant with a thin muslin cloth in the night and cloth was wrung in the morning and secretion was collected; it was considered to cure bronchitis, cutamenia, cholera, constipation, diarrhoea, dysentery, dyspepsia, sun stroke and warts; drinking the secretion is supposed to bring down the blood cholesterol level and serum lipids; fresh plant in hot water is consumed for dysmenorrhoea and nervous affections; it is used externally in paralysis and rheumatism.

Leaf: Abortifacient, alterative, aphrodisiac, astringent, laxative, refrigerant and stomachic; cures anorexia, biliousness, cholera, cutamenia, dyspepsia, dysmenorrhoea, bronchitis, fever, flatulence, nausea, enlarged spleen and retention of urine; boiled leaf poultice is applied to sprains and dislocated joints; fresh juice mixed with potash or carbonate is given in dyspepsia.

Seed: Anthelmintic, appetizer, aphrodisiac, astringent, antibilious, diuretic, expectorant, galactagogue, hypocholesterolemic, nutritive, prophylactic, refrigerant, soporific, spasmolytic, stimulant and tonic; cures bronchitis, burning sensation, hyperdypsia, inflammation, leprosy, pharyngopathy, splenohepatomegaly, skin diseases, thirst and throat complaints; seeds cooked with milk is a nutritive tonic for infants and also controls infantile diarrhoea; seeds eaten in a parched state at night followed by a cup of warm milk gives relief from bronchial catarrh; infusion of seed relieves pain due to urinary calculi; antidote for snake bite.

Unripe seed: Aphrodisiac, stimulant and tonic; cures burning sensation in the body and thirst.

Sprouted seed: Used as prophylactic against scurvy; and to lower serum and liver cholesterol.

Bioactivity: Plant is inhibitor of 5-α-reductase, ACE, aldose-reductase, β-glucuronidase, DOPA-decarboxylase, histidine-kinase, insulinase, lipase, MAO, NADH, ubiquinone-oxidoreductase, ornithine-decarboxylase, PTK, peroxidase, toposiomerase I and II, tyrosine-kinase and xanthine-oxidase; is angiotensin receptor blocker, β-adrenergic receptor blocker, β-blocker, calcium channel blocker, interferon synergist, monoamine precursor, pituitary sensitizer and quinone-reductase inducer.

Ethanol extract of seeds showed anti-implantation activity in albino rats and reduced serum and liver cholesterol levels in rats; triglycerides prevented hyperlipidemia.

Toxicology: Seeds eaten in excess, may cause sterility and effect liver and serum cholesterol as they contain chymotrypsin and trypsin inhibitors;

Uses in Alternative Systems of Medicine

Ayurveda:

Part used: Seed

Uses: Seed is tonic, nutritive; consumed in diabetes, fever, gout, herpes, thirst, tuberculosis and vomiting.

Siddha:

Part used: Plant juice.

Uses: Is astringent; useful in constipation and dyspepsia.

Unani:

Part used: Leaf and seed.

Uses: Anthelmintic, antipyretic, aphrodisiac, deobstruent, diuretic, galactagogue, laxative and tonic; leaf is used in cold and pain; seed improves complexion, resolves tumour, increases sperm count, strengthens hair root, cleanses the uterus, reduces inflammation and sore throat; it cures ear inflammation, ear tumour, fever, leprosy, pimples, rheumatism, scabies, skin diseases, sore throat and is a blood purifier.

Ethnomedicine

Stem: Is used in jaundice.

Seed: Given as tonic.

Non-medicinal Uses

Plant: Improves soil fertility and used as cattle feed.
Seed: Roasted seeds are used as a substitute for coffee; seed flour is used as scrub to remove dead skin on the face; yields starch used in textile sizing and gives a light finish to wool, cotton and silk; is a good adhesive for plywood.
Husk: Used as cattle feed.

23. *Clitoria ternatea* L.

Notes: Three botanical varieties *angustifolia*, *ternatea* and *pleniflora* are recognized.

Common Names: Blue pea, Butterfly pea, Mussel shell creeper and Winged leaf clitoria.

Vernacular Names: Ben: Aparajita; **Hin:** Kajina, Aparajita, Shobhanjan, Vishnukranti; **Kan:** Girikarni balli, Shankapushpa balli, Gokarna; **Mar:** Gokarni, Kajli; **Sans:** Aparajita, Bhoomilagna, Dhandipushpika, Sitapushpa, Supushpi, Vishnukranti; **Tam:** Kakkanam, Kannikkodi, Kavalai, Taruganni; **Tel:** Nellaghentana, Niladintana, Telladintana.

Distribution: Native of Africa, Cambodia, India, Myanmar, Thailand and Vietnam; introduced into Asia, Australia, Brazil, Caribbean Islands, Central America, Europe, Indian Ocean region, Middle East, North America, Pacific Ocean and South America; in India, it grows in Andhra Pradesh, Assam, Bihar, Gujarath, Haryana, Karnataka, Kerala, Madhya Pradesh, Maharashtra, Manipur, Rajasthan, Tamil Nadu and Uttar Pradesh.

Description: Perennial woody climber with pubescent stem; leaves unipinnate; leaflets 5-7, oblong, slightly emarginate or obtuse, short stalked; stipules linear with acute apex; flowers bluish yellow, light blue, white or yellow, solitary on short axillary Pedicels which are resupinate; bracts present; bracteoles orbicular; calyx with upper teeth joined at the base; corolla if blue is tinged with greenish veins on the vexillum or deep blue tinged yellow at the base, wing petal lightly fused to keel petal; stamens 10, diadelphous, vexillary; style hairy inside; pod flat, turgid, linear oblong, beaked, pubescent or glabrous; seeds ellipsoidal, compressed, black, numerous.

Phytochemistry

Plant: β-Methyl glutamic acid, aparajitin, arachidic acid, clitorin, kaempferol and lactone.

Root: Aparajitin, kaempferol-3-O-rhamnosyl (1→6) galactoside, kaempferol-3-O-rhamnosyl (1→6) glucoside, tannin, taraxerol and taraxerone.
Leaf: Kaempferol 3-O-Rhamnosyl (1→6)-glucoside, aparajitin, clitorin, kaempferol, kaempferol-3-O-glucoside, kaempferol-3-O-neohesperidoside, kaempferol-3-0-rhamnosyl (1→6) galactoside, kaempferol-3-O-rutenoside, lactone, linoleic acid, linolenic acid, oleic acid, palmitic acid, sitosterol, stearic acid and triterpenoids.
Flower: (blue variety): 3'-Methyl ether of delphinidin-3-β-glucoside, anthocyanins, delphinidin-3-5-diglucoside, kaempferol, kaempferol glucoside, malvidin-3-β-glucoside, quercetin, robinin and ternalin A and B.

Seed: β-Hydroxycinnamic acid, β-sitosterol, γ-sitosterol, 3,5,7,4'-tetrahydroxyflavone-3-rhamnoglycoside, alanine, anthoxanthin, anthoxanthin glucoside, arginine, aspartic acid, cinnamic acid, ethyl-α-D-galactopyranoside, fixed oils, glutamic acid, glycine, hexacosanol, isoleucine, leucine, p-hydroxycinnamic acid, resin, tannins, tyrosine and valine. **Seed oil:** Behenic acid, cystine, histidine, lignoceric acid, linoleic acid, linolenic acid, lysine, methionine, oleic acid, palmitic acid, phenylalanine¸stearic acid, threonine and valine.

Edibility

Flower: Is used to colour rice to cerulean shade.

Pod: Edible; used as a vegetable.

Medicinal Uses

Plant: Analgesic, diuretic and cathartic; juice is used as nasal drop for nasal congestion; as an antidote for snake bite.

Root: Alexiteric, anthelmintic, aperient, aphrodisiac, cathartic, demulcent, depurative, diuretic, emetic, emmenogogue, laxative, ophthalmic, refrigerant and tonic; useful in anasarca, ascites, asthma, biliousness, burning sensation, bronchitis, dysentery, elephantiasis, fever, goitre, helminthiasis, hemicrania, inflammation, leprosy, leucoderma, ophthalmopathy, pulmonary tuberculosis, strangury, stomach ache, swelling, tubercular glands, ulcer and visceromegaly; it reduces cystic swelling; used by tribals for abortion; root paste is used as poultice for swollen joints; root juice in cold milk removes phlegm in chronic bronchitis; the juice of the root of white flowered variety is blown up the nostrils as a remedy for hemicrania; it also acts as a purgative and useful in ascites and enlargement of abdomen; root mixed with root bark of *Vitex negundo* is given internally as an antidote for snake bite, especially cobra bite.

Root bark: Cathartic, demulcent, diuretic, expectorant, laxative and purgative; useful in cough and pulmonary affections; used as poultice on swollen joints; infusion of root bark is useful in gonorrhoea and given as demulcent in irritation of urethra.

Leaf: Useful in ear ache, eruptions, fever and hemicrania; leaf infusion is used as a lotion for skin eruptions and ulcers; leaf juice mixed with ginger is administered in excessive sweating in fever; leaf juice mixed with salt is applied as dressing around ear for relief from ear ache combined with swelling of the neighbouring glands.

Flower: Blue flowered variety with salt is ground into paste and applied around ear for ear ache; flower decoction relieves dysentery; bark to be taken internally as antidote for snake bite and scorpion sting; it is given with root to relieve pain in cystic swellings and swollen joints.

Seed: Anthelmintic, aperient, cathartic, cooling and purgative; useful in dropsy, skin affections, tumours and visceralgia; improves sight, cures sore throat and mucous disturbance; powdered seeds with ginger is aperient and laxative; powdered seeds fried in ghee gives relief to colic in children and also useful in ascites and constipation.

Toxicology: Medicines containing root extract may cause nausea and vomiting.

Uses in Alternative Systems of Medicine

Ayurveda:

Part used: Plant, root and leaf.

Uses: Plant is aphrodisiac, astringent, brain tonic¸ carminative, nervine tonic and rejuvenating; improves strength and digestive power, relieves pain, improves complexion and voice; cures abdominal disorders, anorexia, cough, diabetes, dyspnoea, dysuria, haematemesis, insomnia, intestinal worms, piles, psychosis, skin diseases¸ sterility, stomatitis and uterine disorders; crushed root paste is applied on parts bitten by animals and used as an antidote to poison; used in treating, burning micturition and hypertension; root and leaf useful in indigestion and skin diseases; root of white flowered variety is alexiteric, anthelmintic, brain tonic, cooling, diuretic and laxative; useful in biliousness, burning sensation, corneal ulcers, elephantiasis, eye diseases, headache, inflammation, leucoderma, pain, snake bite and tuberculous glands; root of blue flower variety is alexiteric, anthelmintic, aphrodisiac, brain tonic, cooling, diuretic and laxative; useful for abdominal enlargement, ascites, asthma, biliousness, bronchitis, constipation, dysentery, elephantiasis, inflammation, leucoderma, pain and ulcers.

Siddha:

Part used: Root, leaf and seed.

Uses: Are diuretic and purgative; useful in constipation, digestive disorders, fever, leucorrhoea, edema, polyurea, retention of urine and worms.

Unani:

Part used: Root and leaf.

Uses: Anthelmintic, antipyretic, biliary, diuretic, emmenogogue, laxative, liver tonic and phlegmatic; root is useful in ascites; leaf paste is applied on eczema, pityriasis and leucoderma; leaf paste with honey applied on wounds helps in healing and also cures ascites, dropsy, jaundice and painful dental caries.

Medicinal Uses Outside india

Indo-China: Root is purgative.

Indonesia: Root infusion of white flowered variety is drunk as a depurative; leaf of blue flowered plants is crushed and applied to abscesses and boils as maturant; water from soaking the leaves and flowers is used as an eye wash for mild form of conjunctivitis.

Koman: The seeds are cathartic and the root is used as diuretic; the powdered seed in combination with ginger powder is laxative.

Madagascar: The root is used as an emetic and the seed as purgative.

Malay Peninsula: Seeds are considered aperient.

Philippines: Root is used as poultice for swollen joints.

Ethnomedicine

Root: Is used as abortifacient and purgative; useful in goiter and leprosy.

Leaf: Useful in swellings.

Seed: Used as laxative.

Non-medicinal Uses

Plant: Grown as garden plant, hedge plant and cover crop.

Root bark: Yields tannin used in tanning industry.

Root juice: Is applied to kill green flies on vegetables and worms on the back of buffaloes.

Flower: During religious ceremonies such as *Shiva Pooja, Durga Pooja* and *Vinayaka Pooja,* flowers are offered to gods; flowers also yield a blue dye; used as a litmus substitute.

24. *Codariocalyx gyroides* (Link) Hassk.

Synonyms: *Desmodium bracteatum sensu* M. Micheli
Desmodium diversifolium sensu auct.
Desmodium gyroides (Link) DC.
Desmodium pseudogyroides Miq.

Distribution: Native of Cambodia, China, Hainan, India, Indonesia, Malaysia, Myanmar, Nepal, Philippines, Srilanka, Thailand and Vietnam; grows in Arunachal Pradesh, Assam, Bihar, Meghalaya, Naagaland, West Bengal and Himalaya in India.

Description: An erect under shrub, with finely glabrous branches; leaves 1-3 foliolate; stipules lanceolate, acuminate; leaflets unequal, terminal one long, oblong lanceolate, obtuse at both ends, glabrous above, silky beneath, lateral leaflets much smaller, obsolete; bracts acute, acuminate; calyx campanulate, teeth deltoid, shorter than tube; flowers pink; pod, pubescent, falcate, flat, joints 6-10, ventral suture indented with dense coating of reddish brown hairs.

Medicinal Uses

Plant: Paste of the aerial parts with leaves of *Oscimum sanctum* is put inside a banana and given to women for conception.

Leaf: Paste is used as poultice for lumbago.

Ethnomedicine

Plant: Is used in lumbago and rheumatism; promotes conception.

Non-medicinal Uses

Plant: Is used as green manure and fodder crop; grown as cover crop.

25. *Codariocalyx motorius* (Houtt.) H. Ohashi

Synonyms: *Codariocalyx gyrans* (L.f.) Hassk.
Desmodium gyrans (L.) DC.
Desmodium motorium (Houtt.) Merr.
Hedysarum gyrans L.f.

Common Names: Moving plant, Telegraph plant.

Vernacular Names: Kan: Elemudure, Naagathagare, Nagargare, Telegraph gida; **Mal:** Naagatakari; **Tam:** Thandicheri, Thozhuganni; **Tel:** Toluganni.

Distribution: Native of Asia including India and Australia; grows throughout Assam, Arunachal Pradesh, Himalaya, Jammu and Kashmir in India.

Description: Erect under shrub; leaves 3-foliolate; leaflets linear, lanceolate, ovate-oblong, obtuse, pubescent beneath; flowers pink, solitary, in the axil of broadly ovate, caducous bracts; calyx teeth very short; wing petal adhering to the keel; style incurved; pod oblong, pubescent, indented along lower suture with 8-10 indistinct septa.

Phytochemistry

Root: Indole-3-alkylamines and N,N-dimethyl tryptophan Nb-oxide.

Leaf: Bufotenine, choline, 5-methoxy-N,N-dimethyl tryptophan-Nb-oxide, β-phenylethylamine.

Medicinal Uses

Root: Used as antidysenteric, emollient and laxative; useful for asthma, cough and fever.

Leaf and Pod: Aphrodisiac, diuretic, febrifuge and tonic; leaf and pod paste is applied to wounds.

Uses in Alternative Systems of Medicine

Siddha

Part used: Whole plant.

Uses: Useful in wound healing and as antidote to snake poison.

Non-medicinal Uses

Plant: Is used as green manure, fodder and grown as cover crop.

26. *Crotalaria albida* Roth.

Synonyms: *Crotalaria deflexa* Benth.
Crotalaria formosana Ito & Matsum
Crotalaria hossei Craib
Crotalaria montana Roxb.

Common Name: Hill rattle wort.

Vernacular Names: Hin: Ban methi; **Kan:** Giligili gida; **Tel:** Konda giligichha.

Distribution: Native of Asia including India and Australasia; it grows throughout India.

Description: Branched, erect, perennial herb with silky pubescence; leaves, linear, oblanceolate, obtuse, retuse, apiculate, gland dotted, glabrous above silky beneath; inflorescence is 6-12 flowered terminal raceme with yellow flowers; stipules absent; bracts and bracteoles linear; calyx with three linear teeth and upper two teeth broad; pod sessile, oblong, glabrous, 2-5 seeded.

Medicinal Uses

Plant: Extract is used in the treatment of impetigo and scabies.

Root: Used as purgative.

Toxicology: Pyrrolizidine alkaloids in the seeds are hepatotoxic and carcinogenic causing liver damage, edema, liver fibrosis, biliary hyperplasia, pulmonary arterial hypertension, ventricular hypertrophy and veno-occlusive disease.

Ethnomedicine

Plant: Is used as purgative.

27. *Crotalaria burhia* Benth.

Vernacular Names: Guj: Ghugaro; **Hin:** Lathia karsan; **Mar:** Ghagari; **Pun:** Bhata, Kauriala, Khippi, Lathia, Pola, Sissi; **Raj:** Sannia.

Distribution: Native of Afghanistan, India, Iran and Pakistan; it grows in Gujarath, Punjab, Rajasthan and Uttar Pradesh in India.

Description: An erect procumbent shrub; old branches end in spines; leaves simple, oblong, lanceolate, rigid, silky, pubescent on both sides; stipules absent; 6-12 yellow flowers on elongated terninal raceme; calyx densely silky, teeth lanceolate; corolla yellow, exerted; pod oblong, villous and 3-4 seeded.

Phytochemistry

Plant: Croburhine, crotalarine, deguelin, elliptone, rotenone, samatrol, tephrosin and toxicarol.

Seed: β-Sitosterol, croburhine, monocrotaline, n-hentriacontane, patulin, squalene, stearic acid and triacontanol.

Medicinal Uses

Plant and Leaf: Used as cooling medicine in burns and swellings; useful in gout and hydrophobia.

Ethnomedicine

Plant: Is refrigerant; stem paste is used as external application on burns.

Non-medicinal Uses

Plant: Yields fiber used for cordage, coarse gunny cloth and canvas; used as fodder.

28. *Crotalaria juncea* L.

Synonyms: *Crotalaria benghalensis* Lam.
Crotalaria fenestrata Sims
Crotalaria porrecta Wallich
Crotalaria sericea Willd.
Crotalaria tenuifolia Roxb.
Crotalaria viminea Wallich

Notes: Ranks next in importance to jute as a fibre crop; a major fiber crop in Punjab; one of the oldest known fiber crops in the Indo-Pakistan subcontinent, mentioned in the ancient Sanskrit literature.

Common Names: Bombay hemp, Brown hemp, False hemp, Indian hemp, Sun hemp.

Vernacular Names: Ben: Shonpat, Ghore; **Hin:** Ghagahi, Patashana, San, Shanahuli, Sunn, Sannai; **Kan:** Apsenabu, Senabu, Sanna senabu, Senabina gobbara, Senabu; **Mal:** Wuckoo nar; **Mar:** Ghagharu; **Sans:** Dhanahaari, Deerghapallava, Deerghashakla, Karnasula, Katuthiktha, Malyapushpa, Nishadana, Sana, Satinaka, Shana, Thvakasara, Vamaka; **Tam:** Chanai, Chanal, Jenaa saanar, Kuttiram, Kuttu, Manji, Nanandam, Sanal, Sannapu, Sanarppai, Uttirabanni, Umadji, Vakkunnar; **Tel:** Gilaka, Gilgichu, Giliginta, Janumu, Shanambo.

Distribution: Native of India; introduced into other countries; grows throughout India; often cultivated.

Description: Herbaceous shrubby annual covered with hairs; stipules minute, pointed, pubescent on both sides; petiole with pulvinous base; leaves simple, linear, elliptic to oblong, dark green; inflorescence a terminal raceme with small linear bracts; flowers showy, deep yellow; calyx hairy, united at the base with pointed lobes; corolla with the standard petal round, streaked with purple on the dorsal surface, keel petal twisted; stamens 10, 5 with short filaments and long narrow anthers, 5 with long filaments and small round anthers; pod inflated, grooved along the upper surface with short pointed beak, brown, soft and hairy; seeds numerous, flat, black or dark grey.

Phytochemistry

Plant: α-Hydroxy norleucine, apigenin-7,4'-O-diglucoside, apigenin-7-glucuronide, cellulose, essential oil, fat, fibre, galactomannan, junceine, oleic acid, pectin, pyrrolizidine alkaloid, resin, riddelline, senecionine, seneciphylline, trichodesmin, β-hydroxy-N-methyl-norvaline and trichodesmine.

Seed: α-Hydroxy-norleucine, apigenin-7,4-O-diglucoside, apigenin-7-glucuronide, galactomannan, galactose, genistein, junceine, linoleate, linoleic acid, linolenic acid, mannose, methyl esters, myrsinate, oleate, oleic acid, palmitate, stearate, trichodesmine and xylose.

Seed oil: Linoleic acid and linolenic acid.

Edibility

Flower: Eaten as vegetable.

Medicinal Uses

Plant: Is abortifacient, antibacterial, anticancer, antifertility, antihaemolytic, antihypertensive, anti-inflammatory, anti-ischaemic, antioxidant, antimicorbial, anti mitogenic, antitussive, cardioprotective, chemopreventive, cytotoxic, demulcent, diuretic, emollient, estrogenic, flatulent, fungicide, haemostatic, hepatotoxic, hypoglycaemic, immunostimulant, laxative, and vasodilator; useful in treating leukaemia, lymphoma, melanoma, neuroblastoma, obesity, osteoporosis, tumour of the breast, colon, ovary, prostate and stomach, ulcer and uterine complaints.

Root: Useful in colic and is astringent in epistasis.

Leaf: Is abortifacient, demulcent, emetic, emmenogogue and purgative; used externally as poultice and given internally in the form of infusion in gastric and bilious fever accompanied by skin diseases, such as impetigo, scabies and psoriasis.

Flower: Used as antifertility agent.

Seed: Used to purify blood in febrile state accompanied by cutaneous eruption such as impetigo and psoriasis; considered cooling and emmenogogue.

Bioactivity: Plant is a inhibitor of α-reductase, catechol-O-methyl-transferase, DOPA-decarboxylase, histidine-kinase, lipase, MAO, NADH-ubiquinone-oxidoreductase, ornithine-decarboxylase, PTK, peroxidase, topoisomerase I and II and tyrosine-kinase; is β-blocker, calcium antagonist, pituitary sensitizer and quinone reductase inducer.

Toxicology: Junceine in the seeds cause biliary arterial hypertension, hyperbilirubinemia, hypoalbuminuria, hepatocarcinogenecity, jaundice, liver damage, liver cancer, liver fibrosis, liver lesions, pulmonary arterial hypertension and neuromuscular blocking; used in large doses it is abortifacient and hepatotoxic; seeds contain trypsin inhibitors.

Uses in Alternative Systems of Medicine

Ayurveda:

Part used: Plant and seed.

Uses: Plant is abortifacient, emetic, emmenogogue and laxative; cures severe body ache; seed is used in impetigo and psoriasis.

Siddha:

Part used: Root, leaf and seed.

Uses: Useful in amenorrhoea, dysentery, fever, poisoning and "vatha".

Ethnomedicine:

Plant: Is a blood purifier, depurative and emmenogogue; used in impetigo, psoriasis and skin diseases.

Flower: Is an antifertility agent.

Seed: Is used in skin diseases and as a blood purifier.

Non-medicinal Uses

Plant: One of the most commonly cultivated fiber crops in India, next in importance to jute as a bast fiber crop; yields a high quality fiber, which is stronger than jute and is used in making twines, sacks, canvas and cordage; used as fodder, green manure and cover crop in coffee, tea, coconut and rubber plantations; the waste, after extraction of the fiber is used in the manufacture of paper.

Bark: Used in making cords, tarpaulins, soles of shoes, sandals, canvas, bags, cot strings, carpets, table linen, rugs, mattresses, fishing nets, mats, tissue and cigarette paper; the sacred thread of *Kshatriya*'s is made from bark fibre.

Leaf and seed: Is used as green manure and fodder.

Seed: Fed to cows for milking.

29. *Crotalaria laburnifolia* L.

Synonyms: *Clavulium pendunculosum* (Desv.) Desv.
Crotalaria pendunculosa Desv.

Notes: Five subspecies, *lab, petiolaris, eldomae, tenuicarpa* and *australis,* are recognized.

Common Names: Bird flower, Large rattle wort.

Vernacular Names: Hin: Muna; **Kan:** Dodda baage kaddi; **Mal:** Keliy elkka, Nellutantalakotte; **Tam:** Kiluguppai, Tamarochi, Tindagam; **Tel:** Chikuragilligiccha, Peddagilligista.

Distribution: Native of Africa, Asia including India and Australasia ; in India it grows in Western Peninsula.

Description: Erect shrub; leaves 3 foliolate; leaflets ovate, elliptic, glabrous above, pubescent beneath; yellow flowers on a terminal raceme; bracts lanceolate; corolla, yellow, glabrous, exerted; pod glabrous, oblong, long stalked; seeds 20-30.

Phytochemistry

Seed: β-Sitosterol, anacrotine, crotalaburmine, flavonoid glycosides, lutexine and necine.

Medicinal Uses

Plant: Plant infusion is used as gargle for inflammation of the mouth and sore throat.

Seed: Anti-inflammatory and antispasmodic.

Ethnomedicine

Plant: Plant is used as gargle for inflammation and sore throat; plant decoction taken internally cures stomatitis.

30. *Crotalaria medicaginea* Lam.

Synonym: *Crotalaria foliosa* Willd.

Common Name: Trefoil rattle pod.

Vernacular Names: Guj: Zensue; **Hin:** Gulabi; **Mar:** Jenjaru; **Pun:** Gulabi.

Distribution: Native of Asia including India and Australasia; in India it grows throughout the plains.

Description: Prostrate herb with filiform stem, sparsely pubescent with divaricate branches; leaves 3-foliolate; leaflets narrow, emarginate, oblanceolate, oval, truncate; bracts minute, linear, subulate, clothed with appressed hairs; stipules minute, filiform; flowers small, yellow on terminal leaf opposed 2-6 flowered raceme; calyx with silky hairs and linear teeth; pod obliquely subglobose, beaked, pubescent and 2 seeded.

Phytochemistry

Plant: β-Sitosterol, n-hentriacontane and patulin.

Root: Prenylated chalcone and 3',5'-di(-prinyl-4,2',4')-trihydroxy chalcone.

Stem bark: Chrysoseriol-7-rutinoside and vitexin-4'-0-xyloside.

Seed: β-Sitosterol, holiotridine, pyrrolizidine, methoxysupinidine, n-hentriacontane and patulin.

Ethnomedicine

Seed: Decoction used for toxaemia.

Non-medicinal Uses:

Plant: Used as fodder.

Seed: Used as cattle feed after cooking with common salt.

31. *Crotalaria pallida* Aiton

Synonyms: *Crotalaria brownei* DC.
Crotalaria falcata DC.
Crotalaria fertilis Del.
Crotalaria hookeri Arn.
Crotalaria mucronata Desv.
Crotalaria siamica F. Williams
Crotalaria striata DC.
Crotalaria tinctoria Baillon

Notes: Two varieties, *pallida* and *obovata,* are recognized.

Common Names: Smooth crotalaria, Streaked rattle pod.

Distribution: Native of Africa, Asia including India and Central America; introduced into Australasia, Australia, Caribbean Islands, Indian Ocean, Pacific Ocean Islands and South America; grows throughout India.

Description: An erect annual with short adpressed hairs; leaves 3-4 foliolate; leaflets elliptic to obovate, glabrous above, thinly puberulous beneath; stipules filiform; numerous flowers on raceme; bracts linear; bracteoles inserted at the base of the calyx, filiform; calyx basally truncate and deflexed, lobes narrow, acuminate; corolla with the standard petal yellow with reddish brown wings, wing petal shorter than the keel, keel round in the middle with a narrow projecting beak; pod linear, oblong, stipitate, incurved, pubescent with 30-40 seeds; seeds oblique, cordiform, smooth, papillose, mottled, ochre or dark grey or brown, glossy or dull.

Phytochemistry

Bark: Apigenin and vitexin-4'-O-xyloside.

Leaf: Crotolarol, vitexin-4'-O-xyloside and vitexin.

Seed: Chrysoseriol-7-rutinoside, crotostriatine, galactomannan, luteolin, mucronatine, mucronatinine, nilgirine, usaramine and vitexin.

Medicinal Uses

Plant: Plant infusion is used as eye wash.

Root: Is used for swelling of joints; tied to wrist and neck of a person suffering from dropsy.

Leaf: Is used as vermifuge.

Bioactivity: Total alkaloid fraction showed antitumour activity against Walker 256 and sarcoma 180 in rats.

Toxicology: Causes convulsions, liver cirrhosis, nausea and asyphyxiation.

Uses in Alternative Systems of Medicine

Ayurveda:

Part used: Whole plant, leaf and seed.

Uses: Plant is useful in bleeding piles, diarrhoea and dysentery; leaf paste is applied locally in leprosy, skin diseases and swellings; seed paste is used to cure ulcers.

Non-medicinal Uses

Plant: Cultivated for its ornamental flowers; grown as pasture, forage and cover crop in tea, rubber and coconut plantations; yields a black dye.

Stem bark: Is a source of fibre used for cordage.

Seed: Used as a substitute for coffee.

32. *Crotalaria retusa* L.

Synonyms: *Crotalaria retusifolia* Stokes
Dolichos cuneifolius Forsskal
Lupinus cochinchinensis Lour.

Notes: Two varieties, *retusa* and *tunguensis*, are recognized.

Common Names: Golden yellow sweet pea, Rattle weed, Shack-Shack, Wedge leaf rattle pod, Wedge leaved crotalaria and Yellow rattle Box.

Vernacular Names: Ben: Bil-jhunjhun; **Hin:** Ghunghunian; **Kan:** Dingala, Gejje gida; **Mal:** Kaatu kirukirakkayi, Kutiri, Tantalakkotti; **Mar:** Ghagri; **Sans:** Sanapushpi, Shanarghandika; **Tam:** Kilukilluppai; **Tel:** Pottigiligichha.

Distribution: Native of Asia including India, Australasia, Indian Ocean Islands, Middle East, Pacific Ocean Islands and Guyana; in India, it grows throughout; often cultivated.

Description: Erect undershrub with furrowed branches covered by white pubescence; leaves simple, obovate, oblanceolate, retuse at the apex, base cuneate; inflorescence is a raceme with 18-25 flowers; flowers yellow tinged with purple; calyx teeth lanceolate, upper teeth ovate, acute, the lower 3 narrow; corolla glabrous; pod stalked, glabrous with 15-20 stipitate seeds.

Phytochemistry

Plant: Isatinecine, monocrotaline, monocrotaline-N-oxide, necine, retronecine, retusamine, retusamine-N-oxide and retusin.

Leaf: Indican.

Seed: Isoquercetin, monocrotaline, quercetin, quercetin-3-O-glucoside, retusanine, retronecine, retusine, spectabiline and vitexin-4'-0-galactoside.

Seed oil: Linoleneic acid, malvalic acid, methyl esters, myristate, oleate, oleic acid, palmitic acid, ricinoleic acid, sterculic acid and tridecanoate.

Medicinal Uses

Plant: Antileukaemic, antitumour, astringent, cardiac depressant, CNS depressant, emetic, expectorant, hypotensive and spasmolytic; useful in cardiac disorders, dyspepsia, fever, impetigo, scabies and "vatha".

Root: Root crushed and mixed with spices is a remedy for colic.

Leaf: Used for curing diarrhoea, fever, impetigo, ptyalism and scabies.

Seed: Depurative, digestive and febrifuge; is useful in fever, flatulence, leprosy and skin diseases; powdered and boiled with milk useful in increasing body strength.

Toxicology: Pyrrolizidine alkoloids present can cause hepatotoxicity, liver lesions and liver damage indirectly through milk of cow which has fed on the plant; retusine can cause veno-occlusive disease, cardiotoxicity and pneumotoxicity.

Medicinal Uses Outside India

Gambia and **Sierra Leone:** The root is crushed and mixed with spices as a remedy for colic; decoction of the leaves is given for fever.

Ethnomedicine

Plant: Is used for fever, haemoptysis, impetigo and scabies.

Non-medicinal Uses

Plant: Used as pesticide; yields fiber; used in making canvas and cordage.

33. *Crotalaria spectabilis* Roth.

Synonyms: *Crotalaria altipes* Raf.
Crotalaria macrophylla Wienmann
Crotalaria sericea Retz.

Notes: The plant is referred as 'pulmonary hypertension plant'.

Common Names: Rattle box, Showy crotalaria, Showy rattle pod.

Vernacular Names: Ben: Pipulijhunjhun, Jhumjhumo, **Hin:** Jhunjhunia; **Kan:** Ghantaarava; **Mal:** Kaatu tandalekotti; **Sans:** Ghantaarava.

Distribution: Native of Asia including India; introduced into other countries; in India, it grows in Karnataka, Punjab and Rajasthan.

Description: Annual herb with angled, subglabrous branches; leaves simple, oblanceolate or obovate, glabrous above, appressed pubescent beneath; stipules obliquely oblong–ovate; bracts cordate, acute or acuminate, foliaceous; bracteoles inserted near the middle of the pedicel and are lanceolate; raceme many flowered; flowers yellow with purplish tinge; calyx tube protracted on lower side, glabrous with triangular lobes longer than the tube; corolla yellow with purplish tinge, glabrous outside, wing petal oblong–obovate, longer than the keel petal, keel petal round in the middle, with a incurved twisted beak; pod short, stipitate, oblong, clavate, glabrous, 20-25 seeded; seeds oblique, cordiform, strongly incurved on the narrow end, smooth and brown.

Phytochemistry

Plant: Monocrotaline, retusine and spectabiline.

Root: α-Carotene, monocrotaline, retusine, sericine, spectabiline and triacontane.

Medicinal Uses

Plant: Is used in the treatment of impetigo and scabies; plant juice is anthelmintic and spasmolytic.

Bioactivity: Fulvin and monocrotaline exhibited prolonged hypotensive effect in dogs and depressant effect on rabbit heart and a stimulant effect on smooth muscle of ileum and uterus; causes pulmonary arterial hypertension and right ventricular hypertrophy in mice and monkey; the compound exhibits activities, which are characteristic of parasympathomimetic drug.

Toxicology: The vegetative parts and seeds contain cronaburmine, which cause liver fibrosis, liver damage, lesions, biliary hyperplasia, pulmonary arterial hypertension, liver cancer, hyperbilirubinemia, hypoalbuminuria, jaundice and edema through consumption of milk and meat of cows and sheep which consume the plant.

Ethnomedicine

Plant: Is anthelmintic; used in rheumatism and scabies.

Non-medicinal Uses

Plant: Grown as an ornamental plant, yields fodder and fiber; grown for erosion control.

34. *Crotalaria verrucosa* L.

Synonyms: *Crotalaria acuminata* (DC.) G. Don.
Crotalaria angulosa Lam.
Crotalaria caerulea Jacq.
Crotalaria coerulea Beddome
Crotalaria flexuosa Moench
Crotalaria mollis J.A. Weinm
Crotalaria semperflorens Vent.
Crotalaria verrucosa L. var., *acuminata* DC.
Crotalaria wallichiana Wight & Arn.

Common Names: Blue rattle box, Blue rattle weed, Purple rattle box.

Vernacular Names: Ben: Thanjana; **Hin:** Banshana, Jhanjhania; **Kan:** Gijigiji gida; **Mar:** Ghagari, Tirat; **Sans:** Dhavanti, Sanapushpi; **Tam:** Killupai, Vattakillu; **Tel:** Ghelegherinta.

Distribution: Native of Asia including India; grows throughout India.

Description: Erect herb with 3-4 angled branches; stipules foliaceous, semilunate; leaves ovate, acute or obtuse, membranous; inflorescence is 8-14 flowered raceme; flowers blue or white; bracts subulate; corolla exerted, glabrous; pod exerted, pubescent, 16-20 seeded.

Phytochemistry

Leaf: Necic acid and propionic acid.

Flower: Acetyl isosenkirkin, flavonolglycoside, kaempferol, vitexin -4'-xyloside and vitexin-O-glycoside, isovitexin and kaempferol-7,4-di-O-α-rhamnopyranoside.

Seed: β-Sitosterol, O-acetylcrotaverrine, apigenin-O-glycoside, crotalburine, crotaverrine, isosenkirkine,

Medicinal Uses

Plant: Is anthelmintic, anti-HIV, antibacterial, anticancer, antiedemic, anti-inflammatory, antinociceptive, antiviral, apoptotic, cardiac depressant, cytotoxic, hepatotoxic, hypotensive and uterine tonic; useful in edema, leukaemia, malaria and melanoma; fresh juice is taken for fever.

Root: Roots crushed and mixed with spices is remedy for colic.

Stalks: Tender stalk is prescribed for impetigo and scabies.

Leaf: Leaf juice is taken internally and applied externally in impetigo and scabies and to diminish salivation; leaves are used as poultice for headache; decoction of leaves is useful in fever.

Bioactivity: Plant is an in inhibitor of phospholipase and prostaglandin synthesis; ethanol extract (50%) of plant is CNS depressant and diuretic in lab animals.

Uses in Alternative Systems of Medicine

Ayurveda:

Part used: Root and leaf.

Uses: Root is useful in blood disorders, diarrhoea, dysentery and leprosy; leaf is expectorant and emetic; cures biliousness, dyspepsia, fever, indigestion, mouth and throat diseases.

Siddha:

Part used: Leaf.

Uses: Useful in skin disorders and as vermifuge.

Ethnomedicine

Plant: Is used for headache, impetigo and scabies.

35. *Cullen corylifolium* (L.) Medikus

Synonyms: *Cullen corylifolia* (L.) Medikus
Cullen corylifolius (L.) Medikus
Psoralea corylifolia L.
Trifolium unifolium Forsskal.

Vernacular Names: Ben: Bavachi, Latakasturi; **Hin:** Babchi, Bakuchi, Baavachi; **Kan:** Somaraji, Bavanchi gida; **Mar:** Bavachya; **Ori:** Bakuchi; **Sans:** Bakuchi, Krishnapala, Kushtanashini, Somaraji; **Tam:** Karpoka arisi, **Tel:** Bavanchalu, Kalaginja; **Urd:** Bakuchi.

Distribution: Native of Africa, India and Middle East; grows through out the plains in India.

Description: An erect annual with grooved stem and branches studded with conspicuous glands and few white hairs; leaves simple, elliptic, round and mucronate at the apex, clothed with white hairs on both the surfaces, base cuneate; petioles hairy and gland dotted; stipules lanceolate, persistent; flowers bluish purple on axillary 10-30 flowered racemes; calyx with teeth, hairy outside, upper teeth linear, lanceolate, the lower ovate and twice as long as the upper; corolla with clawed orbicular standard petal; pod ovoid-oblong, compressed, pitted, mucronate, black, glabrous; seed single.

Phytochemistry

Plant: Angelicin, bakuchiol, corylifolin, coumarin, geranylacetate, limonene, linalool, psoralen and psoralidin.

Root: 8-Methoxypsoralen, angelicin, angelin, bavachalcone, bavachromanol, chalcones, corylifolin, corylifolinin, corylin, methoxychalcone, corylinal, coumesterol, isobavachalcone, flavones, furanocoumarins, isoflavones, isopsoralen, methoxy chalcone, neobavachalcone, psoralen, psoralidin, sitosterol and trilaurin.

Seed: β-Caryophyllene, β-sitosterol, 8-methoxypsoralen, angelicin, bakuchalcone, bakuchicin, bakuchinin, bakuchiol, bavachalcone, bavachromanol, isobavachin, bavachinin, chalcones, corylidin, corylifolin, corylifolinin, corylin, corylinal, coumesterol, essential oil, fixed oil, flavones, flavonoids, furanocoumarins, isobavachalcone, isoflavones, isoneobavachalcone, isopsoralen, isopsoralidin, isopsoralone, linalool, corylin-7-hydroxy-6'-methoxychalcone, methylpsoralenol, monoterpenoid, neobavaisoflavone, parachromanol, psoralen, psoralenol, psoralidin, psoralone, resin, sophorocoumestan, stigmasterol, triacontane, trilaurin and triterpenoid.

Edibility

Seed: Edible.

Medicinal Uses

Leaf: Used in curing diarrhoea.

Pod: Aphrodisiac, bitter, diaphoretic, diuretic and laxative; recommended for anaemia, asthma, bronchitis, difficult micturition, inflammation, leprosy, leucoderma, piles, psoriasis, skin diseases and vomiting; improves complexion and promotes hair growth.

Seed: Acrid, analgesic, anthelmintic, antibacterial, anti-inflammatory, antipyretic, aphrodisiac, bitter, diaphoretic, diuretic, laxative, stimulant, stomachic and tonic; useful in bilious affections, chronic skin diseases, dermatitis, kidney troubles, leprosy, leucoderma, mucous membrane disorders, psoriasis, scabies, ulcers and vitiated conditions of 'pitta'; oleoresinous extract of seeds diluted with chalmurga oil is applied in leucoderma; used as antidote for scorpion sting and snake bite.

Seed oil: Anthelminitic, antifungal, diuretic and stomachic; used in anaemia, anasarca, bacterial skin infections, baldness, leucoderma, psoriasis, respiratory disorders, skin diseases and vomiting.

Bioactivity: Psoralen administered orally for 2-5 months at 30ug/day affected detoxification mechanism in rabbits and caused hypertrophy of liver, kidney and spleen.

Uses in Alternative Systems of Medicine

Ayurveda:

Part used: Pod, seed and seed oil.

Uses: Pod is used in curing anaemia, cough, diabetes, dysuria, edema, epilepsy, fever, inflammation, leprosy, leucoderma, piles, skin diseases, ulcer and urinary disorders; promotes digestion, imparts vigour and vitality, has a cooling effect; seed and seed oil is useful in anorexia, constipation, edema, intestinal worms, leprosy, leucoderma, polyuria and rheumatism.

Chinese:

Part used: Seed.

Uses: Useful as stimulant and tonic; given in impotency, leucorrhoea, polyuria and spermatorrhoea.

Siddha:

Part used: Seed.

Uses: Useful as anthelmintic, carminative¸ deobstruent, diaphoretic, laxative, stimulant, stomachic and tonic; seed paste is used as external application to glandular swellings, leprosy, leucoderma and skin diseases.

Unani:

Part used: Seed.

Uses: Used as anthelmintic, antiflatulant, appetizer, blood purifier, carminative and stomachic; used in cold, cough, impurities of blood, bile and phlegm, phlegmatic fever, pityriasis and leprosy.

Medicinal Uses Outside India

China: The seeds are regarded as aphrodisiac and tonic to reproductive organs; they are used to treat back ache, impotency, menstrual disorders, nocturnal emissions and stomach ache; seed paste is applied in leucoderma.

Non-medicinal Uses

Plant: Used as fodder and manure.

Seed: Used for making perfumed oil.

Seed cake: Used as manure and cattle feed.

36. *Cyamopsis tetragonoloba* (L.) Traub.

Synonym: *Cyamopsis psoralioides* (Lam.) DC.

Notes: This species is considered indigenous to India and Pakistan, but some taxonomists consider that it arose from trans-domestication of the African native *Cyamopsis senegalensis;* seeds yield a gum called guar gum, a valuable commercial product, as it has 5 to 8 times more thickening power than starch.

Common Names: Cluster bean, Siam bean.

Vernacular Names: Ben: Gauri; **Guj:** Guwar; **Hin:** Bharathi, Bhavachi, Gowar; **Kan:** Gorikayi; **Mar:** Bavachi; **Pun:** Guar, Kutti; **Sans:** Bakuchi, Dridabheeja, Goraksha, Gorani, Phalini; **Tam:** Kothaveray, Seeniavarai; **Tel:** Gorichikkudu.

Distribution: Native of India and Sri Lanka; introduced into several countries in Africa, Asia and Middle east; widely cultivated.

Description: An erect annual clothed with appressed grey hairs; stem grooved; leaves 3-foliolate; stipules linear, subulate, persistent; leaflets elliptic, acute, sharply dentate, teeth ending in a weak spine; flowers, small, yellowish white on 6-30 flowered raceme; bracts linear, subulate, persistent; calyx hairy outside, teeth unequal, the lower 2 long, upper 3 short and triangular; corolla longer than the calyx; standard petal orbicular, shortly clawed; pod thick and fleshy with gritty surface found in clusters; seeds 5 or 6, square, slightly compressed, yellowish white.

Phytochemistry

Plant: Galactomannan, D-galactose and D-mannose.

Leaf: Astragalin, calcium, carbohydrates, chlorogenic acid, daidzein-7-0-β-D-glucoside, kaempferol, kaemp ferol-7-0-β-D-glucoside, myricetin, myricetin-7-0-β-D-glucoside, protein, quercetin-3-0-β-D-rutinoside, texais-7-0-β-D-glucoside and trihydroxy benzoic acid.

Pod: Ascorbic acid, β-carotene, calcium, carbohydrates coniferyl-alcohol, fiber, fluoroacetic acid, gallic acid, gentisic acid, iron, juglanin, kaempferol, p-hydroxycinnamyl alcohol, phosphorus, protocatechuic acid and quercetin.

Seed: Astragalin, caffeic acid, chlorogenic acid, ellagic acid, epikatonic acid, gallic acid, kaempferol-3-O-rutinoside, kaempferol-7-O-β-D-glucoside, myricetin and myricetin-7-0-β-D-glucoside.

Seed oil: Linoleic acid, myristic acid, oleic acid, palmitic acid and stearic acid.

Guar gum: Galactomannan and 3-epiketonic acid.

Edibility

Leaf: Used as vegetable; young leaves are eaten like spinach.

Pod and Seed: Used as vegetable.

Medicinal Uses

Plant: Is acidulant, analgesic, antiallergenic, antiandrogenic, antiarrhythmic, and antitumour useful in anaemia, acne, alopecia, anaemia, atherosclerosis, arthritis, asthma, bradycardia, bronchitis, cancer of breast, colon, stomach, liver and skin, candidiasis, cataract, cold, cystitis, depression, dermatitis, diabetes, eczema, edema, fibrosarcoma, fistula, gingivitis, glaucoma, tumour of breast, colon, stomach, liver, lung, mouth, ovary, skin and ulcer; antidote to aluminium, cadmium and lead poisoning; used as bulk forming laxative.

Guar gum: Is appetite suppressor, antidiabetic, hypoglycaemic and hypolipidemic; used in diabetes.

Leaf: Is eaten to cure night blindness.

Pod: Is hypoglycaemic and laxative; useful in biliousness and night blindness.

Seed: Appetizer, cooling, digestive, galactagogue, laxative and tonic; useful in agalactia, anorexia, constipation, dyspepsia¸ nyctalopia and vitiated 'pitta'; seed poultice used for swellings caused by fracture and sprain; seed ash is mixed with coconut oil and applied to boils.

Bioactivity: Plant is inhibitor of 5-α-reductase, 5-lipoxygenase, ACE, ATPase, aldose-reductase, β-glucuronidase, COMT, cyclo-oxygenase, cytochrome-P_{450}, deiodinase, glucosyl-transferase, histamine, iodothyronine-deiodinase, leukotriene, lipoxygenase, lyase, MAO-A, MMP-9, metalloproteinase, NADH-Oxidase, NEP, NO-synthase, ornithine-decarboxylase, PTK, phospho-lipase, prostaglandin synthesis, protein-kinase-C, topoisomerase I and II, tyrosinase, tyrosine-kinase, xanthine-oxidase and CAMP-phosphodiesterase; is angiotensin-receptor blocker, β-adrenergic receptor blocker, β-blocker, calcium channel blocker and quinone-reductase inducer; bean diet caused hypoglycaemia and hypocholesterolemia in albino rats.

Toxicology: Gum is appetite suppressor, hypoglycaemic and hypolipidemic; seeds contain trypsin inhibitors.

Clinical Studies: Addition of bean meal in human diet reduced blood glucose, plasma insulin, serum cholesterol, total lipids and uric acid; it is effective in glycosuria in noninsulin dependent diabetics.

Uses in Alternative Systems of Medicine

Ayurveda:

Part used: Leaf and pod.

Uses: Useful in diseases of "pitta", biliousness, diabetes, indigestion and night blindness.

Ethnomedicine

Plant: Used in hepatomegaly, night blindness, plague, smallpox, swelling and sprain; used as laxative.

Seed: Is used in swelling and sprain.

Non-medicinal Uses

Plant: Is used as fodder and green manure; grown as cover crop.

Guar gum: Is used in food, cosmetic, paper and textile industry; it improves strength of paper; it is used in printing paste, as adhesive for postage stamps; triacetate of galactomannan from the gum is drawn into strong transparent and flexible films; used as binder in tablets; used as thickner in emulsion; used as stabilizer for ice creams, soups, cheese, pharmaceutical jellies, lotions, hair creams, skin creams and tooth paste; as a floculant and coagulant in mining industry; used as salad dressing.

Leaf: Is used as forage.

Pod and Seed: Used as fodder; added as protein supplement to cattle feed.

Seed flour: Is used to strengthen paper; used in ice creams and salad dressing.

Seed cake: After the extraction of gum it is used as protein rich fodder.

37. *Cytisus scoparius* (L.) Link

Synonyms: *Sarothamnus bourgaei* Boiss.
Sarothamnus oxyphyllus Boiss.
Sarothamnus scoparius (L.) Koch
Spartium scoparium L.

Common Names: Broom plant; Scotch broom, Scots broom.

Distribution: Native of Europe, introduced into India; grows in Himachal Pradesh, Punjab and Tamil Nadu in India; naturalized in hill stations.

Description: A shrub of 3 to 5 feet with numerous, long, straight slender bright green branches, which are angled; leaves are alternate, hairy, with 3 small oblong leaflets, the upper ones are sessile; flowers bright yellow, fragrant, borne on axillary stalks either solitary or in pairs; pod flat, oblong, hairy on the edges, smooth on the sides.

Phytochemistry

Plant: Carotenoids, chrysanthemaxanthin, genistein, hydroxytyramine, sparteine and xanthophylls.

Flower: Isorhamnetin-3-β-glucoside, kaempferol, kaempferol glucoside and isoquercetin.

Seed: Sarothamnoside.

Seed oil: Cholesterol, campesterol, stigmasterol and β-sitosterol.

Edibility

Flower and Bud: Pickled and eaten; used as condiment in England; preserved in vinegar it was a dish on every table during coronation festival of James II.

Medicinal Uses

Plant: Antihaemorrhagic, cathartic, depressant, diuretic, emetic, oxytocic and purgative; induces labour; used in kidney affections and tachycardia; constricts peripheral blood vessels in profuse menstruation; regulates the heart and reduces edema of cardiac origin.

Twigs: Before flowering used as diuretic in dropsy and as heart tonic; useful in dysuria and gravel; in large doses emetic and purgative.

Leaf: Diuretic, emetic and purgative.

Toxicology: Cyanaogenic glycoside lupanine in the seeds causes hypertension and hyperglycaemia and toxic in large doses; sparteine is emetic, narcotic, poisonous and purgative in large doses.

Clinical Studies: Increases cardiac output and stimulates neurotransmitter in the brain; exhibits sympathomimetic action on alpha and β-adrenergic receptors.

Uses in Alternative Systems of Medicine

Chinese

Part used: Root and flower.

Uses: Used for cough and cold; decoction of the flower is used to bring out the small pox eruptions.

Homoeopathy

Part used: Tincture of the plant.

Uses: Useful in asthma, cramps, diarrhoea, dysentery, dry cough, inflammation of the eye, hoarseness of voice and swellings.

Ethnomedicine

Plant: Is abortifacient, cardiotonic, diuretic, emetic, purgative, sedative and vasoconstrictor; used for arrhythmia, bruises, cold, cough, dropsy, small pox, tumour and tachycardia.

Non-medicinal Uses

Twigs: Used in making brooms, baskets and veneering.

Bark: Yields fiber; used in paper making; used for tanning leather and thatching.

Leaf: Yields a green dye.

Seed: Used as substitute for coffee.

38. *Dalbergia lanceolaria* L.f.

Synonyms: *Dalbergia lanceolaria* L.f. ssp. *paniculata* (Roxb.) Thoth.
Dalbergia paniculata Roxb.
Dalbergia nigrescense Kurz.

Common Names: Bander siris, Bastard rosewood.

Vernacular Names: Ben: Chakendia; **Guj:** Tantosi; **Hin:** Bithua, Takoli; **Kan:** Belaga, Bettada hasirugoni, Bili beete, Hasirugaani, Hasirugoni, Hassarganni, Kaanaga, Kanagina mara, Kanaaga, Manjula beete; **Mal:** Mannavitti, Ponga, Pulari, Punnu; **Mar:** Dandour, Dandoshi, Dandus, Dandusa, Kaurchi; **Sans:** Goraksha, Kapothavanka; **Tam:** Erigai, Kalvellangu, Kattuppachalai, Nalvellangu, Velangu; **Tel:** Errapachhaari, Kondapachaari, Naagulapachaari, Pachaari, Pasaruganaru, Peddapachaari, Peddasopara, Potupachaari, Thellayarugundu, Vellaari, Yerra pastaru.

Distribution: Native of Asia including India and Pakistan; in India it grows throughout Arunachal Pradesh, Assam, outer Himalaya, Chota Nagpur.

Description: A deciduous tree with glabrous branches; bark greenish grey; leaflets 11-15, ovate, round at the base, emarginate at the apex, glaucous beneath; flowers pink or dull white, unilateral on axillary and terminal panicles; calyx dull purple, silky; corolla pink or white, standard petal broadly obovate, clawed, keel shorter than the wing petal; stamens 10, in 2 bundles of 5 each; pod stalked, glabrous, narrowed at both ends, flexible, brown, 1 seeded, valves reticulate over the seed.

Phytochemistry

Root: Caviunin-7-O-glucoside and triacontane.

Root bark: Ψ-Baptigenin, β-sitosterol, biochanin A, dalbin, dalbinol and lanceolarin.

Stem bark: Biochanin-A, formononetin, isoflavonoids, prunetin and sissotrin.

Heart wood: Biochanin-A-7-O-glucoside and paniculatin.

Leaf and Pod: Ψ-Baptigenin.

Flower: Apigenin, caviunin, dalpanin, luteolin and pinitol.

Seed: Dalpanitin, dalpatin, dalpanin, dalpanol-O-glucoside, milldurone and fatty oil.

Edibility

Leaf: Edible.

Medicinal Uses

Plant: Is antiprotozoal and spasmogenic; used as tooth brush to strengthen gums; useful in dysuria, infection, osteoarthritis, pain in the genital region, pustular eruptions, rheumatism and urinary stones.

Root: Paste is applied to ulcers.

Root bark: Decoction of the bark is used in cutaneous affections, dyspepsia, fever and tongue infections.

Leaf: Is employed in leprosy and other cutaneous diseases; burnt leaf powder is sprinkled on wounds and sores; useful in infection, osteoarthritis and rheumatism.

Flower and Seed oil: Is useful in arthritis and rheumatism.

Bioactivity: Ether extract of leaf exhibited antiarthritic activity in albino rats.

Clinical Studies: Ψ baptigenin from leaf and flower under clinical trials proved to be an anti-inflammatory drug better than butazone for treating arthritis.

Uses in Alternative Systems of Medicine

Ayurveda:

Part used: Plant and seed.

Uses: Useful in anorexia, debility, edema and rheumatoid arthritis.

Tibetan:

Part used: Plant.

Uses: Cures anorexia, diarrhoea, fever and quenches thirst.

Ethnomedicine

Plant: Is used in dyspepsia, desentery, rheumatism and stomach ache.

Stem: Is used for tongue infection.

Stem bark: Used in fever, body pain and skin diseases.

Leaf: Is used in colic.

Non-medicinal Uses

Plant: Grown as shade tree in tea plantations; is a host plant for lac insect.

Wood: Timber is used for tool handles, agricultural implements, scantlings, rafters, packing cases, boards, doors, windows and musical instruments.

39. *Dalbergia latifolia* Roxb.

Synonym: *Dalbergia emarginata* Roxb.

Common Names: Bombay black wood, East India rose wood.

Vernacular Names: Hin: Sheesham, Sheesu, Walayti; **Kan:** Beete mara, Ibidi, Ibri, Karevyadi, Thodegatthi, Thogadatthi, **Mal:** Bitti, Cholavitti, Itti, Kaaritti, Kaarvitti, Ulnalavitti, Vengah, Vitti; **Mar:** Bhotheula, Kalarukh, Sheesham, Sheesu, Sissui, Sisu, Siswa; **Sans:** Krishna-sheemshapa, Sheemshapa, Sheemshipa; **Tam:** Eravadi, Errupootu, Eruvadi, Etti, Itti, Karundorviral, Nukku.

Distribution: Native of India, Indonesia, Nepal, Pakistan, Peninsular Malaysia and Andaman Islands; grows in Bihar, Gujarat, Rajasthan, Uttar Pradesh and West Bengal in India.

Description: A large deciduous tree; branches glabrous; leaflets 7-9, broadly elliptic, orbicular or elliptic-obovate, obtuse, rarely emarginate, glabrous, cuneate at the base, subcoriaceous; petiole long; inflorescence on axillary panicle; pedicel as long as the calyx; calyx teeth obtuse, shorter than the cup; corolla whitish, almost twice as long as the cup; stamens 9; pod oblong-lanceolate, narrowed at both ends, glabrous, 1-4 seeded; seeds reniform.

Phytochemistry

Root bark: β-Sitosterol, γ-sitosterol, acetyl oleanolic acid, campesterol, dalberginene, dalberginone, dalbergi chromene, hentriacontane, latifolin, liquiritigenin, lupeol, methoxydalbergione, methyldalbergin and sisafolin.

Heart wood: 2,7-Dihydroxy-3-methoxy xanthone, 4',7-dihydroxyflavan, dalatenone, dalbergin, dalberginone, dalcriodain, latifolin, latinone and liquiritigenin.

Seed: Coumarin, dalbinol and sisafolin.

Medicinal Uses

Plant: Bitter, stomachic and tonic; used in intestinal worms, leprosy and obesity.

Uses in Alternative Systems of Medicine

Ayurveda:

Part used: Root, bark and leaf.

Uses: Useful in burning sensation, chronic ulcer, edema, leprosy, leucoderma, polyuria, sciatica and urinary disorders.

Ethnomedicine

Plant: Stimulant and vermifuge; used in dysentery, diarrhoea, dyspepsia, leprosy, obesity and stomach ache.
Stem: Is used in eczema and pimples.
Bark: Is used for body ache and wounds.

Non-medicinal Uses

Leaf: Is used as fodder.

Wood: Ranks among the finest of woods; used for cabinets, furniture, carving, ornamental plyboards and veneers; used in making calico printing blocks, mathematical instruments, ammunition boxes, naives, boat knees, agricultural implements, combs, rafters, floor boards, doors and window frames.
Seed oil: Is used for tattooing.

40. *Dalbergia pinnata* (Lour.) Prain

Synonyms: *Dalbergia tamarindifolia* Roxb.
Derris pinnata Lour.

Vernacular Name: Ass: Keti.

Distribution: Native of China, Hainan, Indonesia, Laos, Malaysia, Philippines, Thailand and Vietnam; introduced into India; grows in Arunachal Pradesh, Assam, Bihar, Meghalaya, Orissa and West Bengal in India.

Description: Scandent bushy shrub; leaves with terete petioles; leaflets oblong, base oblique, round or emarginated at the apex, entire along the margins, glabrescent above, slightly pubescent beneath; stipules lanceolate; flowers greenish white on axillary panicles; bracts ovate or suborbicular; calyx pubescent; corolla white; pod oblong, glabrous, flat, reddish brown, stipitate, 1-3 seeded.

Edibility

Root and Bark: Masticatory.

Medicinal Uses

Root: Is used as anthelmintic.

Leaf: Is used as poultice for itching, fever and typhoid.

Uses in Alternative Systems of Medicine

Homoeopathy

Part used: Tincture of the whole plant.

Uses: Useful in asthma, cramps, diarrhoea, dysentery, dry cough, inflammation of the eye, hoarseness of the voice and swelling.

Non-medicinal Uses

Leaf: Is used as fodder.

Root: Is used as masticatory.

41. *Dalbergia sissoides* Wight & Arn.

Synonyms: *Dalbergia emarginata* Prain
Dalbergia javanica Miq.

Common Name: Malabar black wood.

Vernacular Names: Kan: Beete, Bilagada, Bolu beete, Hasiruganneru, Karemuthala, Pachaari, Padaru pachaari, Pottebaage, Vellaga; **Sans:** Kushimshipa; **Tam:** Etti, Iruvadi, Thothagatti, Vel-itti, Vitte; **Tel:** Pachayirugudu.

Distribution: Native of India and Indonesia; it grows in Karnataka, Kerala and Tamil Nadu in India.

Description: Deciduous tree up to 10m; leaf lets 7-9, elliptic, oblong, obovate, acute, shining and glabrous above; calyx glabrous; flowers white, standard petal reflexed; stamens 9, monadelphous; pod acute at apex, stipitate and reticulately veined; seeds 2-3.

Phytochemistry

Plant: Biochanin-A, dalbergin, latifolin, methoxydalbergione, neoflavonoids, tectorigenin and tectorigenin-7-O-gentiobioside.

Wood: β-Sitosterol, isoliquiritigenin, latifolin, liquiritigenin, oleanolic acid and sissotrin.

Leaf: Biochanin A, kaempferol and sissotrin.

Flower: Biochanin A, formononetin and kaempferol.

Medicinal Uses

Bark: Abortifacient, anthelmintic, antipyretic, aphrodisiac, appetizer, expectorant; useful in burning sensation, disease of blood, dysentery, dyspepsia, leucoderma, skin disease, thirst, ulcers and vomiting.

Leaf: Leaf juice is applied on eyes for eye diseases.

Uses in Alternative Systems of Medicine

Ayurveda:

Part used: Root, bark and leaf.

Uses: Useful in chronic ulcers, edema, leprosy, leucoderma, polyuria and sciatica; used as brain tonic.

Non-medicinal Uses

Wood: Is used as timber; used for the same purpose as that of *Dalbergia latifolia.*

42. *Dalbergia sissoo* DC.

Synonym: *Amerimnon sissoo* (Roxb.) Kuntze

Common Names: Rose wood, South Indian red wood, Sissoo.

Vernacular Names: Ben: Sishu; **Hin:** Seesam, Seessai, Seessu, Sheesham; **Kan:** Agara, Agaru, Aguru, Bindi, Birade, Biradi, Biridi, Gette, Ibadi mara, Iragundimaavu, Karimara, Shishma baage, Shishta baage, Sisu; **Mal:** Iruvil;

Sans: Aguru, Agurushinshupa, Dhira, Dhumrika, Kalanusaraya, Kapila, Krishnasaara, Kushimshipa, Mandalapathri, Pichhila, Pipala, Shimshapa, Shyaama, Theevradhumaka, Vira, Yugapathrika; **Tam:** Gette, Itti, Nukku, Pichai, Sissoo, Sisu, Sisu itti, Thaesi maram, Yette; **Tel:** Errasissu, Sinsupa, Sissu.

Distribution: Native of Afghanistan, Bangladesh, Bhutan, India, Iran, Iraq, Pakistan and Oman; introduced into Africa, Asia, Australia, Caribbean islands, China and United States; In India, it is indigenous to sub-Himalayan belt; widely cultivated in other parts of India.

Description: Large tree growing to a height of 10m with drooping branches and rough bark with longitudinal furrows; leaves imparipinnate; leaflets 3-5, obovate, suborbicular, cuspidate, acuminate, glabrous, coriaceous; flowers yellowish white on axillary panicles; calyx teeth ciliate, unequal; bract small, pubescent, caducous; corolla with a flat standard petal, with long claws and slit along the top; stamens 9, monadelphous; pod oblong, long stalked, strap shaped, glabrous, minutely apiculate; seeds 2-4 in number; and are flat.

Phytochemistry

Plant: 3,5-Dihydroxystilbene, 5,7-dihydroxy-4'-methoxyisoflavone, gentiobioside, biochanin, caviunin-7-O-glucoside, cyclitol, dalbergechromene, dalbergenone, dalbergin, dalbergephenol, isocaviudin, isocaviunin, isodalbergin, isotectorigenin, methyldalbergin, nordalbergin and tectorigenin.

Root bark: Dalbergenone, dalbergechromene, dalbergin, isotectorigenin and methoxyisoflavone.

Wood oil: Biochanin-A, bisaboline and nerolidol.

Heart wood: Biochanin-A, bisaboline, dalbergechromene, dalbergenone, dalbergephenol, dalbergin, methoxyisoflavone, methylcoumarin, methyldalbergin, nerolidol, nordalbergin, trans-stilbene and trimethoxyisoflavone.

Leaf: Biochanin-glucoside, dimethoxyisoflavone and isoflavone glycoside.

Flower: Dimethoxyisoflavone, trimethoxyisoflavone, biochanin A, tectoridin and tectorigenin.

Pod: Biochanin A, cavinnin, cavinnin-glucoside, cavinnin-gentiobioside, caviunin, caviunin-glucoside, dalbergenin, isocarididin, isocaviunin, isocaviunin-gentiobioside, methyl coumarin, methyl tectorigenin, tannin, tectoridin, tectorigenin and tectorigenin-glycoside.

Seed: Caviunin-gentiobioside, caviunin-O-glucoside, dalbergin, isocavindin, tectorigenin and tectorigenin-glucoside.

Seed oil: Biochanin-A, bisaboline, dimethoxyisoflavone, methyltectorigenin, methylcoumarin, nerolidol and trimethoxy isoflavone.

Medicinal Uses

Plant: Analgesic, antiarthritic, anti-inflammatory, antipyretic and spasmolytic; corrects dysmenorrhoea, menorrhagia and menopausal disturbances; improves sensory functions.

Root: Astringent and constipating; useful in diarrhoea, dysentery and vomiting; root decoction is applied on eruptive skin affections.

Stem bark: Acrid, abortifacient, alterative, anti-inflammatory, antipyretic, anthelmintic, astringent, aphrodisiac, bitter, depurative, emmenogogue, expectorant, haemostatic, thermogenic and vulnerary; used in curing amenorrhoea, bleeding piles, bronchitis, burning sensation, dyspepsia, dysentery, dysmenorrhoea, hiccup, epistasis, gout, gastropathy, helminthiasis, haemoptysis, intermittent fever, leprosy, leucoderma, menorrhagia, ophthalmopathy, sciatica, scabies, scalding urine, skin diseases, syphilis, ulcers, varicose veins and vomiting.

Wood: Abortifacient, alterative, anthelmintic, antiarthritic, anti-inflammatory, antipyretic, appetizer, aphrodisiac, depurative, emetic, emmenogogue, expectorant, stomachic, thermogenic and vulnerary; useful in amenorrhoea, burning sensation, boils, bronchitis, dyspepsia, dysentery, dysmenorrhoea, gastropathy, gout, helminthiasis, hiccup, inflammation, intermittent fever, leprosy, leucoderma, ophthalmopathy, scabies, sciatica, skin diseases, scalding urine, syphilis, vomiting and ulcers.

Wood oil: Is applied on cutaneous affections.

Leaf: Anthelmintic, astringent, constipating, digestive, diuretic, haemostatic, ophthalmic, stimulant and styptic; useful in burning sensation, colic, diarrhoea, dyspepsia, dysentery, epistasis, excoriation, gonorrhoea, haemorrhage, haematemesis, menorrhagia, ophthalmia, piles, varicose veins and vomiting; leaf mucilage is applied on excoriation; decoction given in acute stage of gonorrhoea.

Bioactivity: Alcoholic extract of aerial parts showed pronounced bronchodilation in rabbits.

Toxicology: Plant contains chymotrypsin and trypsin inhibitors.

Uses in Alternative Systems of Medicine

Ayurveda:

Part used: Heartwood.

Uses: Used in anaemia, chronic ulcers, cutaneous affections, diabetes, obesity, edema, phlegmatic disorders, piles, polyuria and urinary calculi; oil extracted from the wood is used as local application in foul ulcers, leprosy, leucoderma, psoriasis, scabies and sciatica; used as a bath oil for musculoskeletal disorders; given as enema to clean the bowels; given as brain tonic.

Ethnomedicine

Stem: Is used in head ache, leprosy, skin diseases and as stimulant.

Leaf: Is useful in gonorrhoea, scurvy and skin diseases.

Non-medicinal Uses

Plant: Is used as fodder.

Wood: Is used in making high class wooden furniture, cabinets, frames, boats, wheels, carts, railway carriages, yokes, plough, toys, tents, badminton rackets, inlay work, sports bats, drums, country pipes, musical instruments; yields excellent fuel and charcoal; used in paper making.

Seed: Is used in making ear rings.

43. *Dalbergia spinosa* Roxb.

Synonyms: *Amerimnon spinosa* (Roxb.) Kuntze
Dalbergia horrida Graham

Vernacular Name: Tel: Chillanki.

Distribution: Native of Bangladesh, India, Myanmar and Peninsular Malaysia; grows in Andhra Pradesh, Orissa, Tamil Nadu and West Bengal in India.

Description: Stiff erect shrub with numerous short, round horizontal branch lets ending in straight pungent spines; leaves crowded on the nodes of branch lets; leaflets 5-9, obovate-oblong; flowers white on axillary corymbs; calyx teeth short, obtuse; pod brown, falcate, glabrous and reniform; seed one, rarely two.

Phytochemistry

Plant: Caviunin, dalbergin, dalspinin, dalspinosin and isoflavones,

Root: Dalspinosin, dalspinin-7-O-β-D-galactopyranoside and dalspinosin-7-O-β-D glucopyranoside.

Stem bark: Caviunin, dalspinin and dalbergin.

Leaf: Galactoside, methyltectorigenin, prunetin and prunetin-4'-O-β-D-galactoside.

Medicinal Uses

Root: Root powder with water reduces the effects of alcohol.

Root bark: Given in cholera.

Wood: Alterative; useful in boils, eruptions, leprosy and nausea.

Leaf: Decoction given in eruptive diseases and acute stage of gonorrhoea; leaf mucilage is applied in cutaneous affections.

Bioactivity: Ethanol extract of aerial parts is diuretic and spasmogenic.

44. *Dalbergia volubilis* Roxb.

Vernacular Names: Hin: Bhaatiya; **Sans:** Gorakhee, Goraksha, Sirishika; **Tel:** Bandigarjana.

Distribution: Native of India, Laos, Myanmar, Thailand and Vietnam; introduced elsewhere; grows in Assam, Bihar, Gujarat, Meghalaya, Rajasthan, Uttar Pradesh and West Bengal in India.

Description: Large shrubby climber; branches twisted into spiral hooks; leaflets 9-14, oblong, obtuse, glabrous, mucronate at the apex; inflorescence is a raceme with purple or pale blue flowers, standard petal reflexed at the base of orbicular emarginated limb; calyx velvety, teeth lanceolate, lowest longer than the rest; pod short stalked, obtuse and glabrous; seeds 1 to 2.

Phytochemistry

Plant: Dihyroxy-5,7,4'-dihydroxy-3'-methoxy-4-phenyl-24-1-benzopyran-2-one, 7-hydroxy-4-methyl coumarin, P-hydroxycinnamic acid, biochanin A, dalbergin, kaempferol, rotenolone, sishadrin, sitosterol, umbelliferone and volubolin.

Stem bark: (+) Medicarpin, cearoin, dalbergin, methyltectorigenin, tectoridin and tectorigenin-7-gentiobioside.

Leaf: β-Sitosterol, 7-hydroxy-4-methyl coumarin, biochanin A, formononetin, myricyl alcohol, tectorigenin and triacontane.

Flower: Biochanin A, C-rhamnoside and isovolubilin.

Medicinal Uses

Plant: Juice is used as gargle for throat problems and as local application for aphthae.

Root: Juice with cumin cures gonorrhoea.

Bioactivity: Ethanol extract of aerial parts is diuretic.

Uses in Alternative Systems of Medicine

Ayurveda:

Part used: Whole plant.

Uses: Used in anorexia, debility, edema and rheumatoid arthritis.

Ethnomedicine

Plant: Used for apthae, gonorrhoea and sore throat.

Leaf: Is used in sexual debility.

45. *Dendrolobium triangulare* (Retz.) Schindler

Synonyms: *Desmodium triangulare* (Retz.) Merr.
Hedysarum triangulare Retz.

Vernacular Name: Kan: Naaga thagare.

Distribution: Native of Asia including India; grows throughout Eastern Himalaya in India.

Description: Erect under shrub with angled branches; leaves 3-foliolate; leaflets obovate to elliptic, long, acuminate, cuneate, appressed, silky pubescent on both sides; terminal leaflet large, lateral leaf lets small, 20 or more; red or white flowers in subsessile umbel; pod oblong, indented on both sutures, joints 4-5, pubescent, appressed, slightly curved.

Edibility

Pod: Eaten by tribals.

Medicinal Uses

Plant: Is used in blindness, cataract, diarrhoea and stomach pain.

Root: Is useful in rheumatism.

Ethnomedicine

Plant: Is used for abscesses, blindness, cataract, diarrhoea, itch, ophthalmia, rheumatism, sores, splenitis, tumour, whitlow and wound.

Non-medicinal Uses

Plant: Is used as green manure and cover crop.
Leaf: Is used as fodder.

46. *Dendrolobium umbellatum* (L.) Benth.

Synonyms: *Dendrolobium australe* (Willd.) Benth.
Desmodium australe DC.
Desmodium cumingianum (Benth.) Benth.
Desmodium umbellatum (L.) DC.
Desmodium umbellatum (L.) DC. var., *costatum* Carib.
Hedysarum umbellatum L.
Ormocarpum umbellatum (L.) Desv.

Vernacular Names: Kan: Jaenu kaddi chooru, Jaenu kaddi gida, Kome, Murivaal; **Mal:** Naamarallu; **Tel:** Sittenthakura.

Distribution: Native of Kenya, Tanzania, China, India, Indonesia, Malaysia, Myanmar, New Guinea, Philippines, Ryukyu, Sri Lanka, Taiwan, Thailand, Vietnam, Australia and Indian Ocean Islands; grows in Andaman and Nicobar islands and Gujarat in India.

Description: Erect shrub; young branches appressed, pubescent; leaves 3-foliolate; leaf lets ovate, oblong, obtuse, glabrous above, canescent beneath; umbels subsessile, 6-12 flowered; flowers white; pod oblong, indented on both suture joints, 4-5 seeded.

Edibility

Pod: Edible.

Medicinal Uses

Leaf: Is useful in fever.

Uses in Alternative Systems of Medicine

Ayurveda:

Part Used: Whole plant.

Uses: Is used in cough and general debility.

Ethnomedicine

Plant: Is used for constipation and psoriasis.
Leaf: Is used for fever.

Non-medicinal Uses

Plant: Is used as fodder.

47. *Derris elliptica* (Wallich) Benth.

Synonyms: *Galedupa elliptica* Roxb.
Pongamia elliptica Wallich
Pongamia volubilis Zoll. & Moritzi.

Vernacular Names: Hin: Bakal pip; **Kan:** Meenu mari.

Distribution: Native of Bangladesh, Cambodia, China, India, Indonesia, Irian Jaya, Laos, Malaysia, Myanmar, Philippines, Srilanka, Thailand and Vietnam; cultivated in Andaman and Nicobar Islands, Assam, Karnataka, Kerala and Punjab in India.

Description: Scandent shrub or liana, 5-12 m long, velvety pubescent; shoot apex often leafless for several meters; leaves pinnate, 15-30 cm long, mostly with 11-15 leaflets; leaflets narrowly oblong-obovate, rounded at base, entire, rusty-velvety, somewhat glabrescent, glaucous beneath; flowers usually in stalked clusters of 3, on pseudo racemes; calyx, shallowly toothed; petals pink, standard, with 2 auricles at base, softly ferrugeneous, hairy outside; wing and keel with interlocking longitudinal folds; pod elliptic to oblong, flat, leathery, narrowly winged; seeds 1-3, reniform and flat.

Phytochemistry

Plant: 2-Methoxyfuranoflavone, 6-methoxy isopongaglabol, amorphin, hydroxyfuranoflavones, karanjachromene, pongachromene and rotenone.

Root: (–) Maackianin, (+) maackianin, β-tubaic acid, degulein, derrin, elliptinol, ellptinon, rotenone, tephrosin, toxicarol.

Leaf: 4'-Methoxyfuranoflavone, 2,5-dihydroxymethyl-3,4-dihydroxy pyrrolidine, demethoxykaranjin, karanjin, pongapin and rotenone.

Flower: Aurantiamide, gamatin, kaempferol, kaempferol-7-O-glucoside, karanjin, pongaglabol, pongaglabrone, quercetin and quercetin-7-O-glucoside.

Seed: Glabrachalcone, glabrachrome II and pongol.

Seed oil: Isopongachromene.

Medicinal Uses

Leaf: Antimicrobial.

Toxicology: Plant is considered as carcinogenic.

Ethnomedicine

Plant: Is used for hepatosis.

Non-medicinal Uses

Plant: Is used as herbicide and piscicide; extract used as fly spray for horticultural crops.

Root and bark: Insecticidal, larvicidal and piscicidal.

Pod: Is piscicidal.

48. *Derris robusta* (DC.) Benth.

Synonym: *Dalbergia robusta* DC.

Vernacular Name: Ass: Hitkura.

Distribution: Native of Bangladesh, China, India, Myanmar and Thailand; grows in Assam, Eastern Himalaya and Nagaland in India.

Description: Medium sized erect tree with long petioled leaves; leaflets 20, elliptic-oblong, base narrow and unequal, mucronate at the apex, minutely puberulous on both the surfaces; flowers white on axillary racemes; calyx golden coloured, hairy, campanulate; corolla white, vexillum ovate-orbicular, with faint yellow blotch; pod strap shaped, glabrous, reticulately veined, narrow at both ends, winged along the upper suture; seeds 1-5, orbicular, brown to black and compressed.

Phytochemistry

Plant: 4-Hydroxy-3-phenyl coumarin.

Seed: β-Sitosterol, derrone, derrugenin, derrusnin, methoxy isoflavones, methylderrone, robustic acid, robustone, robustone methyl ether, rubone.

Seed shell: Robustigenin-methyl ester and rubone.

Medicinal Uses

Plant: Is used as insecticide.

Seed: Antibacterial and antifungal.

Non-medicinal Uses

Plant: Used as insecticide.

Root: Used as fish poison.

Wood: Is used for making boxes, tea chests, posts and ploughs.

Leaf: Used as fodder.

49. *Derris scandens* (Roxb.) Benth.

Synonyms: *Dalbergia scandens* Roxb.
Derris timoriensis (DC.) Pittier

Common Name: Hog creeper.

Vernacular Names: Ben: Noalota; **Hin:** Gonji; **Kan:** Handi balli; **Mal:** Ponnam valli; **Tam:** Tirani, Thirudankodi; **Tel:** Nallatige.

Distribution: Native of Bangladesh, Cambodia, China, India, Indonesia, Laos, Malaysia, Myanmar, Pakistan, Philippines, Sri Lanka, Thailand and Vietnam; grows in Andaman and Nicobar Islands, Assam, Bengal, Central and South India.

Description: Is a large climber; leaves imparipinnate; leaflets 7-13, oblong to obovate, obtuse, polished and glabrous above, grey and pubescent beneath; flowers white or pink on racemes; bracteoles present under the calyx; calyx campanulate, silky, with obscure teeth; stamens monadelphous; pod oblong, glabrous, winged along one suture with silky hairs, narrow at both ends; seeds 1-4, winged.

Phytochemistry

Plant: β-Sitosterol, lupeol and taraxerol.

Root: Chandrin, coumarin, isoflavone, lonchocarpic acid, nallanin, osajin, robutic acid, scandenin and scandenone.

Seed: Yellow fixed oil.

Medicinal Use

Root bark: Is antimicrobial and antispasmodic; decoction of the bark is given for snake bite.

Non-medicinal Uses

Plant: Is used as fish poison.

Bark: Yields a coarse fibre used in cordage.

50. *Derris trifoliata* Lour

Synonyms: *Derris heterophylla* (Willd.) Baker
Derris uliginosa (Willd.) Benth.
Pongamia madagascariensis Baker
Robinia uliginosa (Roxb.) Benth.

Common Name: Indian tubaroot.

Vernacular Names: Hin: Paanlatha; **Kan:** Vadani gida, Vaduvina gida, Yadathaari; **Mar:** Kaajar bael, Kirthana; **Tel:** Nallathige, Thigekranuga.

Distribution: Native of Africa, Asia including India, Australia, Polynesia, Madagascar, East Africa, Papua New Guinea, Indian Ocean and Pacific Ocean islands; grows in Andaman and Nicobar Islands, Assam, Maharashtra, Tamil Nadu, West Bengal, Himalaya and Meghalaya in India.

Description: Large glabrous, climbing shrub; stem bark deep grey; leaves alternate; leaflets 3-5, ovate to elliptic, acuminate; raceme 3-5 flowered; flowers white or rose red; calyx glabrous; stamens monadelphous; pod orbicular to broadly oblong; glabrous, obtuse at both ends.

Phytochemistry

Plant: α-Amyrin, β-amyrin, β-sitosterol, quercetin-3-0-β-neohesperidoside, rhamnetin-3-O-β-neohesperidoside, 4-phenethylamine, 5-indole-3-alkylamine, betaine, betulin, campesterol, cholesterol, choline, dehydrorotenone, hypaphorine, lupeol, oleanolic acid, rotenone, stachydrine, stigmasterol, trigonelline, tyramine and ursolic acid.

Root: Alkaloids, deguelin, elliptinol, glucosides, ketone, lupeol, rotenoid, rotenone and tephrosin.

Leaf: 4-phenethylamine, 5-indole-3-alkylamines, betaine, choline, hypaphorine, indole-3-acetic acid, N-N-dimethyl tryptophan, quercetin-3-0-β-D-neohesperidoside, rhamnetin-3-O-neohesperidoside, stachydrine, trigonellin and tyramene.

Medicinal Uses

Plant: Is anticarcinogenic, anti-inflammatory, antispasmodic, antiviral, counter irritant, cytotoxic and stimulant; useful in dysmenorrhoea, paralysis and tumour; combined with asafoetida, garlic and *Plumbago*, paste is applied externally for rheumatism.

Leaf: Is used to treat asthma, convulsions, cough, diarrhoea, dysentery and indigestion.

Bioactivity: Plant is inhibitor of prostaglandin and topoisomerase synthesis.

Non-medicinal Uses

Plant: In the British Solomon Islands, the vine is used for catching fish; it is used in making rough cordage.

Root and Bark: Is used as fish poison, insecticide and to kill worms.

Leaf: Leaves are ground with coral and sand and dropped into water to kill fishes.

51. *Desmodium elegans* DC.

Synonyms: *Desmodium cinerascens* Franchet.
Desmodium forrestii Schindler
Desmodium glaucophyllum Pampan.
Desmodium spicatum Rehder

Vernacular Names: Hin: Chamkat, Chamara; **Pun:** Gurkats.

Distribution: Native of Afghanistan, Bhutan, China, India, Nepal and Pakistan; grows in Himalaya in India.

Description: A shrub with densely pubescent branches; leaf trifoliolate, petiolate; leaflets orbicular, ovate to obovate, obtuse to acuminate, glabrous to pubescent on both the sides, hairs silky on the lower side; stipulate; inflorescence is a panicle up to 45 cm long; bracteate; bracteoles linear, small; calyx villous, teeth shorter than the tube; corolla pale-lilac to dark purple; pod sessile, sometimes appearing stipitate by the abortion of basal ovules, upper suture almost straight, lower indented between the seeds and pubescent; seeds many.

Phytochemistry

Plant: Dimethoxyphenethylamine, abrine, hordenine, hypaphorine, salsolidine, salsoline, tryptamine and tyramine.

Root: β-Phenethylamine, betaine, choline, hordenine, hypaphorine, salsolidine, salsoline, tetrahydroisoquinolines, tryptamine and tyramine.

Leaf: Aliphatic alcohol, octacosanoic acid, sitosterol and wax esters.

Medicinal Uses

Plant: Is spasmogenic.

Root: Aperient, blood enricher, carminative, diuretic and tonic; used in amenorrhoea, eczema, edema, lumbago, opthalmia, piles and stuttering.

Non-medicinal Uses

Bark: Yields a fibre employed in making ropes, baskets and paper.

Wood: Yields timber.

Leaf: Is useful as fodder.

52. *Desmodium gangeticum* (L.) DC.

Synonyms: *Desmodium lanceolatum* Walp.
Desmodium maculatum (L.) DC.
Hedysarum gangeticum L.

Common Names: Trifoil

Vernacular Names: Ben: Salpani; **Hin:** Saalpan, Saalwan, Sarivan; **Kan:** Kolaku naaru, Marulu honne, Ondele honne; **Mal:** Orila, Pulatti; **Mar:** Darh, Ranbhal, Saalparni, Saalwan; **Sans:** Amshumathi, Astamathi, Devi, Dhruva, Dheerga moola, Dheerghanghni, Dheerghapatra, Ekamoola, Guha, Keetavinaashini, Kumuda, Nischala, Parnee, Parshniparni, Patini, Pivari, Sarvanukarini, Soumya, Shaalaparni, Shaalini, Shaaliparni, Shaalipatri, Shophaaghni, Shubhapatrika, Sthira, Subhagam, Sudha, Suparni, Thriparni, Vidaarigandha, Vrihiparnika; **Tam:** Pulladi, Sirupulladi; **Tel:** Githanaram, Kolakaponna, Nallanelapariki.

Distribution: Native of Africa, Asia, Australia, Middle East, Palau and Solomon Islands; introduced into Caribbean islands, China, India and Indian Ocean islands.

Description: An erect under shrub with angled branches; leaves, simple, pubescent, ovate, lanceolate, acute, acuminate, glabrous above, pubescent below; stipules scarious; inflorescence is axillary terminal raceme; flowers white or violet in colour; calyx pubescent, teeth longer than the tube; pod broad, subfalcate, moniliform, sticky, upper suture slightly indented, lower deeply indented between the seeds, indehiscent, pubescent, hairs hooked, minute, found in clusters; seeds 4-8.

Phytochemistry:

Plant: β-Carboline, β-phenylethylamine, dimethyltryptamine, β-carbolinium, aminoglycosylated glyceroglycolipid, candicine, cerebroside, desmodin, gangetinin, hypophorine, methyltyramine, methyl-tetrahydroharmen, phenylethylamine, phospholipids and tryptamine.

Root: Alkaloids, candicine, desmodin, gangetinin, phenethylamine and resin.

Leaf: β-Carbolinium, pterocarpan, 2'-hydroxygenistein, dimethyltryptamine, dalbergioidin, desmocarpin, diphysolone and kievitone.

Seed: β-Carboline, β-phenylethylamines, desmodin, gengetinin, carboxylated and decarboxylated tryptamine.

Medicinal Uses

Plant: Is alterative, antiangiogenic, antibacterial, anticancer, antiestrogenic, antifungal, anti-inflammatory, antihaemolytic, antimitogenic, antiviral, apoptotic, bitter, cardioprotective, demulcent, digestive, diuretic, febrifuge, haemostatic, hypotensive, hypocholesterolemic, laxative, nervine tonic, rejuvenator, smooth muscle relaxant, thermogenic and tonic; useful in asthma, atherosclerosis, breast cancer, cough, dysentery, head ache, hazy vision, leukaemia, lymphoma, melanoma, neuroblastoma, osteoporosis, tumour of breast, ovary, prostate and stomach, ulcer and vomiting; plant decoction is used in debility, derangement of humours, flatulence, gout, inflamed chest, intermittent fever, loss of appetite, nausea, piles, seminal weakness and tuberculosis.

Root: Anti-inflammatory, antispasmodic, astringent, diuretic, expectorant, febrifuge and tonic; useful in biliousness, catarrh, cough, diarrhoea, chronic fever and vomiting; used as an antidote for snake bite and scorpion sting.

Root bark: Alterative, anti catarrhal, bitter, digestive, febrifuge and tonic.

Seed: Used in treating asthma, biliousness and cough; used as an antidote for snake bite and scorpion sting.

Bioactivity: Plant is an inhibitor of aldose-reductase, α-reductase, catechol-O-methyl-transferase, histidine-kinase, lipase, MAO, NADH-ubiquinone-oxidoreductase, ornithine-decarboxylase, PTK, peroxidase, topoisomerase I and II and tyrosine-kinase; it is also calcium antagonist, pituitary sensitizer and quinone-reductase inducer.

Glycolipids from ethanolic extract of the whole plant showed *in vitro* and *in vivo* immunostimulant activity against pathogenesis of *Leishmania donovarrin* in Golden hamster model; root extract caused nicotine like effects on dog intestine *in situ* and on carotid blood pressure due to phenylethylamine and candicine; the total alkaloid fraction of stem and leaves exhibited curairiform activity on the rectus muscle of frogs and relaxed the smooth muscles of rabbit, dog and rat uterus.

Uses in Alternative Systems of Medicine

Ayurveda:

Part used: Whole plant.

Uses: Used in ascites, diabetes, diarrhoea¸ dropsy¸ dyspnoea, edema, emaciation, fever, musculoskeletal disorders, pain, piles, polyuria, sprue and vomiting; it is used as oleation catalyst and bowel binder.

Siddha:

Part used: Whole plant.

Uses: Digestive and galactagogue.

Ethnomedicine

Whole Plant: Is used in bronchitis, fever and as tonic.

Root: Is used in asthma, cough, dysentery, vomiting and as antidote to snake poison.

Non-medicinal Uses

Plant: Is used as green manure, fodder and grown as cover crop.

Stem: Is used in paper making; yields fiber.

53. *Desmodium heterocarpon* (L.) DC.

Synonyms: *Desmodium heterophyllum* (Willd.) DC. var., *buergeri* (Miq.) Hosok.
Desmodium polycarpum (Poiret) DC.
Hedysarum heterocarpon L.
Hedysarum polycarpum Poiret
Hedysarum siliquosum Burm.

Vernacular Names: Ori: Krishnupani; **Sans:** Saalaparni; **Tel:** Chepputatta.

Distribution: Native of Bangladesh, Bhutan, Cambodia, China, India, Indonesia, Irian Jaya, Japan, Laos, Malaysia, Nepal, Philippines, Srilanka, Taiwan, Tanzania, Thailand and Vietnam.

Description: Subscandent herb or under shrub, with spreading pubescent branches; leaves 3 foliolate; terminal leaflets obovate to oblanceolate, obtuse, pubescent beneath; inflorescence is a raceme; 1-2 blue flowers per bract; pod linear with 6-8 reticulate joints.

Medicinal Uses

Plant: Spasmogenic; decoction is used in convulsions, cough and fainting; given in stomach ache.

Root: Carminative, diuretic and tonic; useful in fever.

Uses in Alternative Systems of Medicine

Ayurveda:

Part used: Whole plant.

Uses: Rejuvenator and tonic; useful in treating cardiac diseases, diarrhoea, edema, emaciation, dyspnoea, intermittent fever, nausea, pain, piles and polyuria.

Ethnomedicine

Plant: Is used for cough, fainting and as tonic.

Root: Is used in fever.

Non-medicinal Use

Plant: Is used as green manure and fodder; grown as cover crop.

54. *Desmodium heterophyllum* (Willd.) DC.

Synonym: *Desmodium triflorum* (L.) DC. var., *majus* Wight & Arn.

Vernacular Names: Hin: Kudaliya; **Kan:** Kaadu menthe, Kaadu purache, Nela parande; **Mal:** Nilamparanta; **Mar:** Jangli methi, Ranmethi; **Sans:** Hamsapaadi, Tripaadi; **Tam:** Sirupulladi; **Tel:** Muntha mandu, Munuddamoddu.

Distribution: Native of Borneo, China, Hainan, India, Malaysia, Myanmar, Srilanka, Taiwan, Thailand, Vietnam, Australia, Madagascar and Ogasawara-Shoto; grows in Andaman and Nicobar Islands and Eastern Himalayas in India.

Description: Procumbent herb; branches with hairs; leaves 3-foliolate, terminal leaftlets obovate, obtuse, obliquely cordate at base; 3-6 flowered raceme; flowers pink and 2 per bract; pod oblong, indented on lower margin, sparsely pubescent with 2-4 joints.

Medicinal Uses

Plant: Decoction of whole plant is taken for abdominal complaints, dysuria and stomach ache; is a good diuretic and cooling medicine.

Root: Carminative, diuretic and tonic.

Leaf: Galactagogue.

Medicinal Uses Outside India

Malay Peninsula: Poultice of ground plant is applied to boils and sores; plant juice is dropped into the ear for ear ache; decoction of the plant is given for abdominal complaints.

Ethnomedicine

Plant: Diuretic, carminative, galactagogue and tonic; used in dermatosis, ear ache, stomach ache and sores.

Non-medicinal Uses

Plant: Is useful as fodder, green manure and cover crop.

55. *Desmodium oojeinense* (Roxb.) H. Ohashi

Synonyms: *Ougeinia dalbergioides* Benth.
Ougeinia oojeinensis (Roxb.) Hochr.

Common Name: Chariot tree.

Vernacular Names: Hin: Asainda, Arila, Chhaanam, Kalaphulas, Sandan, Syandan, Thimsa, Thinasuna, Thinnas, Thirichha; **Kan:** Betta honne, Huli, Kal honne, Kal mutthuga, Kari honne, Kari mutthuga, Kari mutthala, Male honne, Naayi honne; **Kon:** Kurimutal; **Mal:** Malavenna, Nemi; **Mar:** Kaala palaas, Thanach, Thewas, Thivas, Thinasoo, Syandan; **Sans:** Akshaka, Ashmagarbhaka, Athimuktha, Athimukthaka, Bhasmagharba, Chakri, Chithrakarma, Chithrakritha, Jaladhara, Maeshi, Naemi, Ratha, Rathadru, Rathika, Sarvasara, Shakata, Shatanga, Spandana, Spandanadruma, Spandani, Syandan, Thinashaka, Thineesha, Vanjula; **Tam:** Narivengaai; **Tel:** Adavishikkudu, Athimukthamu, Badanegi, Dargu, Mandamotuku, Nemi chettu, Thellamotuku, Thinisavriksha, Vanjulamu.

Distribution: Native of India and Nepal; grows throughout India in the forests.

Description: Medium sized tree, with a short crooked trunk and deeply cracked brown bark; leaves 3-foliolate; leaflets broadly ovate, glabrous above, sparsely pubescent beneath; flowers white or rose, numerous on a raceme; pod flat, elongated, 2-5 jointed, joints 2-3 times as long as broad; seeds 2-5 per pod.

Phytochemistry

Plant: Betulin, dalbergioidin, ferreirin, genistein, homoferreirin, kaempferol, neophellamuretin, orobol, ougenin, tetrahydroxy-isoflavanone and wedelolactone.

Stem bark: Betulin, hydroxylupeol, lupeol and triterpenes.

Heart wood: Dalbergioidin, homoferreirin, isoflavonoids, ougenin and tetrahydroxyisoflavone.

Leaf: Dalbergioidin, homoferreirin, kaempferol, leucopelargonidin, ougenin and quercetin.

Edibility

Flower: Is edible.

Medicinal Uses

Plant: Is anti-viral, CNS active, febrifuge, hypotensive and spasmolytic; useful in cholera, diarrhoea and dysentery.

Bark: Anthelminitic, anti-inflammatory, astringent, cooling, febrifuge¸ stimulant and sudorific; useful in anaemia, asthma¸ bronchitis, burning sensation, diabetes, diarrhoea, dysentery, general debility, gonorrhoea, haemorrhage, leprosy, leucoderma, ulcers, urorrhagia, verminosis and vitiated conditions of "pitta".

Gum: Astringent; used in diarrhoea and dysentery.

Leaf: Leaf extract cures bronchitis.

Uses in Alternative Systems of Medicine

Ayurveda:

Part used: Stem bark and wood.

Uses: Cures anaemia, biliousness, dysentery, leprosy, leucoderma, obesity, polyuria, pruritus, skin diseases, ulcers, urinary discharges, "vatha" and "kapha", worms and wounds.

Ethnomedicine

Plant: Is used in cholera, diarrhoea, dysentery and fever.

Stem bark: Is used in asthma, bronchitis and dysentery.

Non-medicinal Uses

Plant: Is a host plant for lac insect; used as fodder; used as piscicide.

Bark: Used as fish poison and pestcide; yields fibre, useful in cordage; used in tanning industry.

Wood: Yields timber; used in making bedsteads, shafts, carts, hubs, rafters, cabinets, yokes, axles, spokes, felloes, wheels, carts, tool handles, sugar cane crushers, tent poles, oars, shutters, picking handles, carved furniture, doors and window frames.

Leaf: Is used as pesticide for killing caterpillars.

56. *Desmodium pulchellum* (L.) Benth.

Synonyms: *Dicerma pulchellum* (L.) DC.
Hedysarum pulchellum L.

Vernacular Names: Hin: Jatsalpan; **Kan:** Jenukaddi, Kaaduhuralithe, Kaadulthi, Kaadumuduru, Segate gida, Thigure; **Mal:** Kaattumuthira; **Sans:** Lodhrah, Lodrom; **Tel:** Karranthinta, Kondanthinta, Sarivi.

Distribution: Native of Bhutan, China, India, Indonesia, Irian Jaya, Laos, Malaysia, Myanmar, Philippines, Ryukyu, Srilanka, Taiwan, Thailand, Vietnam and Pacific Ocean Islands; grows in eastern Himalaya, Meghalaya and Dehradun in India.

Description: Erect under shrub with ribbed branches; leaves 3-foliolate; leaflets elliptic, ovate, obtuse, apiculate, tomentose, round at the base, glabrous above, downy beneath, end leaflet largest; stipules lanceolate, acuminate; flowers yellow, white or white tinged with pink in clusters and hidden by leaf like orbicular bracts; calyx incised halfway; pod indented on both sutures, joints 2, glabrous, reticulate.

Phytochemistry

Plant: β-Carboline, carbolinium, bufotenine, bufotenine-0-methyl ether, indole-3-alkylamine, N-methyl-tetrahydroharmon and dimethyltryptamine.

Root: α-Amyrin, β-carboline, β-sitosterol, betulin and indole-3-alkylamine.
Seed: Galactomannan and physcion.

Medicinal Uses

Plant: Is anti HIV, anti-inflammatory, antitumour, antiviral, CNC active, cytotoxic, hypolipidemic, hypotensive and spasmogenic.

Root: Useful in burning sensation in the abdomen.

Root bark: Decoction is used in diarrhoea, eye diseases, haemorrhage, headache, ophthalmia and poisoning.

Flower: Is used in biliousness and related illness.

Bioactivity: Plant is inhibitor prostaglandin-synthesis and topoisomerase.

Medicinal Uses Outside India

Malay Peninsula: Leaf poultice is applied on smallpox pustules and ulcers.

South China: Plant is used for rheumatic fever and tooth ache; it promotes digestion, dissolves blood clots; cures convulsions in infants.

Ethnomedicine

Plant: Is used for diarrhoea, haemorrhage and purturition.

Stem bark: Is used in headache.

Root: Is used for burning sensation in the body.

Non-medicinal Use

Plant: Grown as fodder plant; a branch of the plant is kept in the house to repel bed bugs.

57. *Desmodium styracifolium* (Osbeck) Merr.

Synonyms: *Desmodium capitatum* (Burm. f.) DC.
Desmodium celebicum Schindler
Desmodium retroflexum (L.) DC.
Hedysarum styracifolium Osbeck
Uraria retroflexa Drake

Distribution: Native of Cambodia, China, India, Malaysia, Myanmar, Philippines, Srilanka, Thailand, Vietnam and Palau; grows in Assam, Karnataka, Kerala, Meghalaya and Sikkim in India.

Description: Erect branched herb; leaves 1-3 foliolate; leaflets obovate, broadly elliptic, acute, obtuse, appressed, pubescent beneath; 30-40 flowered raceme; flower purple, two per bract; bract caudate; pod jointed.

Phytochemistry

Leaf: Schaftoside and vicenin-I and III.

Medicinal Use

Root: Aperient, deobstruent, emmenogogue, purgative and stomachic; contracts blood vessels, decreases blood pressure, increases myocardial capillary circulation.

Clinical Studies

Flavonoid extract of the plant increased myocardial capillary circulation, coronary and cerebral blood flow, hypoxia and decreased contraction of blood vessels.

Ethnomedicine

Plant: Diuretic and refrigerant; used for fever, gonorrhoea, hepatitis and urinogenital disorders.

Non-medicinal Use

Plant: Grown for green manure and as cover crop.

58. *Desmodium triflorum* (L.) DC.

Synonyms: *Desmodium triflorum* (L.) DC. var., *minus* Wight & Arn.
Hedysarum triflorum L.

Common Names: Creeping tick, Three flower plant, Beggar weed.

Vernacular Names: Ben: Kudaliya; **Guj:** Jhinopandhio; **Hin:** Motha, Kudaliya; **Kan:** Kaadu menthe, Nelaparande; **Mal:** Munta; **Mar:** Janglimethi, Ranmethi; **Sans:** Hamsapaada, Tripaadi; **Tam:** Sirupulladi; **Tel:** Muntamandu, Munnuddamoddu.

Distribution: Native of Africa, Asia including India, Australasia, Caribbean islands, Indian Ocean Islands, Central America, Middle East, North and South America; grows throughout India in the plains.

Description: A wiry, trailing, spreading herb; branches clothed with fine hairs; leaves 3-foliolate; leaf lets subequal, obovate, obtuse or emarginate, membranous, cuneate at the base, glabrous above, pubescent beneath; stipules lanceolate, persistent; flowers white, pink or purplish blue, standard petal purple, other petals blue; calyx clothed with long white hairs, teeth long, setaceous; pod linear, indented on lower suture with 4-5 reticulate joints.

Phytochemistry

Plant: (+) Pinitol, apigenin, fucosterol, glucosyl vitexin, isovitexin and vitexin.

Root: β-Phenylethylamine, betaine, choline, hypaphorine, indole-3-alkylamines, dimethyl tryptophan, stachydrine and trigonelline.

Stem: β-Phenylethylamine, indole-alkylamines, betaine, choline, indole-3-acetic acid, alkylamines, stachydrine and trigonelline.

Leaf: β-Phenylethylamine, 5-indole-3-alkylamine, betaine, choline, hypaphorine, indole-3-acetic acid, stachydrine, trigonelline and tyramine.

Medicinal Uses

Plant: Acrid, cooling, diuretic, expectorant and galactagogue; useful in abscesses, blindness, body ache, breast pain, bronchitis, burning sensation, cough, erysipelas, diarrhoea, dysentery, flatulence, menorrhagia, pain, pruritus, sores, spleenic disorders, stomach and cardiac troubles, swellings, whitlow and wounds.

Root: Is carminative, CNS stimulant, diuretic, spasmodic and tonic; is useful in asthma, bilious complaints, cough and wounds.

Leaf: Antiseptic and galactagogue; used in abscesses, convulsions, diarrhoea, dysentery, eruptions, tooth ache and wound; fresh leaves applied on sores and wounds.

Uses in Alternative Systems of Medicine

Ayurveda:

Part used: Whole plant, root and leaf.

Uses: Plant paste is useful as application on burns and skin diseases; juice is applied to abscesses, burns and wounds; root is antibilious, carminative, diuretic and tonic; leaf paste in milk is given to children for convulsions and diarrhoea.

Medicinal Uses Outside India

China: A decoction of the plant is given as a remedy for dysentery.

Indonesia: Plant infusion is used in dysentery.

Malay Peninsula: Decoction of the root is taken for stomach ache and at the same time poultice of the leaves is applied on the stomach; leaves are also used in skin diseases.

Philippines: Plant juice is applied to abscesses, ulcers, and wounds.

Taiwan: Plant is used as antipyretic; useful in jaundice and gonorrhoea.

Ethnomedicine

Plant: Used for abscesses, breast pain, colic, dermatosis, diarrhoea, dysentery, eruptions, haematuria, menorrhagia, sores, spleen complaints, stomach and body ache, swellings, tooth ache, whitlow and wounds.

Leaf: Is used in diarrhoea, dysentery, tooth ache and wounds.

Non-medicinal Use

Plant: Used as green manure and fodder; grown as cover crop.

59. *Desmodium velutinum* (Willd.) DC.

Synonyms: *Desmodium lasiocarpum* (P.Beauv.) DC.
Desmodium latifolium (Roxb.) DC.
Hedysarum deltoides Poiret
Hedysarum lasiocarpum P.Beauv.
Hedysarum latifolium Roxb.
Hedysarum velutinum Willd.
Pseudarthria cordata (Klotzsch) C. Mueller

Common Name: Tripura salpani.

Vernacular Names: Kan: Chittaamalli gida; **Mal:** Orila; **Sans:** Prsniparni; **Tam:** Angusabaadi, Kidameri, Kubayam, Kuchattinbaadi, Simbadai, Simbattai, Simanduri, Sirubulladi, Sivamadu; **Tel:** Adiviyanthinta, Gaaba, Mahaalinga, Thellanelapariki.

Distribution: Native of Africa, Asia, Bhutan, Burma, Cambodia, China, India, Indonesia, Laos, Malaysia, Madagascar, Sri Lanka, Taiwan, Thailand, Tropical Africa and Vietnam; grows throughout Himalayas in India.

Description: Erect under shrub with fulvous hairy branches; leaves 1-foliolate; leaflets acute, ovate, obtuse, fulvous, tomentose beneath; 20-30 flowered raceme; flowers violet, 2 per bract; calyx fulvous, hairy, incised half way; pod covered with hooked hairs; upper suture straight, lower indented, 4-7 jointed.

Medicinal Uses

Root: Alterative and tonic; used in bowel complaints, diarrhoea, fever, haematuria, ulcers and vomiting; decoction of root with hot pepper is used as enema to cure blood in the urine.

Non-medicinal Uses

Plant: Is useful as fodder, green manure and cover crop.

Bark: Yields fibre useful in paper making.

60. *Dysolobium pilosum* (Willd.) Marechal

Synonyms: *Vigna pilosa* (Willd.) Baker
Dolichovigna pilosa (Willd.) Niyomdham.

Vernacular Names: Mal: Kaattupayar; **Sans:** Mudgaparni.

Distribution: Native of Cambodia, India, Indonesia, Laos, Myanmar, Nepal, Philippines, Thaiwan, Thailand and Vietnam; grows in Andaman islands, Assam, Bihar, Karnataka, Kerala, Orissa, Sikkim, Tamil Nadu and West Bengal in India.

Description: A slender, extensively spreading hairy twiner; leaves alternate, trifoliate; leaflets large, ovate, acute, terminal leaflet deltoid at the base; flowers red or blue on axillary raceme; pod pendulous, linear, long, pointed; seeds reniform with a dark brown shining seed coat.

Edibility

Seed: Used as vegetable.

Medicinal Use

Root: Aphrodisiac, germicidal and cooling; used for burning in the body, consumption, cough, diarrhoea, dyspepsia, fever, haemorrhoids, opthalmia and vitiated conditions of the 3 humours.

Non-medicinal Uses

Seed and Straw: Used as animal feed.

61. *Erythrina arborescens* Roxb.

Synonym: *Erythrina tienensis* Wang & T. Tang

Distribution: Native of Asia including India; grows in Himalaya, Meghalaya, Rajasthan and Sikkim in India.

Description: A tall tree with brown spines on the branches; leaves 3-foliolate; leaflets broadly triangular, ovoid with broad cuneate base and acuminate tip, glabrous above, hispidulous along the nerves below; stipules lanceolate; flowers bright red on axillary raceme; calyx shallowly bilabiate, sparsely pubescent; standard petal ovate-elliptic, orange red, keel petal ovate, triangular, pale green; pod oblong, ellipsoid, curved, narrow at both ends, 4-6 seeded; seeds oblong and reniform.

Phytochemistry

Plant: Crystamide, erysodine, erysopine, erysorine¸ erysotrine, erythraline and erythrabine.

Bark: Alkylferulates, campesterol, n-alkanes, sitosterol, stigmasterol and wax.

Leaf: 11-hydroxy erysotrine, erybidine, erysodine, erysotramidine, erysotrine, erythrabine, erythratidine, N-nororientaline, tetrahydroxybenzylsuinoline.

Pod: Erysodinophorine and erysopinophorine.

Seed: Alkaloids, erysophorine, erysopinophorine, erythrascine¸ erythrine, fatty oils, hypaphorine, isoerysopinophorine, spiroamine, and tetra-hydroxy-benzyl isoquinoline.

Seed oil: Arachidic acid, behenic acid, eicosanoic acid, linoleic acid, oleic acid, palmitic acid and stearic acid.

Medicinal Uses

Plant: Is spasmogenic and useful in skin diseases.

Bark: Useful in skin diseases.

Leaf: Is ground into paste with ginger and applied externally in skin diseases and for hair fall.

Non-medicinal Uses

Plant: Grown as an ornamental and hedge plant.

Wood: Used for making boards, ladder, wooden jars for storing ghee, drums, boxes, water troughs, packing cases for match sticks, wooden splints, house posts, stakes and railway cross tiers; used as fuel.

62. *Erythrina crista-galli* L.

Synonyms: *Corallodendron crista-galli* (L.) Kuntze
Erythrina crista-galli L. var., *hasskarlii* Backer
Erythrina crista-galli L. var., *leucochlora* A. Lombardo
Erythrina fasciculata Benth.
Erythrina pulcherrima Tod.
Erythrina speciosa Tod.

Notes: It is the national flower of Brazil.

Common Name: Cockspur coral tree.

Distribution: Native of tropical South America, particularly Argentina, Brazil and Bolivia; introduced into India; cultivated throughout India.

Description: Well branched tree of 10 ft height; leaves 3 foliolate, subcoriaceous; odd leaflet rhomboid, lateral leaflets oblong, lanceolate and oblique; raceme erect and terminal; flowers red and showy; calyx tube campanulate with a shallow cleft on one side; corolla red, standard petal reflexed, wing petal oblong, basally auricled, keel oblong lanceolate, basally auricled; pod black with black seeds.

Phytochemistry

Plant: Alanine, arginine, cristamidine, cristacarpin, cristadine, cyanidin-3-glucoside, cyanidin-3-sophoroside, erythrabine, erycristanol, erycristin, erycrystagallin A-C, erydotrine, erythrabyssin II, erythraline, erythramine, glutamic acid, oleanolic acid, oleanoeic acid, pelargonidin-3-glucoside, pelargonidin-3-sophoroside,ursolic acid and vitexin.

Root: Erybidine, erycristagallin, erystamidine, erythrabine, erythraline, erythratine and erytrinine.

Bark and Wood: Cristamidine, daiolzin, daidzein, docosanoic acid, eicosanoic acid, erycristine, erythrabyssin II, genistin, heptadecanoic acid, hexadecanoic acid, octadecanoic acid, pentadecanoic acid and sandwicensin.

Leaf: Cristadine and 8-oxoerythraline.

Flower: β-Methoxyerythraline and 11 methoxyerythratine.

Seed: Alkaloids, arachidic acid, β-erythrodine, cristadine, eicosenoic acid, erysodine, erysonine, erysopine, erysothiopine, erysothiovine, erysovine, erythraline, erythramine, erythratine, erythratinone, hypaphorine, linoleic acid, myristic acid, oleic acid, palmitic acid and protein.

Seed oil: Erythraline, erythratine, erysopine, erysodine and erysovine.

Medicinal Uses

Plant: Is antibacterial, anti-inflammatory, antispasmodic, anticancer, antifibrinolytic, antihistaminic, antioxidant, antiviral, analgesic, anticonvulsant, antiseptic, apoptotic, cardiotonic, CNS active, estrogenic, hepatoprotective, hypnotic, immunomodulator, lactagogue, nervine tonic, neurasthenic, vasodilator and vulnerary; useful in anxiety, asthma, cancer, convulsions, depression, dermatitis, eczema, flatulence, gingivitis, haemolysis, hepatitis, hernia, high blood pressure, high liver enzyme levels, hysteria, inflammation of liver and spleen, insomnia, intermittent fever, leukaemia, liver disorders, lymphoma, melanoma, mental disorders, neuroblastoma, nicotine withdrawl, osteoporosis, tumour of prostate, ovary, breast, stomach, restlessness, sclerosis, shock, sleep disorders, spasms, stress, trauma and ulcer.

Bioactivity: Plant is an inhibitor of 5-α-reductase, aldose-reductase, catechol-O-methyl-transferase, DOPA-decarboxylase, histidine-kinase, lipase, MAO, NADH-ubiquinone-oxidoreductase, ornithine-decarboxylase, PTK, peroxidase, topoisomerase I and II and tyrosine-kinase; is quinone-reductase inducer and pituitary sensitizer.

Toxicology: Due to the presence of erythraline and hypaphorine, curare like paralysis and hyper excitability may result.

Medicinal Uses Outside India

Argentina: Used in diarrhoea, hemorrhoids, respiratory infections, urinary infections and as sedative.

Amazon: Used in hernia, stomach ache and to promote milk flow; is a natural sedative to calm the nervous system.

Brazil: Used in agitation, anxiety, asthma, bacterial infections, convulsions and cough.

Columbia: As diuretic and sedative.

North America: Used in herbal formulation (Kava-Kava) to treat anxiety, asthma, bronchitis, depression, fever, gingivitis, hepatitis, inflammation of liver and nervous cough.

Peru: Used in epilepsy.

USA: Plant is used in epilepsy, hepatitis, hernia, hysteria, insomnia, liver disorders, low blood pressure, shock, stomach ache and trauma; used to regulate heart palpitation and as hypnotic sedative.

Ethnomedicine

Plant: Antiseptic, narcotic and stomachic; useful in cancer and piles.

Non-medicinal Use

Plant: Is grown as ornamental and shade tree in coffee and cocoa plantations; used as insecticide and piscicide.

63. *Erythrina fusca* Lour.

Synonyms: *Corallodendron fuscum* (Lour.) Kuntze
Erythrina glauca Willd.
Erythrina indica sensu R.Vig.

Vernacular Name: Ben: Harikakra.

Distribution: Native of Asia including India, Australasia, Central and South America, Indian and Pacific ocean islands; grows in Andhra Pradesh, Assam, Orissa and West Bengal in India.

Description: Large tree with short trunk and armed branches; leaves appear only during flowering; leaflets elliptic, ovate, coriaceous; inflorescence erect with ferrugeneous hairs; flowers orange or dark red; calyx broadly campanulate, asymmetrically split with a spur on the vexillum side; vexillum round or rhomboid, glabrous, wing petal obliquely obovate, keel petal ovate, falcate; pod woody, linear, torulose with upcurved beak of 2 cm long with dense hairs; seeds 5-10 oblong, ellipsoid, opaque, dark brown with black mottling or black.

Phytochemistry

Stem bark: Erythrinasinate and erythrabyssin I.

Medicinal Uses

Plant: Is analgesic, androgenic, antianginal, antibacterial, anticancer, anticardiospasmic, antihistaminic, anti-inflammatory, antioxidant, antiscorbutic, antiseptic, antispasmodic, antiviral, cardioprotective, detoxicant, diuretic, estrogenic, hepatoprotective, immunostimulant, laxative, neuroparalytic, sedative, vasodilator and vulnerary; used in curing acne, allergy, asthma, anaemia, anorexia, anxiety, arthritis, cancer, diarrhoea, fistula, insomnia, leprosy, lupus, measles, migraine, neuralgia, obesity, orchitis, osteoarthritis, osteoporosis, photophobia, pneumonia, polio, psoriasis, stress, tumour of the lung, ulcer and vertigo.

Bark: Decoction given in painful micturition or discharges.

Leaf: Pounded leaves applied as poultice in bronchitis and ulcers.

Seed: Is used externally and internally for abscesses and cancer.

Bioactivity: Plant is an inhibitor of aldose-reductase, DOPA-decarboxylase, histidine-kinase, lipase, MAO, MDR, NADH-ubiquinone-oxidoreductase, ornithine-decarboxylase, PTK, peroxidase, β-glucuronidase; is estrogen antagonist, pituitary sensitizer, β-blcoker, calcium channel blocker and interferon synergist.

Ethnomedicine

Plant: Is used in beri-beri, cancer and wounds.

Non-medicinal Use

Plant: Is grown as a support plant for betle vine; grown as hedge and cover plant.

64. *Erythrina stricta* Roxb.

Vernacular Names: Kan: Hemmuruku, Hengaara mara, Kaadu haalivaana, Kaadu hongaara, Kaadu paarivaala, Kaelige, Keechige, Maniche; **Mal**: Mullu murikku, Mullumurukku, Murikku; **Sans:** Mura, Paaribhadra; **Tam:** Kinchugam, Mandaaram, Mullumurukku, Murukkai, Murukku; **Tel:** Baditha, Mullumoduga.

Distribution: Native of Bangladesh, Cambodia, China, India, Laos, Myanmar, Nepal, Thailand and Vietnam; in India, it grows in Arunachal Pradesh, Assam, Orissa, Rajasthan and Sikkim.

Description: A tall tree with whitish, conical incurved spines on the branches; leaves pinnately 3-foliolate; leaflets ovate, with round base and acuminate apex, glabrous and shining above, glaucous and minutely hairy along nerves beneath; lateral nerves 6-8 pairs; stipules falcate; flowers coral red at ends of the branches; calyx spathaceous; deeply divided dorsally, brown, pubescent; standard petal scarlet; pod spindle shaped, sharply pointed, glabrous outside, 1-3 seeded; seeds kidney shaped, oblong, red.

Phytochemistry

Plant: Erythrine.

Bark: 7-Methoxy-8-(15-hydroxy pentadecyl)coumarin, alkyl ferulate, coumarin, erysovine, esters, fatty acids, hypaphorine, n-alkanol, sitosterol and stigmasterol.

Seed: Alkyl ferulates, erythraline, erysodine, erysopine, erythrinine, fatty esters and n-alkanols.

Seed oil: Arachidic acid, behenic acid, eicosanoic acid, lignoceric acid, linoleic acid, oleic acid, palmitic acid and stearic acid.

Edibility

Leaf: Used as vegetable.

Medicinal Uses

Root: Root of white flowered variety is given with cold milk as aphrodisiac.

Root bark: Decoction of the bark is given for diabetes.

Stem bark: Antiasthmatic, antiepileptic, antileprotic, antirheumatic, febrifuge and expectorant; useful in biliousness, burning sensation, fever, leprosy, menstrual complaints, rheumatism, snake bite and stomach disorders; the decoction is useful as anthelmintic and antidysenteric; given in opthalmia as collyrium; burnt bark soot is applied to the eye for purulent conjunctivitis; bark juice is given in syphilis; bark decoction with leaves is given to kill worms in sores.

Wood: Washing the face with decoction is useful in reducing facial inflammation.

Leaf: Anthelmintic, antibilious, diuretic, expectorant, febrifuge and laxative; leaf juice is given in dysmenorhoea, to increase breast milk secretion and to relieve painful and difficult micturition; fresh leaves with coconut juice taken internally and applied externally as galactagogue and emmenogogue; the juice with honey is cathartic and vermifuge; leaf juice with castor oil is given in dysentery, dropped into the ear for ear ache and when applied on teeth acts as an anodyne; crushed leaves made hot and applied on joints to relieve rheumatic pain and as poultice it is useful in curing venereal buboes; leaf decoction with decoction of *Emblica officinalis* leaves is a good cathartic; used in chronic dyspepsia with constipation and liver trouble.

Flower: Given as tonic; is an antidote to poison.

Ethnomedicine

Bark: Is used in fever, menstrual complaints and stomach disorders.

Leaf: Used as vermifuge.

Flower: Is used as tonic.

Non-medicinal Uses

Plant: Grown as fence and support plant for pepper, betel and grape vines.

Wood: Used for scabbards, planks, ladders, jars for storing ghee, drums, boxes, water troughs and packing cases for match box.

65. *Erythrina suberosa* Roxb.

Synonyms: *Erythrina alba* Wight & Arn.
Erythrina glabrescens (Prain) R. Parker
Erythrina stricta Roxb. var., *suberosa* (Roxb.) Niyomdham
Erythrina sublobata Roxb.

Common Name: White coral tree.

Vernacular Names: Ass: Mandal; **Guj:** Jhangario; **Hin:** Dauldhak, Kaantepalash, Madara, Nasut, Pangra, Rowanra, Rungra, **Kan:** Mullu hongara, Mullu mutthala, Mullu mutthuga, Mullu paarivaala; **Mar:** Pangra, Kaantepalasa; **Sans:** Kanta palaash; **Tam:** Motaga, Mullumurukku, Mulmurungu, Muni, Munmurukku, Murukku, Vellaikkaliyanamurukku; **Tel:** Baarijama, Baarjapu, Mullumoduga, Munimoduga, Rohi, Rohithakamu.

Distribution: Native of Bhutan, Cambodia, China, India, Myanmar, Nepal, Pakistan, Thailand and Vietnam; in India it grows in Bihar, Himalaya, Jammu and Kashmir and Punjab.

Description: Deciduous tree with yellow or white prickles; leaflets subcoriaceous, matted with grey cottony pubescence beneath, broadly ovate rhomboid, acute with broad deltoid base; calyx bilipped, campanulate; flowers red, standard petal broader than the keel and wing petal, keel petal connate; pod 2-5 seeded, tapering at the ends; seeds black.

Phytochemistry

Bark: Alcohols, ferulates, cholesterol, β-sitosterol and wax esters.

Root: 5,7,3'-4'-tetrahydroxy-2'-(3'-methyl-2"-enyl) flavone.

Leaf: Erysotyrine.

Flower: Diglucosides of delphinidin, cyanidin-3,5-diglucoside and pelargonidin.

Seed: β-Sitosterol, campesterol, erythraline, erysodine, erysotrine, hypaphorine and stigmasterol.

Seed oil: Arachidic acid, behenic acid, myristic acid, oleic acid, palmitic acid and stearic acid.

Medicinal Uses

Root: Is used as antidote for snake bite.

Bark: Anthelmintic, antibilious, diuretic, emmenogogue, expectorant, febrifuge and laxative; decoction is useful in dysentery, syphilis and as vermifuge; is effective as lotion in opthalmia; bark smeared with butter and held over a flame and soot is collected which is used as a salve with coconut oil for purulent conjunctivitis; juice is used for cleansing maggot infested sores and ulcers.

Leaf: Anthelmintic, cathartic¸ discutient, emmenogogue and galactagogue; fresh juice with honey is used as cathartic and vermifuge; cooked leaves useful as emmenogogue; leaf juice relieves painful and difficult micturition, cures syphilis and tooth ache; juice with castor oil is useful in chronic or acute dysentery; leaf poultice is applied on ulcers, kills maggots in sores and act as anodyne; warm poultice is useful to dispense venereal buboes and for rheumatic pains in joints; with jaggery relieves colic when applied on the stomach; is a good antifungal agent; leaf paste or dry powdered leaves ground with coconut relieves joint pain.

Toxicology: β-erythrodine causes neuromuscular blocking.

Ethnomedicine

Bark: Is used in dysentry.

Leaf: Is used to destroy maggots in wounds, to cure tumour and ulcers.

Non-medicinal Uses

Plant: Planted as an ornamental and fence plant.

Stem bark: Is used for manufacturing cork plugs and sheets; useful in making insulation boards and bottle stoppers; is source of fiber used in cordage.

Wood: Is used for making scabbards, boards, planks, ladles, wooden jars, spoons, drums, sieve frames, match boxes, troughs, lacqured boxes, packing cases and toys; wooden splints used for fencing; wood ash and bark is used in dyeing.

66. *Erythrina subumbrans* (Hassk.) Merr.

Synonyms: *Erythrina holosericea* Kurz.
Erythrina sumatrana Miq.

Notes: Normally cultivated under the name *Erythrina indica.*

Distribution: Native of Asia including India and Mauritius; it grows in Andhra Pradesh, Assam, Karnataka, Rajasthan and Tamil Nadu in India.

Description: Evergreen tree often unarmed; leaflets ovate, acuminate, membranous, glabrescent; calyx 2-lipped; flowers red, keel and wing petals subequal; pod flat, broad at the base, turgid, 3-4 seeded, found only at the apex.

Phytochemistry

Plant: Erythratine, erysopine, erysodine, epierythratine and erysoflorinone.

Medicinal Use

Leaf: Juice is used as eye wash; leaf mixed with *Kalanchoe pinnata* is a cure for abdominal distension, bowel complaints, inflammation and internal bleeding.

Medicinal Uses Outside India

Indonesia: Leaf poultice is applied on the body for fever; leaf infusion is taken for bowel complaints, inflammation and internal bleeding; bark and leaf ground together is applied on distended stomach.
Malay Peninsula: Bark decoction is taken for cough.

Ethnomedicine

Plant: Is used in splenomegaly.

Non-medicinal Use

Plant: Is used as fodder and green manure.

67. *Erythrina variegata* L.

Synonyms: *Erythrina corallodendron* Lour.
Erythrina corallodendron L. var., *orientalis* L.
Erythrina mysorensis Gamble
Erythnina orientalis Murray
Erythrina orientalis (L.) Merr.
Erythrina variegata L. var., *orientalis* (L.) Merr.

Common Names: Bastard teak, Indian coral tree.

Vernacular Names: Ben: Palitamandar; **Guj:** Bangaro; **Hin:** Dadap, Mandara, Pangra, Parhad; **Kan:** Bili varijape, Haalivaana, Haarvana, Hongarike, Kempu vaarijapa, Mandara, Mullu murige, Muruku, Nimbatharu, Paalivaana, Paaribhadra, Paarivaana, Pongara, Vaarijapa; **Mal:** Kalyana murikku, Karimurikku, Mandaram, Kulmurikku, Nimbataaru; **Mar:** Mandar, Paangara; **Sans:** Bahupushpa, Kantaku, Kantaki, Mishuka, Krimigna, Krimishatru, Mandara, Nimbathaaru, Palaasha, Paaribhadra, Prabhadraka, Rakthakeshar, Rakthakusuma, Rakhtapushpa, Rohita, Thamrapushpi; **Tam:** Kaliyanamurukku, Kalyana murungai, Kavir, Mullumurukku, Mullu murungai, Paarisadam, Palaasam, Palasu, Savusyam, Sinsugam, Vallai; **Tel:** Baadidachettu, Baadisechettu, Baarjapuchettu, Baridamu, Badida, Mahaameda, Muchikatta, Paaribhadrakamu, Paaribhavyamu, Rohinamu.

Distribution: Native of Tanzania, Asia including India and Australasia; grows throughout India.

Description: A small deciduous tree with black prickles on the branches; bark yellow or grey, peeling off in papery flakes; leaflets 3, broadly deltoid, terminal leaflet long; stipules gland like; calyx spathaceous, oblique; flowers bright red on dense racemes;petals unequal; pod torulose, constricted between the seeds, dehiscent; seeds 6-8.

Phytochemistry

Root: Erycrystagallin, erythrabisin II, isobavachin, phaseollidin, phaseollin and scadenone.

Bark: β-Sitosterol, γ-sitosterol, betaine, campesterol, choline, docosylalcohol, epierythratidine, erysorine, erythraline, erythratidine, hypaphorine and stigmasterol.
Leaf: 11-Hydroxyerysotrine, epierythratidine, erybidine, erysodine, erysotine, erysotramidine, erysotrine, erythrabine, erythratidine, N-nororientaline and orientaline.
Flower: 4-O-Glucosyl-3-O-rhamnoglucoside, caffeic acid, erythratine, ferulic acid, isorhamnetin-3-O-rhamnoglucoside, quercetin and rutin.

Pod: Alkylferualates, betaine, carboxylated indole-3-alkylamines, choline, dihydroxyresveratol, docosyl alcohol, erybidine, erythratine, erythrin A, B & C, erythrosotidienone, erythromotidienone, fatty acids, isoflavones, isoquinoline alkaloids, isorhametin-3-O-rhamnoglucoside, osajin, oxyresveratol, quercetin, quercetin-4-O-glucosyl-3-O-rhamnoglucoside, rutin, saponin, sitosterol, stigmasterol and wighteone.

Seed: (+) Coreximine, caffeic acid, erybidine, erysodienone, erysodine, erysonine, erysophorine, erysopine, erysopitine, erysotine, erysotrine, erysovine, erythraline, erythrasine, erythrinine, ferulic acid, hypaphorine, N,N-dimethyl tryptophan, N-nororientaline, orientaline, reticuline, scoulerine and stachyadine.

Edibility

Young shoots, pod and leaf: Edible.

Seed: Edible after boiling or roasting.

Medicinal Uses

Plant: Acrid, anticancer, anthelmintic, carminative, diuretic, expectorant, febrifuge and galactagogue; useful in convulsions, menorrhagia, paralysis, pimples, snake bite and stomach ache.

Root: Young roots of the white flowered variety is pounded and given with cold milk as aphrodisiac; decoction is given in bronchitis and fever.

Bark: Acrid, antibilious, anticonvulsant, anti-inflammatory, carminative, CNS depressant, diuretic, depurative, emmenogogue, febrifuge, rejuvenating, sedative, stomachic, smooth muscle relaxant, spasmolytic, thermogenic and vulnerary; useful in anorexia, arthritis, colic, cold, cough, conjunctivitis, dysentery, dyspepsia, dysmenorrhoea, flatulence, fever, insomnia, leprosy, liver trouble, ophthalmia, otalgia, edema, rheumatism, skin diseases, strangury and vitiated "kapha" and "vatha"; it is useful in promoting appetite, arresting excessive micturition and destroying pathogenic parasites; decoction of the bark is used in ophthalmia as a wash and taken internally for dysentery; used as an antidote for snake poison. Bark decoction taken orally by women to reduce infertility and induce conception.

Leaf: Anthelmintic, diuretic, emmenogogue, galactagogue, laxative, narcotic and stomachic; leaf paste is applied externally for dispersing venereal buboes and to relieve joint pains; juice of the leaves is cathartic, relieves joint pain, acts as an anodyne in tooth ache and cures chronic dysentery; it is given in cough, inflammations, dysmenorrhoea, gastropathy, odontalgia and otalgia; paste is applied on cuts and wounds for quick healing.

Seed: Emmenogogue and narcotic; crushed seeds boiled in water is regarded as a remedy for abscesses, cancer and snake poison.

Uses in Alternative Systems of Medicine

Ayurveda:

Part used: Bark and leaf.

Uses: Leaf and bark used to give fomentation in neuralgia and the decoction is taken in fever.

Chinese:

Part used: Bark.

Uses: Used as analgesic in arthritis, neuralgia and rheumatism.

Siddha:

Part used: Bark, leaf, flower and seed.

Uses: Used in abdominal disorders, arthritis, diabetes, dysentery, eye diseases, fever, sterility, stomatitis and vomiting.

Medicinal Uses Outside India

Burma: Bark is considered as antipyretic and its decoction is given to treat liver disorders.

Indochina: Bark is used to treat intermittent fever.

Malay Peninsula: Root decoction is given as a remedy for beri-beri and haematuria; bark paste is used as poultice for wounds.

Philippines: Root and leaves are employed as febrifuge; bark is used to treat rheumatism and to relieve painful emissions; leaf and root paste is filled into tooth caries as anodyne; leaf decoction is expectorant.

Solomon Islands: Bark is used to treat broken bones; leaf paste is applied to sores as poultice; as an anodyne in treating ear and tooth ache and as a cough medicine.

Ethnomedicine

Bark: Is used as anthelminitc; used in cold, cough, convulsions, eye complaints, fever, rheumatism, skin diseases and as antidote for snake poison.

Leaf: Is used in cough and ulcers.

Non-medicinal Uses

Plant: Is grown as shade plant for coffee and tea; as support plant for betel and pepper; good hedge and fence plant; is a source of red dye.

Bark: Is used for tanning and dyeing; bark yields fiber suitable for cordage.

Wood: Is used for making boxes, toys, scabboards, planks, boats, trays, lacquer ware, frames for sieves and jars; used as fire wood.

Leaf: Is used as fodder.

Flower: Yields a red dye.

Seed: Is used in making ornamental earrings and necklaces.

68. *Flemingia chappar* Benth.

Synonym: *Moghania chappar* (Benth.) Kuntze

Vernacular Names: Ben: Salpan; **Hin:** Salpan; **Ori:** Singapornno.

Distribution: Native of Cambodia, China, India, Laos, Myanmar, Nepal and Thailand; it grows in Bihar, Eastern Himalaya, South India, Uttar Pradesh and West Bengal in India.

Description: An erect shrub with terete branches, coated with appressed hairs; leaves cordate at the base, acuminate or cuspidate at the apex, glabrous above sub- coriaceous below; inflorescence is a raceme but with straight axis; calyx teeth as long as the tube, teeth lanceolate; flower white, often gregarious; bracts broader than long, deeply emarginate; pod oblong, turgid, 2 seeded.

Phytochemistry

Plant: Dihydroxy chalcone, β-sitosterol and flemichapparin.

Root: β-Sitosterol and 7-dihydroxyflavonone.

Leaf: Flemichapparin-A and hydroxychalcone

Flower: Flemichapparin-A, and 2',4',4-trihydroxychalcones.

Edibility

Tuber: Edible.

Medicinal Uses

Plant: Used in treatment of cancer.

Root: Used in epilepsy, gastritis, hysteria and leucorrhoea; induces sleep and relieves body pain.

Leaf: Used as vermifuge.

Seed: Seed decoction is applied on the eye for eye troubles and cataract.

Ethnomedicine

Seed: Used in eye diseases.

69. *Flemingia grahamiana* Wight & Arn.

Synonyms: *Flemingia pycnantha* Benth.
Flemingia rhodocarpa Baker
Flemingia sericans Kurz
Moghania grahamiana (Wight & Arn.) Kuntze
Moghania rhodocarpa (Baker) Hauman

Notes: Three varieties, *rhodocarpa, glabrescens* and *hockii,* are recognized; Sanjappa (1992) regards *Flemingia sericans* as a synonym of *Flemingia wallichii* Wight & Arn.

Distribution: Native of Africa, Asia including India and Yemen in Middle East; it grows in Karnataka and Tamil Nadu in India.

Description: A small shrub or undershrub with red sticky glands; leaves 3 foliolate; leaflets obovate, obtuse, reticulate, grey or tawny, silky beneath;bracts foliaceous; bracteoles minute; calyx tube short, with subequal teeth; flowers on axillary or terminal raceme; corolla pink, wing petal obliquely oblong, adnate to the keel; stamens diadelphous; pod small, oblique, turgid, 2-seeded; seeds suborbicular.

Phytochemistry

Plant: Deoxy homoflemingin, flemingin A, B and C and homoflemingin.

Medicinal Uses

Plant: Is used as anthelmintic and anti-inflammatory.

Wood: Decoction is a remedy for chill and cough.

Pod: Anthelmintic.

Seed oil: Is used externally for skin diseases and internally as purgative and for cold.

Ethnomedicine

Plant: Is used as vermifuge; used for chill and cough.

Non-medicinal Uses

Plant: Main source of resin known as Warrus dye used in industry.

Wood: Used in cosmetics.

70. *Flemingia macrophylla* (Willd.) Merr.

Synonyms: *Crotalaria macrophylla* Willd.
Flemingia congesta Aiton
Flemingia congesta Aiton var., *latifolia* Baker
Flemingia congesta Aiton var., *semialata* Baker
Flemingia latifolia Benth.
Flemingia philippinensis Merr. & Rolfe.
Flemingia semialata Aiton
Flemingia wallichii Wight & Arn.
Flemingia yunnanensis Franchet
Moghania macrophylla (Willd.) Kuntze
Moghania philippinensis (Merr & Rolfe) H.L. Li
Moghania wallichii (Wight & Arn.) Kuntze

Notes: Treated under the genus *Moghania* in many references.

Vernacular Names: Ass: Sanbasjhatm: **Hin:** Bara, Malikamatteri; **Mar:** Dowdowla; **Ori:** Bibijabdyaki.

Distribution: Native of Asia including India, Pakistan, Nepal and Sri Lanka; grows throughout India.

Description: An erect shrub of 2m height, covered with grey hairs; leaflets 3, lanceolate, acuminate, gland dotted with winged petiole, terminal leaflets cuneate, lateral obliquely rounded at the base, dark green, puberulous above, lower surface with minute ruby coloured glands; flowers pink with purple and orange streaks; pod long, thinly downy; seeds 2, small, round, smooth, black and shining.

Phytochemistry

Plant: Feminone, flemistrictin A,B,C,D,E and F and Flemichin E.

Root: β-Sitosterol, chromoneoflavonones, flemiflavones, flemingin B, C and D, genistein, naringenin and tetrahydroxyisoflavanone.
Stem: Flemiphyllin.
Stem bark: α-Amyrin and lupeol.
Wood: Genistein, isoflavones and 5,7,4'-tetrahydroxy isoflavone.
Leaf: Flemingin C,D,E,F, homoflemingin C and F, kaempferol-3-rhamnoside, myrcetin, myricitrin, neochalcones, quercetin-3-rhamnoside and tamarixetin-3-rhamnoside.
Flower: Chromoneochalcones, flemingin C, D, E and F, flemiwallichin A, B, D, E and F and flemiflavonone D.
Pod: Flemingin A,B and C and homoflemingin.

Edibility

Pod: Used as vegetable.

Medicinal Uses

Plant: Is used in cholera, cold, dysentery, eye troubles, fistula, prolapsed anus, skin diseases, spleen complaints, smallpox and stomach ache; plant decoction is used to wash abscesses and sores and in treating joint pains and paralysis.

Root: Root paste is applied externally for fever, muscle pain, sores and swellings.

Pod: Dried powder is anthelmintic and used for cold.

Medicinal Uses Outside India

China: Plant decoction is used to wash swellings and sores.

Taiwan: Plant is used in treating fever, joint pain and paralysis.

Ethnomedicine

Plant: Decoction of whole plant used in bloody dysentery, cold, fistula, stomach ache, smallpox and spleen complaints.

Root: Paste is used in body pain, fever and muscular pain.

Leaf: Paste is applied on wounds.

Non-medicinal Uses

Plant: Host plant for lac insect; yields a dye called Warrus dye, used in cosmetics and dyeing.

Pod: Is ornamental.

71. *Flemingia strobilifera* (L.) Aiton

Synonyms: *Flemingia bracteata* Wight.
Flemingia fruticulosa Benth.

Flemingia strobilifera (L.) Aiton var., *bracteata* (Roxb.) Baker
Flemingia strobilifera (L.) Aiton.var., *fruticulosa* Baker
Flemingia bracteatum (Roxb.) Wight.
Hedysarum bracteatum Roxb.
Hedysarum strobiliferum L.
Moghania bracteata (Roxb.) H.L. Li
Zornia strobilifera (L.) Pers.

Notes: Treated under *Moghania* in literature.

Common Names: Luck plant, Wild hops.

Vernacular Names: Ass: Makhioti; **Hin:** Kusrunt; **Kan:** Kumaalu; **Mal:** Kumalu; **Mar:** Kanphuti, Bundar; **Tel:** Nallabaddu.

Distribution: Native of Bangladesh, Bhutan, Cambodia, China, Hainan, India and Indonesia; grows throughout India.

Description: An erect shrub of 5-10 feet with terete branches and velvety tip; leaves 1-foliolate, ovate, lanceolate, round at the base; bracts glabrescent, deeply cordate, cuspidate at the apex; calyx teeth lanceolate and pilose; flowers yellow or greenish white on a terminal raceme on rusty pubescent zigzag rachis; corolla with wing petal shorter than the keel petal; pod oblong, turgid.

Phytochemistry

Root: β-Sitosterol, chalcones and n-triacontane.

Stem and Flower: Quercetin and rutin.

Leaf: Flemiculosin, leptosidin, leptosin, naringin, phloridizin, quercetin and rutin.

Medicinal Uses

Plant: Is abortifacient, antiangiogenic, antifertility, antihaemolytic, anti-inflammatory, anti-ischaemic, antimicrobial, antimitogenic, antioxidant, antispasmodic, antiviral, apoptotic, anthelmintic, cardioprotective, cytotoxic, estrogenic, fungicide and immunostimulant; useful in curing atherosclerosis, breast cancer, cough, epilepsy, hysteria, hypocholesterolemia, leukaemia, lymphoma, melanoma, neuroblastoma, osteoporosis, tumour of ovary, prostate and stomach, swellings, skin diseases and ulcer.

Root: Is sedative; useful in body ache, epilepsy, gastric complaints, hysteria and leucorrhoea; induces sleep and relieves pain; pounded root is given in fever.

Leaf: Tonic and vermifuge.

Pod: Anthelmintic.

Seed oil: Used as local application for skin diseases and internally as purgative and in treating cold.

Bioactivity: Plant is inhibitor aldose-reductase, α-reductase, aromatase, catechol-O-methyl transferase, DOPA-decarboxylase, histidine-kinase, lipase, MAO, NADH-ubiquinone-reductase, ornithine-decarboxylase, PTK, peroxidase, topoisomerase I and II and tyrosine-kinase; is calcium antagonist, pituitary sensitizer and quinone-reductase inducer.

Medicinal Uses Outside India

Indonesia: Seeds are used as anthelminitic and in diabetes.

Malay Peninsula: Decoction of the root is tonic; leaf juice is used to treat rheumatism.

Myanmar: Leaf paste is applied to poisonous bites and stings; root is used to treat epilepsy.

Philippines: Roasted seeds are used as emollient and emmenogogue; leaf and flower infusion is given for tuberculosis.

Ethnomedicine

Plant: Is sedative and vermifuge; used in epilepsy, hysteria and rheumatism.

Root: Is used as analgesic; given in body pain, epilepsy, fever, gastric complaints, hysteria and leucorrhoea.

Leaf: Is used as tonic.

Non-medicinal Uses

Plant: Resinous powder obtained from the plant is used in dyeing silk; grown as an ornamental plant and as insecticide.

72. *Flemingia tuberosa* Dalz.

Synonym: *Moghania tuberosa* (Dalz.) Kuntze

Vernacular Names: Mar: Birmova.

Distribution: Nativity not known; grows in India in Goa, Gujarath, Karnataka and Maharashtra; endemic.

Description: Herb of 1 m height; perennial, tuberous; leaflets linear, oblong, elliptic or lanceolate, hairy with acute apex, base round; flowers pink or purple; pod oblong; seeds one or two.

Phytochemistry

Plant: Albuminoids, ash, asparagin, cellulose, starch, sugars, tannin and yellow resin.

Edibility

Root: Raw or roasted tuberous roots are eaten as vegetable.

Medicinal Uses

Plant: Decoction useful in dysentery and leucorrhoea.

Root: Astringent; useful in dysentery and leucorrhoea.

Uses in Alternative Systems of Medicine

Ayurveda:

Part used: Root.

Uses: Used as an astringent in dysentery and vaginal discharge.

Ethnomedicine

Plant: Is used in dysentery and leucorrhoea.

73. *Genista tinctoria* L.

Synonyms: *Genista alpestris* Bertol.
Genista humilis Ten.

Notes: Abundant growth of the plant is an indicator of the poverty of the land; species is highly variable in habit, leaf shape and degree of hairiness; populations with different combinations of these characters are often referred to as subspecies in literature.

Common Names: Dyers green weed, Wood wax plant.

Distribution: Native of Asia including India, Europe and Middle East; cultivated in Nilgiri Hills in Tamil Nadu.

Description: An erect shrub with stripped, glabrous branches; leaves alternate, broad, simple, highly variable in shape, oblong-elliptic to oblong-lanceolate, glabrous; flowers borne singly in the axil of each bract on short racemes

at the ends of the branches; bracts foliaceous; corolla glabrous, bright yellow, standard petal broadly ovate; pod narrow, oblong, pubescent, brown.

Phytochemistry

Plant: Alkaloids, anagyrine, cytisine, daidzein, genisteine, genistein-glucoside, genistin, isoparteine, lupanine, luteolin-diglucoside, luteolin glucopyransoside, N-methyl-cytisine, pachycarpine, retamine, scoparine and sparteine.

Edibility

Flower bud: Used as vegetable; used in sauce and pickled.

Medicinal Uses

Plant: Is antiangiogenic, antiarrhythmic, antibacterial, anticarcinogenic, anticephalagic, antifertility, antihaemolytic, antihepatotoxic, anti-inflammatory, anti-ischaemic, antimicrobial, antimitogenic, antioxidant, antiradicular, antiviral, apoptotic, cancer preventive, cardioprotective, cardiotonic, cathartic, coronary dilator, cytotoxic, demulcent, diuretic, estrogenic, emetic, hypotensive, hallucinogenic, immunostimulant, narcotic, oxytocic, psychotropic, respirostimulant, uterine stimulant and vasoconstrictor; useful in curing atherosclerosis, breast cancer, diarrhoea, dysentery, edema, leukaemia, lymphoma, melanoma, nephritis, osteoporosis, tumour of the breast, ovary, stomach and ulcer; it also cures hypocholesterolemia, hypoglycaemia, hypertension, hydrophobia, rheumatism and tachycardia; poultice of crushed plant is used for abscesses and scrofulous tumour.

Leaf: Cardiotonic, emetic and purgative; useful in dropsy and rheumatism.

Flower and Seed oil: Cathartic, emetic, diuretic, purgative, stimulant, sudorific, narcotic and vasoconstrictor; useful in abscesses, albuminuria, gout, gravel, rheumatism, sciatica, scrofula, stones and tumours; is prophylactic against hydrophobia.

Bioactivity: Plant is inhibitor of aldose-reductase, α-reductase, aromatase, catechol-O-methyltransferase, cyclo-oxygenase, DOPA-decarboxylase, glucosyl-transferase, histidine-kinase, lipase, lipoxygenase, MAO, MDR, NADH-ubiquinone-oxidoreductase, ornithine-decarboxylase, PTK, topoisomerase I and II, tyrosine-kinase and xanthine-oxidase; is calcium antagonist, pituitary sensitizer and quinone-reductase inducer.

Toxicology: Anagyrine in the vegetative parts and seeds is toxic and cause hypertension, hyperglycaemia and tachycardia.

Uses in Alternative Systems of Medicine

Homoeopathy

Part used: Whole plant (Mother tincture).

Uses: Useful in headache.

Ethnomedicine

Plant: Diuretic, emetic, laxative, purgative and sudorific; used in cancer.

Non-medicinal Uses

Plant: Used as fodder, but cattle after eating the plant give bitter tasting milk.

Flower: Yields yellow green dye used for dyeing wool.

Seed: Used as substitute for coffee.

74. *Gliricidia sepium* (Jacq.) Walp.

Synonyms: *Gliricidia pungam* Blanco
Gliricidia lambii Fern
Gliricidia maculata (Kunth) Walp.
Lonchocarpus maculatus (Kunth) DC.
Lonchocarpus rosea (Miller) DC.
Millettia luzonensis A. Gray
Robinia maculata Kunth
Robinia sepium Jacq.

Common Names: Gliricidia, Grow stick, Madre tree, Quick stick.

Distribution: Native of Caribbean islands; introduced into India; cultivated in gardens of Karnataka, Maharastra and Tamil Nadu.

Description: Small tree with grey branches sweeping in arc like fashion; leaves long, imparipinnate, feathery with black spotting glands on the under side; leaflets 7-15, bright green above, pale below; flowers pale pink or purplish pink; pod linear, compressed, with 10 or more seeds.

Phytochemistry

Plant: Erysothiopine, erysothiovine, gliricidin, isomucronulatol, methylsepiol, proteins, sepinol, sepiol and trihydroxyflavanone.

Stem bark: Coumarin.

Wood: Gliricidin, gliricidol, isomucronulatol, methyl sepiol, robinetin, sepinol, sepiol and trihydroxyflavone.

Leaf: β-Carotene, ascorbic acid, calcium, carbohydrates, ceryl alcohol, coumarin, fat, ferulic acid, gallic acid, iron, kaempferol, m-coumaric acid, myricetin, niacin, phosphorus, O-coumaric acid, p-coumaric acid, pinitol, protein, protocatechuic acid, sinapinic acid and thiamine.

Flower: β-Carotene, astragalin, calcium, iron, niacin¸ phosphorus, riboflavin, robinin, thiamine and trifolin.

Seed: Erysodine, erysopine, erysovine, erythraline and fatty oil.

Edibility

Flower: Is used as a vegetable in Central America and Philippines; is a good source of nectar.

Medicinal Uses

Plant: Used as androgenic, antianginal, antibacterial, anticoronary, antihaemorrhagic, antihepatotoxic, antihistaminic, anti-inflammatory, antimutagenic, antipyretic, antiscorbutic, apoptotic, cardioprotective, hepatoprotective, immunostimulant, laxative, sedative, vasodilator and vulnerary; used in acne, anaemia, anorexia, anxiety, arthritis, asthma, atherosclerosis, back ache, beri-beri, cancer, cataract, chilblain, cold, colic, convulsions, delirium, depressions, diabetes, dermatitis, eczema, edema, encephalopathy, epilepsy, fistula, gastritis, gingivitis, glaucoma, heart burn, hepatitis, herpes, hyperactivity, hypertension, insomnia, leprosy, lupus, measles, migraine, malaria, neuralgia, neuropathy, obesity, orchitis, osteoarthritis, osteoporosis, photophobia, pneumonia, polio, psoriasis, stress, tumour of lungs, ulcer and vertigo;as antidote to lead and cadmium poisoning.

Bioactivity: Plant is β-adrenergic-receptor blocker, β-blcoker, calcium antagonist, calcium channel blocker and inhibitor of β-glucuronidase.

Ethnomedicine

Plant: Expectorant, sedative and suppurative; used for alopecia, burn, boils, bruises, cold, cough, debility, eruption, erysipelas, fatigue, fever, gangrene, headache, itch, prickly heat, skin diseases, sores, tumours, ulcers and urticaria.

Non-medicinal Uses

Plant: Cultivated as ornamental and hedge plant; used as green manure and as fire wood; used in house posts, fences, railway cross tiers and stakes.

Bark: Insecticidal and pesticidal.

Wood: Used in making house post, fence, stakes and railway cross ties.

Leaf: Insecticidal and pesticidal; rich source of nitrogen when used as manure and provides good fodder; used as a source of ethylene for advancing ripening of tomato.

Seed: Insecticidal; used as rat poison.

Bark, leaf, pod and seed: Powdered and mixed with rice and used as bait for the destruction of pests.

75. *Glycine max* (L.) Merr.

Synonyms: *Dolichos soja* L.
Glycine hispida (Moench) Maxim.
Glycine soja sensu auct.
Phaseolus max L.
Soja angustifolia Miq.
Soja hispida Moench
Soja japonica Savi
Soja max (L.) Piper

Notes: Two subspecies, *max* and *soja,* are recognized; subsp. *soja,* the ancestor of the cultivated subsp. *max,* yields abundant oil and protein; provides 35 per cent of the world's human consumption of protein; cultivated in China since ancient times; introduced into USA in 1853; unknown in the wild; cultivated widely; genetically engineered varieties for pest resistance with genes from *Bacillus thuringiensis* (Bt) and herbicide resistance "Round up Ready" soybean varieties developed by Monsanto are in cultivation in several countries.

Common Names: Soja, Soya, Soya bean, Soybean.

Vernacular Names: Ben: Garikulay; **Hin:** Bhat; **Pun:** Bhut.

Distribution: Native of China, Laos, Thaiwan, Thailand and Vietnam; introduced into Africa, India, Australia, Caribbean Islands, Central America, Europe, Indian Ocean Islands, North and South America and New Zealand; cultivated in many parts of India.

Description: Bushy, annual herb, terete at the base, angled or subquadrangular above, clothed with grayish brown hairs; leaves trifoliolate with angled, pilose or glabrescent petioles; leaflets membranous, ovate, suborbicular, the lateral leaflets inequilateral, obtuse, mucronate, silky, strigose on both the surfaces; stipules broadly ovate, abruptly acuminate; bracts lanceolate; leafy raceme with 5–8 flowers; calyx teeth unequal, lanceolate, upper pair united; flowers white, pink or violet; corolla with standard petal suborbicular to subovate, wing petal upturned near the apex; pod oblong, subfalcate, pendent with yellowish brown hairs; seeds oval or subspherical, cream or greenish white, olive green or reddish brown to chocolate brown, 2–3 per pod.

Phytochemistry

Plant: β-Carotene, β-indoleacetic acid, acetaldehyde, acetone, acetylgenistin, acid phosphatase, alanine, allantoin, amino-peptidase, aspartic acid, aspartic amino-transferase, astragalin, calcium, carbohydrates, chlorogenic acid, coumesterol, cystine, daidzein, diacetoxyflavone, diacetoxyisoflavone, dimethoxyflavone, dodecene, D-pinitol, esterase, fat, ferulic acid, fibre, formononetin, fructose, genistein, genistin, glucose, glucose-6-phosphate-dehydrogenase, glutamic acid, hexaenylol-acetate, iron, isocitrate dehydrogenase, isoleucine, isomerase, isoquercitrin, kaempferol, kaempferol-rutinoside, leucine, linoleic acid, linolenic acid, lipoxygenase, lysine, malonyldaidzein, mannose-6-phosphate, methanol, methionine, myo-inositol, neochlorogenic acid, oleic acid, oxaloacetic acid, pentanal, pentane, phaseollin, phenylalanine, phosphoglucoisomerase, phosphoglucomutase, pinitol, potassium, proline, propanol, protein, protocatechuic acid, quercetin, quercetin-sophorosides, quercitrin, rutin, serine, sojagol, soyasapogenol, soyasaponin, spermidine, trans-aminase, trans-ocimene, triacetoxyflavone, trigonelline, trimethoxyflavone and tyrosine.

Root: Hydroxybenzoic acid, dihexosyljenkwanin, carlinoside, cis-aconitic acid, coumestrin, coumesterol, daidzein, formic acid, genistein, glycolic acid, isoliquiritigenin, isochaftoside, nodulin, oxalic acid, soyasaponin, trans-aconitic acid, trigonelline, vanillic acid and vitexin.

Leaf: β-Carotene, acetic acid hyxyl ester, acetic acid-cyclohexyl ester, allantoic acid, allantoin, bornesitol, butanoic acid-3-hexenylester, carboxyarabitinol, catechin, coumesterol, crocetin, daidzein, dimethyl-hexanal, D-ononitol, fisetin, flavonol-3-O-β-D-glucoside, gallic acid, glucosyl dihydrozeatin, glucosylzeatin, glyceollin, glycinoprenol, hydroxybenzoic acid, kaempferol, kaempferol-3-O-β-D-neohesperidoside, kaempferol-3-O-β-D-gentiobioside, kaempferol-3-O-β-D-glucosylgentiobioside, methyl glycofuran, myo-inositol, neoxanthin, nicotiflorin, ononitol, phaseol, quercetrin, soyasaponin, trigonelline, trihydroxypterocarpan and violaxanthin.

Pod: Ascorbic acid, β-carotene, calcium, carbohydrates, fat, fibre, iron and niacin.

Seed: (+) Pinitol, α-amylase, β-amylase, α-amyrin, β-amyrin, β-carotene, α-conglycinin, β-conglycinin, α-ketoglutaric acid, α-linolenic acid, β-neocarotene, β-sitosterol, α-tocopherol, acetyldaidzin, acetylgenistin, acetylglycitin, adenine, agmatine, alanine, allantoin, allantoinase, aluminium, amylase, anthocyanin, arabinogalactan, arabinose, arachidonic acid, arginine, ascorbic acid, aspartic acid, barium, betaine, biochanin-A and C, biotin, boron, butyrospermol, cadaverine, caffeic acid, calcium, campesterol, canavanine, carbohydrates, carboxylase, catalase, cephalin, cerabroside, cholesterol, ceramide, choline, ciceritol, citric acid, cobalt, copper, coumesterol, cyanidin-3-glucoside, cysteine, daidzein, daidzin, daizusaponin, dammaradienol, delphinidin-3-monoglucoside, D-galactose, dihydro-β-sitosterol, dimethyl amine, dimethyl texasin, D-pinitol, ergosterol, erucic acid, erythro-neopterin, ethylvinylketone, euphol, fat fibre, flazin, fluorine, formononetin, fumaric acid, galactinol, galactomannan, galactopinitol-A and B, gentisic acid, genistein, genistin, glucosamine, glucose, glucosidase, glucuronic acid, glutamic acid, glutamyl-alanine, glutamyl-tyrosine, glyceocarpin, glyceofuran, glyceollidin, glyceollin, glycerol, glycinin, glycinol, glycitein, glycitein-glucoside, glycitin, glycitin-glucoside, glycolic acid, gossypol, guanidine, hispidol, histidine, homospermidine, hydroxyphaseollin, inositol, inositol-hexaphosphate, iodine, isocoumaric acid, isoferulic acid, isoflavone, isoleucine, isoliquiritigenin, isopropyl-amine, isovalerianic acid, isoxanthopterin, kaempferol, lanosterol, lead, lecithin, lectin, legumelin, leucine, lignoceric acid, linoleic acid, linolenic acid, lipase, lipoxidase, lumazine, lupeol, lysine, magnesium, malic acid, malonylglycitin, malonyldaidzin, malonylgenistin, malonylglycitin, manganese, methionine, methyl amine, methyl genistein, methylagmatine, molybdenum, myristic acid, N-butyl-amine, N-capric acid, N-caprylic acid, neo-β-carotene, neuraminic acid, niacin, nicotianamide, N-methyl-phenethylamine, N-valerianic acid, O-coumaric acid, oleic acid oxalic acid, palmitic acid, palmitoleic acid, pantothenic acid, parkeol, p-coumaric acid, perlolidin, peroxidase, phaseollin, phasin, phenylalanine, phosphatidyl-choline, phosphorus, phosphotidase, p-hydroxybenzoic acid, physeolic acid, phytate, phytic acid, phytin, phytosterol, pinitol, piperidine, potassium, proline, propionic acid, protease, proteins, protocatechuic acid, pseudotaraxasterol, putrescine, pyridoxine, pyroglutamic acid, raffinose, riboflavin, rotenoids, rutin, salicylates, salicylic acid, saponin, selenium, serine, silicon, simiarenol, sinapic acid, sodium, soyasapogenol A,B,C and E, soyasaponin A,C,D and E, spermidine, spermine, squalene, stachyose, starch, stearic acid, stigmasta-4,22-dien-3,6-one, stigmasta-4-6-dien-3-one, stigmasterol, stigmasterone, strontium, succinic acid, sucrose, sugar, sulphur, syringic acid, tannic acid, taraxasterol, taraxerol, tetramethyl pyrazine, thiamine, threoneopterin, threonine, tirucalla-7,24-dien-3-β-ol, tirucallol, titanium, trigonelline, trimethyl-2-hexane, tryptophan, tyrosine, urease, valine, vanillic acid, verbacose, vitamin-D,E,K,B_6, vitexin, xylose, xylogalactomannan, γ-conglycinin, γ-coumaric acid, γ-sitosterol and γ-tocopherol.

Seed oil: Isofucosterol, phylloquinone, stigmasterol, squalene and vitamin E.

Seed sprout: Adenine, adenosine, adenylic acid, α-amyrin, ascorbic acid, β-amyrin, calcium, cycloartenol, glyceollin, hispidol, hypogeic acid, hypoxanthine, inositol, iron, linoleic acid, maltose, niacin, palmitic acid, phosphatidic acid, phosphatidyl ethanolamine, phosphatidyl glycerol, phosphatidyl inositol, phosphorus, potassium, protein, riboflavin, sapogenol-C, sodium, soyasaponin, succinic acid, thiamine and xanthine.

Soya bean lecithin: Lecithin is a protein complex, rich in phosphorus and contains aspartic acid, betaine, choline, cystine, glutamic acid, glycine, histidine, hydrocyanic acid, isoleucine, leucine, lysine, maltose, methionine, oleic acid, oxalic acid, palmitic acid, phenylalanine, saponin, stearic acid, threonine, trigonelline, tryptophan, tyrosine and valine.

Edibility

Plant: Used as vegetable.

Pod: Edible.

Seed: Unripe seeds are eaten as vegetable; dried seeds are eaten whole or split or sprouted after soaking and cooking; ground into flour and consumed; processed to give soya milk which is considered as a rich supplement of protein for infants, old and ailing and is used for making curds, *tofu* and cheese; soya flour is used in ice cream, chocolate bar, meat and in health foods; soya sauce is made from the natural fermented bean which is used in making of many Chinese culinary preparations; it is the main ingredient in most of Chinese food; seed also yields edible semi-drying oil used in cooking, as salad oil and in making of margarine; the seed meal and flour are used to prepare diabetic foods.

Medicinal Uses

Plant: Is analgesic, androgenic, anthelmintic, antianginal, antiapoplectic, antibacterial, anticancer, anticardiospasmic, anticoronary, antidote to mercury, lead, morphine, aluminium and iodine poisoning, antihaemorrhagic, antihepatitic, antihistaminic, anti-inflammatory, anti-ischaemic, antimetastatic, antimitogenic, antinociceptive, antioxidant, antipeptic, antiperoxidant, antiscorbutic, antiseptic, antispasmodic, antitussive, antiviral, aphrodisiac, astringent, bronchodilator, CNS active, cardiotonic, cerebrotonic, cholagogue, cholinergic, cytoprotective, diaphoretic, diuretic, emetic, emmenogogue, emollient, febrifuge and haemolytic; used in curing allergy, anorexia, alcoholism, anxiety, arthritis, atherosclerosis, alopecia, aphthae, apoplexy, arrhythmia, bronchitis, calculi, cancer of cervix, colon, oesophagus, liver, skin, canker, capillary fragility, cardiospasm, cataract, cephalagia, chilblain, colic, cold, convulsions, cystitis, dandruff, delirium, dementia, depression, dermatitis, diabetes, diarrhoea, dysentery, eczema, edema, encephalopathy, epilepsy, fatigue, fibrosarcoma, fistula, flatulence, flu, gastritis, gingivitis, glaucoma, goiter, heart burn, haematuria, haemorrhoids, hepatitis, hepatoma, herpes, hiccup, hydrophobia, hyperactivity, hyperthyroidism, hyperglycaemia, insomnia, leishmaniasis, leprosy, leukaemia, lupus, lymphoma, malaria, measles, melanoma, migraine, myalgia, myocarditis, nausea, nephritis, neuralgia, neuroblastoma, neuropathy, obesity, nyctalopia, opthalmia, orchitis, osteoarthritis, osteoporosis, otitis, photophobia, pneumonia, polio, pruritus, psoriasis, purpura, rabies, retinopathy, rheumatism, rhinitis, senility, sinusitis, sprue, stomatitis, stress, stroke, tachycardia, thrombosis, thyroid, tonsillitis, tuberculosis, tumour of bladder, breast, cervix, colon, oesophagus, liver, lung, mouth, prostate, skin, stomach and brain, tympanitis, ulcer, urinary complaints, vaginitis, varicose veins, vertigo and vitiligo.

Root: Decoction is used as astringent.

Leaf: The bruised leaves are used as local application for snake bite.

Flower: Is used in corneal opacity.

Pod: The green bean hulk smashed and applied to corneal ulcers, smallpox and ulcers.

Seed: Hypoglycaemic and hypocholesterolemic; used as a specific remedy for proper functioning of bowel, stomach, liver and kidney; sitosterol present in the seeds is used as antihypertensive drug; stigmasterol in the seeds is used as antistiffness factor.

Seed oil: Recommended for hypercholesterolemia as oil has high content of unsaturated fatty acids.

Seed lecithin: Commercial grade is used as a potent vasodepressor and lipotropic agent.

Bioactivity: Plant is inhibitor of 5-α-reductase, ACE, ATPase, aldose-reductase, α-amylase, β-glucuronidase, COMT, COX-1 and 2, catechol-O-methyl-transferase, cytochrome-P_{450}, DOPA-decarboxylase, deiodinase, FPTase, glucosyl-transferase; is angiotensin-receptor blocker, β-blocker, calcium channel blocker, circulatory stimulant and ER-β-binder.

Toxicology: Toxic amines such as cadaverine in the seed is a skin irritant; phenolic glycosides cause goiter; isoflavanones such as daidzein, genistein and glycitein has adverse effects on reproduction by disturbing the hormonal balance; antivitamin A & D interferes with vitamin A and D balance and uptake; causes hormonal disturbance; chymotrypsin and trypsin inhibitors reduce proteolysis of dietery proteins, impairment of growth and hypersecretion of pancreatic enzymes; carbohydrates such as stachyose and raffinose produce flatulence; phytates chelate mineral nutrients and render them unavailable to the body.

Uses in Alternative Systems of Medicine

Chinese:

Parts used: Bean, sprout and seed oil.
Uses: Anthelmintic, diaphoretic, diuretic, laxative and resolvent; bean is used in ague, ascites, bone diseases, cold, cold feet, colic, difficult breathing, difficult labour, dysentery, fever, flatulence, haemeturia, haemorrhage, headache, locomotory ataxia, melancholy, nausea, threatened abortion and venereal sores; it is used in the treatment of rheumatism; strengthens the body, lubricates the muscles and skin, improves complexion, tones the muscle; used as antidote for dog bite, horse bite, insect sting and scorpion bite; bean sprout hormonizes the spleen and stomach, removes flatulence; seed oil is used as local application to ulcers and skin diseases; warmed and taken internally or applied externally cures burns, chronic dysentery, eczema, haemorrhages, leprous sores, ophthalmia, scalds and subcutaneous swelling; promotes hair growth.

Non-medicinal Uses

Plant: The vegetative parts provide a good forage, hay, silage and manure; also grown as cover crop.

Straw: Useful as fodder and used in making brown paper.
Seed: Used in making of various food products; seed protein is used in the manufacture of synthetic fibre which is comparable to casein fibre and can blend with rayon or cotton; also used as plywood adhesive; used in textile sizing and fire fighting foams; roasted seeds are used as a substitute for coffee seeds; residue left after the protein extraction is used in the manufacture of phenolic type plastics; the soya bean residue after extraction of oil is a valuable cattle feed.
Seed flour: Used in the brewing industry for improving the body and flavour of beer.
Seed oil: Used in making oil cloth, plastic, leather, linoleum, paint, varnish, enamel, printing ink, soap, disinfectant and insecticide; used in cooking and as salad oil and in making margarine and hydrogenated fats.
Seed lecithin: Lecithin, a byproduct is used as wetting and stabilizing agent in food, cosmetic and pharmaceutical industry.

76. *Glycyrrhiza glabra* L.

Synonyms: *Glycyrrhiza glabra* L. var., *glabra*
Glycyrrhiza glabra L. ssp., *glandulifera* (Waldst & Kit) Ponert

Notes: There are two cultivated varieties of *Glycyrrhiza glabra,* the Oriental and the Spanish; the significance of this species in medicine is due to the glycyrrizin group of compounds (also called liquorice), which are saponins; one of them, the glycyrrhizic acid, is sixty times sweeter than cane sugar and is used as a non-calorie sweetener in numerous medical and nonmedical preparations.

Common Names: Liquorice; also the name of the raw drug, from the underground parts of rhizomes and thick roots.

Vernacular Names: Ben: Jashtimadhum; **Guj:** Jethimadha, **Hin:** Jeshtimadh, Jetimadh, Mulhatti, **Kan:** Atimadhura, Jeshtamadhu, Yashtamadhu; **Mal:** Athimadhuram, Irattimadhuram, Yashtimadhukam; **Mar:** Jeshtamadha, **Sans:** Jalayashti, Klitaka, Lakshmana, Madhoolika, Madhuka, Madhusrava, Madhuyashti, Sthalayashti, Yashtimadhu, Yashtika, Yashtirasakrya, **Tam:** Adimaduram; **Tel:** Athimaduramu, Yashti madhuramu.

Distribution: Native of Afghanistan, China, India, Iraq, Pakistan, former USSR, Yugoslavia, Romania, Sardinia, Sicily, Czech Republic, Slovakia, Cyprus, Greece, Israel, Jordan, Lebanon, Syria and Turkey; introduced for cultivation into subtropical and semi temperate countries; liquorice is cultivated in Northern India; the Oriental liquorice is grown in Syria, Iraq, Russia, Iran and Northern China, while the Spanish liquorice is grown in Spain, Italy and Greece.

Description: Erect branching, perennial herb with a branching rhizome; leaves dark green, alternately pinnate; leaflets 9-17, entire, ovate, pubescent with minute stipules; flowers arise from the axil of leaves on many flowered racemes or spike-like clusters; bracts pubescent and punctate; calyx puberulous to tomentose, punctate; corolla purplish blue, lavender, or violet; pod broad, falcate or coiled, ovate, oblong-linear, compressed, smooth or prickly; seeds small, dark, reniform, 1-6 per pod.

Phytochemistry:

Plant: 2-β-Glucuronosyl glucuronic acid, apigenin, astragalin, bergapten, betulinic acid, galangin, genistein, glycestrone, glycyrrhetinic acid, glycyrrhizic acid, glycyrrhizin, isoliquiritigenin, kaempferol, liquiritigenin-glucoside, liquiritin-diglucoside, lupeol, naringinin, pectin, pinocembrin, polysaccharides, reducing sugars, soya saponins, starch, tannic acid and zanthotoxin.

Root/Rhizome: Acetic acid, acetophenone, acetoxy-methyl isoflavone, acetyl ursolic acid, aluminum, apigenin, ascorbic acid, benzaldehyde, benzoic acid, benzyl alcohol, butan-1-Ol-3-one, butane-2,3-diol, butanoic acid, calcium, caproic acid, carvacrol, choline, chromium, cobalt, cumic alcohol, decanoic acid, deoxyglycyrrhetic acid, dihydro-furanone, dimethyl-pheneyl ethyl alcohol, dodecane, dodecanoic acid, echinatin, eicosane, estragole, estriol, ethyl linolenate, ethyl palmitate, ethyl phenol, ethyl phenyl acetate, eugenol, fenchone, formononetin, fructose, furfural, furfuryl acetate, furfuryl alcohol, furfuryl butyrate, furfuryl butyrolactone, furfuryl formate, furfuryl propionate, geraniol, glabrene, glabric acid, glabridins A and B, glabrol, glabrolide, glabrone, glucose, glycerin, glycocoumarin, glycyglabrin, glycyrin, glycyrol, glycyrrhenitic acid, glycyrrhizic acid, glycyrrhizin, glycarin, hederasaponin-c, henicosane, heptadecane, heptane-1, 2-diol, heptanoic acid, heranol, herniarin, hexadecane, hexadecanoic acid, hexanol, hexyl-formate, hispaglabridin-a and b, hydroxy chalcone, hydroxyglabrol, hydroxyglyyrrhetic acid, hydroxy-methyl isoflovone, isobutyladipate, isoglabrolide, isoliquiritigenin, isoliquiritin, isomucronulatol, isoneoliquiritin, kaempferol, kumatakenin, lavandalol, licochalcone, licoflavonol, licoisoflavanone, licoisoflavones, licoric acid, licoricone, licuraside, linalool, linalool-oxide, liqocoumarin, liquirazide, liquiritic acid, liquiritigenin, magnesium, methoxy-2-methylisoflovone, methyl glabridin, methyl-ethyl-ketone, methyl-hexa-decanoate, methyl-hexanoate, myrtenal, neoliquiritin, n-nonacosane, nonadecane, nonanoic acid, n-tetradecane, o-acetyl salicylic acid, o-cresol, octadecane, octanoic acid, o-methoxyl phenol, palmitic acid, pentadecane, pentadecanoic acid, pentanoic acid, phaseollin, phenethyl alcohol, phenols, phenyl acetaldehyde, phenyl propionic acid, phosphorus, pinocambrin, propionic acid, prunetin, pyrazole, rhamnoisoliquiritin, rhamnoliquiritin, salicylic acid, schaftoside, silica, sodium glycyrrhetinate, stigmasterol, sucrose, tetracosan-1-Ol, tetracosane, tetradecanoic acid, tetramethyl-pyrazine, thujone, tricosane, tridecane, tridecanoic acid, trimethyl pyrazine, umbelliferone, undecane, undecanoic acid, zinc, α-glycyrrhetic acid, α-hydroxyglycyrrhetic acid, α-terpineol, β-lipoproteins, β-sitosterol, γ-butyrolactone, γ-hexalactone, γ-octalactone, ρ-cymenol and ρ-methoxy phenol acetic acid; the root/rhizome also contain anethole, cumic alcohol, eugenol and estragole.

Leaf: Astragalin, bergapten, genistein, kaempferol, mucronulatol, nicotinic acid, prunetin, wighteone and zanthotoxin.

Edibility

Root/rhizome: The underground parts are used without the distinction of root and rhizome; used as flavouring and sweetening agent in foods, chewing tobacco and in beer industry; eaten as chew sticks and used in chewing gum to ensure flexible texture; liquorice is used in cocoa industry as glycyrrhizin has the same flavour as cocoa; utilized in drinks, ice creams, puddings, bakery products, chocolates and candies; It is used as taste corrigent in food and drug industry.

Medicinal Uses

Plant: Plant is demulcent, emollient and soothing; used for cold and cough.

Root/rhizome: The medicinal use of liquorice in both Western and Eastern cultures dates back to several thousand years; it is recommended for all age groups, for male and female alike; it is called the 'grand tonic' of the world; it is a broad spectrum tonic which protects and maintains health and heals injuries; the anti-inflammatory action of liquorice extends to the entire body, both inside and out; the soothing action is on the mucous membranes and the skin; it reinforces the body's ability to withstand bacterial, fungal and viral attacks; it is effective in maintaining the musculoskeletal system.

It is alexipharmic, alexiteric, analgesic, alterative, antiallergic, antiarthritic, antiasthamatic, antibacterial, anticancer, anticholinergic, anticonvulsant, antifungal, antihaemorrhoidal, antihepatotoxic, antihistaminic, antihyperglycaemic, anti-inflammatory, antimalarial, oxytocic, antioxidant, antipruritic, antinephrotic, antitumour, antitussive, antiviral, antiyeast, astringent, aphrodisiac, choleretic, demulcent, depurative, diuretic, emetic, emmenogogue, emollient, expectorant, immuno stimulant, intellect promoter, laxative, memory enhancer, estrogenic, pectoral, smooth muscle relaxant, soothing, suppurative, sudorific, taenifuge, trichogenous and vasodilator; it is used in treating, Addison's disease, anorexia, appendicites, aphthae, biliousness, bronchitis, catarrh, colic, conjunctivitis, consumption, cephalagia, dermatitis, diabetes, duodenal ulcers, dysuria, emaciation, epilepsy, fever, chronic gastritis, hyperdypsia, hypertension, internal inflammations, irritable mucous membrane, muscular pain, ophthalmia, pneumonia, respiratory diseases, rheumatoid arthritis, sore throat, scleroderma, ulcers, vomiting and vitiated "*vatha*"; it cleanses the colon, promotes adrenal gland function, decreases muscle spasms, has estrogen like effects, and stimulates the adrenal cortex after steroid therapy; it has a marked influence on the endocrine system as it has steroid like effect; this explains the use of liquorice in Addison's disease.

Liquorice is used as topical application for skin diseases and scorpion sting; it purifies blood, lowers stomach acid levels thus relieving heart burn; strengthens bones and muscles, relieves indigestion, increases physical strength, helps in retention of sodium and potassium, quenches extreme thirst, increases bile flow, lowers blood cholesterol level and increases extracellular fluids; raw liquorice when chewed, soothens inflammation of mouth, gums, throat, lungs and stomach; it cures abdominal pain, bronchitis, catarrh, constipation, consumption and epilepsy; liquorice powder relieves pain and discomfort caused by acrid matter in the stomach; with jaggery it is used as a mild laxative and with butter or ghee and honey, it is applied on cuts and wounds; liquorice is used as an eye wash in conjunctivitis and to improve sight.

Applied on the head, liquorice prevents hair loss, premature graying and patchy baldness; taken orally it relieves dysuria and pain associated with diarrhoea and urinary ailments.

Hot water extract is aphrodisiac, emmenogogue and galactagogue; is taken internally for tuberculosis and to improve sexual libido; with linseed oil it is useful in dry, irritable cough, laryngitis, peptic and duodenal ulcers and sore throat; hot water extract of the root with *Triticum aestivum* and *Zizyphus jujuba,* acts as a detoxicant; for pyrexia it is taken with *Panax ginseng, Angelica sinensis, Polygala tenuifolia, Euphorbia longum, Paeonia moutan, Zizyphus jujuba, Gardenia jasminoides, Astragalus* species, *Bletilla striata, Agrimonia* species and *Atractylodes macrocephala.* Liquorice root is also taken with *Panax ginseng, Citrus aurantiacum, Astragalus* species, *Angelica sinensis, Altractylodes macrocephala, Paeonia* species, *Rehmannia glutinosa, Bletilla striata, Sanguisorba officinalis, Citrus reticulata, and Equus asinus* hide, to restore vital functions in the body.

Glycyrrhetic acid from liquorice is used as anti-inflammatory compound in various preparations used in curing skin diseases and removal of urethral stones.

Root decoction is used as gargle in gum and teeth disorders. Root infusion taken internally is useful in relieving chronic joint pains, catarrh of the urinary and intestinal tracts, oral inflammations and sore throat; with coriander powder and lemon juice it is a soothing and cooling drink, and a remedy for cold, cough, distressed breathing, extreme thirst, fever and pain; root paste with saffron and milk is applied over bald patches to promote hair growth and to cure dandruff; syrup made from liquorice root is prescribed in gastric and duodenal ulcers.

A mixture of liquorice root with leaves of *Sida spinosa* and *Lycium barbatum*, galls of *Pistacia integerrima,* anthers of *Mesua ferrea*, honey, and cow's milk or ghee, is used as a contraceptive; liquorice root mixed with

Adhatoda zeylanica and *Azadirachta indica,* is taken orally for bronchitis; with *Terminalia arjuna, Sida retusa, Sida spinosa* and ghee it cures heart diseases; root with *Linum, Lobelia* and *Capsicum* cures bronchitis; it is also taken with *Viola pilosa, Adiantum lunulatum* and *Adhatoda zeylonica,* for bronchitis; root with *Cyolonia vulgaris* seeds, *Andropogon muricatus,* camphor, saffron, cinnamon bark, *Cassia fistula* seeds, lettuce seeds, rose petals, water melon seeds and gum tragacanth cures pyrexia and phthisis; root decoction wtih juice of hermophrodite *Amaranth* is taken with honey for leucorrhoea and other uterine complaints; decoction of the root along with *Cyperus rotundus* and coriander seeds, is a remedy for biliousness; root with *Acacia* gum, black pepper, gulacha, saffron and sugar is useful for cough and asthma; liquorice, camphor, asafoetida and gum acacia are together used for influenza; root with *Symphytium, Chamaemelium filipendula* and *Althea* is used in peptic ulcers; root with *Taraxacum* is used in Addison's disease; root with sandal wood paste, maddar, *Andropogon muricatus* paste has a cooling effect; lozenges made of root extract, cubebs, gum arabic, conium and sugar are useful in bronchitis.

Stem: Aphrodisiac, demulcent, emollient, expectorant, laxative, nervine tonic; useful in asthma, catarrh, cough, emaciation, hair loss, anorexia, peptic ulcers, persistent fever, pharyngitis, sore throat, vomiting and urinogenital disorders.

Bark: Given in jaundice.

Leaf: Poultice of leaf paste is a cure for hair loss, and used as a deodorant on stinking armpits and feet.

Bioactivity: When mice were injected with glycyrrhizin (75.0 or 150.0 mg/kg), it increased the weight of spleen and thymus and white blood cell count; it also inhibited increase in levels of serum and liver transaminase, indicating immunostimulant and hepatoprotective properties, against carbon tetrachloride and galactosamine induced cytotoxocity in mouse hepatocytes; at 2.4 mM, glycyrrhizin was effective against *Varicella-zoster* virus (VZV) replication; it inactivated 99 per cent of viral particles within 30 minutes; intravenously injected glycyrrhizin (25 or 50 mg/kg) inhibited passive cutaneous response in mice; it also inhibited concentration dependent contraction of rabbit ileum and guinea pig trachea, induced by histamine or acetylcholine *in vitro;* glycyrrhizin given intraperitonially inhibited granuloma formation induced by intradermal injection of glycyrrhizin and significantly suppressed production of platelet activating factor from rat peritoneal exudate cells stimulated by calcium ionophore 423187; glycyrrhetic acid and glycyrrhetinic acid inhibited tumour promoting activity on skin tumour initiated in mice with 7,12-dimethyl benz (1) anthracene; glycyrrhizic acid inhibited carrageena induced edema in rat paw and leucocyte migration in pleural space induced by detrose injections; it prevented *in vitro* myelo-suppressive effect of *Escherichia coli* endotoxin and hydrocortisone in mice bone marrow cells; isoliquiritigenin inhibited *in vitro* aldose reductase in rat lens; Goliderinin showed pronounced anti-inflammatory activity in mice more than hydrocortisone or amidopyrine; it also showed analgesic, anti-allergic and anti-pyretic properties; sodium glycyrrhetinate and ursolic acid showed hypolipidemic and anti-atherosclerotic activity and decreased cholesterol levels in the aorta; glycyrrhizin significantly prolonged survival time of rats which received a lethal dose of strychnine, shikimin, nicotine, arsenobenzone and caffeic acid; significant diruretic effects was observed in rats and rabbits; glycyrrhizin (150 mg/day) lowered serum TSH response to TSH releasing hormone; it also reduced glycocorticoid catabolism leading to a secondary suppression of pituitary response to TRH; glycyrrhizin (eight per cent) showed spasmolytc activity equivalent of 1/500 of papaverine in rats at 10 mg/kg dosage; glycyrrhizin and 18-β-glycyrrhetinic acid decreased acid phosphatase activity and increased β-N-acetylglucosaminidase activity in mice liver lysomes; crude plant extract administered orally or sub cutaneously to mice, showed loss of blood cell count, increase in thymus weight, atrophic cortex and showed lymphofollicle formation in the medulla of the thymus gland; a protective influence was observed against experimental ulcers; glabranin, pinocembrin, licoflavone and wighteone isolated from crude extract of the rhizome exhibited anti-oxidant activity.

Toxicology: Liquorice can cause metabolic disturbances known as pseudo-aldosteronism due to corticoid effect of glycyrrhizin leading to edema, hypertension and weight gain; excessive usage of liquorice can lead to cardiac dysfunction, hypertension and hypokalaemia; can produce neuro-excilatory symptom in large doses; seeds contain chymotrypsin and trypsin inhibitors.

Clinical Studies: Glycyrrhizin treated human embryonic fibroblast cells inhibited *Varicella-Zoster* virus replication; it also increased proliferation of human embryonic fibroblasts *in vitro;* isoliquiritigenin inhibited sorbitol accumulation in human erythrocytes; glycyrrhizin showed a significant anti-diuretic activity in human volunteers even at high doses of 20 g; rhamnoglucoside from the plant is used as a smooth muscle fibre depressant in many pharmaceutical compositions.

Uses in Alternative Systems of Medicine

Ayurveda:

Part used: Root/rhizome.

Uses: The root/rhizome is used as alexiteric, aphrodisiac, demulcent, diuretic, emmenogogue, expectorant, galactagogue, laxative, refrigerant and tonic; used for abdominal pain, anaemia, asthma, epilepsy, biliousness, catarrh, cough, diseases of the eyelid, fatigue, haemoptysis, headache, hemicrania, hiccup, inflammed respiratory system, irritable mucous membranes of the urinary tract, leprosy¸ ulcers, "vatha", vomiting and wounds; liquorice reduces thirst, purifies blood, improves taste, sight and voice and reduces hoarseness of voice; root paste with ghee (clarified butter) relieves heart burn, burning of stomach and heals ulcers; it promotes growth of healthy tissue after surgery; the dry powder is dusted on operation wounds to promote quick healing; when taken internally, liquorice purifies breast milk; cures dysentery, spleen disorders, fever, hiccup, and subdues deranged humours; when boiled in goat's milk it is used as an eye salve.

Chinese:

Part used: Root, leaf.

Uses: Root is used as alexipharmic, alterative, expectorant and tonic; used to ally cough and distress in breathing, asthma, channel disorders, feverishness, pain, thirst and lung disorders; it is applied to burns, boils and sores; used as emollient in peptic ulcer; is used as antidote for intoxication caused by *Datura metel;* and given for dyspnoea, fever, thirst and uterine cancer; leaf is useful in sore throat, spleen disorders and skin diseases.

Siddha:

Part used: Root/ rhizome.

Uses: The root/rhizome is used as laxative and tonic; is used in acidity, cough, sore throat, ulcers and urino-genital diseases; protects the mucosal membrane.

Tibetan:

Part used: Root/rhizome.

Uses: Root is used to cure lung and channel disorders; controls bleeding from different parts of the body; cleanses and heals ulcers, improves eye sight, voice and complexion and acts as an aphrodisiac.

Unani:

Part used: Root/rhizome and leaf.

Uses: Root/rhizome is as anti-flatulant, demulcent, diuretic, emmenogogue, laxative, maturant and muscular tonic; given in asthma, chest inflammation, chronic fever, cough, diseases of the brain and liver, micturition, paralysis, piles, sore throat and spleenic diseases; with honey liquorice acts as a purgative and helps in the expulsion of the placenta after child birth; given in bronchitis, colic, migraine, head aches, ulcer and vomiting; with *Foeniculum vulgare* and sugar, liquorice improves complexion; when applied to the eye, liquorice root extract improves sight, removes yellow colour in the eye, and reduces cataract; the root/rhizome and the leaf is used to quench thirst; leaf is used for scalds on the head and foul perspiration.

Medicinal Uses Outside India

England: Hot water extract of the root is taken for amenorrhoea and gastric ulcers.

Europe: Root extract is taken for cough, consumption and chest complaints.

France: Root decoction is taken as diuretic, depurative and emollient.

Indochina: Powdered root is taken as diuretic, laxative and sudorific.

Indonesia: Root is chewed for mouth thrush.

Israel: Hot water extract with sugar is taken for kidney stones, lung ailment and ulcers; the fresh leaves are used topically on wounds.

Japan: Root is used as antispasmodic, antitussive, demulcent and diuretic.

Moracco: Root extract is taken orally as cholagogue.

Nepal: Root is used as demulcent, emollient, laxative and tonic; used in cough, sore throat and as antidote for scorpion sting and urino-genital disorders.

South Korea: Hot water extract of root with *Astragalus membranaceus, Panax ginseng, Angelica gigas, Citrus aurantium, Cimicifuga* species and *Bupleurum* species is taken orally to stabilize diagestive functions; when taken along with *Scutellaria baicalensis, Panax ginseng, Zingiber officinale, Zizyphus jujuba, Pinellia tuberosa and Bupleurum falcatum,* cures chill, cold, fever, liver disorders, otitis, tonsillitis and tuberculosis; hot water extract of the root taken daily acts as contraceptive.

Thailand: Hot water extract of the root is taken orally as expectorant.

Tibet: Root/rhizome cures asthma, cough, lung and channel disorders.

Turkey: Used as an alexiteric, alterative, demulcent, depurative, diuretic, emollient, expectorant and pectoral; decoction of the root is taken orally for stomach ache.

USA: Hot water extract of root is taken orally as cathartic, demulcent, expectorant and laxative; root infusion is taken for cystitis and dysmenorrhoea.

Non-medicinal Uses

Plant: Used as sweetener.

Root: Used in the manufacture of toothpaste, cigarettes, cigars, chewing tobacco, moisturizers, conditioners and herbal shampoo; used as a binder for pills; to stabilize foam and to give sparkle and aroma to beer and as a flavouring agent in medicinal syrups; used in fire fighting foams; as sticking, wetting and spreading agent in insecticides; as compost for growing mushrooms.

Wood: Yields a dye employed in calico printing; and used to perfume fabric and as fuel.

Leaf: Added to beverage called *Nabid;* used in medicinal tea as taste corringent; as diabetic sweetener and in cough syrups, lozenges, pastilles and laxative powders.

77. *Indigofera arrecta* A. Rich.

Synonym: *Indigofera madagascariensis* Colla

Common Names: Bengal indigo, Java indigo, Natal indigo.

Vernacular Name: Kan: Jaava neeli.

Distribution: Native of Africa, Asia and Middle East; grows in Andaman and Nicobar islands, Bihar, Karnataka and Uttar Pradesh in India.

Description: Stout, rather woody herb or undershrub, 3m high, often cultivated as annual; stem erect, ridged, densely strigulose, hairs white or brown; stipules subulate, setaceous; leaves bluish green, imparipinnate; leaflets 7-8 pairs with one odd terminal leaflet, elliptic oblong, glabrous above, strigulose beneath; raceme many flowered, axillary, sessile; bracts lanceolate; calyx brown, strigulose, calyx lobes triangular; corolla brown, strigulose outside; pod brown, straight, or slightly tetragonal, curved upwards or deflexed, torulose but not in the plane of sutures; seeds 4-6 per pod, shortly oblong.

Phytochemistry

Plant: Indigotin.

Leaf: Kaempferitrin, kaempferol, nitrogen, potassium, phosphorus and rhamnose.

Medicinal Uses

Plant: Is antiseptic and astringent; useful in bug bites, epilepsy, hysteria, snake bite, spleen and bladder disorders and ulcer.

Non-medicinal Uses

Plant: Grown as cover crop and intercrop in cocoa, coffee, tea and rubber plantations; as green manure crop for rice; recommended as rotation crop for cotton; grown to improve soil fertility and as shade plant; it is the chief source of blue dye-indigo.

78. *Indigofera aspalathoides* DC.

Synonym: *Indigofera aspalathifolia* Roxb.

Common Name: Wiry Indigo.

Vernacular Names: Kan: Koggae, Neela, Neela malli, Shiva mallige, Shivamalli gida, **Mal:** Manali, **Mar:** Shivanimba, Thipoli; **Sans:** Rathakohomba, Shevaniarba, Shivanimba, Shivanti; **Tam:** Anthritthan poondu, Iraivanvembu, Shivamalli, Shivanaarvembu.

Distribution: Grows in the coastal areas of Andhra Pradesh, Karnataka, Kerala, Orissa and Tamil Nadu in India.

Description: Silvery undershrub with silvery pubescence; leaves simple or trifoliolate, oblanceolate, folded, entire; stipules small and shortly adnate to the petioles; flowers red on axillary racemes; calyx campanulate with subequal teeth; standard petal ovate or orbicular, wing petal oblong, keel petal erect; stamens with connate filaments; pod cylindrical.

Medicinal Uses

Plant: Useful in liver and spleen enlargement; promotes appetite.

Leaf, flower and tender shoots: Demulcent; leaf decoction is used in abscesses, cancerous affections and leprosy.

Uses in Alternative Systems of Medicine

Siddha:

Part used: Whole plant.

Uses: Useful in abscesses, cancer, edema, leprosy, psoriasis, skin diseases and syphilis.

79. *Indigofera cassioides* DC.

Synonyms: *Indigofera elliptica* Roxb.
Indigofera pulchella Roxb.

Vernacular Names: Hin: Hakna, Neel, Sakena; **Kan:** Chennaata, Gogge; **Mal:** Manali; **Mar:** Baroli, Chirmati, Nirda; **Tam:** Naarinchi, Naarina, Naarinji; **Tel:** Karkandi, Siralli.

Distribution: Native of Asia including India; it grows throughout hills in India.

Description: Deciduous, grey, pubescent shrub; leaflets 13-23, opposite, sometimes alternate, elliptic-oblong, obtuse or emarginate; 15-20 yellow flowered raceme; pod straight, cylindrical, 10-12 seeded.

Edibility

Leaf, flower bud and pod: Used as vegetable.

Medicinal Uses

Plant: Decoction is useful in cough, epilepsy, menorrhagia and mucous in urine.

Root: Anti-viral and spasmolytic; decoction useful in cough; root powder with oil is applied externally for pain in the chest and swellings.

Leaf and Flower: Useful in cough.

Ethnomedicine

Plant: Used in epilepsy, menorrhagia and mucous in urine.

Leaf: Used in cough and inflammation of the face.

Non-medicinal Use

Plant: Used as fodder.

80. *Indigofera hendecaphylla* Jacq.

Synonyms: *Indigofera endecaphylla* Jacq.
Indigofera pectinata Baker
Indigofera spicata sensu auct. mult.
Indigofera stricata sensu auct. mult.

Common Names: Trailing Indigo.

Vernacular Names: Mal: Cherupullate; **Mar:** Bhuiguli; **San:** Vasuka; **Tam:** Cheppunerunji; **Tel:** Chalapachi, Cherragaddalu; Yerrapalleru.

Distribution: Native of Africa and Asia including India and Indian Ocean Islands; grows throughout the plains in India.

Description: Prostrate or ascending herb with strigulose, ridged stem; stipules broadly scarious at base, tapering to a subulate tip, glabrescent along the margins; leaf rachis prolongs up to 3 mm beyond the lateral leaftlets; leaves pinnate with 5-11 leaflets which are cuneate, obovate, apiculate; inflorescence is raceme with many purple flowers; bracts brown, lanceolate; corolla strigose outside; pod reflexed or downward curved, apiculate, tetragonal, strigulose; seeds 5-8 per pod, subglobular, yellow and smooth.

Phytochemistry

Plant: Ash, carbohydrates, fibre, indospicine and proteins.

Medicinal Uses

Plant: Antiscorbutic, alterative and diuretic; juice of plant is used in old venereal affections; decoction given in epilepsy and insanity.

Toxicology: Not recommended as a forage crop; causes abortion in cattle and sheep; toxicity is due to indospicine which is a hepatotoxic amino acid which interferes with synthesis and utilization of arginine and inhibits DNA synthesis.

Non-medicinal Uses

Plant: Grown as cover and green manure crop.

Leaf: A source of indigo dye.

81. *Indigofera heterantha* Brandis

Synonyms: *Indigofera gerardiana* Baker
Indigofera gerardiana Baker var., *heterantha* (Brandis) Baker

Common Name: Indigo bush.

Vernacular Name: Pun: Kati, Simhalese, Kattu.

Distribution: Native of Asia including India; in India it grows throughout Himalaya, Jammu and Kashmir and Arunachal Pradesh.

Description: Ornamental shrub with flowers having vanilla scent; young parts greyish, pubescent; leaves 5-7 foliolate; leaflets obovate, oblanceolate, base cuneate, mucronate at the apex, pubescent on both the surface; flowers pink or purple on raceme; calyx hairy with lanceolate teeth; standard petal obovate; pod linear, pubescent.

Phytochemistry

Plant: β-Sitosterol, acacetin¸ apigenin, isoquercetin and ursolic acid.

Edibility

Flower: Used as vegetable.

Medicinal Uses

Plant: Anticancer.

Root: Anthelmintic; useful in cough, muscular pain and urinary complaints.

Leaf: Purgative; leaf given in powder form for inflammation of the liver.

Ethnomedicine

Root: Is used in urinary complaints.

Non-medicinal Uses

Plant: Used as fodder; grown as ornamental plant.

Twig: Useful in basket work and making rope bridges; also used as fuel.

Wood: White in colour with an irregular dark colour, used in making decorative articles.

82. *Indigofera hirsuta* L.

Synonyms: *Indigofera ferruginea* Schum. & Thonn.
Indigofera fusca G. Don
Indigofera hirsuta L., var., *pumila* Baker
Indigofera hirta Bojer

Common Name: Hairy indigo.

Vernacular Names: Tel: Peruvempali.

Distribution: Native of Africa and Asia including India; in India, it grows in Assam, Goa, Karnataka, Kerala, Tamil Nadu, West Bengal and West Himalaya.

Description: Summer annual or biennial herb; stem erect with grey or brown pubescence; leaves pinnate with 5-11 obovate leaflets; leaflets opposite, membranous, greyish green above, glaucous beneath; flowers pink or red on axillary racemes; calyx densely pubescent, teeth long, setaceous, plumose; corolla exerted; pod narrow, 4-angled, brown, pubescent, in clusters; hairs on the pod point downwards; seeds 4-6.

Medicinal Uses

Plant: Decoction is given in cerebral disorders.
Leaf: Decoction is used for diarrhoea and as stomachic.
Pod: Crushed pod is applied as poultice for eye diseases.

Medicinal Uses Outside India

Goldcost: Leaf decoction is used as lotion for yaws.
Philippines: Plant decoction is given in cerebral disorders.
Senegal: Crushed pod is used for eye diseases.

Ethnomedicine

Plant: Is used as antidysenteric and stomachic; used in diarrhoea and dysentery.

Non-medicinal Uses

Plant: Grown as cover crop to control soil erosion and as fodder; yields a dye.
Leaf: Is a source of indigo dye.

83. *Indigofera linifolia* (L.f.) Retz.

Synonym: *Hedysarum linifolium* L.f.
Vernacular Names: Ben: Bhangra; **Guj:** Jhinkigali, Nahanigali; **Hin:** Torki; **Mar:** Bhangra.
Distribution: Native of Ethiopia, Sudan, Asia including India, Australia, Saudi Arabia and Yemen.
Description: A slender, procumbent or trailing herb covered by silvery pubescence; leaves simple, linear, subsessile, lanceolate; flower brilliant red on a 1-8 flowered subsessile raceme; pod, silvery, globose, minute, one seeded.

Phytochemistry

Plant: β-Sitosterol, linifolin, linoleic acid, linolenic acid, oleic acid and palmitic acid.

Medicinal Use

Plant: Used in amenorrhoea, febrile eruptions and as vermifuge.

Ethnomedicine

Plant: Used in amenorrhoea, sores and wounds.

84. *Indigofera spicata* Forsskal

Synonyms: *Indigofera endecaphylla sensu auct.*
Indigofera neglecta N.E. Br.
Indigofera parkeri Baker
Indigofera hendecaphylla sensu auct.

Common Names: Creeping indigo, Spicate indigo.
Vernacular Name: Kan: Maalamandi.
Distribution: Native of Africa, China, Madagascar, Mauritius and Yemen; introduced into India; in India, it grows in Andhra Pradesh, Bihar, Gujarath, Karnataka, Kerala, Maharashtra, Manipur, Orissa, Tamil Nadu and West Bengal.
Description: Erect straggling herb; leaves alternate; leaf lets 9-11, alternate, pink, elliptic, oblong, acute, obtuse; stipules, linear, narrow and acute; calyx incised to more than half way, lobes lanceolate; raceme 15-18 flowered; flowers pink, standard petal ovate with a short claw, keel petal spurred; pod tetragonous, deflexed, 6-10 seeded.

Phytochemistry

Plant: 3-Nitropropionic acid and endecaphyllin.

Seed: Indospicine and hepatotoxic amino acids.

Toxicology: Teratogenic; causes cleft palate in rats.

Non-medicinal Uses

Plant: Cultivated as cover crop and for green manure and fodder.

Leaf: Source of indigo dye.

85. *Indigofera suffruticosa* Miller

Synonyms: *Indigofera argentea sensu* Baker
Indigofera divaricata Jacq.
Indigofera micrantha Desv.
Indigofera oligophylla Lam.
Indigofera tinctoria L., var., *brachyacarpa* DC.

Notes: Cultivated as dye yielding plant.

Common Names: Guatemala Indigo, Wild Indigo.

Vernacular Names: Hin: Vilaayati nil; **Kan:** Shimeneeli; **Tam:** Shimaiyaviri; **Tel:** Shimanil.

Distribution: Native of Asia including India, Caribbean Islands, Central America and Indian Ocean Islands; introduced else where; grows in Assam, Bihar, Andaman and Nicobar Islands and West Bengal in India.

Description: Shrub, 3 m high; stem pubescent; leaves imparipinnate, 9-17 foliolate; leaflets opposite, oblong, elliptic, apiculate at the apex, acute at the base, slightly pubescent above, densely glabrous below; stipules filiform; flower brick red or pink on axillary racemes; bracts filiform to lanceolate; calyx pubescent with subulate deltoid lobes; pod subterete, auriculate with thick margins, pubescent, 3-8 seeded.

Phytochemistry

Plant: β-Sitosterol, (+) pinitol, α-D-glucopyranose and lauisfieserone.

Root: α-D-Glucopyranose and nitropropanoyl-glycopyranoside.

Medicinal Uses

Plant: Diuretic, febrifuge, stomachic and vulnerary; used in back ache, bronchitis, rheumatic pain and syphilis; extract of the plant given in enlargement of spleen, liver, epilepsy and nervous disorders and to improve appetite; an ointment made from the plant is applied on haemorrhoids, sores and ulcers.

Root and Seed: Decoction is used for destroying vermin in ulcers and for urinary infections.

Leaf: Juice is used for hydrophobia and menorrhagia.

Ethnomedicine

Plant: Is used as abortifacient, analgesic, anodyne, ecbolic, depurative, diueretic, purgative, stomachic, sudorific and vulnerary; used for back ache, cholera, epilepsy, fever, headache, herpes, inflammation, neurosis, rheumatism, sores, spasm, stomach ache, syphilis, ulcer and urinogenitial disorders; used as an antidote for bee sting and snake bite.

Non-medicinal Use

Plant: Grown as green manure crop; also grown as cover crop in coffee plantations; yields a dye.

86. *Indigofera tinctoria* L.

Synonyms: *Anila tinctoria* (L.) Kuntze var., *normalis* Kuntze
Indigofera indica Lam.
Indigofera oligophylla Baker.

Common Names: Ceylon Indigo, Dyer's Indigo, Indian Indigo, True Indigo.

Distribution: Native of Malaysian Archipelago; introduced into India; grows throughout India.

Vernacular Names: Guj: Gali; **Hin:** Gouli, Neel, **Kan:** Bangaali neeli, Hennuneeli, Neeli gida, Olle neeli; **Mal:** Amari, Amil, Dronica, Grymina, Madhuparnika, Neel, Neelaamari, Neelam, Neeli, Ranjani; **Mar:** Nali, Neela, Neeli, Nuli, Gulee; **Sans:** Aklika, Asita, Bhadra, Bharavahi, Charathika, Dola, Dronika, Duli, Dulika, Gandhapushpa, Gandhapushi, Grameena, Gramya, Kaala, Kaeshi, Klitaka, Kutsala, Madhuparnika, Mahaabala, Mahaaphala, Meghavarna, Mochakrishna, Neelaa, Neelapushpika, Neeli, Neelika, Neelini, Rangapathri, Rangapushpi, Ranjani, Ranjini, Shaarathee, Shodhini, Shyaama, Shyaamalika, Sriphali, Sthiraranga; **Tam:** Asidai, Atthipurashadaam, Averi, Aviri, Avuri, Chaamundi, Irasani, Karundoli, Madubarunigai, Neelakkaali, Neeli, Neelum, Toli, Vannanavuri; **Tel:** Aviri, Neeli chettu, Syaama.

Description: Grey, pubescent undershrub with angular branches clothed with white hairs; leaflets 5-13, opposite, obovate, oblong, obtuse, apiculate, bluish green, smooth above, silvery pubescent beneath; flowers red, inner side greenish yellow with radiating purple streaks on 15-20 flowered subsessile raceme; calyx incised half way, silvery pubescent; standard petal greenish yellow, wing petal pink; pod linear curved, glabrescent, tapering towards apex with 6-10 seeds.

Phytochemistry

Plant: Apigenin, indican, indirubin, kaempferol, luteolin and quercetin.

Leaf: Indigotin and potash.

Seed: Galactomannan.

Medicinal Uses

Plant: Is antihepatotoxic, antineoplastic, hypoglycaemic and CNS depressant, useful in amennorrhoea, cholera, convulsions, hysteria and nervous disorders; ointment made out of the plant is applied on piles, sores and old ulcers.

Root: Anthelmintic, antitumour, bitter, diuretic, emetocathartic, expectorant, febrifuge, laxative, thermogenic, tonic and trichogenous; useful in bronchitis, cardiopathy, colic, consumption, epilepsy, giddiness, gonorrhoea, hepatitis, insanity, scorpion sting and sores; root decoction is useful in calculi, fever, inflammation and convulsions; used as antidote for arsenic poisoning.

Stem: Expectorant and febrifuge; useful in cardiopathy, chronic bronchitis, consumption, dental caries, hair growth and insanity; antidote to snake bite.

Leaf: Decoction of leaves used as anti-inflammatory, anthelmintic, diuretic, expectorant, laxative, thermogenic, tonic and trichogenous; useful in asthma, cardiopathy, chronic bronchitis, dropsy, epilepsy, heart palpitation, hydrophobia, kidney complaints, leucoderma, liver enlargement, lung diseases, nervous disorders, piles, spleen and urinary complaints; poultice of the leaves is recommended in burns, haemorrhoids, old ulcers and various skin affections, scalding of skin and wounds; applied to bites of venomous snakes and insects to relieve the burning pain.

Seed: Seed powder mixed with alcohol is used to kill lice.

Uses in Alternative Systems of Medicine

Ayurveda:

Part used: Whole plant, root and leaf.

Uses: Plant is used as a remedy against abdominal and spleen enlargement, ascites, calculi, constipation, dog bite, edema, fainting, giddiness, gout, hepatomegaly, intestinal obstructions, poisonous affections and urinary disorders; root is useful in bladder stones and epilepsy; leaf poultice is applied on wounds and ulcers; leaf juice is given for asthma, kidney complaints, rapid heart beat and whooping cough.

Chinese:

Part used: Whole plant, leaf and pod.

Uses: Useful in bruises, dysentery, fever, infantile disorders, swellings, tumours and typhoid; bruised leaves used as antidote to medicinal poison, arrow poison, cantheridial blister, insect stings and wolf bite; pod is antifebrile, astringent, anthelmintic and restorative.

Homoeopathy:

Part used: Tincture of plant juice.

Uses: Useful in acne, amenorrhoea, constipation, cough, diarrhoea, epilepsy, flatulence, head ache, hysteria, inflammation of stomach and kidney, itching, nausea, pimples, prolapsed anus, renal colic, sciatica, skin diseases, tooth ache and urethral pain.

Siddha:

Part used: Root and leaf.

Uses: Useful in anaemia, arthritis, epilepsy, fever, haemorrhoids, hepatitis, jaundice, leucorrhoea, nervous disorders, sores and ulcers.

Unani:

Part used: Whole plant, root and leaf.

Uses: Plant is useful in bronchitis, epilepsy and nervous disorders; root is used in hepatitis and hydrophobia; ointment made from the plant is used locally on haemorrhoids, sores and old ulcers; leaf juice is given for hydrophobia.

Medicinal Uses Outside India

Indo-China: The plant is used to treat haemorrhages and tetanus; leaf paste with turmeric is applied on contusions, inflammations and scabies; root is used in blennorrhagia and as diuretic.

Malay Peninsula: The plant is ground into a paste and applied on boils and yaws; taken internally for fever.

Philippines: Root decoction is taken internally for arsenic poisoning and kidney stones.

South China: Plant is used in treating dysentery and liver disorders; used as antidote for bee and wasp sting.

Ethnomedicine

Plant: Alexiteric, anodyne, antidote to arrow poison and cantheridial sting, antiseptic, astringent, deobstruent, emetic, purgative, restorative, stimulant, vermifuge and nervine tonic; used in bronchitis, cardiac ailments, dropsy, epilepsy, kidney and spleenic disorders, sores, ulcers, whooping cough, adenopathy, bug bite, blenorrhagia, bruise, cancer, convulsion, depression, dysentery, dysmenorrhoea, fever, hepatomegaly, mumps, piles, scabies, spasm, splenomegaly, trauma, swelling, tumour of abdomen, typhoid and urinogenital disorders.

Root: Root is used for fever and to heal wounds.

Leaf: Used for eye diseases, fractures and nervous disorders; promotes hair growth.

Non-medicinal Uses

Plant: Yields indigo dye used in painting, textile and tattooing; grown as a green manure crop in coffee plantations.

Leaf: Yields a blue colored indigo dye used in dyeing industry.

87. *Indigofera trifoliata* L.

Synonym: *Indigofera glandulifera* Hayata

Notes: Two varieties, *trifoliata* and *duthei,* which are endemic to India are recognized.

Common Name: Indigo.

Vernacular Names: Hin: Jangli methi; **Mar:** Lal Methi; **Tam:** Shimaiyaviri; **Tel:** Baragadama.

Distribution: Native of Asia including India and Australasia; grows through out India.

Description: Herb; leaflets 3, oblong, oblanceolate, emarginate, end ones almost or sessile, faintly hairy; obtuse with brown glands beneath; 6-10 red flowered raceme; calyx incised to more than halfway; pod straight, tetragonous, deflexed, narrowly winged on both sides of the suture, 3-6 seeded.

Medicinal Uses

Plant: Extract is useful in bronchitis, epilepsy, nervous disorders and as ointment applied on haemorrhoids, sores and ulcers.

Leaf: Juice is used as prophylactic agent against hydrophobia.

Seed: Alterative, aphrodisiac, astringent, restorative and tonic; useful in leucorrhoea and rheumatism; decoction useful in relieving back pain.

Ethnomedicine

Plant: Is used as tonic.

Non-medicinal Use

Plant: Is used as fodder and green manure; grown as cover crop.

88. *Indigofera trita* L.f.

Synonyms: *Indigofera carinata* De Wild
Indigofera heterophylla Presl.
Indigofera laevis Rydb.
Indigofera oxycarpa Desv.
Indigofera subulata Griseb.
Indigofera tephrosioides M. Micheli
Tephrosia frutescens (Miller) DC.

Notes: Four varieties, *subulata, maffei, scabra* and *trita* are recognized.

Common Name: Three leaved indigo.

Vernacular Names: Guj: Vekhario; **Kan:** Tore menthi; **Tam:** Kaattuyaviuri, Kandaram, Kutthukaran chammatti, Punalmurangai, Saubanjam; **Tel:** Nakkanaru.

Distribution: Native of Africa, Asia including India, Australasia, Madagascar; grows in Gujarath and Maharashtra in India.

Description: Grey pubescent undershrub; leaves 3 foliolate, oblong, obovate, slightly emarginate; leaflets obovate, oblong, terminal leaftlet stalked and larger than the other two; raceme spike like, 6-12 flowered; flowers salmon red; calyx teeth setaceous; pod subtetragonous, pointed, 4-angled and hairy; seeds cylindrical, truncate at both ends, glabrous, smooth and olive-green.

Medicinal Uses

Plant: Used to strengthen gums.

Seed: Used as nutritive tonic.

Uses in Alternative Systems of Medicine

Siddha:

Part used: Root.

Uses: Stimulant; useful in dropsy, fainting spells, itch and venereal disorders.

Ethnomedicine

Plant: Is used as restorative.

Twigs: Is used as toothbrush to strengthen gums and teeth.

Non-medicinal Uses

Plant: Grown as forage and green manure crop.

89. *Lablab purpureus* (L.) Sweet

Synonyms: *Dolichos lablab* L.
Dolichos purpureus L.
Lablab niger Medikus
Lablab vulgaris (L.) Savi
Vigna aristata Piper

Notes: Three subspecies, *purpureus, bengalensis* and *uncinatus* are taxonomically recognized.

Common Names: Banner bean, Black bean, Bonavist bean, Hyacinth bean, Lablab bean.

Vernacular Names: Hin: Bhata vaasu, Bhatavans, Betharasu, Borboti, Kaala lobia, Katajang, Lobia, Makhansim, Raajashimbi, Sem, Valm; **Kan:** Avare, Avare baele, Avare balli, Avare kaalu, Avare kaayi, Bili maniavare, Holadavare, Maniavare; **Mar:** Anvare, Kadavebaala, Pandharepavate, Parvate, Paote, Tambadepavade, Val; **Sans:** Nespava, Nishpava, Raajashimbi, Shimbi, Shvethashimbika, Simbi, Vallaka; **Tam:** Avarai, Avarai-motchai, Thattaam payiru; **Tel:** Adavichikkudu, Anapachikkudu, Annapa, Anumulu, Boberulu, Chikkudu, Thellachikkudu, Thellachikurkai.

Distribution: Native of Africa, China and Madagascar; introduced into India.

Description: Annual, creeping or twining herb; leaves trifoliolate; stipules reflexed; flowers reddish purple or white, borne on axillary raceme; calyx bilipped, upper entire or emarginate, lower 3-lobed; vexillum round, auriculate with 2 callosities or appendages on the inner face, keel with incurved beak at a right angle; style laterally flattened, slightly curved at the apex, hairy on the inner side; pod flat or inflated, linear, incurved with remnants of style; seeds 5-7, globose, ovate or flat, white or black.

Phytochemistry

Seed oil: α-Amino-n-butyric acid, 3-O-β-D-glucopyranosyl, D(+) pinitol, galactinol, linoleic acid, linolenic acid, myoinositol, oleic acid and palmitic acid.

Edibility

Pod and Seed: Edible.

Medicinal Uses

Pod: Used in nasal bleeding.

Seed: Aphrodisiac, febrifuge and stomachic.

Uses in Alternative Systems of Medicine

Ayurveda:

Part used: Seed.

Uses: Alexiteric, aphrodisiac, diuretic, fattening, galactagogue and laxative; useful in burning sensation and diseases of blood, inflammation and flatulence.

Unani:

Part used: Leaf and pod.

Uses: Used in cholera, diarrhoea, fever, flatulence, leucorrhoea and menorrhagia; reduces heat from sunstroke and quenches excessive thirst; used as antidote to fish and snake poison.

Medicinal Uses Outside India

China: Flowers are used for bloody stools, dysentery with pus, heat disorders, leucorrhoea and menorrhagia; seeds are used as antispasmodic, febrifuge, sedative, stomachic and tonic and to quench thirst in diabetics; used in cholera, colic, diarrhoea, dysentery and rheumatism; leaf poultice is used as an antidote for fish and snake poison.

Malay Peninsula: Leaf poultice is applied on eczema and leaf juice is dropped into the ear for ear ache.

90. *Lablab purpureus* (L.) Sweet var., *lignosus* Prain

Notes: In India, var., *lignosus* Prain is widely cultivated; var., *typicus* Prain with long and narrow pods is uncommon.

Common Names: Lablab bean, Poor man's bean, Tonga bean.

Vernacular Names: Guj: Val; **Hin:** Ballar; **Kan:** Avare; **Mal:** Mochakotta; **Tam:** Mochai; **Tel:** Anumulu.

Distribution: Native of Africa, widely cultivated throughout the tropics.

Description: Twining herbaceous annual; leaves trifoliolate; leaflets broad, ovate to rhomboid, acute at the tip, smooth above, short haired below; inflorescence is a many flowered raceme; flowers white, blue or purple; pod scimitar shaped; seeds 2-4, brown, ovoid and laterally compressed.

Phytochemistry

Plant: Gibberellin, myristic acid and pantothenic acid.

Pod: Arabinose, ascorbic acid, fat, fibre, galactose, galacturonic acid, glucose, iron, phosphorus and protein.

Seed: β-Carotene, β-D-ribofuranosyl-cis-zeatin, α-linolenic acid, β-D-ribofuranosyl-trans-zeatin, alanine, arginine, arsenic, ascorbic acid, aspartic acid, brassinolide, calcium, carbohydrates, copper, cystine, dolicholide, dolichosterone, fat, fibre, galactose, glucose, glutamic acid, glycine, histidine, homodolichosterone, iron, isoleucine, lauric acid, leucine, linoleic acid, lysine, magnesium, manganese, methionine, niacin, oleic acid, palmitic acid, palmitoleic acid, phenylalanine, phosphorus, potassium, proline, riboflavin, serine, sodium, stearic acid, thiamine, threonine, trans-zeatin, tryptophan, tyrosine, valine and zinc.

Edibility

Pod: Edible, young pods are salted, steamed and consumed.

Seed: Eaten after cooking.

Medicinal Uses

Plant: Is antiadenomic, antiallergenic, antiandrogenic, antibacterial, anticancer, anticoronary, antifibrinolytic, antihepatotoxic, antihistaminic, antihypertensive, anti-inflammatory, antilymphocytic, antimutagenic, antioxidant, antiperiodontic, antipyretic, antiseptic, antispasmodic, antiviral, apoptotic, cardioprotective, choleretic, diuretic, haemolytic, hepatoprotective, immunostimulant, laxative, tonic, vasodilator and vulnerary; useful in acne, alcoholism, anaemia, anorexia, antidote to aluminium, cadmium and lead poison, anxiety, arrhythmia, arthritis, asthma, atherosclerosis, back ache, beri-beri, canker, cataract, chilblain, cold, convulsions, delirium, dementia, depression, diabetes, eczema, edema, epilepsy, fatigue, gastritis, gingivitis, glaucoma, heart burn, hepatitis, haemorrhages, herpes, hyperactivity, hypoglycaemia, impotency, insomnia, leprosy, lupus, measles, migraine, myocardia, neuralgia, neuropathy, obesity, orchitis, osteoarthritis, osteoporosis, phthisis, pneumonia and polio.

Leaf: Emmenogogue; given in colic; used as poultice for snake bite.

Flower: Emmenogogue; used in treating leucorrhoea and menorrhagia.

Pod: Juice mixed with salt and applied in inflammation of the ear and throat.

Seed: Antispasmodic, aphrodisiac, febrifuge, sedative, stomachic and tonic; used in treating colic, cholera, diarrhoea, dysentery, flatulence and rheumatism; to reduce heat from sun stroke and to nullify liquor intoxication; as antidote for fish poisoning; to quench excessive thirst in diabetic patients.

Bioactivity: Plant is inhibitor of 5-α-reductase, ACE, aldose-reductose, β-glocuronidase, COX 1 and 2, insulinase and prostaglandin synthesis; is angiotensin-receptor blocker, β-adrenergic receptor blocker, β-blocker, CNS active, calcium channel blocker, interferon synergist and pituitary stimulant.

Toxicology: Ration of lablab to cows imparts "feedy" flavour to milk and makes milk unpalatable; chymotrypsin and trypsin inhibitors present.

Uses in Alternative Systems of Medicine

Ayurveda:

Part used: Leaf and seed.

Uses: Leaf and seed useful in anaemia, burning sensation, flatulence, haemorrhages, hiccup and inflammations; seed is alexiteric, aphrodisiac, diuretic, galactagogue and laxative.

Chinese:

Part used: Leaf, flower and pod.

Uses: Leaf is used as poultice in snake bite; flower is useful in leucorrhoea and menorrhagia; pod is useful as antidote to fish poison; cures cholera, diarrhoea, excess thirst, fever and flatulence.

Siddha:

Part used: Leaf and pod.

Uses: Aphrodisiac, febrifuge and stomachic; useful in cholera, chronic dysentery, headache and ulcers.

Medicinal Uses Outside India

China: Infusion of the dried flowers is taken in summer for heat disorders, for cooling the body, dysentery with pus and bloody stools.

Indonesia: Leaf juice with fennel is taken as a cooling drink.

Malay Peninsula: Leaf poultice is applied for eczema.

Ethnomedicine

Plant: Antivenous, carminative and vermicide; used for ear ache, blindness, cholera, diarrhoea, epilepsy, fever, leucorrhoea, menorrhagia, splenomegaly, sun stroke, swelling, thirst, throat troubles, tumour and ringworm; prevents grey hairs.

Root: Is vermicide.

Leaf: Is used for ringworm and cough; is used as antidote for plant poisons.

Non-medicinal Uses

Plant: Is used as green manure and fodder.

Leaf: Yields a green dye used for tattooing.

Pod and Seed: Is used as cattle feed.

91. *Lathyrus aphaca* L.

Synonym: *Orobus aphaca* (L.) Doll

Common Names: Yellow pea, Yellow vetch, Yellow vetchling, Yellow pod.

Vernacular Names: Ben: Jangli matar, Masur channa; **Hin:** Jangli mattar; **Pun:** Rawari.

Distribution: Native of Africa, Asia including India, Europe and Middle East; grows in West India, Kashmir, Himachal Pradesh in India.

Description: Trailing or scrambling annual; stem not winged, glabrous; stipules foliaceous, broadly ovate, rest of the leaf reduced to a tendril; inflorescence a 1-2 flowered, axillary raceme; pendunde 1-4 times as long as the stipule; calyx teeth unequal, 3 times as long as the tube; corolla bright to pale yellow; pod glabrous, 4-6 seeded.

Edibility

Seed: Unripe seeds are eaten but ripe ones have narcotic effects.

Medicinal Uses

Flower: Resolvent.

Toxicology: Seeds cause lathyrism.

Ethnomedicine

Plant: Is narcotic and resolvent.

Non-medicinal Uses

Plant: Is used as fodder.

92. *Lathyrus odoratus* L.

Common Name: Sweet pea.

Distribution: Native of Italy, Sicily and Greece; introduced into India; grows in most of North India and Karnataka in India.

Description: Climbing herb; leaflets ovate oblong, pinnately compound, upper leaflets modified into tendrils; flowers white, rose or purple; pod flat; seeds many, yellow to brown.

Phytochemistry

Plant: Glucopyranosyl 3-isoxazoline and 2-carboxymethyl 3-isoxazolin.

Leaf: Lathodoratin and methyl ester.

Flower: Allantoin, arbutin, cyanidin, kaempferol, delphinidin diglucoside, dirhamnoside, kaempferol, kaempferol glycoside, malvidin-5-glucosido-3-rhamnoside, myricetin-3-rhammoside, pelargonidin-3-rhamnoside, peonidin-3-rhamnoside, petunidin, quercetin-3-rhamnoside, 3-xyloglucoside and 3,5-xylosylglucoside.

Pod: Lathodoratin.

Seed: 2-(2-carboxyethy l-3-isoxazoline) 5-one.

Edibility

Pod and seed: Are cooked and eaten.

Bioactivity: Seed diet caused skeletal and vascular changes, growth disturbance, lesions in small blood vessels, osteoporosis and degenerative arthritis in rats.

Toxicology: Seed causes angiolathyrism; toxic principle has been identified as β-(L-glutamyl)-amino propionitrile; contains chymotrypsin and trypsin inhibitors.

Ethnomedicine

Plant: Is used as insect repellant.

Non-medicinal Uses

Plant: Most popular garden plant; used as insecticide.

Flower: Yields a sweet and delicate perfume resembling scent of orange flowers, used in perfumery.

93. *Lathyrus pratensis* L.

Common Names: Meadow pea, Meadow vetchling

Distribution: Native of Africa, Asia including India, Nepal, Pakistan, Europe and Middle East; grows in Western Himalayas in India.

Description: Perennial, scrambling; stem angled, subglabrous or pubescent; leaves paripinnate; each leaf has a pair or narrow leaflets and a terminal tendril; flower yellow in groups of 5-12; stipules as long as the leaflets; calyx with unequal teeth, lowest tooth 2 times a long as the tube; pod broad, glabrous, 4-8 seeded.

Phytochemistry

Plant: Luteolin, pentahydroxyflavone-3'-β-glucoside and quercetin.

Leaf: Apigenin-7'-diglucoside, arabinoside, apigenin, kaempferol 3'-glucoside, luteolin 4'-glucoside, luteolin-3'-glucoside, myricetin, myricetin-3'-arabinoside, myricetin-3'-diglucoside, quercetin, tricetin-3'-glucoside and tricetin-7'-diglucoside.

Flower: 3'-Methyl myricetin, apigenin and 5-deoxykaempferol.

Medicinal Use

Seed: Used as resolvent.

Ethnomedicine

Plant: Used as resolvent.

94. *Lathyrus sativus* L.

Synonym: *Lathyrus asiaticus* (Zalk.) Kudrj.

Notes: An ancient old world cultigen; unknown wild origin; one of the first crops to be domesticated in Europe; intraspecific and interspecific crossing very common.

Common Names: Birdseed, Blue vetchling, Chickling, Chickling pea, Chickling vetch, Grass pea, Indian pea.

Vernacular Names: Ass: Khesari; **Ben:** Khesari; **Guj:** Langi; **Hin:** Khesari dhal; **Kan:** Khesari bele; **Mar:** Lakh; **Ori:** Khesara; **Pun:** Chural, Kesari; **San:** Triputa; **Tel:** Lamka.

Distribution: Nativity not known; introduced into Africa, Asia including India, Australia, Europe, Middle East and Pacific Ocean Islands; grows wild in Western Himalayas; cultivated throughout Central, North and South India.

Description: Much branched sub-erect, annual herb with winged stem; leaves pinnately compound; leaflets generally 2, linear, lanceolate, upper leaflets modified into tendrils; flowers axillary solitary, blue, purple or white; pod flat, oblong, dorsally 2 winged, slightly curved; seeds 4-5, spotted or mottled, brown or yellow.

Phytochemistry

Plant: Cyanin, homoagamatine, lathycarpin, pelargonin and pisatin.

Leaf: Lathycarpin and pterocarpin.

Seed: β-Alanine, β-N-oxalyl-α-β-diaminopropionic acid, aflatoxin, agmatine, alkaloids, aspartic acid, cadaverine, citric acid, cyanin, cyanogenic glucosides, cystine, fructose, glucose, glutamic acid, glycine, isoleucine, lathycarpin, lectin, leucine, L-homoarginine, malic acid, mannose, methionine, O-acetyl ethanolamine, phenylalanine, poly-β-hydroxybutyric acid, proline, putrescine, saponin, serine, spermidine, spermine, tartaric acid and tryptophan.

Seedling: Guanidine, guanidine amine and omoagamatine.

Edibility

Plant: Is used as food in times of famine.

Leaf: Eaten as potherb.

Pod: Boiled and eaten as vegetable.

Seed: Seed flour made into chapathi or seeds are boiled and eaten.

Medicinal Uses

Leaf: Cures biliousness, "kapha" and is astringent to bowels.

Seed oil: Is a powerful cathartic; useful as application in burning piles, inflammations and as cooling agent for sores and wounds.

Clinical Studies: Used as a tool in connective tissue research.

Toxicology: The vegetative parts and seeds contain cadaverine causing skin irritation; propionic acid, β-N-oxalylamine-L-alanine (BOAA), β-diamino propionic acid, L-3-oxalylamino-2-aminopropionic acid in the seeds causes neurolathyrism; β-amino propionitrile causes osteolathyrism, Zelweger's syndrome and infantile Refsum's disease; contains chymotrypsin and trypsin inhibitors.

Uses in Alternative Systems of Medicine

Ayurveda:

Part used: Leaf and seed.

Uses: Leaf removes biliousness and "kapha"; seed is astringent to bowels, cooling and tonic; improves taste and appetite; seed is used in burning piles, pain and inflammation.

Homoeopathy:

Part used: Tincture of the seed.

Uses: Useful in beri-beri, impotency, locomotary ataxy, lumbago, paralysis, paraplegia, stiffness of lower limbs, rheumatism and spinal sclerosis.

Ethnomedicine

Plant: Is used for dysentery, jaundice, leg pain, liver complaints, skin diseases, sore throat and stomach disorders.

Non-medicinal Uses

Plant: Is used as fodder and green manure crop.

Seed: Is used as cattle feed; ground seeds are used as plywood adhesive.

95. *Lens culinaris* Medikus

Synonym: *Ervum lens* L.

Common Name: Lentil.

Vernacular Names: Hin: Masoor; **Kan:** Channangi baele, Lentil, Masoor baele; **Mar:** Masoori; **San:** Gabholika, Gurubeeja, Halasaka, Kalyaanabeeja, Mangaliya, Masoora, Masooree, Masooraka, Masoorika, Prithubeejaka, Ragadali, Sura, Thamboolaraaga, Vrihikanchana; **Tam:** Misurpurpu; **Tel:** Chirisenagalu, Masuripappu.

Distribution: Native of Asia, including India and Pakistan and Middle East; cultivated throughout India.

Description: Annual, bushy herb, much branched, with soft hairs; stems slender, angular; leaves pinnate ending in a tendril; leaflets 4–7 pairs, alternate or opposite, oval; stipules small, entire; flowers small, pale blue, purple, white or pink on axillary 1–4 flowered raceme; pod oblong flat, smooth, 1–2 seeded; seeds biconvex, small, greenish brown or light red, speckled black.

Phytochemistry

Plant: Aflatoxin, ciceritol, deoxykaempferol, diglycosyl delphinidin, dihydroxyflavone, 4',7-dihydroxy-3'-methoxyflavone, ferulic acid, indolyl acrylic acid, kaempferol glucoside, lenticin, luteolin, oxosteroids, proanthocyanidin, syringic acid, trans-ferulic acid, trans-p-coumaric acid, tricetin, trihydroxy flavone, triterpene alcohols and variabilin.

Seed: β-Carotene, alanine, amylase, arginine, ascorbic acid, aspartic acid, biotin, calcium, carbohydrates, chlorine, choline, copper, cystine, daidzein, diglycosyl delphinidin, esculin, ethanolamine, fat, ferulic acid, fibre, folic acid, gadoleic acid, genistein, glutamic acid, hemicellulose, histidine, γ-hydroxy–L-homoarginine, imidazole, inositol, iodine, iron, isoleucine, kaempferol, kaempferol–3-O-L-glucoside, kaempferol-3-O-L-rhamnoside, β-D-glucosyl-7-O-α-L-rhamnoside, leucine, linoleic acid, linolenic acid, luteolin, lysine, magnesium, myristic acid, niacin, oleic acid, oxalic acid, palmitic acid, palmitoleic acid, pantothenic acid, paragalactoaraban, p–coumaric acid, phenylalanine, phosphorus, phytate, potassium, proanthocyanidin, proline, proteins, pyridoxine, riboflavin, 3-O-rutinosyl-7-O-α-L-rhamnoside, saponin, serine, sodium, soyasaponin, stachyose, starch, stearic acid, sulphur, thiamine, threonine, tocopherol, tricetin, 3,4,7-trihydroxy-flavone, tryptophan, tyrosine, valine, variabilin, vitamin A and B_6, wyerone and zinc.

Seed sprout: Alanine, arginine, ascorbic acid, asparagin, aspartic acid, β-carotene, calcium, carbohydrate, copper, cystine, fat, fibre, glutamic acid, glycine, histidine, iron, leucine, linolenic acid, lysine, magnesium, manganese, methionine, niacin, oleic acid, palmitic acid, pantothenic acid, phenylalanine, phosphorus, potassium, proline, riboflavin, serine, sodium, stearic acid, thiamine, threonine, tyrosine, valine, vitamin B_6 and zinc.

Edibility

Pod: Young pods are eaten as vegetable.

Seed: Edible; used in soup flavoured with spices and condiments; lentil flour is mixed with barley and other cereal flours; used in cakes and as food for children.

Medicinal Uses

Plant: Is antialcoholic, antiandrogenic, antianginal, antibacterial, antidepressant, antiemetic, antiestrogenic, antifertility, antifibrinolytic, antihaemorrhagic, anti-hepatotoxic, antihistaminic, antihypertensive, anti-inflammatory, antimicrobial, antimitotic, antimyocarditic, antinephritic, antineuralgic, antinociceptive, antioxidant, antiperiodontal, anti-prostaglandin, antipsychotic, antipyretic, antiradicular, antiscorbutic, antiseptic, antitumour, antiviral, aphrodisiac, apoptotic, arteriodilator, astringent, CNS-depressant, cardiodepressant, cardioprotective, cardiotonic, cerebroprotective, cerebrotonic, cholagogue, choleretic, contraceptive, cytotoxic, diuretic, emetic, emollient, estrogenic, haemolytic, haemostatic, hepatoprotective, hypercholesterolemic, hypotensive, hypnotic, immunomodulator, sedative, uterorelaxant, vasodilator and vulnerary; used in treating acne, allergy, anaemia, anorexia, anxiety, arrhythmia, atherosclerosis, arthritis, asthma, back ache, beri-beri, bronchitis, cancer of breast, colon, liver, skin, canker, capillary fragility, cardiospasms, cataract, chilblain, colic, cold, convulsions, dandruff, delirium, depression, dementia, dermatitis, diabetes, eczema, edema, epilepsy, fatigue, gastritis, gingivitis, glaucoma, goiter, gout, heartburn, haemorrhoids, hepatitis, herpes, hyperactivity, hypoglycaemia, impotency, insomnia, keratitis, leprosy, leukaemia, lupus, lymphoma, measles, melanoma, menorrhagia, metaplasia, neoplasia, nephritis, neuralgia, neuropathy, neuroblastoma, obesity, ophthalmopathy, orchitis, osteoarthritis, osteoporosis, photophobia, pneumonia, polyp, psoriasis, retinopathy, rheumatism, schizophrenia, senility, scabies, spasms, stomatitis, stress, stroke, sunburn, tremors, thrombosis, thyroid, tumour of breast, colon, cervix, stomach, ovary and skin, ulcer, vertigo, vitiligo and uterine complaints; antidote to aluminium, cadmium, iodine and lead poisoning,

Seed: Astringent, cooling, diuretic and laxative; improves appetite, enriches blood, removes "kapha" and biliousness, cures bronchitis, constipation, dysentery, eye and chest disorders, intestinal affections, skin diseases, small pox, strangury, stomatitis and tumours; made into paste and applied on foul ulcers, sores and slow healing ulcers.

Bioactivity: Plant is inhibitor of 5-α-reductase, ACE, aldose-reductase, β-glucuronidase, catechol-O-methyl-transferase, decarboxylase, histidine-kinase, insulinase, lipase, lipoxygenase, MAO, MDR, NADH–ubiquinone-oxidoreductase, ornithine-decarboxylase, PTK, peroxidase, phospholipase-2, protein-kinase, prostaglandin synthesis, topoisomerase, tyrosinase and xanthine-oxidase; is angiotensin receptor blocker, β-adrenergic receptor blocker, β-blocker, CNS-depressant, calcium antagonist, calcium channel blocker, pituitary stimulant and monoamine precursor.

Toxicology: Plant contains chymotrypsin and trypsin inhibitors; excess use may cause night mares and melancholia.

Uses in Alternative Systems of Medicine

Ayurveda:

Part used: Leaf and seed.

Uses: Leaf and seed astringent and cooling to bowels, improves appetite; cures biliousness, dysentery, eye and skin diseases, strangury and tumours; removes "kapha".

Tibetan:

Part used : Seed

Uses: Seed heals wounds and cures lung diseases.

Unani:

Part used: Seed.

Uses: Astringent, deobstruent, emmenogogue, laxative and tonic; useful in breast inflammation, bronchitis, cough, diphtheria, eye complaints, palsy, paralysis and stomatitis.

Ethnomedicine

Plant: Used for abdominal tumour, jaw cancer, scurvy and ulcer.

Non-medicinal Uses

Plant: Used as fodder and green manure; is a source of commercial starch which is used in calico printing and textile industry.
Leaf: Used as fodder.
Seed: Employed in textile and calico printing.

96. *Lespedeza juncea* (L.f.) Pers.

Synonyms: *Hedysarum junceum* L.f.
Lespedeza aitchisonii Ricker
Lespedeza cytisoides (Pallas) Nakai
Lespedeza hedysaroides (Pallas) Kitagawa
Trifolium cytisoides Pallas
Trifolium hedysaroides Pallas

Common Name: Japanese clover.

Distribution: Native of Asia including India, Pakistan, Nepal and Australasia; grows throughout Arunachal Pradesh, Bihar, Central and North East India, Himalaya, Kashmir and Rajasthan in India

Description: An erect densely pubescent, under shrub, 2-4 ft high; leaves ascending, trifoliolate, crowded; leaflets linear, cuneate, emarginate, glabrous above, clothed with grey or white silky hairs beneath; stipules linear; flowers pale white, tipped with purple tinge; 2-4 flowers in the leaf axils; calyx pubescent with long linear subulate teeth; keel petal much incurved, pale yellow or pink; style persistent, twice as long as the pod; pod silky.

Phytochemistry

Plant: β-Sitosterol, avicularin, juglanin, kaempferol, pinitol, quercetin, succinic acid, triacontan-1-Ol and trifolin.
Root: Cinnamic acid, ferulic acid, gallic acid, gentisic acid and syringic acid.
Leaf: Isoorientin, isovitexin, lucenin-2 and vicenin-2.

Medicinal Uses Outside India

China: Plant decoction is used as stomachic; used in anaemia, dysentery, haemorrhages and nocturnal emissions; root decoction is given in malaria.

Non-medicinal Uses

Stem: Dried and used for paper pulp.
Leaf: Used as manure and fodder.
Seed: Yields semi-drying oil.

97. *Lonchocarpus sericeus* (Poiret) DC.

Synonyms: *Derris sericea* (Poiret) Ducke
Lonchocarpus cruentus Lundell
Lonchocarpus formosanus DC.
Robinia sericea Poiret

Notes: Two varieties, *sericeus* and *palmeri*, are recognized.

Distribution: Native of Africa, India, Caribbean islands, Central America and South America; cultivated in Botanical gardens in West Bengal, India.

Description: Large shrub or small tree, 8-10m high; branches hanging; leaves 3 to 5 pairs, dark green above, silky pubescent below,; leaflets 6 to 10 cm; flowers on crowded axillary raceme, silky pubescent when young; corolla purplish; pod ovate, compressed; seeds 1 or 2.

Phytochemistry

Root, Leaf and Seed: Lonchocarpine.

Leaf: p-Coumaric acid, quercetin, rutin and β-sitosterol.

Medicinal Uses

Root bark: Is employed as laxative and stomachic; useful in back ache and convulsions; applied locally on skin eruptions.

Ethnomedicine

Plant: Diuretic, laxative and stomachic; used for back ache, convulsion, skin eruption, scurvy, skin diseases and wound healing.

Non-medicinal Uses

Plant: Used as insecticide and parasiticide.

98. *Lotus corniculatus* L.

Synonyms: *Lotus ambiguus* Sprengel
Lotus arvensis Pers.
Lotus balticus Miniaev
Lotus carpentanus Lacaita
Lotus major Scop.

Notes: Very variable; recognition of subspecies and varieties is disputed.

Common Name: Bird's foot trifoil.

Distribution: Native of Africa and Asia including India and Pakistan, Europe and Middle East; grows in Western Himalaya in India.

Description: A prostrate herb with pentafoliolate leaves; leaflets 2, asymmetrical, pointed and glabrous; flowers yellow on a umbel like cyme with a tinge of red; keel petal abruptly incurved and pointed; calyx bell shaped with a dentate tube with five distinct teeth; upper stamen free, others united, 5 alternate ones longer than the other five and thickened upwards; pod cylindrical, straight; seeds 15-20, round, greenish yellow or dark brown in colour.

Phytochemistry

Plant: Kaempferol-7-rhamnoglucoside, kaempferol-3-glucoside, kaempferol-3-glucosyl-7-rhamnoside and kaempferol-3-rhamnoside.

Leaf: Ferulic acid, kaempferol, leucocyanidin, leucodelphinidin, p-coumaric acid and quercetin.

Flower: Quercetagetin-3-galactoside, quercetin-3,7-dirhamnoside.

Seed: Kaempferol-7-glucoside, kaempferol-3-glucosyl-7-dirhamnoside, kaempferol-3-glucosyl-7-rhamnoside, kaempferol-3-rhamnoside, quercetin-3,7-dirhamnoside, quercetin-3-glucosyl-7-rhamnoside and quercetin-7-rhamnoside.

Edibility

Plant: Is an important source of nectar.

Medicinal Use

Plant: Is vulnerary.

Toxicology: Cyanogenic glycosides such as linamarin and lotaustralin in the seeds cause ambylopia, endocrinological and neurological disorders, tropical ataxy, neuropathy, inhibition of respiratory enzymes, alteration in cardiac physiology and paralysis, due to thiocyanate formed in the body; trypsin inhibitors present in the plant.

Uses in Alternative Systems of Medicine

Chinese:

Part used: Whole palnt.
Uses: Used as antifebrile, carminative, restorative and tonic.

Tibetan:

Uses: Heals wounds and cures lung diseases.

Ethnomedicine:

Plant: Carminative, cyanogenic, restorative and tonic; used for fever, nervous disorders and thirst.

Non-medicinal Uses

Plant: Used as manure, pasture and fodder plant; yields good silage and hay; good soil binder and ground cover.
Flower and Leaf: Yields colouring matter imparting yellow or orange colour to cotton fabric; it is an important bee plant.

99. *Lupinus albus* L.

Common Names: Egyptian lupine, Lupine, White lupin.
Vernacular Names: Ben: Turmuz; **Hin:** Turmas.
Distribution: Native of Europe and Middle East; introduced into India; grows in some parts of South India.
Description: Short hairy annual; leaflets obovate, cuneate, mucronate, glabrous above, villose beneath; stipules setaceous; flowers white to blue, on long racemes; calyx lips shallowly dentate; pod flat, broad, longitudinally rugulose, glabrous, yellow; seeds 3–6, orbicular or quandrangular, compressed, smooth, yellow, sometimes with dark variegation.

Phytochemistry

Plant: Albine, dehydroalbene and asparagine.
Root: Hydroxylupalbigenin, coumaronochromones, dihydrofurano isoflavones, erythrin–C, licoisoflavones A and B, lupin isoflavones A, B, C, D, E and F, lupinalbin A, B, C, D and E, lupalbigenin, luteone, parisoflavone, prenylated isoflavones and wighteone.
Pod: Apigenin-di-C-glycoside.
Seed: (–)Asparteine, β-carotene, β-galactan, α-isolupanine, 13-(angeloyloxy)-4-hydroxylupanine, 13-(benzoyloxy) lupanine, 13-(trigloyloxy) lupanine, 13-angeloyloxylupanine, 13–dihydroxylupanine, 13-hydroxymultiflorone, 4-hydroxylupaine, alanine, alkaloids, ammodendrine, angustifoline, arginine, asparagine, aspartic acid, benzoic acid, calcium, carbohydrates, cephalin, cinnamic acid, citric acid, copper, cystine, dihydrostigmasterol, epigenin-di-C-glycoside, epilupinine, fat, ferulic acid, gadoleic acid, glutamic acid, glycine, glycolic acid, histidine, iron, isoleucine, lauric acid, lecithin, leucine, linoleic acid, lupanine, lupeol, lupinine, lysine, magnesium, malic acid, methionine, methyl albine, multiflorine, myristic acid, niacin, oleic acid, oxalic acid, palmitic acid, palmitoleic acid, phenylalanine, phosphorus, potassium, proline, riboflavin, serine, sodium, sparteine, stearic acid, succinic acid, tetrahydrorhombifolin, thiamine, threonine, trans–13-(cinnamoyloxy) lupanine, tryptophan, tyrosine, valine, vanillin, vernine, zeaxanthin and zinc.
Seed sprout: Allantoic acid, allantoin, homogentisinic acid and trigonelline.

Edibility

Seed: Edible; used in making bread; yields edible fatty oil.

Seed and Flour: Lupine flour is added to biscuits, bread, noodles, sauces, soups, margarine, oil, cheese and coffee.

Medicinal Uses

Plant: Is abortifacient, analgesic, antiallergenic, antiandrogenic, antianginal, antibacterial, anticancer, antidysmenorrheic, antifibrinolytic, antihaemolytic, antihistaminic, anti-inflammatory, antimicrobial, antimitogenic, antinociceptive, antioxidant, antiperiodontic, antiperoxidant, antiprostaglandin, antiseptic, antitumour, anthelmintic, aphrodisiac, apoptotic, cardioprotective, cholinergic, cytotoxic, diuretic, emetic, estrogenic, haemolytic, haemostatic, hepatotonic, hypercholesterolemic, hypertensive, hypouricemic, immunostimulant, insulinogenic, keratolytic, laxative, neuroexcitant, pectoral, refrigerant, sedative, uterorelaxant, vasodilator, vulnerary and tonic; cures acne, alopecia, anaemia, anorexia, anxiety, arteriosclerosis, arrhythmia, arthritis, asthma, atherosclerosis, ataxia, back ache, beri-beri, calculi, cancer of the breast, cervix, liver, canker, cardiospasms, cataract, chilblain, cirrhosis, cold, colitis, convulsions, delirium, dementia, depression, dermatitis, diabetes, eczema, encephalopathy, epilepsy, gastritis, glaucoma, glossitis, herpes, hyperglycaemia, hepatitis, insomnia, leprosy, leukaemia, lymphoma, malaria, mastalgia, melanoma, menorrhagia, myocarditis, nephritis, neuralgia, neurasthenia, neuroblastoma, obesity, polio, pruritus, retinopathy, rheumatism, sprue, stomatitis, tumour of lung, ovary and stomach, ulcer and vertigo; antidote to aluminium, cadmium, lead and paracetamol poisoning.

Root: Discutient and increases digestive power.

Seed: Alterative, anthelmintic, carminative and deobstruent; given in internal heat and leprosy; used to counter act the effects of alcohol; reduces blood pressure; increases sugar tolerance in diabetic patients.

Bioactivity: Plant is a inhibitor of 5-α-reductase, ACE, aldose-reductase, aromatase, COX-I, catechol–0– methyl transferase, DOPA – decarboxylase, FPTase, histidine, α-amylase, kinase, insulinase, lipase, MAO, MDR, NADH–ubiquinone-oxidoreductase, ornithine-decarboxylase, peroxidase, topoisomerase and tyrosine-kinase; is angiotensin–receptor blocker, β-blocker, calcium channel blocker, estrogen antagonist, monoamine precursor, pituitary stimulant and quinone – reductase inducer; seed extract showed contractile effect on rat duodenum and isotropic effect on rat heart muscles.

Toxicology: Quinolizidine alkaloids such as anagyrine, lupanine and lupinine in the vegetative parts and seeds cause ataxia, coma, diaphoresis, nausea, respiratory arrest, visual disturbances and tachycardia and is harmful to stomach; multiflorine in the vegetative parts and seeds causes central nervous system depression; the pyrimidine glycosides such as convicine and divicine cause favism; (–) sparteine causes hypoglycaemia and diuresis; contains trypsin inhibitors.

Uses in Alternative Systems of Medicine

Unani:

Part used: Bark and seed.

Uses: Anthelmintic, antiflatulant, appetizer, cleansor, deobstruent, diuretic, emmenogogue, laxative and lithotropic; useful in asthma, biliousness, bronchitis, catarrh, headache, leucoderma, liver tumour, nausea, phlegmatic fever, pityriasis, rheumatism, sciatica, scrofula and skin diseases.

Ethnomedicine

Plant: Is alterative, carminative, deobstruent, depurative, digestive, discutient, diuretic, emmenogogue, pectoral and vermifuge; used for tumours.

Non-medicinal Uses

Plant: Used as green manure; lupine fiber useful in making cloth, towel and envelopes with lupine adhesives.

Seed: Boiled and used as fodder; seed paste is employed in culture media in the commercial production of tuberculin.

Seed oil: Used in soap making.

100. *Lupinus luteus* L.

Synonym: *Lupinus luteolus sensu* Clewell

Common Name: Yellow lupine.

Distribution: Native of Portugal and Greek Islands; introduced into India; grows in hill stations of Tamil Nadu.

Description: Hairy annual; leaflets obovate, oblong, mucronate, sparsely villous; stipules dimorphic, those of the lower leaves subulate, those of the upper leaves linear-obovate; flowers verticillate, scented; petals bright yellow; pod densely villous, black; seeds 4-6 per pod, orbicular-quadrangular, compressed, smooth, with a white curved line on each side or all white.

Phytochemistry

Plant: Genistein-7-O-glucosylglucoside and lupanine.

Root: 5-O-methyl genistein.

Leaf: Genistein.

Edibility

Seed: Yields fatty oil after refining and is edible.

Medicinal Uses

Plant: Is abortificient, antiangiogenic, antiestrogenic, antifertility, antihaemolytic, anti-implantation, anti-inflammatory, anti-ischaemic, antimicrobial, antimitogenic, antioxidant, antispasmodic, antiviral, apoptotic, cytotoxic, fungicide, hypocholesterolemic and immunostimulant; used in atherosclerosis, cancer of the breast, flatulence, inflammation, leukaemia, lymphoma, melanoma, neuroblastoma, osteoporosis, tumour of breast, ovary, prostate and stomach.

Bioactivity: Plant is inhibitor of aldose-reductase, α-reductase, catechol-O-methyl-transferase, DOPA-decarboxylase, histidine-kinase, lipase, MAO, NADH-ubiquinone-oxidoreductase, ornithine-decarboxylase, PTK, peroxidase, topoisomerase I and II and tyrosine-kinase; is calcium antagonist, pituitary sensitizer and quinone-reductase inducer.

Toxicology: Quinolizidine alkaloids in the vegetative parts and seeds cause ataxia, coma, diaphoresis, nausea, respiratory arrest and visual disturbance; cyanogenic glycoside anagyrine is toxic, teratogenic, induces tachycardia; lupanine causes hypertension and hypoglycaemia and is toxic in large doses.

Non-medicinal Uses

Plant: Yields fodder and silage.

101. *Lupinus mutabilis* Sweet

Synonym: *Lupinus cruckshanksii* Hook.

Common name: Sweet lupine

Distribution: Native of South America; introduced into India.

Description: A stout showy annual herb; leaflets lanceolate, oblong, obtuse; flowers white, but standard petal yellow or violet; pod pubescent; seed white and flat.

Phytochemistry

Pod: β-Carotene, calcium, carbohydrates, fat, fiber, iron, niacin, phosphorus, riboflavin and thiamine.

Seed: (–)-Sparteine, (+) lupanine, 4,13-(angeloyloxy)–4-hydroxylupanine, 13-(angeloyloxy) –lupanine, 13-dihydroxy lupanine, 4-hydroxy lupanine, β-isolupanine, β-carotene, ammodendrine, apigenin-diglucoside, calcium, carbohydrates, fat, fiber, iron, multiflorine, niacin, phosphorus, protein, riboflavin, tetra hydrorhombifoline, thiamine, trans-13-(cinnomoyloxy) lupanine and zeaxanthin.

Medicinal Uses

Plant: Is abortifacient, analgesic, androgenic, antiallergenic, antianginal, antiarrhythmic, anticancer, anticardiospasmic, anticolitic, anticoronary, antidysphagic, antiencephalopathic, antiglossitic, antihistaminic, antihypertensive, antilupus, antimenorrhagic, antimutagenic, antineuralgic, antioxidant, antiperiodontic, antiradicular, antispasmodic, antitumour, antiulcer, cancer preventive, cardioprotective, diuretic, hepatoprotective, hypocholesterolemic, hypoglycaemic, hypolipidemic, hypotensive, immunostimulant, laxative, oxytocic and phagocytotic; useful in acne, alcoholism, alopecia, anaemia, anorexia, anxiety, asthma, atherosclerosis, bach ache, beri-beri, cataract, chilblain, convulsions, delirium, dermatitis, depression, diabetes, epilepsy, fatigue, gastritis, heart burn, herpes, insomnia, migraine, myocardia, neuropathy, obesity, osteoporosis, psoriasis, spasms, stress, tumour, ulcer and vertigo; antidote to aluminum and lead poison.

Bioactivity: Plant is angiotensin-receptor blocker, β-blocker, calcium channel blocker and interferon synergist.

Toxicology: The quinolizidine alkaloids in the vegetative parts and seeds may cause nausea, ataxia, respiratory arrest, visual disturbances, diaphoresis and central nervous system depression.

102. *Macrotyloma uniflorum* (Lam.) Verdc.

Synonyms: *Dolichos biflorus sensu auct.*
Dolichos uniflorus Lam.

Notes: Three varieties, *stenocarpum, verrucosum* and *benadirianum* are recognized.

Common Name: Horse gram.

Vernacular Names: Ben: Kurtikalai; **Hin:** Kutthi; **Kan:** Hurali; **Mal:** Muthiva, Muthera; **Mar:** Kulith; **Tam:** Kollu; **Tel:** Ulavalu.

Distribution: Native of Africa, India, Pakistan, Sri Lanka and other Asian countries and Australia; grows in the sub-Himalayan tract, Dehradun, Punjab and South India.

Description: Annual; branches twining, glabrescent: stipules, oblong, basifixed; leaflets broadly lanceolate, membranous; bracts lanceolate, one at the base of each pedicel and two lateral at the base of each flower; flowers yellow; calyx teeth lanceolate, setaceous, keel petal narrow, obtuse, shorter than the standard; pod scimitar shaped, compressed, recurved, tipped with persistent style: seed 5–6, compressed, reniform, grey or reddish brown.

Phytochemistry

Stem: Coumesterol, psoralidin and pterocarpon.

Leaf: Coumesterol, psoralidin and pterocarpon.

Seed: β-Sitosterol, allantoinase, coumesterol, dalbergiodin, dolichin A, dolichin B, genistein, hydroxygenistein, isoferreirin, kievitone, L-rhamnopyranosyl-(1→2)–0-β-D-glucopyranoside, neohesperidoside, phaseollidin and psoralidin.

Edibility

Seed: Edible after cooking.

Medicinal Uses

Plant: Is antiangiogenic, anticancer, antiestrogenic, antihaemolytic, anti-inflammatory, antimicrobial, antimitogenic, antioxidant, antispasmodic, antitumour, antiviral, cardioprotective, cytotoxic, hypocholesterolemic, immunostimulant, sedative and vasodilator; used in adenitis, burns, dysuria, dysentery, fistula, ischaemia, leprosy, leukaemia, lymphoma, measles, melanoma, neuralgia, neuroblastoma, osteoporosis, pleurisy, pneumonia, prolapsed anus, small pox, sores, stomach disorders, tumour of breast, ovary and ulcer.

Seed: Aphrodisiac, astringent, diuretic and tonic; decoction is given in enlarged liver and spleen, leucorrhoea and menstrual problems.

Bioactivity: Plant is inhibitor of aldose-reductase, α-reductase, catechol–O–methyl transferase, DOPA-decarboxylase, histidine-kinase, lipase, MAO, ornithine-decarboxylase, PTK, peroxidase, topoisomerase I and II and tyrosine-kinase; is also calcium antagonist, estrogen antagonist, pituitary sensitizer and quinone-reductase-inducer.

Globulin fraction showed hypolipidemic effect in rats; the seed lectins are used to identify human A1, and A1B blood groups; ethanol extract is spasmolytic and affected respiration in rats.

Toxicology: Plant contains chymotrypsin and trypsin inhibitors.

Uses in Alternative Systems of Medicine

Ayurveda:

Part used: Seed.

Uses: Is antipyretic, astringent, anthelmintic, diuretic and tonic; cures abdominal complaints, asthma, bronchitis, constipation, cough, dyspnoea, edema, epistasis, eye diseases, flatulence, hiccup, inflammations, leucorrhoea, liver trouble, menstrual disorders, piles, poisoning by animal bites, strangury, tumours, ulcer, urinary calculi, urinary discharges and wounds.

Siddha:

Part used: Whole plant.

Uses: Astringent, diuretic and tonic; useful in cough, diarrhoea, dysentery, dyspnoea, eye diseases, fever, filariasis, leucorrhoea and menstrual disorders.

Unani:

Part used: Seed.

Uses: Antidiaphoretic, deobstruent, diuretic, emmenogogue, expectorant and lithotriptic; useful in amenorrhoea, biliousness, colic, hiccup, leucorrhoea, menstrual disorders, piles, renal calculi and spleenic obstructions.

Ethnomedicine

Plant: Is used in abdominal tumour, measles, pleurisy, pneumonia and small pox.

Seed: Used for kidney stones, menstrual complaints and tooth ache.

Non-medicinal Use

Seed: Used as cattle feed.

103. *Medicago lupulina* L.

Common Names: Black medic, Hope clover, Yellow trefoil.

Distribution: Native of Africa, Asia including India, Europe and Middle East; grows in Bihar, Punjab and West Bengal in India.

Description: Annual or perennial; stem prostrate or ascending, pubescent, hairs simple or glandular; leaflets obovate, cuneate, retuse to cordate, apiculate, serrate in the upper-half; stipules cordate, dentate; inflorescence an axillary, pedunculate raceme; calyx pubescent, teeth almost equal to or slightly longer than the tube; corolla yellow; pod curved through 180°, sparsely pubescent to glabrescent, black, 1-seeded.

Phytochemistry

Plant: Arabinose, galactose, glucose, glucuronic acid, medicagenic acid, rhamnose, soya sapogenol B, C, D, E and F and xylose.

Root: Hederagenin-3-O-β-D-glucopyranoside, medicagenic acid and soyasaponin.

Seed: β-D-mannopyranosyl (1→4)-O-[α-D-galactopyranosy]-(1→6)-D-mannose, 4-O-β-mannopyranosyl-D-mannose, 6-O-β-D-galactopyranosyl(1→4)-D-mannose, galactomannan, medicagenic acid, medicagenic acid-3-O-β-D-glucopyranoside and soya sapogenol.

Edibility

Seed: Edible, used to impart colour and flavour to butter.

Medicinal Uses

Plant: Antibacterial and galactagogue.

Non-medicinal Uses

Plant: Used as green manure, fodder and forage.

Seed: Used as cattle feed.

104. *Medicago monantha* (C. Meyer) Trautv.

Synonyms: *Trigonella incisa* Benth.
Trigonella monantha C.A. Meyer
Trigonella monantha C.A. Meyer ssp., *noeana* (Boiss.) Huber-Morath.

Vernacular Names: Hin: Chanihari; **Pun:** Chini, Sainji.

Distribution: Native of Asia and Middle East; introduced into India; grows in Himachal Pradesh, Punjab, Rajasthan, Uttar Pradesh in India.

Description: Prostrate to ascending annual; leaflets oblong to cordate-cuneate, dentate or pinatifid; stipules semi sagittate, denticulate; inflorescence pedunculate, 1-3 flowered, in leaf axils; calyx campanulate, teeth about as long as cup; corolla yellow; pod reticulately veined, many seeded.

Phytochemistry

Seed: Diosgenin, gitogenin, sapogenins and trigonin.

Edibility

Plant and Pod: Eaten as vegetable.

Medicinal Uses

Seed: Antidysenteric and antidiarrhoeic.

Non-medicinal Uses

Plant: Is used as fodder.

105. *Medicago polymorpha* L.

Synonyms: *Medicago apiculata* Willd.
Medicago denticulata Willd.
Medicago nigra (L.) Krocker
Medicago nigra (L.) Krocker var., *nigra.,* (L.) Krocker
Medicago pentacycla DC.
Medicago reticulata Benth.

Notes: A very variable species.

Common Names: Bur medic, California bur clover, Toothed bur clover, Toothed medic.

Vernacular Names: Ben: Maina; **Hin:** Chandausi; **Pun:** Maina.

Distribution: Native of Africa, Asia including India, Europe and Middle East; grows in Bihar, North West India, West Bengal and Western Himalaya in India.

Description: Annual, somewhat spreading; petiolate; leaflets obovate to cuneate, obtuse, truncate to retuse, slightly toothed; stipules laciniate; inflorescence is 2-8 flowered peduncled raceme; calyx teeth subequal to the tube; corolla yellow; pod spiral, glabrous, spiny; spines hooked, in 2-rows, oriented almost parallel to the surface of the disc.

Phytochemistry

Plant: Medipolymorphol.

Edibility

Plant and Leaf: Edible.

Medicinal Use

Plant: Depurative, deobstruent, diuretic and stomachic; used in intestinal and kidney disorders; infusion is an antidote for poison.

Medicinal Uses Outside India

China: Plant is used as deobstruent, depurative and diuretic; used for intestinal and kidney disorders.

Non-medicinal Use

Plant: Makes a good pasture, green fodder and green manure; also grown to prevent soil erosion; grown as hedge plant.

106. *Medicago sativa* L.

Synonyms: *Medicago grandiflora* (Grossh.) Vassilcz.
Medicago ladak Vassilcz.
Medicago mesopotamica Vassilcz.
Medicago orientalis Vassilcz.
Medicago praesativa Sinsk.
Medicago sogdiana (Brand.) Vassilcz.

Notes: Alfalfa has been used for centuries for strength, vitality and as rejuvenator. Four subspecies, *ambigua, microcarpa, sativa* and *varia* are recognized; a complicated polymorphic species.

Common Name : Alfalfa

Vernacular Names: Kan: Kudure masaale soppu, Vilaayathi hullu; **Sans:** Ashvabhala.

Distribution: Introduced into most of the countries and naturalized; origin not known; it grows in North India, West India and North West India.

Description: Erect, ascending perennial plant with alternate trifoliolate leaves; leaflets obovate, toothed in the upper third with mucronate tip; stipules linear-lanceolate, serrate; inflorescence is a raceme; flowers purple; pod spirally coiled, green to brown, glabrous; seeds 2-5, kidney shaped, yellow or brown.

Phytochemistry

Plant: Adenine, aesculetin, alfalfone, amylase, anthocyanin, apigenin, apigenin-glucuronide, arabinose, arginine, ascorbic acid, azukisaponin, benzoyl mesotartaric acid, biochanin-A, biotin, calcium, campesterol, carbohydrates, cerebroside, cholesterol, choline, chromium, chrysoeriol, chrysoeriol diglucuronide, chrysoseriol-triglucuronide, citric acid, citrullin, coumesterol, cryptoxanthin¸ cycloartenol, cystine, cytidine, daidzein, daphnoretin, dehydro soyasaponin, dihydroxyflavone, esculetin, ferredoxin, foliacin, folic acid, formononetin, fructose¸ fumaric acid, galactose, genistein, glucose, glucuronic acid¸ glycerolipids, glycolipids, guanine, guanosine, hederagenin, hentriacontane, histidine, hydroxycoumestin, hydroxyformononetin, hypoxanthine, inosine, inositol, invertase, iron, isocytosine, isoleucine, L-arginine, lauric acid, leucine, liquiritigenin, L-stachydrine, lucernol, lutein, lysine, magnesium, malic acid, malonic acid, manganese, medicagenic acid, medicagol, medicoside, methionine, methoxycoumesterol, methoxy medicarpin, methylcoumesterol, molybdenum, myristic acid, myristone, myrsetinol, neoxanthin, niacin, nonocosane, norspermidine, oleic acid, oxalic acid, palmitic acid, pantothenic acid, pectin, pectinase, peroxidase, phenylalanine, phospholipids, phosphorus, phylloquinone, pinitol, potassium, rhamnose, riboflavin, ribose, salicylic acid, saponin, scopoletin, shikimic acid, sodium, soyasapogenol, soyasaponin, sphingosine, stachydrine, stachyose, succinic acid, sucrose, thiamine, threonine, trans-cinnamic acid, triacontane, triacontanol, tricin, tricin-7-diglucuronide, trifoliol, trigonelline, trihydroxyflavone, trimethoxy isoflavan, tryglycerides, tryptophan, valine, vestitol, violaxanthin, xanthophylls, xylose, zeaxanthin, zenoxanthin, β-sitosterol, β-spinasterol and zinc.

Root: Aminobutyric acid, autoside, caulosaponin-β, coumesterol, dihydroxyflavone, formononetin, formononetin-7-O-6'-O-malonylglucoside, glycerolipids, medicagenic acid, medicagenic acid-3-O-β-D-glucopyranoside, medicarpin, medicarpin-β-D-glucoside, medicoside, medicoside-3-O-glycoside and soyasapogenol.

Leaf: Ascorbic acid, auroxanthin, bayogenin, biochanin-A, carboxyarabinitol, coumesterol, coumesterol-O-glucoside, cyanogenic glycosides, daidzein, dihydroxyflavone, ferulic acid, flavoxanthin, formononetin, formononetin-7-O-glucoside, genistein, geraniol, glycine, hederagenin, hydroxyproline, kaempferol rhamnoside, kaempferol-7-O-β-D-glucuronide, L-canavanine, lucernic acid, lucernol, lutein, lutein-5,6-epoxide, luteoxanthin, medicagenic acid, medicarpin, methoxycoumesterol, molybdenum, neoxanthin, ononin, p-coumaric acid, phytofluene, pipecolic acid, populnin, potassium, proline, sativol, stigmasterol, soyasaponin, soyasapogenol, spinasterol, trifoliol, violaxanthin, wairon, zanthic acid, zeaxanthin, β-carotene, β-cryptoxanthin, β-ionone.

Flower: β-Carotene, β-cryptoxanthin, auroxanthin, kaempferol, laricytrin, lutein, myricetin, quercetin and zeaxanthin.

Pod: β-Amyrin, β-spinasterol, β-sitosterol, esculetin and myrsellinol.

Seed: β-Amyrin, β-sitosterol, asparagines, bayogenin, betaine, canaline, canavanine, caproic acid, glucose, glucuronic acid, hederagenin, homostachydrine, hyperoside, isopalmitic acid, isopentadecanoic acid, lauric acid, L-canavanine, L-homostachydrine, medicagenic acid, oleic acid, palmitic acid, purines, saponin, soyasapogenol A,B,C and E, soyasaponin, stachydrine, stachyose, stearic acid and trigonelline.

Sprout: Coumesterol, L-canavanine, medicanine, medicarpin, saponin and vestitol.

Seed oil: β-Sitosterol, linoleic acid, linolenic acid, oleic acid, palmitic acid and stearic acid.

Edibility

Flower: Edible.

Medicinal Uses

Plant: Abortifacient, analgesic, androgenic, aneasthetic, anticancer, anticardiospasmic, anticholinergic, anticoronary, antiemtic, antiencephalopathic, antifibrinolytic, antigonadotrophic, antihaemolytic, antihaemorrhagic, antihepatotoxic, antihistaminic, antihypercholesterolemic, antihyperthyroid, anti-inflammatory, antikeratitic, antimastalgic, antimetastatic, antimicrobial, antimitogenic, antineurasthenic, antinociceptive, antioxidant, antiperiodontal, antiprostaglandin, antipsychotic, antirhinitic, antiscorbutic, antiseptic, antispasmodic, antitussive, antiviral, aphrodisiac, apoptotic, arteriodilator, astringent, bronchoprotectant, capillary protective, cardioprotective, cerebrotonic, chemopreventive, cholagogue, cytoprotective, demulcent, diaphoretic, diuretic, emetic, emmenogogue, emollient, energizer, febrifuge, hepatoprotective, hyperuricemic, immunostimulant, lactagogue, laxative, lithogenic, memory-enhancer, narcotic, neuroprotective, oxytocic, phagocytotic, radioprotective, refrigerant, sedative, stimulant, systolic-depressant, thermogenic, tranquilizer, uterorelaxant, vasodilator, vermifuge and vulnerary; used in allergy, acne, alcoholism, anaemia, anorexia, anxiety, antidote to aluminium, cadmium, camphor, iodine, lead, mercury and morphine poisoning, atherosclerosis, arthritis, asthma, back ache, beri-beri, calculi, cancer of breast, cervix, colon, oesophagus, liver, pancreas and skin, canker, capillary fragility, cataract, cervical palsy, chilblain, colic, cold, convulsions, cystitis, dandruff, deafness, delirium, dementia, depression, dermatitis, diabetes, dypsia, diarrhoea, dysentery, dysmenorrhoea, eczema, edema, epilepsy, fatigue, fibrosarcoma, gastritis, gingivitis, glaucoma, glycosuria, gout, heart burn, hepatitis, hepatoadenoma, hepetocarcinoma, herpes, hiccup, hyperactivity, hydrophobia, hypertension, impotency, inflamed bladder, insomnia, ischaemia, leukaemia, lymphoma, malaria, measles, melanoma, metaplasia, migraine, myalgia, myocardia, myopia, nausea, nephritis, neuralgia, neuroblastoma, neuropathy, obesity, ophthalmia, osteoarthritis, osteoporosis, pharyngitis, photophobia, pneumonia, polio, polyp, psoriasis, retinopathy, rheumatism, schizophrenia, senility, sinusitis, spasms, stress, tachycardia, thalassimia, thrombosis, tumour of bladder, breast, cervix, colon, oesophagus, stomach, liver, lung, ovary, pancreas, prostate, skin and brain, tymphinitis, ulcer, varicose veins, vertigo and vitiligo; useful in treating cold, depression, fistula, flatulence, goiter, hypertension, hypoglycaemia, tremors and urinary complaints.

Alfalfa is very high in minerals and vitamins and is easily digested and assimilated by man; it is a good source of α-carotene and chlorophyll; it is used as prostatic, heart, reproductive and musculoskeletal tonic; it stabilizes blood sugar, balances the pituitary gland function and detoxifies blood and kidneys, helps to increase estrogen production;

it is used for anaemia, colon disorders, haemorrhage, liver and stomach dysfunction, it helps in reconstituting bone and useful in rickets; it inactivates dietery chemical carcinogens; it is a effective appetizer and nervous relaxant.

Bioactivity: Plant is inhibitor of β-glucuronidase, catechol-O-methyl transferase, DOPA-decarboxylase, deiodinase, COMP, COMT, COX I and II, glucosyl-transferase, histamine, histamine-kinase, insulinase, iodothyronine-deiodinase, leukotriene, lipase, lipoxygenase, lyase, MAO, MDR, MMP-9, metalloproteinase, NADH-oxidase, NADH-ubiquinone-oxidoreductase, NO, ornithine-decarboxylase, oxidase, peroxidase, phospholipase–A_2, polyamine synthesis, prostaglandin synthesis, protein-kinase, topoisomerase I and II, tyrosinase, tyrosine-kinase, xanthine-oxidase, CAMP, phosphodiesterase; is β-adrenergic receptor blocker, β-blocker, CNS active, calcium antagonist, copper chelator, cytochrome–P_{450}-inducer, calcium channel blocker, ER-β-binder, estrogen antagonist, mast cell stabilizer, metal chelator, monoamine precursor, P_{21}-inducer, pituitary stimulant and quinone-reductase inducer.

Uses in Alternative Systems of Medicine

Chinese:

Part used: Whole plant and root.

Uses: Cooling, depurative and emetic; prescribed in fever and gravel.

Homoeopathy:

Part used: Tincture of the plant.

Uses: Useful in diabetes, general debility, hypertrophy, insomnia, nervous disorders and neurasthenia.

Tibetan:

Part used: Plant

Uses: Heals wounds and cures lung diseases.

Clinical Studies: Chalcones in the plant exhibited cytoxicity to leukaemic cells.

Medicinal Uses Outside India

Iraq: Used in arthritis and rheumatism.

Turkey: Used as appetizer, cardiotonic, intellect promoter and stimulant.

United States: Used as anticancer and cyanogenic.

Non-medicinal Uses

Plant: Used as fodder; cultivated for pasturage, hay and silage; a commercial source of chlorophyll; useful for bee pasturage.

Seed: Yields a drying oil suitable for use in paints and varnishes; also yields a yellow dye.

107. *Melilotus albus* Medikus

Synonyms: *Melilotus alba* Medikus
Melilotus leucanthus DC.
Melilotus melanospermus Ser.
Melilotus vulgaris Willd.

Common Names: Honey clover, Siberean melilot, Sweet clover, White melilot, White sweet clover.

Vernacular Names: Ben: Safed banmethi; **Ori:** Safed banmethi, **Pun:** Safed senji.

Distribution: Native of Libya and Asia including India, Europe and Middle East; grows throughout India as a weed.

Description: Erect or ascending biennial, up to 1.5 m, with trifoliolate leaves; leaflets obovate, dentate in upper third and with terminal leaflet on longer petiole than lateral leaflets (a feature used to differentiate the plant from Lucerne, since both species have a similar appearance in the vegetative state); single well-branched stem in establishment year but in following year, several stems originate from crown buds; the succulent stems become fibrous with age; deeply-tap-rooted; inflorescence is elongated one-sided raceme of white flowers, 4-5 mm, borne on long axillary stalks; indeterminate flowering; pod has a reticulated, ridged coat which turns black with ripening and contains a single smooth, kidney-shaped and brownish-yellow to brown seed.

Phytochemistry

Plant: Clovin, melilotin, melilotocarpan, melitin, 4-O-methyl glucuronoxylan, oligosaccharides, terpenoid glycosides and triterpenoidal saponins.

Flower: Melilotin.

Seed: Apigenin, clovin, isoorientin, melilotin, melitin, orientin, quercetin-3-rhamnogalactoside, resin, robonoside and taumarin.

Edibility

Young shoot and Leaf: Edible.

Medicinal Uses

Plant: Antidiabetic, antiedemic, antihepatotoxic, antibacterial, aromatic, astringent, diuretic, emollient, hypotensive, laxative, memory enhancer, narcotic nutritive, stimulant and vulnerary; used in curing hypercholesterolema, hyperglycaemia, hypouricemia, obesity, tumour of lung and ovary, ulcer and varicose veins; ointment made from the plant promotes healing of skin eruptions; paste is used as poultice or plaster for rheumatic and abdominal pain.

Bioactivity: Plant is acetylcholinergic and angiotensin-receptor-blocker.

Toxicology: Chymotrypsin and trypsin inhibitors present.

Uses in Alternative Systems of Medicine

Homoeopathy:

Part used: Tincture of the fresh plant during flowering.

Uses: Useful in cough, dysentery, epilepsy, fever, headache, leucorrhoea, neuralgia and pneumonia.

Ethnomedicine

Plant: Expectorant; used for sores, spasm and sclerosis.

Non-medicinal Uses

Plant: Is used as green manure, fodder and cover crop.

Seed oil: Suitable in paint and varnish industry.

Seed meal: Used as cattle feed.

108. *Melilotus indicus* (L.) All.

Synonyms: *Melilotus indica* (L.) All.
Melilotus officinalis sensu Bojer
Melilotus parviflora Desf.
Trifolium indica L.
Trifolium indicum L.

Common Names: Annual melilot, Indian melilot, King Island melilot, Small melilot, Sour clover.

Vernacular Names: Hin: Ban methi; **Kan:** Kaadu loosnar soppu; **Sans:** Vanamethika.

Distribution: Native of Africa, Asia including India, Europe and Middle East.

Description: Erect annual herb; stem pubescent; leaves 3-foliolate; leaflets obovate, or oblong, dentate, retuse, emarginate, glabrous; flowers yellow on pedunculate raceme; bracteate; pedicel reflexed after flowering; pod ellipsoid, olive green; seeds 6-10, oblong, ellipsoid, compressed and brown.

Phytochemistry

Plant: Coumarin, fraxidin, herniarin, umbelliferone, scopoletin, clovin, azukisaponin II, robinin and soyasapogenol B.

Leaf: Alkaloids, amino acids, azukisaponin II, robinin, clovin, saponin, soyasapogenol B, sterols, sugars, tannin and triterpenoids.

Flower: Azukisaponin II, clovin, robinin, saponin and soyasapogenol.

Edibility

Plant and Seed: Is used as vegetable.

Medicinal Uses

Plant: Is antiallergenic, antiandrogenic, antibacterial, antidepressant, antifertility, antihypertensive, anti-inflammatory, antimutagenic, antioxidant, antiperiodontic, antiviral, cancer preventive, cardioprotective, diuretic, febrifuge, hepatoprotective, immunostimulant, laxative and memory enhancer; useful in atherosclerosis, breast cancer, depression, diabetes, hypocholesterolemia and hypoglycaemia; also used in edema, insomnia, leukaemia, lymphoma, nausea, obesity, osteoporosis, swellings, tumour of the breast, cervix, lung and ulcer.

Seed: Useful in bowel complaints, dysmenorrhoea, infantile diarrhoea, rheumatism and scrofula; given to purify blood.

Bioactivity: Plant is acetylcholinergic, angiotensin receptor blocker and calcium-channel blocker.

Uses in Alternative Systems of Medicine

Ayurveda:

Part used: Seed.

Uses: Is antipyretic, anthelminitic and tonic; increases appetite, astringent to bowels; cures bronchitis, leprosy, piles, "vatha" and vomiting.

Homoeopathy:

Part used: Tincture of the fresh plant.

Uses: Useful in congestion, constipation, convulsions, cough, delirium, dysmenorrhoea, epilepsy, fever, haemoptysis, haemorrhages, headache, leucorrhoea, neuralgia, palpitation and pneumonia.

Tibetan:

Part used: Plant and seed.

Uses: Cures chronic fever caused by poisoning.

Unani:

Part used: Plant and pod.

Uses: Anti-inflammatory, astringent, analgesic, demulcent, diuretic, emmenogogue and galactagogue; useful in abdominal pain; pod extract is applied for earache, palsy and paralysis; taken internally for melancholia.

Non-medicinal Uses

Plant: Grown as cover crop and cultivated for fodder, green manure and to improve alkaline/saline soil.

109. *Melilotus officinalis* (L.) Pallas

Synonyms: *Medicago officinalis* (L.) E.H.L. Krause
Melilotus arvensis Wallr.
Melilotus neglectus Ten.
Melilotus pallidus Ser.
Trifolium officinale L.

Common Names: Common yellow melilot, Medicinal sweet clover, Ribbed melilot, Yellow sweet clover.

Distribution: Native of Asia and Europe; introduced into India.

Description: A annual herb with erect or ascending branches tinged with red; leaves trifoliolate; leaflets of lower and middle leaves broadly oval, lanceolate to rhomboid ovate, rounded at the tip; those of the upper leaves oblong lanceolate, truncate at the tip and irregularly dentate; stipules entire, lanceolate or subulate; raceme 30-80 flowered; flowers yellow; calyx teeth as long as the tube; standard petal and wing petal equal or longer than the keel petal; pod compressed, grey to straw coloured, obtuse at the tip; seed one, rarely two, ovoid elliptic, yellow or greenish yellow, sometimes with purple spots.

Medicinal Uses

Plant: Analgesic, antiallergenic, antiandrogenic, antibacterial, anticancer, antihaemolytic, antihistaminic, anti-inflammatory, antimetastatic, antimitotic, antimyocarditic, antinociceptive, antioxidant, antiperiodontic, antiprostaglandin, antispasmodic, antitussive, antiviral, astringent, bronchodilator, cardioprotective, cardiotonic, cholagogue, cytotoxic, diaphoretic, diuretic, hepatoprotective, hypnotic, immunomodulator, laxative, lymphocytogenic, narcotic, phagocytotic, sedative, styptic, tranquilizer, uterotonic, vasodilator and vulnerary; useful in arthiritis, asthma, atherosclerosis, bronchitis, cancer of breast, colon, stomach, kidney, liver, skin, cystitis, dandruff, dermatitis, diabetes, eczema, edema, fatigue, fibrosarcoma, gastritis, gingivitis, hepatoadenoma, herpes, hepatoma, hydrophobia, hypoglycaemia, inflammation, leukaemia, malaria, melanoma, neuralgia, pharyngitis, polio, psoriasis, pyrosis, rheumatism, stomatitis, stress, tumour of bladder, breast, colon, kidney, liver, lung, mouth, ovary, prostate and skin and ulcer.

Bioactivity: Plant is inhibitor of 5-lipoxygenase, ACE, ATPase, aldose-reductase, β-glucuronidase, COMP, cyclo-oxygenase, cytochrome-P_{450}, deiodinase, elastase, glucosyl-transferase, iodothyronine-deiodinase, leukotriene, lipoxygenase, lyase, MAO, MM, metallo-proteinase, NADH-oxidase, NEP, ornithine-decarboxylase, oxidase, PTK, phospholipase, prostaglandin synthesis, protein-kinase, topoisomerase I and II, tyrosine, tyrosine-kinase, xanthine-oxidase, CAMP and phospho-diesterase; is CNS active, calcium antagonist, copper chelator, DNA protective, mast cell stabilizer and quinone-reductase inducer.

Edibility

Herb, Flower and Seed: Edible.

Uses in Alternative Systems of Medicine

Unani:

Part used: Pod.

Uses: Analgesic, anti-inflammatory, aphrodisiac, astringent, carminative, demulcent, diuretic, emmenogogue, galactogogue, maturant and tonic; useful for ear ache, leucoderma, melancholia, palsy, paralysis and stomach pain.

Non-medicinal Uses

Plant: Is cultivated for fodder; yields silage rich in protein.

Seed: Yields fatty oil, which is used in paints and varnishes; seed meal is used as protein supplement in cattle feed.

110. *Melilotus suaveolens* Ledeb.

Synonyms: *Melilotus altissimus sensu auct.*
Melilotus officinalis sensu auct.

Notes: *Melilotus officinalis* (L.) Pallas is a valid species, which is a native of several Asian countries and Europe but introduced into India; there is confusion between the two valid specific names in literature.

Common Names: Heart wort, King's crown, Plaster clover,

Vernacular Name: Ben: Banpiring.

Distribution: Native of Asia including India; grows in Jammu, Kashmir and Himachal Pradesh in India.

Description: A short, pubescent or nearly glabrous herb with upright, spreading, terete branched stem; stipules entire or subulate, dilated at the base; leaves unequally denticulate with toothed margins, lower leaves obovate, orbicular, the upper ones oblanceolate, obtuse; leaflets narrowly elliptic to broadly lanceolate; bracts linear, subulate; flowers yellow; calyx teeth narrowly triangular; standard, wing and keel petal are equal; pod broadly ovate, greenish yellow, sometimes flecked with purple.

Phytochemistry

Plant: Coumarin, coumaric acid and melilotic acid.
Leaf: Coumarin.
Flower: Essential oil.
Seed: Fatty oil and galactomannan.

Edibility

Root, Leaf and Seed: Edible; seed is used as a substitute for Tonka beans.

Medicinal Uses

Plant: Useful in ophthalmia.

Uses in Alternative Systems of Medicine

Tibetan:
Part used:
Uses: Cures fever due to poisoning.

111. *Millettia extensa* (Benth.) Baker

Synonyms: *Millettia auriculata* Brandis
Millettia macrophylla Kurz.

Vernacular Names: Hin: Gauj, Gonj; **Mal:** Valli muritali; **Ori:** Arkawla, Kissi, Rekorlo; **Tel:** Kondatangedutige.

Distribution: Native of Bhutan, Cambodia, India, Nepal, Pakistan and Vietnam; grows in Assam, Bihar, Himalaya, Kashmir, Madhya Pradesh and Orissa in India.

Description: A woody climber; branches finely pubescent; leaves imparipinnate; leaflets 9-11, ovate, oblong, obtuse, glabrous above, silky pubescent beneath, shortly stipitate; flowers pinkish red on short dense axillary and terminal raceme; calyx finely pubescent; corolla denseiy silky; pod woody, glabrescent and strap shaped.

Phytochemistry

Plant: Auriculatin, milletin, scandonone and sumatrol.
Root: Auriculatin, isoauriculatin, auriculin, milletin, scandonone and rotenone.
Pod: Isoaurimillone.
Seed: Aurimillone and milletin.

Medicinal Uses

Root: Useful in stomach ache; helps in conception.

Ethnomedicine

Plant: Used for spleen diseases and tooth ache.
Root: Is used to kill lice and ticks; heals sores.

Non-medicinal Uses

Plant: Is insect repellent, piscicidal and larvicidal.
Root: Is used as fish poison; applied on sores of cattle to kill vermin; used as insecticide.
Bark: Yields fibre used for rough cordage; used as brushes.
Leaf: Is used as fodder and hay.

112. *Mucuna bracteata* DC.

Synonyms: *Mucuna brevipes* Craib.
Mucuna venulosa (Piper) Merr. & Metcalf
Stizolobium venulosum Piper

Distribution: Native of China, Laos, Myanmar, Thailand and Vietnam; introduced into India; grows in Assam, Eastern Himalayas and Meghalaya in India.

Description: A climber with young parts pubescent; leaves 3-foliolate; lateral leaflets obliquely ovate or deltoid, base rounded, truncate, acute at the apex, glabrous above, appressed pubescent beneath, terminal leaflets rhomboid; flowers blackish purple on axillary raceme; bracts ovate, pubescent with scattered bristles; calyx and corolla covered with scattered bristly hairs; wing petal longer than the standard petal; keel petal exceeds the wing; pod terete, reddish brown, tomentose, abruptly recurved at the ends; 5-6 seeded; seeds covered with reddish bristles, ovoid and compressed.

Edibility

Seed: Edible.

Medicinal Uses

Pod: Medicinal uses similar to those of *Mucuna pruriens.*

Seed: Used as aphrodisiac and tonic.

Toxicology: Pod hairs cause itching and dermatitis when in contact with skin.

Ethnomedicine

Plant: Astringent, diuretic, emmenogogue, haemostatic, purgative, stimulant and vermifuge.

Pod: Hairs used as vermifuge.

Seed: Seed is used in impotency; used for bladder trouble, dropsy, elephantiasis, evil eye, hepatitis, nervous disorders, rhinitis, sores and cuts.

113. *Mucuna gigantea* (Willd.) DC.

Synonyms: *Dolichos gigantea* Willd.
Dolichos giganteus Willd.
Mucuna grevei Drake
Stizolobium giganteum (Willd.) Sprengel

Common Names: Elephant cow itch, Sea bean.

Vernacular Names: Kan: Thuribilangi; **Mal:** Kaa kavalli; **Tam:** Kalgaayvalli; **Tel:** Enugadoolagondi.

Distribution: Native of Africa, Asia including India, Australasia, Indian Ocean and Pacific Ocean countries.

Description: A large, woody twiner; stipules linear; leaflets ovate or elliptic, acuminate, terminal leaflets equal sided and cuneate, lateral leaflets equilateral and rounded at the base, glabrous on both the surfaces; calyx with a few irritant bristles, bilipped, upper lip subtruncate, the 3 lower teeth small and triangular; flowers greenish yellow on a corymb; keel petal not inflexed at the tip; pod broadly winged but not plaited across the face, clothed with yellowish brown irritant bristles; seeds 2-6.

Phytochemistry

Seed: L-dihydrophenylalanine (L-DOPA).

Edibility

Seed: Edible.

Medicinal Uses

Bark: Paste is applied for rheumatism.

Seed: Powdered and used as aphrodisiac and purgative.

Toxicology: Pod hairs cause intense skin irritation and dermatitis.

Ethnomedicine

Plant: Aphrodisiac and laxative.

114. *Mucuna monosperma* Wight

Synonym: *Stizolobium monospermum* (Wight) Kuntze

Common Names: Elephant's scratch wort, Negro bean.

Vernacular Names: Guj: Adadaveliya, Kagadolia; **Kan:** Anipeballi, **Mal:** Malanthelli; **Mar:** Mothikuhili, Sonagaravi; **Ori:** Bai donka, Sarni; **Sans:** Khatavangi; **Tam:** Periyattalargai, Thelu kodi; **Tel:** Enugadulagondi, Peddadulagondi.

Distribution: Native of Bangladesh, India, Myanmar, Sri Lanka and Thailand; in India, it grows in Assam, Himalaya, Meghalaya, Orissa, Tripura and Western Ghats in Karnataka.

Description: A large woody climber with glabrescent branches; leaves 3-foliolate; leaflets ovate oblong, lateral leaflets very oblique, base rounded, abruptly acuminate at apex, subcoriaceous, glabrous above, pubescent beneath; flowers 6-12, purple, on axillary raceme; calyx bowl shaped, with bristle like reddish hairs on both outside and inside; pod ovate-ovoid, reddish brown, bristly, 1-seeded.

Phytochemistry

Seed: Amino acids, oleic acid¸ oleanolic acid¸ palmitic acid and stearic acid.

Edibility

Seed: Edible.

Medicinal Uses

Stem: Watery sap from stem is used for asthma, cough, fever and tongue infection.

Pod: Vermifuge.

Seed: Used as expectorant, restorative and sedative; useful in asthma, cough, fever, throat pain and tongue infections; pod hairs used as counter irritant.

Toxicology: Pod hairs cause dermatitis.

Uses in Alternative Systems of Medicine

Ayurveda:

Part used: Root, pod hairs and seed.

Uses: Root and pod hairs, used as aphrodisiac and tonic; useful in diseases of "pitta" and wasting diseases; seed is aphrodisiac, cardiotonic and refrigerant; astringent to bowels; cures "*tridosha*" and biliousness.

Ethnomedicine

Plant: Antispasmodic, expectorant and sedative.

Seed: Is used in asthma, cold, cough and sores.

115. *Mucuna pruriens* (L.) DC. var., *pruriens*

Synonyms: *Dolichos pruriens* L.
Mucuna prurita Hook

Mucuna prurita (L.) Hook

Stizolobium niveum (Roxb.) Kuntze

Notes: Two varieties, var., *pruriens* and var., *utilis* are recognized; var., *pruriens* has straw yellow irritant hairs on pods, and the seeds are known by the name "Nescafe" in Central America and Brazil.

Common Names: Coffee bean, Cat bean, Common cowitch, Cowage, Cowhage, Cowitch, Velvet bean.

Vernacular Names: Ben: Alkushi, Bichochoti; **Guj:** Kivanch, Kavatch; **Hin**: Goncha, Kaevanch, Kajohara, Kamaach, Kawanch, Kivanch; **Kan:** Hasaguni, Chinaki beeja, Kaadu thavare, Markata, Koogari, Nayisonagu balli, Nasagunni, Sonagee bheeja, Thuruchi avare; **Mal**: Naaykorana, Shorivalli, Shoriyanam; **Mar:** Kaajkuhilee, Kavacha, Kuhili; **Sans:** Aathmaguptha, Adhyanda, Ajalomi, Ajavha, Arshabi, Badari, Chanda, Durabhigraha, Dusparsha, Gatrabhanga, Guru, Kaachumathi, Kandura, Kapikacchu, Kapiprabha, Kapiromaphala, Kashiroma, Kundali, Markati, Maharshabhilangali, Rishabhajatha, Rishyaprokata, Romavalli, Sadhyashotha, Shimbi, Shukapindi, Shukashimbika, Shukravathi, Sugupta, Svayamgupta, Theekshna, Vaanari, Varahika, Vyaaghra; **Tam:** Amudaari, Arugrattam, Kaadigandu, Kavi, Naikurini, Maaguttam, Ponnepoonduku averaikody; Poonaikkali, Poonaippidukkan, Sugasimbi; **Tel:** Dodagondi, Dooradagondi, Nayikkorana, Pilliyadagu, Praavrishayani, Theeta kogial, Thelladooradagondi, Thotadoolagondi.

Distribution: Native of Africa, Caribbean Islands. Eastern Asia, Indian subcontinent, Madagascar and Mexico; introduced into Central and South America and United States; in India, it grows throughout the plains of Assam, Himalayas and Meghalaya.

Description: An annual twiner with slender, glabrescent branches; leaves 3 foliolate, petioles silky; leaflets membranous, obovate, elliptic, terminal leaf let slightly small, rhomboid, ovate with cuneate base, lateral leaflet with truncate base, lower side greatly dilated, mucronate, pubescent above, densely clothed with silvery grey hairs beneath; stipules present; flowers purple, solitary or 2-3 on racemes; pedicels hairy; bracts and bracteoles hairy, caducous; calyx silky with irritant bristles outside, calyx tube campanulate, upper teeth completely connate into a triangular lip, lateral and lower teeth lanceolate; corolla with incurved keel petal, wing petal adherent to keel petal; stamens 9+1; pod turgid with longitudinal rib running the length of each valve, falcately curved on both ends, like the letter S, densely clothed with persistent irritant bristles, which are yellow to steel grey; seeds 5-6, small, reddish orange, reniform and glossy.

Phytochemistry

Plant: Mucuadine, mucuadinine, nicotine, prurienidine, N,N-dimethyl tryptamine, bufotenine, 5-methoxy-N, N-dimethyl tryptamine, β-carboline, 5-oxy-indole-3-alkylamine.

Root: Choline.

Leaf: Bufotenine, carboline, choline, dimethyl-hydroxytryptamine, dimethyl tryptamine, methoxy harman, methoxy tryptamine and oxindole-3-alkylamine.

Pod: Bufotenine, carboline, choline, dimethyl tryptamine, dimethyl-hydroxy tryptamine, methoxy tryptamine and oxindole-alkylamine.

Pod hairs: 5-Hydroxy tryptamine, histamine, mucunain and serotonin.

Seed: Alanine, arabinose, arachidic acid, arginine, aspartic acid, behenic acid, bufotenine, calcium, carbohydrates, carboline, choline, cis-12, 13-epoxyocta-trans-9-enoic acid, cystine, fat, fatty oil, fibre, gallic acid, glucose, glutathione, glycine, histidine, iron, isoleucine, L-dopa (L-3, 4-dihydroxy phenylalanine), lecithin, leucine, linoleic acid, linolenic acid, lipids, lysine, methionine, mucuadine, mucuadinine, mucunadine, mucunine, myristic acid, N, N-dimethyl tryptamine, niacin, nicotine, oleic acid, oxindole-alkylamine, palmitic acid, palmitoleic acid, phosphorus, proline, proteins, prurienidine, prurienine, prurienvirine, riboflavin, saponin, seratonin, serine, sitosterol, stearic acid, sterols, sucrose, thiamine, threonine, tryptamine, tyrosine and valine.

Seed oil: Arachidic acid, cis-12, 13-epoxy octadeca-trans-9-enoic acid, lecithin, linoleic acid, myristic acid, oleic acid, palmitic acid, stearic acid and vernolic acid.

Edibility

Leaf, Pod and Seed: In Central America the seeds (commonly called as mucuna beans) are roasted and ground as coffee substitute for decades; seeds are also cooked and eaten; tender leaves and pods are also cooked and eaten.

Medicinal Uses

Plant: Anabolic, anti-inflammatory, anthelmintic, aphrodisiac, hypocholesterolemic and hypoglycaemic; useful in anasarca, bone fracture, cancer, dropsy, ringworm, sores and syphilis; *Mucuna* with camphor, mace, *Argyreia speciosa, Acorus calamus* and sugar is given for general debility; *Mucuna* with *Plumbago zeylanica*, dry ginger, long pepper, cinnamomum and clove is given in colic, dyspepsia worms and to improve libido and to cure sexual dysfunction; the juice of the stem is useful in eye diseases.

Root: Anthelmintic, aphrodisiac, astringent, diuretic, emmenogogue, emollient, febrifuge, purgative, stimulant, thermogenic and tonic; useful in amenorrhoea, cough, constipation, delirium, dysmenorrhoea, dystentery, dropsy, fever, hemiplegia, nephropathy, facial paralysis, strangury and ulcers; smoke from burnt root is given during child birth to reduce pain; root decoction is antipyretic and diuretic; root infusion with honey is given in cholera; root paste is given in dropsy and bloody dysentery; an ointment of the root is applied on elephantiasis.

Leaf: Anthelmintic, aphrodisiac, blood enricher and tonic; useful in blisters, boils, cephalagia, general debility, inflammation, piles and ulcers; leaf infusion is given as anticoagulant; leaf decoction is antihypercholesterolemic and antipyretic.

Flower: Useful as vermifuge.

Pod: Anthelmintic, antispasmodic, aphrodisiac, hypoglycaemic and tonic; used against abdominal tumour, cancer, indolent ulcers and malignant ulcers; useful in ringworm.

Pod hairs: Used as vermifuge; an ointment made from hairs is applied externally as a local stimulant in paralysis and as a vesicant.

Seed: Anabolic, androgenic, analgesic, anti-inflammatory, aphrodisiac, anti-parkisonian, antispasmodic, blood cleanser, CNS stimulant, digestive, febrifuge, hypotensive, hormonal, hypocholesterolemic, hypoglycaeimic, immunomodulator, nervine, neurasthenic, uterine stimulant and vermifuge; as a carminative to expel intestinal gas and to eliminate kidney stones; it is useful in cancer, consumption, cough, cholera, delirium, dropsy, debility, gonorrhoea, hemiplegia, impotency, muscular disorders, paralysis, sterility, vaginal flaccidity and urinary complaints; seed decoction is useful in leucorrhoea, spermatorrhoea, menstrual disorder and urinary complaints; seed paste is applied on the body for dropsy; a nutritive tonic prepared by mixing *Mucuna* seeds with *Tribulus terrestris* seeds, *Eriodendron afractuosum* root, *Asparagus adescendens, Emblica,* myrobalan, *Tinospora,* ghee and sugar, is given to enhance libido and as a cure for spermatorrhoea.

Bioactivity: Plant is inhibitor of 5-β-reductase, ACE, aldose-reductase, α-reductase, aromatase, catechol-O-methyl transferase, cyclo-oxygenase, DOPA-decarboxylase, histidine-kinase, NADH-ubiquinone oxido reductase, ornithine-decarboxylase, PTK, peroxidase, succinic-dehydrogenase, topoisomerase I and II, tyrosine-kinase and xanthine-oxidase; is CNS stimulant, angiotensin receptor blocker, calcium-channel blocker, monoamine precursor, pituitary stimulant and quinone-reductase inducer.

Seed extract exerted antiparkinsonian, hypoglycaemic, hypocholesterolemic activity in rats; it produced an striking sexual activity in male rats; alkaloids from the extract increased sperm count, testes and seminal vesicle size and weight and caused neuromuscular block on frog rectus and abdomen; the total indole alkylamines produced marked behavioural changes, antagonized pentabarbitone induced hypnosis, inhibited reserpine induced ptosis, hypothermia; chloropromazine induced catatonia and enhanced amphetamine toxicity in rats.

Toxicology: Leaf and seed contain 3,N,N-dimethyl tryptamine which cause hypertension, papillary dilation, anxiety, hallucination and perceptual disorders; altering the levels of dopamine in the brain may effect the balance of other hormones and enzymes and thus its continuous use may be dangerous; it may cause abortion and should not be used by pregnant women; the powerful biological actions of this plant should not be abused and should be taken under medical supervision.

Clinical Studies: Mucuna seeds are clinically studied and found to have anabolic and growth hormone stimulant properties; the anabolic effect is due to its ability to increase testosterone level in the body; in 2002, U.S patent was filed on the use of beans to stimulate the release of growth hormones in humans; L-Dopa is converted to dopamine in the body which stimulates the pituitary gland to secrete the hormones; dopamine is also inhibitor of prolactin; increased level of prolactin in the body causes erectile dysfunction and decreases sperm production; velvet bean is gaining popularity in sports nutrition as it increases muscle mass due to increased level of testosterone and is a common ingredient in products and preparations used for body building; due to high level of L-Dopa in the seeds, it has been studied for its possible use in Parkinson's desease, which is a age related neurodegenerative diseases associated with progressive degeneration of dopaminergic neurons in the brain.

Uses in Alternative Systems of Medicine

Ayurveda:

Part used: Root, leaf, pod, pod hairs and seed.

Uses: Root is emmenogogue; cures cough, facial paralysis, general weakness, internal tumours, muscular stiffness and respiratory disorders; the smoke from dried and burnt root accelerates delivery and lessens the pain, promotes wound healing and is uterine stimulant; leaf is aphrodisiac, anthelmintic and tonic; lessens inflammation; improves blood; juice is given for headache. Seed and pod given in bilious diseases, consumption, diseases due to derangement of "vata" and "pitta", dysentery, indolent ulcers, uterine trouble, wasting diseases, worms and wounds; pod hairs are aphrodisiac, diuretic, purgative and nervine tonic; promotes strength; used in cholera, delirium, diarrhoea, impotency, internal tumours, intestinal worms, leucorrhoea, malignant ulcers, muscular pain, parkinsonism, respiratory trouble, sexual debility, spermatorrhoea, tuberculosis and urinary disorders; seed is alexipharmic, aphrodisiac, laxative and tonic; used in dystentery, diarrhoea, gonorrhoea, respiratory troubles, tuberculosis and muscular pain; seed paste is applied on the body in dropsy, and in wound healing; antidote to scorpion sting;

Homoeopathy

Part used: Tincture of the pod.

Uses: Used in asthmatic cough, constipation, cough, haemorrhoids, herpes, itching, jaundice, neuralgia of gums, pain, pruritus and sore throat.

Siddha:

Part used: Root, pod hairs and seed.

Uses: Aphrodisac, laxative and nervine tonic; useful in cholera, diarrhoea, dropsy, eczema, filaria, hemiplegia, leucorrhoea, menorrhagia and worm infestation.

Unani:

Part used: Root, and seed.

Uses: Alexipharmic, anthelmintic, anti-inflammatory, aphrodisiac, emmenogogue, laxative and tonic; root juice is given in headache; seeds are used as antidote to scorpion sting and used in gonorrhoea.

Medicinal Uses Outside India

Brazil: Seed is used as nervine tonic; taken for edema, gastritis, impotency, Parkinson's disease and worms.

Columbia: Root infusion with honey is prescribed for cholera; leaf infusion is given for piles; pod hairs are used as vermifuge when administered with honey.

French Guiana: Root used as purgative.

Germany and Guadeloupe: Seeds given to children as vermifuge.

Guinea Bissau: Plant juice is given as emmenogogue; seeds as aphrodisiac.

Haiti: Pod decoction is taken orally for intestinal parasites.

Ivory coast: Hot water extract of seeds is taken orally as emmenogogue.

Madagascar: Decoction of seed is taken as aphrodisiac.

Mozambique: Hot water extract of seed is aphrodisiac.

Nepal: Hot water extract of seed is aphrodisiac; seed useful as nervine tonic and antidote to scorpion sting; root is used as purgative; useful in delirium and dropsy; pod is anthelmintic; infusion of pod with honey is given in cholera; plant is used as vermifuge.

Nigeria: Leaf extract is taken orally for snakebite.

Pakisthan: Stem juice is applied on burns, sores and wounds.

Philippines: Fresh stem juice is used to treat burns, sores and wounds.

Senegal: Pod hairs used as anthelmintic with honey.

Thailand: Dried and burnt leaves and stem are used as application for cuts and burns; it is also mixed with *Oroxylum indicum* bark and leaves for cuts and burns.

Virgin Islands: Hot water extract is taken orally as vermifuge.

Trinidad: Crushed seeds are eaten as vermifuge.

West Indies: Root decoction is taken as a powerful diuretic and cleanser; leaves are applied to ulcers; pod infusion is a remedy for dropsy; pod hairs taken as vermifuge, followed by purge.

Ethnomedicine

Plant: Aphrodisiac, anodyne, diuretic, resolvent, rubefacient and nervine tonic; useful in asthma, boils, blisters, cancer, cholera, cough, diarrhoea, dysentery, dropsy, dysuria, fracture, mumps, pruritus, ringworm, sores, syphilis, tumour of the abdomen and ulcers; used as antidote for dog and snake bite.

Root: Is used in dysentery and cough.

Stem juice: Is used in eye diseases.

Leaf: Is useful in boils and blisters.

Pod and pod hairs: Used in ringworm and as vermifuge.

Seed: Aphrodisiac, appetizer and digestive; used in impotency, night blindness, sores, syphilitic wounds and urinary complaints; used as antidote for scorpion sting and snake bite.

Non-medicinal Uses

Plant: Grown as hedge plant, cover crop and green manure crop; used as pesticide.

Leaf: Is used as fodder.

Pod and Seed: Roasted and used as substitute for coffee.

116. *Mucuna pruriens* (L.) DC., var., *utilis* (Wight) Burck

Synonym: *Mucuna utilis* Wight

Notes: The pods of this variety have soft blackish brown hairs while those of var., *pruriens* have stiff straw-coloured hairs.

Common Names: Bengal bean, Mauritius bean, Portugese bean

Vernacular Names: Ben: Kamach, Mundari, Kursi; **Hin:** Kiwach.

Distribution: Native of Caribbean Islands; introduced into India; grows in Bihar, Haryana, Himalaya, Madhya Pradesh, West Bengal and Sikkim in India.

Description: A climbing twiner; leaves 3 foliolate; leaflets membranous, obovate, elliptical; terminal leaflet small, rhomboid with cuneate base, lateral leaflet with truncate base, pubescent above, clothed with silvery grey hairs below; flowers purple with hairy pedicles; bracts and bracteoles hairy; calyx with irritant bristles; corolla with wing petal adherent to keel petal; pod "s" shaped, similar to that of *Mucuna pruriens* var.*pruriens*; pod hairs in var. *pruriens* are stiff, irritating and those in var. *utilis* are dark brown and soft.

Edibility

Young leaf and Pod: Edible.

Medicinal Uses

Plant: Is abortifacient, antiangiogenic, anticancer, antiestrogenic, antihaemolytic, anti-inflammatory, antimicrobial, antimitogenic, antioxidant, antispasmodic, antiviral, cytotoxic and immunostimulant; useful in ischaemia, leukaemia, lymphoma, melanoma, neuroblastoma, osteoporosis, tumours and ulcer.

Bioactivity: Plant is inhibitor of aldose-reductase, α-reductase, catechol-O-methyl transferase, DOPA-decarboxylase, histidine-kinase, lipase, MAO, NADH-ubiquinone-oxidoreductase, ornithine-decarboxylase, PTK, peroxidase, topoisomerase I and II and tyrosine-kinase; is calcium antagonist, estrogen antagonist, pituitary sensitizer and quinone-reductase inducer.

Uses in Alternative Systems of Medicine

Chinese:

Part used: Whole plant.

Uses: Used in respiratory affections.

Unani:

Part used: Root, leaf and seed.

Uses: Abortifacient, aphrodisiac, astringent, diuretic, emmenogogue, laxative, nervine tonic and vermifuge; useful in leucorrhoea, paralysis and spermatorrhoea.

117. *Mundulea sericea* (Willd.) A. Chev.

Synonyms: *Cytisus sericeus* Willd.
Dalbergia sericea Bojer
Mundulea suberosa (DC.) Benth.
Tephrosia suberosa DC.

Common Names: Bastard satin wood, Madagascar bean.

Vernacular Names: Kan: Bettadahurali, Kaadubettadahurali, Kaadu thuvari, Henukuruvina gida; **Mal:** Kattutuvara; **Tam:** Pilavaram; **Tel:** Kondavempali, Palasaram, Verri billudu.

Distribution: Native of Africa, India, Sri Lanka and Madagascar and Indian Ocean Islands.

Description: Silky pubescent shrub; leaves numerous; leaflets 17-23, opposite or subopposite, ovate, oblong, lanceolate, glabrous above silvery silky beneath; flowers pinkish violet on terminal raceme; calyx clothed with silky hairs and short deltoid teeth; standard petal silky outside; pod flat, densely velvety with golden brown hairs, thickened at the sutures; seeds 3-8, yellowish brown.

Phytochemistry

Plant: Hydrocyanic acid, lupinifolin, mundulin, mundulinol and sericone.

Root bark: Mundulone, and rotenone.

Stem bark: Mundulin, mundulinol, mundulone, munduserone, pyranoflavonoids, rotenone, sericetin and sericone II.

Leaf: Isorhamnetin.

Seed: 2-Aminoimidazole and mundulone.

Toxicology: All parts of the plant are toxic; bark and seeds are highly toxic.

Ethnomedicine

Plant: Is used for tuberculosis.

Non-medicinal Uses

Stem bark: Insecticide and piscicide.

Pod: Used as fish poison; toxic to house flies and cockroaches.

Seed: Used as fish poison.

118. *Myroxylon balsamum* (L.) Harms

Synonyms: *Myrospermum toluiferum* DC.
Toluifera balsamum L.

Notes: Balsam of Tolu exudes from the trunk as a result of injury; yields a volatile oil, oil of Tolu balsam, which is commercially important; the generic epithet is derived from the Greek word "myron" meaning perfume or sweet oil and "xylon" wood.

Common Names: Balsamum of Tolu, Tolu balsam tree.

Distribution: Native of El Salvador, Mexico, Panama and South America; introduced into India; cultivated in India, cultivated in Lalbagh, Bangalore.

Description: Tall tree with spreading crown; bark grey, more or less spotted with yellow rough areas; leaves alternate, evergreen, oddly pinnate; leaflets without stipules, ovate oblong, acuminate, subcoriaceous, lustrous; racemes closely tomentulose; flowers white; pod broad, 8-13 cm long; seed with 2 globose balsam pits between the corky mesocarp and endocarp.

Phytochemistry

Plant: α-Bournonene, α-cadinene, α-calacorene, α-capaene, α-curcumene, α-murolene, α-pinene, benzaldehyde, benzoic acid, benzyl alcohol, benzyl-benzoate, benzyl-cinnamate, β-baerbonene, β-elemene, benzyl-ferulate, β-isoferulate, cadalene, cadenine, calamenene, cinnamein, caryophyllene, cinnamaldehyde, cinnamic acid, cinnamic alcohol, cinnamyl-benzoate, cinnamyl-cinnamate, cis-ocimene, coumarin, dihydro benzoic acid, dihydro cinnamic acid, dihydromandelic acid, ethyl benzoate, eugenol, farnesol, ferrulic acid, hydroxy hopanone, murolene, methyl cinnamate, nerolidol, oleonolic acid, peruviol, styrene, sumaresinolic acid, tannin and vanillin.

Seed: Afromosin, carbreuvin, (+) 7-hydroxy-4'-methoxyisoflavonone.

Medicinal Uses

Plant: Analgesic, anaesthetic, anthelmintic, anti-HIV, antibacterial, anticancer, anticonvulsant, antidiuretic, antiedemic, antihepatotoxic, anti-inflammatory, antimelanomic, antimetastatic, hypocholesterolemic; used in acne, asthma, cancer of cervix, kidney, pancreas, colon, lung, convulsions, cough, diarrhoea, dysentery, dsmenorrhoea, edema, haemorrhage, hypoglycaemia, flu, herpes, hypertension, leukaemia, melanoma, nephritis, odantopathy, pneumonia, polio, psoriasis, rheumatism, thrombosis, scabies and sleep disorders.

Oil of Tolu: Is antiseptic, expectorant and stimulant; used as an inhalent in catarrh; ingredient of cough mixture.

Gum: Is antiseptic, anti-inflammatory, antibacterial, antifungal and emollient; as documented in the U.S.pharmacopeia 1820, it is used for bronchitis, cough, laryngitis, dysmenorrhoea, diarrhoea, dysentery and leucorrhoea; it is used as topical application in eczema, scabies, skin parasites, sores, ulcers and wounds; it is used in cough syrups and lozenges for sore throat and as vapour inhalant for respiratory ailments; is used in hair tonics and anti dandruff preparation, in aromatherapy, and considered as warming and comforting and used to relieve nervous tension and strain.

Bioactivity: Plant is a inhibitor of catechol-O-methyl transferase, DOPA- decarboxylase, histidine-kinase, lipase, MAO, MDR, NADH-ubiquinone-oxidoreductase, ornithine-decarboxylase, PTK, peroxidase, topoisomerase I and II and tyrosine-kinase; is calcium antagonist, estrogen antagonist, pituitary sensitizer, quinone-reductase-inducer, CNS active, circulatory stimulant, motor depressant, testosterone inducer, trypsin enhancer and cytochrome-P_{450} inducer.

Medicinal Uses Outside India

Amazon: Gum is used for abscesses, asthma, bronchitis, catarrh, headache, flu, rheumatism, sores, sprains, tuberculosis, venereal diseases and wound.

Britain: Balsam is used topically for eczema, prurigo and scabies; taken internally for asthma, bronchitis and to check excess mucous secretion.

Caledona: Gum is used for bronchitis, cough, skin sores and wounds.

Central America: The leaf and pod is used for asthma, cold, flu, rheumatism and wound.

Dominican Republic: Gum is used for excessive mucous, indigestion, sores and wounds.

Mexico: Gum is used for asthma, bronchitis, colic, flu, freckles, gout and itch.

United States: Gum is used for bronchitis, cough, dandruff, diarrhoea and dysmenorrhoea.

Ethnomedicine

Plant: Is used as antiseptic, deodorant, expectorant, stimulant, stomachic, pectoral, vulnerary and tonic; used for asthma, bronchitis, catarrh, cold, cough, headache, rheumatism, sclerosis, sores, swelling, tuberculosis, venereal diseases and wounds.

Non-medicinal Uses

Wood: Is used in decorative woodwork.

Gum: Is used to give fragrance to soaps, detergents, creams, lotions, perfumes and sprays; used in confectionary and chewing gum.

Oil: Is used in cosmetics, perfumes and soaps.

119. *Myroxylon peruiferum* L.f.

Synonym: *Myrospermum pedicellatum* Lam.

Notes: Peru balsam is a soft yellow, brown or black, bitter, viscous, acrid substance with vanilla like flavour, which is obtained by incising the stem.

Common Names: Black balsam, Honduras balsam, Peru balsam, Peruvian balsam, Surnam balsam.

Distribution: Native of South America; introduced into India; cultivated in Lalbagh, Bangalore.

Description: Tall tree with terete branches, ash to ash brown in colour; leaves alternate, evergreen, compound; leaflets 5-11, alternate, oblong, ovate, slightly tapering at the base, apex abruptly contracted into an emarginate point; petioles and midrib with short glossy hairs; leaflets marked by round and linear pellucid spots; flowers white on axillary racemes; pod stalked, winged, indehiscent; seed thin, single, white, membranous with a bitter almond taste and golden colored juice.

Phytochemistry

Gum: Balsamic esters of benzyl cinnamate, benzoic acid, benzyl benzoate, benzyl alcohol, cinnamic acid, cinnamyl cinnamate, farnesol, peruviol and vanillin.

Edibility

Gum: Edible.

Medicinal Uses

Plant: Is a source of balsam of Peru, which is used in ointments for bed sores, chilblain, pediculosis, ringworm, scabies and ulcers; used as a dressing for cuts, indolent ulcers and wounds; used as suppository for haemorrhoids; given internally as antacid, antiseptic, expectorant, diueretic, stimulant and vulnerary; useful in urinary troubles.

Ethnomedicine

Plant: Is used for cancer.

Non-medicinal Uses

Oil: Oil is used in soaps and perfumes.

Wood: Is used in furniture making.

120. *Neonotonia wightii* (Wight & Arn.) Lackey

Synonym: *Glycine wightii* (Wight & Arn.) Verdc.

Notes: Three subspecies, *wightii*, *pseudojavanica* and *petittiana,* are recognized.

Distribution: Native of Africa, Asia including India and Middle East; introduced into Australasia, Indian Ocean and Pacific Ocean Islands.

Description: Herbaceous perennial with strong taproot and trailing or climbing and twining stem; stem slender and well branched; leaves, pinnately trifoliolate with ovate leaflets 5 to 10 cm long and 3 to 6 cm wide, with short hairs on both surfaces and small triangular stipules; elongated raceme bear clusters of white or violet flowers; in some varieties the flowers become yellow or orange-yellow; pod hairy, straight or slightly curved, 1 to 4 cm long, about 3 mm wide, with three to eight seeds; seeds vary in size, shape and colour, depending on variety.

Edibility

Leaf: Edible.

Medicinal Uses

Plant: Is analgesic, antiangiogenic, antihaemolytic, anti-inflammatory, anti ischaemic, antimicrobial, antimitogenic, antioxidant, antispasmodic, antitumour, antiviral, apoptotic, cardioprotective, estrogenic, flatulent, fungicide, immunostimulant and uterotropic; used in atherosclerosis, cancer of breast, leukaemia, lymphoma, melanoma, neuroblastoma, osteoporosis, tumour of breast, ovary, prostrate, stomach and ulcer.

Bioactivity: Plant is inhibitor of aldose-reductase, α-reductase, catechol-O-methyl transferase, DOPA-decarboxylase, histidine-kinase, lipase, MAO, ornithine-decarboxylase, PTK, peroxidase, toposisomerase I and II, tyrosine-kinase and xanthine-oxidase; is calcium antagonist and quinone-reductase inducer.

121. *Ononis spinosa* L., ssp., *hircina* (Jacq.) Gams.

Synonyms: *Ononis arvensis* L.
Ononis hircina Jacq.
Ononis intermedia sensu auct.
Ononis pseudohircina Schur.
Ononis repens L. ssp., *arvensis* (L.) Greuter
Ononis spinescens Ledeb.
Ononis spinosa L. ssp., *arvensis* (L.) Greuter & Burdet
Ononis spinosiformis Simonkai
Ononis vulgaris Rouy

Common Name: Wild liquorice.

Distribution: Native of Asia including India and Pakistan, former USSR, Europe and Turkey; grows in Western Himalayas in India.

Description: Shrubby perennial; leaf trifoliolate, elliptic to ovate, denticulate; stipules large, foliaceous; inflorescence terminal raceme; flowers in pairs; calyx teeth exceed the tube; corolla pink; pod 1 to 3 seeded.

Edibility

Shoot: Young shoots boiled or eaten raw in salads.

Medicinal Uses

Root and Bark: Dissolves kidney stones.

Ethnomedicine

Plant: Aperient, depurative, diuretic and expectorant; used for cancer and gout.

122. *Ormocarpum cochinchinense* (Lour.) Merr.

Synonyms: *Diphaca cochinchinensis* Lour.
Ormocarpum glabrum Teijsm & Binn.
Ormocarpum orientale (Sprengel) Merr.
Ormocarpum sennoides DC.

Common Name: Jungle moringa.

Vernacular Names: Kan: Kaadunugge; **Mal:** Kattumurinna, Punamurinna; **Sans:** Kananashekhara; **Tam:** Kattumuringai, Thavasumurungai; **Tel:** Adavimunaga, Gunnangi, Nallakashina.

Distribution: Native of Asia including India, Australasia and Pacific Ocean Islands; grows in Western India.

Description: Straggling shrub with warty branchlets; leaves imparipinnate; leaflets 6 pairs, obovate, emarginate; flowers yellow, axillary on 3-5 flowered raceme; pod prickly and warty; seeds narrowly oblong.

Medicinal Uses

Root: Used as stimulant and tonic; cures lumbago.

Root bark: Paste mixed with oil is rubbed on the paralysed parts.

Ethnomedicine

Plant: Stimulant and tonic; used for lumbago and paralysis.

Non-medicinal Uses

Plant: Grown as support plant for pepper and shade plant for coffee and cocoa; used as a fish poison.

123. *Pachyrhizus erosus* (L.) Urban

Synonyms: *Dolichos articulatus* Lam.
Dolichos erosus L.
Pachyrhizus angulatus DC.
Pachyrhizus articulatus Walp.

Common Name: Yam bean.

Vernacular Name: Ben: Sankalu.

Distribution: Native of Central America; introduced into India; cultivated throughout India.

Description: Strong, climbing vine; stem terete, pubescent; leaf rachis with ascending hairs; stipules lanceolate or acute; leaflets with sparse pubescence, blunt or acute apically, truncate basally; the lateral leaflet ovate, shallowly lobed, inequilateral; inflorescence axillary, many flowered; bracteoles linear, lanceolate; flowers bluish white; calyx tube with canescent or fulvous hairs and acute lateral and lower lobes; petals membranous, glabrous, auricle incurved, acuminate with a claw, vexillum obovate, elliptic with semicircular ridge connecting the auricles above the claw; pod long, hairy with ferrugineous or fulvous hairs; seeds 10, brown and shining.

Phytochemistry

Plant: Hydroxy rotenone and rotenone.

Leaf: Pachyrrhizin A and B.

Seed: Erosone, pachysaponin A and B, pachyrrhizin and pachyrrhizone.

Edibility

Tuber: Eaten raw or cooked; sliced, dried and made into chips.

Pod: Young pods eaten.

Medicinal Uses

Root: Antipyretic, anticephalagic and haemostatic; is a recommended cure for cholera and smallpox.

Seed: Anthelmintic, laxative and vermifuge; powdered seeds used for skin affections.

Uses in Alternative Systems of Medicine

Chinese:

Part used : Tuber

The fleshy tubers are given as a cooling food in fever.

Medicinal Uses Outside India

Java: Powdered leaves mixed with sulphur applied on skin eruptions, which spread by scratching.

Non-medicinal Uses

Plant: Is used as green manure.

Tuber: Is used to feed cattle and yields starch of superior quality.

Stem: Used in making fishing nets.

Leaf: Toxic to mosquito larvae; insecticidal.

Seed: Powdered seeds used as insecticide and piscicide.

124. *Paracalyx scariosus* (Roxb.) Ali

Synonym: *Cylista scariosa* Roxb.

Vernacular Names: Guj: Kamalawel; **Kan:** Kadalenaaru; **Mar:** Ranghevada; **Sans:** Nadinishpara; **Tel:** Karuchikkudu.

Distribution: Native of India, Pakistan and Nepal; introduced into other countries; in India it grows in Andhra Pradesh, Bihar, Goa, Gujarat, Karnataka, Kerala, Madhya Pradesh, Orissa, Tamil Nadu, Uttar Pradesh and West Bengal.

Description: Woody twiner with tomentose branches; leaves 3 foliolate; terminal leaflets rhomboid ovate, acuminate, distinctly nerved, tomentose beneath, lateral leaflets obliquely ovate, acute, pubescent on both sides; inflorescence is a raceme with yellow flowers; calyx pubescent, lower lip boat shaped, prominently veined; corolla yellow, concealed by the calyx; pod small, pubescent, 1 seeded, enclosed in the calyx.

Phytochemistry

Plant: Sparteine.

Root: Tannin.

Leaf: Isorhyncospermin, kaempferol, kaempferol-3-O-rutinoside, quercetin, rutin and scariosin.

Medicinal Uses

Plant: Astringent, diuretic, cathartic, CNS stimulant, cardiac depressant and oxytocic; used to induce and stimulate labour; useful in dysentery, leucorrhoea and tumours.

Root: Astringent; decoction given for dysentery and leucorrhoea; applied externally to tumours.

Leaf: Juice is taken to kill intestinal worms and as an antidote to snake bite.

Uses in Alternative Systems of Medicine

Ayurveda:

Part used: Pod.

Uses: Acrid, appetizer and alexiteric; astringent to bowels; enriches the blood; cures biliousness and improves taste.

125. *Phaseolus acutifolius* A. Gray

Notes: Two varieties, *acutifolius* and *latifolius* are recognized.

Common Names: Tapery bean.

Vernacular Names: Ben: Banmudgakheri; **Hin:** Mothe; **Pun:** Moth; **Sans:** Makushthaka; **Tam:** Tulkapyrai; **Tel:** Minumulu.

Distribution: Nativity unknown; introduced into India; grows in Himalayas in India.

Description: Plant is 10 feet long vine which climbs on shrubs; cultivated varieties are semi-viny or bushy; leaves pointed, trifoliolate; flowers pea like, white or pinkish on axillary racemes; calyx 4 lobed; standard petal reflexed, keel petal narrow with 2-3 turns to form a spiral; pod straw coloured with 5-6 seeds per pod; seeds flat and buff coloured, some times freckled.

Phytochemistry

Pod: Carbohydrates, fibre and protein.

Seed: Ascorbic acid, calcium, chromium, cobalt, copper, cysteine, fat, fibre, iron, isoleucine, leucine, lysine, magnesium, manganese, methionine, niacin, phenylalanine, phosphorus, phytate, phytic acid, potassium, raffinose, riboflavin, sodium, stachyose, sucrose, thiamine, threonine, tyrosine, valine and zinc.

Edibility

Seed: Edible; boiled, ground and made into paste and eaten; seed flour is used in making cakes.

Medicinal Uses

Root: Is narcotic.
Seed: Boiled and given in fever.

Non-medicinal Uses

Plant: Is used as fodder and grown as cover crop.

Seed: Fed to cattle and horse to fatten them.

126. *Phaseolus coccineus* L.

Synonyms: *Phaseolus multiflorus* Lam.
Phaseolus multiflorus Willd.

Notes: Occurs as relict of cultivation.

Common Names: Haricot bean, Runner bean, Scarlet bean, Scarlet runner bean.

Distribution: Native of tropical Central America; introduced into India; cultivated in Assam, Himachal Pradesh, Sikkim and Uttar Pradesh.

Description: Dwarf plant with trifoliolate leaves, with purple tinged veins; leaflets ovate; stipules triangular; flowers, large, bright scarlet, occasionally white; petals 4, free, (a large standard petal, 2 lateral wing petals and a single boat shaped keel); pod broad, stout, variable in colour from shades of pink, spotted or blotched with black, sometimes entirely black; seeds 6-10, broad, oblong, convex, flat, dark purple or pink with red marking or some times white, cream, brown, violet, black, black with a large, white, flat incurved scar.

Phytochemistry

Plant: Absicic acid, cyanidin-glucoside, cyanidin-3-sophoroside, cyclokievitone, demethyl vestitol, daidzein, dihydroxykaurenolide, dimethoxyisoflavone, 3'-O-β-D-glucopyranosyl gibberelin, glycuol, 2'- hydroxy isoprunetin, hydroxygenistein, isoprunetin, tetrahydroxydihydrodaidzein and trihydroxy kaurenoic acid.

Leaf: Dalbergioidin, demethylvestitol, hydroxygenistein, kievitone, phaseollidin and phaseollin.

Pod: Ascorbic acid, β-carotene, calcium, carbohydrates, 7,β-1,3-dihydroxykaurenolide, fat, fibre, iron, niacin, phosphorus, phylloquinone, riboflavin and thiamine.

Seed: Ascorbic acid, β-carotene, calcium, carbohydrates, chromium, cobalt, copper, fat, fibre, gibberellinene, iron, magnesium, manganese, niacin, phaseic acid, phosphorus, potassium, protein, riboflavin, sodium, thiamine and zinc.

Edibility

Pod: Edible.

Medicinal Uses

Plant: Abortifacient, antiangiogenic, antibacterial, anticancer, anticephalalgic, anticoronary, antiestrogenic, antifertility, antihaemolytic, anti-inflammatory, antimitogenic, antioxidant, antiradicular, antiseptic, antispasmodic, antitumour, antiviral, apoptotic, cancer preventive, coronary dilator, hypocholesterolemic, hypotensive, immunostimulant and laxative; useful in arrhythmia, atherosclerosis, breast cancer, cephalagia, diarrhoea, ischaemia, leukaemia, lymphoma, melanoma, neuroblastoma, tumour of breast, ovary and stomach.

Toxicology: Malonic acid in the plant causes damage to the mucous membrane and skin; chymotrypsin and trypsin inhibitors present.

Medicinal Uses Outside India

China: Plant is used as a remedy for diabetes.

127. *Phaseolus lunatus* L.

Synonyms: *Phaseolus bipunctatus* Jacq.
Phaseolus limensis Macfad.
Phaseolus puberulus Kunth
Phaseolus saccharatus Macfad.

Notes: Widely cultivated and naturalized.

Common Names: Butter bean, Lima bean, Pole bean, Rangoon bean and Sugar bean.

Vernacular Names: Ben: Bunbur, Bunburbuthy; **Kan:** Dabbale beans, **Mal:** Khasi, **Pun:** Lobiya; **Sans:** Akshipeedaka; **Tam:** Binusaveraikodi.

Distribution: Native of Bahamas; introduced into India and Pakistan; cultivated in most parts in India.

Description: A small annual or perennial climber with glabrous or pubescent stem; leaves trifoliolate; leaflets 2-8, entire or dentate; lateral leaflets oblique, ovate to lanceolate; leaf rachis ends in a branched tendril; stipules obliquely ovate, toothed below, semiamplexicaul at the base; inflorescence is a peduncled raceme; bracteate and bracteolate; calyx pubescent, toothed, upper 2 teeth joined, lower teeth longest; corolla white, lilac or reddish purple, or pale rose, standard petal sparsely pubescent externally, keel petal twisted in one and half turns; pod oblong, falcate, 3-4 seeded, glabrous.

Phytochemistry

Plant: Methoxyflavone, garbanzol, methyl cysteine and irigenin.

Leaf: γ-Dimethyl allyl kievitone, γ-dimethyl allyl phaseollidin, carbohydrates, coumesterol, cyclokievitone hydrate, dalbergioidin, deoxykievitol, deoxykievitone, di-γ-dimethylallylglycinol, dihydrocyclokievitone, dihydrokievitol, dihydrokievitone, dihydroxygenistein, fat, genistein, hydroxydaidzein, hydroxygenistein, isoferreirin and methoxygenistein.

Seed: Alanine, α-linoleic acid, aluminium, amygdalin, arginine, ascorbic acid, β-carotene, calcium, carbohydrates, chromium, cobalt, copper, coumesterol, cystine, daidzein, fat, genistein, glutamic acid, glycine, histidine, hydrocyanic acid, iron, isoleucine, leucine, lignin, linoleic acid, lysine, magnesium, manganese, methionine, molybdenum, myristic acid, niacin, nickel, oleic acid, palmitic acid, pantothenic acid, phaseolunatin, phenylalanine, phosphorus, potassium, proline, riboflavin, serine, sodium, stearic acid, thiamine, threonine, tryptophan, tyrosine, valine, vanadium, zinc and zirconium.

Sprout: Ascorbic acid, ash, β-carotene, calcium, carbohydrates, fat, iron, niacin, phosphorus, protein, riboflavin and thiamine.

Edibility

Pod: Used as green vegetable and eaten alone or mixed with other vegetables.

Seed: Dried or boiled or baked and eaten.

Medicinal Uses

Root: Used to stop excessive salivation.

Stem and Pod: Decoction is given in diabetes and dropsy.

Seed: Astringent; decoction of green seeds is given in diabetes and dropsy; seed paste is applied over the abdomen for stomach ache.

Toxicology: Cyanogenic glycosides such as linamarin, lotaustralin and phaseolunatin in the seeds cause ambylopia, endocrinological and neurological disorders, tropical ataxy, neuropathy, inhibition of respiratory enzymes, alteration in cardiac physiology and paralysis; chymotrypsin and trypsin inhibitors present.

Ethnomedicine

Plant: Nutritive; used in head ache, fever, stomach ache, tumour and warts.

Non-medicinal Uses

Plant: Grown as cover crop; used as fodder.

128. *Phaseolus vulgaris* L.

Synonyms: *Phaseolus communis* Pritzel.
Phaseolus esculentus Salisb.

Notes: There are over several hundred varieties that are cultivated in the world; often occurs as relict of cultivation; highly polymorphic species.

Common Names: Common bean, Dwarf bean, French bean, Field bean, Garden bean, Green bean, Kidney bean, Wax bean.

Vernacular Names: Guj: Phanasi; **Hin:** Baklasem, Lobia, Vilaitisem; **Kan:** Huralikayi; **Mal:** Barigalu; **Mar:** Shravanghevda; **Sans:** Brihat shimbe, Raajamaasha; **Tam:** Seemai averai kody; **Tel:** Chikkudi kaayi.

Distribution: Native of Mexico, introduced into India; cultivated throughout India.

Description: Annual herb, erect and bushy; leaves alternate, green or purple, trifoliolate, stipulate, petiolate, with a marked pulvinus base; leaflets ovate, entire, acuminate; flowers axillary, on racemes, zygomorphic, variegated, white, pink, or purplish; pods slender, green, yellow, black, or purple, cylindrical or flat; seeds 4-6, usually glabrous, sometimes puberulent, oblong with prominent beak, white, red, tan, purple, grey or black, often variegated, reniform or oblong.

Phytochemistry

Plant: Caffeic acid, chlorogenic acid, coumarin, dalbergioidin, deoxykievitone, dihydrokievitone, esculetin, esculin, ethanolamine, ferulic acid, gluconic acid, hydroxymethylpentonic acid, inositol, lysophosphatidylcholine, methylerythronic acid, N-acetyl derivative of phosphatidylethanolamine, neochlorogenic acid, phaseollidin, phaseollin, phaseoloside A and E, phaseolutenone, phosphatidic acid, phosphatidyl derivative of choline, saponin, scopoletin, umbelliferon and vomifoliol.

Root: Glycinoeclepin A, B, C, 4',7'-dihydroxy-2',3', dimethoxyisoflavan.

Leaf: Allantonin, esculetin, esculin, 6-hydroxy-7-methoxy coumarin, phaseolic acid, sphingolipid and umbelliferone.

Pod: β-Sitosterol, aspargine, daidzein, dehydroxy kievitone, deoxykievitone, dihydroxypterocarpon, fatty oil, glucoquinones, glucosides, glutamic acid, glutamine, hydroxyphaseollin, isoflavanone, isoflavone, lipids, methionine, oestriol, oestrone, phaseollidin, phaseollin, proline, quercetin, rubidium, rutin, serine, sterol, trihydroxy isoflavan and trihydroxyisoflavone .

Seed: Aceyl-phosphatidyl-ethanolamine, brassinolide, castasterone, choline, clerosterol, deoxocastasterone, deoxydihydrocastasterone, deoxydihydrohomodolichosterone, deoxydihydrolichosterone, deoxydolichosterone, dolicholide, dolichosterone, ethanolamine, ethyl desmosterol, ethylidene, gramisterol, indole-3-acetic acid, inositol, isocitrostadienol, isoquercetin, kaempferol, lichosterone, lysophosphatidyl choline, methyl fucosterol, methyl gramisterol, methylcholesterol, methyldolichosterone, methylfucosterol, oestriol, oestrone, ostradiol, quercetin, quercetin-3-glucorunide, robinin, rutin, soyasapogenol B, soyasaponin and terpenoids.

Edibility

Leaf: Young leaves are eaten raw in salads.

Pod: Green immature pods are cooked and eaten; mature pods are baked and used in soups in Spanish and Mexican dishes.

Seed: Edible, processed in different ways, and consumed.

Medicinal Uses

Plant: Used as emollient.

Leaf: Antibilious, cooling, sedative and tonic.

Pod: Is carminative, depurative, diuretic and resolvent; used in acne, bladder disorders, burns, cardiac diseases, diarrhoea, dysentery, dropsy, eczema, hiccup, itch, rheumatism, sciatica and tetanus; it lowers the blood sugar level in diabetics; used as emollient in cataplasms.

Green pod shells: Diuretic; used in kidney and heart diseases and as an adjuvant in mild diarrhoea.

Toxicology: Toxic amino acids in the seed cause neurolathyrism; chymotrypsin and trypsin inhibitors present.

Uses in Alternative Systems of Medicine

Ayurveda:

Part used: Seed.

Uses: Useful in hyperacidity.

Homoeopathy:

Part used: Dried bean and pod.

Uses: Useful in albuminuria, breast tumour, diabetes, dropsy, hernia, headache, impotency, pleurisy, haemorrhage and pericarditis.

Medicinal Uses Outside India

Indochina and Europe: Plant is used as emollient, purgative and resolvent.

Ethnomedicine

Plant: Carminative, depurative, diaphoretic, digestive, diuretic, emmenogogue and resolvent; used for albuminuria, burns, chest pain, cold, diabetes, dysentery, eye complaints, fever, headache, hiccup, rheumatism, scurvy and tumour.

Non-medicinal Uses

Plant: After the harvest of beans, straw is used as fodder.

Pod: Used to feed livestock; bean extract is used as a supplement for culture medium in chromosomal analysis as PHA initiates mitosis in leucocytes.

129. *Physostigma venenosum* Balf.

Common Names: Calabar bean, Ordeal bean.

Distribution: Native of Africa, introduced into India.

Description: A woody climber; leaves trifoliolate; leaflets ovate, acuminate; flowers purple, shell shaped, on axillary drooping racemes; pod yellowish brown, narrowed at ends; seeds, few, ellipsoid, one side nearly straight, the other grooved by depressed hilum.

Phytochemistry

Seed: Alkaloids, behenic acid, β-sitosterol, calabacine, calabarine, calabarol, calabatine, eseramine, eseridine, eserine, geneserine, isophysostigmine, linoleic acid, N-8-norphysostigmine, oleic acid, palmitic acid, physosterine, physovenine, physostigmine, protein, stearic acid, stigmasterol and trifolianol.

Medicinal Uses

Plant: Is antiadenomic, antiallergic, antiandrogenic, antibacterial, anticancer, anticoronary, antifibrinolytic, antihepatotoxic, antihistaminic, anti-inflammatory, antilymphocytic, antimutagenic, antioxidant, antiperiodontic, antipyretic, antiseptic, antispasmodic, antiviral, apoptotic, cardioprotective, choleretic, diuretic, haemolytic, hepatoprotective, immunostimulant, laxative, memory enhancer, narcotic, sedative, tonic, vasodilator and vulnerary; is used in treating anorexia, alopecia, anxiety, arthritis, atherosclerosis, depression, diabetes, diarrhoea, gingivitis, herpes, hypertension, hypoglycaemia, insomnia, leukaemia, lupus, lymphoma, menorrhagia, nephritis, obesity, odontalgia, osteoporosis, tumour of breast, cervix, lung, varicose veins, paralysis and ulcer; useful as a stimulant; acts on the nervous system.

Seed: A powerful sedative of spinal cord; useful in atony of the urinary bladder and intestine after operation; cures, cholera, epilepsy, flatulence, nervous affections, edema, poisoning, rheumatism and tetanus; crushed and applied on skin diseases; application of paste with oil is a remedy for lice on head; physostigmine which is an alkaloid present in the seed is used in ophthalmology as an ointment; used as local application on eye to contract the pupil; physostigmine salicylate is given intravenously to reverse the intoxication caused by tricyclic antidepressants, antihistamines and sedatives; used as an antispasmodic agent in rheumatoid arthritis, bursitis, colic and fibrositis; eseridine is given as purgative and used in dyspepsia and as eye drops in glaucoma; calabarine has stimulating effect on spinal cord; physoreine is strongly myotic.

Bioactivity: Plant is inhibitor of 5-α-reductase, 5-lipoxygenase, ACE, aldose-reductase, cyclo-oxygenase, glucosyl-transferase, iodothyronine-deiodinase, lipoxygenase, MAO, ornithine-decarboxylase, PAF, topoisomerase I and II, tyrosinase, xanthine-oxidase, CAMP, phosphodiesterase; is angiotensin receptor blocker, β-blocker, calcium channel blocker, haemoglobin inducer, memory enhancer and quinone-reductase-inducer.

Toxicology: Physostigmine-a indole alkaloid in the seeds is highly toxic, large doses are fatal; causes paralysis of the lower limbs, heart and central nervous system and death by asphyxia; affects the smooth and striated muscles, contracts the entire digestive tract, bladder, spleen, uterus and bronchial muscle; but the poisoning principle in the seed is destroyed by boiling and scorching; contains chymotrypsin inhibitors.

Clinical Studies: Used in medical research for treating chronic dermatosis, gastrointestinal disorders and *myasthenia gravis*.

Uses in Alternative Systems of Medicine

Homoeopathy:

Part used: Tincture of bean.

Uses: Useful in astigmatism, constipation, diarrhoea, dysentery, epilepsy, glaucoma, haemorrhoids, headache, hemiplegia, herpes, hysteria, gastralgia, corneal ulcers, inflammation, leucorrhoea, muscular atrophy, paralysis, sore throat, spinal sclerosis and tetanus.

Ethnomedicine

Plant: Is used as antidote for atropine and strychnine poisoining; used for arthritis, bursitis, colic, fibrositis, rheumatism, spasm and tetanus; used as sedative.

Non-medicinal Uses

Leaf: Yields a black dye.

Stem: Split into layers and stitched into mats used for drying cocoa.

Seed: Is used to kill mice.

130. *Piscidia piscipula* (L.) Sargent

Synonyms: *Erythrina piscipula* L.
Piscidia erythrina L.
Ichthyomethia piscipula (L.) Hitchc.

Common Names: Dogwood, Fish poison tree, Jamaica dogwood, Jamaica wood.

Distribution: Native of Caribbean islands, Central America and USA; introduced into India and cultivated.

Description: A large shrub or a small tree; leaves imparipinnate; leaflets 7-9; inflorescense is a much branched panicle; flowers deep pink borne on long stalk; pod with 4 longitudinal wings.

Phytochemistry

Root: Isoflavone, 8-prenylpinocembrin and rotenone.

Root bark: 2'-Deoxy piscerythrone, 6-prenylpiscerythrone and piscidone.

Stem bark: 5,7,2',4'-tetrahydroxy-5'-methoxy-3',6'-disoprenyl isoflavone, 5,7,4'-trihydroxy-3'-5-di isopenta-2-enyl isoflavone and rotenone.

Medicinal Uses

Root bark: Anodyne, sedative and uterine depressant; useful in dysmenorrhoea, insomnia, migraine and neuralgia.

Toxicology: Rotenone is toxic to fish.

Uses in Alternative Systems of Medicine

Homoeopathy:

Part used: Tincture of bark.

Uses: Useful in fever, sleeplessness and profuse sweating.

Ethnomedicine

Plant: Used as analgesic, anaesthetic, anodyne, narcotic, sedative, sialogogue and sudorific; useful in dysmenorrhoea, fracture, headache, neuralgia, spasm and tooth ache.

Non-medicinal Uses

Plant, Root and Leaf: Used as fish poison and insecticide.

Wood: Used in heavy construction and boat building; used as fuel; yields charcoal; is a good substitute for mahagony.

131. *Pisum sativum* L.

Synonyms: *Lathyrus oleraceus* Lam.

Pisum vulgare Jundz.

Notes: Two sub species, *sativum* and *elatius* are recognized; peas are the fourth most important pulse.

Common Names: Field pea, Garden pea, Green pea and Sugar pea.

Vernacular Names

Ben: Mara-vatara; **Hin:** Bahtaana, Baramattar, Battaanichola, Buttani, Golmattar, Kerava, Mattar; **Kan:** Bataani; **Mal:** Pattani, Pattaani payaru; **Mar:** Bataan, Vatane, Watana; **Sans:** Athivarthula, Shaman, Harenu, Kalaaya, Kanti, Khindaka, Mundhachandaka, Neelaka, Raenuka, Satila, Satilaka, Satinaka, Shoman Thriputa; **Tam:** Pattaani, Vellappattani; **Tel:** Bataani, Bataanilu, Gundu senigheui, Pataanilu.

Distribution: Native of Africa, Afghanistan, Iran, Europe and Middle East; introduced and cultivated throughout India.

Description: Annual herbaceous climber with tendrils and hollow stem; leaves pinnately compound; stipules auricled; flowers white, or light blue or purple pink; pod straight or curved; seeds smooth round, green, white, grey, 6–8 per pod, sometimes wrinkled.

Phytochemistry

Plant: Acetyl homoserine, aminobutyric acid, alanine, allantoin, aspartic acid, betaine, caffeic acid, choline, cobalt, coumaric acid, 4,4'-dihydroxy-2'-methoxychalcone, dihydromaleimide-β-D-glucoside, genistein, glutamic acid, histamine, homoserine, indole–3–acetic acid, indole–3-actaldehyde, kaempferol–3-glucoside, homoserine, acetyl homoserine, leucine, lutein, magnesium, manganese, mevalonic acid, obtustyrene, phosphorus, phylloquinone, potassium, proline, protein, quercetin-3-triglucoside, sulphur, trigonelline, valine and xenognosin.

Root: Betaine, choline, myo-inositol, ononitol, p-hydroxybenzoic acid, proline, syringic acid, trigonelline, tyramine and vanillic acid.

Leaf: Cadaverine, cyanidin–3-sophoroside–5–glucoside, kaempferol, kaempferol-3-coumaroyl-triglucoside, kaempferol-3-feruloyl-triglucoside, kaempferol-3-glucoside, lutein, putrescine, quercetin-3-coumaroyl-triglucoside, quercetin-3-feruloyl-triglucoside, spermidine, spermine, trigonelline and tyramine.

Flower: Cyanidin, delphinidin, delphinidin-3-glucoside, malvidin-3-rhamnoside, paeonidin and pelargonidin.

Pod: Alanine, α-linolenic acid, arginine, ascorbic acid, aspartic acid, β-carotene, caffeic acid, caffeic acid-4-O-β-D-glucoside, calcium, calcium oxalate, carbohydrate, choline, cystine, fat, glutamic acid, glycine, iron, isoleucine, jasmonic acid, leucine, linoleic acid, lysine, magnesium, methionine, myristic acid, niacin, oxalic acid, palmitoleic acid, pantothenic acid, L-pipecolic acid, phenylalanine, pisatin, pisatoside, potassium, proline, protein, riboflavin, saponin, serine, sodium, stearic acid, thiamine, threonine, trigonelline, tryptophan, tyrosine, valine and vitamin B_6.

Seed: Abscicic acid, alanine, α-amino adipic acid, α-amyrin, α-carotene, α-linolenic acid, aluminium, allantoin, anthocyanins, arachidic acid, argenine, arsenic, ascorbic acid, aspartic acid, β-amyrin, β-carotene, β-cryptoxanthin, β-sitosterol, biotin, boron, bromine, cadmium, calcium, campesterol, carbohydrates, cellulose, cephalin, cerebroside, chlorine, cholesterol, choline, chromium, cis-sinapic acid, citric acid, cobalt, copper, coumesterol, cryptoxanthin, cystine, daidzein, dehydroxy ascorbic acid, ferulic acid, fluorine, folacin, genistein, gentisic acid, gibberellin, glucosyl transferase, glutamic acid, glycine, glycolic acid, histidine, inositol, iodine, iron, isoflavonoids, isoleucine, kaempferol triglucoside, lecithin, legumin, leguminelin, leucine, linolenic acid, lysine, m-xylohydroquinone, magnesium, maleic acid, manganese, mercury, methionine, methyl-4-chloro-indoyl-3-acetate, molybdenum, myo-inositol, neo-β-carotene, niacin, nicotinic acid, nitrogen, o-coumaric acid, oleic acid, p-coumaric acid, p-hydroxy benzoic acid, palmtic acid, pantothenic acid, paragalactan, pectin, phenylalanine, phosphorus, phytic acid, phytopricipitin, pinitol, pisatin, pisatoside, pipecolic acid, potassium, protein, pyridoxine, quercetrin, quercetin triglucoside, quinic acid, riboflavin, rubidium, selenium, serine, silicon, sodium, sphingolipids, soyasaponin, , sphingenine, stachyose, starch, stearic acid, stigmasterol, succinic acid, sucrose, sugar, sulphur, syringic acid, thiamine, threonine, tocopherol, trans-sinapic acid, trigonelline, tryptophan, tyrosine, valine, vanillic acid, verbascose, vicilin and vitamin B-6.

Seed sprout: 5’-0-(6-0-malonyl-β-D-glucopyranosyl)–pyridoxine, 5’-0[(6-0-13-hydroxy-3-methyl-4-carboxy-buyanoyl) β-D-glucopyranosyl]pyridoxine, ascorbic acid, cycloartenol, hydroxy-homoarginine, isoxazolin-5-one-2-β-D-glucoside, homoserine, lupeol, mesoxalic acid, nicotinamide, nicotinic acid, pisatin, pyrroline, squalene and trigonelline.

Seed oil: Metaxylohydroquinone.

Wax: 14-Hentriacontanol and hentriacontane.

Edibility

Pod and Seed: Used as vegetable; eaten cooked.

Medicinal Uses

Seed: Astringent, depurative, emollient, laxative, nutritious, refrigerant, resolvent and sweet; applied as cataplasm on cuts, wounds and burning of the skin; cures bronchitis, cough, diarrhoea, leprosy, phlegm; seed extract is given to children with measles.

Seed flour: Emollient, ecbolic, fungistatic, resolvent and spermicidal.

Bioactivity: Plant is inhibitor of 5-α-reductase, 5-lipoxygenase, ACE, aldose-reductase, α-amylase, α-reductase, β-glucuronidase, COX-2, catechol-0-methyl transferase, DOPA-decarboxylase, FPTase, histamine, histidine-kinase, insulinase, iodothyronine-deiodinase, leukotriene, lipase, lipoxygenase, lyase, MAO, NADH-ubiquinone-oxidoreductase, ornithine-decarboxylase, PTK, peroxidase, phospholipase-A_2, polyamine synthesis, prostaglandin synthesis, protein-kinase–C, topoisomerase I and II, tyrosinase, tyrosine-kinase, xanthine-oxidase, CAMP and phosphodiesterase; is angiotensin receptor blocker, β-blocker, CNS active, calcium channel blocker, copper chelator, DNA protective, P_{21}-inducer, pituitary stimulant, quinone-reductase-inducer and cyclo-oxygenase activator.

Metaxylohydroquinone (MXHQ) in the seed oil showed anti vitamin E effect in immature female rats; injection of MXHQ (5mg) to pregnant rats in the first trimester caused foetal resorption of abortion or still birth; it also showed antiprogesterone effect; exhibited stimulatory effect on the responsiveness of the accessory genital organs to the homologous sex hormones; reduced spermatogenesis in male rats; depolymerised DNA and blocked the glycolytic cycle at G-6-PD state, thus showing anticancer effect; it produced CNS depression in mice. Methemoglobinemia induced by MXHQ was possibly a factor responsible for CNS depression; it caused hypertension in cats, dogs, rats and Guinea pigs and showed hypocholesterolemic effect in rats when fed with seed diet for 4 weeks; in studies on ovariectomized rats, the seed oil prevented the peripheral action of progesterone on the endometrium.

Toxicology: Toxic amino acids in the seeds cause neurolathyrism; pyrimidine glycosides in the seeds cause favism; isoflavonones cause haemolysis and inhibition of respiratory enzymes; phenolic glycosides cause goiter.

Uses in Alternative Systems of Medicine

Ayurveda:

Part used: Seed.

Uses: Cooling, contraceptive and laxative; useful in biliousness, bronchitis, burning sensation, flatulence and haemorrhage; improves appetite and purifies blood.

Chinese:

Part used: Seed.

Uses: Cooling; prescribed for fever, nausea and urinary disorders; promotes secretion of milk.

Siddha:

Part used: Plant.

Uses: Plant is antifertile in activity.

Medicinal Uses Outside India

China: Cooling; recommended for fever and to increase body tissue.

Ethnomedicine

Plant: Used as contraceptive, diuretic, emmollient, lactagogue, refrigerant and resolvent; used in dysentery, nausea, testicular tumour, urinogenital disorders and warts.

Seed: Used in diabetes.

Non-medicinal Uses

Plant: Is a valuable fodder.

Seed: Fed to livestock.

132. *Pongamia pinnata* (L.) Pierre

Synonyms: *Cytisus pinnatus* L.
Pongamia glabra Vent.
Derris indica (Lam.) Bennet

Notes: Seed oil is extensively studied as a source of biodiesel.

Common Names: Indian beech tree, Pongam, Pongam oil tree.

Vernacular Names: Ass: Karcha; **Ben:** Karanja, Dahur, Karana, Herran, Khawari; **Guj:** Karanja; **Hin:** Karanja, Karanji; **Kan:** Honge; **Mal:** Pungu, Punnu, Pongam, Ungu, Unnu; **Mar:** Ghanera, Karanji, Karanja, Keedamaar; **Sans:** Angaravalli, Badhapala, Chirabilva, Dhana, Gaura, Ghrethaparnaka, Gucchapushpaka, Hasthija, Hasthivaruni, Kakaghni, Kalimara, Karabhandika, Karanja, Karaanjika, Maathrey, Nakthaahva, Nakhtamaala, Prakriya, Puthika, Puthi karanja, Puthiparna, Puthipathraka, Rochana, Shadagrantha, Shamgashta, Snigda patra, Thapasvi, Udakriya, Uthundika, Vishari, Vritthaparni; **Tam:** Agirunanandam, Kaanjanam, Kolliyam, Naguttam, Nanandam, Nattamalam, Neerpungu, Pungam, Thattaippungu, Udagu; **Tel:** Kaagu, Kaadepathige, Kaanuga chettu, Kaggera, Kraanuga, Vyaaghranakamu.

Distribution: Native of Asia including India, Australasia and Pacific Ocean Islands; grows throughout India.

Description: A medium sized glabrous tree with a short bole and spreading crown; branches glabrous, spreading; bark soft, greyish green or brown and covered with tubercles; leaves imparipinnately compound; leaflets opposite, 5-9, ovate oblong, acute, shortly acuminate, glabrous, base acute or rounded, main nerves 6-8 pairs, prominent beneath; stipules small, oblong, obtuse, caducous; flowers lilac or white, tinged with pink or violet on axillary raceme with glabrous or puberulous rachis; bracteoles ovate, acute; calyx purplish brown, pubescent, truncate or obscurely toothed; corolla with suborbicular, broad, emarginated, standard petal with appendiculate base and short claw; pod oblong, or slightly obovoid, compressed, woody, narrow at the base and with a short decurved beak at the apex, glabrous, indehiscent, yellowish grey when ripe; seeds elliptical or reniform, wrinkled with reddish brown leathery testa; usually one seed per pod, rarely two.

Phytochemistry

Root: Demethoxy kanugin, kanugin, pongachromene and tetra-O-methyl fisetin.

Stem bark: β-Sitosterol, desmethoxykanujin, gamatin, kanujin, karanjin, lanceolatin β, ovalitenone, pinnatin, pongachromene, pongaflavone, pongaglabrone, pongapin and tetra-O-methylfisetin.

Heartwood: β-Sitosterol, demethoxykanugin, glabrachromene, lonchocarpin, pongachalcone I and II, pongachromene and pongaglabrone.

Leaf: 3-Methoxypongapin, glabrachalcone, glabrachromene II, karanjin and kanjone.

Flower: Aurantiamide, desmethoxykanugin, fisetin, friedelin, (–) isolonchocarpene, gamatin, glabrosaponin, 6-methoxy isopongaglabol, kaempferol, kanjone, kanugin, karanjin, lanceolatin β, neoglabrin, ovalichromene β, ovalitenone, pinnatin, pongaglabol, pongaglabrone, pongamin, pongamol, pongone, quercetin and γ-sistosterol.

Seed: Glabrin and glabrachromene II.

Seed shell: Desmethoxykanugin, isopongaflavone, kanjone, kanugin, karanjin, lanceolatin-B, pinnatin, pongamol and pongapin.

Seed oil: β-Sitosterol, 6-methoxyfurano (3',2',6,7)-flavone, 7-methoxyfurano (2',3',5,6)-flavone, 8-methoxyfurano(2',3',5,6)-flavone, desmethoxykanugin, glabrachalcone, isolonchocarpin, isopongaflavone, isopongachromene, kanjone, karanjachromene, karanjin, lanceolatin-β, linoleic acid, methyl fisetin, oleic acid, ovalitenone, pongaglabrone, pongamol, pongapin.

Edibility

Flower: Edible.

Medicinal Uses

Plant: Antibacterial, anticholinergic, antipyretic, CNS depressant, hypoglycaemic, hypotensive, sedative and wound healing; useful in abdominal enlargement, eczema, epilepsy, fracture, leprosy, scabies and ulcers; used as antidote for scorpion sting and snake bite.

Root: Anthelmintic, antiparasitic, antiseptic and wound healing; useful in ascites, biliousness, diarrhoea, gonorrhoea, itching, piles, spleen enlargement, tumours, urinary discharges and wounds; root paste with rice water is used as local application for scrotal and scrofulous enlargement; fresh root with human urine is ground and applied to the eyes and a small quantity is dropped into the nostrils in stupor and coma; root juice applied for eczema; root juice mixed with lemon juice is used as liniment for fistulous sores, foul ulcers, pityriasis, psoriasis and rheumatism; root juice with neem leaves is ground into a paste and applied on foul sores infested with worms; root juice with coconut milk, *Hydnocarpus* oil, lime water, camphor and sulphur is a external remedy for gonorrhoea; root juice is taken internally for curing urethritis.

Stem: Juice of the stem useful in diarrhoea, cough and gonorrhoea; with lemon juice it is used in rheumatism; young branches are used to clean teeth and to strengthen gums.

Bark: Alexipharmic, alexiteric, antibacterial and astringent; useful in ascites, beriberi, biliousness, dermatopathy, haemorrhoids, itching, malaria, ophthalmopathy, spleen enlargement, tumours, ulcers, vaginopathy, wounds and urinary discharges; fresh decoction is taken internally for piles; bark extract is antipyretic, antispasmodic, sedative and smooth muscle relaxant.

Leaf: Antifungal, antigonorrheoic, antiparasitic, antipyretic, antirheumatic, antispasmodic, anthelmintic, carminative, cholagogue, digestive, expectorant, galactagogue, laxative, nematicidal and styptic; used in treating dyspepsia, fever, flatulence, inflammation, liver affections, leprosy, piles and skin diseases; tender leaves are useful in bleeding piles; leaf juice is prescribed for cough, diarrhoea, dyspepsia, flatulence, gonorrhoea and leprosy; leaf juice has antiseptic and cleansing properties; hot infusion is good for cleansing sores, ulcers and wounds; leaf paste is used for whitlow; leaves of *Pongamia* and *Plumbago* mixed with pepper is applied for leprosy and with lemon juice it is used for rheumatism: broken rice and leaves, salt, *Pongamia, Morinda citrifolia* is fed to children to cure glandular enlargement of the abdomen; leaf and seed paste is used to arrest surgical bleeding, bursting wounds and leprosy; taken internally in abdominal gland, acute constipation, fever, intestinal absceses, jaundice, migraine, paralysis, piles and urinary diseases; reduces obesity; leaf poultice is applied to worm infested ulcers.

Flower: Useful in quenching excessive thirst.

Pod: Anthelmintic, febrifuge and tonic; cures bronchitis, eye problems, keratitis, piles, skin diseases and urinary discharges.

Seed: Acrid, anthelmintic, antiparasitic, blood purifier, carminative, expectorant and tonic; useful in anaemia, bronchitis, cough, ear complaints, fever, hydrocele, haemorrhoids, inflammation, otalgia, pectoral diseases and whooping cough; seed paste is applied for boils, burns, itch, eczema, leprotic wounds, piles, ringworm, rheumatic pain, scabies, septic wounds, urinary discharges and scorpion sting; the seed powder is febrifuge and tonic; useful in bronchitis and whooping cough; seeds with seeds of *Cassia tora* and root of *Aplotaxis auriculata* is ground with cow's urine and applied to eruptive skin diseases.

Seed sprout: Alexiteric, anthelmintic and stomachic; improves apptetite, cures "kapha", piles, skin diseases and inflammation.

Seed oil: Anthelmintic, antiparasitic, antiseptic, cholagogue, depurative, laxative with cleansing properties, stimulant and styptic; used in abdominal enlargement, baldness, cold, dyspepsia, cutaneous affections, eczema, eye diseases, fever, haemorrhages, haemorrhoids, herpes, intestinal worms, itch, leprosy, leucoderma, liver disorders, lumbago, paralysis, piles, ptyriasis, ringworm, scabies, stomach ache and ulcers; seed powder is used in asthma, bronchitis, ear ache, fever, herpes, hydrocele, leucoderma, lumbago, rheumatism, scabies, whooping cough and wounds; seed oil is applied for leucoderma, lumbago, pain in the liver, skin diseases, sores and ulcer.

Bioactivity: Plant extract exhibited hypoglycaemic effect in normal and alloxan diabetic rabbits; pongapin, kanugin and desmethoxykanugin isolated from the stem bark were devoid of anticonvulsant and analgesic activity; in higher doses, however, kanugin and desmethoxykanugin led to convulsions in rats; the refined oil (free from toxic flavonoids, bitterness and unpleasant odour) from seeds when given at 10 per cent level in a 20 per cent protein diet for 12 weeks in rats showed poor growth performance, altered lipid metabolism and fat infiltration in liver.

Toxicology: In an attempt to study the nature of toxic substance(s) in various portions of the karanj oil, its acidic, basic and neutral fractions were found toxic, the severity being more in the basic and neutral fractions than in the acid fraction.

Clinical Studies: Clinical trials conducted on patients suffering from herpetic lesions, impetigo, ringworm, pityriasis and eczema were treated with pongamia seed oil and root paste topically and remarkable healing effect was observed; pongamia seed oil was effective on herpetic lesions and seed paste on Shingles Zoster and impetigo; leaf paste was very useful in whitlow during clinical trials; herbal ointment containing aqueous extract of plant was effective against fungal, protozoal and vaginal infections.

Uses in Alternative Systems of Medicine

Ayurveda:

Part used: Leaf, pod and seed.

Uses: Used in epilepsy, dysuria, helminthiasis and musculoskeletal disorders, dropsy, fever, itching, leprosy, skin diseases and piles; it is also used in ascites, tumours, ulcers, vaginal diseases and wounds.

Siddha:

Part used: Root, leaf, flower and seed oil.

Uses: Useful in abdominal disorders, cough, eczema, glandular swellings, helminthiasis, pain, poisoning, skin eruptions and venereal diseases.

Tibetan:

Part used: Plant.

Uses: Used to increase digestive heat and to cure kidney disorders.

Unani:

Part used: Leaf, seed and seed oil.

Uses: Leaf poultice is applied to ringworm infection and ulcers; seed is anti-inflammatory, blood purifier, carminative and expectorant; seed paste is used as external application for leprous wounds, painful rheumatic joints and skin diseases; seed oil is used as massage oil in herpes, leucoderma, rheumatism, scabies, skin diseases and sores;

Medicinal Uses Outside India

Ceylon: Root juice is applied to sores for cleansing the teeth and strengthening the gums; seeds and root is ground with human urine and applied on the bitten part and small quantity is administered in stupor and coma.

Nepal: Seed paste is used as external application in herpes, rheumatism, scabies and skin diseases; pod skin is used in bleeding piles; leaf poultice is applied on worm infested sores and ulcers; root juice is applied for healing fistulous sores; taken with coconut milk and lemon, it is useful for gonorrohoea.

Polynesia and South China: Seed oil is applied on skin diseases and rheumatism; used as antiseptic and cleanser; young shoots used in rheumatism; decoction of root bark is abortifacient.

Ethnomedicine

Root: Used in fistula.

Bark: Is used in malaria, piles, skin diseases and wounds.

Leaf: Is used in skin diseases.

Flower: Is used in diabetes.

Seed: Is used in bronchitis, burn, ear complaints, joint pain, rheumatism, skin diseases and whooping cough.

Seed oil: Useful in cold, eczema, fever, itch and scabies.

Non-medicinal Uses

Plant: All plant parts are used as insecticide and piscicide.

Root: Used as fish poison.

Stem: Used as tooth brush.

Bark: Yields a coarse fibre used for cordage.

Wood: Used for making ploughs, combs, yokes, oil mills, cart wheels, rafters, turnery articles, plywood and packing wood; wood ash is used for dyeing.

Leaf: Used as manure and preservative for paper, books and stored grains; used as fish poison.

Seed: Used as fish poison.

Seed oil: Used as illuminant; used for leather dressing, in tanning industry; in making washing soap and candles; can be used as lubricant for heavy lathes, chains, gas engines and heavy machinery.

Seed cake: Used as manure and cattle feed.

133. *Pseudarthria viscida* (L.) Wight & Arn.

Synonyms: *Desmodium timoriense* DC.
Desmodium viscidum auct. non DC.

Vernacular Names: Guj: Chapakno; **Kan:** Antuparni; **Mal:** Moovila, Muvvila; **Sans:** Saaliparni, Salaparni, Sanaparni; **Tam:** Kodiottai, Moovilai, Nirmalli; **Tel:** Muyyakuponna, Naayaakuponna.

Distribution: Native of Asia including India; grows in Bihar, Gujarat and West Bengal in India.

Description: A perennial, viscid, pubescent, semi-erect under shrub, clothed with whitish hairs; leaves 3 foliolate; terminal leaflets rhomboid-ovate, lateral ones obliquely ovate, oblong, subrhomboid, hairy above, silky beneath; flowers purple to pink, smal!, numerous on terminal racemes, clothed with fine viscous hooked hairs; calyx hairy, teeth 5, the lower 3 teeth linear, subulate, upper 2 short and connate; pod linear, oblong, flat, covered with sticky hairs; seeds 4-6, brownish black and sub-reniform.

Medicinal Uses

Plant: Used in asthma, biliousness, diarrhoea, excessive heat, fever, heart disease, piles, rheumatism and worms.

Root: Anthelmintic, anti-inflammatory, aphrodisiac, astringent, bitter, cardiotonic, digestive, diuretic, emollient, febrifuge, nervine tonic, rejuvenating and thermogenic; useful in asthma, bronchitis, cough, diabetes, dyspepsia, emaciation, general debility, gout, haemorrhoids, hyperthermia, inflammation, neurasthenia, strangury and tuberculosis; decoction given in asthma, biliousness, cardiac problems, diarrhoea, excessive heat, fever, intestinal worms, piles and rheumatism.

Ethnomedicine

Plant: Is used as astringent, antibilious and vermifuge; used in asthma, fever, body heat, piles and rheumatism.

134. *Psophocarpus tetragonolobus* (L.) DC.

Synonym: *Dolichos tetragonolobus* L.

Common Names: Asparagus pea, Four angle bean, Goa bean, Manila bean, Princess pea, Winged bean.

Vernacular Names: Ben: Chara-konisem; **Hin:** Chaudhaari phali; **Kan:** Shambe kayi; **Tam:** Morisuavarii.

Distribution: Country of origin unknown, may be Asia; cultivated in India.

Description: A perennial twining, glabrous vine with tuberous roots and ridged stem; leaves trifoliolate; leaflets broadly ovate, acute, glaucous beneath; flowers 2-10, in axillary racemes; corolla large, exerted, with broad standard petal and much flexed auricle at base, emarginated, pale green on the back, white or pale blue within; pod with 4 longitudinal jagged ridges, 2-valved; seeds 4-16 per pod, globular, smooth, white, black, brown or yellow, with different markings.

Phytochemistry

Plant: Methoxypterocarpan, isopentenyl pterocarpan and phasseollidin.

Root: Alanine, α-linolenic acid, arginine, aspartic acid, calcium, carbohydrates, cystine, erucic acid, fat, glutamic acid, glycine, histidine, iron, isoleucine, leucine, linoleic acid, lysine, methionine, oleic acid, palmitic acid, phenylalanine, phosphorus, proline, protein, serine, stearic acid, threonine, tryptophan and valine.

Leaf: Alanine, α-linolenic acid, arginine, aspartic acid, calcium, carbohydrates, cystine, erucic acid, fat, glutamic acid, histidine, iron, leucine, linoleic acid, lysine, magnesium, methionine, myristic acid, oleic acid, palmitic acid, phenylalanine, phosphorus, potassium, proline, protein, serine, stearic acid, threonine, tryptophan, tyrosine, valine, vitamin A and C.

Pod: Ascorbic acid and β-carotene.

Seed: Alanine, α-linolenic acid, arginine, ascorbic acid, aspartic acid, β-carotene, calcium, carbohydrates, copper, cystine, erucic acid, fat, glutamic acid, glycine, iron, isoleucine, leucine, linoleic acid, lysine, magnesium, manganese, methionine, myristic acid, niacin, oleic acid, palmitic acid, phenylalanine, phosphorus, potassium, proline, riboflavin, serine, sodium, stearic acid, thiamine, threonine, tryptophan, tyrosine, valine, vitamin B_6 and zinc.

Edibility

Shoot: When fungus *Synchitrium psophocarpum* infects the young shoots, the shoot swells and swollen parts are steamed as a delicacy.

Leaf: Young leaves are steamed and eaten; also used in soups.

Flower: Fried flowers taste like mushroom; eaten alone or added to various dishes, salads and soups to colour them.

Pod: Tender pods are eaten raw or cooked.

Seed: Roasted and eaten.

Medicinal Uses

Plant: Is analgesic, anthelmintic, antialopecic, antiandrogenic, antianginal, antiarrhythmic, antibacterial, antiberiberi, anticancer, antihaemorrhagic, antihistaminic, anti-infertility, anti-inflammatory, antilithic, antineurotic, antinociceptive, antioxidant, antiperiodontitic, antiphotophobic, antiprostaglandin, antipyretic, antiradicular, antiscorbutic, antiseptic, antispasmodic, antitussive, antivertigo, antiviral, aphrodisiac, apoptotic, asthma-preventive, astringent, bronchosedative, cardioprotective, contraceptive, cytotoxic, diuretic, emetic, haemolytic, hepatoprotective, hypertensive, hypnotic, hypolipidemic, hypotensive, immunomodulator, laxative, neuroexcitant, pituitary stimulant, sedative, vasodilator, vasomotor stimulant and vulnerary; used in acne, allergy, anaemia, anorexia, anxiety, arteriosclerosis, arthritis, asthma, atherosclerosis, bach ache, canker, cataract, cervical dyspalsy, chilblain, cold, convulsions, dandruff, delirium, dermatitis, depression, diabetes, antidote to aluminium, cadmium and lead poisoning, menorrhagia, eczema, edema, encephalopathy, epilepsy, fatigue, gastritis, gingivitis, glaucoma, heart burn, hepatitis, herpes, hyperactivity, hyperglycaemia, impotency, insomnia, keratitis, leprosy, leukaemia, malaria, measles, menopausal complaints, migraine, nephritis, neuralgia, neuropathy, orchitis, obesity, osteoarthritis, osteoporosis, pthisis, plague, pneumonia, polio, psoriasis, retinopathy, rheumatism, stomatitis, stress, stroke, tumour of lung, tremor, ulcer and vitiligo.

Bioactivity: Plant is a inhibitor of 5-α-reductase, aldose-reductase, β-glucuronidase, phospholipase A_2, prostaglandin and xanthine-oxidase; it is angiotensin receptor blocker, β-adrenergic receptor blocker, β-blocker, CNS depressant, calcium channel blocker, copper antagonist, insulinogenic, interferon synergist, lymphocytogenic and monoamine precursor.

Ethnomedicine

Plant: Used for boils, conjunctivitis, smallpox and vertigo.

Non-medicinal Uses

Leaf: Used as fodder.

Seed: Yields fatty oil used for cooking and soap making.

Seed cake: Used as livestock feed.

135. *Pterocarpus dalbergioides* DC.

Synonyms: *Lingoum dalbergioides* (Roxb.) Pierre.
Pterocarpus indicus auct. non Willd.

Common Names: Burmese red wood.

Vernacular Names: Tam: Vengai; **Tel:** Yerravigisa.

Distribution: Native of Andaman Islands; endemic; introduced into mainland India.

Description: A very large semi deciduous, evergreen, buttressed tree; leaves imparipinnate; leaflets 5-9, ovate-lanceolate, gradually narrowing into a point, with prominent nerves beneath; flowers golden yellow on panicles; pod orbicular, flat, winged, with 1-2 seeds; seeds smooth and shining.

Phytochemistry

Plant: Carpusin, erythrodio-3-monoacetate, erythrodiol, eudesmol, marsupin, marsupol and pterostilbene.

Root: Garbenzol, hydroxybenzaldehyde, isoliquiritigenin, liquiritigenin, pseudobaptigenin and pterosupin.

Heart wood: Eudesmol, hydroxy flavanone, irisolidone, isoliquiritigenin, liquiritigenin, marsupin, p-hydroxy benzaldehyde, propterol, pterostilbene, pterosupin, retusin, santalin, sesquiterpene alcohol and yellow colouring matter-santal.

Bioactivity: A nondialysable polyphenolic compound isolated from bark inhibited plasmin esterolytic activity at 2.5 mg/ml and showed carcinostatic effect on mice bearing Escites Ehrlich carcinoma at a dose of 2mg/kg.

Non-medicinal Uses

Wood: Is used in making decorative furniture, articles, panelling, cabinets, railway carriages, bridges, turnery, staircase, flooring, counters, book cases, musical instruments, wardrobes, cots, gun carriages, ammunition boxes, boats, carts, door frames, beams and poles; it can be sliced or peeled for veneer work and attractive plywood sheets; earlier it was used in making aircraft propellers and air screws; Heart wood yields reddish yellow dye.

136. *Pterocarpus indicus* Willd.

Synonyms: *Pterocarpus indica* Willd.
Pterocarpus wallichii Wight & Arn.
Pterocarpus zollingeri Miq.

Notes: Wood has a scent similar to sandalwood; yields a red dye and kino gum.

Common Names: Blood wood, Malaya padauk, Sang dragon.

Vernacular Names: Kan: Honne; **Tam:** Vengai; **Tel:** Erravegisa, Gandhamriga punetturu, Seema gadda.

Distribution: Native of Asia, including India, Australasia and Pacific Ocean islands; grows in Eastern and Western peninsular India.

Description: A large buttressed tree up to 36m in height; leaves ovate, round; leaf rachis sparsely hairy; leaflets 5-13, subcoriaceous, greyish brown, sometimes greenish, slightly shiny, glabrous, sparsely hairy beneath; stipules linear, caducous; inflorescence on lax branched axillary panicles; flowers numerous; calyx hairy; corolla yellow to brown, densely hairy, wing petal membranaceous; pod woody, orbicular; seeds 1-2, smooth, dark brown.

Phytochemistry

Wood: α-Eudesmol, β-eudesmol, angolesin, formononetin, isoliquiritigenin, narrin, p-hydroxy hydrotropic acid, pterocarpol, pterofuran and santalin.

Bark: Gum kino, homopterocarpin and pterostilbene.

Edibility

Leaf and Flower: Edible.

Medicinal Uses

Plant: Decoction of the twigs is given in dropsy and stones in the bladder.

Bark: Antidiarrhoeal.

Kino: Decoction is taken to quench thirst and useful as a wash to heal sores and in diarrhoea.

Wood: Antidysenteric, antilithic and diuretic; decoction is given in dropsy, stones in the bladder and diarrhoea.

Leaf: Infusion taken for liver complaints; applied as lotion for bilious headache.

Seed: Is emetic.

Ethnomedicine

Plant: Emetic; used for abdominal tumours, bladder disorders, boils, diarrhoea, dropsy, headache, nephritis, prickly heat, sore, stomatitis, syphilis, thrush and urinary stones.

Non-medicinal Uses

Wood: Valuable as ornamental wood for decorative articles and furniture; used in railway coaches, ship saloon, cabinets, billiard tables, counter tables in offices; used in preparation of red dye used in staining light colored woods.

Leaf: Ground leaves used as hair wash.

137. *Pterocarpus marsupium* Roxb.

Notes: Yields Indian Kino gum; the timber is highly priced.

Common Names: East India kino, Indian kino, Malabar kino tree.

Vernacular Names: Hin: Aasan, Banda, Beejasal, Beeja, Beejasaar, Biya, Hiradokhi, Piasaal, Peisar, Pitshola; **Kan:** Bange, Benga mara, Bengada mara, Bibbla, Honne mara, Olle honne, Rakhta honne, Nettharu honne; **Mal:** Karintakara Malantakara; **Mar:** Asan, Asaaha, Bibla, Dhorbenla, Honi; **Sans:** Asana, Bandhukapushpa, Beejaka, Beejavriksha, Mahaakutaja, Neelaka, Peethasalaka, Peethasaara, Paramaayudha, Priyaka, Priyasalaka, Rakhtabeejaka; **Tam:** Asanam, Kaani, Kurinji, Pidasaralam, Pirasaaram, Pidogaragam, Sarvasaadagam, Sarudagam, Thamisam, Thannini, Thimil, Thimisu, Udiravengai, Vandunarmalar, Vengai, Visaga; **Tel:** Beeja saramu, Pedda vegisa, Pedegu, Vengisa, Yagisa, Yegi.

Distribution: Native of India and Sri Lanka and introduced into Taiwan and Madagascar; grows in Bihar, South India, Uttar Pradesh and Western Peninsula.

Description: Large deciduous tree growing to a height of 30m; bark thick, grey with vertical cracks; leaves imparipinnate, elliptic, ovate-oblong, obtuse or emarginate, coriaceous, glabrous above, pubescent beneath; leaflets 5-7, elliptic ovate, emarginate at the apex, glabrous, above, pubescent below; flowers yellow, fragrant; calyx white, silky pubescent; stamens monadelphous; style filiform; pod convex, pubescent, orbicular, flat, winged; seeds 2, bony, convex.

Phytochemistry

Plant: β-Eudesmol, garbenzol, carpusin, chalcone, irisolidol glucopyranoside, isoliquiritigenin, liquiritigenin, marsupinol, marsupol, marsupin, mucilage, non-glucosidal tannin, pentosan, propterol, protocatechuic acid, pseudobaptigenin, pterosupin and sesquiterpene alcohol.

Root: β-Eudesmol, β-hydroxy dihydrochalcone, retusin-8-O-α-L-arabinopyranoside, carpusin, chalcones, deoxykaempferol, erythrodiol-3-monoacetate, flavone glycosides, garbenzol, isoliquiritrigenin, kaempferol, liquiritrigenin, lupeol, naringinin, p-hydroxbenzaldehyde, pseudobaptigenin, pterostilbene, pterosupin and retusin.

Bark: Epicatechin, oleanolic acid, pterostilbene, tannic acid and trihydroxychalcone.

Heartwood: β-Eudesmol, alkaloids, aukrone glycosides, benzofuranone, carpusin, catechol, colouring matter, deoxykaempferol-glucoside,5,7-dihydroxy-6-methoxyisoflavone, epicatechin, erythrodiol, essential oil, fisetin-8-C-glucoside, fixed oil, gallic acid, hydroxyflavanone, irisolidone-7-rhamnoside, isoflavone glycoside, isoliquiritigenin, lactic acid, liquiritigenin, marsupin, marsupinol, marsupol, oleanolic acid, pectin, phenolic compounds, phlobaphene, propterol-A and B, protocatechuic acid, pterostilbene, resin, retusin-7-O-β-glucopyranoside, retusin-7-glucoside, tannins, terpenoids and trihydroxy flavone.

Leaf: Nitrogen, phosphoric acid and potash.

Flower: Aukrone glycosides.

Gum Kino: Gallic acid, kinotannic acid, pectin, protocatechol, pyrocatechin, pyrocatechuic acid and resin.

Edibility

Flower and Seed: Used as vegetable.

Medicinal Uses

Plant: Is alterative, anticancer, antifibrinolytic, antigonadotropic, anthelmintic, antihepatotoxic, antiherpetic, anti-inflammatory, anti-ischaemic, antimutagenic, antioxidant, astringent, hepatoprotective, hyperglycaemic, hypocholesterolemic, immunomodulator, nephrotoxic, phagocytotic, sedative and styptic; useful in "kapha", allergy, arrythmia, arthritis, asthma, bronchitis, burns, cancer of breast, colon, cholera, depression, diabetes, diarrhoea, dysentery, elephantiasis, erysipelas, eye trouble, fever, flu, haematuria, heart burn, hepatopathy, herpes, inflammation, leprosy, leucoderma, leukaemia, menorrhagia, mouth and skin diseases, nephritis, obesity, odontalgia, opthalmia, psoriasis, pyrosis, skin eruptions, spasms, stomach ache, syphilis, tooth ache, tumour of breast and ulcers; plant extract is used as wash for leucorrhoea and passive haemorrhages; it is applied topically on indolent ulcers; used as gargle for sore throat; topical application of the plant juice is a remedy for burns, boils, skin eruptions and ulcers.

Root bark: Astringent and hypoglycaemic; useful in diabetes, diarrhoea, dysentery and urinary complaints.

Heart wood: Acrid, alterant, anodyne, anthelmintic, antibiotic, anti-inflammatory, astringent, bitter, constipating, cooling, depurative, haemostatic, hypoglycaemic and rejuvenating; used in anaemia, asthma, body pain, bronchitis, bruises, cough, diarrhoea, disorders due to deranged fat, dysentery, elephantiasis, erysipelas, fractures, gout, grey hair, haemophilic disorders, haemorrhages, hypoglycaemia, leprosy, leucoderma, odontalgia, opthalmopathy, rectalgia, rectitis, rheumatoid arthritis, skin diseases, urethrorrhoea and verminosis; decoction is given to lower blood sugar; containers made from the wood are used to store water over night and the water which given to diabetic patients to drink early in the morning, is considered to act as antidiabetic agent; ointment made of wood is antifungal.

Kino gum: Is alterative, anthelminitic, antipyretic, astringent, laxative, liver tonic, styptic and vulnerary; useful in boils, dysentery, diarrhoea, elephantiasis, fever, erysipelas, gastralgia, haematuria, heart burn, haemorrhages, hepatopathy, helminthiasis, intermittent fever, leprosy, leucoderma, leucorrhoea, odontalgia, ophthalmia, psoriasis, vitiated conditions of "pitta", "kapha" and 'vatha", pyrosis, skin eruptions, tooth ache, urinary discharges, ulcers, wounds and urethrorhoea.

Leaf: Crushed leaves are applied to boils, sores and other skin diseases.

Flower: Appetizer, febrifuge, cooling and sweet; useful in anorexia, fever and vitiated conditions of "pitta".

Bioactivity: Plant is inhibitor of ACE, aldose-reductase, cyclo-oxygenase, MAO, topoisomerase and xanthine-oxidase; is CNS active, calcium antagonist, haemoglobin inducer, insulin sparing and quinone-reductase inducer.

Alcoholic extract of stem significantly lowered blood sugar and improved glucose tolerance in rabbits; extract of heart wood showed significant hypoglycaemic action in fasting rabbits, 3 to 5 hours after oral administration; no harmful effects were noticed.

Uses in Alternative Systems of Medicine

Ayurveda:

Part used: Heart wood, gum and leaf.

Uses: Used in erysipelas, leprosy, leucoderma, polyuria, worm infestation and wounds; aqueous extract of wood known as "Bijasar" is used in Ayurveda for diabetes; active principle being (–) epicatechin.

Siddha:

Part Used: Stem bark, leaf and gum.

Uses: Useful in cough, dysentery, dental diseases, diseases of "vatam", cough, leucorrhoea, skin diseases and vomiting.

Unani:

Part used: Leaf and gum.

Uses: Leaf paste is used in boils, sores and skin diseases; gum is used in diarrhoea, pyrosis and tooth ache.

Ethnomedicine

Plant: Astringent; used in asthma, boils, burns, chest pain, cholera, diabetes, diarrhoea, dysentery, fever, hyperglycaemia, leucorrhoea, skin diseases, sores, stomach ache, syphilis, tooth ache and urinary complaints.

Bark: Is used as tonic; useful in dysentery, tongue diseases and urinary complaints.

Wood: Used in asthma, body pain, chest pain and diabetes.

Resin: Is useful as cooling agent and in diarrhoea.

Leaf: Is used in boils, burns, skin diseases and sores.

Flower: Is febrifuge.

Non-medicinal Uses

Plant: Yields kino gum which is used in preparation of wine.

Bark: Is employed for dyeing; used in tanning, printing and papermaking.

Wood: Utilized in making pulp which is used for making wrapping paper, boats, bridges, carts, electrical poles, doors, window frames, furniture, rafters, bent beams, agricultural implements, railway wagons, shafts, axels, yokes, spokes, oars, toys, drums, picture frames, combs, felloes, mathematical instruments, textile looms, sports rifles, bridges, electric transmission poles, tool handles, sleepers and ships. building; is a good substitute for teak; used in railway carriages;

Kino gum: Is used in dyeing, tanning, printing and in paper industry.

Leaf: Is used as fodder; is a excellent manure for arecanut plants.

138. *Pterocarpus santalinus* L. f.

Common Names: Red sandalwood, Red sanders, Ruby wood and Sanders red.

Vernacular Names: Guj: Ratanjali; **Hin:** Lalchandana, Ragath chandan, Raktha chandan, Rukhtochandan, Undum; **Kan:** Agaru, Kempu honne, Rakthachandana, Kempugandha, Kempugandhada chakke, Kshudra chandana, Pathranga; **Mal:** Patrangam, Raktha shandanam, Senshendanam, Thilaparnni; **Mar:** Rakthachandan, Rathaanjili, Thambada chandana; **Sans:** Agarugandha, Bhaaskarapriya, Harichandana, Kumoda, Kushikam, Lohitha, Lohitha chandana, Malayaja, Pathranga, Pravaalphala, Rakthabeeja, Rakthachandana, Rakthaktha, Rakthanga, Rakthasaara, Ranjana, Thaamrabhra, Thaamrasara, Thaamravriksha, Thilaparni; **Tam:** Atti, Chandana vengai, Chenchandanam, Kusandanam, Picanam, Pidagattam, Salliyam, Segappusandanam, Semmaram, Sensandanakkattai, Sensandanam, Shandanam, Shandanavengai, Sivappuchandanam; **Tel:** Agarugandhamu, Erragandhamu, Kuchandanamu, Raktha chandanamu, Rakthagandhamu, Shandanamu.

Distribution: Native of India; grows in Andhra Pradesh, Karnataka and Tamil Nadu in India; endemic.

Description: Tree of 12mts height; leaves 3 foliolate; leaflets ovate, orbicular, round at both ends; flowers yellow; pod silky, beaked, narrowly winged; seeds reddish brown, smooth and leathery.

Phytochemistry

Plant: β-Amyrin, β-amyrone, β-sitosterol, erythrodiol, liquiritigenin and stigmasterol.

Shoot: Fresh shoot yields colouring matter and glucoside.

Stem bark: β-Amyrone, betulin, epilupeol, homopterocarpin, lupenone, lupeol, lupendeol, pterocarpin, santalic acid, sitosterol and triterpenoids.

Sapwood: Acetyloleanolic acid, acetyloleanolic aldehyde and erythrodiol.

Heart wood: β-Eudesmol, acetyl oleanaldehyde, acetyloleonolic acid, caffeic acid, carptriol, cedrol, cryptomeridiol, deoxysantalin, erythrodiol, eudesmol, gallic acid, homopterocarpin, isopterocarpin, isopterocarpolone, pterocarpodiolone, pterocarpin, pterocarpol, pterocarptriol, pterostilbene, quercetin, santalin A and B, and stigmasterol.

Leaf: β-Amyrin, betulin, epilupeol, erythrodiol, lupenone, lupeol, sitosterol and stigmasterol.

Edibility

Wood: Yields a dye, which is used as food colour.

Medicinal Uses

Plant: Is antiandrogenic, antibacterial, antiestrogenic, antihepatotoxic, anti-inflammatory, antilithic, antinociceptive, antipeptic, antiperoxidant, antipyretic, antispasmodic, antiviral, apoptotic, astringent, bronchodilator, CNS active, cancer preventive, cytotoxic, hepatoprotective, hypercholesterolemic, immunostimulant, nephrotoxic and sedative; it is used for anorexia, allergy, asthma, bronchitis, cancer of breast, depression, diabetes, edema, flu, herpes, hyperglycaemia, leukaemia, lymphoma, malaria, rheumatism, tumour of breast, cervix, lung and ulcer.

Heart wood: Acrid, alexiteric, anthelmintic, anti-inflammatory, aphrodisiac, bitter, cooling, depurative, diaphoretic, febrifuge, haemostatic, ophthalmic and tonic, useful in bilious affections, boils, burning sensation, cephalagia, diarrhoea, dysentery, dyspepsia, fever, general debility, haemoptysis, haemorrhage, haemorrhoids, headache, hemicrania, inflammation, leprosy, mental depression, odontalgia, opthalmia, skin diseases, ulcers, verminosis, vitiated conditions of "pitta" and vomiting; wood paste is applied on head in headache; used as an antidote for scorpion sting and snake bite.

Pod: Astringent and tonic; useful in chronic dysentery and urorrhagia.

Seed: Decoction is astringent and tonic; given in chronic dysentery.

Bioactivity: Plant is inhibitor of cyclo-oxygenase, FPTase, MAO, prostaglandin synthesis, topoisomerase – I and II and xanthine-oxidase; it is calcium antagonist, insulin sparing and haemoglobin synthesis inducer.

Toxicology: Excessive usage may be harmful to sexual vigour.

Uses in Alternative Systems of Medicine

Ayurveda:

Part used: Wood.

Uses: Antipyretic, alexiteric, anthelmintic, aphrodisiac, astringent, cooling, diaphoretic and tonic; used in abscesses, biliousness, boils, colic, edema, eye diseases, head ache, inflammation, leprosy, poisonous affections, skin diseases, thirst, ulcers, vomiting, "vatha" and "kapha".

Chinese:

Part used: Wood.

Uses: Considered as blood remedy; useful in bleeding ulcers, haemorrhages and wounds; useful in ascites, chest and abdominal pain, cholera and dysentery.

Siddha:

Part used: Heartwood.

Uses: Useful in diseases of "pittam", dropsy, dysentery, fever, herpes, piles, skin eruptions and venereal diseases.

Unani:

Part used: Bark, wood and seed.

Uses: Bark is anti-inflammatory, antiseptic, antipyretic, blood purifier; cures bilious affections, excessive perspiration, fever, headache due to heat and as astringent to bleeding bowels; bark decoction is gargled for sore throat, checks haemorrhage and heals wounds; seeds are used in burning micturition, dysentery, haemorrhages, hemicrania, tooth ache, wounds; wood is used as astringent and refrigerant.

Medicinal Uses Outside India

Indonesia: Wood scrapings soaked in rice water and the water is taken internally to reduce haemorrhages and is a specific remedy for poisoning.

Taiwan: Wood infusion is used to reduce swellings.

Ethnomedicine

Plant: Is antibilious, astringent, diuretic, emetic, suppurative and tonic; used for arrhythmia, diarrhoea, fever, haemoptysis, haemorrhage, haematuria, inflammation, leucorrhoea, rheumatism, skin diseases, sprue, stomach ache and yaws.

Non-medicinal Uses

Fresh shoot: Yields colouring matter used in dyeing; used as fodder.

Wood: Yields a red dye used in dyeing leather, paper pulp and in calico printing; a histological stain is prepared by alcoholic fraction of the wood; used for construction of cart shafts, yokes, spokes, felloes of wheels, plough, agricultural implements, ornamental boxes, picture frames and idols.

Leaf: Used as fodder.

139. *Pueraria montana* (Lour.) Merr. var., *chinensis* (Ohwi) Sanjappa and Pradeep

Synonyms: *Dolichos trilobus* Lour.
Pachyrhizus trilobus DC.
Pueraria chinensis Ohwi
Pueraria lobata (Willd) Ohwi ssp., *chinensis* Ohwi
Pueraria thomsonii Benth.

Distribution: Native of Bhutan and Hainan; introduced into India; grows in Arunachal Pradesh, Gujarath, Meghalaya, Manipur, Nagaland, Sikkim and West Bengal in India.

Description: Large climber with brown hairy deflexed branches; leaflets ovate, rhomboid, subcuneate or round at the base, acuminate; flowers bluish purple or red; calyx purplish green; pod flat, oblong, densely hairy; 8-12 seeded.

Phytochemistry

Root: Daidzein-4',7-diglucoside, daidzein-4,7-formononetin, kudzusapogenol A, B and C, cantoniensistriol, puerarin and sophoradiol.

Leaf: Irisolidone-7-glucoside and robinin.

Edibility

Stem and Leaf: Edible.

Medicinal Uses

Plant: Used as diaphoretic, febrifuge and lactagogue.

Root: Decoction is used in cold, dysentery and fever.

Uses in Alternative Systems of Medicine

Chinese:

Part used: Root.

Uses: Antipyretic and refrigerant; useful in reducing soreness and stiffness in the neck.

Medicinal Uses Outside India

Indochina, Japan and Korea: The roots are used as antifebrile, antiemetic, diaphoretic and refreshing; used in treating alcoholic intoxication, chicken pox, diarrhoea, dysentery, enteritis, fever, flatulence, gastritis, influenza and vomiting; flowers are given in dysentery and distended intestine; decoction of the stem without leaves and flowers is given as tonic and as a remedy for cough.

Ethnomedicine

Plant: Is antispasmodic, diaphoretic, emetic, febrifuge, lactagogue, refrigerant and styptic; useful in aphthae, boils, cold, deafness, dysentery, fever, flu, food poisoning, hypertension, measles, nausea, skin diseases, spasms, stiffness in the body, stomatitis, thirst, tumours and wound; is used as antidote for alcoholic intoxication and crotalaria poisoning.

Non-medicinal Uses

Plant: Is grown as cover crop and pasture; used as green manure; yields hay and silage; is an ideal crop to control soil erosion.

140. *Pueraria montana* (Lour.) Merr. var., *lobata* (Willd.) Sanjappa and Pradeep

Synonyms: *Dolichos lobatus* Willd.
Pueraria lobata (Willd.) Ohwi. ssp., *lobata* (Willd.) Ohwi.
Pueraria thunbergiana (Siebold & Zucc.) Benth.
Pueraria triloba (Houtt.) T.Makino

Common Name: Kudzu.

Vernacular Names: Hin: Mugani; **Sans:** Mudgaparni, Vanamudga; **Tam:** Narippayaru.

Distribution: Native of Bhutan and Hainan; introduced into India; grows in Meghalaya in India.

Description: Semi woody perennial vine with whitish spreading or reflexed brown hispid hairs; leaves compound, subovate to subrotund; leaflets hirsute on the upper surface, densely puberulent beneath, 2-3 lobed, terminal lobe rhomboid, abruptly acuminate, lateral lobes bifid; flower reddish purple on a short peduncled many flowered raceme; flowers grape scented; bracts linear, pilose, caducous; pod flat, dark brown, hispid with brown hairs; seeds 8-20.

Phytochemistry

Plant: (–) Glycinol, dihydroxybenzoic acid, dihydroxybenzoic acid-methyl ester, dihydroxyisoflavone, malonyldaidzein, malonylgenistein, prenyl daidzein, prenyl genistein, β-sitosterol, calcium, carotene, coumesterol, daidzein, daidzein-diglucoside, daidzin, genistein, genistin, kaempferol, linoleic acid, linolenic acid, lupiwighteone, magnesium, p-hydroxybenizoic acid, p-hydroxy benzoic acid methyl-ester, palmitic acid, phosphorus, potassium, protein, puerarin, stearic acid, stigmasterol and tectoridin.

Root: (+) Pinitol, (+)D-pinitol, (+)L-lactic acid, magnesium lactate, β-sitosterol, acetylcholine, alantoin, apigenin, arachidic acid, biochanin-A, calcium, carbohydrates, choline, coumarin, coumesterol, daidzein, daidzein-glucoside, daidzein-methyl ether, daidzin, daucosterol, dihydroxy isoflavone, dimethoxycoumarin, fat, fibre, flavones, fluoride, formononetin, genistein, genistein-glucoside, geranyl-dihydroxycoumestan, glucosyl-genistein, hydroxypuerarin, iron, isoflavones, isoliquiritigenin, kudzubutenolide, kudzusapogenol, kudzusapogenol-B-methylester, kudzusaponins, lupenone, malonyl daidzin, malonyl genistin, malonyl puerarin, methoxydaidzein, methoxypuerarin, methylhydantoin, methylpuerarin, phosphorus, protein, pueraria glycoside, puerarin, puerarin-apioside, puerarin-glucoside, puerarinxyloside, puerarol, pueroside A and B, quercetin, scoparone, sitosterol, sophoradiol, soyasapogenol, soyasaponin, starch, stigmasterol, succinic acid and xylosylpueratin.

Leaf: Daidzein, damascenone, genistein, genistin, glycinol, irisolidone, irisolidonexylosylglucoside, nicotiflorin, p-coumaric acid, phaseollidin, quercetin, tuberosin and wighteone.

Flower: β-Sitosterol, biochanin-A-7-O-glucoside, daidzein, daucosterol, delphinidin-3-diglucoside, delphinidin-3-O-β-D-glucoside, formononetin, genistein, glycitein, irisolidone, irisolidone-7-O-β-D-glucoside, irisolidone-xylosylglucoside, kaikasaponin, kakkalide, kakkanin, ononin, p-coumaric acid, quercetin, robinin, shekanin and tectorigenin.

Seed oil: (–)-Linalool, benzoic acid, benzyl alcohol, caproic acid, isovaleric acid, phenylethanol and propionic acid.

Edibility: The dried roots yield starch, which resembles cassava starch and is used in food; in China and Japan, flour called ko-fen is made from the roots, which is used in soups.

Medicinal Uses

Plant: Febrifuge and lactagogue; is used for bradycardia, haematuria, haemorrhoids, herpes, hydrophobia, hypertension, hyperglycaemia, insomnia, keratitis, leukaemia, lymphoma, malaria, melanoma, menorrhagia, migraine, nephritis, neuroblastoma, obesity, odontalgia, osteoporosis, osteosclerosis, pharyngitis, plague, polio, pruritis, psoriasias, purpura, retinopathy, spasms, stroke, sun burn, tachycardia, thrombosis, tumour of bladder, cervix, colon, lung, ovary, skin, ulcer and varicose veins.

Root: Decoction of the root given as febrifuge and in cold and dysentery.

Leaf: Antibilious, cooling, sedative and tonic; useful in fever; leaf poultice is applied in the form of paste to the eyes to improve eyesight and for haemorrhoids.

Pod: Used as an antidote for scorpion sting.

Bioactivity: Plant is inhibitor of 12-lipoxygenase, 5-α-reductase, 5-lipoxygenase, COMT, COX-I, catechol-0-methyl transferase, cyclohexogenase, cytochrome-P_{450}, DOPA-decarboxylase, glucosyl-transferase, histidine-kinase, lipase, lipoxygenase, MAO, metalloproteinase, NADH-oxidase, NADH-ubiquinone-oxidoreductase, NEP, ornithine-decarboxylase, P_{450}, PTK, peroxidase, phospholipase, polyamine-synthesis, prostaglandin synthesis, protein-kinase, topoisomerase I and II, tyrosinase and tyrosine-kinase; is CNS depressant, calcium channel blocker, β-binder, mast cell stabilizer, metal chelator and quinone-reductase inducer.

Non-medicinal Uses

Plant: Grown as cover crop and hay crop and for green manure; used as fish poison; yields crude fibre which is used in the manufacture of cordage.

141. *Pueraria phaseoloides* (Roxb.) Benth.

Synonym: *Dolichos phaseoloides* Roxb.

Notes: Three varieties are recognized: var., *jawanica*, *phaseoloides* and *subspicata.*

Commoin Name: Tropical kudzu.

Vernacular Name: Ass: Jermei-kyn-saw.

Distribution: Native of Asia including India, Andaman and Pacific Ocean islands; cultivated in Arunachal Pradesh, Assam, Manipur, Meghalaya, Nagaland, Sikkim and West Bengal in India.

Description: Annual twining herb; young stems covered with brown hairs; leaves pinnate, 3-foliolate; stipules basifixed, lanceolate, ciliate with long brown hairs; leaflets rhomboid to broadly ovate, lateral leaflets oblique, obtuse or acute; racemes long peduncled; calyx hairy outside; corolla purplish; pod linear, black, clothed with stiff hairs, 10-15 seeded; seeds black, transversely oblong.

Phytochemistry

Plant: Arabinosylapigenin, calcium, carbohydrates, fat, glucosylapigenin, glucosyl-xylosylapigenin, hexosylapigenin, hexosyl-pentosylapigenin and phosphorus.

Leaf: Genistein, hydroxygenistein and glycosylflavones.

Seed: Fat, gallic acid, hydroquinone diacetate, prodelphinidin, protein and protocatechuic acid.

Edibility

Root: Is used as vegetable.

Medicinal Uses

Plant: Is abortifacient, antiangiogenic, anticancer, antiestrogenic, antihaemolytic, anti-inflammatory, anti-ischaemic, antimitogenic, antioxidant, antispasmodic, antiviral, cancer preventive, cardioprotective, cytotoxic and immunostimulant; useful in atherosclerosis, boils, cancer of breast, inflammation, leukaemia, lymphoma, melanoma, neuroblastoma, osteoporosis, spasms, tumour of breast, ovary and stomach and ulcer.

Leaf: Used in tootache.

Bioactivity: Plant is inhibitor of aldose-reductase, α-reductase, aromatase, catechol-o-methyl transferase, DOPA-decarboxylase, histidine-kinase, lipase, MAO, MDR, NADH-ubiquinone-oxidoreductase, ornithine-decarboxylase, PTK, topoisomerase I and II and tyrosine-kinase; is calcium channel blocker, pituitary sensitizer and quinone-reductase inducer.

Non-medicinal Uses

Plant: Grown as cover crop; used as green manure and pasture plant; grown to control soil erosion.

Stem: Yields fibre useful in making ropes and twines.

142. *Pueraria tuberosa* Willd.

Synonm: *Hedysarum tuberosum* Willd.

Common Name: Tuberous honeysuckle.

Vernacular Names: Ben: Shimiabatraji; **Guj:** Vidaarikand; **Hin:** Badar, Bedarikand, Bhooyikooshamand, Bidarikand, Bilaikand, Billi, Bodar, Bolaikand, Dedarikand, Paathaalakand, Paathaalkohnada, Pona, Safed vidaari, Siaali, Sural, Surur, Thirra, Vidaarikand; **Kan:** Bhoochakra gedde, Bhoosakkare gedde, Gummadi balli, Gummadi gida; **Mal:** Paalmutukku, **Mar.:** Badra, Bharda, Bhendar vaela, Bhooyi kohlaa, Godha bael, Pithana, **Pun:** Silai; **Sans:** Bhukooshmandi, Bhumikooshamand, Gajavajipriya, Gajeshta, Gandhaphala, Ikshugandha, Kandapalash, Kroshtri, Ksheerashukla, Ksheeravalli, Kundapalasha, Payasvini, Shrigalika, Shukla, Seetha, Svadhukanda, Svadulatha, Thriparna, Vaajivallabha, Vidaali, Vidaari, Vadaarika, Vidaarikanda, Vrikshavalli, Vrishyakanda, Vrishyavallika, Vrishyavardhini, **Tam:** Nelapoosani; **Tel:** Dhaarigumadi, Kubaayathige, Vidhaarikanda.

Distribution: Native of India, Nepal and Pakistan; grows throughout India.

Description: A large twiner with tuberous root; leaves 3 foliolate; stipules ovate oblong, cordate; leaflets subcoriaceous, terminal leaflets broadly ovate, acuminate, equal sided, cuneate at the base, lateral leaflets inequilateral, ovate oblong, truncate at the base, glabrescent above, silky beneath; flowers on lax leafless raceme; bracteoles oblong, silky; calyx densely silky, teeth shorter than the tube, oblong, obtuse, ciliate, upper teeth connate; corolla blue, standard petal spurred, auricled at the base; pod membranous, flat, constricted between the seeds, clothed with long, silky, bristly brown hairs; seeds 3-6.

Phytochemistry

Root: β-Sitosterol, 3-O-methyl anhydrotuberosin, 4',6'-di-O-acetylpuerarin, amino acid, cantoniensistriol, daidzein, glycosides, 3-O-methylanhydrotuberosin, hydroxytuberosone, isoflavonepuerarone, coumestan, kudzusapogenols A & B, pterocarpone, puerarin, puerarone, puerarostan, sophoradiol, stigmasterol, tuberosin and tuberostan.

Edibility

Root and Leaf: Boiled and eaten.

Medicinal Uses

Plant: Useful in excessive lactation, menorrhagia, rheumatism, syphilis, tumours and ulcers.

Root: Alterant, aphrodisiac, cardiotonic, demulcent, diuretic, emetic, emollient, expectorant, febrifuge, galactagogue, lactagogue, laxative, refrigerant, rejuvenating, sweet and tonic; useful as cataplasm on swollen joints.

Uses in Alternative Systems of Medicine

Ayurveda:

Part used: Tuber and flower.

Uses: Alterative, aphrodisiac, cooling, demulcent, galactagogue, rejuvenator, restorative and tonic; used in biliousness, bronchial asthma, emaciation, fever, general debility, haemorrhage, leprosy, rheumatism and urinary discharges; overcomes vitiated "pitta" and "vatha"; increases breast milk and semen quantity.

Unani:

Part used: Tuber and flower.

Uses: Anthelmintic, cholagogue, diuretic, blood purifier, galactagogue and muscular tonic.

Ethnomedicine

Plant: Is lactagogue and tonic; used in arthritis, asthma, body ache, chest pain, cholera, diarrhoea, fever, gravel, menorrhagia, rheumatism, skin diseases, swelling, syphilis, tumours and ulcers.

Root: Used in asthma, body ache, chest pain, cholera, fever and rheumatism.

Seed: Is used in skin diseases and swellings.

Non-medicinal Uses

Plant: Grown as soil cover and for restoring soil fertility.

Root: Yields starch.

Leaf: Used as fodder.

143. *Rhynchosia minima* (L.) DC.

Synonyms: *Dolichos minimus* L.
Rhynchosia minima (L.) DC. var., *diminifolia* Walraven

Vernacular Names: Guj: Nahanikamalavel; **Kan:** Ghattavare; **Mar:** Dhaktaranghevada; **Tam:** Chittavari; **Tel:** Gadi chikkudu, Nellatumma.

Distribution: Native of Africa, Asia including India, Australasia, Caribbean islands, Central, North and South America and Middle East; grows throughout plains of India and Himalayas.

Description: Twining glabrescent herb; branches slender; stipules lanceolate; leaflets rhomboid, gland dotted beneath, acute; the end one is distinctly stalked; flowers yellow on 7-9 flowered raceme; bracts linear, minute; calyx pubescent, lower tooth setaceous; pod oblong, slightly bent upwards; sparsely pubescent; seeds 2.

Phytochemistry

Leaf: C-Glycosides of vitexin, isoorientin, isovitexin, lucenin, orientin, schaftoside and vicenin.

Seed: Gallic acid, hydroquinone, proanthocyanidins, prodelphinidin and protocatecheuic acid.

Medicinal Uses

Plant: Is analgesic, antiangiogenic, antiarrhythmic, antibacterial, anticancer, antifibrinolytic, antihepatotoxic, antihypertensive, anti-inflammatory, antioxidant, antiperoxidant, antiseptic, antispasmodic, antiviral, apoptotic, astringent, bronchodilator, cancer preventive, cardioprotective, choleretic, diuretic, estrogenic, hypotensive, immunostimulant, nephrotoxic, phagocytotic and vasodilator; useful in allergy, anxiety, asthma, atherosclerosis, bronchitis, cancer of breast, depression, diabetes, flu, herpes, insomnia, leukaemia, melanoma, neuroblastoma, obesity, osteoporosis, polio, tumour of breast, colon, mouth, ovary, prostrate, skin and stomach and ulcer; antidote to aluminium and lead poison.

Leaf: Used as abortifacient and in piles.

Bioactivity: Plant is inhibitor of aldose-reductase, α-reductase, aromatase, catechol-0-methyl transferase, cyclo-oxygenase, DOPA-decarboxylase, histidine-kinase, lipase, MAO, MDR, NADH-ubiquinone-oxidoreductase, aromatase, cyclo-oxygenase, glucosyl-transferase, iodothyronine-deiodinase, lipoxygenase, MAO, ornithine-decarboxylase, oxidase, topoisomerase, xanthine-oxidase and CAMP-phosphodiesterase; is angiotensin receptor blocker, calcium channel blocker, motor depressant and quinone-reductase inducer.

Ethnomedicine

Plant: Used as abortifacient.

Leaf: Used in asthma and piles.

Non-medicinal Uses

Plant: Is used as fodder and manure.

144. *Rhynchosia rufescens* (Willd.) DC.

Synonym: *Glycine rufescens* Willd.

Vernacular Names: Kan: Kaadu thogari, **Tam:** Kaatu kollu, Malaikollu, **Tel:** Chirvu lava.

Distribution: Native of Bangladesh, Cambodia, China, India, Indonesia, Java, Sri Lanka and Thailand; grows in Bihar, Meghalaya, Orissa and West Bengal in India.

Description: Trailing or twining, pubescent plant; leaflets elliptic-ovate, subacute, velvety; flowers yellow; raceme 5-7 flowered; calyx accrescent, lobes oblong, obtuse; seed single.

Phytochemistry

Plant: C-Glycosides of isoorientin, isovitexin, lucenin-2, orientin, pinitol and vitexin.

Leaf: Kaempferol-3-mythyl ester and quercetin-3-methylether.

Medicinal Uses

Plant: Is analgesic, antihypertensive, anti-inflammatory, antimicrobial, antioxidant, antiseptic, antispasmodic, antitussive, astringent, chemopreventive, choleretic, cytotoxic, diuretic, estrogenic, hepatotoxic, hypocholesterolemic, immune modulator, nephrotoxic, uterotrophic and vasodilator; useful in allergy, arrhythmia, atherosclerosis, bronchitis, cancer, depression, diabetes, ischaemia and uterine complaints; antidote to aluminium and lead poisoning.

Bioactivity: Plant is a inhibitor of ACE, aldose-reductase, α-reductase; is angiotensin-receptor blocker.

Non-medicinal Uses

Plant: Used as green manure and fodder.

145. *Robinia pseudoacacia* L.

Common Names: Black locust, False acacia.

Distribution: Native of Sicily in Europe, Canada and United States; introduced into India; grows in Jammu and Kashmir, Himachal Pradesh, Punjab and Tamil Nadu in India.

Description: Thorny medium sized tree with thick brown, longitudinally furrowed bark; leaves alternate, branches armed with setaceous stipules or stipular spines at the base of petioles; leaflets alternate, 7-23, elliptic, thin and ovate; flowers, fragrant, white, yellow, on axillary pendulous raceme; pod flat, shiny, smooth, narrow and brown; seeds reniform, dark brown.

Phytochemistry

Root: β-Sitosterol, lupenone and lupeol.

Root bark: Robinin.

Stem bark: Phasin, phytosterols, pyrocatechin, resin, robinin, robinetin, syringin, tannin and urease.

Wood: Butein, cellulose, dihydrorobinitin, 2,4'dihydroxy benzoic acid, leucorobinitinidine, lignin, liquiritrigenin, pentahydroxy chalcone, robinin, robinitin, trihydroxy chalcone and trihydroxyflavan.

Heart wood: Absicic acid, butein, dihydrorobinetin, liquiritigenin and robinitin.

Leaf: Acacetin, acacin, apigenin-7-trioside, apigeninglycoside, arachidic acid, aspargines, calcium, carbohydrates, carotene, coumarin, glucose, indican, L-aspargine, lauric acid, linarin, methoxyflavone, phosphorus, rhamnose, robinin, scopoletin, trioside, volatile oil and xylose .

Flower: β-Sitosterol, benzaldehyde, benzyl alcohol, caffeic acid, chlorogenic acid, heliotropine, kaempferol, linalool, kaempferol-L-rhamnopyranoside, terpineol, rhamnorobin, robinin, robinitin and umbelliferone.

Pod: Acacin, bioquercetin, biorobin and robinin.

Seed: Abscicic acid, arachidic acid, behenic acid, calcium, canavanine, carbohydrates, fat, lignoceric acid, linarin, linoleic acid, linolenic acid, oleic acid, palmitic acid, protein, raffinose, stachyose; dihydrorobinitin, robinin and robinitin.

Seed oil: Linoleic acid, linolenic acid, palmitic acid, stearic acid and sucrose.

Edibility

Leaf and flower: Used as flavouring and seasoning agent.

Seed: Edible after cooking.

Medicinal Uses

Plant: Is antiadenomic, antiandrogenic, antibacterial, anticancer, anticoronary, antifibrinolytic, antihepatotoxic, antihistaminic, antihypertensive, anti-inflammatory, antilymphocytic, antimutagenic, antioxidant, antiperiodontic, antipyretic, antiseptic, antispasmodic, antiviral, apoptotic, cardioprotective, diuretic, haemolytic, hepatoprotective, immunostimulant, laxative, memory enhancer, narcotic, sedative, tonic, vasodilator and vulnerary; is used in treating anorexia, allergy, alopecia, anxiety, arthritis, atherosclerosis, depression, diabetes, diarrhoea, gingivitis, herpes, hypertension, hypoglycaemia, insomnia, leukaemia, lymphoma, lupus, menorrhagia, nephritis, obesity, odontalgia, osteoporosis, paralysis, tumour of breast, cervix and lung, varicose veins and ulcer.

Root: Emetic and purgative.

Wood: Dihydrorobinitin from wood is antifungal.

Leaf: Antispasmodic, diuretic, emollient and laxative; leaf infusion is prescribed in digestive disorders.

Flower: Antispasmodic, diuretic and emollient.

Bioactivity: Plant is inhibitor of 5-α-reductase, 5-lipoxygenase, ACE, aldose-reductase, cyclo-oxygenase, glucosyl-transferase, iodothyronine-deiodinase, lipoxygenase, MAO, ornithine-decarboxylase, PAF, topoisomerase I and II, tyrosine, xanthine-oxidase, CAMP and phosphodiesterase; is angiotensin receptor blocker, β-blocker, calcium channel blocker, haemoglobin inducer, memory enhancer and quinone-reductase inducer; robinin showed anti-inflammatory activity in rats and rabbits at a dose of 10 g/kg.

Toxicology: Bark and young shoots are poisonous to live stock.

Uses in Alternative Systems of Medicine

Homoeopathy:

Part used: Tincture of fresh root bark, stem bark and seed.

Uses: Useful in acidity, constipation, convulsions, coryza, diarrhoea, dyspepsia, flatulence, gastritis, head ache, herpes, indigestion, leucorrhoea, nausea, neuralgia, nymphomania, urticaria and vertigo.

Ethnomedicine

Plant: Used as antispasmodic, astringent, cholagogue, diuretic, emetic, emollient, laxative, purgative and sedative.

Non-medicinal Uses

Plant: Grown for soil conservation and afforestation; used as fodder and timber.

Wood: Used in making agricultural implements, sails, tool handles, insulator pins, boat ribs, brackets, sleepers, fence post, gates, hubs, cart wheel, ship building, furniture, turnery, table tops and cabinets; used as fuel; yields strong dye used in dyeing cotton to brown or orange; yields pulp suitable for paper making.

Leaf and Flower: Essential oil extracted from flower is used in preparation of sherbets and aromatic waters.

Seed cake: Seed cake is used as animal feed.

146. *Rothia indica* (L.) Druce

Synonyms: *Rothia trifoliata* (Roth) Pers.
Trigonella indica L.

Vernacular Names: Tam: Nurrepitten keerai; **Tel:** Nuchakoora.

Distribution: Native of India, Indonesia, Laos, Sri Lanka, Vietnam and Australia; grows throughout India.

Description: Annual with digitate 3-foliolate leaves; leaflets obovate to oblanceolate, acute or obtuse, membranous; flowers small, solitary or in short racemes from the axil of the leaves; calyx teeth subequal, lanceolate, the 2 upper broad and falcate; corolla with long clawed petals, standard petal ovate or oblong, wing petal narrow; stamens monadelphous, filaments united; pod linear, slender, compressed, straight, many seeded; seeds reniform.

Phytochemistry

Plant: β-Sitosterol, ammodendrine, hentriacontanol, amyrin acetate, L-isolupanine, L-isoparteine, lupanine, lupeol acetate, nuttalline, p-coumaric acid, pseudobaptigenin, rothindin and soyasaponin I.

Edibility

Leaf: Is boiled and eaten as vegetable during famine.

Pod: Is used as vegetable.

Medicinal Uses

Plant: Useful in fever.

Ethnomedicine

Plant: Is used in fever.

Non-medicinal Uses

Plant: Is used as green manure; grown to reclaim saline and alkaline soils.

147. *Sesbania bispinosa* (Jacq.) W. Wight.

Synonyms: *Aeschynomene aculeata* Schreber
Aeschynomene bispinosa Jacq.
Sesbaina aculeata (Willd.) Pers.
Sesbania aegyptiaca sensu Bojer.

Notes: Widely introduced; native range uncertain; two varieties, *bispinosa* and *micrantha* are recognized; *Sesbania aculeata* (Willd.) Pers., is incorrectly treated as a separate species in literature.

Common Name: Prickly sesban.

Vernacular Names: Hin: Brihatchakramed, Dhunchi, **Kan:** Jayanthi, Vanajayanthi **Mal:** Kitannu, **Mar:** Bhuiavali, Chinchani, Kansevari, Ranshevari; **Sans:** Itakata, Jayanthi, Solo, **Tam:** Mudchembai, Nirchembai, Netti, **Tel:** Errajiluga.

Distribution: Native of Africa, Asia, including India and Pakistan, Indian ocean islands and introduced elsewhere; cultivated throughout the plains in India and in Western Himalaya.

Description: Annual or biennial shrub; stem erect, glabrous or sparsely pubescent when young; leaves paripinnate, rachis aculeate; leaflets oblong, obtuse, mucronate, sparsely pilose on margins and midrib below; stipules membranous, acuminate; inflorescence is a 3-12 flowered raceme and pedicellate; calyx membranous, margin pubescent; corolla pale yellow, red or violet flecked; pod glabrous erect, long, 35-40 seeded.

Phytochemistry

Plant: Richinaringinine, histidine and gum.

Gum: Galactose and mannose.

Edibility

Seed: Edible.

Medicinal Uses

Plant: Used in healing wounds.

Root: Alterative, alexiteric, anthelmintic, anti-inflammatory, bitter, carminative, diuretic and galactagogue; useful in biliousness, diabetes, eye diseases, fever, "kapha", leucoderma, tuberculous glands, throat trouble and ulcers; powdered root is used as emetic after snake bite.

Leaf: Anthelmintic, demulcent, maturant and purgative; useful in hydrocele.

Seed: Astringent, emmenogogue and stimulant; used in chronic ulcers, diarrhoea, diseases of spleen, excessive menstrual flow, ringworm, skin diseases and smallpox marks.

Uses in Alternative Systems of Medicine

Ayurveda:

Part used: Root and seed.

Uses: Root is alexiteric, anthelmintic, diuretic and galactagogue; seed is used in treating ringworm and other skin diseases.

Non-medicinal Uses

Plant: Used as green manure for rice, sugarcane, cotton and coconut; cultivated for fibre used in sails, fishing nets and ropes as it is durable in sea water and is superior in strength and durability to jute; used as fodder and grown to reclaim saline and alkaline soils.

Leaf: Used as forage.

148. *Sesbania cannabina* (Retz.) Pers.

Synonyms: *Aeschynomene cannabina* Retz.
Sesbania aculeata Pers. var., *cannabina* (Retz.) Baker
Sesbania cannabina (Retz.) Pers. var., *floribunda* Gagnepain

Vernacular Names: Guj: Sasi-ikad; **Hin:** Dhaincha; **Kan:** Mullujeerangi; **Mal:** Kitannu; **Mar:** Ranshevari; **Ori:** Dhaincha; **Sans:** Jayanthi; **Tam:** Mullagathi; **Tel:** Errajiluga.

Distribution: Native of Irian-Jaya in Asia and Australasia; introduced into India; naturalized in Gujarat and West Bengal in India.

Description: Glaucous erect subshrub, up to about 1m tall; leaves up to 30 cm long; leaflets 20-40 pairs, small oblong, about 1cm long and 3.5 mm wide, shortly petiolulate, opposite, greyish-green, blunt; flowers rather small, pale yellow; pod slender, curved, pendent, slightly constricted between the seeds; seeds numerous.

Edibility

Seed: Edible.

Non-medicinal Uses

Plant: Yields fiber, used for fishing nets, ropes and sails; grown as shade plant in tea and coffee plantations and as wind break and green manure crop for rice, wheat, jowar, tea, cotton and coconut; reclaimer for saline and alkaline soil; used as fodder.

Leaf: Used as fodder.

Stem: Pith from stem employed in making paper.

Seed: Yields gum which is similar to guar gum; seed meal is used in the medium for culturing micro organisms in laboratories.

149. *Sesbania grandiflora* (L.) Pers.

Synonyms: *Aeschynomene grandiflora* L.
Agati grandiflora (l.) Desv.

Common Names: Sesban, Swamp pea, Tree bean, West Indian sesban.

Vernacular Names: Ben: Agati, Baagphal; **Guj:** Agatoio; **Hin:** Agasthya, Basna, Hathiya, Vaasana, Hadaga; **Kan:** Agache, Agase, Chogache, Kempu agase, Vakrapushpi; **Mal:** Agatti, **Mar:** Agasthi, Chopechini, Haadga, Shevari; **Sans:** Agasthi, Agasthya, Dheerghaphalaka, Dheerghashimbi, Kanali, Kharadhvansi, Munidruma, Munipriya, Munipushpa, Munitharu, Pavithra, Rakthapushpa, Shighrapushpa, Shuklapushpa, Sthoolapushpa, Surapriya, Vaka, Vakrapushpa, Vangasena, Vranapaha, Vranaari; **Tam:** Acham, Agatthi, Kariram, Muni, Peragatthi, Sevvagatthi; **Tel:** Agise, Avise, Bakapushpam, Erragise, Sukanaasamu, Thellayavise.

Distribution: Native of Indonesia and introduced into India and other tropical countries; cultivated throughout India.

Description: Soft wooded tree of 20-30 feet high; leaves paripinnate, leaflets 30-60 pairs, glabrous; stipules minute, deciduous; flowers white or pink on an axillary raceme; calyx glabrous, 2-lipped, truncate; pod long, falcate with swollen, margins and thick sutures, valves constricted between the seeds; seeds 25-35.

Phytochemistry

Plant: All parts of the plant contain carotenoids, along with alanine, aspargine, aspartic acid, cyanidin-3-glucoside, cystine, galactose, glucuronic acid, grandiflorin, histidine, isoleucine, kaempferol-3,7-diglucoside, (+) leucocyanidin, oleanolic acid, phenylalanine, rhamnose, threonine, tryptophan and valine.

Wood: Gum and tannin.

Gum: Agathine, bassorine, resin, tannin and xanthoagathine.

Leaf: Aliphatic alcohol, galactose, glucuronic acid, oleanolic acid, pectin, rhamnose, sapogenin, saponins and vitamin C.

Flower: Kaempferol-3-rutinoside, oleanolic acid, and vitamin B.

Seed: (+) Leucocyanidin, cyanidin-3-glucoside, galactomannan, kaempferol-3,7-diglucoside, oleanolic acid and proteins.

Edibility

Tender leaf and Flower: Edible.

Pod: Edible; cooked and eaten as vegetable or raw as salad.

Medicinal Uses

Plant: Diuretic, hypothermic and spasmolytic; useful in anaemia, arthritis, epilepsy, fever and nicotine poisoning.

Root: Removes "vata", "kapha" and reduces abscesses, calculi, inflammation and tumour; the root paste of red flowered variety used as poultice reduces swelling and pain in rheumatism; the juice of the leaves is given with honey as an expectorant; the root paste along with root paste of *Stromonium* is applied for painful swellings.

Root bark: Anthelmintic, astringent, bitter, cooling, febrifuge and tonic; as a paste, it is useful as an external application for scabies and ulceration of tongue and taken internally is useful in ulceration of alimentary canal and tongue; bark juice or decoction is useful in dysentery, diarrhoea, dyspepsia, gastralgia and in small doses is aperient and emetic; bark infusion is given in first stage of malaria, eruptive fevers and small pox; decoction with salt and asafoetida relieves colic; the bark of the white flowered variety is made into a paste and administered internally as an antidote for snake bite.

Leaf: Acrid, aperient, bitter, cooling, diuretic, laxative and tonic; useful in abscesses, calculi, fever, headache, mumps and tumours; leaf juice is used in cephalagia, epilepsy, headache, nasal catarrh, night blindness and nyctalopia; blown up into the nostrils causes a copious discharge of fluid, thus relieving the pain and sense of weight in the frontal sinus; leaves chewed as an antiseptic and disinfectant for mouth and throat infections and is useful in fever, leprous eruptions and stomachalgia; leaf poultice is applied on bruises; and leaf juice dropped into the nostrils cures epilepsy.

Flower: Acrid, antipyretic, astringent, bitter, cooling, emollient, haemolytic and laxative; flower juice is used as eye wash, for vision disturbance and to improve eye sight; useful in cough, headache, nasal catarrh and sinus; taken orally it cures broncho-pneumonia and is an expectorant, and used as an "douche" for leucorrhoea; fresh juice and turmeric with warm water is given as urethral injection for gonorrhoea in males and leucorrhoea in females.

Pod: Alexiteric, bitter, cooling and laxative; useful in anaemia, colic, emaciation, flatulence, jaundice, poisoning, sinus, vitiated condition of *tridosha* and tumours; used as an appetizer, memory enhancer and nutritive tonic.

Seed: Is emmenogogue; used in abscess, calculi and tumour.

Bioactivity: Crude plant extract caused central nervous system depression in mice, hypotensive effects in cats, histamine like stimulation in Guinea pig ileum which was blocked by mepyramine shortened phenobarbitol sleeping time in white mice, protected against edema and had diuretic effects; aqueous extract of flowers caused haemolysis of human and sheep erythrocytes which is due to methyl ester of oleanolic acid.

Uses in Alternative Systems of Medicine

Ayurveda:

Part used: Root, bark, leaf, flower, pod and seed.

Uses: Root is astringent, antihistaminic and tonic; reduces "vatha", "kapha" and inflammation;bark is astringent, imparts body strength, cures intermittent fever, discoloration of the skin and diseases caused due to discordance of "pitta" and "kapha", promotes memory and is anthelmintic; it is useful in alcoholism, diabetes, epilepsy, fever, itching, gout, loss of body tissue, night blindness, ophthalmia, poisoning, spleenic and bile diseases, toxicosis, uterine disorders and wasting chronic diseases; leaf and flower are useful in body pain, boils, catarrh, coryza, cough, emaciation, epilepsy, fever, night blindness, tumour, worms and haemorrhage; flower is used as astringent and cooling; cures biliousness, bronchitis, catarrh, consumption, cough, fever, gout and night blindness; improves appetite; pod is alexiteric, laxative and memory enhancer; cures anaemia, bronchitis, consumption, fever and pain; reduces tumour growth and glandular enlargement; allays thirst and improves taste; seed is useful in diabetes, fever, poisoning and bile disorders; is used as uterine tonic and taken internally cures calculi, internal abscesses and migraine, hastens suppuration and heals ulcers; it also reduces obesity.

Siddha:

Part used: Root, bark, leaf and flower.

Uses: Astringent, cooling and tonic; cures chicken pox, eye diseases, headache, intermittent fever, sinusitis and wasting diseases.

Unani:

Part used: Leaf.

Uses: Leaf is antipyretic and tonic; cures biliousness and night blindness.

Medicinal Uses Outside India

Amboyna: Leaf poultice is a popular remedy for bruises.

Cambodia: Bark decoction is given in diarrhoea, dysentery and malaria; flowers are used as emollient and laxative.

Indochina: Bark is used as astringent, antipyretic and tonic and is a remedy for gastric trouble, colic in diarrhoea and dysentery.

Indonesia: Crushed leaves are used as resolvent; leaf juice as gargle cleanses the mouth and throat; bark in small doses is used to treat dysentery and sprue; in large doses it is used as laxative; in very large doses it is emetic; the pounded bark is applied in scabies.

Java: The bark is used for infantile disorders of the stomach and thrush; gum is used as an astringent.

Malay Peninsula: Crushed leaves are applied to bruises and sprains and associated pain; leaf juice is used as gargle to cleanse mouth and throat; bark is useful in dysentery and sprue; pounded bark is applied to scabies; in small doses bark decoction is laxative but in large doses it is an emetic.

Nepal: Bark is used as astringent and tonic; infusion is given in small pox; leaf is a remedy for headache and nasal catarrh; flower juice is a remedy for headache and nasal catarrh.

Philippines: Bark decoction is given for haemoptysis; flowers used as emollient and laxative; gum is used as astringent.

Tibet: The plant is recommended as a supreme emetic.

Ethnomedicine

Plant: Nutritive.

Leaf: Is used in headache and mumps.

Non-medicinal Uses

Plant: Grown as ornamental tree and as a support for betel vine, vanilla and pepper; used as posts to support the roof of houses as a substitute for bamboo; grown as shade plant for coconut seedlings and wind break in banana plantations.

Tender shoots: Are used as livestock feed; wood is white and soft; charcoal of the wood is used in gunpowder; valued for making toys and is a source of paper pulp.

Plant juice: Is used to make fishing nets tough.

Bark: Yields good fiber. bark is also used in tanning and colouring mats. Dried and powdered bark is used as a cosmetic in Java.

Red gum: Is used as a substitute for gum arabic

Leaf: Is useful as forage and fodder.

Flower: Is considered sacred and is offered to Lord Shiva in temples in the month of Karthik.

Seed: Is used as filler for casein and glue; as an adhesive for plywood.

150. *Sesbania sesban* (L.) Merr.

Synonyms: *Aeschynomene sesban* L.
Sesbania aegyptiaca Poiret
Sesbania pubescens sensu auct.

Notes: Two subspecies, *sesban* and *punctata* are recognized.

Vernacular Names: Hin: Dhandiain, Jait, Janjhan, Jayanthi, Jet, Jhijam, Jhijan, Jothara seena, Jouth, Rasin, **Kan:** Arisina jeenangi, Jayanthi, Jeenangi, Karee jeenangi; **Mal:** Kedangu, Kitannu, Nellittaali, Shempa, **Mar:** Jaanjana, Jayath, Joth, Sevari, Shewari; **Sans:** Janthi, Jaya, Jayanthi, Jayaya, Kaesharuha, Naadeyi, Raaja, Sookshma moola, Sookshma pathra, Vaijayanthi; **Tam:** Asnapanni, Chithagatthi, Chittakatti, Karumsembai, Karunchempai, Karunjembi, Sagudai, Sembai, Sirragatthi; **Tel:** Jalugu, Lingashimma, Mallasominta, Sominta.

Distribution: Native of Africa, Asia including India and Australia; grows throughout Himalayas.

Description: Tall shrub; leaves paripinnate; leaflets linear, oblong, entire, mucronate, glabrous, acuminate; flowers 3-20, yellow with purple streaks on axillary raceme; calyx glabrous, broadly triangular; pod slender, cylindrical, slightly curved with 10-40 seeds.

Phytochemistry

Plant: 3-O-α-L-rhamnopyranosyl(1'!3)-β-D-glucuronopyranosyl oleonolic acid, α-ketoglutaric acid, β-sitosterol, campesterol, cholesterol, galactomannan, lignoceric acid, linoleic acid, linolenic acid, oleic acid, oxaloacetic acid, palmitic acid, pyruvic acid and stearic acid.

Flower: Cyanidin, delphinidin, flavonols and glycosides.

Pod: β-Sitosterol, campesterol and cholesterol.

Seed: Stigmasta-5,24(28)-diene-3β-O-β-D-galactopyranoside, canavanine, fatty acids and galactomannan.

Edibility

Leaf, Flower and Seed: Used as vegetable.

Medicinal Uses

Plant: Useful in colic, goiter, pleurisy, pneumonia and urinary complaints.

Root: Paste is applied for scorpion sting as an antidote.

Root bark: Astringent; juice is taken internally for itch and other cutaneous eruptions.

Leaf: Poultice of leaves promotes suppuraton of boils and ulcers; useful in hydrocele, inflammatory and rheumatic swellings; juice is anthelmintic.

Flower: Boiled in gingelly oil and used to massage the body as stimulant and tonic.

Seed: Antibacterial, astringent, antidepressant, blood purifier, emmenogogue, hypoglycaemic and stimulant; useful in diarrhoea, excessive menstrual flow, itch, skin eruptions and spleen enlargement; used as an antidote for scorpion sting.

Toxicology: Chymotrypsin and trypsin inhibitors present in the plant.

Uses in Alternative Systems of Medicine

Ayurveda:

Part used: Root, bark, leaf, flower and seed.

Uses: Alterative, anthelmintic, bitter and carminative; useful in dysuria, intermittent fever, inflammations, leprosy, leucoderma, polyuria, scorpion sting, throat affections, tuberculous glands and ulcers.

Siddha:

Part used: Root, leaf, flower and seed.

Uses: Astringent; useful in coryza, diarrhoea, eczema, head ache, leucorrhoea, menorrhagia, skin diseases, ulcer and wounds.

Unani:

Part used: Leaf and seed.

Uses: Anthelmintic, astringent, demulcent, emmenogogue, maturant, purgative and stimulant; leaf is useful in chronic ulcers, diarrhoea, hydrocele, inflammation and skin eruptions; checks excess of menstrual flow.

Ethnomedicine

Plant: Is used as anodyne, astringent, emmenogogue, lactagogue, stimulant, suppurative and vermifuge; used for abscesses, catarrh, colic, coryza, dog bite, inflammation, itch, neuralgia, pleurisy, pain, pneumonia, rheumatism, skin diseases, urinogential disorders and tumour.

Leaf: Is used in ulcers.

Non-medicinal Uses

Plant: Is used as green manure and cover crop; grown as shade tree for coffee and as support plant for betel and pepper.

Stem: Pith of the stem is plaited into mats and used as fishing floats.

Bark: Yields fiber used for making ropes.

Wood: Is used as substitute for bamboo; used for roofing huts; making of toys; as fuel and charcoal for gunpowder.

Leaf: Used as fodder.

151. *Smithia conferta* Sm.

Synonyms: *Smithia conferta* Smith var., *geminiflora* (Roth) Cooke
Smithia geminiflora Roth
Smithia geminiflora Roth var., *conferta* Baker

Vernacular Names: Mal: Elakanni, Thiruthali; **Mar:** Naichibha; **Tam:** Elakanni.

Distribution: Native of China, India, Indonesia, Vietnam and Australasia; grows throughout India.

Description: Annual with long smooth branches; leaves small, abruptly pinnate; stipules large, acute with acuminate auricles; leaflets 2-8 pairs, linear, densely bristly on the margins and on the midrib beneath; flowers yellow, solitary or 2 in the axil of the upper leaves which are crowded at the ends of the branches to form a congested head; bracteoles with a long bristled apex; calyx teeth equal, oblong, acute with bristles on top and at the back; corolla yellow; pod 3-6 jointed, papillose.

Edibility

Leaf: Used as vegetable.

Medicinal Uses

Plant: Anti inflammatory, CNS active, diuretic, laxative, spasmolytic and tonic; useful in biliousness and rheumatism; removes wrinkles.

Ethnomedicine

Plant: Bilious, laxative and tonic; used for rheumatism and sterility.

152. *Smithia sensitiva* Aiton

Synonyms: *Aeschynomene tribuloides* Baillon
Smithia javanica Benth.

Vernacular Names: Ben: Nullakashina; **Hin:** Odabrini; **Mar:** Kaola; **Pun:** Odobrini.

Distribution: Native of Asia including India and Australia; grows throughout Himalayas.

Description: Dwarf semi woody, much branched herb; stiff and wiry branches with spreading hairs; leaves 3 foliolate; leaflets 5-12 pairs, oblong; calyx rigid, covered with spreading hairs, entire, acute; flowers yellow; pod is a lomentum with 6 joints, papillose, concealed by persistent bracts.

Edibility

Leaf: Edible.

Medicinal Uses

Plant: Boiled and given for micturition and gravel; as lotion, it is used for headache.

Ethnomedicine

Plant: Lactagogue and refrigerant; used for dysuria, gravel and headache.

Non-medicinal Uses

Plant: Good cattle feed and yields good hay.

153. *Sophora flavescens* Aiton

Synonyms: *Sophora angustifolia* Siebold & Zucc.
Sophora flavescens Aiton var., *angustifolia* (Siebold & Zucc.) Kitag.

Distribution: Native of China, Japan, Korea, Taiwan and USSR; introduced into India.

Description: Deciduous shrub 1.5 m high with annual shoots; leaves very variable in shape, pinnate, 15-25 cm long, with numerous ovate to oblong leaflets; lamina dull green with scattered hairs above, paler and more hairy beneath; flowers greenish-yellow, on terminal erect racemes; pod brown, narrowed between the seeds.

Phytochemistry

Plant: Anagyrine, baptifoline, methylcytisine, matrine, matrine-N-oxide and sofranol,

Root: (+) Kurarinone, anagyrine, formononetin, kosamol-A, kuraninol, kuraridin, kurarinone, kushenol, leachianone-A, matrine, methoxykurinone, methyl sophoranol, N-methylcytisine, norkuraninol, oxymatrine, sophocarpine and sophoraflavonone.

Pod: Rhamnose, rutin, sophoretin, sophoricoside and sophorabioside.

Medicinal Uses

Plant: Abortifacient, alexiteric, antidiarrhoeic, antiphlogistic, antiseptic, astringent, cancer preventive, carminative, cytotoxic, diuretic, stomachic and vermifuge; useful in hypercholesterolemia, hypolipidemia and ulcer.

Root: Used as stomachic; cures dysentery, jaundice and scrophula.

Bioactivity: Subcutaneous injection of matrine from roots produced intense convulsions and respiratory paralysis in rabbits; methonolic extract of the roots show cytotoxicity against cultured human tumour cell lines.

Toxicology: Ammothamnine and cytisine cause ataxia, nausea, respiratory arrest, diaphoresis and visual disturbance.

Uses in Alternative Systems of Medicine

Chinese:

Part used: Root.

Uses: Stomachic; used as an astringent in dysentery and enterorrhagia.

Ethnomedicine

Plant: Useful in cancer, diarrhoea, dysentery, enterorrhagia, enteritis, fever, haemorrhage, jaundice, metrorrhagia, piles, rash, scabies, scrofula and vaginitis.

154. *Sophora mollis* (Royle) Baker

Synonyms: *Edwardsia hortensis* Boiss. & Buhse
Edwardsia mollis Royle
Keyserlingia buxbaumii Boiss.
Keyserlingia hortensis (Boiss & Buhse) Yakovlev

Vernacular Name: Pun: Kohen, Malan.

Distribution: Native of Afghanistan, China, India, Iran, Nepal, Pakistan and USSR; grows in Jammu, Kashmir, Punjab and Himachal Pradesh in India.

Description: Erect deciduous shrub, young shoots finely pubescent; leaves stipulate, stipules triangular, acute, caducous; leaflets opposite or alternate, ovate or elliptic, entire; obtuse or emarginate, sparsely pilose on both sides; flower yellow, pedicellate, on axillary racemes; bract concave; calyx oblique, finely pilose; ovary densely hairy; pod moniliform, 5-10 jointed, 4 winged, glabrous when mature.

Phytochemistry

Plant: (–) Buptifoline, 5,6-dehydrolupanine, anagyrine, cytisine, formylcytisine, lupanine, N-methyl cytosine, rhombifoline and sparteine.

Leaf: Cytisine and pachycarpin.

Seed: Cytisine.

Medicinal Uses

Plant: Juice is used to wash sore eyes.

Root: Warm decoction is applied on the head for headache; promotes hair growth.

Seed: Powder mixed with oil is applied on the head to kill lice.

Ethnomedicine

Plant: Is used for alopecia and as antidote to poison.

Non-medicinal Uses

Plant: Is used as fodder; used to colour the hair black.

Wood: Is used as fuel.

155. *Sophora secundiflora* (Ortega) DC.

Synonyms: *Broussonetia secundiflora* Ortega
Calia erythrosperma Teran & Berland
Sophora sempervirens Engelm.
Sophora speciosa Benth.

Common Names: Coral bean, Mescal bean, Texas mountain bean.

Distribution: Native of Mexico and United States; introduced into India; cultivated in the gardens in India.

Description: An evergreen shrub or a small tree; leaves pinnately compound; leaflets 7-9, notched, glossy green on the upper side, silky below; flowers pea like violet blue, fragrant on a pendulous raceme; pod is woody with soft velvety coating with 1-5 bright red or orange seeds, which are toxic.

Phytochemistry

Plant: Cytisine, m-sparteine and isoflavone.

Leaf: 11-Oxycytisine, 13-hydroxysparteine, 3-methoxyquercetin, 5-deoxykaempferol, anagyrine, apigenin, argentine, baptifoline, cytisine, fisetin, kaempferol, kaempferol-3-O-rhamnoglucoside, kaempferol-3-O-rhamnoside, lupine, lupinine, N-acetylcytisine, N-formyl cytisine, N-methylcytisine, quercetin, rutin and sparteine.

Pod: Allylcytisine, lupanine and rhombifoline.

Seed : Cytisine.

Medicinal Uses

Plant: Is abortifacient, antiarrthymic, anti-inflammatory, antioxidant, antiperoxidant, antispasmodic, antitumour, antiulcer, cardioprotective, cathartic, diuretic, hallucinogenic, hypotensive, immunostimulant, laxative, osteogenic, oxytocic, psychoactive, respiratory depressant, toxic and vasodilator; useful in allergy, anxiety, atherosclerosis, bradycardia, depression, diabetes fatigue, hypertension, hypoglycaemia, insomnia, obesity, osteoporosis, spasm, tachycardia, tumour and ulcer; antidote to aluminium and lead poisoning.

Bioactivity: Plant is angiotensin receptor blocker and calcium channel blocker.

Toxicology: Seed is highly toxic due to cytisine in the seed; causes nausea, convulsions and death by asphyxia.

Ethnomedicine

Plant: Is used as intoxicant and narcotic; used for ear ache and as antidote to poison.

Non-medicinal Uses

Plant: Grown for its fragrant violet flowers.

Seed: Insecticidal.

156. *Sophora tomentosa* L.

Synonym: *Sophora fometosa* L.

Notes: Two subspecies, *tomentosa* and *occidentalis,* are recognized; widespread in tropics and subtropics.

Common Names: Seacoast laburnum, Silver bush.

Distribution: Native of Asia including India, Australasia, Caribbean Islands, Central America, Indian Ocean, United States, Pacific Ocean Islands and South America; grows in Andaman and Nicobar Islands in India.

Description: Shrub with imparipinnate leaves; leaflets opposite, 15-17, broadly elliptic; stipules lanceolate, deciduous; flowers on terminal racemes or panicles; bracts linear, caducous; calyx tube campanulate with short deltoid teeth; corolla yellow, exerted, petals clawed; style incurved; pod terete, grey, velvety downy, 3-8 jointed; seeds many, obovoid.

Phytochemistry

Plant: Benzofurans, epilamprolobine, formononetin, isobavachin, isosophoranol, isosophoranone, L-maackianin, quinolizidine-N-oxide, medicagol, methylene dioxybenzofuran, sophoracarpan, sophoro isoflavanone A and B, sophoronol, stearyl ester, trans caffeic acid and trihydroxy chalcone.

Root: Baptifoline, cystine, isosophoranone, oxymatrine, sophoraflavanone and sophoronol.

Leaf, Stem and Pod: Alkaloids, aminomethyl trans quinolizidine-N-oxide, ammodendrine, anagyrine, baptifoline, cytisine, epilamprolobine, epilamprolobine-N-oxide, matrine, matrine-N-oxide, N-acetyl derivatives and sophocarpine N-oxide.

Seed: Cytisine and poiosonous alkaloid sophorine.

Medicinal Uses

Root: Specific for biliousness and cholera.

Leaf: Cathartic and purgative.

Seed: Astringent, febrifuge, purgative and stomachic; useful in biliousness; ground seeds given in cholera and dysentery; used as an antidote to seafood poisoning; seed decoction cures biliousness.

Seed oil: Expectorant.

Ethnomedicine

Plant: Used in cholera, colic, diarrhoea, dropsy, dysentery, dysuria, enterosis, food poisoning, gonorrhoea, haematuria, haemoptysis, nausea and swelling.

Non-medicinal Uses

Plant: Is used as fish poison.

Root: Used as insecticide.

Wood: Yields timber used for making furniture.

157. *Spartium junceum* L.

Notes: "Absolute" (essential oil) is obtained from dried flowers, which is used in aromatherapy; produced mainly in France.

Common Names: Spanish broom, Weavers broom.

Distribution: Native of Africa, Asia including India and Pakistan, Europe and Middle East; grows in Northern India and Tamil Nadu.

Description: A shrub with sparingly leafy or almost leafless, cylindrical, striate glabrous branches; leaves alternate, simple, oblong linear to lanceolate, entire, glabrous above, appressed pubescent beneath, subsessile, glaucous; flowers on lax terminal many flowered raceme; calyx is split above, hence one lipped with 5 minute teeth; sweet scented, yellow, standard petal, showy, the lower edge of keel petal hairy with a small bract at the base and 2 bracteoles; the standard and keel petals are longer than the wing; stamens monadelphous; pod flat, strigose, pubescent; seeds black, 10-18, with a basal appendage.

Phytochemistry

Plant: Caffeic acid, calcium, carbohydrates, chrysin, chrysin-7-gentiobioside, chrysin-7-glucopyranoside, essential oil, fat, flavonoids, formononetin, genistein, isoquercetin, luteolin, methyl genistein, ononin, orientin, quercetin, rutin, spireoside and vitexin.

Leaf: Sparticarpin and hydroxygenistein.

Flower: β-Amyrin, anagyrine, caffeic acid, caprylic acid, chrysin, chrysin gentiobioside, chrysin glucoside, cytosine, cytosine-11-oxide, hexacosanediol, isolupanine, juncein, lupeol, luteolin glucoside, N-nonacosane, octadecanediol, orientin, quercetin, scoparin, sitosterol, sparteine and thermopsine.

Pod: Cystine and volatile alkaloids.

Seed: Linoleic acid, linolenic acid, myristic acid, oleic acid, palmitic acid, protein and stearic acid.

Medicinal Uses

Plant: Is abortifacient, analgesic, antiandrogenic, antibacterial, anticarcinogenic, antihepatotoxic, anti-inflammatory, antimitogenic, antimicrobial, antimyocarditic, antinociceptive, antioxidant, antiperoxidant, antiseptic, antispasmodic, antithrombic, cancer preventive, cardioprotective, cardiotonic, cathartic, choleretic, cytoprotective, diaphoretic, hallucinogenic, haemolytic, hepatocarcinogenic, hypouricemic, insulinogenic, laxative, neuroprotective, psychoactive, sedative, utero contractant, vasodilator and vulnerary; useful in allergy, anaemia, aloepecia, anxiety, atherosclerosis, arthritis, asthma, bradycardia, cancer of breast, cataract, colic, cystitis, depression, dermatitis, diabetes, eczema, edema, fibrosarcoma, flu, ganglion, gastritis, herpes, hepatoma, hydrophobia, hypertension, hyperthyroidism, hypourecemia, insomnia, leukaemia, lymphoma, malaria, melanoma, neuroblastoma, obesity, osteoporosis, pharyngitis, polio, psoriasis, rheumatism, stomatitis, sun burn, tachycardia, tumour of bladder, breast, colon, lung, ovary, prostate, skin, stomach and ulcer; antidote to aluminium and lead poison.

Flower and Seed: Diuretic, laxative and narcotic.

Bioactivity: Plant is an inhibitor of 5-α-reductase, 5-lipoxygenase, aldose-reductase, DOPA-decarboxylase, deiodinase, glucosyl-transferase, histidine-kinsae, lipase, lipoxygenase, lyase, MAO, MDR, MMP-9, metalloproteinase, NADH-oxidase, NADH-ubiquinone-oxidoreductase, NEP, NO-P_{450}, PTK, peroxidase, prostaglandin synthesis, phospholipase, protein-kinase-C, topoisomerase I and II, tyrosinase, tyrosine-kinase and CAMP-phosphodiesterase; is angiotensin receptor blocker, DNA protective, estrogen antagonist, mast cell stabilizer, pituitary sensitizer and quinone-reductase-inducer.

Toxicology: Sparteine, the active constituent in flowers in highly toxic.

Non-medicinal Uses

Plant: Yields fiber used in the manufacture of mats, baskets, canvas, ropes, paper, rubberized conveyer belts in mines, pillows, mattresses; is considered superior to cotton or flax as it does not rot or loose strength; dry stem yields high grade cellulose; grown as hedge plant.

Flower: Source of raw material for perfumery.

158. *Spatholobus parviflorus* (DC.) Kuntze

Synonyms: *Butea parviflora* DC.
Spatholobus roxburghii Benth.

Vernacular Names: Hin: Maula; **Kan:** Mukkate balli; **Mal:** Athampuvalli, Athambu; **Mar:** Phalsam; **Ori:** Porso; **Punj:** Mula; **Tam:** Plashivalli, Pilachivalli; **Tel:** Botegateega.

Distribution: Native of Asia including India; grows throughout India.

Description: Large ever green woody climber; leaves 3-foliolate; terminal leaflets broadly elliptic, ovate-obovoid, lateral leaf lets oblique, rhomboid, orbicular, with a round or cuneate base with short acuminate apex, glabrous above silky tomentose beneath; flowers creamy white on terminal tomentose panicles; calyx pubescent; petals subequal; pod erect, coriaceous, flat, wing like, brown and velvety.

Phytochemistry:

Root: Rotenone.

Stem: Gum and tannin.

Seed: β-Sitosterol and fiedelin.

Edibility

Seed: Is cooked and eaten.

Medicinal Uses

Bark: Decoction of the bark given in bowel complaints, dropsy and as vermifuge; used as an antidote to snake poison.

Gum: Is used in colic and hysteria.

Non-medicinal Uses

Bark: Yields a fibre used in making ropes.

Seed: Yields fatty oil used for cooking and for anointing.

159. *Styphnolobium japonicum* (L.) Schott

Synonym: *Sophora japonica* L.

Common Names: Chinese scholar tree, Japanese pagoda tree, Umbrella tree.

Distribution: Native of China, Japan and Korea; introduced into India; cultivated in gardens.

Description: A tall tree with spreading branches; leaflets ovate to ovate-lanceolate, acute, glossy above pubescent below; petiolulate; inflorescence a loose panicle; flower yellowish white; pod stalked, glabrous, terete.

Phytochemistry

Plant: (+) Medicarpin, β-sitosterol, 5,7-dihydroxy-3'methylene dioxyisoflavone, anhydropisatin, biochanin A, caffeic acid, cytisine, ferulic acid, genistein, irisolidine-glucoside, kaempferol, kaempferol-3-7-glycoside, p-coumaric acid, pisatin, pterocarpan, quercetin and quercetrin.

Root: Maackianin, rutin, sophojaponicin, sinapic acid, sopharol and sophocarpine.

Wood: Biochanin, biochanin-7-β-D-gentiobioside, irisolidine, pratensein and sissotrin.

Flower: Betulin, rutin and sophoradiol.

Pod: Arginine, aspargine, aspartic acid, cystine, genestein, genestein-4'-O-glucoside, genestein-4'-0-neohesperidoside, glutamic acid, glycine, isoleucine, kaempferol-L-rhamnoside, kaempferol-3,7-diglucoside, neohesperidoside, N-methyl cystine, proline, rutin, serine, sopharicoside, sophocarnine, sophorabiose, sophorabioside, sophoretin, sophoricoside, sophoraflavonoids, sophorin, stizolamine, tyrosine and valine.

Seed: Calcium, carbohydrates, cystine, fat, galactomannan, linolenic acid, linoleic acid, matrine, oleic acid, phosphorus, potassium, protein, rutin and stizolamine.

Medicinal Uses

Plant: Is abortifacient, analgesic, androgenic, antibacterial, anticancer, anticapillaryfragility, anticoronary, anticystitic, antifibrinolytic, antihaemolytic, antihepatotoxic, antihypercholesterolemic, antihyperthyroid, anti-inflammatory, antimitotic, antimyocarditic, antinociceptive, antiperiodontal, antipyretic, antiscorbutic, antiseptic, diaphoretic, diuretic, hepatoprotective, hypolipidemic, hypothermic, immunomodulator, laxative, neuroprotective, psychoactive, vasodilator and vulnerary; useful in anxiety, arthritis, arrhythmia, asthma, atherosclerosis, bradycardia, cataract, cephalagia, cold, colic, convulsions, dementia, depression, dermatitis, diabetes, dysmenorrhoea, eczema, edema, fatigue, fibrosarcoma, flu, gastritis, gingivitis, glaucoma, haematuria, haemorrhages, haemorrhoids, hepatitis, hepatoadenoma, herpes, hydrophobia, insomnia, ischaemia, leprosy, leukaemia, lymphoma, malaria, measles, melanoma, menorrhagia, migraine, nephritis, neuroblastoma, obesity, orchitis, osteoarthritis, osteoporosis, pharyngitis, plague, pneumonia, polio, psoriasis, purpura, tumour of bladder, skin, stomach, ulcer and varicose veins.

Flower: Antibacterial, antihaemorrhagic, astringent, haemostatic and styptic.

Pod: Antibaterial and vermicide.

Toxicology: Trypsin inhibitors present in the plant.

Uses in Alternative Systems of Medicine

Chinese:

Part used: Flower and pod.

Uses: Flower is astringent, anthelmintic and haemostatic; pod is astringent, haemostatic, styptic and tonic; useful in fever, haemorrhoids, profuse salivation, venereal sores and wounds.

Ethnomedicine:

Plant: Abortifacient, anodyne, astringent, haemostatic, purgative, styptic and tonic; used for conjunctivitis, convulsion, dysentery, epilepsy, epistasis, eye troubles, fever, gonorrhoea, haematemesis, hypertension, leprosy, menorrhagia, metrorrhagia, orchitis, purpura, scabies, skin diseases, tumour, venereal diseases, vertigo and wound.

Non-medicinal Uses

Plant: Grown as ornamental tree.

Wood: Used in making ornamental doors and pillars.

Leaf: Used as fodder.

Flower: Yields a yellow dye used in batik work.

Pod: Yields a yellow dye.

Seed: Used as livestock feed; the endosperm yields mucilage which is a substitute for carob mucilage.

Seed cake: Used as livestock feed.

160. *Sutherlandia frutescens* (L.) R. Br.

Common Names: Bladder senna, Cancer wort, Cape balloon pea.

Distribution: Native of Africa; introduced into India.

Description: Plant is an attractive, small, soft wooded shrublet; leaves pinnately compound; leaflets are 4-10 mm long, grey-green, giving the bush a silvery appearance; flowers orange-red on short racemes in the leaf axils at the tips of the branches; the wing petal small and concealed in the calyx and the standard petal is much shorter than the keel; pod large, bladder-like, papery, inflated and is almost transparent.

Medicinal Uses

Root: Powdered root in water is useful as eye wash in eye troubles.

Leaf: Decoction taken for back ache, dropsy, haemorrhoids, influenza, intestinal and uterine troubles, liver complaints and rheumatism.

Ethnomedicine

Plant: Febrifuge; used for back ache, cancer, cough, dropsy, eye troubles, flu, intestine and liver disorders, neuralgia, piles, rheumatism and stomach ache.

Non-medicinal Uses

Plant: Is used as a fodder.

Leaf: Is dried and smoked like tobacco.

161. *Tadehagi triquetrum* (L.) H. Ohashi

Synonyms: *Desmodium triquetrum* (L.) DC.
Hedysarum triquetrum L.
Pteroloma triquetrum (L) Benth.

Vernacular Names: Kan: Doddotte, Molada gida, Nariyale gida, **Mal:** Kaattarali, Sattukai, **Mar:** Ranabhaala, **Sans:** Srushtaparni.

Distribution: Native of Asia including India, Australasia, Indian Ocean and Pacific Ocean Islands; grows in Andaman and Nicobar Islands, Bihar, Karnataka, Meghalaya, Orissa, Punjab, Tamil Nadu and Uttar Pradesh in India.

Description: Erect under shrub with grooved triangular branches with appressed pubescence along the angles; leaves 1-foliolate, petiole winged; leaflets ovate, lanceolate, acuminate, truncate at base, pubescent along the nerves; 15-20 pairs of lateral nerves; raceme 50-80 flowered, with bluish purple flowers in the axil of stipule like bracts; calyx sparsely hairy, upper teeth deltoid, lower linear, incised more than half way; pod oblong with appressed pubescence and 6-8 joints.

Edibility

Root: Is used as masticatory.

Medicinal Uses

Plant: Is antiviral; extract used in fever and piles.

Root: Useful as antidote for snake bite.

Leaf: Extract used in abdominal pain, cold and piles.

Medicinal Uses Outside India

China: Plant is used in infantile spasms, as tonic for dyspepsia and as external application for abscesses and haemorrhoids.

Indonesia: Leaf infusion is used in gravel, lumbago and as astringent and diuretic.

Myanmar: Plant is used as anthelmintic; used in dyspepsia, infantile spasms and tonic.

Ethnomedicine

Plant: Used for abdominal disorders, fever and piles.

Root: As antidote to snake bite.

Non-medicinal Uses

Plant: Is useful as green manure.

Leaf: Is used as substitute for tea.

162. *Tephrosia candida* (Roxb.) DC.

Synonyms: *Cracca candida* (Roxb.) Kuntze
Robinia candida C.Presl.

Common Name: White tephrosia.

Vernacular Names: Hin: Kulthi, Lashtai, Masethi; **Kan:** Kaashi agasthe, Koggi gida, Sanna agasthe; **Sans:** Sharapunkha.

Distribution: Native of Asia including India; grows in India in Assam, Bihar, Guajrat, Himalaya, Karnataka, Manipur, Meghalaya, Rajasthan, Sikkim, Tamil Nadu and West Bengal in India.

Description: A low shrub with slender grooved branches with brown or grey, persistent velvety pubescence; leaflets 19-25, ligulate, acute, glabrous and green above, grey and silky below; flowers white; calyx-teeth shorter than the calyx tube; standard petal densely silky; pod recurved with brown silky hairs; style flat and silky.

Phytochemistry

Plant: 6-Hydroxykaempferol-4'-methylether, 5-methoxy-2',2'-dimethylpyranoflavone, boron, candidin, candidone, dehydrorotenone, flemichapparin B, iron, kaempferol-6-hydroxy-4-methoxy-3,7-dirhamnoside, manganese, molybdenum, nitrogen, obovatin-methyl ether, ovalichalcone, rotenoid, rotenone and zinc.

Root: Tephrosin.

Root bark: Rotenone and tephrosin.

Leaf: Candidone, dehydrorotenone, flemichapparin B and ovalichalcone.

Seed: Caffeic acid, candidin, candidol, dehydrodeguelin, pongachalcone, pongachin, rotenone, sitosterol, tephrone and tephrosin.

Medicinal Uses

Plant: Anticoronary, anti-inflammatory, antinociceptive, antitumour, antiviral, astringent, cytotoxic, estrogenic, immunostimulant and vulnerary; used in acne, alopecia, anaemia, anorexia, antidote to cadmium poisoning, arthritis, cancer of oesophagus, canker, cataract, cold, colic, convulsions, diabetes, eczema, epilepsy, herpes, impotency, leprosy, menorrhagia, obesity, osteoarthritis, osteoporosis, plague, rheumatism, stomatitis, scabies, tumour of oesophagus and ulcer; plant decoction is used to wash sores and wounds.

Bioactivity: Plant is ACE and topoisomerase inhibitor and copper antagonist.

Ethnomedicine

Leaf: Paste is applied on sores.

Non-medicinal Uses

Plant: Grown as ornamental plant; provides good manure and serves as shade plant and windbreak; used as insecticide.

Bark and Leaf: Used as fish poison.

163. *Tephrosia maxima* (L.) Pers.

Synonym: *Tephrosia purpurea* (L.) Pers., var., *maxima* (L.) Baker.

Distribution: Native of Tanzania, India and Srilanka; grows in the plains of India.

Description: A slender, usually prostrate undershrub; leaves imparipinnate; leaflets numerous, parallel-nerved, silky beneath; stipules free from the petiole, lanceolate or subulate; flowers pale pink on terminal or leaf opposed racemes; calyx teeth unequal, the lower teeth longer, upper two are connate; petals clawed, standard petal suborbicular, wing petal adherent to incurved keel petal; style incurved, bearded; pod linear, flat, obscurely septate; seeds many, reniform, elliptic.

Phytochemistry

Plant: Isoflavones, A, C, D, E and F, methylene and dioxyisoflavone

Root: Dimethoxyisoflavone and tetramethoxyisoflavone.

Pod: Purpuranin.

Seed: Rotenoids.

Bioactivity: Maxima isoflavones in the plant showed dose dependent, nonspecific, antispasmodic activity in Guinea pig ileum; maxima isoflavone G was the most potent.

Uses in Alternative Systems of Medicine

Siddha

Part used: Whole plant.

Uses: Useful in asthma and cough; used as diuretic.

Non-medicinal Use

Plant: Grown as shade plant and used as green manure; yields a dye.

164. *Tephrosia pumila* (Lam.) Pers.

Synonyms: *Tephrosia dichotoma* Desv.
Tephrosia procumbens sensu auct.
Tephrosia purpurea (L.) Pers., var., *pumila sensu* Cronq.

Notes: Pantropical; three varieties, *aldabrensis, ciliata* and *pumila,* are recognized.

Distribution: Native of Africa, Asia including India, Australasia, Madagascar, Seychelles and Middle East; Grows throughout India.

Description: Annual or short lived perennial; branches procumbent; stem pilose; leaves imparipinnate, petiolate; leaflets oblong or oblanceolate, truncate to retuse, glabrous or pubescent above, pilose below; stipulate; inflorescence terminal or leaf-opposed, 1-3 flowered raceme; bracteate; pedicillate; calyx hispid; corolla white, pale-pink or purplish; pod pubescent; curved towards the tip; seeds 8-14.

Phytochemistry

Root: β-Diketone, fisetin-ether, fisetin-7-ethylether, 7-4'-dihydroxy-3'-methoxyisoflavone, obovatin, precansone-A, and -B, pumila isoflavone A, B, rotenone, sumatrol and 7-ethoxy-3,3,4-trihydroxy-flavone.

Pod: Precansone and pumila isoflavone A, B, C and D.

Medicinal Uses

Plant: Used as stomachic.

Non-medicinal Uses

Plant: Insecticidal.

165. *Tephrosia purpurea* (L.) Pers.

Synonym: *Cracca purpurea* L.

Common Names: Purple tephrosia, Wild Indigo.

Vernacular Names: Ben: Bannilgacch, Sarphonka; **Guj:** Ghodakan; **Hin:** Sarphoka, Sarphanka; **Kan:** Adavineeli, Kogge, Koggli, Phanike, Vajradanthi, Vajra neeli; **Mal:** Kalinil, Korineel; **Mar:** Sharapunkha, Udhadi; **Ori:** Kolathiyapokha; **Sans:** Banapunkha, Ishupunkhika, Kalika, Krithi, Kolingi, Kollukai, Neelavriksha, Sarapunka; **Tam:** Kattu kolingi, Mollukhya, Kollukkayvelai, Paavali; **Tel:** Bonta vempali, Neela vempali, Pampara chettu, Vempali.

Distribution: Native of Asia, Australasia, Indian and Pacific Ocean Islands and Middle East; introduced into Africa and North America; grows throughout India in particular the coastal areas.

Description: Perennial under shrub, erect, glabrous or sparsely pubescent, ash coloured; leaf imparipinnate; leaflets 11-21, elliptic, oblong, truncate, mucronate, glabrous above, appressed silky below; stipules narrowly triangular; bracts present; inflorescence is a leaf opposed raceme; flowers reddish purple or bright pink; calyx pubescent; standard petal suborbicular, clawed, wing petal oblong, slightly adnate to the keel; stamens 9 + 1; style incurved, glabrescent; pod pubescent, with curved tip; seeds 6-8, smooth, greenish grey.

Phytochemistry

Plant: β-Sitosterol, albumin, anthocyanidin, anthocyanin A & B, delphinidin, gum, isolonchocarpin, karanjin, lanceolatin A, B & C, maackianin, manganese, pongamol, purpurin, purpurenone, rutin, spinosterol, tephrinose, ursolic acid, villinol, villol, villosin and villosinol.

Root: (–) Isolonchocarpin, (–) maackianin, (+) purpurin, β-hydroxychalcone, β-sitosterol, apollinine, 2-methoxy-3,9-dihydroxy coumestone, deguelin, dehydroisoderricin, 5,7-dimethoxy-8(2,3-epoxy-3-methyl butyl) flavanone, epoxyflavone, flavanone, flavonoids, flemichapparins B & C, isolonchocarpin, isotephrosin, kanjone, karanjin, lanceolatin A & B, maackianin, methyl karanjic acid, methyl pongamol, pongamol, purpurenone, purpurin, quercetin, rotenone, rutin, semiglabrin, semiglabrinol, spinasterol, tephrosin, ursolic acid and wallichin.

Leaf: β-Sitosterol, deguelin, lupeol, quercetin, rotenoid, rutin, tephrosin and wallichin.

Flower: Cyanidin chloride and delphinidin chloride.

Pod: Lanceolatin, n-triacontanol, purpurin A & B and tephrone.

Seed: Caffeic acid, deguelin, flavonone, furanoflavones, isolonchocarpin, kanjone, karanjin, lanceolatin A & B, linolenin, linolin, oleic acid, palmitic acid, pongamol, purpurin, purpuritenin A & B, quercetin, sitosterol, stearic acid and tephrosin.

Edibility

Root: Used to flavour milk.

Seed: Edible.

Medicinal Uses

Plant: Is anthelmintic, alterant, alexiteric, antihepatotoxic, anti-inflammatory, astringent, blood purifier, digestive, diuretic, depurative, deobstruent, laxative, styptic, thermogenic and tonic; is used in bilious febrile attacks, bronchitis, bleeding piles, boils, cough, enlargement of liver and spleen, kidney, liver and spleen obstructions, tightness in the chest; plant extract is useful in increasing the functional efficiency of liver; decoction is given in Bright's disease, dropsy and specific for nose bleed; the plant given with pepper is diuretic and is useful in gonorrhoea; the plant in combination with *Cannabis indica* leaves is a remedy for bleeding piles; dried and powdered, the plant is given for bronchitis, biliousness and cough; sprinkled on boils and pimples for healing.

Root: Is abortifacient, anthelminitic, diuretic, laxative and tonic; useful in abscesses, aphthae, boils, carbuncles, colic, diarrhoea, haemorrhoids, inflammations, liver complaints, pimples, stomach ache, tooth ache and ulcers; is an effective remedy for adenoids, anaemia, bronchitis, dysmenorrhoea, dysentery, elephantiasis, flatulence, gingivitis, hydrocele, hepatosplenomegaly, inflammed tonsil, odontalgia, scrofula, skin diseases, strangury, stomachalgia, tympanitis and verminosis; root powder is used as snuff for asthma and cough; dried root is smoked for relief from asthma and cough; root decoction is given in asthma, cough, dyspepsia, fever, kidney and liver disorder, rheumatism and as vermifuge and mouth wash; root with pepper is given in colic; liniment prepared from root is employed in elephantiasis; root ground in butter milk is given for hepatic dropsy; root powder with honey is applied to ulcers; external application of root paste is a remedy for hydrocele; root boiled in milk and is applied to leprous wounds; root paste with turmeric and cow's milk is applied to scrofulous glands; root is chewed in stomach pain and as antidote for bites of venomous animals.

Root bark: Is antigonorrhoeic, antisyphilitic, hypotensive and tonic; useful in bruises, dyspepsia, jaundice, lung and chest affections, pectoral diseases and piles.

Leaf: Is antigonorrhoeic, antisyphilitic and hypotensive; useful in bruises, chest complaints, chronic fever, dysmenorrhoea, dyspepsia, elephantiasis, flatulence, gingivitis, haemorrhoids, hepatosplenomegaly, hydrocele, inflammation of tonsils, jaundice, lung affections, odontalgia, pectoral diseases, scrofula, skin diseases, strangury, stomachalgia and verminosis; used in treating adenoids, anaemia, asthma, bronchitis, colic and diarrhoea; leaves with *Cannabis indica* leaves ground with curds is useful as poultice for bleeding piles; with black pepper it acts as diuretic and useful in gonorrhoea; dried leaves are smoked for relief from asthma and cough; leaf powder with rhizome of *Zingiber officinale* and *Curcuma longa* is taken for imparting strength and vigour to women after child birth; leaf juice is beneficial in dropsy, diabetes and mad dog bite.

Pod: Pod extract is a cure for asthma, diarrhoea, dyspepsia, inflammation, pain, rheumatism and urinary disorders; pod decoction is used as vermifuge and to stop vomiting.

Seed: Useful in rat poisoning, skin diseases and employed as anthelmintic; seed infusion is a cooling medicine and lowers blood glucose level.

Seed oil: Used as an external application in eczema, skin eruptions and scabies.

Uses in Alternative Systems of Medicine

Ayurveda:

Part used: Plant, root and seed.

Uses: Plant is anthelmintic, alexiteric, alterative, alexipharmic, antipyretic and digestive; cures asthma, bronchitis, dyspepsia, fever, leprosy, liver and spleen disorders, pain, piles, teeth carries and ulcers; root is anthelmintic, antipyretic and laxative; seed anthelminitic; used as antidote for rat bite.

Siddha:

Part used: Whole plant, root and seed.

Uses: Is anthelmintic and diuretic; employed in abdominal disorders, asthma, chronic diarrhoea, cough, dyspepsia, dyspnoea, fever, gingivitis, hepatomegaly, hydrocele, inflammation and enlargement of spleen and liver, leprosy, obstinate colic, skin diseases, splenomegaly, thirst, ulcers and wounds; used as antidote to snake poison; root and seed is anthelmintic and antidote for rat poisoning.

Unani:

Part used: Plant, root, leaf and seed.

Uses: Plant is alterative, anthelmintic, antidiabetic, anti-inflammatory, blood purifier, cholagogue, deobstruent, diuretic, emmenogogue, febrifuge, laxative and tonic; cures asthma, pimples, biliary fevers, bronchitis, cough, diseases of bladder and kidney, polypepsia, tympinitis and ulcers; root is digestive and exhilarant; is used in chronic diarrhoea, dyspepsia and leprotic fits; decoction is antidote to poisons especially that of *Strychnos nux-vomica;* paste is applied in breast tumours; root with pepper is used in obstinate colic; leaf useful in bleeding piles, gonorrhoea and syphilis; improves the appetite; useful in diseases of lungs and chest; seed used in diabetes.

Medicinal Uses Outside India

Ceylon: Root is employed as an anthelmitic.

Indochina: Root is deobstruent and emmenogogue.

Myanmar: Root administered as cardiac tonic and is stomachic.

Ethnomedicine

Plant: Is abortifacient, anodyne, colic, deobstruent, diuretic, laxative and tonic; useful in asthma, bronchitis, cough, diarrhoea, dyspepsia, eczema, elephantiasis, epistasis, fever, gonorrhoea, inflammation, itch, kidney and liver disorders, piles, rheumatism, scabies, sclerosis, skin diseases, ulcers, tumour and urinogenital complaints.

Root: Is used in aphthae, colic, dyspepsia, fever, impotency, piles, stomach ache and tooth ache; used as blood purifier and antidote to snake poison.

Non-medicinal Uses

Plant: Useful as green manure, fodder and cover crop.

Leaf: Used as manure and fodder; the leaves yield a colouring matter, which imparts excellent and comparatively fast colours used in dyeing industry.

Root, Pod and Seed: Insecticidal and piscicidal.

166. *Tephrosia sinapou* (Buc'hoz) A. Chev.

Synonyms: *Galega sinapou* Buc'hoz
Tephrosia emarginata Kunth
Tephrosia toxicaria (Sw.) Pers.

Common Name: Surinam poison.

Distribution: Native of Caribbean islands, Central America and South America; introduced into India.

Description: Erect perennial of 1m height; leaves alternate, 25-37 foliolate, oblong-elliptic, apex retuse to emarginate; inflorescence terminal; anthers 10, diadelphous, style barbed; stigma glabrous; pod with 8-13 black seeds.

Phytochemistry

Root: Contains highly toxic toxicarol.

Medicinal Use

Plant: Sap is used as purgative; applied for skin affections.

Ethnomedicine

Plant: Purgative; used for skin diseases.

Non-medicinal Uses

Plant: Used as fish poison.

Root: Insecticidal.

167. *Tephrosia spinosa* (L.f.) Pers.

Common Name: Spiny wild indigo.

Vernacular Names: Kan: Mullu kolinji, **Sans:** Kantapunkha, **Tam:** Ghekkvuri, Karchurai, Mulkolingi, **Tel:** Vempalimullu.

Distribution: Native of India, Java and Sri Lanka; grows throughout the plains in India.

Description: Rigid, spiny shrub; leaves imparipinnate; stipules spinose; leaflets 5-7, linear, obovate, cuneate, oblanceolate; flower red, 1-2, or 3 on axillary raceme; pod sickle shaped, 5-6 seeded.

Phytochemistry

Root: Prenylated chalcone.

Medicinal Uses

Plant: Analgesic and anthelmintic; used for swellings.

Root: Decoction is antidiarrhoeal, antirheumatic, febrifuge and stomachic; used to control excessive thirst and as local application, reduces swellings.

Root bark: Decoction is given in diarrhoea, excessive thirst, fever, indigestion, rheumatism and swellings.

Leaf: Antisyphilitic.

Uses in Alternative Systems of Medicine

Ayurveda:

Part used: Whole plant and root.

Uses: Useful in chronic ulcers, cough, dyspnoea, fever, liver disorders and pain.

Ethnomedicine:

Plant: Is used for diarrhoea, dyspepsia, fever, rheumatism and thirst.

168. *Tephrosia uniflora* Pers.

Synonym: *Tephrosia quartiniana* Cufod.

Vernacular Names: Guj: Bishoni; **Hin:** Bishonee; **Sans:** Kanthapunkha, Kantthaalu.

Distribution: Native of Africa, India and Middle East; grows in Gujarat, Karnataka, Maharashtra, Punjab and Rajasthan in India.

Description: Semi erect perennial, with pubescent stem covered with silky hairs; leaves imparipinnate; leaflets 5-9, oblanceolate, elliptic, obtuse or acute, glabrous or pubescent above, pilose below; flowers pink, 2-3, in the leaf axil; calyx pubescent; standard petal long and pubescent; pod pubescent, 7-8 seeded; seeds round and greenish yellow.

Medicinal Uses

Plant: Deobstruent, diuretic, laxative and tonic; used in biliousness, bleeding piles, boils, bronchitis, febrile affections and pimple.

Root: Decoction is given in asthma, diarrhoea, dyspepsia, elephantiasis, rheumatism and urinary disorders; with milk and pepper it is used for colic; pulverized and dried, roots are smoked for relief from asthma.

Leaf: Boiled leaves used in syphilis.

Pod: Decoction is used as antiemetic and vermifuge.

Seed oil: Specific against eczema, itch and scabies.

Ethnomedicine

Plant: Used in syphilis.

Non-medicinal Uses

Root and seed: Used as insecticidal and piscicidal.

169. *Tephrosia villosa* (L.) Pers.

Synonym: *Cracca villosa* L.

Notes: Two subspecies, *villosa* and *ehrenbergiana* are recognized.

Vernacular Names: Guj: Runchalisarpankho; **Kan:** Kaadu thogari; **Ori:** Srotobthiya; **Sans:** Sharapunkha, **Tam:** Punaikkaivettai, Punaikkaivezhai, **Tel:** Nuguvempali.

Distribution: Native of Africa, India, Indonesia, Pakistan, Srilanka, Vietnam, Madagascar, Saudi Arabia and Yemen; grows throughout the plains in India.

Description: Erect, tomentose, undershrub; branches pubescent; leaves imparipinnate; leaflets oblanceolate, obovate, obtuse, mucronate, sparsely villose above; flowers white or pink, on lax racemes; pod wooly and curved, seeds 6-8.

Phytochemistry

Plant: Deguelin, elliptone, methylglabranin, prenylated flavonone, rotenone, tephrinone, villinol, villol, villosin, villosinol, villosol and villosone.

Root: Coumestone, dehydrorotenone, flavonine, rotenone, rotenoids, stigmasterol, tephrinine, tephrinone and tephrosol.

Pod: Rotenoid, villinol, villosone, villol, villosinol and villosol.

Seed: Prenylated dehydro rotenoids.

Medicinal Uses

Plant: Used in cold and cough.

Root: Used in hypoglycaemia.

Leaf: Juice is given in diabetes, dropsy and hypoglycaemia.

Uses in Alternative Systems of Medicine

Ayurveda:

Part used: Whole plant and Root.

Uses: Useful in cough, dyspnoea, fever, liver disorders, pain, poisoning, spleenic disorders and ulcer.

Ethnomedicine

Plant: Used for cough, dropsy and hyperglycemia.

Root: Is useful in scrotal pain.

Non-medicinal Use

Plant: Provides good manure.

170. *Tephrosia vogelii* Hook. f.

Synonym: *Cracca vogelii* (Hook. f.) Kuntze

Common Names: Fish poison bean, Fish bean, Vogel tephrosia.

Distribution: Native of Africa, Asia including India and introduced into Australia, Central and South America, Caribbean and Indian Ocean islands; in India it grows in Tamil Nadu.

Description: A soft woody herb or shrub; branches tomentose with white or fulvous hairs; leaves compound; leaflets 11-29, elliptic oblong to elliptic lanceolate, silky tomentose above, densely tomentose beneath; stipules lanceolate, tomentose; flowers white or purple on dense, terminal or axillary pseudoraceme; bracts ovate, acuminate; calyx brown, tomentose, 2 upper lobes oblong and united, lateral lobes oblong, free, lower lobe narrow, acute; standard petal white, silky, keel petal pubescent at the margins; style pubescent, incurved at the tip; pod straight, rusty tomentose with slightly curved beak; seeds 10-20, black, smooth, obliquely transverse with a well marked annular aril.

Phytochemistry

Plant: Rotenoid, vogeletin and vogeloside.

Root: 5-Methoxyisolonchocharpine, deguelin, pongamol, rotenone and tephrosin.

Leaf: Deguelin, dehydroxydeguelin, isoquercetrin, methoxyflavone, quercertin-3-arabinoside, rotenoids, rotenone, rutin, tephrosin and tephrosol.

Seed: Deguelin, dehydroxydeguelin, fatty oil, protein, rotenone, rotenoid, tephrosin, vogeletin and vogeloside.

Pod: Deguelin and tephrosin.

Edibility

Plant: Eaten as pot herb.

Medicinal Uses

Root: Used as purgative and paste is used to dress wounds and sores.

Leaf: Used as abortifacient; used in skin diseases, sores and wounds; leaf infusion is emetic and purgative.

Bark and Pod: Decoction is used as abortifacient and for skin diseases.

Ethnomedicine

Plant: Abortitacient, emetic and purgative; used for skin diseases and sores.

Non-medicinal Uses

Plant: Used as green manure and fodder; grown as wind break and fence plant; used as insecticide and piscicide.

Root: Used as fish poison.

Leaf, Pod and Seed: Insecticidal and piscicidal.

171. *Teramnus labialis* (L.f.) Sprengel

Synonym: *Glycine labialis* L.f.

Notes: Extremely variable; two subspecies *labialis* and *arabicus,* the former with 5 varieties are recognized. Sanjappa, 1992 recorded *Teramnus mollis* (*T. labialis*) var., *mollis* (Wight and Arnold) (Baker) from Myanmar.

Vernacular Names: Hin: Banurdee, Maasparnee, Mashonee, Mashvan, **Kan:** Adavi uddu, Soya beeja, Vilaayithi uddu, **Mal:** Kaattu lunnu, **Mar:** Ram-udida, **Sans:** Ardramaasha, Ashvapuchhi, Athmodbhava, Bahuphala, Ghana, Hansamaasha, Hayapuchhi, Hayapuchhika, Kalyaani, Kaamboji, Krishnavantha, Krishnavrintha, Mahaasahaa,

Mangalya, Mansamaasha, Maashaaparni, Maashaaparnika, Maashaakhyaparni, Panduloma, Pandulomaasha, Pandulomaashaparni, Pandura, Parnini, Shaliparni, Sinhamukhi, Sinhapuchhi, Sinhapuchhika, Sinhavina, Sooryaparni, Sulabha, Surpaparni, Svayambhu, Thrashiprokta, Vajramooli, Vishambika, **Tam:** Kaatu uzhundu, **Tel:** Karu minumullu.

Distribution: Native of Africa, Asia including India, Australasia, Indian Ocean Islands, Middle East and South America; grows throughout Western Himalayas to South of India.

Description: A twining herb with hairy, slender stem; leaves 3-foliolate; stipules lanceolate and acute; leaflets membranous, oblong, lanceolate, subacute, apiculate, glabrous, with round base; flowers solitary, red; bracts linear, lanceolate; calyx silky, teeth lanceolate, as long as the calyx tube; corolla red; pod linear, straight or slightly incurved, glabrous with a short stout beak bent upwards; seeds 8-12, oblong, truncate, smooth and dark brown.

Edibility

Leaf: Used as vegetable.

Medicinal Uses

Plant: Astringent, bitter, cooling, febrifuge and stomachic; used in catarrh, consumption, disorders of blood and urine, fever, haemoptysis, nervous affections, rheumatism and tuberculosis.

Leaf: Young leaves are chewed for relief from tooth ache.

Pod: Antibilious, antipyretic, aphrodisiac, astringent, stomachic and tonic; used in bronchitis, gout, haemoptysis, nervous diseases, paralysis, rheumatism and tuberculosis.

Seed oil: Anticancer and hypoglycaemic.

Uses in Alternative System of Medicine

Ayurveda:

Part used: Whole plant, root and pod.

Uses: Aphrodisiac, antipyretic, galactagogue and tonic; cures biliousness, bronchitis, burning sensation, consumption, fever, gout, inflammation, paralysis, rheumatism, thirst and *tridosha.*

Siddha:

Part used: Seed.

Uses: Useful in fever, pain and as aphrodisiac.

Ethnomedicine

Plant: Astringent, depurative and stomachic; used for catarrh, fever, haemoptysis, malaria, nervous disorders, polyps, rheumatism and tuberculosis.

Non-medicinal Uses

Plant: Grown as green manure and cover crop.

172. *Teramnus mollis* Benth.

Synonym: *Teramnus labialis* (L.f.) Spreng. var., *mollis* Baker.

Vernacular Name: Sans: Maashaparni.

Distribution: Native of Bangladesh, India, Indonesia, Myanmar and Sri Lanka; occurs throughout India from Himalaya southwards.

Description: A twining herb with long stem and numerous spreading hairs; leaves 3-foliolate; stipules lanceolate and acute; leaflets membranous, oblong, lanceolate, apiculate, densely pubescent; flowers red; calyx silky, teeth shorter than the tube; pod linear, straight or slightly incurved, densely hairy with a short stout beak; seeds 8-12, oblong, truncate, smooth and dark brown.

Medicinal Uses

Plant: Used as antiperiodic and blood purifier.

Root and Pod: Used to cure ringworm.

Uses in Alternative Systems of Medicine

Ayurveda

Part used: Whole plant and root.

Uses: Aphrodisiac and tonic; useful in burning sensation, diseases of "pitta", edema and fever.

173. *Trifolium alexandrium* L.

Notes: Not known in the wild.

Common Name: Egyptian clover.

Distribution: Nativity uncertain; recorded as native of Israel, Lebanon, Turkey and Syria; introduced into India; cultivated in many parts in India.

Description: Sparingly appressed, pubescent herb with glabrous branches; petioles of the lower leaves equaling the leaflets, those of the upper leaves very short; leaflets oblong to broadly elliptic or oblong lanceolate, obtuse; stipules lanceolate, free part subulate to setaceous, villous; bracts linear, caducous; calyx with hirsute, triangular subulate teeth; corolla yellow, twice as long as the calyx; pod oblong, ovoid, one seeded; seed similar to red clover but larger and less distinctly lobed.

Phytochemistry

Seed: Caproic acid, kaempferol, lauric acid, oleic acid, palmitic acid and velutin.

Medicinal Uses

Plant: Is antiangiogenic, anticancer, antifertility, antihaemolytic, anti-inflammatory, anti ischaemic, antimicrobial, antimutagenic, antispasmodic, cardioprotective, cytotoxic and immunostimulant; is used in curing atheroclerosis, cancer of breast, flatulence, leukaemia, lymphoma, melanoma, neuroblastoma, osteoporosis, tumour of breast, ovary, stomach and ulcer.

Bioactivity: Plant is an inhibitor of aldose-reductase, α-reductase, aromatase, catechol-o-methyl transferase, DOPA-decarboxylase, histidine-kinase, lipase, MAO, MDR, NADH-ubiquinone-oxidoreductase, ornithine-decarboxylase, PTK, peroxidase, topoisomerase I and II, tyrosine-kinase; is calcium and estrogen antagonist and pituitary sensitizer.

Non-medicinal Uses

Plant: Used as fodder, hay a.ıd green manure; grown as cover crop in orchards; to reclaim alkaline and saline soil.

174. *Trifolium pratense* L.

Synonyms: *Trifolium bracteatum* Schousb.
Trifolium pratense L., var., *lenkorianicum* Grossh

Common Names: Trifoil , Wild clover.

Distribution: Native of Africa, Asia including India and Europe; grows in Himalaya and Tamil Nadu in India.

Description: A small herb with erect pilose stem; leaves at the base are long petioled, those on the main stem nearly sessile; leaflets oval, or elliptic, cuneate, obovate; stipules oblong oval to triangular, abruptly tapering to an erect setaceous tip; flowers rosy purple, to creamy white on terminal heads subtended by stipules of the upper pair of leaves; calyx tube pubescent with filiform teeth from a triangular base, sparsely hirsute; pod oblong, ovoid; seeds ovoid, asymmetrical, yellow.

Phytochemistry

Plant: α-Ketoglutaric acid, α-tocopherol, 7-O-β-glucoside-3-methoxy quercetin, allantoic acid, arabinose, arginine, asparaginic acid, benzyl formate, biochanin A, biochanin-A-7-O-β-D-glucoside-6"-malonate, boron, calcium, calycosin, calycosin-7- glycoside, calycosin-7-galactoside, campesterol, chlorogenic acid, choline, citrulline, clovamide, coumarins, coumesterol, cyanogenic glycosides, cyanidin, daidzein, demethylpterocarpan, flavonoids, formononetin-7-O-β-D-glucoside-6"-malonate, formononetin, furfurol, genistein, glucuronic acid, glucose, glutamic acid, histidine, homopisatin, homopterocarpin, homoserine, hydroxypterocarpan, hyperoside, irolone, iron, isoflavones, isoflavonoids, isoleucine, isorhamnetin, isotrifolin, L-aspargine, leucine, lysine, maackianin, maakianin-3-O-β-D-galactopyranoside, medicagol, melilotic acid, methionine, methoxy isoflavone, molybdenum, ononin, orbol-4'-methyl ether, oxalic acid, phenolic glycosides, phenylalanine, pinitol, pipecolic acid, potassium, pratenol, pratensein, protein, prunitrin, pseudobaptigenin, rhamnose, salicylates, selenium, sissotrin, soyasapogenol, threonine, trifolianol, trifolin, tryptophan, tyrosine and valine.

Root: (–) Maackianin-3-O-glucoside, 2-methyl-butanol, 2-phenylethanol, 3-methoxy quercetin-7-0-β-D-gluco pyranoside, 4'5-dihydroxy-6,7-methylene dioxyisoflavone, acetic acid, acetoin, arabinose, benzoic acid-ethyl ester, benzoic acid-methyl ester, benzyl alcohol, biochanin-A, calycosin, calycosin glycoside-7, 3'-dihydroxy-4-methoxy flavone-7-O-β-D-galactopyranoside, caproic acid, coumaric acid, decanoic acid-ethyl ester, dodecan-1-ol, estragole, eugenol methylester, formononetin, formononetin-7-O-β-D-glucoside-6"-malonate, furfurol, genistein, glucose, glucuronic acid, glutamic acid, hexadecanoic acid-ethyl ester, homopterocarpin, inermin, inermin glucoside, irilone, irilone-4'-O-glucoside, lauric acid, lauric acid ethyl ester, maackianin-3-0-β-D-galactopyranoside, maackianin-3-O-β-D-galactoside, medicarpin, medicarpin glucoside, methoxy medicarpin, myristic acid, ononin, pseudobaptigenin, rhamnose, sissortrin, soyasapogenols, tetradecane-1-Ol, trifolin, trifolirhizin and xylose.

Leaf: 6-α-7-Dihydroxy medicarpin, 6-α-7-dihydroxymaackianin, 6-α-methoxy maackianin, biochanin-A, biochanin-A-7-O-β-D-glucoside-6"-malonate, boron, calycosin, calycosin-7-O-β-D-glucoside, caryophyllene, cis-β-ocimene, cisclovamide, coumesterol, daidzein, formononetin, formononetin-7-O-glucoside-6"-malonate, hyperoside, irilone, irilone-4'-O-β-D-glucoside, irilone-4"-O-β-D-glucoside-6"-O-malonate, isochanin-A, isoflavones, isoquercetrin, isoquercetrin-6"-O-malonate, kaempferol, kaempferol-3-O-galactoside, maackianin, maltose, medicarpin, methoxy isoflavone, methyl ester, ononin, pisatin, pratensein, pratensein-7-O-β-D-glucoside, prunetin, prunetin-4'-O-glucoside, pseudobaptigenin, rothindin, salicylic acid, selenium, sissotrin, trans-β-farnesene, trans-β-ocimene, trans-clovamide, trifolin, trifoside-6"-O-malonate and variabilin.

Flower: β-Carotene, β-caryophyllene, β-sitosterol, 2-ethyl hexanal, 2-phenylethyl alcohol, 3-methyl quercetin, 3-methylquercetin-7-O-β-D-glucopyranoside, acetoin, acetophenone, afromorsin, afromorsin-7-O-β-D-glucoside, aluminium, arabinose, ascorbic acid, benzaldehyde, benzyl acetate, benzyl alcohol, biochanin-A, calcium, calycosin, caryophyllene, chromium, cinnamic acid, cinnamic acid methyl ester, cinnamic alcohol, cinnamic aldehyde, cinnamyl hexonate, cis-β-ocimene, cis-clovamide, cobalt, coumarin, coumesterol, diadzein, ethyl cinnamate, formononetin, formononetin-7-O-β-D-glucoside-6"-O-malonate, furfural, genistein, genistin-6"-O-malonate, geraniol, glucose, glucuronic acid, hentriacontane, heptacosane, hyperoside, irilone-4'-O-β-D-glucoside-6"-O-malonate, iron, isoflavone, isoquercitrin, isoquercitrin-6"-O-malonate, isorhamnetin, isotrifolin, kaempferol, L-DOPA, limonene, linalool, linalool oxide, linalyl-acetate, magnesium, malvidin, manganese, medicagol, methyl anthralinate, methyl cinnamate, methyl salicylate, myricyl alcohol, N-heptacosane, niacin, ocimene, ononin, paeonidin, peictolinarigenin, phenol, phenylethyl alcohol, phenylethyl acetate, phosphorus, pictolinarin, phaseolic acid, potassium, pratensein, pratensein-7-O-β-D-glucoside-6"-O-malonate, pratensol, pratol, protein, prunetin, pseudobaptigenin, quercetin, quercitrin, rhamnose, riboflavin, rothindin, salicylic acid, selenium, silicon, sissortrin, slaframine, sodium, soyasapogenol A, B, C and D, thiamine, trans-β-ocimene, trans-clovamide, trifolin, trifolionol, trifolirhizin, trifoside, trifoside-6"-O-malonate, xylose and zinc.

Medicinal Uses

Plant: Is antiadenomic, antiandrogenic, antibacterial, anticancer, anticoronary, antifibrinolytic, antihepatotoxic, antihistaminic, antihypertensive, anti-inflammatory, antilymphocytic, antimutagenic, antioxidant, cardioprotective, choleretic, carminative, cerebrotonic, chemopreventive, cholinergic, cytoprotective, diuretic, haemolytic, hepatoprotective, hypocholesterolemic, haemostat, hypnotic, hyperlipidemic, immunostimulant, lithogenic, laxative, memory enhancer, narcotic, sedative, tonic, vasodilator and vulnerary; used in treating allergy, anorexia, cystitis, ache, allergy, anaemia, anxiety, arrhythmia, atherosclerosis, asthma, back ache, bronchitis, beri-beri, bradycardia, cancer of breast, oesophagus, kidney, liver, pancreas and skin, capillary fragility, canker, cardiospasms, cataract, cephalagia, chilblain, cold, colic, constipation, convulsions, cystitis, dandruff, dementia, delirium, depression, dermatitis, diabetes, diarrhoea, dysmenorrhoea, eczema, edema, encephalopathy, fatigue, fibrosis, fibrosarcoma, flu, gastritis, gingivitis, glaucoma, glycosuria, gout, heart burn, haemorrhages, haemorrhoids, hepatitis, herpes, hydrocele, hyperactivity, keratitis, hyperthyroidism, hypoglycaemia, impotency, insomnia, leprosy, leukaemia, lupus, lymphoma, malaria, measles, melanoma, menopausal disturbance, menorrhagia, migraine, myalgia, myocardites, nephritis, neuralgia, neuroblastoma, neuropathy, obesity, odontalgia, orchitis, osteoarthritis, osteoporosis, pharyngitis, plague, pneumonia, polio, psoriasis, purpura, rheumatism, tumour of bladder, breast, cervix, colon, oesophagus, kidney, liver, lung, ovary, pancreas, prostate, skin, stomach and brain, ulcer, uremia, vaginitis, varicose veins, vertigo, vitilago and wrinkles; antidote to lead and mercury poison.

Flower: Antispasmodic, antifungal, depurative, expectorant, sedative and vulnerary; useful in asthma, bronchitis, cough, corns and cancerous ulcers; it improves appetite and regulates digestion.

Uses in Alternative Systems of Medicine

Homoeopathy:

Part used: Tincture of flower head.

Uses: Useful in cancer, constipation, cough, fever, head ache, mumps and sore throat.

Ethnomedicine

Plant: Is used as alterative, depurative, diuretic, expectorant, sedative and tonic; used in asthma, bronchitis, burn, bowel cancer, catarrh, corn, cough, dyspepsia, eye problems, pruritus, scrofula, sores, spasm, tumour and whitlow.

Non-medicinal Use

Plant: Grown for manure, fodder, hay and pasturage; used in the preparation of anti asthmatic cigarettes.

175. *Trifolium repens* L.

Common Names: Dutch clover, White clover.

Vernacular Name: Pun: Shaftal.

Distribution: Native of Africa, Asia including India and Europe; grows in Himalaya, Rajasthan and Tamil Nadu in India.

Description: A long lived perennial with long petioled leaves; leaflets cuneate, obovate to broadly oblong, emarginate to obtuse, dull green, occasionally marked with white "V" and sometimes with dark red flecks; stipules abruptly narrowed to a point; flowers fragrant, white or rose tinged; bracts membranous; calyx teeth unequal, the upper shorter than the tube, the lower two-thirds the length; pod sessile, linear, 3-4 seeded; seeds ovoid, truncate, smooth, yellow, some times reddish.

Phytochemistry

Plant: Azukisaponin, calcium, coumarin, coumesterol, daidzein, daphnoretin, formononetin, genistein, linamarin, lotaustralin, melilotic acid, phosphorus, pinitol, soyasapogenol, soyasaponin, trihydroxyflavone and umbelliferone.

Leaf: Cyanoglucosides such as linamarin and lotaustralin.

Flower: Quercetin glucoside.

Seed: Dihydroxycyclopentadienone.

Medicinal Uses

Plant: Antirheumatic, depurative, diuretic, prophylactic, tonic, antiangiogenic, antifertility, anti-inflammatory, antimicrobial, antimitogenic, antioxidant, antispasmodic, antiviral, apoptotic, astringent, cytotoxic, estrogenic, expectorant, hypocholesterolemic, immunostimulant, tonic and uterotrophic; used in adenopathy, atherosclerosis, cancer of the breast, flatulence, gout, leukaemia, lymphoma, melanoma, mumps, neuroblastoma, osteoporosis, rheumatism, scrofula, sore, spasm, swelling, tumour of the breast, ovary, prostate and stomach ulcer and whitlow.

Resin: Astringent, diuretic and emmenogogue; used as antidote for scorpion sting.

Leaf: Depurative; tincture is applied for gout.

Flower: Astringent; useful in gout.

Bioactivity: Plant is inhibitor of aldose-reductase, α-reductase, aromatase, catechol-O-methyl transferase, DOPA-decarboxylase, histidine-kinase, lipase, NADH-ubiquinone-oxidoreductase, ornithine-decarboxylase, peroxidase, topoisomerase I and II, and tyrosine-kinase; is calcium antagonist, pituitary sensitizer and quinone-reductase-inducer.

Toxicology: Plant contains chymotrypsin and trypsin inhibitors.

Uses in Alternative System of Medicine

Homoeopathy

Part Used: Tincture of flower head.

Uses: Useful in mumps and excess salivation.

Ethnomedicine

Plant: Astringent, depurative, estrogenic and expectorant; used for adenopathy, cancer, gout, poison, mumps, parotitis, rheumatism, scrofula, sore, spasm, swelling and whitlow.

Non-medicinal Uses

Plant: Used as forage, fodder, hay, silage, and green manure; grown as cover crop to reduce soil erosion and as pasture crop.

176. *Trifolium subterraneum* L.

Common Names: Subclover, Subterranean clover.

Distribution: Native of Africa, Asia, Europe and Middle East; introduced into India; naturalized in Hill stations of Tamil Nadu.

Description: Softly pubescent annual with slender procumbent stem forming circular clumps; leaves long petioled; leaflets broadly cordate, entire, except for the shallowly crenate apex; stipules ovate, oblong, the lower most acuminate; inflorescence is a few flowered globose head; fertile flowers 2-5, white, striped pink; sterile flowers develop after anthesis, numerous, apetalous and completely enclose the pod; calyx of fertile flowers nerveless, tube glabrous, with subequal, ciliate flexuous teeth; calyx of sterile flowers with unequal, narrow, irregularly bent calyx teeth; seed one, large, broadly ellipsoid, purplish black.

Phytochemistry

Root: (–)-Maackianin-3-O-glucoside, acetic acid, acetoin, isoflavone, methoxy medicarpin.

Leaf: Genistein and its methyl ester.

Medicinal Uses

Plant: Is antiangiogenic, anticancer, antihaemolytic, anti-inflammatory, antimicrobial, antimitogenic, antioxidant, antispasmodic, antiviral, apoptotic, cardioprotective, cytotoxic, immunostimulant and uterotrophic; is used in atherosclerosis, cancer of breast, flatulence, leukaemia, lymphoma, melanoma, neuroblastoma, osteoporosis, tumour of breast, ovary and stomach and ulcer.

Bioactivity: Plant is an inhibitor of aldose-reductase, α-reductase, aromatase, catechol-O-methyl transferase, DOPA-decarboxylase, histidine-kinase, MAO, MDR, NADH-ubiquinone-oxidoreductase, ornithine-decarboxylase, peroxidase, topoisomerase I and II, and tyrosine-kinase; is calcium and estrogen antagonist, pituitary sensitizer, and quinone-reductase-inducer.

Non-medicinal Uses

Plant: Primarily grown for pasture and as soil binder.

177. *Trigonella balansae* Boiss. & Reuter

Synonym: *Trigonella corniculata* (L.) L.

Vernacular Names: Hin: Champa methi, Kasturi methi; **Kan:** Kasthuri menthya; **Mar:** Pirang; **Sans:** Malaya.

Distribution: Native of Crete, Greece in Europe, and Turkey in Asia; introduced into India; grows in Gujarath, Punjab, Sikkim, Uttar Pradesh, West Bengal and Western Himalaya in India.

Description: Erect to procumbent annual; leaves 3-foliolate; leaflets obovate to oblong cuneate, obtuse to emarginate, glabrous above, sparsely pilose beneath along the nerves, upper half toothed, lower half entire; stipules lanceolate, dentate or incised; inflorescence is raceme with 8-20 flowers; peduncle extends beyond the flower in a awn-like point; calyx teeth subequal, shorter than tube; corolla yellow, wings shorter than the keel petal; pod linear, acuminate, curved, glabrous with transverse anastomosing veins; 4-8 seeded.

Phytochemistry

Plant: Diosgenin and vitexin.

Leaf: Diosgenin, gitogenin and vitexin.

Flower: Gitogenin and titogenin.

Seed: β-Sitosterol, betaine, apigenin, cholesterol, choline, digalactopyranoside, dihydrostigmasterol, diosgenin, fatty oil, flavone-c-glycosides, glucopyranosyl acacetin, linoleic acid, oleic acid, palmitic acid, stearic acid, triacontane, vitexin and yuccagenin.

Edibility

Plant: Eaten as pot herb; used for flavouring.

Medicinal Uses

Pod: Astringent and styptic; applied to bruises and swellings.

Ethnomedicine

Plant: Astringent and styptic; used for bruises and swellings.

178. *Trigonella foenum-graecum* L.

Common Name: Fenugreek.

Vernacular Names: Guj: Methi; **Hin:** Methi; **Kan:** Menthe, Menthya; **Mal:** Venthayam, Uluva; **Pun:** Methi; **Sans:** Bahuparni, Bahupathrika, Chandrika, Deepani, Gandhabeeja, Gandhaphala, Jyothi, Kunchika, Methika, Methini, Misrapushpa, Munindrika, Peetha beeja, Vallari, Vedhani; **Tam:** Vendiyam, Venthayam; **Tel:** Menthikoora, Menthulu.

Distribution: Native of Afghanisthan, Iran, Iraq, Kuwait and Yemen; introduced into India; cultivated throughout India.

Description: A smooth erect annual; leaves trifoliolate; leaflets toothed; stipules not toothed; flowers axillary, sessile, yellow; calyx teeth linear; corolla much exerted; pod long, straight or curved with a long persistent beak, often falcate; 10-20 seeded, with out transverse reticulations; seed brown, small with striae.

Phytochemistry

Plant: γ-Aminobutyric acid, acetyl choline, acetoxy-methyl coumarin, apigenin, ascorbic acid, behenic acid, coumarin, D-galactosyl myoinositol, 4',7-dihydroxy flavone, dioscin, diosgenin, D-mannose, fenugreekin, folic acid, formononetin, fructose, galactinol, galactomannan, gitogenin, graecunin-B, homo-orientin, isoquercitrin, kaempferol, kaempferol-3,7-diglycoside, kaempferol-3-glycoside, lecithin, lignin, luteolin, 4-methyl-7-acetoxy coumarin, myoinositol, neogitogenin, palmitic acid, p-coumaric acid, pectin, pyridoxine, quercetin, sapogenins, saponins, sulphur, tigogenin, trigocoumarin, trigofoenoside, trigoforin, trigonelline, trigonelloside, 3',4',7-trihydroxy flavone, trimethyl coumarin, verbascose, vitamin-K, vitexin, xylose, β-galactosidase, β-mannosidase, α-mannan and α-sitosterol.

Root: Diosgenin, tigogenin and yamogenin.

Leaf: Arginine, calcium, choline, copper, digogenin, diosgenin, gitogenin, graecunin, homoorientin, isoleucine, kaempferol, leucine, lysine, methionine, methoxy-quercetin, neotigogenin, phynylalanine, quercetin, saponaretin, sarsasapogenin, smilagenin, spirasterol, threonine, transaminase glutamate, trigonelloside-c, trigonenin, tryptophan, valine, vicenin, xanthophylls, yamogenin, β-carotene and β-sitosterol.

Flower: Kaempferol and quercetin.

Pod: Catechol, flavonol, phenol, phytate, raffinose, saponin, stachyose and sucrose.

Seed: α-Carotene, α-sitosterol, α-trigonellic acid, 4-hydroxy isoleucine, 7-acetoxy-4-methyl coumarin, alanine, alkaloid, aluminum, arabinose, arachidic acid, arginine, ascorbic acid, aspartic acid, calcium, campesterol, carbohydrates, carpaine, catechol, cholesterol, choline, chromium, cobalt, colouring matter, copper, coumarin, cystine, dihydroactinidiolide, dihydrobenzofuran, dihydroxyflavone, dimethyl pyrazine, dimethyl-furanone, dioscin, diosgenin, elemene, epimilagenin, ethylmethylpyrazine, ethyl-pyrazine, fenugreekin, fenugrin, fixed oil, flavonoid glycosides, flavonols, fenugraecin, fructose, furostanol, saponins, galactomannan, galactose, genistein, gentianine, gitogenin, glucose, glutamic acid, glycine, graecunin A to M, grigonelloside, hexadecane, histidine, homorientin, hydroxyisoleucine, iron, isoleucine, isoorientin, isoorientinarabinoside, iso-propylpyrazine, isovitexin, jamogenin, L-arginine, leucine, L-histidine, luteolin, lysine, magnesium, manganese, mannose, methionine, methoxymethyl pyrazine, methoxy-quercetin, methylcaprolactone, mucilage, mucine, neodiosgenin, neotigogenin, neo-α-carotene, niacin, nicotinamide, nicotinic acid, oleic acid, orientin, phenol, phenylalanine, phosphatase, phosphorus, phytate, phytic acid, phytosterols, polysaccharides, potassium, proline, protein, pyrazine, quercetin, raffinose, rhamnose , riboflavin, saponaretin, saponin, sarsasapogenin, selenine, selenium, serine, silicon, smilagenin, sodium, sotolone, stachyose, starch, stearic acid, steroidal saponins, stigmasterol, sucrose, tannic acid, thiamine, threonine, tigogenin, trigofoenoside B-1 and C-1, trigonella-glycoside, trigonelline, trigonelloside-c, trihydroxyflavone, trillin, trimethyl coumarin, tryptophan, tyrosine, undecane, valine, vicenin I and II, vitamins, vitexin, vitexincoumarate, vitexinglucoside, vitextin-cinnamate, xylose, yamogenin and yamogenin tetroside B anc C.

Sprout: Biotin, cyanocobalamine, diosgenin, peroxidase and uridine-diphosphogalactose-4-epimerase.

Seed oil: Behenic acid, linoleic acid, linolenic acid, octadecatrienoic acid, oleic acid, palmitic acid, palmitoleic acid and stearic acid.

Edibility

Plant, Leaf and Pod: Plant is eaten as pot herb; leaf and pod edible.

Seed: Edible; it is used as condiment to flavour rice and curries; in Egypt, ground seeds are mixed with wheat flour for making bread; in Switzerland, it is used in flavouring cheese; a good substitute for coffee.

Medicinal Uses

Plant: Is anthelmintic, antiandrogenic, antiestrogenic, antifertility, antihepatotoxic, anti-inflammatory, antineuralgic, antinociceptive, antiperoxidant, antithyroid, antitussive, antiviral, astringent, cardioprotective, cerebrotonic, cytotoxic, demulcent, diuretic, emetic, goitrogenic, haemolytic, hepatoprotective, immunosuppressant, laxative, neurotoxic, sedative, spermicide, tranquilizer, vasodilator and vulnerary; is used in acne, allergy, anaemia, anorexia, anxiety, arrhythmia, atherosclerosis, arthritis, asthma, back ache, beri beri, cancer of the breast, cervix, kidney, liver, canker, capillary fragility, cardiospasms, cataract, cephalagia, cervical dyspalsy, chilblain, cold, cough, colic, convulsions, cystitis, delirium, dementia, depression, dermatitis, diabetes, dropsy, dysentery, menorrhagia, eczema, edema, encephalopathy, enlargement of liver and spleen, fatigue, fibrosarcoma, gingivitis, glaucoma, goiter, glycosuria, gout, heart burn, haematuria, haemorrhages, haemorrhoids, hepatitis, hiccup, hydrophobia, hyperactivity, hyperglycaemia, impotency, insomnia, ischaemia, keratitis, leprosy, lupus, lymphoma, malaria, measles, melanoma, meningitis, menorrhagia, menopausal disorders, migraine, mouth ulcers, myalgia, nausea, nephritis, neuralgia, neuroblastoma, neuropathy, obesity, ophthalmia, orchitis, osteoarthritis, osteoporosis, pharyngitis, photophobia, pneumonia, polio, polyp, psoriasis, purpura, rheumatism, retinopathy, sprue, stress, stroke, sunburn, tachycardia, tumour of bladder, breast, cervix, colon, kidney, liver, lung, pancreas and stomach, ulcer, varicose veins and vertigo; pessary made out of the herb is introduced into the vagina to check leucorrhoea; is also applied inside the mouth for ulcers, antidote to cadmium, aluminium, lead and mercury poisoning and pesticides.

Leaf: Aperient, aromatic, carminative, cooling, diuretic and refrigerant; useful in indigestion, flatulence, sluggish liver, febrile conditions and vitiated "pitta"; leaf poultice is applied for burns and swelling; leaf paste is applied on the head to prevent hair fall and dandruff; leaf paste applied on face is a cure for pimples, black heads, acne, dryness; prevents early appearance of wrinkles; boiled and fried in butter it cures biliousness; leaf infusion is used as gargle for recurrent ulcers.

Seed: Anti-inflammatory, aphrodisiac, aromatic, astringent, bechic, carminative, demulcent, diuretic, emmenogogue, emollient, galactagogue, hypoglycaemic and nutritive tonic; seed paste is used as poultice for abscesses, boils and ulcers; seed infusion is used as gargle for sore throat and mouth ulcers; regular intake of seeds soaked in water is highly recommended for reducing blood sugar in diabetics; seed fried in ghee and eaten with wheat flour and sugar, helps in normalizing the system after child birth; tea made from the seeds soothens inflammed stomach and intestine; it cleanses the stomach, bowel, kidney and respiratory tract; it is beneficial in healing peptic ulcers; it helps in detoxifying the body and shortens gestation period of fever; seed decoction is useful in removing the toxins from the nasal and oral passages, gastrointestinal tract, stomach, kidney, respiratory and urinary tract, blood, uterus and vagina; it is beneficial in reducing unpleasant odour from the nose, mouth, body and vagina; it is an astringent for leucorrhoea; reduces back pain and excess bleeding during menustration; it prevents respiratory infections like bronchitis, catarrh, diphtheria, influenza, pneumonia and sinusitis; it helps in normal perspiration; powdered seed is taken to stop vomiting; it is given as a cooling drink.

Bioactivity: Plant is a inhibitor of 5-α-reductase, 5-lipoxygenase, ATPase, aromatase, aldose-reductase, catechol-O-methyl transferase, histidine-kinase, MAO, MDR, NADH-ubiquinone-oxidoreductase, ornithine-decarboxylase, PTK, insulinase, iodothyronine-deiodinase; it is angiotensin receptor blocker, calcium antagonist, quinone-reductase inducer and pituitary sensitizer.

Ether extract of seeds showed anti-inflammatory and antiulcer activity when administered to rats with induced inflammation by formalin or carragenan exposure; glucose tolerance test conducted on normal rats showed that the oral administration of powdered seeds prior to glucose load resulted in the inhibition of glucose induced hyperglycaemia; blood glucose level significantly reduced by 13 per cent when compared to control group; trigonelline in seeds counter acts the hyperglycaemic effect of cortisone given three hours before or simultaneously; hydroxy isoleucine an insulinotropic amino acid from seeds causes significant increase in the plasma insulin level accompanied by progressive decrease in blood glucose, it also promotes healing of gastric ulcer and is smooth muscle relaxant; intragastric administration of alcoholic extract of seeds possess anti-inflammatory effect on carrageenan induced paw edema in rats; oral administration of alcoholic extract of seeds before and after inoculation of EAC cells in mice produced 70 per cent inhibition of tumour cell growth, the effect was both on peritoneal exudate cells and intraperitoneal macrophages indicating that inflammation is associated with cancer development.

Toxicology: Plant contains chymotrypsin and trypsin inhibitors.

Uses in Alternative Systems of Medicine

Ayurveda:

Part used: Leaf and seed.

Uses: Used as antidiabetic, antispasmodic constipating, cooling.

Siddha:

Part used: Seed.

Uses: Used as aphrodisiac, antidysenteric, carminative, refrigerant and tonic.

Unani:

Part used: Seed.

Uses: Analgesic, anti-inflammatory, aphrodisiac, diuretic, emmenogogue, laxative and resolvent; cures cough, dropsy, gout, piles, spleen enlargement, tumours and uterine disorders.

Ethnomedicine

Plant: Anodyne, aphrodisiac, astringent, carminative, demulcent, deobstruent, diuretic, emmenogogue, lactagogue and stimulant; useful in arthritis, beri-beri, bladder complaints, boils, cancer, chancre, cystitis, diabetes, diarrhoea, dysentery, dyspepsia, edema, frigidity, gastritis, gonorrhoea, hernia, hydrocele, impotency, piles, sores, syphilis, tumour, uterine disorders and wounds.

Leaf: Paste applied to prevent hair loss.

Seed: Is cooling, carminative and tonic; useful in gastric complaints, piles, rheumatism, sores and syphilis.

Non-medicinal Uses

Plant: Used as fodder and green manure and as a good soil rejuvenator.

Seed: Mucilage in the seed is used as an adjuvant in pharmaceutical preparations as a tablet disintegrator; yields yellow dye.

179. *Ulex europaeus* L.

Synonym: *Ulex armoricanus* Mabille
Ulex compositus Moench

Notes: Contrary to popular belief, plant does not cause haematuria.

Distribution: Native of Corsica, France, Germany, Great Britain, Ireland, Italy, Netherlands, Portugal, Spain and Switzerland; naturalized in hill stations of Tamil Nadu, West Bengal and Sikkim in India.

Description: Densely branched shrub; branchlets striate, green, hirsute with reddish brown hairs; short branchlets on the main branches have numerous spines; leaves 3 foliolate with pointed petioles; bracts conical; bracteoles boat shaped, pubescent; calyx deeply 2 lobed, yellow, boat shaped, lobes subequal, adaxial short, 2 toothed, abaxial 3 toothed; corolla, yellow, clawed, standard petal obovate with a deep notch, keel petal basally pouched; stamens 10, filaments alternately shorter and longer; style apically curved; pod oblong, densely pubescent; seeds 2-4.

Phytochemistry

Plant: Good source of β-carotene, anagyrine, calcium, protein and sodium.

Medicinal Uses

Plant: Antiedenomic, antiandrogenic, antiarrhythmic, antibacterial, anticoronary, antiestrogenic, antihypertensive, anti-inflammatory, antinociceptive, antioxidant, antiperoxidant, antiprostaglandin, antipyretic, antitumour, antiviral, cardioprotective, cytotoxic, hallucinogenic, hypoglycaemic, hepatoprotective, hypolipidemic, hypouricemic, immunostimulant, laxative, oxytocic, phagocytotic, respirostimulant and vasodilator; is useful in allergy, anorexia, acne, asthma, cancer of breast, cervix and liver, diabetes, edema, flu, gastritis, hyperglycaemia, leukaemia, lupus, lymphoma, malaria, migraine, obesity, photophobia, psoriasis, rheumatism, stress, tumour of breast, liver, lung and ulcer.

Bioactivity: Plant is angiotensin-receptor blocker, β-blocker, ATPase-inhibitor, interferon synergist, quinone-reductase-inducer and TOPO-2-inhibitor.

Toxicology: Anagyrine in the plant causes "crooked calf disease" in cows, which ingest the plant as feed.

Ethnomedicine

Plant: Cardiotonic and diuretic.

Non-medicinal Uses

Plant: Used as fodder; yields a yellow dye; dried plants used as fuel.

Leaf: Used as substitute for tea.

Flower: Yields an essential oil used in perfumery.

180. *Uraria crinita* (L.) DC.

Synonyms: *Uraria crinita* (L) DC. var., *macrostachya* (Wallich) Schindler
Uraria macrostachya Wallich

Distribution: Native of Cambodia, China, India, Indonesia, Laos, Malaysia, Philippines, Taiwan, Thailand and Vietnam; grows in Assam, Manipur, Meghalaya, Mizoram and Tripura in India.

Description: A shrub of 2m high; leaves 3-7 foliolate; leaflets oblong, base round, apex acuminate, glabrous above, strongly veined and hairy along the nervules beneath; flowers purple on simple, hairy, terminal raceme; bracts ciliate; calyx segments densely plumose; pod compressed, divided transversely into 4-6 opaque segments each folded over the other; seed black, orbicular and single.

Medicinal Uses

Plant: Useful in diarrhoea, dysentery, enlarged spleen, fistula, liver disorders, pustules and tumours; herb extract heals fractures quickly.

Root: Used for treating chills, colic, diarrhoea and fever.

Pod: Employed in treatment of prolapsed anus and cures mouth sores.

Ethnomedicine

Plant: Used for diarrhoea, dysentery. fever, hepatomegaly, hepatosis, pimple, small pox, splenomegaly and tumour.

Non-medicinal Uses

Plant: Grown as cover crop; used as manure.

Root: Used as fish poison.

Leaf: Used to kill lice.

181. *Uraria lagopodioides* DC.

Synonym: *Uraria alopecuroides* (Roxb.) Sweet

Common Name: Tick trifoil.

Vernacular Names: Ben: Chakulia, **Guj:** Ranaganga, **Hin:** Daula, Davada, Davala, Peethavana, **Kan:** Laageli, Naribaalada gida, Narihonne, Nariyaala honne, **Mal:** Muvila, Orila, **Mar:** Davala, Peethavana, **Sans:** Anghriparni, Anghrivallika, Athiguha, Bramhaparni, Chakrakulya, Chakraparni, Chithraparni, Chithraparnika, Deergha, Deerghapathra, Dhamani, Dhavani, Dheergha, Ghastila, Guha, Kadala, Kalashi, Kalasi, Kankashathru, Kroshtukamekhala, Kroshtukapuchhika, Kroshtupuchhi, Kroshtuvinna, Laangali, Laangulika, Mahaguha, Mekhala, Pishtaparni, Prishniparni, Pristhiparni, Prithakparni, Poornaparni, Saaliparni, Shirnamala, Shrigalvinna, Sinhalanguli, Sinhapuchhi, Sinhapuchhika, Sinhapushpi, Srigalavantha, Thanvi, Thriparni, Vishnuparni, **Tam:** Orilai, **Tel:** Anghriparnika, Jibilike chettu, Kolaponna, Nakkathoka ponna.

Distribution: Native of Bhutan, West China, India, Myanmar, Nepal, Polynesia, Ryukyu, Taiwan, Australia, Andaman and Pacific Ocean Islands; grows in Andaman and Nicobar Islands, Andhra Pradesh, Assam, Himalaya, Himachal Pradesh, Karnataka, Maharashtra, Manipur, Orissa, Punjab, Tamil Nadu and West Bengal in India.

Description: An erect under shrub or herb with shortly pubescent woody branches; leaves 3-foliolate, simple, broadly oblong, or roundish; petiole pubescent; leaflets oblong, rhomboid or truncate; stipules lanceolate-acuminate; calyx bristly, calyx teeth subequal; lower teeth elongated; flowers white or blue or purple on a dense raceme; pod oblong, glabrous with 2-6 joints.

Phytochemistry

Plant: 3-hydroxy-7,4'dimethoxyflavone.

Medicinal Uses

Plant: Alterative and tonic; given with milk as abortifacient; used in catarrh, chest inflammation and intermittent fever.

Root: Decoction is used in bleeding piles, body ache, catarrh, fever, and rheumatism; is used as antidote for snake bite.

Leaf: Decoction given in diarrhoea.

Uses in Alternative Systems of Medicine

Ayurveda

Part used: Root.

Uses: Hot, bitter and aphrodisiac; useful in curing edema, intermittent fever, respiratory problems and tumours; promotes growth of healthy tissues.

Siddha

Part used: Whole plant and root.

Uses: Abortifacient, alterative, aphrodisiac, carminative, stomachic and tonic; useful in amenorrhoea, catarrh, general debility.

Medicinal Uses Outside India

China: Plant decoction is used to wash haemorrhoids.

Indonesia: Leaves are used in enlarged liver and spleen; flower paste with coconut oil is applied for pimples and small pox eruptions.

Malay Peninsula: Leaf decoction is given in dysentery, colic and fever; crushed leaves are applied on the head to remove lice.

Ethnomedicine

Plant: Abortifacient, oxytocic and spasmogenic; used for abdominal tumour, chest complaints, diarrhoea, fever, inflammation and malaria.

Root: Given in body ache; to prevent miscarriage; as antidote to snake poison.

Non-medicinal Uses

Plant: Used as green manure.

182. *Uraria picta* (Jacq.) DC.

Synonyms: *Uraria aphrodisiaca* Welw.
Uraria leucantha Span.
Uraria linearis Hassk.

Vernacular Names: Ben: Sankarjate; **Hin:** Dabra, Peethavan, Pransiparni; **Mal:** Muvila; **Mar:** Peethavan, **Sans:** Angriparni, Chithraparni, Dhaavanee, Guha, Kalashee, Prayakooparni, Prishniparni, Saalaparni, Sargaalavinna; **Tam:** Kalaponna, Sitthirappaaldai, **Tel:** Kolkuponna.

Distribution: Native of Africa, Asia including India, Australia and Papua New Guinea; grows throughout India.

Description: Suffrurticose perennial, with stout stem clothed with fine hooked hairs; leaves imparipinnate; stipules triangular, long, acuminate; leaflets on the upper part of the stem 5-7, linear, oblong, acute, blotched with white hairs, minutely pubescent beneath, base rounded; leaflets on the lower part of the stem 1-3 foliolate, suborbicular; flowers on cylindrical raceme; pedicels long, curved upwards after flowering; bracts large, striate, ciliate, upper ones lanceolate, acuminate, lower ovate-acuminate; calyx teeth plumose, subulate; corolla purple; pod glabrous, lead coloured with 3-6 smooth polished joints; seeds 1-2, red, reniform.

Medicinal Uses

Plant: Used in anasarca, bleeding piles, blood diseases, body cramps, bone setting, colitis, cough, delirium, difficult micturition, dropsy, fever, heart troubles, herpes, musculoskeletal disorders, piles, purpura, respiratory disorders and wounds; used as antidote for snake bite.

Root: Aphrodisiac, diuretic, expectorant, febrifuge and nervine tonic; is given for abdominal pain, body ache, chill, cough, fever and prolapsed anus; used as antidote for snake bite.

Leaf: Pulverised leaves used as a remedy for fever and gonorrhoea.

Pod: Decoction is given in cold, fever and rectal prolapse; applied to mouth sores.

Uses in Alternative Systems of Medicine

Ayurveda:

Part used: Inflorescence.

Uses: Promotes strength and semen quality; cures fever.

Ethnomedicine

Plant: Antiseptic and aphrodisiac; used for aphthae, chill, cough, fever, fracture, gonorrhoea and stomatitis.

Root: Given for body ache, fever and vomiting; used as antidote for snake bite.

Leaf: Is antipyretic.

183. *Uraria rufescens* (DC.) Schindler

Synonym: *Uraria hamosa* Wight & Arn.

Vernacular Name: Sans: Salaparni.

Distribution: Native of Asia including India; grows through out India.

Description: Straggling under shrub; leaf 3 foliolate; leaflets ovate or elliptic ovate; flowers purple on terminal branched racemes; pod with 4-6 hispid joints.

Medicinal Uses

Plant: Is used in diarrhoea, dyspnoea, emaciation, intermittent fever, edema, polyuria, vomiting and as rejuvenator.

Uses in Alternative Systems of Medicine

Ayurveda:

Part used: Whole plant.

Uses: Useful in cardiac problems, diarrhoea, dyspnoea, edema, emaciation, intermittent fever, pain, piles, polyuria and vomiting.

184. *Vicia faba* L.

Synonyms: *Faba bona* Medik.
Faba equina Medik.

Common Name: Broad bean,.

Vernacular Names: Hin: Bakla anhuri, Kala matar; **Kan:** Kaadu huralikayi; **Pun:** Chastang, Kali Bakla.

Distribution: Native region uncertain, probably West Africa; known only under cultivation throughout the world, with a large number of varieties; cultivated in Bihar, Himachal Pradesh, Punjab, Rajasthan and Uttar Pradesh in India.

Description: An upright annual herb with compound leaves; leaflets usually 6, broad, oval; flowers large, white with dark purple markings born in clusters of 1-6 in the axil of leaves; pod, greenish black, brown, glabrous, broad, terete, oblong, obliquely acuminate at both ends; seeds 3-4, oblong or oval, flat or round, smooth, bright reddish brown or greenish brown or dark purple, obscurely mottled or dotted with colours similar to base colours.

Phytochemistry

Plant: α-Terpineol, antoside, bioquercetin, cinaroside, cosmosin, ethanone, eugenol, ferulic acid, heptadecanoic acid, hexadecanoic acid, isoeugenol, isojasmonic acid, isoquercetrin, kaempferol, linalool, 1-(4-methoxyphenyl)ethanone, methoxyeugenol, octadecadienoic acid, octadecanoic acid, octadecotrienol, p-hydroxy benzoic acid, rutin, sinapic acid, tetradecanoic acid, trimethyl–entadecanone and wyerone.

Leaf: Brassinone, castasterone, cholesterol, 23-dehydrocampesterol, 23-dehydrocholesterol, 24-dehydrositosterol, epibrassinolide, isofucosterol, kaempferol, 24-methyl desmosterol, methylene cholesterol, nigellic acid and sitosterol.

Flower: N-[(–)jasmonosyl] (s) tryptophan I, N-[(+) cucurbinoyl (s) tryptophan II and epibrassinolide.

Flower oil: α-Terpineol, antoside, bioquercetin, cinaroside, cosmosin, eugenol, heptadecanoic acid, hexadecanoic acid, isoeugenol, isoquercetin, linalool, methoxyeugenol, octadecadienoic acid, octadecanoic acid, octadecatrienol, rutin, tetradecanoic acid and 14-trimethy l-2-pentadecanone.

Pod: β-Carotene, ascorbic acid, butein, calcium, chlorine, choline, citric acid, convicine, dihydroxyflavone, divicine, geraldone, glyceric acid, kaempferol, kaempferol-3-O-glucoside, kaempferol–3-galactoside-7-rhamnoside, kaempferol–3-rhamnosyl-(6''-acetyl)galactoside-7-rhamnoside, L-dioxyphenylalanine, L-tyrosine, magnesium, malic acid, niacin, oxalic acid, phosphorus, potassium, quercetin–3-galactoside-7-rhamnoside, quercetin–triglycoside, sulphur, tetrahydroxyflavone, thiamine, trihydroxyflavone and vicine.

Seed: (–)-Jasmonic acid, alanine, arginine, ascorbic acid, β-carotene, β-cyanoalanine, β-sitosterol, biotin, calcium, campesterol, carbohydrates, coumesterol, cystine, D-hydroxylysine, favin, gallocatechin, genistein, glutamic acid, glycine, hexosamine, histidine, iron, isoleucine, leucine, leucocyanidin, leucodelphinidin, linoleic acid, linolenic acid, lysine, L-DOPA, magnessium, matiresinol, medicarpin, methionine, myristic acid, niacin, oleic acid, palmitic acid, palmitoleic acid, pantothenic acid, phenylalanine, phosphorus, phylloquinone, phytate, phytosterol, pipecolic acid, potassium, proanthocyanidin, proline, raffinose, riboflavin, salicylate, saponin, serine, sinapic acid, sodium, stearic acid, stigmasterol, stizolamine, thiamine, threonine, tryptophan, tyrosine, valine, vicine and wyerone.

Edibility

Pod and Seed: Edible.

Medicinal Uses

Plant: Is analgesic, androgenic, antibacterial, anticoronary, antiestrogenic, antifibrinolytic, antigonadotropic, antihaemolytic, antihaemorrhagic, antihepatotoxic, antihistaminic, anti-inflammatory, antimastalgic, antimicrobial, antimitogenic, antineurotic, antinociceptive, antioxidant, antiperoxidant, antiprostaglandin, antipsychotic, antipyretic, antiradicular, antiseptic, antispasmodic, antitussive, antiviral, aphrodisiac, CNS active, cancer preventive, capillary protective, cardiospasmic, cardiotonic, cerebrotonic, cholagogue, choleretic, cytoprotective, dermatogenic, diuretic, emetic, estrogenic, hallucinogen, hepatoprotective, hypercholesterolemic, hyperglycaemic, hypnotic, hyperlipidemic, immunomodulator, insulinotonic, laxative, lithogenic, neuroprotector, phagocytotic, refrigerant, respirostimulant, sedative, spermicide, tranquilizer, uterosedative, vasodilatory and vulnenary; useful in anorexia, acne, anaemia, anorexia, arrhythmia, atherosclerosis, arthritis, asthma, beri-beri, calculi, cancer of breast, colon, liver, skin, cataract, cephalalgia, chilblain, colitis, convulsions, delirium, dementia, depression, dermatitis, diabetes, dypsomania, antidote to aluminium, cadmium, lead, paracetamol poisoining, dysmenorrhoea, eczema, edema, epilepsy, gastritis, gingivitis, glaucoma, glossitis, goiter, hepatitis, herpes, hyperglycaemia, insomnia, impotency, ischaemia, keratitis, leprosy, leukaemia, lupus, lymphoma, measles, melanoma, menopausal disturbance, menorrhagia, migraine, myocardites, nephritis, neuralgia, neuroblastoma, neuropathy, odontalgia, orchitis, osteoarthritis, osteoporosis, pneumonia, polio, pruritus, psoriasis, retinopathy, rheumatism, scabies, thyroid disturbance, tumour of the breast, colon, fore stomach, liver, lung, ovary prostate, ulcer, vertigo and vitiligo.

Root: Diuretic and laxative.

Leaf: Diuretic.

Seed: Aphrodisiac, diuretic, emollient and resolvent; fresh seeds antifungal and estrogenic.

Bioactivity: Plant is inhibitor of 5-a-reductase, 5–lipoxygenase, ACE, aldose-reductase, α-amylase, COX-2, catechol–0-methyl transferase, DOPA-decarboxylase, histamine, histidine-kinase, insulinase, iodothyronine-deiodinase, lipase, lipoxygenase, lyase, MAO, NADH-ubiquinone-oxidoreductase, ornithine-decarboxylase, PAF, PTK, peroxidase and prostaglandin–synthesis; it is angiotensin receptor blocker, CNS depressant, calcium channel blocker, interferon synergist, metal chelator, monoamine precursor, pituitary sensitizer and quinone-reductase inducer.

Injected intravenously into rabbits, the seed extract produced haemoglobinuria and death; ethanol extract of seeds showed estrogenic activity; the seeds also contain chymotrypsin and trypsin inhibitors.

Toxicology: Seeds cause favism; inducing toxins are aglycone moieties of vicine and convicine; the toxic amino acids and pyrimidine glycosides in the leaf and seed also cause favism resulting in haemolytic anaemia, disturbance of functioning of RBC, oxidative stress, damaged erythrocytes, altered fat and mitochondrial metabolism, haemolysis and abnormal flow of intercellular electrolytes; seeds contain chymotrypsin and trypsin inhibitors.

Uses in Alternative Systems of Medicine

Unani:

Part used: Pod and seed.

Uses: Aphrodisiac, antiemetic, anti-inflammatory, expectorant and demulcent; useful in bronchitis, cough, diarrhoea, hoarseness of voice, inflammation of chest, sore throat, tonsillitis and vomiting; checks gum bleeding; pod pulp is dried and powdered and is applied to the eyes as eye tonic.

Ethnomedicine

Plant: Is aphrodisiac, diuretic, estrogenic, expectorant, stomachic and tonic; used for chest complaints, pneumonia, sclerosis, tumour and warts.

Non-medicinal Uses

Plant: Grown for fodder, green manure, hay, silage and as cover crop.

185. *Vicia hirsuta* (L.) S.F. Gray

Synonyms: *Cracca hirsuta* (L.) Gennari
Cracca minor Gren. & Godr.

Common Names: Hairy vetch, Tiny vetch.

Vernacular Names: Ben: Masur chana; **Hin:** Munmuna, Jhunjhuni ankari.

Distribution: Native of Africa, Asia including India, Europe and Middle East; grows in Bihar, Central Himalaya, Tamil Nadu, North West India and West Bengal in India.

Description: A climbing annual with pubescent stem; leaves pinnate, ending in a tendril; leaflets 5-10 pairs, truncate; flower white, tinged with pale blue on a raceme; pod oblong, oblique with 2 black seeds.

Edibility

Leaf: Eaten as pot herb.

Seed: Edible.

Medicinal Uses

Pod and seed: Blood enricher and promotes strength and vigour; useful in chronic malaria.

Ethnomedicine

Plant: Is depurative and tonic; useful in malaria.

Non-medicinal Uses

Plant: Used as fodder and green manure.

186. *Vicia sativa* L.

Synonyms: *Vicia communis* Rouy
Vicia cornigera St.- Amans
Vicia vulgaris Uspensky

Notes: Four subspecies, *amphicarpa, macrocarpa, nigra* and *sativa,* are recognized.

Common Name: Spring vetch.

Vernacular Names: Ben: Ankari; **Hin:** Akra, Akta; **Ori:** Rothi.

Distribution: Native of Africa, Asia, including India, Europe and Middle East; it grows in Gujarat, North West India, West Bengal, Tamil Nadu and North Eastern India.

Description: Erect herb; leaves pinnate with a terminal tendril; leaflets 4-8 pairs, oblong, oblanceolate, truncate, apiculate; flowers reddish or purplish, solitary or paired in leaf axils; pod 8-10 seeded; seeds brown.

Phytochemistry

Plant: Carotene, cryptoxanthin, lutein, neoxanthin, vicianine, violaxanthin and zeaxanthin.

Leaf: Sterol and sterol glucosides.

Pod: γ-L-glutamyl-aminobutyric acid and γ-L-glutamyl-β-alanine.

Seed: Betaine, cholesterol, choline, citronic acid, convicine, lecithin and vicine.

Edibility

Seed: Used as food.

Medicinal Uses

Seed: Used as astringent and remedy for diarrhoea; decoction of seed used in treatment of small pox and measles; seed floor is used as poultice.

Toxicology: The amino acids, β-cyano-L-alanine and γ-glutamyl-β-cyano-L-alanine in the seeds cause neurolathyrism and favism, which results in haemolytic anaemia, RBC dysfunction, oxidative stress, damaged erythrocytes, altered fat and mitochondrial metabolism, abnormal flow of intercellular electrolytes; the cyanogenic glycosides such as convicine, divicine and vicianin, cause endocrinological disorders, ambylopia, alteration in cardiac physiology, inhibition of respiratory enzymes and paralysis.

Ethnomedicine

Plant: Astringent; used for cancer, diarrhoea and tumour.

Non-medicinal Uses

Plant: Yields fodder, green manure and silage; grown as cover crop and hedge plant.

187. *Vigna aconitifolia* (Jacq.) Marechal

Synonyms: *Phaseolus aconitifolius* Jacq.
Vigna aconitifolius (Jacq.) Marechal

Common Names: Aconite bean, Dew bean, Mat bean and Moth bean.

Vernacular Names: Ben: Kheri; **Guj:** Math; **Hin:** Bhringga, Methkalai, Mot, Moth, Mothi, Urad; **Kan:** Madaki, Thuraka hesaru; **Mar:** Mat, Math, Matha, Matki; **Pun:** Bhionia; **Sans:** Amritha, Aranyamudga, Bassunta, Khandi, Krimilaka, Kulinaka, Madyaka, Makushtha, Makushthaka, Mapashtha, Mayastha, Mudjashtaka, Nigudhaka, Raajamudga, Vallimudga, Vanamudga, Varaka, Vasunta; **Tam:** Tulkapayir; **Tel:** Kunkuma pesalu, Minumulu and Pijjipesaru.

Distribution: Native of China; introduced into India; grows throughout Himalaya and southwards in India.

Description: Prostrate or diffused hairy herb with ascending branches; stipules lanceolate; leaves 3-foliolate; leaflets deeply lobed, oblanceolate, apiculate; flowers yellow on long peduncled raceme; bracteoles linear, twice as long as calyx; pod cylindrical, beaked, sparsely pubescent.

Phytochemistry

Plant: Calcium, carbohydrate, glutamic acid, dehydroxylase and phosphorus.

Pod: Carbohydrates, fat, fibre and proteins.

Seed: α-Linolenic acid, ascorbic acid, calcium, carbohydrates, chlorine, copper, cystine, fat, histidine, iron, isoleucine, leucine, linoleic acid, lysine, magnesium, manganese, methionine, niacin, oleic acid, palmitic acid, phenylalanine, phosphorus, phytin, potassium, protein, riboflavin, sodium, starch, stearic acid, sulphur, thiamine, tryptophan, valine and zinc.

Edibility

Pod: Used as vegetable.

Seed: Eaten as pulse; boiled and mixed with vegetables and consumed; ground into flour to make bread; seeds ground into flour and mixed with other flours for making unleavened bread.

Medicinal Uses

Plant: Is antiandrogenic, antianginal, antibacterial, anticancer, antifibrinolytic, antihepatotoxic, antihistaminic, antifertility, anti-inflammatory, antinociceptive, antiperiodontic, antipyretic, antiradicular, antiscorbutic, antiseptic, antispasmodic, antitumour, antiviral, apoptotic, astringent, cardioprotective, CNS depressant, diuretic, emetic, emollient, haemolytic, hepatoprotective, hypoglycaemic, hypotensive, hypolipidemic, immunostimulant, laxative, metastatic, sedative, uterorelaxant, vasodilatory and vulnerary; useful in curing acne, allergy, alopecia, anaemia, anorexia, anxiety, arrhythmia, arteriosclerosis, arthritis, asthma, back ache, beri-beri, cancer, canker, cardiospasms, cataract, chilblain, cold, colitis, convulsions, dandruff, delirium, dementia, depression, dermatitis, diabetes, dysmenorrhoea, eczema, edema, encephalopathy, epilepsy, gastritis, gingivitis, glaucoma, glycosuria, haemorrhage, hepatitis, herpes, hypertension, hyperglycaemia, impotency, insomnia, keratitis, lupus, leprosy, mastalgia, measles, menopausal disorders, menorrhagia, migraine, myocarditis, neuralgia, nephritis, pneumonia, polio, retinopathy, rheumatism, scabies, stomatitis, tumour, ulcer, vertigo, vitiligo and tumours.

Root: Narcotic.

Seed: Given as diet in flatulence and fever.

Bioactivity: Plant is a inhibitor of 5-α-reductase, ACE, aldose-reductase, insulinase and prostaglandin synthesis; is angiotensin receptor blocker, β-blocker, β-adrenergic receptor blocker, calcium channel blocker, copper antagonist, interferon synergist and monoamine precursor.

Toxicology: Seeds contain chymotryspin and trypsin inhibitors.

Uses in Alternative Systems of Medicine

Ayurveda:

Part used: Seed.

Uses: Fattening, stomachic and tonic; useful in abdominal glands, burning sensation, fever, "kapha", leprosy, respiratory troubles, tuberculosis, tumours, "vatha", vomiting and wasting diseases.

Siddha:

Part used: Seed.

Uses: Useful as a nutritive tonic.

Non-medicinal Uses

Plant: Cultivated for hey and green manure, checks erosion; highly esteemed as fodder.

Leaf: Used as fodder.

Seed: Good animal feed; starch obtained is used in textile sizing and calico printing.

188. *Vigna adenantha* (G. F. Meyer) Marechal & *al*.

Synonym: *Phaseolus adenanthus* G.F. Meyer

Vernacular Names: Sans: Aranyamudga; **Tam:** Kattupayru.

Distribution: Native of Africa, Asia and Pacific Ocean Islands; introduced into India.

Description: Herb with 3-foliolate leaves; leaflets entire, subacute, stipules basifixed; flowers showy; calyx tube campanulate, teeth distinct; corolla exerted, standard petal orbicular, auricled at the base, wing petal slightly adnate to the keel petal, incurved; pod recurved, glabrous, acuminate, septate between the seeds; seeds reniform.

Edibility

Root: Eaten at times of scarcity.

Seed: Edible.

Medicinal Uses

Plant: Decoction is used in bowel complaints and gonorrhoea.

Root: Used to stop excessive salivation; with sandalwood, fresh butter and sugar, it is used as liniment for sore eyes.

189. *Vigna angularis* (Willd.) Ohwi & H. Ohashi

Synonym: *Phaseolus angularis* (Willd.) Wight

Notes: The species is next in importance to soybean in China and Japan.

Common Name: Azuki bean.

Vernacular Name: Kan: Ghattada uddu.

Distribution: Native of Vietnam; introduced into India; in India, it is cultivated in Assam, Gujarat, Karnataka, Punjab, Sikkim and Tamil Nadu.

Description: Bushy, erect, hairy annual; leaves trifoliolate, leaflets entire or 3 lobed; flower bright yellow in clusters; pod cylindrical, glabrous, constricted between the seeds; seed oblong, white to blue black, mottled, hilum with a protruding ridge on the long side.

Phytochemistry

Plant: 2-Hydroxybehenic acid, 2-hydroxylignoceric acid, 2-hydroxypalmitic acid, ceramide, cerebroside, palmitic acid and sphingolipids.

Leaf: Kaempferol and Robinin.

Seed: (+) Catechin-7-β-D-glucopyranoside, 3-O-[β-D-glucopyranosyl(1→2)-β-D-glucuronopyranosyl] azukisapogenol, α-L-glutamyl-L-β-phenylalanine, 3-O-[β-D-glucopyranosyl (1→2)-β-D-glucuronopyranosyl] sapogenol, 3-O-[β-D-glucopyranosyl(1→2)-β-D-glucuronopyranosyl]sophoradiol, β-carotene, α-globulin, anhydrosophoradiol, arginine, ascorbic acid, aspartic acid, azukisaponin, calcium, carbohydrates, carnaubic acid, copper, cystine, daidzein, delphinidin-3-monoglucoside, furanomethanol-β-D-glucopyranoside, genistein, glutamic acid, glycine, gypsogenic acid; alanine, histidine, iron, isoleucine, leucine, lysine, magnesium, manganese, methionine, ononitol, palmitic acid, phenylalanine, phosphorus, phytosterol, pipecolic acid, potassium, proline, prosapogenin, protease, protein, raffinose, riboflavin, robinin, serine, sodium, soyasapogenol B, stachyose, stearic acid, stigmasterol, sucrose, thiamine, threonine, tryptophan, tyrosine, valine and zinc.

Edibility

Pod and Seed: Eaten boiled or mixed with rice; used for making confections, cakes and soup.

Medicinal Uses

Plant: Is antiadenomic, antiandrogenic, antibacterial, anticancer, anticoronary, antifibrinolytic, antihepatotoxic, antihistaminic, antihypertensive, anti-inflammatory, antilymphocytic, antimutagenic, antioxidant, antiperiodontic, antipyretic, antiseptic, antispasmodic, antiviral, apoptotic, cardioprotective, choleretic, diuretic, haemolytic, hepatoprotective, immunostimulant, laxative, memory enhancer, narcotic, sedative, tonic, vasodilator and vulnerary; is used in treating anorexia, allergy, alopecia, anxiety, arthritis, atherosclerosis, depression, diabetes, diarrhoea, gingivitis, herpes, hypertension, hypoglycaemia, insomnia, leukaemia, lupus, lymphoma, menorrhagia, nephritis, obesity, odontalgia, osteoporosis, paralysis, tumour of breast, cervix and lung, varicose veins and ulcer.

Root: Aphrodisiac, bitter and cold; cures bleeding wounds, burning sensation, discordance of *tridosha*, cough, diarrhoea, eye diseases, fever, indigestion, piles, rheumatism and swelling.

Bioactivity: Plant is inhibitor α-reductase, ACE, aldose-reductase, cyclo-oxygenase, glucosyl-transferase, iodothyronine-deiodinase, lipoxygenase, MAO, ornithine-decarboxylase, PAF, topoisomerase I and II, tyrosine, xanthine-oxidase, CAMP and phosphodiesterase; is angiotensin receptor blocker, β-blocker, calcium channel blocker, copper antagonist, interferon synergist, monoamine precursor, haemoglobin inducer, memory enhancer and quinone-reductase inducer.

Toxicology: Plant contains chymotrypsin and trypsin inhibitors.

Uses in Alternative Systems of Medicine

Ayurveda:

Part used: Whole plant and root.

Uses: Aphrodisiac; useful in burning sensation, consumption, cough, diarrhoea, eye diseases, fever, indigestion, piles, rheumatism and swelling; effective in diseases caused due to discordance of *tridosha*; arrests bleeding from wounds.

Medicinal Uses Outside India

China: Root pulp is used as a dressing for abscesses and boils.

Japan: Seeds are used in treating beri-beri.

Malay Peninsula: The leaf paste is applied on the stomach for stomach ache.

Ethnomedicine

Plant: Lactagogue; used in boils, constipation and fever; prevents miscarriage.

190. *Vigna dalzelliana* (O. Kuntze) Verdc.

Synonyms: *Phaseolus dalzellii* Cooke
Phaseolus pauciflorus Dalz.

Vernacular Names: Guj: Magavaine, **Mal:** Kaattu lunnu; **Mar:** Mugavel; **Sans:** Maasaparni, Mudgaavalli.

Distribution: Native of Cambodia, India, Laos, Philippines, Thailand and Vietnam; introduced into India; grows in Bihar, Gujarat, Karnataka, Kerala, Madhya Pradesh, Maharashtra, Orissa and Rajasthan in India.

Description: A slender climber with slender and glabrous leaves; leaflets entire or faintly lobed, green above, pale beneath; flowers pale yellow on capitate raceme; pod glabrous, subcylindrical, slightly recurved; seeds 6-10, subcylindrical, truncate, smooth and dark brown.

Edibility

Pod and Seed: Edible.

Medicinal Uses

Seed: Boiled seeds are very nutritious for patients recovering from acute illness and for those with enlarged liver and spleen; useful in relieving heat and burning of the eyes; seed powder is used to clean the body; when applied it reduces distention of the mammary glands and checks secretion of milk; used as an antidote for scorpion sting.

Non-medicinal Uses

Plant: Used as fodder.

191. *Vigna mungo* (L.) Hepper

Synonym: *Phaseolus mungo* L.

Common Name: Black gram.

Vernacular Names: Hin: Urad dhal; **Kan:** Uddu; **Mal:** Ulunnu; **Sans**: Mudga, Soopashresta; **Tam:** Pacchai payaru, Ulundu; **Tel:** Kandipappu, Minapappu.

Distribution: Native of India, cultivated through out plains, mainly grows in tropics.

Description: An erect hairy subscandent, annual with twining branches covered with reddish brown hairs; leaves trifoliolate; leaflets ovate, entire, deltoid or round at the base, membranous; flower small and yellow on elongated peduncles; pod hairy, subcylindrical, unseptate between the seeds with short hooked beak; seeds 1-4 per pod, black, mottled, grey or green with protruding white hilum.

Phytochemistry

Pod: Calcium, carbohydrate, fat, phosphorus and protein.

Seed: β-Carotene, 2'-hydroxygenistein, 2'-hydroxy daidzein, 4'-O-methyl kievitone, 5-deoxy kievitone, allantoin, allantoinase, arsenic, ascorbic acid, aspargines, behenic acid, calcium, carbohydrates, cellobiose, choline, copper, cyclokievitone, dalbergioidin, diethylstilbesterol, eurenol, genistein, glycinol, genistin, glutathione, hemicellulose, iron, isoferreirin, kievitone, lactose, lead, lecithin, lignoceric acid, linoleic acid, magnesium, maltose, melibiose, mercury, niacin, oleic acid, palmitic acid, pectin, phosphorus, phytin, potassium, protein, raffinose, riboflavin, sodium, stachyose, starch, stearic acid, sucrose, sulphur, thiamine, verbascose, vestitol and vitamin B_{12}.

Seed sprout: β-Carotene, coniferin, dehydro ascorbic acid, hydroxylamine, indole-3-acetamide, indole-3-carboxyaldehyde, p-hydroxycinnamic acid, phenylalanine and shikimic acid.

Edibility

Green pod: Edible.

Seed: Chief constituent of papads, idli, vada, sweet meats and pancakes.

Medicinal Uses

Plant: Is antiadenomic, antiandrogenic, antibacterial, anticancer, anticoronary, antifibrinolytic, antihepatotoxic, antihistaminic, anti-inflammatory, antilymphocytic, antimutagenic, antioxidant, antiperiodontic, antipyretic, antiseptic, antispasmodic, antiviral, apoptotic, cardio-protective, choleretic, diuretic, haemolytic, hepatoprotective, immunostimulant, laxative, memory enhancer, narcotic, sedative, tonic, vasodilator and vulnerary; is used in treating anorexia, allergy, alopecia, anxiety, arthritis, atherosclerosis, depression, diabetes, diarrhoea, gingivitis, herpes, hypertension, hyperglycaemia, insomnia, leukaemia, lupus, lymphoma, menorrhagia, nephritis, obesity, odontalgia, osteoporosis, paralysis, tumour of breast, cervix and lung, varicose veins and ulcer.

Root: Narcotic; used as poultice for abscesses and inflammations.

Seed: Is aphrodisiac, astringent, cooling, digestive, diuretic, fattening, galactagogue, laxative, and tonic; strengthens the eye, cures biliousness, blood diseases, cephalagia, dropsy, hepatic trouble, nervous disorders and rheumatism; baked gram is a nutritive light diet; useful in affections of liver and nervous system, burning of the eyes, cystitis, diarrhoea, dysentery, fistula, catarrh, leucorrhoea, paralysis, piles, seminal debility and uterine disorders; reduces distension of mammary glands and lowers cholesterol; seed paste is externally applied for contracted knee, rheumatic pain and stiff shoulder.

Bioactivity: Plant is inhibitor α-reductase, 5-lipoxygenase, ACE, aldose-reductase, cyclo-oxygenase, glucosyl-transferase, iodothyronine-deiodinase, lipoxygenase, MAO, ornithine-decarboxylase, topoisomerase I and II, tyrosine, xanthine-oxidase, CAMP and phosphodiesterase; is angiotensin receptor blocker, β-blocker, calcium channel blocker, haemoglobin inducer, memory enhancer and quinone-reductase inducer.

Toxicology: Seeds contain chymotrypsin and trypsin inhibitors.

Uses in Alternative Systems of Medicine

Ayurveda:

Part used: Seed.

Uses: Aphrodisiac, galactagogue and tonic; used in diseases of nervous system, rheumatism and ulcers; retards loss of vision and hastens suppuration.

Chinese:

Part used: Leaf, flower, pod and seed.

Uses: Leaf is used in cholera; flower is used to reduce alcoholic intoxication; pod is used in obstinate dysentery; seed is used as antifebrile, carminative and resolvent; useful in obstinate dysentery and bladder disorders in aged people; used as a poultice for abscesses, boils and in corneal opacity.

Siddha:

Part used: Root and seed.

Uses: Aphrodisiac and narcotic; useful in diseases of "vatam", arthritis, cough, fever, nervous disorders, paralysis, piles and rheumatism.

Unani

Part used: Seed.

Uses: Astringent and tonic; enriches the blood; useful in biliousness, bronchitis, eye diseases, fever, headache, inflammation, kidney diseases and throat inflammation.

Ethnomedicine

Plant: Used as carminative, colic, diuretic, narcotic and resolvent; used for abscesses, bladder complaints, boils, cholera, dropsy, dysentery, fever, fistula, gravel, hepatitis, impotency, rheumatism, scabies, smallpox, splenitis, tumour of the abdomen and lachrymal glands.

Medicinal Uses Outside India

China: Seeds are used as carminative, depurative and resolvent; used to treat febrile disorders, obstinate dysentery, small pox and urinary trouble.

Indochina: Seeds are used as diuretic; used in treating blisters on the skin caused by centipede.

Indonesia: Seeds are used as tonic; useful in dyspepsia, enteritis, smal pox eruptions and tuberculosis.

Malay Peninsula: Used in treating vertigo.

Non-medicinal Uses

Plant: Grown as cover crop; used as fodder.

192. *Vigna radiata* (L.) R. Wilczek

Synonym: *Phaseolus radiatus* L.

Notes: Three varieties, *radiata, sublobata* and *glabra* are recognized; it ranks high among the pulse crops of India; rich in protein.

Common Names: Green gram, Golden gram.

Vernacular Names: Hin: Moong; **Kan:** Hesaru; **Mal:** Cherupayaru; **Tam:** Pasipayaru; **Tel:** Pacchapesaru, pesarupappu.

Distribution: Native of Pakistan; introduced into India; cultivated through out India.

Description: Erect or semi erect, hairy annual; diffusely branched with tendency to twine at the tips; leaves alternate, trifoliolate, dark green with scattered hairs on both sides; leaflets large, ovate, entire, rarely trilobed; stipules peltate; flowers pale yellow, crowded, on axillary or terminal racemes in clusters of 10-25; pod black, grey or brown, reflexed with short hairs; seeds 10-20, globose or oblong, slightly flat, green but may be olive, yellow, brown or purplish brown, marbled or mottled, smooth or marked with crenulate lines.

Phytochemistry

Plant: 3-O-β-D-Glucuronide, aspartic acid, biochanin-A, chlorogenic acid, daidzein, formononetin, genistein, homoglutathione, isoleucine, isoquercetin, isoscopoletin, kaempferol, kaempferol-3-O-β-D-glucuronic acid, leucine, naringenin, neochlorogenic acid, ononin, p-coumaric acid, phenylalanine, quercetin-3-O-β-dophoroside, robinin, scopoletin, umbelliferone and γ-glutamic acid.

Leaf: β-Sitosterol, choline, cystine, glycerol, isoleucine, linolenic acid, L-pipecolic acid, lysine, threonine, tryptophan, tyrosine¸ valine and vitexin.

Seed: Alanine, α-linolenic acid, arabinose, arachidic acid, ascorbic acid, ascorbigen, aspargine, aspartic acid, behenic acid, β-carotene, β-sitosterol, cadmium, calcium, campesterol, cerotic acid, choline, chlorogenic acid, copper, cyanidin-3-glucoside, cystine, daidzein, delphinidin-3,5-diglucoside, folacin, folic acid, fucose, galactose, galactokinase, genistein, glucose, glucuronic acid, glutamic acid, glycine, hemicellulose, homoglutathione, histidine, inositol, iodine, iron, isoleucine, leucine, linoleic acid, linolenic acid, lysine, magnesium, manganese, mannose, matiresinol, methionine, niacin, oleic acid, palmitic acid, pantothenic acid, pectin, phenylalanine, phosphorus, phytin, phosphoglucomutase, phytosterol, potassium, protein, pyridoxine, quercetin, raffinose, rhamnose, riboflavin, ribose, robinin, rutin, salicylate, saponin, scopoletin, isoscopoletin, serine, shikimic acid, sodium, soyasapogenol-C, stachyose, stearic acid, stigmasterol, sucrose, sulphur, thiamine, threonine, tryptophan, tyrosine, umbelliferone, valine, verbascose, vitamin-B_{12}, xylose and zinc.

Sprout: Adenosine-5'-monophosphate, adenosine diphosphate, alanine, α-linolenic acid, arginine, ascorbic acid, β-carotene, β-sitosterol, calcium, campesterol, coumesterol, cystine, daidzein, foliacin, glutamic acid, glutathione, glycine, guanosine-5'-monophosphate, histidine, homoglutathione, imidazole-acrylic acid, iron, isoleucine, isoquercitrin, kaempferol, kaempferol-7-O-rhamnoside, leucine, linoleic acid, magnesium, manganese, methionine, myristic acid, niacin, oleic acid, palmitic acid, pantothenic acid, phenylalanine, phosphorus, phytosterol, potassium, protein, quercetin, riboflavin, robinin, rutin, serine, sodium, stearic acid, stigmasterol, thiamine, threonine, tryptophan, tyrosine, uridine-monophosphate and uridine-diphosphate.

Edibility

Pod: Edible.

Seed: Parched seeds made into porridge or ground into flour in bread making; it is sprouted, fried, salted and eaten.

Medicinal Uses

Plant: Is antiadenomic, antiandrogenic, antibacterial, anticancer, anticoronary, antifibrinolytic, antihepatotoxic, antihistaminic, anti-inflammatory, antilymphocytic, antimutagenic, antioxidant, antiperiodontic, antipyretic, antiseptic, antispasmodic, antiviral, apoptotic, cardioprotective, choleretic, diuretic, haemolytic, hepatoprotective, hypoglycaemic, immunostimulant, laxative, memory enhancer, narcotic, sedative, tonic, vasodilator and vulnerary; is used in treating anorexia, allergy, alopecia, anxiety, arthritis, atherosclerosis, depression, diabetes, diarrhoea, gingivitis, herpes, hypertension, insomnia, leukaemia, lymphoma, menorrhagia, nephritis, obesity, odontalgia, osteoporosis, paralysis, tumour of breast, cervix and lung, lupus, ulcer and varicose veins.

Root: Narcotic.

Seed: Acrid, aphrodisiac, astringent, constipating, cooling, digestive, diuretic, galactagogue, styptic and tonic; used in affections of nervous system, agalactia, asthma, cephalagia, cough, dropsy, epilepsy, epistasis, eye diseases, fatigue, fever, general debility, gout, polyuria and wound; decoction of the seed given in beri beri, gonorrhoea, haemorrhoids, leprosy, leucoderma, neuralgia, ophthalmia, paralysis, rheumatism, scabies, skin diseases, strangury, vertigo and vitiated conditions of "pitta" and "kapha".

Bioactivity: Plant is inhibitor α-reductase, 5-lipoxygenase, ACE, aldose-reductase, cyclo-oxygenase, glucosyl-transferase, iodothyronine-deiodinase, lipoxygenase, MAO, ornithine-decarboxylase, PAF, topoisomerase I and II, tyrosine, xanthine-oxidase, CAMP and phosphodiesterase; is angiotensin receptor blocker, β-blocker, calcium channel blocker, haemoglobin inducer, memory enhancer and quinone-reductase inducer.

Toxicology: Seeds contain chymotrypsin and trypsin inhibitors.

Uses in Alternative Systems of Medicine

Ayurveda:

Part used: Seed.

Uses: Aphrodisiac, appetizer, diuretic, galactagogue, laxative and tonic; used in asthma, diseases of "kapha" and "pitta", fatigue, epilepsy, fever, gout, obesity, piles, polyuria and wounds.

Chinese:

Part used: Leaf and seed.

Uses: Used in abdominal pain, carcionomatous and purulent swellings, difficult child birth, dropsy, fever, spleenic disorders, threatened abortion, urinary disorders and wounds.

Siddha:

Part used: Seed.

Uses: Galactagogue; useful in inflammations and diseases of "pittam".

Unani

Part used: Seed.

Uses: Aphrodisiac, diuretic, galactagogue and styptic; useful in epistasis, gonorrhoea, leucoderma and scabies.

Medicinal Uses Outside India

China: Seeds are used in beri-beri, carcinomatous swellings, dropsy, obstetrical difficulties, and to dissolve blood clots; seed sprouts are fed against threatened abortions; the seeds are used as febrifuge, diuretic and laxative.

Indochina: Beans are given to cure beri-beri.

Japan: Beans are given in beri-beri.

Korea: Seeds are used as a remedy for dropsy and diarrhoea.

Ethnomedicine

Plant: Used as abortifacient, carminative, colic, diuretic, lactagogue, narcotic, restorative and resolvent; used for abscesses, beri-beri, bladder disorders, boil, carcinoma, cholera, dysentery, fever, purpura, smallpox, swelling, tumour, urinogenital complaints and vertigo.

Non-medicinal Uses

Plant: Grown for fodder, green manure and as cover crop.

Seed: Used as fodder; germinated seed is a substitute for making corn steep liquor; used as supplement in the culture medium for growing *Streptomyces*.

193. *Vigna sublobata* (Roxb.) Babu & Sharma

Distribution: In India, it grows in Assam, Bihar, Karnataka, Kerala, Madhya Pradesh, Maharashtra, Orissa, Rajasthan, Tamil Nadu and Uttar Pradesh.

Description: A suberect twining annual with short slender stem clothed with reddish brown hairs; leaves 3-foliolate; leaflets ovate, lobed, silky, hairy on both sides, prominently 3 nerved at the base; flowers yellow on 6-12 flowered raceme; pod cylindrical, beaked, 2 valved, septate between the seeds, sparsely tomentose; seeds 4-6 and black.

Edibility

Pod and Seed: Edible.

Medicinal Uses

Root: Anti-inflammatory, aphrodisiac, constipating, depurative, digestive, emollient, febrifuge, galactagogue, haemostatic, refrigerant and tonic; useful in agalactia, burning sensation, colic, consumption, cough, diarrhoea, dyspepsia, emaciation, facial paralysis, fever, flatulence, general debility, haemorhage, haemoptysis, hemiplegia, inflammation, leprosy, pyrexia, seminal weakness and skin diseases.

Uses in Alternative Systems of Medicine

Ayurveda:

Part used: Seed.

Uses: Aphrodisiac and constipating; useful in cough, emaciation, haematemesis and pyrexia; improves colour of the skin and promotes hair growth and growth of lymphatic tissue; removes impurities in the blood; cures diseases caused by disturbed "vatha" and "pitha" and strengthens the body.

Non-medicinal Uses

Plant: Used as fodder and forage; grown to increase the soil fertility and to control soil erosion.

194. *Vigna trilobata* (L.) Verdc.

Synonyms: *Dolichos trilobatus* L.
Phaseolus triolobus sensu auct.
Phaseolus trilobatus (L.) Schreber

Common Name: Wild gram.

Vernacular Names: Hin: Mugani, Mugawana, Mungani, Thrianguli; **Kan:** Kohesaru, Mudgaparni, Nari hesaru; **Mal:** Kaattupayar; **Mar:** Arkamath, Janglimath, Ranmath, Ranmuga; **Sans:** Aranyamudga, Hrasva, Kaakamudga, Karanjika, Koshila, Kshudrasaha, Kurangika, Marjaragandhika, Mudgaparni, Saha, Shimbiparni, Shimbiparnika, Shurpaparni, Vanajaringini, Vanamudga, Vanodbhava, Vanya; **Tam:** Naripayar, Panipayir; **Tel:** Pillippersara.

Distribution: Native of Bhutan, China, India, Indonesia, Myanmar, Pakistan, Srilanka, Taiwan and Vietnam; grows in Andhra Pradesh and Himachal Pradesh in India.

Description: Diffuse, trailing pubescent herb; stipules, peltate, spurred; leaflets ovate, deltoid, acute; the middle leaflet is largest, spathulate, lateral leaflets with subacute or obtuse apex, hairy on both sides; flowers yellow in subcapitate raceme; pod oblong, slightly curved, hispid; seeds 2-5, brownish black.

Phytochemistry

Root: Doliroside A and medicagenic acid.

Leaf: Alantoin.

Seed: β-Sitosterol, epifriedelin, friedelin, galacturonic acid, leucine, lysine, medicagenic acid, methionine, phenylalanine, stigmasterol, tannin, tryptophan, tyrosine, uridine diphosphate, valine, vitexin and γ-glutamylphenylalanine.

Edibility

Pod and Seed: Edible.

Medicinal Uses

Plant: Aphrodisiac, bitter, blood purifier, constipating and rejuvenating; cures cold, cough, edema, eye troubles, haematemesis, haemorrhage, pyrexia and tuberculosis; is prescribed in fever.

Root: Used in constipation, ophthalmia, skin diseases and tumour.

Leaf: Antibilious, cooling, laxative, sedative and tonic; used as cataplasm for weak eyes and fever; leaf poultice is used as an antidote to scorpion sting.

Pod: Anthelmintic, aphrodisiac, astringent, cooling and styptic; useful in biliousness, burning sensation, consumption, cough, dysentery, fever, gout, inflammation, piles and thirst.

Seed: Astringent and cooling; used as diet in fever to relieve thirst; strengthens weak eyes; relieves heat and burning of the eyes; used as an external application in haemorrhoids; seed soup is nutritious.

Uses in Alternative Systems of Medicine

Ayurveda:

Part used: Whole plant, root and pod.

Uses: Anthelmintic, aphrodisiac, astringent, blood purifier, cooling and styptic; useful in biliousness, consumption, cough, dermatosis, dysentery, edema, eye complaints, fever, gout, haemorrhages, inflammation, piles, rheumatic conditions, rat bite poisoning, sprue, thirst, *tridosha* and tuberculosis; increases secretion of semen and breast milk.

Siddha:

Part used: Leaf and seed.

Uses: Expectorant and refrigerant; useful in eczema, leucorrhoea and micturition.

Non-medicinal Uses

Plant: Used as green fodder.

Root: Source of arrowroot like starch.

195. *Vigna umbellata* (Thunb.) Ohwi & H. Ohashi

Synonym: *Dolichos umbellatus* Thunb.

Common Names: Climbing mountain bean, Oriental bean, Rice bean.

Vernacular Names: Hin: Sutri, Ghurush; **Mal:** Kaattulunnu; **Pun:** Ghurush; **Sans:** Maasaparni.

Distribution: Native of Asia including India; grows in Assam, Gujarat, Himachal Pradesh, Kerala, Manipur, Meghalaya, Nagaland, Rajasthan, Sikkim, Uttar Pradesh and West Bengal in India.

Description: Annual or short lived perennial with suberect stem with vining grooved branches, covered with short white hairs; leaves trifoliolate; stipules conspicuous, ovate-lanceolate; leaflets ovate, entire, sometimes 3-lobed; inflorescence erect axillary raceme with 5-20 flowers; flowers 2-3 together at nodes, bright yellow with linear-lanceolate bracteoles; pod long, slender, 6-12 seeded, glabrous, falcate; seeds oblong, subtrapezoidal, smooth, green, yellow, brown, black, speckled or mottled.

Phytochemistry

Seed: Arabinose, arachidic acid, behenic acid, galactose, galacturonic acid, lignoceric acid, linoleic acid, myristic acid, oleic acid, palmitic acid, stearic acid, volatile oil and xylose.

Edibility

Leaf and Pod: Eaten as vegetable.

Seed: Ripe seeds are boiled or fried and eaten.

Medicinal Uses

Seed: Aperient and nutrient.

Toxicology: Seeds contain chymotrypsin and trypsin inhibitors.

Uses in Alternative Systems of Medicine

Chinese:

Part used: Pod.

Uses: Used to restore kidney strength, to harmonize abdominal organs, to quench excess thirst; given in diarrhoea, nausea and polyurea.

Non-medicinal Uses

Plant: Useful as fodder and green manure.

196. *Vigna unguiculata* (L.) Walp.

Synonym: *Vigna catjang* Savi.

Notes: Five subspecies, *unguiculata, cylindrica, sesquipedalis, dekindtiana* and *mensensis* are recognized; there are many cultivated forms in this species resulting in disputed nomenclature; *sesquipedalis* and *cylindrica* are cultivar groups.

Common Names: Asparagus bean, Black eyed bean, Black-eyed pea, Cow pea, Long bean, Yard long bean.

Vernacular Names: Hin: Lobiyaa; **Kan:** Alasande, Alasandi, Alasandige, Halasande, Kaaraamani, Kursana pairu, Kursandi, Tadagani, Taranikaayi; **Mar:** Alasunda, Chavali; **Sans:** Raajmaasha; **Tam:** Caramunnipayara, Karamani, Thattapayir; **Tel:** Alasandalu, Kaaraamanulu.

Distribution: Native of Africa; introduced and cultivated throughout India.

Description: Annual herb, erect or suberect, spreading, glabrous; the first formed leaves above the cotyledons are simple and opposite, subsequent leaves are alternate, the terminal leaflet often bigger and longer than the two asymmetrical laterals; leaflets ovoid, rhomboid, entire or slightly lobed, apex acute, lateral leaflets oblique; inflorescence is 2-4 flowered axillary raceme on short curved peduncles; calyx campanulate with triangular teeth, upper 2 teeth connate and longer than the rest; corolla dull white, or yellow; keel petal truncate; pod curved or straight or coiled; seed globular to reniform, smooth or wrinkled, red, black, brown, buff or white, spotted, marbled, speckled or blotched.

Phytochemistry

Plant: (–) Phaseollidin, (–) phaseollin, β-carotene, α-galactosidase, β-galactosidase, α-galactoside, β-glucoside, β-linolenic acid, α-mannoside, acetylglucosaminidase, alanine, ascorbic acid, aspartic acid, coumesterol, cysteine, dalbergioidin, D-arabinose, D-glucose, dolichin A and B, erucic acid, genistein, glycine, hydroxygenistein, iron, isoferreirin, kievitone, L-ascorbic acid, linoleic acid, linolenic acid, L-rhamnose, magnesium, methylisoflavone, myristic acid, neohesperidoside, niacin, oleic acid, palmitic acid, palmitoleic acid, phosphorus, potassium, protein, psoralidin, riboflavin, serine, sodium, stearic acid, streptogenin, thiamine and urease.

Pod: α-Linolenic acid, arginine, ascorbic acid, β-carotene, calcium, carbohydrates, cystine, fat, histidine, iron, isoleucine, leucine, linoleic acid, lysine, magnesium, methionine, myristic acid, niacin, oleic acid, palmitic acid, palmitoleic acid, phenylalanine, phosphorus, potassium, protein, riboflavin, sodium, stearic acid, thiamine, threonine, tryptophan, tyrosine and valine.

Seed: α and β Pinene, β-amylase, β-carotene, α-linolenic acid, β-ocimene, alanine, albuminoids, amylose, arachidic acid, arginine, ascorbic acid, aspartic acid, behenic acid, boron, calcium, carbohydrates, cedrene, choline, chromium, cobalt, copper, coumesterol, cycloartenol, cystine, daidzein, daucosterol, fats, folic acid, genistein, glucose-6-phosphatase, glutamic acid, glutamyl-methyl-cysteine, glycine, hexyl acetate, histidine, homopterocarpin, iron, isoleucine, kievitone, lead, leucine, lignoceric acid, limonene, linoleic acid, lysine, magnesium, manganese, matiresinol, methionine, myristic acid, niacin, oleic acid, ononitol, palmitic acid, palmitoleic acid, pantothenic acid, phenylalanine, phenylphosphatase, phosphorus, phosphorylase, phytate, phytic acid, potassium, proline, pyrophosphatase, raffinose, riboflavin, rubidium, salicylate, selenium, serine, sodium, stachyose, stearic acid, stigmasterol, strontium, sucrose, thiamine, threonine, tryptophan, tyrosine, urease, valine, verbascose, vignafuran, vignalin, zinc and zirconium.

Edibility:

Shoot, Leaf, Pod, Seed: Edible; used in salads; seed flour is cooked to make sweet meats, cakes and porridge.

Medicinal Uses

Plant: Is analgesic, androgenic, antianginal, anticolic, anticonvulsant, antihistaminic, antihypercholesterolemic, anti-inflammatory, anti-ischaemic, antilipidemic, antimastalgic, antimicrobial, antimyocarditic, neuralgic, antinociceptive, antioxidant, antipelargic, antipyretic, antiseptic, antispasmodic, antitumour, antiviral, aphrodisiac, astringent, cardioprotective, cerebrotonic, cholagogue, cytotoxic, demulcent, diuretic, emetic, emollient, estrogenic, haemolytic, hepatoprotective, hypnotic, hypourecemic, immunomodulator, laxative, metastatic, phagocytotic, sedative, tranquilizer, uterorelaxant and vulnerary; used in anorexia, acidity, acne, dermatitis, allergy, alcoholism, anaemia, anxiety, arrhythmia, atherosclerosis, arthritis, asthma, back ache, beri-beri, cancer of breast, oesophagus, canker, cardiospasm, cataract, cephalagia, cervical dysplasia, chilblain, convulsions, delirium, dementia, depression, dermatitis, diabetes, dysmenorrhoea, eczema, edema, encephalopathy, epilepsy, fistula, glaucoma, herpes, hypertension, hyperglycaemia, impotency, insomnia, keratitis, leprosy, leukaemia, lupus, lymphoma, measles, melanoma, menorrhagia, migraine, myalgia, nephritis, neuroblastoma, neuropathy, obesity, ophthalmia, orchitis, osteoarthritis, osteoporosis, photophobia, plague, pneumonia, polio, polyp, psoriasis, retinopathy, rheumatism, stomatitis, stress, tumour of breast, oesophagus, lung, ovary, stomach and brain, ulcer, vertigo and vitiligo.

Seed: Antibilious, aphrodisiac, appetizer, astringent, diuretic, galactagogue, laxative, tonic and vermifuge; useful in liver complaints and jaundice; used as pessary for haemorrhoids and in the preparation of steroidal hormones; as a part of the diet it helps in reducing blood sugar and cholesterol.

Bioactivity: Plant is a inhibitor of 5-α-reductase, ACE, ATPase, aldose-reductase, aromatase, DOPA-decarboxylase, histidine-kinase, insulinase, lipase, MAO, MDH, NADH-ubiquinone-reductase, ornithine-decarboxylase, PTK, polyamine synthesis, prostaglandin synthesis, protein-kinase, topoisomerase I and II, tyrosine-kinase; is angiotensin receptor blocker, adrenergic receptor blocker, β-blocker, calcium antagonist, calcium channel blocker, copper antagonist, monoamine precursor, quinone-reductase-inducer.

Toxicology: Seeds contain chymotrypsin and trypsin inhibitors; cause flatulence, indigestion and nausea.

Uses in Alternative Systems of Medicine

Ayurveda:

Part used: Seed.

Uses: Aphrodisiac, appetizer, diuretic, galactagogue, laxative and tonic; useful in hyperacidity.

Siddha:

Part used: Root and seed.

Uses: Aphrodisiac; useful in diseases of "vatam", cough, diarrhoea, dysentery, dyspnoea, fever, filariasis and hypertension.

Unani:

Part used: Seed.

Uses: Is analgesic, demulcent, diuretic, emmenogogue, galactagogue and resolvent; increases sperm count and cleanses the uterus and is an analgesic in kidney pain.

Medicinal Uses Outside India

Indonesia: Pulverized leaves with alum are applied on the breast to stop milk flow.

Malay Peninsula: Stem and leaf juice is dropped into the ears for ear ache.

Ethnomedicine

Plant: Used as astringent, cyanogenic and diuretic; used for adenopathy, biliousness, diarrhoea, dysentery, dysuria, ear pain, fistula, jaundice, kidney disorders, leprosy, liver complaints, measles, nausea, neuralgia, pleurisy, pneumonia, polyuria, smallpox, sore throat and tumour.

Non-medicinal Uses

Plant: Used as green manure and fodder.

Stem: Yields fibre.

Leaf: Yields a green dye when processed with turmeric.

Seed: Starch from seed is used in textile sizing and as thickening agent in calico printing; roasted seeds are used as substitute for coffee; seed powder is a nutrient supplement to culture media for *Streptomyces*.

197. *Wisteria sinensis* (Sims) Sweet

Synonym: *Wisteria chinensis* DC.

Common Names: Blue acacia, Chinese wisteria.

Distribution: Native of China; introduced into India; grows in North Eastern India in the hills.

Description: A woody climbing shrub, flowers profuse, pale purple or bluish violet, sweet scented.

Phytochemistry

Plant: Afrormosin.

Stem bark: α-Amyrin, hydroxytriacontane, kaempferol, protosterol, triacontanol and wistarin.

Leaf: Apigenol-7-rhamnoglucoside, flavanols, luteolol-7-glucorhamnoside and luteolol-7-rhamnoglucoside.

Edibility:

Flower: Edible.

Medicinal Uses

Seed: Diuretic.

Toxicology: Wisterin and resin in the bark, cause diarrhoea and vomiting.

Uses in Alternative Systems of Medicine

Chinese:

Part used: Seed.

Uses: Useful as cardiotonic and diuretic.

Ethnomedicine

Plant: Diuretic.

Non-medicinal Uses

Plant: Grown as an ornamental and hedge plant.

Leaf: Used as substitute for tea.

198. *Zornia gibbosa* Span

Synonyms: *Zornia angustifolia sensu* Bojer
Zornia cantonensis Mohl.
Zornia diphylla sensu auct.
Zornia graminea Span.
Zornia pratensis sensu Mohl.

Vernacular Names: Guj: Samrapani; **Kan:** Nellujollusoppu; **Mal:** Nelammari; **Tam:** Nelammari.

Distribution: Native of Asia including India and Australasia; introduced into Indian Ocean and Pacific Ocean Islands; grows in Bihar, Himachal Pradesh, Karnataka, Madhya Pradesh, Maharashtra, Punjab, Rajasthan, Tamil Nadu and West Bengal in India.

Description: A diffuse annual with zigzag wiry branches; leaflets 2, lanceolate, acute, glabrous, dotted with black glands; stipules lanceolate with a long spur; flowers yellow on 3-12 flowered raceme; bracts ovate, acute, dotted; pod 1-6 jointed, pubescent, prickly.

Medicinal Uses

Plant: Useful in dysentery.

Root: Soporific; given to induce sleep in children.

Ethnomedicine

Plant: Soporific; used in burns, convulsion, dysentery, fever, sores and spasm.

Non-medicinal Use

Plant: Used as fodder and green manure.

Fig. 1. *Abrus precatorius flower*

Fig. 2. *Abrus precatorius pod with seeds*

Fig. 3. *Abrus pulchellus*

Fig. 4. *Arachis hypogaea*

Fig. 5. *Butea monosperma*

Fig. 6. *Butea monosperma*

Fig. 7. *Cajanus cajan*

Fig. 8. Pods and seeds of: *Castanospernum australe*

Fig. 9. *Clitoria ternatea* (white variety)

Fig. 10. *Clitoria ternatea* (Blue variety)

Fig. 11. *Crotalaria juncea*

Fig. 12. *Cytisus scoparius*

Fig. 13. *Dalbergia sissoo*

Fig. 14. Pods of *Dalbergia sissoo*

Fig. 15. *Erythrina crista-galli*

Fig. 16. *Erythrina*

Fig. 17. Rhizome of *Glycyrrhiza glabra*

Fig. 18. *Gliricidia sepium*

Fig. 19. *Lotus corniculatus*

Fig. 20. *Millettia peguensis*

Fig. 21. Pods of *Mucuna monosperma*

Fig. 22. *Mucuna pruriens*

Fig. 23. Pods of *Mucuna pruriens* var *pruriens*

Fig. 24. *Mucuna pruriens* var. *utilis*

Fig. 25. Tender pods of *Mucuna pruriens* var *utilis*

Fig. 26. Dry pods of *Mucuna pruriens* var *utilis*

Fig. 27. Seeds of *Mucuna pruriens* var *utilis*

Fig. 28. Flower buds of *Pongamia pinnata*

Fig. 29. Flowers of *Pongamia pinnata*

Fig. 30. Pods of *Pongamia pinnata*

Fig. 31. *Pterocarpus marsupium*

Fig. 32. *Trifolium dubium*

Fig. 33. *Trifolium pratense*

Fig. 34. *Trifolium repens*

Fig. 35. *Trigonella foenum-graecum*

Fig. 36. *Trigonella foenum-graecum*

Fig. 37. *Ulex europaeus*

Fig. 38. *Vicia sativa*

Fig. 39. *Wisteria sinensis*

❑❑❑

CHAPTER-26

Mimosaceae

1. *Acacia auriculiformis* Benth.

Synonym: *Acacia auriculaeformis* Benth.

Common Name: Ear pod wattle.

Distribution: Native of Australasia, Irian Jaya and Moluccas; introduced into India; cultivated throughout India.

Description: Unarmed tree with white bark; branches glabrous; leaves phyllodic; phyllode narrow at both ends; inflorescence an axillary spike with fragrant yellow flowers; pod spirally twisted and woody.

Phytochemistry

Plant: Auriculoside and α-spinasterol.
Stem bark: Flavones.
Wood: Chalcone, flavanone, flavonol, isotetracacidin, mollisacacidin, tetracacidin and α-dihydroflavonol.
Seed: Fatty oil.

Medicinal Uses

Plant: Is hypotensive and CNS depressant.
Toxicology: Plant is a central nervous system depressant.

Non – Medicinal Uses

Plant: Is a good soil binder to control erosion; grown as an ornamental plant and shade tree and avenue tree.
Bark: Used in tanning industry.
Wood: Pulp is used for making wrapping paper, writing and printing paper; yields charcoal and fire wood; used for furnitures.

2. *Acacia caesia* (L.) Willd.

Synonyms: *Acacia intsia sensu auct.*
Acacia intsia Willd., var., *caesia* (L.) Baker
Mimosa caesia (L.) Willd.

Notes: Several varieties are recognized; var. *caesia* grows in India and Srilanka.
Common Name: Black catechu.
Vernacular Names: Ben: Shiahkanta; **Hin**: Aila, Shiahkanta; **Kan**: Antarike, Chende mullu, Kaadu seege; **Mal:** Attushini, Inna, Inja, Kattasinikka; **Mar**: Chilari; **Pun**: Arlu; **Sans**: Nikunji, Nikunjika; **Tam:** Ingi, Kaari indu, Ingikai, Kaatusikka; **Tel:** Korinta, Korinda, Thellakorinda.

Distribution: Native of Asia, including India; grows throughout India.

Description: A large woody climber with grayish brown rough bark; the branches are glabrous with numerous hooked thorns; young branches and leaves minutely tomentellous; stipules ovate and acute; leaf rachis with small hooked decurved thorns and large, oval disc like glands near the base; the pinnae are in 5–6 pairs with ring like glands between the bases of the upper pinnae; leaf lets 15–17 pairs, oblong, linear, mucronate, unequal with a truncate base, midrib eccentric; umbels are usually 3 together with small hooked thorns; buds brownish red; flowers greenish white; calyx funnel shaped, pubescent; pod glabrous, flat, strap shaped, reddish brown, stalked with acute apex; seeds 5–7.

Phytochemistry

Root: β-amyrin, β-sitosterol and Friedelin.

Stem bark: Acacinoic acid, flavonoids, leucofisetinidin, leucoorbinitinidin, lupeol and stigmasterol.

Seed: Fatty oils.

Medicinal Uses

Plant: Is hypothermic and CNS active; useful in bronchitis, consumption, cough, fistula, measles, pleurisy, small pox and tuberculosis; crushed plant is applied on the head to kill vermins and lice.

Root: Useful in general debility.

Stem bark: Bark infusion is used to cleanse the skin and to protect the skin from infections.

Flower: Is used to treat menstrual disorders.

Ethnomedicine

Plant: Is useful in bronchitis, cough, fistula, measles, pleurisy, small pox and tuberculosis.

Non-medicinal Uses

Plant: Is used as a substitute for soap.

Bark: Infusion of bark is used in washing hair; is used as a protective coating for fishing nets; used in dyeing with lac to give luster.

Leaf: Fresh leaves used for dyeing.

3. *Acacia catechu* (L.f.) Willd.

Synonyms*:* *Acacia sundra* (Roxb.) Bedd.
Mimosa catechu L.f.

Notes: The most important product, catechu of this species, is obtained from the Heart wood of 20 to 30 year old trees and is marketed in the trade name Kaachu, Katha or Khadira.

Common Names: Black catechu, Black cutch, Cutch tree.

Vernacular Names: Hin: Katha, Khair, Khairbaabul; **Kan:** Kaachu, Kaggali, Kanti, Kempu Jaali, Kugli, Khaira, Tare; **Mal:** Kadaram, Karingaali, Senkarinnaali; **Mar:** Khaderi, Khair, Khairkatha, Laalkhair; **Sans:** Bahushalya, Balapathra, Balatanaya, Danthadhaavana, Gaayatri, Karkati, Khadira, Kadyapatri, Kshithikshana, Kushthari, Medhya, Mridupathrika, Pathidruma, Prasakha, Rakhthasara, Saradruma, Sushalya, Vakrakantaka, Yajnika; **Tam:** Kaaschukkatti maram, Kadiram, Karangali, Kodam, Kudakkini, Odalai, Podalam, Sengarun gaali, Sirumarodam; **Tel:** Kaachu, Khadiramu, Podalamaanu, Pongamu, Sundra, Tellathumma.

Distribution: Native of Bhutan, China, India, Nepal, Pakistan, Sri Lanka and Thailand; introduced into China, Java, Mauritius, Ryukyu Islands and Taiwan; grows through out India.

Description: A moderate sized tree of 30–40 feet; bark rough, dark coloured; young shoots purple or dark brown, glabrous; leaves 2 pinnate, main rachis pubescent with glands between the pairs of pinnae and a large gland near the middle of the petiole; leaf rachis pubescent; leaflets 30–50 pairs, linear, subacute, ciliate; stipular spines hooked and arise from triangular base; flowers pale yellow on axillary spikes; calyx campanulate, long, hairy outside, teeth deltoid, ciliate; corolla 2–3 times as long as the calyx with ovate, oblong, subacute, pubescent lobes; pod stalked, flat, thin, brown, shining with triangular beak at the apex, narrowed at the base into a stalk; seeds 3–10.

Phytochemistry

Plant: (–) Acacatechin, (–) epicatechol, (+) afzelechin, (+) afzelin, 7,3',4'– trihydroxy-8-methoxyflavonol, 8-methoxy flavone-3,4-dion, 8-methoxyfisetin, aldobiuronic acid, arabinose, catechol, catechuic acid, epicatechin, fisetin, flavotannin, galactose, gallic acid, gallotannin, glucose, glucosyluronic acid, gum, isocatechin, isorhamnetin, kaempferol, L–arabinose, machaerinic acid, magnesium, manganese, methoxy flavones, peroxidase, phosphorus, procyanidin, propionic acid, quercetin, quercitrin, taxifolin, uronic acid, xylose, α-catechin and β-catechin.

Gum: Aldobiuronic acid, arabic acid, calcium, catechin, D–galactose, D–rhamnose, L–arabinose, L–glucoronic acid, magnesium, malic acid, potassium, sugar and tannic acid.

Heartwood: Afzelechin, catechin, cyanidanol, dihydrokaempferol, epicatechin, fisetin, gossypetin, isorhamnetin, kaempferol, procyanidin, quercetin and taxifolin.

Edibility

Heart wood: Powder from heartwood is used as a flavouring agent in ice cream, beverages, candy, baked food and chewing gum; catechu from heart wood produces a good colour when used as a food additive; catechu is marketed in the form of "katha"; occupies an important position in Indian culture by its use in betel chew (*paan*).

Medicinal Uses

Plant: Anaesthetic, analgesic, antiallergenic, antianginal, antiatherogenic, antibacterial, anticancer, anticoronary, antidiabetic, antiestrogenic, antifungal, antiherpetic, antihypertensive, anti-inflammatory, antimastalgic, antimutagenic, antioxidant, antispasmodic, antitumour, antiulcer, antiviral, apoptotic, cardioprotective, cytotoxic, diaphoretic, diuretic, haemostat, hepatoprotective, hypotensive, immunostimulant, neuroprotective, radioprotective, respiroparalytic, spasmolytic, styptic, uterorelaxant and vasodilator; it is used in treating asthma, bronchitis, cancer, colic, diarrhoea, dysentery, leprosy, pain in the chest, stomatitis and ulceration; juice along with asafoetida cures haemoptysis.

Heartwood: Acrid, antifungal, anthelmintic, antidysenteric, anti-inflammatory, antihyperlipidemic, antipyretic, antiseptic, astringent, cooling, depurative, haemostatic, hypoglycaemic, hypotensive and tonic; useful in anaemia, anorexia, cancer, catarrh, cough, diabetes, itch, helminthiasis, leprosy, leucoderma, skin diseases, vitiated "kapha" and "pitha"; used in healing ulcers and wounds; it is an useful remedy for diarrhoea, dysentery, haemorrhages, haematemesis, haemoptysis, inflammations, intermittent fever, pharyngodynia, odontopathy, splenomegaly, tumour of the abdomen and tooth ache; smoke of burnt heart wood is inhaled in nose bleed; decoction of heartwood with tobacco and cow dung is administered as nasal drop in nasal bleeding; the antioxidants in the heart wood protects lungs, kidney and spleen from oxidative damage and increases the resistance to bacterial and viral infections.

Catechu: Acrid, antidiabetic, antidysenteric, antileprotic, antipyretic, antiseptic, aphrodisiac, appetizer, astringent, cooling, depurative, digestive, haematinic, hepatoprotective, stomachic, thermogenic, tonic and vulnerary; useful in anorexia, bronchitis, conjunctivitis, colporrhagia, diseases of the gum, diarrhoea, dysentery, erysipelas, flatulence, gingivitis, gonorrhoea, haemoptysis, heteropathy of the tonsil, helminthiasis, leucorrhoea, laryngopathy, leucoderma, melancholia, odontopathy pharyngites, stomatitis, uterine haemorrhage, relaxed uvula, urorrhoea, urinary and vaginal discharges; it is recommended as a mouth wash in diseases of gum and mouth; as gargle for sore throat, spongy gums and relaxed throat; useful as a cooling agent when applied externally on boils, sores, ulcers, wounds and other skin eruptions; ointment made of catechu is a local applicant in burns, chronic ulcerations with foetid discharges, itch and syphilis; the tincture is excellent for bed sores; decoction is used for washing foul sores and cracked nipples; an injection of catechu infusion is useful in treatment of gonorrhoea, otitis, otorrhoea, leucorrhoea, menorrhagia and acute haemorrhage after child birth; catechu with cinnamon or opium is used in diarrhoea, haemorrhages, loss of voice, sore throat and tooth ache; as lozenge it is used in hoarseness of voice, relapsed sore throat and spongy gums; catechu with asafoetida, opium, carbonates of sodium and potassium is made into pills and given with betel leaf juice in chronic dysentery; catechu with myrobalans, neem bark, root of *Cocculus villosus* and *Cocculus cordifolius,* leaves of *Adathoda vasika* in the form of decoction is useful in prurigo and other skin diseases; catechu with nutmeg, camphor, cardamom and gum of *Acacia nilotica* is made into a bolus and chewed for affections of gum, teeth, tongue and palate; catechu with opium is applied on prolapsed anus and protruding piles; catechu with coconut oil is kept in the mouth and water slowly swallowed is a cure for hoarseness of voice; in combination with burnt charcoal, myrr and areca nut it is used as styptic in haemorrhages and sprinkled on wounds; powdered and mixed with water, it is used as eye wash in conjunctivitis; catechu with *Commiphora wightii* is given as tonic to women after child birth to promote secretion of milk; with *Commiphora* resin and *Hamamelis* it is used as a gargle for pharyngitis, gingivitis and tonsillitis; it is combined with *Acorus, Mentha piperita, Agrimonia* and *Quercus* and used in bowel disorders.

Kheersal: Is a white powder found in the cavities in the wood; it is used in the treatment of chest affections, cough and sore throat.

Bark: Is anthelmintic, antidysenteric, anti-inflammatory, antileprotic, antipyretic, astringent and cooling; it is used in anaemia, bronchitis, conjunctivitis, diarrhoea, elephantiasis, gum trouble, haemoptysis, menstrual complaints, psoriasis, skin diseases and ulcers; bark juice with asafoetida is useful in haemoptysis; bark of *Acacia* with conessi bark, margosa bark, sweet flag root, triphala, root of *Ipomoea turpethum* is made into a decoction and given in gonorrhoea, rheumatism and tooth ache; bark powder with cumin is used in stomatitis; applied on itching, leprosy, leucoderma, scabies and wounds; bark decoction is administered for gum trouble and chronic diarrhoea; used as eye wash for conjunctivitis.

Leaf: Astringent; used for boils, bloody dysentery, cough and throat affections.

Flowering tops: Juice of the flowering tops with cumin, milk and sugar is useful in body heat, gonorrhoea and syphilis.

Seed: Hypoglycaemic.

Bioactivity: Plant is inhibitor of 5-lipoxygenase, ACE, ATPase, aldose-reductase, COMT, cyclo-oxygenase, cytochrome-P_{450}, deiodinase, glucosyl-transferase, HIV-RT, lipoxygenase, MAO, MMP, metalloproteinase, NADH-oxidase, ornithine-decarboxylase, PTK, prostaglandin synthesis, protein-kinase, reverse-transcriptase, xanthine-oxidase, CAMP-phosphodiesterase and topoisomerase; is angiotensin receptor blocker, CNS depressant, calcium antagonist, calcium channel blocker, mast cell stabilizer and quinone-reductase inducer.

Toxicology: Contains serine protein inhibitors which inhibits the action of digestive proteinases which reduce the proteolysis of dietery proteins and cause growth impairement; though used for centuries in ayurveda and considered safe, excess usage may cause haemolysis.

Uses in Alternative Systems of Medicine

Ayurveda:

Part used: Bark, wood, flower.

Uses: Bark is anthelmintic, antidysenteric, antipyretic, aphrodisiac, astringent to bowels, cooling and geriatric; bark and wood useful in anaemia, asthma, boils, bronchial asthma, distaste, indigestion, inflammation, itch, leprosy, leucoderma, pruritus, psoriasis, sore throat, stomatis, ulcer and wound; given in "kapha" and "vata", elephantiasis, erysipelas, leucorrhoea, piles, strengthening the teeth, throat diseases, ulcers, urinary and vaginal discharges; flower taken internally or external application of paste is used in herpes, purpura, skin diseases and urethral discharges.

Chinese:

Part used: Gum.

Uses: Is used as astringent, antiphlogistic, stimulant and styptic.

Siddha:

Part used: Root, bark and heartwood.

Uses: Anthelmintic, astringent and liver tonic; useful in abdominal disorders and leprosy

Tibetan:

Part used: Heartwood.

Uses: It dries lymph accumulation.

Unani:

Part used: Heartwood.

Uses: Used as an external application in skin diseases.

Medicinal Uses Outside India

China: Catechu is used to stimulate salivary gland secretion and to reduce fever, phlegm and excessive thirst; it is used as astringent, antiphlogistic and styptic.

Myanmar: Wood extract is used to treat chest affections and ulcers.

Nepal: Bark is used as astringent, cooling and digestive; useful in cough, diarrhoea, gum pain, mouth ulcers and sore throat; it is applied on boils, skin diseases and sores.

Taiwan: Heart wood decoction is given as astringent, styptic and for diarrhoea.

Tibet: Heartwood is used to reduce lymph accumulation.

Ethnomedicine

Plant: Is useful in asthma, boils, bronchitis, cancer, colic, consumption cough, malaria, phthisis, sores, stomatitis, throat infections, tuberculosis and tumour of the abdomen.

Root: Is useful in toothache.

Bark: Is astringent; useful in diarrhoea, leucorrhoea and menstrual complaints.

Leaf: Used in dysentery.

Heartwood: Smoke from burnt heartwood is inhaled for nosebleed.

Non-Medicinal Uses

Plant: Grown for controlling soil erosion; is a host plant for lac insect; yields a pale yellow dye and khair gum.

Wood: Used in construction of carts, axles, house posts, shafts, yokes, spokes, felloes, boats, wells, bed post, tool handles, tent pegs, furniture, spokes of spinning wheels, railway sleepers, combs, pestles, oil and sugar cane crushers, agricultural implements, bowls, spears, shafts, musical instruments and sword handles; used as fire wood and charcoal; yields tannin used in dyeing and tanning industry; tannin from cutch is used as a coating to protect metals from corrosion; as a fast dye to give shades of olive and brown colour with different mordants; as a main ingredient in hair dye; as water softner, in ore floatation and to remove mercapton impurities to give odourless gasoline; in dyeing fishing nets, canvas and ropes; as an antifungal agent.

Gum: Known as Khair gum; used as plywood adhesive, as mud additive in petroleum refinery and in drilling operations in oil industry to reduce the viscosity of mud.

Bark: Yields fibre used in cordage.

Sap wood: Is ground and two by products "Jharua" and "Gondi" are obtained; Jharua is used in foundaries as binder and Gondi in galavanizing industry; it is used as an auxillary in dyeing cotton and silk, canvas for boat sails and calico printing; it acts as a mordant and its antioxidant property is used to arrest free radicals reaction of auto oxidation resulting in rancidity of fats and oils and to give more shelf life.

4. *Acacia chundra* (Rottler) Willd.

Synonyms: *Acacia sundra* (Roxb.) DC.
Acacia catechu (L.f.) Willd. var., *sundra* (Roxb.) Prain
Mimosa chundra Rottler

Common Names: Red cutch, Red ebony.

Vernacular Names: Hin: Katha, Laal kheir; **Kan**: Kempu jaali, Tharegida; **Mal**: Kannali; **Mar**: Laal khair; **Sans**: Khadira, Rakhta chandana; **Tam:** Karangali, Parambai; **Tel**: Sundra, Nallachandra.

Distribution: Native of India, Myanmar and Srilanka; grows in Western India.

Description: A tree growing upto 8 m height with dark brown rough bark; branchlets smooth, glabrous, pale purplish brown in colour; thorns hooked, solitary and found below the stipules; leaves with small, circular, concave glands between the bases of the two terminal pinnae; pinnae 9–18 pairs, sparsely pilose; leaf lets subsessile, 20–30 pairs, overlapping, small, linear, obtuse, glabrous with oblique base; flowers white to pale yellow, sessile or short stalked on a spike; calyx campanulate with short teeth; corolla lobes linear lanceolate and three times as long as the calyx; pod strap shaped, flat, thin, straight, tapering at both ends, dark reddish brown; seeds 3 – 6, flat, greenish brown.

Phytochemistry

Gum: Arabinose, aldobiuronic acid, galactose and rhamnose.

Medicinal Uses

Plant: Antifertility and spasmolytic; tender branches are used as tooth brushes for spongy and infected gums; stem paste is applied to heal abscesses.

Stem bark: Antiseptic and astringent; bark paste or decoction is applied on boils, sores, tooth ache and ulcers.

Wood: Paste is applied on leucoderma patches to bring back the skin colour.

Leaf: Paste is applied on abscesses.

Bioactivity: Ethanol extract of aerial parts is spermicidal and spasmolytic.

Non-medicinal Uses

Wood: Is used for making beams, posts, agricultural implements, steamship propellers and tail shaft bearings.

5. *Acacia concinna* (Willd.) DC.

Synonyms: *Acacia concinna* (Willd.) DC., var., *rugata* (Benth.) Baker
Acacia rugata (Lam.) Merr.
Mimosa concinna Willd.

Notes: The soap nut tree or the soap pod tree, which is in extensive economic use in India, has been known all along as *Acacia concinna* DC; following a controversy on the correct name for this species, ILDIS has accepted *Acacia sinuata* (Lour.) Merr., which makes the binomial *Acacia concinna* irrelevant, even as a synonym; however, ILDIS has also accepted the binomial *Acacia concinna* (Willd.) DC., as the correct name for a different species; both *Acacia sinuata* and *Acacia concinna* are native to the Asian region including India, but the former has a much wider distribution and economic applications, than the latter; on account of the use of *Acacia concinna* for the soap nut tree for a very long time, there is overlap and confusion in the information on this species in literature and in spite of taking care, possibly in the profiles given here; it is, therefore, necessary to be cautious.

Distribution: Native of Asia including India; the plant grows throughout India and is often planted in gardens.

Description: A large prickly climbing shrub; branchlets, petioles and peduncles covered with fine grey hairs and numerous recurved prickles; leaf rachis with a large gland near the base and one between the upper most pinnae; stipules cordate, ovate; pinnae 3–6 pairs; leaflets 6–20 pairs, linear, obtuse, unequal at the base, glabrous; an odd leaf let is usually found near the base of each pinna; flowers pink, in panicles; bracts obliquely cordate ovate; calyx tubular, crimson in colour; corolla white, veined with red; pod straight, thick, succulent with 6–10 seeds.

Phytochemistry

Stem bark: Acacic acid, hexacosanol, lupeol, α-spinasterol and α-spinasterone.

Pod: Acacic acid, acacic acid ester, acacidiol, acacinin C, D and E, arabinose, fucose, lactone, machaerinic acid, raffinose, rhamnose, sapogenin–B, sonunin I and II, stachyose, sucrose, verbascose and xylose.

Seed: Acacinin A and B, arabinose, concinnin, fucose, glucose, rhamnose and xylose.

Medicinal Uses

Leaf: Cathartic; cures biliousness.

Pod: Anthelmintic, antidiarrhoeal, cardiotonic, cooling, deobstruent, digestive and purgative; improves appetite; cures ascites, burning sensation, biliousness, diseases of blood, erysipelas, leucoderma, piles and stomatitis.

Bioactivity: Saponins from leaves, bark, pod and seed exhibited spasmolytic activity in Guinea pig ileum and also showed antihistaminic activity; the triterpenoidal moiety was more effective than the steroidal moiety.

Uses in Alternative Systems of Medicine

Ayurveda:

Part used: Leaf and pod.

Uses: Leaf is cathartic; cures biliousness; pod is anthelmintic, cardiotonic, cooling, deobstruent, digestive and purgative; improves appetite; cures ascites, diarrhoea, erysipelas, "kapha" and leucoderma.

Siddha:

Part used: Leaf and pod.

Uses: Emetic and purgative; useful in constipation, fever, jaundice and ulcer.

Medicinal Uses Out Side India

China and Japan: The pods are considered as emetic and laxative; they are used in constipation and kidney troubles; externally pod paste is applied to absecesses, eczema, leprous patches and prurigo; the consumption of seeds is said to hasten childbirth.

Ethnomedicine

Plant: Cures tumour.

Non-medicinal Uses:

Plant: Is used as insecticide, pesticide and piscicide.

6. *Acacia decurrens* Willd.

Synonym: *Acacia decurrens* Willd. var., *angulata* (Desv.) Benth.

Notes: The plant is a rich source of tannin and valued in tanning industry.

Common Names: Common wattle, Green wattle.

Vernacular Names: Kan: Peeka jaali, Thambati jaali; **Tam:** Velam pattai.

Distribution: Native of Australia; introduced intc India (cultivated in Darjeeling, West Bengal and Tamil Nadu).

Description: Tall tree with angled branches; leaves alternate with densely pubescent, grooved rachis with a gland below each pair of pinnae; pinnae 10–15 pairs; leaflets 30–40 pairs, linear, oblong, obtuse at the apex; flowers cream coloured in compound raceme, bracteoles one for each flower; calyx lobes 5, obtuse; corolla lobes 5, twice as long as the calyx, acute; stamens numerous.

Phytochemistry

Plant: (+) Catechin, 7,3',4',5'-tetrahydroxy flavan-3-ol, catechin, 3,3',4',5',7- pentahydroxy-2-phenylchromone, 3,3',4,4',7'-pentahydroxy flavan, 3–O–methyl–L–rhamnose, acetic acid, aldobiuronic acid, anthocyanidin, anthocyanilidin, carbohydrates, cellulose, D–galacturonic acid, D–glucose, D–mannose, D–pinitol, fisetinidin, fructose, gallocatechin, gallic acid, indole acetic acid, L–arabinose, lignin, L–rhamnose, mearnsitrin, methyl salicylic ester, pelargonidin, pentahydroxy flavan, phenylchromone, phlobaphene, phlobaphene anhydride, phloroglucinol, proanthocyanidin, protocatechuic acid, robinetin, rutin, tetrahydroxy flavan–3–Ol–catechin, xanthophylls, α-cellulose and β-carotene.

Stem bark: Leucofisetinidin, leucoorbinitinidin and tannin.

Seed: Fatty oil.

Medicinal Uses

Plant: Is allergenic, analgesic, antiatherogenic, antibacterial, anticancer, anticapillary fragility, anticoronary, antifibrinolytic, antihepatotoxic, antihypertensive, antinociceptive, antioxidant, antiperoxidant, antiseptic, antispasmodic, antitussive, antiviral, bronchodilator, cardioprotective, chemopreventive, cytotoxic, diuretic, expectorant, haemostat, hepatotoxic, immunostimulant, laxative, nephrotoxic, phagocytotic, styptic and vasodilator; cures acne, arrhythmia, asthma, bronchitis, cataract, convulsions, dermatitis, diabetes, diarrhoea, edema, glaucoma, haematuria, haemorrhoids, hyper cholesterolaemia, ischaemia, leukaemia, lupus, malaria, nausea, nephritis, polio, purpura, tumour of the colon, mouth and skin , ulcer, vaginitis and varicose veins.

Stem bark: Astringent; decoction given in severe dysentery.

Bioactivity: The plant is inhibitor of ACE, aldose-reductase, cyclo-oxygenase, topoisomerase I and II, xanthine-oxidase, CAMP and phosphodiesterases; is β-adrenergic receptor and angiotensin receptor blocker.

Non-medicinal Uses

Plant: Grown as a windbreak and for manure; tender branches are used as reinforcing material for mud walls.

Wood: Used as fuel; wood pulp is useful in paper making.

Flower: A perfume of good quality is obtained from flowers.

7. *Acacia farnesiana* (L.) Willd.

Synonyms: *Acacia farnesiana* (L.) Willd. var., *lenticellata* (F.Muell.) Bailey
Acacia lenticellata F. Muell.
Acacia minuta (M.E. Jones) Beauch
Mimosa farnesiana L.
Pithecellobium minutum M.E. Jones
Pithecellobium acuminatum M.E. Jones

Notes: Origin uncertain, probably South American; source of "cassie" perfume.

Common Names: Cassie, Ellington curse, Fragrant acacia, Mimosa wattle, Needle bush, Prickly moss, Sponge wattle, Sweet acacia, Thorny acacia, Thorny feather wattle and Wild briar.

Vernacular Names: Ass: Tarwakadam; **Ben:** Guyababul; **Guj:** Jabbaval; **Hin:** Vilayatikikar, Gandhbabul; **Kan:** Kasthuri jaali; **Mal:** Kariveelam, Arivelam; **Mar:** Guibabhool, Kankar; **Pun:** Vilayathi kikar; **Sans:** Arimedah, Girimedha, Kaalaskandha, Marudruma; **Tam:** Pavalam, Pikkaruvil, Veddavala; **Tel:** Kempucumma, Nugatumma.

Distribution: Introduced into Asia including India, Africa, Australia, Europe, China, America, Indian Ocean Islands; planted and naturalized throughout India.

Description: Shrub with zig zag branches with dots; stipular spines straight; leaf rachis with petiolar glands; leaves with 3–4 pairs of pinnae; leaflets 10–20 pairs, linear, oblong, greyish green, coriaceous; flowers bright yellow, fragrant; pod oblique, cylindrical, dull brown, 2–seeded.

Phytochemistry

Plant: (+) Catechol, (+) gallocatechol, 3-methyl-3-E-nonenoic acid, 3-methyl-3-Z-decen-1-Ol, 3-methyl-3-Z-nonenoic acid, 4-hydroxy pipecolic acid, aromadendrin, aspartic acid, benzaldehyde, benzoic acid, butyric acid, campesterol, catechol, cholesterol, coumarin, cresol, cuminaldehyde, decyl aldehyde, diketone ethyl gallate, eicosane, ellagic acid, farnesol, gum, homoterpene, hydroxy acetophenone, isorhamnetin-3-rutinoside, kaempferol–7-galloylglucose, kaempferol-7-glucoside, lactones, ethyl phenol, linalool, linalyl acetate, linamarin, lotaustralin, lynalyl-acetate, m-digallic acid, myricetin-4'-methyl ether-3-rhamnoside, nerolidol, N-methyl-β-phenethylamine, nonacosane, nonenoic acid, palmitic acid, pinitol, 4-hydroxypipecolic acid, prunin–O-6'-gallate, quercetin-3-O-rutinoside, salicylic acid, sitosterol, stigmasterol, tannins, trans-3-methyl-dec-4-enoic acid, tyramine, wax, α-spinasterol, α-terpineol, β-geraniol and β-coumarin.

Stem bark: Ethyl gallate and gallic acid.

Leaf: 3-7-Glucorhamnoside, alanine, apigenin, apigenin-6-8-bis-C-glucopyranoside, arginine, ascorbic acid, calcium, campesterol, carbohydrates, catechol, cholesterol, fat, fibre, glutamic acid, glycine, histidine, hydrocyanic acid, iron, isoleucine, leucine, linamarin, lotustralin, lysine, methionine, niacin, nitrogen, phenylalanine, phosphorus, proline, riboflavin, rutin, serine, sitosterol, stigmasterol, tannin, thiamine, threonine, trans-3-methyl-dec-4-enoic acid, triacontanol, tyrosine, β-carotene and β- diketone.

Flower: 3-Methyl-3-Z-decen-1-Ol, 3-methyl-3-Z-nonenoic acid, 5-7',4'-trihydroxyflavanone, 7-O-β-D-6'-gallolyl-glucopyranoside, 7-glucorhamnoside, anisaldehyde, anisic dealdehyde, aromadendrin, benzaldehyde, benzoic acid, benzyl alcohol, butyric acid, catechol, cis-3-methyl-dec-3-enoic acid, coumarin, cresol, cumin alcohol, cumin aldehyde, cuminicdealdehyde, digallic acid, dihydroactinidiolide, eicosane, ellagic acid, enoic acid, ethyl phenol, eugenol, farnesol, gallic acid, geraniol, geranyl acetate, glucorhamnoside, homoterpene, isorhamnetin, kaempferol-7-gallolylglucose, kaempferol-7-glucoside, lactone, linalool, linalyl acetate, methyl eugenol, methyl salicylate, naringenin-7-diglucoside, naringenin-7-glucoside, nerolidol, N-methyl-β-phenethylamine, nonacosane, nonenoic acid, palmitic acid, p–cresol, pinitol, salicylic acid, trans-3-methyl-dec-4-enoic acid, α-ionone, β-geraniol and β-ionone.

Pod: Aromadendrin, ellagic acid, gallic acid, glucopyranoside, kaempferol, kaempferol-7-galloylglucose, kaempferol-7-glucoside, m-digallic acid, methyl gallate, naringenin, naringenin-7-diglucoside, naringenin-7-glucoside, naringenin-7-rhamnoglucoside, tannin and trihydroxy flavanone.

Seed: Calcium, campesterol, carbohydrates, cholesterol, cystine, fat, fatty oil, isoleucine, leucine, linoleic acid, lysine, methionine, N-acetyl-L-dijenkolic acid, oleic acid, palmitic acid, phenylalanine, phosphorus, sitosterol, stearic acid, stigmasterol, threonine, tryptophan, valine and α-spinasterol.

Edibility

Gum: Used in confectionary.

Leaf: Used as a substitute for tamarind in chutneys and curries.

Medicinal Uses

Plant: Is abortifacient, allergenic, anaesthetic, analgesic, anthelmintic, antiandrogenic, antibacterial, anticardiospasmic, antidote to aluminium, cadmium, lead, mercury and strychynine poisoning, antiadenomic, antifertility, antifibrinolytic, antihistaminic, antioxidant, antiseptic, antispasmodic, aphrodisiac, apoptotic, bronchodilator, cardiotonic, carminative, chemopreventive, demulcent, diaphoretic, diuretic, emetic, haemostatic, hypolipidemic, hypotensive, laxative, motordepressant, nephrotoxic, radioprotective, respirostimulant, sedative, styptic and thermogenic; cures acne, arteriosclerosis, arthritis, asthma, atherosclerosis, beri-beri, bronchitis, cancer of the kidney, cardiospasms, cataract, cold, colitis, convulsions, delirium, dementia, depression, dermatitis, diabetes, eczema, edema, gastritis, gingivitis, glaucoma, haematuria, haemorrhage, hemorrhoids, hepatitis, herpes, impotency, keratitis, menorrhagia, nephritis, neuralgia, neuropathy, pityriasis, pneumonia, pruritus, tumours, varicose veins and vertigo.

Root: Alterative, antidiarrhoetic, antirheumatic, antispasmodic, aphrodisiac, astringent, demulcent, febrifuge and stimulant; the roots are chewed for sore throat.

Stem bark: Anti-inflammatory, antidysenteric, alterative, anthelmintic, astringent, aphrodisiac and demulcent; useful in the treatment of bronchitis, dental caries, erysipelas, leucoderma, polyuria, spermatorrhoea, stomatitis, skin diseases, sores and ulcers; a decoction of the stem bark is given in leucorrhoea and prolapsed rectum; decoction of the bark is used with ginger as a mouth wash for bleeding gums as haemostatic.

Gum: Astringent, demulcent, emollient and nutrient; decoction is taken internally in diarrhoea and dysentery.

Leaf: Tender leaf paste eaten for gonorrhoea; tender leaves are boiled, crushed and applied as cataplasm on ulcers and wounds along with washing with leaf decoction; useful in washing inflamed eyes.

Flower: Stimulant; useful in venereal diseases; perfume extracted is used in aromatherapy.

Pod: Alterative and astringent; pod lotion is used to wash sores and skin diseases; given internally for diarrhoea; crushed pods are used in conjunctivitis.

Pod oil: The essential oil from pod is a muscle relaxant, cardic depressant and sedative when taken internally.

Bioactivity : Plant is inhibitor of 5-a-reductase, 5-lipoxygenase, ACE, aldose-reductase, a-reductase, b-glucoronidase, cyclo-oxygenase, cytochrome P_{450}, DME, decarboxylase, glucosyl-transferase, insulinase, iodothionine deiodinase, lipoxygenase, prostaglandin synthesis, quinone-reductase, reverse-transcriptase, serotonin, tyrosinase, xanthine-oxygenase and CAMP phosphodiesterase; is blocker of angiotensin and b-adrenergic receptors, along with quinone-reductase inducers.

Glycosidal fraction of plant extract (2–10 mg) showed vasodilator action in hind limb perfusion of dog by increased femoral venous out flow; it inhibited carrageenin and formaldehyde induced inflammation and exhibited smooth muscle relaxant activity.

Toxicology: Contains trypsin inhibitors.

Uses in Alternative Systems of Medicine

Ayurveda:

Part used: Stem bark, heartwood and gum.

Uses: Stem bark is alexiteric, anthelmintic and antidysenteric; cures bronchitis, dental caries, diseases of blood, erysipelas, inflammations, leucoderma, stomatitis and ulcers; wood is useful in bronchial asthma, leprosy, pruritus, stomatitis and wounds; gum is aphrodisiac and tonic.

Siddha:

Part used: Stem bark, gum, leaf, flower and pod.

Uses: Useful in gingivitis and leucorrhoea.

Medicinal Uses Outside India

Gold coast and Nigeria: The gum is chewed for sore throat; bark and leaf decoction is applied as a lotion for sores and skin diseases and taken internally for diarrhoea.

Indochina: Leaf poultice is applied on ulcers and sores.

Indonesia: Plant is used as emetic; leaf juice is given for lumbar pains.

Malay Peninsula: Pounded roots are applied to swellings.

Philippines: A decoction of the stem bark is given for prolapsed anus and leucorrhoea; the tender leaves are boiled and applied as cataplasm to wounds and ulcers, which are also cleaned with the decoction of leaves.

West Indies: Bark is tied around the joints to alleviate rheumatism.

Ethnomedicine

Plant: Is used as alterative, aphrodisiac, astringent, anticancer, stimulant and stomachic; various parts of the plant is used in curing arthritis, cancer of stomach, carbuncle, cholera, convulsions, conjunctivitis, delirium, dysentery, dyspepsia, epilepsy, fever, inflammation, insanity, mucous membrane disturbance, nausea, rabies, rheumatism, skin diseases, sores, spasms, sterility, tuberulosis, tumour and wounds.

Root: Is antifertility in action.

Bark: Used as astringent and in malaria.

Leaf: Is used in eye complaints, headache and gonorrhoea.

Non-medicinal Uses

Plant: Grown as an ornamental, fence and hedge plant; grown to check soil erosion; is a host tree for lac insect; it is largely cultivated in France for the blossoms which yield perfume; used as nematicide and piscicide; yields a gum marketed with other *Acacia* gums.

Tender branches: Used as tooth brush.

Wood: Yields timber, which is hard and durable; used for tent pegs, posts, beams, agricultural implements, knees of ships; yields a gum which is a substitute for *Acacia* gum.

Gum: Used in tanning and dyeing industry in combination with iron and in making ink; used as an adhesive for mending broken crockery.

Flower: Is a source of "Cassie" perfume and is a good substitute for expensive perfumes; used in cosmetic industry.

8. *Acacia ferruginea* DC.

Synonym: *Mimosa ferruginea* Roxb.

Common Names: Anasandra bark tree, Cat claw acacia, Tamarind cutch.

Vernacular Names: Guj: Kaigu; **Hin:** Gandh babul, Divaana babul, Sonkhair; **Kan:** Bannimara, Kiribanni; **Mal:** Thimavelvelam, Vanni; **Mar:** Pandhra khair; **Sans:** Brahmashalya, Dvijapriya, Kantadhya, Karumuka, Kadiroparna, Kujakantaka, Mahavriksha, Memivriksha, Shevathasara, Shyamasara, Somasara, Somavallika, Somavriksha; **Tam:** Perambai, Peyakkarungadi, Seemaivellel, Vanni, Velvelam; **Tel:** Ansandra, Vannesandra, Thellatumma and Vuni.

Distribution: Native of India and Sri Lanka; in India it grows in Rajasthan and West Bengal.

Description: Tree growing upto 8 m height with rough, dark rusty brown, deeply fissured bark; branchlets glabrous, smooth and purplish brown; thorns stipular, paired, hooked and sharp; leaves glabrous with a circular or oval gland found on the petiole and small glands between the bases of one or two apical pinnae; pinnae 3–6 pairs; leaflets upto 20 pairs, glabrous, linear oblong, obtuse with a oblique base, the mid rib slightly off the center; flowers glabrous, pale yellow (after anthesis dark yellow), sessile and found in spikes; calyx thin and campanulate; corolla three times as long as the calyx with oblong lobes; pod strap shaped, thin, flat, stalked, dark brown with 1–6 flat, oval seeds.

Edibility

Leaf: Edible.

Medicinal Uses

Stem bark: Astringent, hypotensive and spasmolytic; decoction is used as gargle for sore throat and mouth.

Wood: Is useful in scabies.

Gum: Astringent, demulcent, emollient and nutrient; useful in catarrh, diarrhoea, dysentery, genitourinary diseases, pulmonary affections and gonorrhoea; gum is chewed for cough and throat irritation.

Leaf: Is astringent, styptic and tonic; it enriches blood; useful in alimentary canal disorders, diarrhoea, dysentery, gonorrhoea, liver complaints and urinary tract infections; leaf paste is applied to burns and scalds.

Pod: Astringent and demulcent; useful in treating genitourinary diseases, diarrhoea, dysentery, gonorrhoea and cystitis.

Bioactivity: Ethanol extract of stem bark is hypotensive and spasmolytic.

Uses in Alternative Systems of Medicine

Ayurveda:

Part used: Stem bark and pod.

Uses: Is anthelmintic; cures itching, leucoderma, stomatitis and ulcers; useful in "vatha" and "kapha", cough, diarrhoea, diseases of blood, dysentery, dyspnoea, giddiness, haemorrhage, leucoderma, piles, skin diseases, stomatitis and ulcers.

Unani:

Part used: Leaf.

Uses: Astringent, styptic and tonic; useful in alimentary and urinary tract diseases, burns, diarrhoea, dysentery, gonorrhoea, liver complaints and scalds; it enriches blood and promotes suppuration.

Ethnomedicine

Plant: Is used as an astringent and in treating tumour.

Non-medicinal Uses

Plant: Young branches are used as cattle feed.

Wood: Is used for making agricultural implements, cart wheels, beams, framework and tent pegs.

9. *Acacia leucophloea* (Roxb.) Willd.

Synonym: *Mimosa leucophloea* Roxb.

Common Names: Panicled acacia, White babool, White bark acacia.

Vernacular Names: Ben: Safed babul; **Hin:** Safed babul, Sarab kikikar, Safed Kikar, Karir, Nimbar, Ringa, Rinj, Rohini; **Guj:** Haribaval; **Kan:** Bili jaali; **Mal:** Pattocharaya maram, Velvelam; **Mar:** Hewar, Pancharya, Babbuliche jada; **Ori:** Goira, Gulara; **Pun:** Safed kikar; **Sans:** Hemagaura, Irimeda, Kinkirata, Peethabhadra, Peethaka, Shatpadananda, Somavalka, Shwethabara, Shewethakhadira, Vipralambi, Vipralobhi; **Tam:** Pattaiccaraya maram, Velvelan; **Tel:** Tellatumma.

Distribution: Native of India, Indonesia, Java, Lesser Sunda, Malaysia, Pakistan, Srilanka, Thailand and Vietnam; in India it occurs in Gujarat, Madhya Pradesh, Maharashtra, Orissa, Punjab, Rajasthan, Tamil Nadu and Uttar Pradesh.

Description: A large tree with smooth pale yellow bark with white areas; stipular thorns paired, straight, black and glossy, pilose and connate at the base found on the lower part of the trunk and branches; a large, round, subsessile gland is found between the bases of the lower pinnae and smaller glands are found between the 2–3 pairs of apical pinnae; pinnae 5–12 pairs; leaflets sessile, 15–30 pairs, oblong, linear, the base truncate, oblique with slightly eccentric midrib; flowers white, pubescent and on globose heads on a terminal panicle; calyx funnel shaped; corolla lobes exserted from the calyx; pod strap shaped, somewhat fleshy, slightly curved, flat, short stalked with a round apex and minutely tomentose; seeds numerous, oval and flat.

Phytochemistry

Plant: Betunic acid-3-O-maltoside, leucophleol, leucophleoxol, leucoxol, myricetin, N-octacosanol, tannin and β-amyrin.

Root: 1,3-Dihydroxy-5-methoxy-2-methylanthraquinone, 8-O-L-rhamnopyranoside, betulinic acid-3-O-β-maltoside, galangin, galangin-3-O-α-L-rhamnopyranoside, leucokol, leucophleol, leucophleoxol, maltoside and saponins.

Root bark: Cyanin, isokannin, leucodelphinidin-3-O-α-L-rhamnopyranoside, leucophenol, leucophleol, leucophleoxol and leucoxol.

Stem bark: Cyanin, isokannin, leucodelphinidin-3-O-α-L-rhamnopyranoside, leucodelphinidin-8-O-L-rhamnopyranoside, n-hexacosanol, tannin, β-amyrin and β-sitosterol.

Wood: N-Hexacosanol, octacosanol, pinitol, β-amyrin and β-sitostrol.

Heart wood: N-Hexacosanol, β-amyrin and β-sitosterol.

Flower: (+) Pinitol, 3'-hydroxy-3'-methoxy isoflavone, apigenin, aliphatic acid, behenic acid ester, apigenin-8-C-glucoside, gallic acid, isoflavone, kaempferol, mannitol, myricetin-3'-hydroxy-3'-methoxy isoflavone, quercetin–3–glucoside, methyl gallate, methoxy isoflavone, myricetin, octacosane, p–hydroxy benzoic acid, quercetin-3-glucoside, quercetin, rutin, triacontane, triacontanol and β-sitosterol.

Pod: Hydrocyanic acid, leuteolin, mannitol and β-sitosterol.

Seed: Galactomannan.

Seed oil: Coranaric acid.

Edibility

Pod: Edible; ground and made into flour at times of famine in India.

Seed: Edible.

Medicinal Uses

Plant: Is antiallergenic, antibacterial, anticancer, antihypertensive, anti-inflammatory, antinociceptive, antioxidant, antiperiodontic, antiradicular, antiseptic, antitussive, antiviral, apoptotic, bronchosedative, diuretic, hepatoprotective, respirostimulant, vasodilator and vasomotor stimulant; useful in diarrhoea, dysentery, edema, gastritis, gingivitis, hepatotoxicity, hypoglycaemia, nephritis and ulcer.

Stem bark: Alexiteric, anthelmintic, antipyretic, aphrodsiac, astringent, bitter, constipating, demulcent, expectorant, styptic, thermogenic and vulnerary; it is used in biliousness, bronchial asthma, burning sensation, cough, dental caries, diarrhoea, dysentery, erysipelas, inflammation, internal and external haemorrhages, leprosy, leucoderma, oral ulcers, protopsis, pruritus, psychosomatic disorders, rectal prolapse, scabies, skin diseases, stomatitis, ulcers, vitiated "kapha", vomiting and wounds.

Gum: Aphrodisiac, demulcent and tonic; useful in bronchitis, biliousness and general debility.

Bioactivity: Plant is inhibitor of cyclo-oxygenase, glucosyl-transferase, hypoxygenase, ornithine-decarboxylase, quinone-reductase, topoisomerase I and II, tyrosine-kinase and xanthine-oxidase.

Ethanol extract of aerial parts is hypotensive and CNS depressant.

Uses in Alternative Systems of Medicine

Ayurveda:

Part used: Bark.

Uses: Acrid, alexiteric, anthelmintic, antipyretic, bitter and cooling; useful in biliousness, bronchitis, burning sensation, eczema, herpes, inflammation, leprosy, pustular patches, ringworm, scabies, stomatitis, thirst and vomiting.

Siddha:

Part used: Bark, gum, leaf, root and seed.

Uses: Used in treating arthritis, fainting spells, fever and rheumatism.

Medicinal Uses Outside India

Indochina: Leaf and pod used as diuretic, emetic and laxative; used as an external application for buboes, eczema, leprosy, prurigo and tumours.

Myanmar: The bark is used as astringent.

Ethnomedicine

Root: Is abortifacient.

Bark: Is useful as astringent and in curing ulcer.

Non-medicinal Uses

Bark: The yellowish white bark, known as velam bark is used in arrack distillaries as a clarifyng and flavouring agent; it yields a coarse fibre; in Ceylon it is used for dyeing sails, mats, ropes, fishing nets and for tanning; it gives black color used in tanning as a substitute for wattle bark; useful in making liquor from sugarcane.

Gum: Is used in dyeing, gives a black colour.

Wood: Is used for making agricultural implements, oil wells, roof, cart wheels, beams, posts, turnery and tool handles; employed as fuel and fire wood.

Leaf: Yields a black dye used in dyeing; used as fodder.

Pod: Used as fodder.

10. *Acacia mearnsii* de Wild.

Common Name: Black wattle.

Vernacular Names: Ben: Sirish; **Guj:** Pilo sarshio; **Hin:** Siris; **Kan:** Bangemara; **Mal:** Vaga; **Mar:** Sirisha; **Tam:** Chavukku, Vagi.

Distribution: Native of Australia and Tasmania; introduced into India; cultivated in Tamil Nadu.

Description: A tall tree of 6–20 m height with conical or round crown; all parts of the plant except flowers are pubescent; stem without spines or prickles; leaves bipinnate with a gland above; leaf rachis with numerous raised glands all along the upper side; pinnae 8–30 pairs; leaflets 16–20 pairs, linear oblong; flowers yellow, fragrant and borne on raceme; pod grey, puberulous, moniliform, dehiscent with 3–12 joints; seed black, smooth, elliptic, compressed, ovoid with a areole.

Phytochemistry

Plant: Gum.

Stem bark: (+) Catechin, (–) robinetinidol, biflavonoids, fisetinidol, gallocatechin, tannin and triflavonoids.

Heartwood: Biflavonoids, butein, fisetin, fisetinidol, fustin, leucofisetinidin, recorcinol and tannin.

Seed: Fatty oil.

Edibility

Seed: Yields oil, which is used for cooking.

Medicinal Uses

Plant: Astringent, diuretic, hypotensive and styptic.

Wood: Recorcinol content is used in lotions to cure skin disease.

Bioactivity: Ethanol extract of stem bark is hypotensive and diuretic.

Non-medicinal Uses

Plant: Yields timber; used as manure.

Bark: Adhesive is prepared from the bark and used as flotation agent; used in making straw board and wrapping paper; used in tanning industry.

Wood: Used in paper and board manufacture; used as tent poles in building construction; distillation of wood tar yields high content of acetic acid; yields charcoal.

Seed: Yields oil used in leather industry.

11. *Acacia melanoxylon* R. Br.

Synonyms: *Acacia melanoxylon* R. Br. var., *obtusifolia* Ser.
Mimosa melanoxylon R. Br.

Common Names: Australian black wood, Black acacia, Black wood, Sally wattle, Tasmanian lightwood.

Vernacular Names: Kan: Jaali, Seeme gobli; **Tam:** Simppaagam, Therchavukku, Velam.

Distribution: Native of Australia and Tasmania; introduced into India; in India it grows in the hill stations of Tamil Nadu.

Description: Small tree, 5–7 m in height; leaves reduced to elliptic, oblong, flattened, lanceolate, leathery, glabrous, usually curved phyllodes; flower heads on short axillary or terminal racemes; flowers white; pod narrow and much curved.

Phytochemistry

Plant: 2-Methyl-6-methoxy–furanobenzoquinone, 2,6-dimethoxy-1,4-benzoquinone, acamelin, anthocyanidin, cellulose, dihydroxyflavonols, flavan-3,4-diol, gibberellic acid, hyperoside, leucoanthocyanins, phlobaphene, tannin and tetra hydroxy dihydro flavonol.

Heartwood: 2,3-Cis-dihydroflavonol-3,7,8,4',5'-pentahydroxyflavanone, ethylmelacacidin, isomelacacidin, leucoanthocyanidins, stigma-7-enol and β-spinasterol.

Flower: Quercetin-3-galactoside.

Medicinal Uses

Plant: Antibacterial, antioxidant, anti HIV, antihepatotoxic, anticancer and antiviral.

Seed: Hyperglycaemic and hypercholesterolemic.

Bioactivity: Plant is inhibitor of collagenase, cyclo-oxygenase, glucosyl-transferase, lipoxygenase, MAO, ornithine-decarboxylase, xanthine-oxidase and CAMP.

Toxicology: Quinonoid principle in the plant causes allergy, contact dermatitis and bronchial asthma.

Non-Medicinal Uses

Plant: Yields fuel.

Wood: Used for making agricultural implements, furniture, cabinet, coaches and carriages; yields good firewood.

12. *Acacia nilotica* (L.) Del.

Synonyms: *Acacia arabica* (Lam.) Willd.
Mimosa nilotica L.

Notes: Seven subspecies, *adstringens, indica, kraussiana, leiocarpa, nilotica, subalata* and *tomentosa* are recognized.

Common Names: Black babool, Egyptian mimosa, Indian gum arabic tree.

Vernacular Names: Ben: Babul; **Guj:** Banul; **Hin**: Babul kikar; **Kan**: Jaali, Gobbli; **Mal**: Karivelam; **Punj**: Kikar; **Sans:** Abhashatpada, Ajabaksha, Babboola, Babboolaka, Barboora, Dheerghakantaka, Dridhabheeja, Dridharuha, Goshringa,

Kanta, Kaphantaka, Kashaaya, Kashtaki, Kinkiraatha, Kinkirata, Maalaphala, Modini, Sapitaka, Sukshmapatra, Theekshnakantaka, Vavara, Yugalaksha and Yugmakanta; **Tam:** Iramangadam, Karuvelum; **Tel:** Barbooramu, Nallathumma, Thummachettu.

Distribution: Native of Africa, Asia, India, Iran, Iraq, Pakistan, Israel, Oman, Syria, Yemen; introduced into Australia, Caribbean islands, China, Madagascar, Mauritius, Rodrigues and United States; in India it is found throughout.

Description: Small tree growing to a height of 2–14m; bark thin, rough, fissured, deep reddish brown; branch lets purple brown with dense grey pubescence and lenticels; stipular spines grey, slightly recurved; leaves 10–17 pairs with one or two petiolar glands between all or only the upper most pair of pinnae; leaflets 10–30 pairs, glabrous or pubescent, with obtuse apex; flowers yellow on axillary heads; an involucel present from the base to half way up the peduncle; calyx subglabrous; corolla densely ciliate; pod variable, linear, stalked, indehiscent, straight or curved, glabrous, or velvety, turgid, black, 10-12 seeded; seeds blackish brown, smooth, suborbicular, compressed.

Phytochemistry

Plant: Yields gum arabic, which is similar to true gum arabic along with aldobiouronic acid, 6-hydroxy quinoline, 7-coumarinyl acid, 4-aldobiouronic acid, amylase, arabic acid, ascorbic acid, ascorbic acid oxidase, betulin, calcium, carbohydrates, cellulose, chloragenic acid, D–catechol, D–galactose, D–glucuronic acid, diastase, dicatechin, digitonin, ellagic acid, epicatechol, epicatecholgallate, epigallocatechin, fructose, gallocatechol, hydroquinone, indole acetic acid, isoquercitrin, isorhamnetin–3–rutinoside, kaempferol, kaempferol–3–O–glucoside, kaurenic acid, L– arabinose, leucocyanidin, lignin, linoleic acid, lysolecithin, magnesium, malic acid, maltase, manganese, monobutyrinase, mucilage, myrecetin–4'–methylether–3– rhamnoside, oleic acid, oxalic acid, peroxidase, phlobaphene, phloroglucinols, polyphenol oxidase, potassium, pyrocatechol, pyrogallol, 3-quinolyl acid, quercetin, riboflavin, robidandiol, saponin, tannase, tannin, thiamine, trihydroxy flavan-3-Ol, uronic acid; (+) catechin, β-sitosterol and β-carotene.

Root: Betulin, hentriacontane, lapachol, n-hentriacontanol, octacosanol, paulownin, α-sitosterol and α-amyrin.

Stem bark: (+) Dicatechin, catechin, catechol, epicatechin, epigallocatechin-5,7-digallate, gallic acid, leucocyanidin, m–digallic acid, octacosanol, pyrogallol, quercetin, stearic acid, sucrose, tannin and β-amyrin.

Gum: 2-O-β-L-arabinofuranosyl–L–arabinose, galactose, gallic acid, rhamnose and tannic acid.

Leaf: Gallic acid and tannin.

Flower: Isoquercitol, kaempferol-3-glucoside, leucocyanidin and stearic acid.

Pod: 7–Digallate quercetin, arabinose, catechin, catechol, chlorogenic acid, epicatechin, epicatechol, 5-epigallocatechin, gallic acid, m-digallic acid, rabidandiol, tannic acid, tannin, trihydroxy flavan and tetrahydroxy flavan-3-Ol.

Seed: Amino acids, arachidic acid, capric acid, fatty oil, lauric acid, linoleic acid, lecithin, myristic acid, oleic acid, palmitic acid and stearic acid.

Edibility

Gum: Yields gum, which is similar to gum-arabic used in confectionary.

Pod and seed: Edible, used as masticatories in sweet meats; seeds are roasted and eaten.

Medicinal Uses

Plant: Alexiteric, alexipharmic, alterative, anaesthetic, analgesic, antiandrogenic, antiasthmatic, antiatherosclerotic, antibactierial, anticarcinogenic, anticholinergic, anticonvulsant, anticoronary, antidiabetic, antifungal, antihaemolytic, antihepatotoxic, antiherpetic, antihistaminic, anti-inflammatory, antimetastatic, antimutagenic, antinociceptive, antioxidant, antiperiodontic, antiperoxidant, antiprostaglandin, antiprotozoal, antiseptic, antispasmodic, antitumour, antitussive, antiulcer, antiviral, aphrodisiac, apoptotic, astringent, autotoxic, cardioprotective, cholagogue, CNS depressant, cytotoxic, demulcent, depurative, diuretic, emetic, emollient, emmenogogue, expectorant, febrifuge, haemostatic, hepatoprotective, hypoglycaemic, hypolipidemic, hypotensive, immunostimulant, intellect promoting, laxative, memory enhancer, nephrotoxic, neuroprotective, estrogenic, pectoral, refrigerant, sedative, smooth muscle relaxant, spasmolytic, stimulant, styptic, sudorific, suppurative, tonic, transquilizer, trichogenous, vasodilator and vulnerary; used in treating aluminium, lead and cadmium poisoning, arrhythmia, arthritis, arteriosclerosis, back ache, breast and skin cancer, bronchitis, cancer of the colon, cataract, cholera, cold, colitis, cystitis, delirium, dermatitis, diabetes, diarrhoea, dysentery, ear ache, eczema, edema, encephalopathy, epilepsy, gastritis, gingivitis, glaucoma, glossitis, gonorrhoea, haemorrhage, hyperchloresterolaemia, hyperthyroidism, hypoglycaemia, leprosy, sore throat, syphilis and tooth ache.

Tender tops are ground into paste and used as demulcent; water extract of the tops is astringent and used as douche in leucorrhoea; useful in Addison's disease, anorexia, aphthae, appendicitis, biliousness, bronchitis, catarrh, colic, conjunctivitis, consumption, diabetes, duodenal and stomach ulcers, dysuria, emaciation, epilepsy, gastralgia, urinogenital diseases, hyperdipsia, hyperprolactinaemia, hypertension, respiratory affection, sore throat, thirst, vitiated conditions of "vatha" and vomiting.

Stem: Antifungal, analgesic, antiprotozoal, CNS active, hypoglycaemic and spasmolytic.

Bark: Acrid, alexipharmic, alexiteric, analgesic, antibilious, antiseptic, antiscorbutic, aphrodisiac, astringent, cooling, constipating, demulcent, depurative, diuretic, emetic, emollient, expectorant, hypoglycaemic, nutritive, stimulant, styptic and vulnerary; useful in ascites, asthma, burning sensation, bronchitis, cold, cough, diabetes, diarrhoea, dysentery, gonorrhoea, haemorrhages, helminthiasis, haemorrhoids, leprosy, leucorrhoea, leucoderma, odontopathy, ophthalmia, oral ulcers, pharyngodynia, proctoptosis, pulmonary bronchitis, skin diseases, seminal weakness, sore throat, small pox, syphilitic ulcers, strangury, tuberculosis, tooth ache, ulcers, uterine disorders, urinary diseases, vitiated conditions of "vatha" and wounds; used as gargle or mouth wash in aphthous conditions of mouth, excessive salivation, sore throat, spongy and bleeding gums; bark extract with milk is used as eye lotion in conjunctivitis; bark is burnt and ash is used as tooth powder along with shell powder of almonds and salt; chewing fresh bark strengthens loose teeth and bleeding gums; bark boiled with mango bark in water is used as treatment for eczema; bark decoction used as vaginal douche for leucorrhoea; bark decoction with salt is used as gargle for tonsillitis; bark powder is dusted over sores and ulcers for quick healing.

Gum: Antiasthmatic, alexipharmic, anthelmintic, antibilious, antipyretic, astringent, constipating, cooling, emollient, expectorant, haemostatic, liver tonic and styptic; useful in asthma, bronchitis, burns, chest complaints, catarrh, colic, cough, debility, diabetes, diarrhoea, dental carries, dysentery, gonorrhea, haemorrhoids, intermitent fever, leprosy, leucoderma, piles, pharyngodynia, prolapsed rectum, pneumonia, pulmonary disorders, scalds, seminal weakness, urinary and uterine diseases; gum fried with sugar and ghee is a nutritive tonic and aphrodisiac in sexual debility; gum with latex of *Calotropis procera* is given for asthma; powdered gum is given in diarrhoea and dysentery.

Leaf: Alexipharmic, anthelmintic, antiasthmatic, antibilious, antiscorbutic, aphrodisiac, astringent, demulcent, stimulant and tonic; used in cough, diarrhoea, diabetes, dysentery, gonorrhoea, leucoderma, pulmonary bronchitis, piles, prolapsed rectum, small pox, sore throat, tuberculosis, uterine and urinary diseases; paste of burnt leaves with coconut oil is applied for itching skin; decoction is used as gargle for sore throat, mouth ulcers and wounds; applied to sore and watering eyes along with breast milk; young leaves with sugar is eaten in the morning and evening to allay irritation in gonorrhoea; leaves with sugar and cumin is eaten as a cure for bloody seminal discharge, prolapsed rectum and pulmonary diseases.

Flower: Flower juice taken from newly opened flowers early in the morning is used as effective eye wash for sore eyes.

Pod: Antiscorbutic, astringent, expectorant, demulcent and stimulant; used in treating cold, cough, diarrhoea, dysentery, diseases of mucous membrane, leucorrhoea, ophthalmia, sore eyes, syphilitic ulcers, tuberculosis and urinogenital discharges; eating fresh pods removes phlegm from bronchial tubes; shade dried pods with sugar and milk is a effective remedy for spermatorrhoea; pod decoction with ginger is used as mouth wash; chewing fresh pod cures cough; pod juice applied on cuts and wounds arrests bleeding; it is applied on syphilitic ulcers to hasten suppuration.

Pod pulp: Astringent and stimulant; used as poultice for sore eyes and ulcers with discharges.

Bioactivity: Plant is inhibitor of 5-α-reductase, 5–lipoxygenase, ACE, ATPase, aldose-reductase, ATPase, β-glucuronidase, COMP, COMT, cyclo–oxygenase, cytochrome-P_{450}, deiodinase, glucosyl-transferase, HIV–RAT, histamine, leukotrine, MAO, MMP–9, metalloproteinase, NADH–oxidase, NEP, protein kinase–C, CAMP-phosphodiesterase, prostaglandin synthesis, topoisomerase I and II, tyrosine, tyrosine-kinase and xanthine-oxidase; is blocker of angiotensin receptor, calcium channel, β-adrenergic receptor, mast cell stabilizer and P_{450} inducer.

Toxicology: Plant contains chymotrypsin and trypsin inhibitors.

Uses in Alternative Systems of Medicine

Ayurveda:

Part used: Stem bark, leaf, pod and gum.

Uses: Bark is astringent to the bowels and alexipharmic; cures ascites, biliousness, bronchitis, cough, diarrhoea, dysentery, haemorrhages, leucoderma, piles, skin diseases and urinary discharges; pacifies "kapha" and "pitta"; leaf cures bronchitis, piles and "vatha"; heals fractures; pod is cooling and astringent to bowels; cures "kapha" and biliousness; gum is astringent, antidysenteric and styptic; cures biliousness, diarrhoea, dysentery, haemorrhages, leprosy, skin diseases and vaginal discharge.

Siddha:

Parts used: Root, bark, gum, leaf, seed and pod.

Uses: Useful in cough, diarrhoea, dysentery, polyuria, pyrexia, ulcer and urinogenital irritations.

Unani:

Parts used: Whole plant, stem bark, gum, leaf, flower, seed.

Uses: Antiflatulant, antipyretic, aphrodisiac, astringent, demulcent, deobstruent, emetic, expectorant, haemostatic, liver tonic and vitaliser; useful in burning sensation, dysuria, gonorrhoea, leucorrhoea, micturition and stomatitis; bark is emetic; strengthens the gums; gum is used as astringent in diarrhoea and dysentery, cures burns and colic; leaf cures cough, diabetes, gonorrhoea, jaundice, leucoderma, ophthalmia, phlegmatic liver, piles, sore throat, strangury and urethral discharge; flower is used as brain tonic and for insanity; seed cures haemorrhages and stomach pain.

Ethnomedicine

Plant: Is used in cancer and sclerosis.

Root: Is used in diarrhoea, liver complaints, skin diseases and syphilis.

Stem bark: Is used in asthma, bronchitis and diabetes.

Gum: Used locally on burns; and taken internally in diabetes and diarrhoea.

Leaf: Is used in eye complaints, hair fall, headache, skin diseases, syphilis, throat infections and urinary complaints.

Pod: Is used in diabetes and diarrhoea.

Non-medicinal Uses

Plant: Host plant for lac insect; source of gum.

Twigs: Used as tooth brush and chewing sticks; green, fresh twigs yields fibre used for basket making and fishing nets; used in the manufacture of paper and coarse ropes.

Spines: Used as fishing hooks and paper pins.

Bark: Source of fibre used in cordage; source of tanning material; used for dyeing fabrics and paper pulp; used in pharmaceuticals as emulsion for tablets, pills, lozenges and pastilles.

Wood: Useful in making beams, rafters, posts, door frames, crushers, furnitures, agricultural implements, tent pegs, knees of boat, oars, railway wagon, turnery, walking sticks, ploughs, yokes, carts, axles, spokes, felloes, wheels, blocks, gymnasium equipments, boats and oars; provides goad fuel and charcoal.

Gum: Used in calico printing; in sizing silk and cotton; in manufacture of paper, matches and ink; as a fixative for paint that is used for clay toys.

Leaf: Used as fodder; used for dyeing and manufacture of ink.

Pod: Green pods used as fodder; used for dyeing and manufacture of ink; yields black coloured dye.

Seed: Seeds are pounded and mixed with dates and fermented to make a intoxicating drink.

13. *Acacia pennata* (L.) Willd.

Synonym: *Mimosa pennata* L.

Notes: Name often misapplied to other species of the genus; there are several subspecies.

Common Names: Pineland wattle, Rusty mimosa.

Vernacular Names: Ben: Kuchui; **Hin:** Aari, Aila, Bisbul, Khairbel; **Kan:** Kaaadu seege, Mullu seege; **Mal:** Kaattusinikka, Sunna; **Mar:** Araati, Shemba vaela, Shemberti, Shembi; **Sans:** Ari, Ballikhadira, Khadirapathrika, Sandanika, Silikhadira, Svadikhallari, Uddala, Valli khadira; **Tam:** Idambubai, Idamullu, Indu, Kaattindu, Mullu chingai, Peychiyakkai, Singai, Singaimullu, Vellaiindu; **Tel:** Giddukorinda, Kaarushikaaya, Karrachecka, Mullukorinda.

Distribution: Native of Asia and Indian Ocean Islands; in India it grows in Assam, Bihar, Central and East Himalaya, Gujarat, Meghalaya, Rajasthan, Uttar Pradesh and West Bengal.

Description: A woody climber; branchlets, rachises and inflorescence densely tomentellous; the branches and rachises have recurved, hooked, smooth prickles; petiole with large, flat, oval gland near the base; more glands which are round are formed between the bases of upper 1–3 pairs of pinnae; pinnae 7–18 pairs; leaflets 30–50 pairs, sessile, overlapping, glabrous with fimbriate margins, narrow, linear, base oblique, mid rib eccentric almost near margin; flowers 1–3, subsessile in the axils of tiny spathulate bracts; buds brownish red; calyx thin, campanulate, glabrous, teeth distinct, brownish red; corolla light green, tips of the lobes green or reddish brown; pod strap shaped, flat with raised sutures, reddish brown, dehiscent, distinctly stalked with acute apex; seeds 6–12, flat, round; pod slightly bullate over the seeds.

Phytochemistry

Stem: β-Sitosterol.

Stem bark: β-Spinasterol, lupeol and tannin.

Edibility

Leaf: Edible.

Medicinal Uses

Plant: Is aphrodisiac and hypotensive; useful in burns, coryza, cough, digestive disorders, head ache, leucorrhoea and pain.

Root: Aphrodisiac, evacuant and digestive; powdered root is soaked in wine and plugged onto carious tooth before extraction to minimize the pain.

Bark: Acrid, bitter, hot, sour; it cures asthma, biliousness and bronchitis; bark juice is a cure for the poison of Russel viper's bite.

Leaf: Leaf juice is mixed with milk and given to infants suffering from indigestion with black stools; chewing the leaves with cumin and sugar, cures bleeding gums; leaf paste with sugar, cumin seeds and cow's milk is a remedy for scalding urine; decoction of young leaves is given in body pain, headache and fever and is a antidote to snake bite; pounded leaves are used as dressing for wounds.

Uses in Alternative Systems of Medicine

Ayurveda:

Part used: Stem and heartwood.

Uses: Useful in asthma, biliousness, bronchitis, haemorrhages, leprosy and wounds.

Siddha:

Part used: Whole plant.

Uses: Useful in burns, coryza, cough, digestive disorders, head ache, leucorrhoea and pain; leaf juice is expectorant, stimulant, stomachic and tonic; useful in bronchitis, colic, convulsions, respiratory affections in children, spasms, menstrual disorders, rheumatism, sciatica and sinusitis.

Medicinal Uses Outside India

Gold coast: The bark of white flowered variety is given with pepper as an enema for stomach complaints.

Senegal and Sudan: The vapour from the hot decoction of the plant is inhaled as a remedy for chest complaints; ground leaves are applied as dressing to wounds.

Sierra Leone: Carious tooth is plugged with root or bark soaked in palm wine as a preparatory procedure for extraction.

Ethnomedicine

Plant: Is used in cough, rheumatism and small pox.

Root: Is used in digestive complaints.

Bark: Is useful in dandruff and as antidote to snake bite.

Leaf: Is useful in headache and body pain; used as febrifuge.

Non-Medicinal Uses

Plant: Useful as fodder; host for lac insect.

Root: Serve as toothbrush and to clean ornaments.

Stem bark: Is used as substitute for soap; tannin in the bark is used in tanning industry; yields fibre used in making fishing nets and cordage; ash of charred fibre is used in cleaning mirrors and as fish poison.

Pod pulp: Is used for tanning; as insecticidal and fish poison.

14. *Acacia polyacantha* Willd.

Synonyms: *Acacia suma* (Roxb.) J.Voigt.
Mimosa suma Roxb.

Common Names: White cutch, White sundra.

Vernacular Names: Ben: Sai kanta; **Kan:** Bilijaali, Buguli, Mugali, Mugali Soppu, Mukkili; **Mal:** Venkarinnali; **Mar:** Kamitya, Sonekhair; **Sans:** Kadara, Shamee, Samavaika; **Tam:** Kovalai, Mullu maram, Selai, Selaivunjai, Silai; **Tel:** Tella sundra, Tellachandra.

Distribution: Native of Africa and Asia including India; in India it grows in Assam, Bihar and West Bengal.

Description: A medium sized tree with white bark exfoliating in papery flakes; the trunk is marked by horizontal patches of dark colour; stipular prickles found below each pinna; pinnae 10–12 pairs; leaf lets numerous, linear; flowers white or pale yellow; seeds 6–8 in number.

Phytochemistry

Stem bark: (–) Epicatechin and proanthocyanidin.

Heartwood: Quercetin and 5,4'–dihydroxy–7,3'–dimethoxyflavone–3–O–D–galactopyranoside.

Leaf: N, N-Dimethyl tryptamine.

Medicinal Uses

Plant: Acrid, anticancer, astringent and refrigerant; used in cough, diabetes, diarrhoea, epistasis, haemorrhagic diseases, leprosy, leucoderma, obstinate urinary disorders, piles, pthisis and ulcers.

Stem bark: Astringent; used as a gargle for sore throat and as mouth wash for tooth ache; bark powder is sprinkled on ulcers to promote healing.

Wood: Acrid, anthelmintic, astringent, bitter and thermogenic; useful in diabetes, epilepsy, helminthiasis, insanity, leprosy, leucoderma, obesity, pruritis, rheumatism, skin diseases, ulcers and vitiated conditions of "kapha".

Gum: Obtained from heartwood is used in dental preparations.

Leaf: Infusion of tender leaves is astringent and is given in diarrhoea and dysentery.

Pod: Decoction of the pod is given for urinogenital diseases.

Seed: Hypoglycaemic.

Toxicology: Excessive use causes hair loss.

Uses in Alternative Systems of Medicine

Ayurveda:

The medicinal uses are similar to *Acacia catechu,* which is sometimes used as a substitute.

Tibetan:

Part used: Plant.

Uses: Cures anaemia, diabetes, leukoderma, skin diseases and ulcers.

Non-medicinal Uses

Plant: Grown as avenue tree and hedge tree; yields gum.

Bark: Is a source of tannin used in tanning industry.

Wood: Used as fuel.

Gum: Is used in dental preparations and as tanning material.

15. *Acacia pycnantha* Benth.

Synonyms: *Acacia falcinella* Meissner
Acacia leiophylla Benth.
Acacia petiolaris Lehm.
Acacia pycnantha Benth., var., *petiolaris* H. Vilm
Acacia westonii Maiden.

Notes: The plant is the official floral emblem of Australia.

Common Names: Broad leaved wattle, Golden wattle, Sydney golden wattle.

Distribution: Native of Australia; introduced into India; cultivated in hill stations of Tamil Nadu.

Description: A shrub or a small tree with more or less pendulous branches; leaves with a gland slightly away from the base; leaflets oblong lanceolate to broadly obovate; flowers large, golden yellow, showy, numerous; pod contracted and contricted between the seeds; seeds club shaped with aril.

Phytochemistry

Plant: Albizzine, carboxyethyl cysteine, phenethylamine, tannin, β-acetylpropionic acid and β- diaminopropionic acid.

Gum: Arabinose, arabogalactin and tannin.

Flower: Essential oil.

Non-Medicinal Uses

Plant: Grown as hedge plant, soil binder and ornamental; yields gum called wattle gum, which is red in colour.

Bark: Used in tanning industry; rich in tannin (50%).

Gum: Is equal in quality to catechu gum with similar uses.

Flower: Used in perfumery.

16. *Acacia senegal* (L.) Willd.

Notes: Very variable; five intergrading varieties are recognized by Ross; it yields Senegal gum or gum arabic.

Common Names: Gum arabic, Senegal gum.

Vernacular Names: Guj: Goradio babul; **Hin:** Kumta, Kheira, Kumbhat, Kumbhariya; **Mar:** Khair; **Pun:** Khor, Kumata; **Sans:** Shvetakhadira.

Distribution: Native of Africa, India, Pakistan and Middle East; in India it grows in Gujarat, Punjab, Rajasthan and Uttar Pradesh.

Description: A flat topped tree of 30 ft height with short trunk and pale bark, fissured and flaking with age; a translucent gum exudes from the cracks on the tree; the branches bear stipular spines which are found in groups of three on the enlarged nodes; the central thorn is strongly recurved; bipinnately compound, 3–6 pairs of pinnae; the leaflets are grayish green, oblong, leathery and 8–18 pairs in number; flower buds are red but flowers are ivory white on dense spikes, solitary or in groups of 2 or 3; pod flat, brown, papery, pubescent, oblong with 1–6 greenish brown disc like seeds; the tree is said to produce seeds at five year intervals.

Phytochemistry

Plant: 4–Methoxy glucuronic acid, arabic acid, arabinose, ascorbic acid, aspartic acid, cysteine, D–galactose, dimethyltryptamine, erythrodiol, galactoglucuronic acid, glucuronic acid, hentriacontane, hentriacontanol, hydrocyanic acid, kaempferol, L–arabinose, L–rhamnose, leucine, magnesium, octacosanol, peroxidase, quercetin, rhamnose, serine, sitosterol, sodium, sucrose, tannin, uronic acid, valine, and β-sitosterol;

Root: Betulin, ceryl alcohol and sitosterol.

Stem bark: Erythrodiol, octacosan–1–Ol, octacosanol, β-amyrin and β-sitosterol.

Wood: Hentriacontanol, nonacosane, octacosanol, pinitol and sitosterol.

Leaf: Calcium, carbohydrates, D–pinitol, dimethyl tryptamine, fibre, N–hentriacontane.

Flower: Betulin, octacosan–1–Ol and β-sitosterol.

Pod: Carbohydrates, fat, fibre, kaempferol, quercetin, β-sitosterol and β-amyrin.

Seed: Carbohydrates, fat and fibre.

Edibility

Gum: Yields a gum called "gum arabic" which is used in confectionary, chewing gums, spray dried instant foods, sweet meats and bakery products and to preserve flavour of the soft drinks.

Seed: Used as vegetable.

Medicinal Uses

Plant: Is analgesic, antiandrogenic, antiallergenic, anthelmintic, antibacterial, antihaemorrhagic, antihistaminic, antihypertensive, antihypoglycaemic, antifertility, anti-inflammatory, antimastalgic, antinociceptive, antioxidant, antiperiodontic, antiprostaglandin, antipyretic and antispasmodic; cures adenoma, anorexia, arrhythmia, arthritis, asthma, atherosclerosis, cancer of the breast, dermatitis, diabetes, diarrhoea, dysentery, dysmenorrhoea, eczema, edema, encephalopathy, epilepsy, fibrosarcoma, gastritis, gingivitis, glaucoma, hepatotoxicity, herpes, leprosy, leukaemia, lymphoma, malaria, myocarditis, nephritis, ophthalmia, orchitis, osteoporosis, pharyngitis and pneumonia; antidote to morphine poison.

Stem bark: Is anti-inflammatory, spasmolytic and CNS active.

Gum: Antihaemorrhagic, antitussive, astringent, bechic, demulcent, emollient, expectorant and nutritive; useful in bleeding gums, burns, catarrh, cold, cough, diarrhoea, dysentery, fever, nodular leprosy, throat irritation and typhoid; it is also useful in reducing inflammation of mucous membranes of intestine, irritated bladder and urethra, edema, haemorrhages and nephrosis; substitute for amylase in diabetic food; in modern pharmacy it is employed as demulcent in preparations to treat, diarrhoea, dysentery, cough, throat irritations and fever; it is a valuable remedy for gonorrhoea, inflammations, nodular leprosy and sore nipples; it was formerly given intravenously to counter act low blood pressure, haemorrhages and edema associated with nephrosis but this practice is now abandoned as it caused kidney and liver damage and allergy.

Bioactivity: Plant is inhibitor of 5–lipoxygenase, ATPase, aldose-reductase, β-glucoronidase, cyclo-oxygenase, cytochrome P_{450}, deiodinase, glucosyl-transferase, lipoxygenase, MAO, MMP–9, metalloproteinase, NADH–oxidase, ornithine-decarboxylase, phospholipase, prostaglandin synthesis, protein-kinase, topoisomerase I and II, tyrosine-kinase, xanthine-oxidase, CAMP phosphodiesterase along with blocker of angiotensin receptor and calcium channel and inducer of quinone reductase.

Toxicology: A solution of the gum given intravenously to arrest haemorrhage (a practice abandoned now) causes kidney and liver damage and allergic reactions; contains serine proteinase inhibitors which inhibit the action of digestive proteinases; cause reduction of proteolysis of dietery proteins and growth impairment.

Medicinal Uses Outside India

Nigeria: Gum is used for local inflammations and nodular leprosy.

Non-Medicinal Uses

Plant: Grown as an ornamental tree and wind break; it enriches the soil and reclaims sandy soil; used as pesticide; yields gum arabic.

Root: Has a tough tap root used as tool handles; yields a strong fibre used in making ropes, fishing nets and horse girdle.

Stem bark: Yields fibre used in making ropes and fishing nets; the tough stem is used as tool handles.

Wood: Yields charcoal; used in making tool handles, wheels, sugar cane crushers, weaver's shuttle, fencing posts and agricultural implements.

Gum: The gum arabic obtained from the stem bark is used in manufacture of inks, paints, wax polish, cosmetics and crayons; it is used to give luster to silk and crepe and as a mordant in calico printing and in dyeing fabrics; useful as an adhesive for postage stamps and envelope flaps; is used in the pharmaceutical industry as it serves as emulsifier and gives viscocity to powdered drugs; used as a binding agent in pills, tablets, cough syrups and lozenges; used in wax polishes and hair sprays; it is used in lithography, pottery pigment, water colours as a thickening agent and mordant and as a substitute for amylase in diabetic food.

Leaf: Grazed by animals and also used as green manure.

17. *Acacia seyal* Del.

Synonyms: *Acacia fistula* Schweinf.
Acacia flava (Forsskal) Schweinf. var., *seyal* (Del.) Roberty
Acacia stenocarpa A. Rich

Notes: Two varieties, *seyal* and *fistula* are recognized; the wood of *Acacia seyal* was used in making coffins in ancient Egypt and the Ark of old testament.

Common Names: Shittim wod, Whistling wood.

Distribution: Native of Africa and Middle East; introduced into India.

Description: A flat topped tree with white to greenish yellow or orange red bark; sparsely branched, covered by sparse hairs with numerous sessile reddish glands; leaves with large glands on petiole and between the top 1–2 pairs of pinnae; stipules spinescent; pinnae 3–7 pairs; leaf lets 11–20 pairs, sparingly glabrous; flowers bright yellow, axillary; pod dehiscent, falcate, constricted between the seeds, glabrous, 6–9 seeded; seeds elliptic, compressed, olive brown.

Phytochemistry

Plant: Tannin.

Edibility

Gum: Is edible.

Medicinal Uses

Stem bark: Decoction is used for dysentery and leprosy.

Gum: Aphrodisiac, astringent and emollient; used for cold, cough, diarrhoea, haemorrhage and ophthalmia.

Non-Medicinal Uses

Plant: Yields a good quality gum and tannin; used in tanning industry.

Wood: Used as fuel.

Gum: Is used in textile sizing and as flavouring agent.

Leaf: Used as forage.

Pod: Fed to sheep to fatten them.

18. *Acacia sinuata* (Lour.) Merr.

Synonyms: *Acacia rugata* (Lam.) Voigt
Acacia rugata Benth.
Mimosa simuata Lour.

Notes: The soap nut tree or the soap pod tree, which is in extensive economic use in India, has been known all along as *Acacia concinna* DC; following a controversy on the correct name for this species, ILDIS has accepted *Acacia sinuata* (Lour.) Merr., which makes the binomial *Acacia concinna* irrelevant, even as a synonym; however, ILDIS has also accepted the binomial *Acacia concinna* (Willd.) DC., as the correct name for a different species; both *Acacia sinuata* and *Acacia concinna* are native to the Asian region (including India), but the former has a much wider distribution and economic applications, than the latter; on account of the use of *Acacia concinna* for the soap nut tree for a very long time, there is overlap and confusion in the information on this species in literature and in spite of taking care, possibly in the profiles given here; it is, therefore, necessary to be cautious.

Common Names: Soap pod tree, Soap pod wattle, Washing pod tree.

Vernacular Names: Ben: Banritha; **Guj:** Chikaki; **Hin:** Ritha, Kochi, Shikakai; **Kan**: Seege kayi, Olleseege; **Mal:** Chikaka, Shikai, Cheeyakayi, Innaa, Shinikkaaya; Shinkkayya; **Mar:** Shika, Shikakai; **Sans:** Bhooriphena, Charmakaasha, Deeptha, Pathragha, Phenila, Saathala, Sapthala, Sara, Sarmakasa, Shreevalli, Svarnapushpi, Vidala, Vimala, Vishanika **Tam:** Shikai, Seekai, Sigakai, Nallachiyakkay, Shivakkai; **Tel**: Cheekaya, Manjisikaya, Shikaya, Sheeka kaayi.

Distribution: Native of Asia; grows in China, India and Indian Ocean Islands; it grows throughout India.

Description: Prickly climbing shrub, rarely a tree; branches pubescent; leaves with a petiolar gland below the middle; rachis with a gland between upper most pair of pinnae; pinnae 4–7 pairs; leaf lets 20–25 pairs, pubescent; flowers white; calyx red, finely 5–lobed, glabrous; pod fleshy, sinuate, wrinkled when dry; 10–14 seeds.

Phytochemistry

Plant: Acacinin, hexacosanol, lupeol, saponins and α-spinasterol.
Stem bark: Hexacosanol, lupeol and α-spinasterol.
Leaf: Amino acid, ascorbic acid, citric acid, nicotine, oxalic acid, proteins, rutin, saponin, succinic acid, tannin, tartaric acid and triterpenoids.
Pod: Acacic acid, acacidiol, acacigenin–B, acacinin C, D and E, lactone, sapogenin B, α-amyrin, β-sitosterol and β-amyrin.
Seed: Acacic acid, acacinin A, arabinose, concinin, coronaric acid, fucose, macherinic acid, rhamnose, sapogenin B, xylose and β-sitosterol.

Edibility

Leaf: Used in culinary preparation like curries and chutneys as a substitute for tamarind.

Medicinal Uses

Plant: Aperient, astringent, CNS active, diuretic, emetic, expectorant, germicidal, hypotensive and spasmolytic.

Stem bark: Saponins from the bark is haemolytic, spasmolytic and spermicidal; the juice of the bark taken daily in a small dose, is an effective remedy for asthma and leprosy.

Leaf: Aparient, cathartic and febrifuge; leaf infusion is useful in biliousness and malarial fever; prevents flatulence and is a mild laxative; chutney of leaves with salt and pepper is eaten during jaundice.

Pod: Anthelmintic, aperient, astringent, bitter, cardiotonic, cooling, deobstruent, depurative, diuretic, emetic, expectorant, febrifuge and purgative; useful in abscesses, burning sensation, biliousness, constipation, diseases of blood, eczema, erysipelas, haemorrhoids, jaundice, leprosy, leucoderma, prurigo, renal calculi, stomatitis, vitiated conditions of "kapha" and "pitta"; an ointment of the pod is useful in skin diseases; decoction of the pod used regularly as hair wash promotes hair growth and removes dandurff and lice in the hair; taken internally it relieves biliousness in jaundice; emulsified in water it acts as collyrium; used as antidote for snake bite.

Uses in Alternative Systems of Medicine

Ayurveda:

Part used: Pod.

Uses: Useful in diseases of the blood, edema and regurgitation.

Siddha:

Part used: Leaf and pod.

Uses: Emetic and purgative; useful in constipation, fever, jaundice and ulcers.

Tibetan:

Part used: Pod.

Uses: Cures dysuria, heart diseases and phantom tumours.

Non-medicinal Uses

Plant: Host plant for lac insect.

Bark: Used in tanning and dyeing; infusion is used as domestic germicide.

Leaf: Because of the tannin content, leaf paste mixed with turmeric gives a green dye used in tanning industry.

Pod: Employed as fish poison; decoction of the pod applied on the hair removes lice; also used as a detergent for washing silk and woollen fabrics; used as hair wash.

Seed: Fish poison.

19. *Acacia tomentosa* Willd.

Synonyms: *Acacia chrysocoma* Miq.
Mimosa tomentosa (Willd.) Roxb.

Common Names: Elephant thorn, Jungle nail tree.

Vernacular Names: Kan: Aane mullu; **Mar:** Anjar; **Tam:** Aanai mullu.

Distribution: Native of India, Indonesia, Thailand and Vietnam.

Description: A small tree with branch lets and leaf rachis covered with fine grey hairs; leaf rachis with several glands; thorns spindle shaped, black; pinnae crowded; leaf lets 8–16 pairs, sessile, oblong, linear, subcoriaceous, glabrescent; flowers white in axillary pedunculate heads; pod thin, flat, ligulate, short stalked, falcate; seeds 6–10, flat, oval, areolate, brownish green.

Medicinal Uses

Plant: Cures dysuria, subdues deranged "kapha" and excites "pitta".

Stem bark: Decoction used as gargle for sore throat.

Leaf: Astringent; useful in dysentery and diarrhoea.

Pod: Useful in urinary diseases.

Non-Medicinal Uses

Plant: Yields gum.

Stem bark: Rich in tannin and used in tanning industry.

Wood: Suitable for making fencing posts and agricultural implements.

Pod and leaf: Eaten by cattle.

20. *Acacia torta* (Roxb.) Craib

Synonym: *Mimosa torta* Roxb.

Vernacular Names: Hin: Aila; **Kan:** Antarike, Chande mullu, Kempu seege; **Mal:** Alterincha, Inna; **Mar:** Chilar; **Ori:** Dentari; **Pun:** Arhai; **Tam:** Kariyundu; **Tel:** Tellakorinda, Konda korinda, Korinta, Seengaikkodi.

Distribution: Native of India and Thailand; in India it grows in Bihar, Madhya Pradesh, Punjab, Rajasthan, Uttar Pradesh and West Bengal.

Description: Straggler with yellowish, pubescent branchlets; internodal thorns straight; pinnae alternate, 10–14 pairs; leaflets 25–40 pairs, elliptic, glabrous with truncate base and subacute apex; petioles sparsely prickled with a gland at the base; rachis, grooved, pubescent, with prickles on the underside along with glands, opposite to 4 uppermost pinnae; flowers white with 5 petals and lanceolate bracts; calyx tube 5 lobed, pubescent; stamens numerous, filaments basally connate; pod flat, thin, prominently nerved, with attenuate base and obtuse apex; seeds 6.

Medicinal Uses

Plant: Is used in the treatment of bronchitis, consumption, cough, measles, menstrual disorders and tubercular fistula; used as hypotensive.
Bark: Astringent.

Phytochemistry

Stem bark: Acacinoic acid, aliphatic hydroxy compounds, lupeol and wax.

Non-Medicinal Uses

Plant: Useful as manure.

Bark: Soaked bark is used to wash hair and clothes as a substitute for soap; its extract is used as protective coat for boats and fishing nets; used in dyeing industry; contains 17% tannin used in tanning industry.

Leaf: Used for dyeing.

21. *Acacia tortilis* (Forsskal) Hayne

Synonym: *Mimosa tortilis* Forsskal.

Common Name: Umbrella thorn.

Distribution: Native of Africa; introduced into India; in India it grows in Rajasthan.

Description: Tree reaching a height of 20 m with umbrella shaped canopy; branches are reddish brown or dark brown with grey lenticels and white, straight, slender spines with black or brown, sharply curved tips; leaves 3–10 pairs; leaf lets 6–15 pairs; flowers axillary, pale yellow, fragrant, set on globose heads; pod spiral, glabrous, pale orange to light brown at maturity; seeds 10, dark with a oval depression on lateral face.

Phytochemistry

Plant: (+) Catechol, (+) gallocatechol, betulin, gallic acid, hexacosanol, α-amyrin and β-amyrin.

Stem bark: Betulin and tannin.

Heartwood: Betulin and 3–acetyl-β-sitosterol.

Leaf: Apigenin-6,8-bis-glucoside, calcium, carbohydrates, fibre, hydrocyanic acid, isorhamnetin-3-rutinoside, luteolin–7–glucoside, myricetin, pipecolic acid, proteins, quercetin, quercetin-3'-glucoside, quercetin-3'-O-rutinoside and tannin.

Gum: Aldobiouronic acid.

Pod: Calcium, carbohydrates, fibre and protein.

Seed: 4–Hydroxy pipecolic acid, carbohydrates, N-aetyl-D–jenkolic acid and phosphorous.

Medicinal Uses

Plant: Is antiallergenic, antibacterial, anticancer, antifibrinolytic, antihepatotoxic, antihypertensive, anti-inflammatory, antioxidant, antiradicular, antiseptic, antitumour, antitussive, antiviral, apoptotic, astringent, bronchodilator, cytotoxic, cancer preventive, cardioprotective, diuretic, haemostat, hepatoprotective, hypouricemic, immunostimulant, laxative, psychotropic, respirostimulant, styptic and vasodilator; useful in asthma, atherosclerosis, bronchitis, dermatitis, diabetes, diarrhoea, dysentery, hepatitis, herpes, nephritis, osteoporosis and ulcer.

Leaf: Vermifuge; useful in allergic dermatitis, edema and skin infections.

Bioactivity: Plant is inhibitor of ACE, aldose-reductase, cyclo-oxygenase, glucosyl-transferase, ornithine-decarboxylase, prostaglandin synthesis, topoisomerase I and II and xanthine-oxidase; is blocker of angiotensin receptors and calcium channel; quinone-reductase inducer.

Non-Medicinal Uses

Plant: Yields gum.

Bark: Yields tannin used in tanning industry.

Wood: Is used for fencing posts and agricultural implements; yields charcoal and firewood.

Leaf: Used as livestock feed.

22. *Adenanthera pavonina* L.

Common Names: Acacia coral, Agati, Bead tree, Coral wood, Curly bean, Jumbi bead, Peacock flower fence, Red bead tree, Red sandalwood.

Vernacular Names: Hin: Barigumchi, Kuchandan, Raktha chandan, Ranjana; **Kan:** Aanegulaganji, Dodda gulaganji, Manda theeya, Manjaadi, Manjitte; **Mar:** Thorlaguni, Thorligunji; **Sans:** Kaamboji, Ksharaka, Kuchandana, Ranjaka, Thaamraka; **Tam:** Aanaikundumani, Manjaadi, Thilam, Thilagam; **Tel:** Bandigurivenda, Bandiguruginja, Yenugaguruginji, Gurivenda, Mensenikotta, Peddaguriginja.

Distribution: Native of Asia including India and Australasia; it grows in Andaman and Nicobar Islands, Assam, Meghalaya, Tripura and Uttar Pradesh.

Description: A large tree with grey smooth bark; branchlets sparsely to densely glabrous or tomentellous; stipules small; pinnae 2–6 pairs, opposite; leaf lets 3–9 pairs, alternate, elliptic to ovate, base slightly oblique, broadly cuneate; flowers white, turns yellow with age on solitary raceme; calyx glabrous; petals glabrous, linear – lanceolate, acute: pod straight to falcate, slightly twisted, tapered at the base and acute at the apex; seeds 6–15, elliptic, lenticular and glossy red.

Phytochemistry

Plant: Chalcone, butein, dihydromyricetin, dihydrorobinetin and robinetin.

Root: Echinocystic acid, methyl echinocystate, methyl oleolate, oleanolic acid and stigmasterol.

Bark: 2–4–Dihydro benzoic acid, butein, chalcone, dihydromyricetin, dihydrorobinetin, echinocystate, echinocystic acid, methyloleonolate, oleonic acid, saponins and stigmasterol.

Wood: 2,4-dihydrobenzoic acid, butein, chalcone, dihydrorobinetin and robinetin.

Leaf: Dulcitol, octacosanol, stigmasterol and β-sitosterol.

Seed: ψ-Methylglutamic acid, butein, chalcone, dihydrorobinetin, dulcetol, flavonoids, γ-ethilidene glutamic acid, polysaccharide, robinetin, sitosterol, stigmasterol, γ–ethylene glutamic acid, γ–methylene glutamic acid and γ–methylene glutamine.

Edibility

Seed: Roasted and eaten.

Medicinal Uses

Plant: Astringent; useful in boils, cholera, inflammation, paralysis, rheumatism and sore throat; ground with honey, the paste is applied on abcesses and sores to hasten suppuration.

Root: Useful as emetic.

Bark: Aphrodisiac, astringent and vulnerary; useful in colonorrhoea, gout, haematuria, pharyngopathy, ulcers and vitiated conditions of "vata"; bark decoction is taken internally for haematemesis, haemorrhage from bowels and chronic rheumatism.

Wood: Aphrodisiac, astringent, emetic, tonic and haemostatic; useful in diarrhoea, haemorrhages, orchitis and vitiated conditions of "vatha"; decoction is used internally in pulmonary affections and is applied externally in chronic ophthalmia; wood paste is applied on the body for prickly heat and on the fore head for headache.

Leaf: Aphrodisiac, astringent and vulnerary; useful in colonorrhoea, haematuria, haematemesis, haemorrhage from bowels, pharyngopathy and vitiated conditions of "vatha"; macerated leaves are used in gout and rheumatism.

Seed: Aphrodisiac, astringent, antiemetic, anti-inflammatory, anti-cephalalgic, cooling, febrifuge and suppurative; useful in boils, burning sensation, cholera, colic, fever, giddiness, gonorrhoea, gout, hyperdipsia, paralysis, prickly heat, rheumatism, vitiated conditions of "vatha" and "pitta" and vomiting; decoction is taken internally for haemorrhage from bowels, haematuria and pulmonary affections; powdered seeds hasten suppuration of boils and cures prickly heat; decoction of the leaf given in chronic gout and rheumatism; seed decoction is used as eye wash for chronic ophthalmia.

Toxicology: Contains serine proteinase inhibitors, which inhibit the action of digestive proteinases by reducing the proteolysis of dietery proteins and resulting in growth impairment.

Uses in Alternative Systems of Medicine

Siddha:

Part used: Root, leaf and seed.

Uses: Emetic; useful in joint pains and warts.

Medicinal Uses Outside India

La Reunion: The plant is considered astringent; it is used for rheumatism and sore throat.

Non-Medicinal Uses

Plant: Is used as preservative of milk and confectionery; used in paint, leather and cosmetic industry; in the manufacture of plastic, tooth paste, matches and perfumes; as adhesive powder for artificial teeth; for polishing sewing threads, ropes and twines; for lubricating surgical instruments; in textile sizing, printing, printing inks and dyeing.

Bark: Used for washing hair and clothes.

Wood: Powdered wood yields a red dye which is used for dyeing fabrics and used by Hindus for making mark on forehead; used for making furnitures, cabinet and general construction.

Gum: Wood yields a gum used in pharmaceuticals for its thickening power and stabilizing action in the manufacture of pills, tablets, medicinal jellies, insect repellants and paraffin oil.

Seed: Used in making rosaries, necklaces and other ornaments; powdered seeds with borax and water can be used as cement.

23. *Albizia amara* (Roxb.) Boivin

Common Names: Dark siris, Oil cake tree.

Vernacular Names: Ben: Krishnasirish; **Guj:** Mota sariso; **Hin:** Kaalishirish; **Kan:** Bilikambi, Bilatti, Chigare, Chiguru, Kaadu seege, Thuggligida; **Mal:** Sulivaaka, Vaaraasi; **Mar:** Lali, Lulai, Thugli; **Sans:** Krishna sirisha; **Tam:** Arappu, Karuvaagai, Sayalvaagai, Sikkaram, Thurinjil, Unjai, Unjil, Ushilai, Ushilam, Wunja; **Tel:** Chikkareni, Nallangi, Nalla regu, Seekireni, Sigara, Sirikenu.

Distribution: Native of Africa, India and Srilanka; it occurs in Peninsular India; often planted.

Description: A moderate sized tree with a crooked bole; branches and leaf rachis densely pubescent; leaves with a small gland on middle of petiole and between bases of upper pair of pinnae; pinnae 6–16 pairs, densely pubescent; stipules minute, lanceolate; leaf lets 10–30 pairs, sessile, linear, very small; flowers yellow, fragrant and borne on axillary umbels; calyx funnel shaped with short teeth; corolla funnel shaped; pod thin, flat, oblong, apiculate, veiny, pubescent and shortly stalked; seeds 10–12, ovate, orbicular and grey.

Phytochemistry

Plant: Yields gum.

Stem bark: Methyl esters of fatty acids, saponins, triterpenes and β-sitosterol.

Wood: 4'–O–methyl quercetin–3–rutinoside, arabinose, echinocystic acid, glucose, methyl esters of fatty acid, oleonic acid, rhamnose, saponins, triterpenes and β-sitosterol.

Leaf: 4'–O–methyl quercetin-3–rutinoside, echinocystic acid, arabinose, flavonolglycoside, glucose, methyl esters of fatty acids, oleanolic acid, rhamnose, rutin, saponins, triterpene saponin, unsaturated carboxylic acid and β-sitosterol.

Seed: Yields dark yellow fixed oil along with 10–demethyl–protoemitinol, arabinose, echinocystic acid, glucose, oleanolic acid, protoberberine, rhamnose, taxifolin and β-sitosterol.

Medicinal Uses

Plant: Spasmogenic.

Root: Useful in cough and dyspnoea.

Root bark and wood: Useful in cough and dyspnoea.

Gum: Astringent; used as topical application in ulcers.

Leaf: Useful in ophthalmia; paste used as local application on boils and erysipelas; the dried leaves or dried leaf powder is taken orally during famine to kill appetite with out loss of debility.

Flower: Cooling; paste is applied externally on boils, eruptions, erysipelas, inflammations, swellings and ulcers.

Seed: Used as astringent; given in diarrhoea, gonorrhoea and piles.

Seed oil: Applied locally on leucoderma and leprosy.

Uses in Alternative Systems of Medicine

Siddha:

Part used: Root, root bark and heartwood.

Uses: Used in cough and dyspnoea.

Non-Medicinal Uses

Plant: Used as green manure; grown as shade tree.

Wood: Is a source of gum; wood is heavy and durable and used in making posts, beams, rafters, carts, shafts, axles, yokes, ploughs, furniture, tool handles, agricultural implements, cabinets and turnery; yields fire wood.

Leaf: Used as fodder and green manure; also used as a substitute for soap and as adulterant of tea.

24. *Albizia chinensis* (Osbeck.) Merr.

Synonyms: *Acacia stipulata* DC.
Albizia marginata (Lam.) Merr.
Albizia stipulata Boivin
Mimosa chinensis Osbeck
Mimosa marginata Lam.
Mimosa stipulacea Roxb.

Common Names: Stone siris, Stipulate siris.

Vernacular Names: Ben: Amalaki, Chakura; **Hin:** Kanujera, Siran, Pattia, Samsundra; **Kan:** Baage, Bettadabaage, Hottebaage, Kottabaage; **Mal:** Mottavaaka, Pottuvaaka; **Mar:** Laeli, Udala; **San:** Bhandi, Sairisam; **Tam:** Kaatturinjil, Pilivaagai, Silaivaagai; **Tel:** Chinduga, Kondachiragu.

Distribution: Native of Asia including India; occurs throughout India.

Description: An unarmed ever green tree with a flat crown upto 40 m high; bark dark grey, smooth, hoop ringed, lenticillate; stipules very large, caducous, obliquely cordate, pinkish orange, pubescent, with filiform tail with a dilated base at one side; leaf rachis lenticillate with protruding circular gland near the base and smaller glands between the bases of most pinnae; pinnae 6–20 pairs; leaf lets 20–45 pairs, sessile, falcate, lanceolate, oblique with a obtuse base; leaf surface pitted; flowers white, upper half light red and arranged on pedunculate globose head; calyx funnel shaped; corolla funnel shaped, pubescent; staminal tube yellow, glossy; pod thin, flat, strap shaped, reddish brown, glossy, reticulate; seeds 8–12, ovate, dark brown, with out aril.

Phytochemistry

Plant: Yields gum and saponin.

Stem bark: Arabinose, glucose, oleanolic acid, quercetin, echinocystic acid and β-sitosterol.

Heartwood: (+) Pinitol, norepinephrine, seretonin, α-spinasterol and α-spinasterone.

Leaf: 4'–O–Methylquercetin–3–rutinoside, arabinose, echinocystic acid, glucose, oleanolic acid, rhamnose and β-sitosterol.

Seed: Albiside, arabinose, echinocystic acid, fucose, rhamnose and xylose.

Medicinal Uses

Plant: Diuretic and spasmogenic.

Stem bark: Decoction is taken to contract uterine muscles; infusion is used as lotion for cuts, scabies and skin diseases.

Flower: Externally applied to boils, inflammatory swellings and ulcers.

Leaf: Infusion of leaves used in conjunctivitis.

Seed: Astringent and anti-inflammatory; useful in diarrhoea, gonorrhoea and piles.

Non-Medicinal Uses

Plant: Grown as shade tree in coffee and tea plantations; is a rare host for lac insect; yields tannin used in tanning industry; used as insecticide and fish poison; branches used as fodder.

Wood: Used in making tea boxes, packing cases, agricultural implements, furniture, planks, turnery, combs, spoons, yokes, wheels and cabinets; yields fire wood and charcoal; wood pulp used in making writing paper and printing paper.

Leaf: Is used as fodder and manure.

Gum: Used for sizing paper.

25. *Albizia julibrissin* Durazz.

Synonyms: *Acacia julibrissin* (Durazz.) Willd.
Mimosa julibrissin (Durazz.) Scop.
Mimosa speciosa Thunb.

Common Names: Pink siris, Silk tree, Silk Albizzia.

Vernacular Names: Hin: Bararu, Barulia, Bhokra, Karmaru, Laalsiris; **Kan:** Chelabaage, Elesujjalu baage; **Tam:** Sela vaagai; **Tel:** Chanchu, Kondagnam, Nalla sinduga, Sinduga.

Distribution: Native of Asia including India; in India it occurs in Himalaya upto 2500 m and in Meghalaya.

Description: A middle sized unarmed tree; leaf rachis with a small gland at the base of petiole and between upper pinnae; leaflets 8–24 in number, membranous, sessile, sensitive, cuspidate, straight on the upper side, round at the base; flowers white; calyx funnel shaped, shortly toothed; anther filaments rose red in colour; pod glabrous, membranous, indehiscent, grey or pale brown with a short stalk and narrow beak; seeds 8–12.

Phytochemistry

Plant: Acacic acid, acacigenin B, albicide, lactone, machaerinic acid, quercetin–3–O–galactoside and quercetin–3–O–rhamnoside.

Stem bark: 3',4',7–trihydroxy flavone, acacigenin, calcium, copper, macherinic acid, magnesium, manganese, saponin, sodium and α-spinasterol.

Flower: Cis and trans linalool oxide, isopentanol and α-ocimene.

Seed: Albizzin, fat, proteins and β-carboxy ethyl cysteine.

Edibility

Leaf: Aromatic, edible; used as flavouring agent in food industry.

Medicinal Uses

Plant: Is anthelmintic, antiallergenic, antianginal, antibacterial, anticoronary, anti-inflammatory, antimastalgic, antinociceptive, antioxidant, antiperiodontic, antispasmodic, antistomatitic, antitumor, antiviral, astringent, diuretic, hepatoprotective, hypotensive, immunosuppressant, nephrotoxic, vasodilator and vulnerary; useful in anaemia, anorexia, arrhythmia, arthritis, asthma, atherosclerosis, cancer, canker, cataract, colitis, diabetes, diarrhoea, dysentery, eczema, epilepsy, glaucoma, herpes, hypoglycaemia, leprosy, menorrhagia, mastalgia, nephritis, osteoporosis, retinopathy and rheumatism; every part of the plant is prescribed for the treatment of snake bite and scorpion sting.

Root: Is used in hemicrania.

Root bark: Used as tooth brush to strengthen spongy gums and to cure gum ulcers; powdered root bark also is used for gum ulcers.

Stem bark: Anthelmintic, antileprotic, antiparalytic and astringent; useful in bronchitis, diarrhoea, dysentery, eye injuries and piles.

Leaf: Stimulant; useful in leprosy, ophthalmia, syphilis and to stimulate hair growth.

Flower: Cooling and emollient; externally applied on boils, eruptions, carbuncles and swellings; as antidote for snake bite.

Seed: Antigonorrhoeic and astringent; useful in diarrhoea and piles; powdered seeds taken in enlargement of scrofulous glands; paste is applied at the same time.

Seed oil: Useful as application in leprosy.

Bioactivity: Plant is inhibitor of ACE, cyclo-oxygenase, glucosyl-transferase, lipoxygenase, MAO, ornithine-decarboxylase and xanthine-oxidase; is angiotensin receptor blocker, calcium channel blocker, β-blocker and copper antagonist.

Uses in Alternative Systems of Medicine

Chinese:

Part used: Stem bark and gum.

Uses: Bark is useful as anthelmintic, astringent, sedative, stimulant, tonic and vulnerary; gum is used as poultice for carbuncles, fractures, sprain and swellings.

Medicinal Uses Out Side India

China: Antidiaphoretic, resolvent, vulnerary and vermifuge; dried stem is powdered and is prescribed for broken bones, haemorrhoids, insomnia; the flowers are given as digestive and tonic.

Korea: Bark is considered as sedative and tonic.

Non-Medicinal Uses

Plant: Used as fish poison.

Wood: Is used in flooring of railway carriages and making furniture.

26. *Albizia lebbeck* (L.) Benth.

Common Names: East Indian walnut, Lebbeck tree, Parrot tree, Sizzling tree, Woman's tongue tree.

Vernacular Names: Ben: Sirish; **Guj:** Pilo sarishio; **Hin:** Badaashirish, Sirar Shirish, Tanita; **Kan:** Baagemara, Dodda sirisha, Goddu hunise, Hombage, Sibigida, Kaadumanjeete; **Mal:** Kaattuvaka, Nenmani; **Mar:** Chichola, Kaalashiris, Motashiris; **Sans:** Barhapushpa, Bhandira, Kalinga, Kapeethaka, Karnashirista, Karnapoora, Madhupushpa, Murdhapushpa, Shankinipala, Shukadruma, Shukapriya, Shukapushpa, Shukeshta, Shyamala, Shyamavarna, Supushpaka, Swarnapushpa; **Tam:** Adukkavaagai, Kaalindi, Kattuvagai, Sugathram, Uddanaka, Vishaghati, Vishahanta, Vishanaashana; **Tel:** Baagichettu, Billavara, Dirisanamu, Girisamu, Kapithanamu, Peddadirisanamu.

Distribution: Native of India, Guyana, Laos, Myanmar and Surinam; introduced into Africa, Australia, Afghanistan, Argentina, Brazil, Bolivia, Cambodia, Caribbean Islands, China, Columbia, Central America, French Guiana, India, Indonesia, Iran, Iraq, Malaysia, Middle East, Mexico, Pakistan, Papua New Guinea, Peru, United States and Venenzula; grows throughout India.

Description: An unarmed deciduous tree of 40–70 feet hight; bark pale; young shoots glabrous; leaves abruptly 2–pinnate, main rachis pubescent or glabrous, furnished with large gland on the petiole above the base and one below the upper most pairs of pinnae; leaves 2–3 (rarely 4) pairs, rachis glabrous or pubescent, swollen at the base; leaflets 5–9 pairs with glands between their bases; lateral leaflets elliptic oblong, the terminal two leaf lets obovate oblong, unequal sided, very obtuse, glabrous above, pubescent and reticulately veined beneath, with obliquely round or truncate base; petioles swollen and hairy at the base; flowers white or greenish yellow, fragrant, on globose umbellate heads; peduncles long, pubescent; pedicels pubescent; calyx pubescent, teeth short, deltoid; corolla tube glabrous with triangular lobes, pubescent outside; stamens longer than the corolla; filaments connate at the base forming short tube; pod linear, oblong, bluntly pointed, thin, pale yellow, reticulately veined above the seeds, smooth, shining; seeds 4–12, ellipsoid, oblong, compressed, foveolate on the faces, pale brown.

Phytochemistry

Plant: Albigenic acid, albigenin, albiziagenin, biochanin, cyclitols, echinocystic acid, flavones, oleanolic acid, saponins, tannin and triterpenes; yields a gum called "Gum siris".

Root: 3–O–L–rhamnopyranosyl(1→5)-β-D–xylofuranosyl (1→5)–β-D–arabino furanosyl (1→4)-β-D-glucopyranoside, echinocystic acid, olbitosin, oleanic acid, stigma sterol and β-sitosterol.

Stem bark: 3'–O–methylmelanoxetin, acacic acid, echinocystic acid, fridelin, glucose, lebbecacidin, lebbekanin–C, leucocyanidin, melacacidin, melanoxetin, methylmelacacidin, neoleucoanthocyanidin, rhamnose, tannin, γ–sitosterol, α-catechin and β-sitosterol.

Heartwood: (–) Melacacidin, (+) pinitol, 3'–O–methylmelanoxetin, acacic acid, D–pinitol, echinocystic acid, lebbecacidin, leucopelargonidin, melanoxetin, methyl melacacidin, okanin and oleonic acid.

Flower: Benzoic acid, benzyl acetate, benzyl alcohol, benzyl benzoate, crocetin, echinocystic acid, lebbekanin, lupeol, quercetin glycosides, sapogenins, saponins and β-amyrin.

Leaf: Albigenin, alkaloids, caffeic acid, echinocystic acid, glyoxalate, kaempferol, keto acids, myricitin, oxalacetate, phosphoenol pyruvate, quercetin, reynoutrin, robinin, rutin, saponins, vicenin–2, α-oxoglutarate and β-sitosterol.

Pod: Albigenic acid, albigenin, arabinose, decasanoic acid, echinocystic acid, fucose, galactose, glucose, lebbekanin–C, lupeol, melacacidin, melanoxetin, oleanolic acid, rhamnose, saponins, xylose, γ-sitosterol and β-sitosterol.

Seed: Acacic acid, albigenic acid, albigenin, arabinose, echinocystic acid, galactose, glucose, lebbekanin A and D, oleanolic acid, rhamnose, sapogenins, xylose and β-sitosterol.

Medicinal Uses

Plant: Abortfacient, alexipharmic, analgesic, antiallergic, antiasthmatic, antibacterial, anticancer, anticarcinogenic, antidepressant, antidiabetic, antigingivitic, antihepatotoxic, antiherpatic, anti–HIV, antihypertensive, anti-inflammatory, antileukaemic, antimalarial, antimutagenic, antiparkinsonian, antiprotozoal, antiulcer, antiviral, bronchodilator, CNS depressant, cytotoxic, diuretic, estrogenic, fungicidal, hepatoprotective, hypocholesterolaemic, hypoglycaemic, hypouricaemic, immunostimulant, laxative, neuroprotective, phagocytotic, sedative, spermicidal, uterotonic, vasodilator and vulnerary; cures anaemia, arteriosclerosis, asthma, arthritis, atherosclerosis, breast cancer, cataract, chilblain, colitis, diabetes, diarrhoea, dysentery, eczema, encephalopathy, epilepsy, gastritis, gingivitis, glaucoma, haemorrhage, hyperglycaemia, hyperthyroidism, leprosy, leukaemia, lupus, migraine, menorrhagia, melanoma and odantalgia; every part of the plant is recommended for snake bite and scorpion sting.

Root: Abortifacient, anticancer, hypoglycaemic and spermicidal; useful in hemicrania and ulcers; powdered root bark is used for brushing the teeth as it strengthens the gums; decoction of root with leaves and bark of *Diospyros embryopteris* is used as an eye wash; used as an antidote for snake bite.

Stem bark: Acrid, analgesic, anodyne, anthelmintic, anti-inflammatory, aphrodisiac, astringent, bitter, cooling, depurative, detergent, expectorant, restorative, thermogenic and tonic; useful in all types of poisoning, allergy, asthma, boils, bronchitis, catarrh, cough, diarrhoea, dysentery, enlarged cervical glands, erysipelas, itching, leprosy, leucoderma, neuralgia, night blindness, nyctalopia, ophthalmopathy, piles, respiratory disorders, scrofula, skin eruptions, sprains, strengthening of gums, ulcers, vitiated conditions of "kapha", "pitta", "vata" and wounds; decoction of the bark is used as mouth wash in relaxed gums and as eye wash in eye injury.

Wood: Abortifacient and spermicidal.

Gum: Useful in eye troubles.

Leaf: Useful in night blindness, ophthalmia and syphilis; crushed leaves applied on the parts bitten by snake.

Flower: Used as cooling medicine; useful in asthma, bronchitis, cough and as an external application for boils, carbuncle, eruptions and swellings; as antidote to poison.

Pod: Abortifacient and spermicidal; useful as as antidote to snake bite.

Seed: Abortifacient, anticancer, anti-inflammatory, astringent, hypoglycaemic, restorative and spermicidal; useful in catarrh, diarrhoea, dysentery, gonorrhoea, leprosy, leucoderma, ophthalmopathy, piles, poisoning, scrofula, skin diseases and seminal weakness; powdered seeds useful in enlargement of cervical glands and seed paste is used as local application at the same time.

Seed oil: Used in leprosy.

Bioactivity: Plant is inhibitor of lipoxygenase, aldose-reductase, aromatase, β-glucoronidase, COMT, COX–2, cytochrome P_{450}, deiodinase, elastase, FPTase, glucosyl-transferase, HIV–RT, histamine, lipoxygenase, leukotriene, lyase, MAO, MMP9, metalloproteinase, NADH oxidase, NEP, nitrous oxide-synthetase, ornithine-decarboxylase, phospholipase, prostaglandin synthesis, topoisomerase I and II, tyrosinase, tyrosine-kinase, xanthine-oxidase and CAMP phosphodiesterase; is blocker of angiotensin receptor, β-adrenergic receptor, calcium channel and mast cell stabilizer.

Decoction of bark and flower protected Guinea pigs against histamine and acetylcholine induced bronchospasm and also protected the sensitized Guinea pigs against antigen challenge and showed anaphylactic action; seed extract and saponin fraction of plant reduced number of rupturing mast cells in both mesenteric and peritoneal fluids.

Toxicology: Plant contains chymotrypsin and trypsin inhibitors; albitocin present in the plant is anaphylactic.

Clinical Studies: Clinical trials conducted showed that patients suffering from tropical pulmonary oesnophilia showed excellent response to flower powder.

Uses in Alternative Systems of Medicine

Ayurveda:

Part used: Root, bark, leaf, flower and seed.

Uses: Root is used in treating cough, dyspnoea, erysipelas, food poisoning, hemicrania, leprosy, rat bite and wounds; bark is alexiteric, anthelmintic, astringent and cooling; cures bronchitis, excessive perspiration, granulated wounds, herpes, inflammation, insect bites, itching, leucoderma, piles, pruritus, scabies and skin diseases; leaf is used in ophthalmia; flower is given in asthma and as an antipoison drug; seed powder strengthens gums, cures musculoskeletal disorders, piles and scrophulous ulcers.

Siddha:

Part used: Root, bark, leaf, flower and seed.

Uses: Useful in bleeding piles, dental diseases, diarrhoea, leprosy, leucorrhoea, poisoning, stomatitis, thirst and venereal diseases.

Unani:

Part used: Root, bark, leaf, flower, seed and seed oil.

Uses: Anthelmintic, aphrodisiac, astringent, emollient, maturant and tonic; root is used in opthalmia; bark used in boils, leprosy, paralysis, scabies, syphilis, tooth ache and weakness; strengthens gums and teeth; leaf used in night blindness; flower useful in hemicrania; seed used for gonorrhoea and tuberculous glands; seed oil is applied for leucoderma.

Medicinal Uses Outside India

Indochina: Bark and seeds are used as astringent; prescribed for diarrhoea, dysentery and piles; flower is used as emollient and its paste is applied to boils and carbuncles.

Madagascar: Leaf paste is used in syphilis.

Myanmar: Leaves and seeds are used to treat ophthalmia; bark paste is used to treat boils.

Ethnomedicine

Bark: Is useful in diarrhoea, dysentery, eyesore and gonorrhoea.

Flower: Is used in swellings.

Pod: Is used in boils, carbuncles, mouth ulcers and night blindness.

Seed: Is used in diarrhoea, dysentery and gonorrhoea.

Non-Medicinal Uses

Plant: Grown as an avenue tree, wind break, cover crop and shade tree; used also as fodder; is a host for lac insect.

Bark: Used for tanning fishing nets and in dyeing industry.

Wood: Used in house building, furnitures, agricultural implements, sugar cane crushers, shafts, canoes, oil mills, boats, axles, carts, carriages, stakes, felloes, combs, walking sticks, toys, picture frames, turnery, interior decoration, panelling, flooring, bridges, drums, railway carriages and house beams; used as fuel.

Wood pulp: Used in writing and printing paper.

Gum: Tree yields gum, which is used as an adhesive in furniture industry and in textile industry in calico printing; as an adulterant for gum arabic.

Twigs and leaf: Used as fodder and manure.

Flower: Used by tribal women to decorate the ears.

27. *Albizia odoratissima* (L.f.) Benth.

Synonyms: *Acacia odoratissima* (L.f.) Willd.
Albizia odoratissima (L.f.) Benth. var., *mollis* Baker

Common Names: Fragrant albizia, Fragrant siris.

Vernacular Names: Ben: Kakur siris; **Guj:** Kala sarasio; **Hin:** Bansa, Bersa, Bhandir, Kaalasiris, Shirisha, Siris; **Kan:** Aenu baage, Bilikambi, Bilwaara, Bilvaarada mara, Goddu Hunase, Kaadu Baage, Kadache, Kalangi; **Mal:** Karintakara, Karuvaaka, Nellivaaka, Pulivaaka; **Mar:** Chichunda, Kaalashirish, Pandurashirish, Shiras, Siris; **Sans:** Shvethashirisha, Shirisha, Sireesha, Sittilaivaagai; **Tam:** Karuvaagai, Kaliturinjil, Kaliyanjai, Purusilai, Silai, Silaiyunjai, Sittila vaagai; **Tel:** Bandisinduga, Chinduga, Chinta yelaga, Konda sigara, Thelasu.

Distribution: Native of Asia including India; in India it occurs throughout North India and in Tamil Nadu, Sikkim and Himalaya.

Description: Unarmed tree with a smooth yellowish gray soft bark; branchlets covered with yellow brown hairs; leaf rachis glabrous with a large oval gland near the base and a small round gland between the bases of the upper two pinnae; pinnae 2–8 pairs, glabrous; leaflets 6–24 pairs, sessile, obliquely oblong, obtuse with a oblique base; flowers sessile, greenish white on globose heads arranged at the bracteate nodes of corymbose brachlets; calyx campanulate, pilose with a small teeth; corolla funnel shaped, minutely puberulous; pod strap shaped, flat, thin, oblong, reddish brown with mucronate or obtuse tip containing 6–12 flat oblong-orbicular, aereolate seeds.

Phytochemistry

Plant: Gum, machaerinic acid, odoratissimin and tannin.

Wood: Acacic acid, methyl ester of dihydroflavonol, penta–O–methyl–dihydromelanoxetin and tannin.

Leaf: Tannin.

Seed: Arabinose, echinocystic acid, glucose, machaerinic acid, odoratissimin, rhamnose and xylose.

Edibility

Leaf: Edible; used as vegetable.

Medicinal Uses

Plant: Is anthelmintic, antibacterial, anticancer, anticapillary fragility, antihypertensive, anti-inflammatory, antioxidant, antiviral, cancer preventive, diuretic, expectorant, haemostat, hepatoprotective and immunosuppressant; useful in dermatitis, diarrhoea, dysentery, hepatotoxicity, nephritis, ulcer and vaginitis.

Stem bark: Acrid, alterative, astringent, cooling, depurative and expectorant; useful in bronchitis, burning sensation, diabetes, erysipelas and skin diseases; bark powder with butter taken daily, is an excellent tonic; bark decoction is used as a mouth wash to strengthen the gums; the bark paste is applied externally in leprosy and ulcers.

Leaf: Leaves boiled in ghee is taken internally for cough; leaf juice is given for night blindness.

Seed: Semen coagulant and is diuretic in action; seed paste is useful in reducing enlarged cervical glands.

Bioactivity: Ethanol extract of seeds is semen coagulant and diuretic; it is hypoglycaemic on normal albino rats.

Uses in Alternative System of Medicine

Ayurveda:

Part used: Bark, leaf, flower and seed.

Uses: Useful in cough, dyspnoea, erysipelas, food poisoning and other poisons, leprosy and wounds.

Ethnomedicine

Leaf: Is useful in leucorrhoea and menorrhagia.

Non-Medicinal Uses

Plant: Grown as an avenue tree; yields gum; is a host for lac insect.

Bark: Yields a dye, which is brown in colour; is a source of fermented drink called Basi in Philippines.

Wood: Used for construction work; it is mainly employed for making carts, shafts, axles, spokes, felloes, plough, furniture, decorative panelling and tool handles; is suitable for bridge construction and railway sleepers.

Leaf: Used as fodder; yields a brown colour dye.

28. *Albizia procera* (Roxb.) Benth.

Common Names: Hill siris, Silver bark rain tree, Tall albizia, White siris.

Vernacular Names: Hin: Baro, Garso, Garar, Gurbari, Gurkur, Karanji, Karhar, Karo, Karolu, Karro, Safed siris; **Kan:** Baagi, Bilibaage, Salabage; **Mal:** Chaalavaaka, Karuntakara, Kutamvaaka, Vaaka, Vellavaaka; **Mar:** Kinai, Kinnigurai; **Sans:** Katabhi, Kinihi, Mahaa shvetha, Shirisha; **Tam:** Kandaivaagai, Kondavaaghe, Nallavaagai, Saalaiyunjil, Velvaagai; **Tel:** Chigara, Dirisanamu, Ganaru, Kondadirisanamu, Peddapaccharu, Thellachinduga, Thellasapara, Thelladinisanamu.

Distribution: Native of Asia including India, Papua New Guinea, Bismarck Archipelago; in India it occurs in Andaman Islands, Arunachal Pradesh, Bihar, Jammu, Gujarat, Meghalaya, Nagaland, Uttar Pradesh and South India.

Description: A tall unarmed tree with yellowish or greenish white bark; leaves evenly 2 pinnate with a large gland near the base; pinnae 2–6 pairs; leaflets 6–12 pairs, obliquely ovate to oblong, obliquely cuneate on the upper margin with an ovate gland between the last pair; flowers many in corymbose branches of terminal panicles; flowers sessile, yellowish white with yellow stamens; calyx glabrous, teeth triangular; corolla lobes lanceolate, densely hairy outside; pod thin, flexible, reddish brown, shining; seeds 6–12, pale brown.

Phytochemistry

Root: Oleanolic acid and α-spinasterol.

Bark: Biochanin A, daidzein, demethylmedicarpin, formononetin and genistein.

Gum: 4–O–methyl–D–glucuronic acid, galactose, glucuronic acid, mannose and proceranin A.

Leaf: Hentriacontane, hexacosanol and α-spinasterol.

Seed: Arabinose, fucose, galactose, genin, glucose, machaerinic acid, proceraginin, proceranin A, proceric acid, rhamnose, sapogenins, xylose and β-sitosterol.

Seed oil: Arachidic acid, behenic acid, linoleic acid, oleic acid, palmitic acid and stearic acid.

Edibility

Pod and seed: Edible.

Medicinal Uses

Plant: Is abortifacient, antiangiogenic, anticarcinomic, antifertility, antihaemolytic, anti-inflammatory, antimicrobial, antioxidant, antispasmodic, antiviral, antitumour, cytotoxic, fungistatic and immunostimulant; useful in ischaemia, leukaemia, lymphoma, melanoma, neuroblastoma, osteoporosis, tumour of the breast, ovary and stomach and ulcer.

Stem bark: Paste is applied in backache.

Leaf: Is made into paste and applied as poultice to ulcers.

Pod: Anticancer and antiprotozoal.

Bioactivity: Plant is inhibitor of aldose reductases, aromatase, α-reductase, catechol–O–methyl–transferase, DOPA–decarboxylase, histidine-kinase, lipase, MAO, MDR, NADH–ubiquinone–oxidoreductase, ornithine-decarboxylase, PTK, peroxidase, topoisomerase I and II and tyrosine-kinase and quinone-reductase inducer.

Ethnomedicine

Stem bark: Is used in backache.
Seed: Useful in body pain.

Non-Medicinal Uses

Plant: Is grown as avenue and hedge tree; host tree for lac insect.

Stem bark: Used for tanning and dyeing, and is piscicidal.

Wood: Used in making furniture, large panels, carts, carriages, oars, agricultural implements, cart wheels, scantling, rice pounder, carving, cane crushers, plywoods, railway sleepers, tool handles, house posts, panelling, tea boxes; yields charcoal; wood pulp is used for making writing and printing papers.

Gum: Is used in sizing paper.

Leaf: Used as fodder and insecticide.

29. *Dichrostachys cinerea* (L.) Wight & Arn.

Synonyms: *Dichrostachys glomerata* (Forsskal) Chiov.
Dichrostachys platycarpa Welw.
Mimosa glomerata Forsskal
Mimosa cinerea (L.)

Notes: Eight subspecies and seven varieties are recognized.
Common Names: Saint Dominique tree, Sore eye plant.
Vernacular Names: Hin: Baelanthar, Barbael, Beertharu, Kaanraai, Kheri, Varthuli; **Kan:** Edathari, Odathari, Odavina, Shamee vriksha, Vadavaarada gida, Vaduvaara, Yedathari; **Mal:** Veeravriksham; **Mar:** Bellathoor, Segumkati, Sigamkaati, Vaelathoora; **Sans:** Bahuvriksha, Deerghamoola, Kricchari, Kshudhaku Shaalasandnaka, Mahaakapittha, Vellanthara, Veeratharu, Veeravriksha; **Tam:** Anatter, Mavilandam, Vadatharam, Varittulla, Vedathala, Vidathalan, Veluthullimaram; **Tel:** Nallavenuturu, Nallajammi, Samiramu, Sarathumma, Veluthuruchettu, Venuthuru.

Distribution: Native of Africa, Asia including India, Australasia, Indian Ocean Islands and Middle East; in India it grows in Karnataka and Tamil Nadu.

Description: Shrub or tree of 6m hight with numerous, stiff, spreading branches; bark yellowish grey, fibrous, longitudinally fissured; branchlets end in a spine; leaves 8–12 pairs; leaf rachis with erect, cylindrical concave gland between the bases of the pinnae; leaf lets 12–20 pairs, sessile, closely packed, oblique, linear to lanceolate, ciliate, base truncate; stipules subulate from triangular base; spike solitary; lower flowers pink to violet, becoming white after anthesis, upper flowers yellow; calyx thin and pubescent; staminodes of barren flowers filiform, red; pod indehiscent, linear, glabrous, dark brown, swollen over the 6–10 ellipsoid seeds.

Phytochemistry

Root: Friedelan–3–Ol, friedelan-3-one, n–octacosanol, β-amyrin and β-sitosterol.

Stem bark: Fridelin, sitosterol and α-amyrin.

Wood: Octacosanol and sitosterol.

Leaf: Hentriacontanol.

Flower: Cyanidin and quercetin.

Medicinal Uses

Plant: Useful in diseases of female genital tract, disorders of "vatha", dysuria, gravel, ophthalmia and retention of urine.

Root: Astringent; useful for rheumatism, renal troubles, urinary calculi and vaginal diseases.

Young shoots: Ground and applied on eyes in ophthalmia.

Toxicology: Trypsin and chymotrypsin inhibitors are reported.

Uses in Alternative Systems of Medicine

Ayurveda:

Part used: Whole plant and gum.

Uses: Used in diarrhoea, dysentery, eczema, eye diseases, leucorrhoea and piles.

Medicinal Uses Outside India

East Africa: Root powder and leaf powder used in asthma, bone pain, hernia, joint pain and snake bite.

Ethnomedicine

Plant: Is astringent and vermifuge; useful in leucorrhoea, ophthalmia, rheumatism, urinogenital disorders; is antidote to scorpion sting.

Non-Medicinal Uses

Plant: Is a host for lac insect.

Stem bark: Yields tannin used in tanning industry; also yields a yellowish white fibre.

Wood: Is very tough and is used for making walking sticks, tentpoles and pegs; employed as fuel.

Leaf: Used as fodder.

30. *Entada gigas* (L.) Fawcett & Rendle

Synonyms: *Entada scandens* (L.) Benth.
Mimosa gigas L.
Mimosa scandens L.

Notes: The application of the binomials *Entada rheedei, Entada gigas, Entada phaseoloides, Entada pursaetha* and *Entada scandens* is in controversy; consequently, there is a considerable amount of overlap and confusion regarding the information on the species with the now synonymized binomials; the position taken by ILDIS is adopted here, recognizing *Entada rheedei* and *Entada gigas;* care was taken to avoid overlap in the profiles presented here; yet some caution is needed.

Common Names: Garbee bean, Mackay bean, Water wine.

Vernacular Names: Ass: Gilalewa; **Hin:** Gila; **Kan:** Dodda kampi, Hallekayi balli; **Mal:** Kakkavalli, Nalamanchadi; **Mar:** Garambi, Garbe; **Ori:** Geredi.

Distribution: Native of Africa, Caribbean Islands, Central and South America; introduced into India; grows throughout India.

Description: Gigantic climber, unarmed, tendril bearing; leaves bipinnate; flowers small, yellowish, on long spikes; calyx companulate; petals slightly connate; stamens 10, free, anthers, gland crested; pod huge, flat, woody with many 1-seeded discoid joints; seeds large, circular, compressed.

Phytochemistry

Plant: Echinocystic acid, entadamide A, lupeol, oleanolic acid, saponins ES II, ES III and α-spinosterol.

Stem bark: α-Spinasterol, echinocystic acid, entagenic acid, lupeol, oleanolic acid and saponins.

Leaf: Echinocystic acid and entadamide A and C.

Seed: α-Amyrin, arachidic acid, behenic acid, cyanidin chloride, cysteine, dilinolcin, diocin, dopamine glucoside, entadamide A and B, entagenic acid, fixed oil, L–tyrosine, myristic acid, oleodilinolein, phaseoloidin, prosapogenin A, gallic acid, raffinose, resins, saponins, steroids, trilinolein, triolcin, trypsin, tyrosine and β-sitosterol.

Uses in Alternative Systems of Medicine

Unani:

Part used: Seed.

Uses: Anthelmintic, emetic, haemolytic, irritant and tonic; paste of seeds applied to glandular swelling, swollen joints, hands and feet and painful loin.

Medicinal Uses Outside India

Indonesia: Pounded root is rubbed on the stomach and stem juice is drunk for dysentery; roasted seeds are eaten as depurative, emetic and given for stomach ache.

Mangolia: Plant is used to lessen heat in the spleen.

Myanmar: Seeds are used as febrifuge; used to treat haemorrhoids.

Phillippines: Seed paste is used as poultice for abdominal complaints.

Non-Medicinal Uses

Plant: Is used as fish poison.

Bark: Yields fibre used for cordage and fishing nets; used as soap substitute.

Leaf: Juice of the leaves is used as fish poison.

Seed: Roasted seeds are coffee and soap substitutes.

Seed oil: Used as illuminant.

31. *Entada rheedei* Sprengel

Synonyms: *Entada gigas* G. Gilbert & Boutique
Entada phaseoloides sensu auct.
Entada pursaetha DC.
Entada scandens sensu auct.
Enterolobium cyclocarpum (Jacq.) Griseb.

Notes: The application of the binomials *Entada rheedei, Entada gigas, Entada phaseoloides, Entada pursaetha* and *Entada scandens* is in controversy; consequently, there is a considerable amount of overlap and confusion regarding the information on the species with the now synonymized binomials; the position taken by ILDIS is adopted here, recognizing *Entada rheedei* and *Entada gigas;* care was taken to avoid overlap in the profiles presented here; yet some caution is needed.

Common Names: Elephant's creeper, Giants rattle, Indian cocoon, Lady nut, Mackaybean, Matchbox bean.

Vernacular Names: Hin: Chian, Gaarveej, Gila, Katbael; **Kan:** Doddaganapi balli, Dodda kampi, Ganape kayi, Hale balli, Halekaai balli; **Mal:** Ahkkaatla, Anathata, Kakkavalli, Makkanaka, Paranta, Perinkakkuvalli, Vattalvalli; **Mar:** Gaarambi, Gaarabeej; **Sans:** Gilla; **Tam:** Chillu, Irikki, Sillu, Thellukkai, Vattavalli; **Tel:** Gilatheege, Kondachinta, Peddamadupu, Yenuguchinta.

Distribution: Native of Africa, Asia including India and Australasia; in India it grows in Andaman and Nicobar islands, Bihar, Karnataka, Kerala, Orissa, Tamil Nadu, Eastern Himalayas and North East India.

Description: An unarmed liana with glabrescent branches; leaves alternate, 2-pinnate, glabrous; pinnae 2-pairs; leaflets 4 pairs, ovate, obovate with cuneate base and obtuse apex; stipules setaceous; tendril forked; flowers cream coloured, 5– merous, with 5 oblong lobes; stamens 10, filaments subequal; anthers gland crested; pod woody, flat; seeds ovoid and 6-15 in number.

Phytochemistry

Plant: Entadamide A, echinocystic acid, entada saponins II and III, entagenic acid, lupeol, oleanolic acid, phaseolidin and α-spinasterol.

Leaf: Entadamide A and C.

Seed: 2–Hydroxy-α-L–rhamnopyranosyl-oxybenzene, arabinose, arachidic acid, behenic acid, linoleic acid, cyanidin chloride, cysteine, dilinolein, diolin, dioleolinolein, dopamine–3–O–glucoside, echinocystic acid, entadamide A and B, entagenic acid, gallic acid, glucose, L–tyrosine, linoleic acid, lupeol, myristic acid, oleodilinolein, oleolinolein, palmitic acid, phaseoloidin, prosapogenin–A, quercetin, raffinose, sapogenins, stearic acid, trilinolein, triolein, trypsin, tyrosine, xylose, α-amyrin and β-sitosterol.

Seed oil: Arachidic acid, behenic acid, linoleic acid, linolenic acid, myristic acid, oleic acid, palmitic acid and stearic acid.

Edibility

Pod and seed: Roasted and eaten; seeds used as coffee substitute.

Medicinal Uses

Plant: Is anthelmintic, antiperiodic, CNS active and emetic; useful in anasarca, inflammatory glandular swellings, cancer, dropsy, epilepsy, constipation, joint pain and ulcers.

Stem bark: Useful in anasarca, cancer, dropsy, epilepsy and constipation; bark paste is applied on ulcers and loin pain.

Wood: Decoction is applied for ulcers; is taken internally for anasarca, cancer, constipation, dropsy, epilepsy and loin pain.

Leaf: Ground and applied on wounds as vulnerary.

Pod: Roasted pods are purgative.

Seed: Alexiteric, anodyne, anthelmintic, anti-inflammatory, antiperiodic, antirheumatic, bechic, carminative, emetic, febrifuge, laxative, narcotic, stomachic and tonic; useful in diabetes, eye diseases, fever, head ache, liver complaints, ophthalmia and phantom tumours; seed paste is applied on inflammatory glandular swellings, mumps, joint pains, loin pain and swollen hands and feet; raw seeds are used as oral contraceptive; seeds boiled in water and hot water vapour is taken for watering of the eye.

Seed kernel: Anodyne, carminative, emetic, febrifuge and stomachic; eaten by women for some days after delivery for allaying body pain and to ward off cold.

Seed oil: Used as massage oil for glandular swelling, joint pains and paralysis; given internally as emetic and febrifuge.

Uses in Alternative Systems of Medicine

Ayurveda:

Part used: Seed.

Uses: Used as emetic; seed paste is applied on inflammatory swellings, painful joints, loin, hands and feet.

Chinese:

Part used: Stem.

Uses: Cures blood stasis and rheumatism.

Medicinal Uses Outside India

Indochina: Seeds are used as alexiteric, emetic, febrifuge and narcotic.

Myanmar: Seed paste is used as febrifuge.

New Guinea: Bark decoction is taken as astringent.

Philippines: Spongy fibres are soaked in water and used for various skin diseases.

Sikkim: Kernel paste is applied to cure mumps.

South Africa: Seeds are given to infants to bite during teething and also as a remedy for cerebral haemorrhage; is purgative.

Ethnomedicine

Plant: Is useful in anasarca, cancer, inflammation, malaria, sores and wounds.

Leaf: Useful in cuts and wounds.

Seed: Useful in eye diseases, liver complaints and throat swelling; used as antidote for snake bite.

Non–Medicinal Uses

Plant: Used as fish poison; used as a substitute for soap; used as pesticide.

Stem bark: Yields a coarse fibre used for cordage and fishing net.

Pod: Used as fish poison.

Seed: Used as a substitute for soap for washing hair; ground and used for crinkling clothes, for burnishing pottery, polishing hand made paper; roasted and used as a substitute for coffee; used as fish poison.

Seed shell: Is used in making snuff boxes.

Seed oil: Used as illuminant.

32. *Leucaena leucocephala* (Lam.) de Wit.

Synonyms: *Acacia glauca* (L.) Willd.
Acacia leucocephala (Lam.) Link.
Leucaena glauca Benth.
Leucaena glauca (Willd.) Benth.
Mimosa leucocephala Lam.

Notes: Economically the most important species of the genus; the species arose as a hybrid of *Leucaena pulverulenta* and *Leucaena lanceolata* in Veracurz; three subspecies, *leucocephala*, *glabrata* and *ixtahuacana* are recognized.

Common Names: Aroma blanca, Jumbie bean, Jump and go tree, White popinac.

Vernacular Names: Guj: Lasobaval, Vilayatibaral; **Kan:** Subaabul, Tankeshvara, Vilayathi bili gobli; **Mal:** Takaranniram; **Mar:** Kubaabhala; **Ori:** Rajokasundiri; **Tam:** Tagarai, Nattu, Naattu cavindaal, Periyadagarai, Thagaraanchiram, Thagaraimaram; **Tel:** Kaniti, Jilangi.

Distribution: There is much uncertainity over the origin; probably a native of Cyprus; introduced into India; grows throughout India.

Description: Low, erect shrub or small tree, without spines; bark brown, warty, more or less striate; leaves bipinnate with a cupular gland at the base; pinnae 4–8 pairs; leaf lets 10–15 pairs, linear, oblong, obovate, base oblique, acute at the apex, membranous; flowers white on dense, axillary heads; calyx tubular, campanulate, teeth short; petals oblong, free, pubescent; pod straight, glabrous, strap shaped with acute apex on 3cm long stalk; seeds brown and glossy, 10-20 in number.

Phytochemistry

Plant: Asparagine, aspartic acid, histidine, leucine, selenium, tannin, tyrosine, valine and β-sitosterol.

Stem bark: Selenium and tannin.

Flower: Quercetagetin, quercetin and quercetin–3–O galactoside.

Seed: (–) Mimosine, galactomannoglycan, proteins and selenium.

Seed oil: Arachidic acid, lauric acid, lignoceric acid, linoleic acid, oleic acid, palmitic acid and stearic acid.

Edibility

Young shoots and pod: Are used as vegetable.

Seed: Parched and eaten.

Toxicology: The plant has the capacity to absorb selenium from the soil and concentrate in the plant body–more in seeds; the excess use of the seeds as livestock feed causes selenium poisoning in animals; the plant has a toxic principle leucanol which may cause temporary sterility in animals; chymotrypsin and trypsin inhibitors are present in the plant.

Medicinal Uses Outside India

Indonesia: Seeds are used as anthelmintic and as a remedy for diabetes; roasted seeds are used as emollient and emmenogogue.

Myanmar: Paste of the leaves is applied to poisonous bites and stings.

Ethnomedicine

Plant: Used as abortifacient, anodyne, carminative, depilatory, depurative, emmenogogue and laxative; useful in fever, lung infections, typhoid and intestinal worms.

Non-Medicinal Uses

Plant: Useful as fodder.

Wood: Is a source of paper pulp.

Leaf: Used as a supplement to alfalfa meal for chickens.

Seed: Used as cattle feed.

33. *Lysiloma latisiliquum* (L.) Benth.

Synonyms: *Acacia formosa* A. Rich.
Leucaena latisiliqua (L.) Gillis & Stearn

Common Names: Horse tamarind, Singing bean.

Vernacular Names: Guj: Lasobaval, Vilayathi baral; **Hin:** Subabul, Vilayathi baval; **Kan:** Subaabul, Tankeshvara, Vilayathi bili gobli; **Mal:** Takaranniram; **Mar:** Kubaabhala; **Ori:** Rajakosundari, Mattuccavundal; **Tam:** Naattucavindaal, Periyadagara; Thagaramaram, Thagaraanchiram; **Tel:** Jiluga, Kaniti.

Distribution: Native of Caribbean Islands, Central America and North America; introduced into India; it is cultivated and naturalized throughout India.

Description: A large erect shrub or a small tree with unarmed trunk; pinnae 3–6 pairs; leaf lets 10–15 pairs, membranous, linear–oblong, acute; flowers white in colour on globose heads; calyx membranous, teeth short, triangular; corolla spathulate–oblong, indurated at the tip; stamens 10, exserted; pod stalked, slightly pubescent, flat; seeds 15–20.

Phytochemistry

Plant: Kaempferol–3–arabinoside, kaempferol–3–xyloside, leucenine, leucenol, mimosine, quercetin–3–arabinoside, quercetin–3–galactoside and quercetin–3–rhamnoside.

Seed: Mucilage composed of mannans, galactans, xylans and stachyose along with dark green fatty oil.

Edibility

Shoot: Young shoots are eaten fresh or cooked.

Pod: Immature pods are eaten as vegetable.

Seed: Young seeds are eaten fresh or dried; mature seeds are eaten after parching or cooking.

Medicinal Uses

Plant: Is a worm repellant and antifertility agent.

Root: A decoction of the root is given as contraceptive, ecbolic and emmenogogue.

Stem bark: Is useful as contraceptive, ecbolic and emmenogogue; eaten for any kind of hidden internal pain.

Pod and seed: Eaten for any internal pain.

Non-Medicinal Uses

Plant: Grown as hedge plant and grown for afforesting grass lands and soil reclaimation; as green manure crop; as wind break; as a cover crop, for checking soil erosion; shade tree in coffee, tea and rubber plantation; used as insecticide and fish poison; it is valued as nutritive, protein rich fodder.

Wood: Is hard and heavy; is used as raw material for paper pulp; used for construction of furniture and heavy wooden articles; used as fuel; yields charcoal.

Leaf: Used as manure and fodder.

Pod: Used as animal feed.

Seed: Powdered seeds are mixed with soil to increase soil fertility; seeds are used in making jewellary such as earrings and necklaces, fancy bags and basket; ripe seeds are used as a substitute for coffee.

34. *Mimosa pudica* L.

Notes: Probably introduced into the old world; five varieties, *hispida, pudica, setosa, tetrandra* and *unijuga* are recognized .

Common Names: Bowing plant, Humble plant, Sensitive plant, Shame face, Shame lady, Shame plant, Shame weed and Touch me not plant.

Vernacular Names: Assam: Nilary ban; **Ben:** Lajak, Lajapati; **Hin:** Choyimuyi, Lajjavathi, Lajawanti, Sharam patta; **Kan:** Lajjavathi, Mududu daavare, Muttidaremuni, Naachike gida, Naachike mullu; **Mal:** Thindaar mani, Total vaadi; **Mar:** Lajjari; **Sans:** Anjalikaaraka, Gandamalika, Khandiraka, Khadhira patrika, Lajjavati, Lajjika, Mahaabhita, Namaskari, Prasaarini, Rakhthamoola, Sankochini, Sapthaparni, Sparshalajja, Svagupta, Thamramoola, Varshini, Varaahakrantha; **Tam:** Karappuindirikody, Samangai, Totta churangi, Thotaal vaadi, Tottar sinugi; **Tel:** Atthapatti, Mudathadamara, Munugudu, Peddanidrakanti, Sigakku, Thottalasingi.

Distribution: Native of tropical America, but widely established as a common weed in the tropical and warm temperate countries; grows throughout India.

Description: A diffuse under shrub of 2–3 feet height; stem and branches sparingly prickly and clothed with long weak bristles from bulbous base; leaves imparipinnate, sensitive, digitate, bristly, rachis clothed with ascending bristles; leaf lets 10–20 pairs, sessile, coriaceous, linear oblong, acute, glabrous above, clothed with appressed bristles beneath, base obliquely rounded; stipules triangular; flowers pink, tetramerous in globose heads; peduncles prickly, in axillary pairs; bracteole solitary, linear, acute, ciliolate; calyx minute, pubescent; corolla divided about half the way down with 4 lobes, lobes ovate, oblong, obtuse; stamens 4, much exerted; style flat, glabrescent, stigma penicillate to glabrous; pod flat, slightly recurved with 3–5 one seeded joints, sutures clothed with spreading yellow weak bristles; seeds 3–5.

Phytochemistry

Plant: 4-3,5-Dihydroxy benzoic acid, crocetin dimethyl ester, D–pinitol, friedelin, mimosine, tannin, turgorin, β-amyrin and β-sitosterol.

Root: Mimosine.

Leaf: Flavone glycosides, leucoanthocyanidin, mimosine, norepinephrine and stigmasterol.

Seed: D–glucuronic acid, D–xylose, greenish yellow fatty oil, mimosine, mimosine–O-β-D–glucoside and mucilage.

Seed oil: Behenic acid, D–pinitol, linoleic acid, oleic acid, palmitic acid and stearic acid.

Edibility

Seed: Edible.

Medicinal Uses

Plant: Alterative, analgesic, antibacterial, antihepatotoxic, antiherpatic, antiseptic, antiviral, astringent, blood purifier, cardioprotective, carminative, cooling, demulcent, detoxicant, diuretic, hepatoprolective, immunostimulant, laxative, metastatic, neuroprotective and resolvent; used for biliousness, burning sensation in the body, diarrhoea, dysentery, eczema, fistulous sores, haemophilic conditions, morbid vagina, myalgia, scabies, small pox, uterine tumours and vesicular calculi.

Root: Alexipharmic, alterative, antiseptic, antispasmodic, aphrodisiac, astringent, blood purifier, carminative, constipating, cooling, diuretic, emetic, febrifuge, resolvent, and vulnerary; useful in asthma, burning sensation of the body, dysentery, epilepsy, fever, fistula, gravel, glandular tumours, gum trouble, haemorrhoids, head ache, inflammation, jaundice, "kapha", leucoderma, leprosy, metropathy, small pox, tumours, ulcers, vitiated "pitta", uterine and vaginal complaints; powdered root with milk is prescribed for fistula and piles; boiled root is taken to aid easy passage of urine; root juice is alterative, antiseptic and blood purifier; applied as lotion for piles, sores and scorpion sting.

Stem: Ground and made into paste and used as poultice for healing wounds; also used as a antidote for snake bite and scorpion sting.

Leaf: Alterative, carminative, resolvent, sudorific and tonic; leaf decoction is employed as a hip bath in hip and kidney pain, fistula, glandular tumours, gravel, boils, conjunctivitis, haemorrhoids, hydrocele, scrofula, swellings and wounds; ground leaves applied externally for hydrocele, fistula, glandular swellings and piles; juice of the leaves demulcent; used in mouth ulcers and spongy gums; used for dressing all forms of sinus; applied for burns and hydrocele; leaf juice with horse urine is dropped in the eye for removing the film on cornea by setting upon of artificial inflammation; as an antidote for snake bite and scorpion sting.

Seed: Emetic and blood purifier.

Bioactivity: Plant is inhibitor of 5-α-reductase, aldose-reductase, α-reductase, β-glucoronidase, cyclo oxygenase, glucosyl-transferase, lipoxygenase, MAO, ornithine-decarboxylase and xanthine-oxidase; is angiotensin receptor blocker, andrenergic receptor and calcium channel blocker.

Ethanolic extract of the whole plant showed antiviral activity against *Vaccinia* virus; it also showed spasmogenic activity in Guinea pig ileum; when the plant extract was given to rats parenterally (1.6 mg / 100g) for 120 days, it regenerated the experimentally injured sciatic nerve and the regeneration was 30–40% higher than hydrocortisone; leaf decoction showed moderate diuretic response in albino rats.

Toxicology: In large doses seeds are irritant, poisonous and can result in hair loss; has goitrogenic effect resulting in reproductive disorders and cataract; contains chymotrypsin and trypsin inhibitors.

Clinical Studies: Clinical study of female patients with heavy menstrual blood loss and dysfunctional uterine bleeding were given distilled extract of the plant during the menstrual cycle and were also given dired powder of root (1000 mg) and the results were promising.

Uses in Alternative Systems of Medicine

Ayurveda:

Part used: Whole plant and root.

Uses: Root is alexipharmic, alterative, antiseptic, astringent, blood purifier, bowel binder, cooling and vulnerary; useful in asthma, biliousness, bleeding piles, burning sensation, diseases of buccal cavity and sensory organs, diarrhoea, dysentery, dyspnoea, edema, fatigue, haemophilic conditions, haemorrhage, inflammation, leprosy, leucoderma, leucorrhoea, musculoskeletal disorders, skin diseases, uterine disorders and wounds; as an ingredient of ointment it is used to heal wounds quickly, to promote adhesion of fractured bones, to cure erysipelas, to heal ulcers and wounds; paste or decoction of root is useful in diarrhoea, haemothermia and piles; applied on abscesses; as a pessary in bleeding piles and vaginal discharges; plant cures diseases of head and sensory organs, erysipelas, fever, gynaecological disorders, prolapsed rectum, purpura, ulcer, vitiated blood and wounds.

Homoeopathy:

Part used: Leaf.

Uses: Useful in conjunctivitis, rheumatism and swollen ankles.

Siddha:

Part used: Root and leaf.

Uses: Useful in diabetes, diseases of "vatha", enuresis, eye complaints, sinus, urinary calculi and wounds.

Unani:

Part used: Whole plant, leaf, root, stem, flower and seed.

Uses: Alterative, phlegmatic, laxative and resolvent; decoction of root is used in bilious fever, gravel, inflammation, jaundice, leprosy, small pox and ulcer; leaf and root is used in fistula and piles; leaf paste applied to hydrocele; leaf and stem is used as an antidote for scorpion sting; flower paste is used for bleeding piles; seeds useful in treating fractures.

Medicinal Uses Outside India

Brazil: Root is used as emetic; leaves given in scrofula.

Cambodia: Plant is recommended for edema, myalgia, rheumatism, tumour of the uterus and vesicular calculi.

Gold coast: Leaves used as a cure for Guinea worm; decoction of the root is a remedy for gravel; leaf infusion is given in dysentery.

Guiana: Leaves are sudorific and tonic; seeds and root in small doses emetic.

Indo-China: Leaf infusion is tonic.

Indonesia and Malay Peninsula: Bathing in the decoction of the plant induces sleep; pounded leaves applied as poultice to swellings.

LaReunion: Astringent, antispasmodic and diuretic; used for convulsions, diphtheria, worms, painful and difficult menstruation; in abnormal cessation of periods, the leaves with turmeric as infusion is given as foot bath.

Martinique: Seeds used as emetic and leaf infusion as tonic.

Nepal: Root decoction useful in gravel; leaf and root used for piles; leaf paste applied to hydrocele; leaf and stem used in the treatment of scorpion sting.

Philippines: Leaves soaked in coconut oil and applied on wounds and ulcers or crushed and mixed with oil is applied as poultice on the abdomen to treat hernia or enlarged abdomen; root is diuretic and used in dysentery and dysmenorrhoea; plant decoction is alterant and antiasthmatic.

Ethnomedicine

Plant: Is emetic, sedative and tonic; is useful in adenopathy, alopecia, gravel, heat, insomnia, oligosuria, piles, sinus and tumour.

Root: Used in epilepsy, fever, fistula, gum trouble and tooth ache; used as an antidote to snake bite.

Leaf: Useful in boils, fistula, headache, hydrocele, kidney diseases, piles, swelling and wounds.

Non-Medicinal Uses

Plant: Grown as hedge plant.
Seed oil: Is used for dimerization and production of coating materials.

35. *Mimosa rubicaulis* Lam.

Vernacular Names: Ben: Shiah kanta; **Hin:** Shiah kanta; **Kan:** Uriseege; **Mal:** Kattusinikka; **Mar:** Huziru garhwal khinkari; **Pun:** Arulu; **Tam:** Chemba, Igai, Ingi, Ingu, Iraittu, Karindu, Karunjundi, Kattuppyindiriyakodi, Kattuchiyakkay, Kattuseekkai, Kundai, Nakkiyakkeeram, Singili, Thuppirasam; Thulobam; **Tel:** Boda chandra, Kodimundusu, Korinda, Sandrathige, Vallikataru, Ventra.

Distribution: Native of Asia; grows throughout India.

Description: A prickly shrub with glabrous branches; branches armed with small, copious, hooked spines; bark smooth; leaves with prickly rachis; stipules subulate; leaves 4–12 pairs with bristle like gland; leaf lets 12–15 pairs, oblong, cuspidate, mucronate at the apex, obliquely truncate at the base, membranous; flowers pink or red at first and later turn white, 4–merous; stamens 8 in number; pod glabrous without any prickles, linear, oblong; seeds 6–10 with distant prickles.

Phytochemistry

Plant: β-Amyrin, β-sitosterol and friedelin.

Medicinal Uses:

Root: Root powder is given to check vomiting.

Leaf: Leaf infusion is used as douche for piles and applied on burns.

Ethnomedicine

Plant: Is used in bronchitis, cholera, cough, dyspepsia, dysuria, fever, jaundice, measles, pleurisy, small pox, sores, syphilis and tuberculosis; as antidote to snake bite.

Non-Medicinal Uses

Plant: Grown as hedge plant.

Wood: Is used in making gun powder charcoal.

36. *Mimosa rubicaulis* Lam. ssp., *himalayana* (Gamble) H. Ohashi

Synonyms: *Mimosa himalayana* Gamble
Mimosa rubicaulis auct, non Lam.

Common Name: Slow mimosa.

Vernacular Names: Ass: Kuchoi, Kaitkusia kanta, Kauri kanta; **Ben:** Shaiak kanta; **Hin:** Shaiak kanta; **Mal:** Kattu sinikka; **Pun:** Deokhadia; **Tam:** Ingai; **Tel:** Kodimudusu, Undra, Ventra.

Distribution: Native of Afghanisthan; is introduced into India; in India, it grows in Arunachal Pradesh, Assam, Central India, Delhi, Jammu and Kashmir.

Description: An aquatic herb with stout, creeping or floating stems, rooting at the nodes; leaves bipinnate; pinnae 4–6; leaflets sensitive, 5–15 pairs, oblong, obtuse, glabrous; flowers minute on dense heads; pod oblique, oblong, beaked with 6–8 seeds.

Medicinal Uses

Plant: Useful in bronchitis, cholera, constipation, cough, fever, indigestion, measles, pleurisy, small pox, sores, syphilis, throat infections and urinary complaints; as an anitaodte to snake bite.

Root: Powder is used to stop vomiting of food due to weak intestine.

Leaf: Useful in epilepsy and hysteria; infusion is useful in piles; leaf paste is applied on burns.

Non-Medicinal Uses

Plant: Grown as ornamental or hedge plant.

Wood: Heart wood is useful in making tent pegs and gun powder; yields charcoal.

37. *Neptunia oleracea* Lour.

Synonyms: *Mimosa natans* L.f.
Mimosa prostrata Lam.
Neptunia stolonifera Guillemin

Vernacular Names: Ben: Pani-najak; **Hin:** Lajalu; **Kan:** Nidraaganti; **Mal:** Nittitoddavaddi; **Sans:** Alambusa; **Tam:** Sadai, Sundaikkirai, Neersondaikody; **Tel:** Neeruthalavapu nidryam, Nidravathi.

Distribution: Native of Africa, Asia including India, Central America, Caribbean Islands, Madagascar, Brazil, Columbia, Guyana and Peru; grows throughout India.

Description: Annual free floating aquatic herb; stem terete, glabrous, rarely branched; pinnae in 3 pairs with out any glands between them; leaf lets 12–15 pairs, oblong, glabrous; stipules obliquely cordate; bracts small, ovate, subobtuse; flowers yellow, minute, in dense heads, the lower replaced by numerous ligulate yellow staminodes; pod stalked, obliquely oblong, deflexed and beaked with 6–10 seeds.

Edibility

Plant: Edible; used as vegetable.

Stem: Young stem is eaten as pot herb.

Pod and seed: Eaten as vegetable.

Medicinal Use

Plant: Is anodyne, anthelmintic, antiodiarrhoeal, astringent, diuretic and refrigerant; useful in burning sensation, cephalalgia, diarrhoea, dypsia, helminthiasis, otalgia, strangury, syphilis and vitiated conditions of "pitta".

Root: Useful in syphilis.

Stem: Stem juice is dropped into the ear for earache.

Leaf: Is antitumour.

Ethnomedicine

Plant: Is used as astringent and refrigerant; used in treating ear ache, syphilis and tumour.

38. *Parkia biglandulosa* Wight & Arn.

Synonym: *Mimosa pedunculata* Roxb.

Common Name: Mexican palo-verde.

Vernacular Names: Kan: Byandkaayi mara, Shivalingada mara; **Tel:** Bantithumma, Saambraani maramu.

Distribution: Native country unknown; introduced into India; in India it is cultivated in most parts.

Description: Tree, 8–10 m high; leaves bipinnate; pinnae 20–40 pairs; leaf lets 60–100 pairs; flowers white to light yellow, in dense subspherical heads; pod long.

Edibility

Pod pulp and seedlings: Are sweet and edible.

Medicinal Uses

Stem bark: Is astringent.

Leaf: Anti-inflammatory and antiulcerogenic.

Medicinal Uses Outside India

Java: Seeds are given as a remedy for cholera, diabetes, liver diseases and worms.

Indonesia: The seed paste is applied to ulcers and wounds.

Malay Peninsula: Seeds are used as carminative; decoction is given as tonic and for colic and distended stomach; bark paste is applied as poultice for absecesses and boils; the leaf and bark decoction is applied on ulcers.

Ethnomedicine

Plant: Is used as astringent and refrigerant.

Non-Medicinal Uses

Plant: Grown as avenue and shade tree; is grown in gardens as an ornamental tree for its attractive feathery foliage.

Bark: Used in tanning.

Wood: Is hard and yields fuel and charcoal; also used in construction.

Flower: Flower and pollen mixed in water makes a refreshing drink.

39. *Parkia timoriana* (DC.) Merr.

Synonyms: *Inga timoriana* DC.
Parkia roxburghii. G. Don

Vernacular Names: Ass: Khorial, Yongchak, Zongta; **Hin:** Kihorial, Supata.

Distribution: Native of Asia including India; in India it grows in Assam, Meghalaya and Mizoram.

Description: Large tree upto 25m height with spreading branches; bark grayish brown; leaves bipinnate; rachis with a gland below the lower pair of pinnae and a few between the upper pinnae; leaf lets 40–80 pairs, oblique, oblong, acute, base obliquely truncate, glabrous; flowers yellow or dull white, narrow, tubular on dense clavate peduncled head; pod flat, dark brown, drooping.

Edibility

Pod: Tender pods are used for flavouring food; used as vegetable.

Seed: Eaten raw or after roasting.

Medicinal Uses

Root: Cyanophoric.

Stem bark and leaf: Cyanophoric; used for preparing lotions for sores, skin diseases and ulcers.

Pod and seed: Used in colic and stomach disorders.

Medicinal Uses Outside India

Java: Seed decoction is given in colic, cholera, and as general tonic; poultice of seed paste is applied on the stomach as a remedy for distended stomach; the bark is applied to absecesses and boils; the leaf and bark paste used as dressing for ulcers.

Malay Peninsula: Seeds are used as carminative.

Philippines: Seed decoction is used to wash ulcers and wounds.

Non-Medicinal Uses

Wood: Is used for temporary construction, soles of shoes, bowls and utensils; is a source of paper; used as fuel.

40. *Pithecellobium dulce* (Roxb.) Benth.

Synonym: *Mimosa dulcis* Roxb.

Common Names: Deccan babool, Madras thorn, Manila tamarind, Sappan fruit tree, Sweet babool.

Vernacular Names: Ben: Dekhani babul; **Hin:** Dakhani babul, Vilayathi imli; **Kan:** Chakkuli mara, Madras mullu, Manila hunase, Seeme hunase; **Mal:** Korukka puli, Kodapuli, Korakapuli; **Mar:** Hatichinch, Vilaythi chinch; **Tam:** Karapilli, Karkapilli, Kodukka puli, Kattuppilli, Korukapuli; **Tel:** Seemachinduga, Seemechinta.

Distribution: Native of Central America, United States and South America; introduced into India and naturalized; cultivated throughout India.

Description: A large tree with grey, smooth or lenticillate, hoop ringed bark with fissures; ultimate branch lets with stipular thorns; one pair of pinnae with a small cup-like gland between the bases of the pinnae and with similar gland between the bases of the petiolules; leaflets glabrous, oblique, obovate oblong with rachis and rachillae forming an acute point; flowers white, in dense heads on puberulous peduncles in the axils of small, lanceolate bracts, on the branchlets; the bracts of the lower two third pairs are accompanied by minute, stipular thorns; calyx subcampanulate, shortly toothed; corolla densely puberulous, funnel shaped with acute lobes; stamens in a staminal tube; pod turgid, curled into a circle, falcate, slightly lobed between the seeds, glabrous, valves coriaceous, claret red; seeds obovate, oblong with glossy testa and a firm, edible, sweet, pink to white aril.

Phytochemistry

Leaf: Afzelin, behenic acid, kaempferol, lignoceric acid, octacosanol, quercetin and α-spinasterol.

Flower: Asparagines, glucose, hexacosane, hexacosanol, kaempferol, L-glycine, L-leucine, L-lysine, quercetin, xanthophylls and α-spinasterol.

Seed: Echinocystic acid, flavone, hexacosanol, kaempferol, lecithin, leucine, oleanolic acid, pithogenin, proline, quercetin and saponins.

Mesocarp: Asparagines, glucose, hexacosanol, leucine, proline, sterol glucoside and valine.

Edibility

Seed: Aril on the seed is edible; in Mexico the aril is used to make lemonade like drink.

Seed oil: Edible.

Medicinal Uses

Stem bark: Is antidysenteric, astringent, febrifuge and spasmolytic; decoction of stem bark is given as enema;

Leaf: Abortifacient and emollient; useful in leprosy; promotes hair growth.

Seed: Anti-inflammatory and spermicidal.

Toxicology: Continuous contact with the bark in tanning industries can cause eye infection and swelling of the eye lids.

Ethnomedicine

Root: Is useful in leprosy and to promote hair growth.

Leaf: Used in rabies; cures earache, convulsions, dysentery, dyspepsia, fever, sores, tooth ache and venereal diseases; acts as astringent and anodyne.

Non-Medicinal Uses

Plant: Grown as avenue, hedge and ornamental plant; yields gum, which is reddish brown in colour.

Bark: In Philippines, the bark is used for tanning leather.

Wood: Used for making carts, packing boxes and panelling; used as fuel.

Leaf: Used as fodder.

41. *Prosopis cineraria* (L.) Druce

Synonyms: *Mimosa cineraria* L.
Prosopis spicigera L.

Notes: In the Indian epic *Mahabharatha*, the Pandavas concealed their arms on this tree during their exile.

Common Names: Indian mesquit, Sponge tree.

Vernacular Names: Hin: Chinkur, Chokara, Chownkra, Jaant, Jhand, Safed kikkar, Sejada, Shame; **Kan:** Banni mara, Perumbe; **Mal:** Marampu, Parambu, Thavivanni; **Mar:** Saunder, Savandad, Shambaree; **Sans:** Bhadra, Dhurithadamani, Havirgandha, Ishana, Ishta, Kachariphala, Kaeshahanthri, Keshamathani, Kananari, Lakshmi, Mangalya, Medhya, Paapanashini, Paapashamani, Pavitra, Priyadarshini, Sakthuphala, Samira, Samudra, Shamee, Shankari, Shantha, Shiva, Shivaphala, Shubhadha, Subhadra, Shubakaari, Sukhada, Surabhi, Thapanatanaya, Thunga, Vanhigarbha; **Tam:** Jambu, Kalisam, Parambai, Sami, Seemai mullu, Siva, Thamali; **Tel:** Jammichettu, Priyadarshini.

Distribution: Native of Asia including India and Middle East; in India it grows in Gujarat and Punjab.

Description: A moderate sized tree armed with scattered, broad, conical, straight prickles; pinnae usually 4, opposite, with a gland between each pair; leaflets 8–12 pairs, obliquely oblong, cuspidate, glabrous; flowers yellow on axillary panicles; pod pendulous, torulose; seeds oblong, brown, many, in sweetish pulp.

Phytochemistry

Plant: Albumin, amino acids, arabinose, campesterol, galactose, glucose, glutelin, mannose, octacosanol, paulitricin, prosogerin A and C, B, C and E, rhamnose, spicigerin, stigmasterol, ursolic acid and β-sitosterol.

Wood: Esters of lauric acid, linolenic acid, myristic acid, n–decanol, palmitic acid, stearic acid and β-sitosterol.

Leaf: Alkaloids.

Flower: Flavones, glucosides, patulitrin, prosogerin A and B.

Pod: Carobin, carobone and carobonic acid.

Seed: Gallic acid, isoprosogerin, luteolin, patuletin, patulitrin, prosaponol D and E, prosogerin A and B, prosogerinic acid, rutin and spicigerin.

Edibility

Bark: Is sweet and edible; eaten by poor during famine.

Gum, leaf, pod and seed: Edible.

Medicinal Uses

Plant: Is recommended for the treatment of snake bite.

Young shoot: Astringent, demulcent and pectoral; used as tooth brush.

Root: Is antidysenteric.

Stem: Crushed and applied on the skin removes the hairs.

Stem bark: Anti-inflammatory, anthelmintic, cooling and tonic; cures asthma, bronchitis, cough, diarrhoea, dysentery, dyspnoea, giddiness, haemorrhage, leucoderma, piles, rheumatism, skin diseases and tremors; used for removing the hairs on the skin.

Leaf: Crushed and applied on the skin removes hairs.

Flower: Eaten during pregnancy to check miscarriage; burnt and ash is rubbed over to remove hair on the skin.

Pod: Astringent, demulcent and pectoral.

Toxicology: Causes hair fall.

Uses in Alternative Systems of Medicine

Ayurveda:

Part used: Bark and leaf.

Uses: Useful in cough, diarrhoea, dysentery, dyspnoea, giddiness, haemorrhage, piles and skin diseases; stem bark is anthelmintic, cooling and tonic; cures asthma, bronchitis, dysentery, leprosy, leucoderma, piles and tremors; smoke of dried and burnt leaves is good for eye complaints; dries up excessive accumulation of mucous and fluid, fumigation of dry leaves is given in piles.

Siddha:

Part used: Bark, leaf and pod.

Uses: Useful in delirium, diseases of "kapham" and poisoning.

Medicinal Uses Outside India

Central Province: The bark is used as a remedy for rheumatism.

Las Bela: Ash of the plant is used as a depilator.

Saruma: The flowers are pounded and eaten against miscarriage.

Non–Medicinal Uses

Plant: Grown as a hedge plant; considered as a sacred tree in North India by Sikhs as Gurunanak had planted many trees.

Bark: Is used for tanning.

Wood: Is used as fuel.

Leaf: Used as green manure and fodder.

Pod: Used as cattle feed.

42. *Prosopis glandulosa* Torrey.

Synonyms: *Porsopis chilensis sensu auct.*
Prosopis chilensis (Molina) Stuntz. var., *glandulosa* (Torrey) Standley

Prosopis juliflora sensu auct.
Prosopis juliflora (Sw.) DC. var., *constricta* Sarg.
Prosopis juliflora (Sw.) DC. var., *glandulosa* (Torrey) Cockerell

Notes: Three varieties, *glandulosa, prostrata* and *torreyana* are recognized.

Common Name: Honey mesquite.

Vernacular Names: Hin: Vilaayati kikkar, Kabuli kikhar, Vilayati khejra.

Distribution: Native of Mexico and United States; introduced into India, Iran, Iraq, Australia, Caribbean Islands, South Yemen and African countries; naturalized in Delhi, Punjab and Rajasthan.

Description: Small tree with glabrous zig zag shoots, covered with stipular thorns; leaves bipinnate; rachis with a gland at the apex; pinnae 2; rachilla with glands between bases of leaflets from the tip to 8th pair of leaflets; leaflets 20 pairs with obliquely truncate base; flowers on cylindrical spikes; calyx 5, toothed, ciliate; corolla 5, acute, white; stamens10, glabrous; pod flat, glabrous; seeds ovoid, compressed.

Phytochemistry

Plant: Juliprosopine, oleanolic acid, prosopol, triacontanol and β-sitosterol.

Root: 3, 3'–Di–O–methyl ellagic acid–4–O–rhamnoside, procyanidin and tannin.

Root bark: Kaempferol–4'–O–methyl–3–O-β-D–galactopyranoside and retusin–7–O–neohesperidoside.

Stem bark: Hexacosan–25–one–1–ol and leucoanthacyanins.

Pod: Ellagic acid and rutinoside.

Edibility

Pod: Edible after removal of seeds.

Medicinal Uses

Plant: Antiandrogenic, antibacterial, antihepatotoxic, anti-inflammatory, antioxidant, antiviral, anti-HIV, antihistaminic, CNS depressant, diuretic, estrogenic, hypoglycaemic, hepatoprotective, hypocholesterolaemic, immunomodulator and sedative; cures arthritis, edema, infections, inflammation and ulcers.

Toxicology: Plant contains trypsin inhibitors.

Non-Medicinal Uses

Plant: Used as pesticide and insecticide.

Leaf: Used as forage.

43. *Prosopis juliflora* (Sw.) DC.

Notes: Three varieties, *juliflora, horrida* and *inermis* are recognized.

Common Names: Cashew of Texas, Texas mesquit bean, Velvet mesquit.

Vernacular Names: Hin: Vilayathi kikar, Vilayati babul; **Tam:** Vaelikkaruvai, Vanni; **Tel:** Circar kampa, Circar mullu.

Distribution: Native of Caribbean Islands, Central America, United States and South America; introduced into India and also naturalized; grows throughout arid areas.

Description: Tree or shrub of 12 m hight; bark fissured; spines straight, solitary or paired, divergent; leaves glabrous with 2 or 4 pinnae, with a small orbicular gland near the base of the lower pinnae; pinnae ending in a short stipe; leaflets opposite, sessile, 6–30 pairs, oblong–linear, both ends obtuse or apex, emarginate, midrib slightly eccentric; flowers greenish yellow, fragrant, subessile on axillary racemes; calyx campnulate, glabrous with small, triangular teeth; pod pale yellow, glossy, smooth, flat, with straight parallel sutures or irregularly submoniliformous; seeds oval, brown in colour, 10–20 in number.

Phytochemistry

Plant: 3,3',5,5'-Tetrahydroxy-4'-dimethoxy diphenic acid, procyanidin, 3,3-di-O-methyl ellagic acid-4-O-rhamnoside.

Root: Ellagic acid, juliflorine and julifloridine.

Stem bark: Juliprosopine, kaempferol–4'–O–methyl–3–O-β- D–galactopyranoside, ketoalcohol, hexacosan-25-one-1-ol, leucoanthocyanin S, leucodelphinidin–3–O-α-L– rhamnopyranoside and retusin.

Leaf: Juliflorinine, juliprosin, juliprosinene and prosopidione.

Pod: Ellagic acid, julifloridine, juliflorine, O–rutinoside and patulitrin.

Medicinal Uses

Plant: Is antiallergenic, antiatherosclerotic, antibacterial, anticancer, antidepressant, antidiabetic, antiherpatic, antihypotensive, anti-inflammatory, antimutagenic, antioxidant, antitumour, antiulcer, antiviral, cytotoxic, diuretic, hepatoprotective and vasodilator.

Edibility

Pod and seed: Edible.

Bioactivity: Plant is inhibitor of 5–lipoxygenase, ATPase, aldose-reductase, aromatase, cytochrome-oxidase, deiodianase, glucosyl-transferase, HIV–RT, lipoxygenase, MAO, MMPI, metalloprotein, NADH–oxidase, NEP, ornithine-decarboxylase, PKC, PTK, phospholipase, polyamine synthesis, prostaglandin synthesis, protein-kinase, topoisomerase, tyrosine-kinase, xanthine-oxidase, CAMP phosphodiesterase; is blocker of angiotensin receptor, calcium channel and calmodulin; inducer of quinone-reductase.

Toxicology: Plant contains chymotrypsin and trypsin inhibitors.

Uses in Alternative Systems of Medicine

Siddha:

Part used: Stem bark, leaf and pod.

Uses: Useful in delirium, diseases of "kapham" and poisoning.

Ethnomedicine

Plant: Used as cathartic, cyanogenic, discutient, emetic, stomachic and vulnerary; useful in catarrh, cold, diarrhoea, dysentery, flu, inflammation, itch, measles, stomach ache, throat trouble and wounds; is used as antidote to poison.

Non-Medicinal Uses

Plant: Is grown as windbreak and hedge plant; used as pesticide.

Wood: Used in paper making; yields fire wood and charcoal.

Pod: Is used as fodder; pod shell is used in dyeing.

44. *Samanea saman* (Jacq.) Merr.

Synonyms: *Enterolobium Saman* (Jacq.) Prain.
Mimosa saman Jacq.
Pithecellobium saman (Jacq.) Benth.

Notes: Commonly known in Inida as *Enterolobium saman* (Jacq.) Prain, this species may have to be named *Albizia saman* (Jacq.) Merr.

Common Names: Coco tamarind, Cow tamarind, East Indian walnut, French tamarind, Monkey pod, Rain tree.

Vernacular Names: Kan: Male mara; **Tam:** Thoongu moonji mara; **Tel:** Tella thurai.

Distribution: Native of Jamaica (Caribbean Islands), Belize, Costa Rica (Central America) and South America; introduced into India; grows throughout India.

Description: A large tree with rough, deeply fissured yellowish brown bark; stipules small, lanceolate, densely pubescent; petiole of the leaf without gland, swollen at the base; pinnae 4 pairs; glands found between the bases of pinnae; leaf lets sessile, 12–16 pairs on the upper pinnae, 6–10 pairs in the lower pinnae, glossy above, softly pubescent below with eccentric midrib, ovate oblong in shape with truncate or attenuate ends; flowers in dense heads in the axil of upper leaves, pink, with a lanceolate, pilose bract; calyx funnel shaped, teeth short and triangular; corolla pink, funnel shaped, with ovate lobes; anther filament pink and connate at the base; pod sessile, fleshy, straight, indehiscent with a glossy brown epicarp and sweet, sticky mesocarp and endocarp forms a continuous septa between the seeds; seeds 16–20, transversally ovate, smooth, brown and glossy.

Phytochemistry

Plant: Acacic acid and β-smyrin.

Stem bark: Gallic acid, glucose, lupenone, lupeol, octacosanoic acid, phytosterol, samarin, α-spinasterol and β-spinasterone.

Wood: Arabinose, fucose, glucose, rhamnose, samanin–C and xylose.

Flower: Arabinose, glucose, rhamnose, samanin–D and xylose.

Medicinal Uses

Plant: Spermicidal.

Non-Medicinal Uses

Plant: Is a host plant for lac insect; grown as shade tree in coffee, cocoa and vanilla plantations; yields gum of inferior quality.

Wood: Gives charcoal, which is used as fuel in railways; wood shavings used for making hats.

Leaf: Is used as fodder and green manure.

Pod: Used as fodder for livestock.

45. *Tetrapleura tetraptera* (Schum. & Thonn.) Taubert

Synonym: *Adenanthera tetraptera* Schum. & Thonn.

Common Name: Aidan tree.

Distribution: Native of Benin, Burkina Faso, Cambodia, Gambia, Ghana, Guinea, Liberia, Mali, Nigeria, Senegal, Sierra Leone and Uganda; introduced into India; cultivated in India in Botanical Gardens.

Description: A deciduous tree of 20–25 m height; bark smooth, grey brown; branches and leaves are glabrous; leaves sessile, channeled on the upper surface; pinnae 5–9 pairs, opposite, sometimes alternate; leaf lets 6–12 in number, alternate, elliptic, rounded at both ends, apex slightly notched, base unequal, glabrous, with slender stalks; flowers pinkish cream turning to orange and crowded on racemes; stamens 10 with a gland at the apex; pod hanging from the ends of branches on stout stalk, shiny, glabrous, purple brown, slightly curved with 4 longitudinal wing like ridges, two of the wings are woody, other 2 filled with soft sugary pulp, oily and aromatic; seeds are small, black, hard, flat and rattle in the pod.

Phytochemistry

Root bark: Saponins, saponoside and tannin.

Pod: Coumarin, oleonolic triglycosides, saponosides, scopoletin and tetramethyl pyrazine.

Medicinal Uses

Bark: Used as emetic and purgative.

Pod: An infusion of the pod is used as stimulant and tonic; pod is used in making anticonvulsant drink.

Bioactivity: Alcoholic axtract of the pod exhibit marked tranquilizing properties in male albino mice and lowers body temperature; oral dose of alcoholic extract of the pod sedated mice within 30 minutes; antibroncho constrictive and anti-arrhythmic effects of scopoletin in the pod was demonstrated *in vivo* and *in vitro*; it protected the Guinea pigs from induced arrhythmia and increased the functional refractory period of myocardium in the same way as quinidine; it also relaxed histamine induced contractions of tracheal muscles in Guinea pigs.

Non-Medicinal Uses

Wood: Yields timber and tannin.
Flower: Used in perfumes and in pomades from palm oil.
Pod: Is rich in sugars; is used in flavouring food and in perfumery; used as insecticide.

46. *Xylia xylocarpa* (Roxb.) Taubert

Synonyms: *Xylia dolabriformis* Benth.
Mimosa xylocarpa Roxb.

Common Name: Burma iron wood.

Vernacular Names: Hin: Jambu, Suria; **Kan:** Ambaarke, Arasuthega, Bettadavarike, Hommavarike, Jambe, Jambemara, Shivve, Takku, Thiruva; **Mal:** Erubogam, Irimpullam, Irumulla, Kadamorum; **Mar:** Jamba, Suria, Tambia; **Sans:** Simsapa; **Tam:** Aruvapallam, Irul, Iruvel; **Tel:** Channangi, Eravalu, Errachennanangi, Kondatangedu.

Distribution: Native of Asia including India; grows in Karnataka and Bihar in India.

Description: Unarmed deciduous tree with reddish grey bark; leaves bipinnate; leaflets 3–5 pairs, elliptic, lanceolate, acuminate with a gland in between, lower leaflets smaller than the upper; flowers sessile on racemes; corolla yellowish white, lanceolate; calyx campanulate; stamens 10, free; pod brown, pubescent, woody, axe–shaped; seeds 6–10, oblong.

Phytochemistry

Stem bark: Oil and triterpenes.

Heart wood: Diterpenes, manoyl oxide and sandaracopimaradien-3-one.

Leaf: β-Sitosterol and trans–5–hydroxy pipecolic acid.

Pod: Fatty oil and proanthocyanidin.

Seed: Oil.

Edibility

Pod and seed: Edible.

Medicinal Uses

Stem bark: Decoction is used as anthelmintic, antidiarrhoeal, antileprotic and emetic; used in gonorrhoea, ulcers and vomiting.
Seed oil: Is applied for leprosy, piles and rheumatism.

Ethnomedicine

Plant: Is used in diarrhoea, gonorrhoea, leprosy, nausea, piles, rheumatism, sores, swellings and as vermifuge.

Non-Medicinal Uses

Plant: Useful as fodder; is a host plant for lac insect.

Bark: Yields tannin used in tanning industry.

Wood: Is water resistant and resistant to white ants; is an important timber used for making railway sleepers, boats, canoes, knees, beams, scantling in house construction, wooden crubs of wells, ladders, bridges, agricultural implements, telegraph posts, oil presses, yokes, wheels of timber carts, spokes and felloes of wheels, bowls, tool handles, turnery and flooring; yields a good tanning material; used in making, writing, wrapping and printing paper; is a good substitute for sal and teak wood; used as fuel and yields charcoal.

Leaf: Used as manure.

Fig. 1. Pods of *Acacia catechu*

Fig. 2. *Acacia dealbata*

Fig. 3. *Acacia planiformis*

Fig. 4. *Acacia podalyriifolia*

Fig. 5. Pods of *Albizia lebbeck*

Fig. 6. *Mimosa pudica*

Fig. 7. *Samanea saman*

❒❒❒

Glossary

This glossary contains explanation of the medical terms used both in the Chapters and the species profiles.

Abortifacient : A drug that induces abortion or miscarriage.

Abscess : A localized collection of pus anywhere in the body, surrounded and walled off by damaged and inflamed tissues.

Acidity : The state of being acid or sour.

Acidosis : Depletion of the body's alkali reserve, resulting in disturbance of the acid-base balance.

Acidulant : Having an acid quality.

Acne : A skin disorder in which the sebaceous glands become inflamed.

Actinomycosis : A disease caused by the species of the fungus *Actinomyces*, the sites most affected being the lung, jaw and intestine; granulomatous tumours form which usually suppurate, discharge a thick, oily pus containing yellowish granules.

Addison's disease : A form of anaemia that results when bone marrow fails to produce adequate number of RBC's due to vitamin B_{12} deficiency.

Adenitis : Inflammation of the gland or group of glands.

Adenocarcinoma : Malignant tumour of the lymphoid tissue of the nasopharynx.

Adenoma : A non-malignant tumour of glandular tissue.

Adenopathy : A general term for glandular disease.

Agalactia : Absence or abnormally low production of milk in a woman who has just given birth.

Ague : An infectious disease caused by presence of parasitic protozoa; also called malaria.

Albuminuria : Presence of serum albumin, serum globulin or other serum proteins in the urine.

Alexiteric : Protective against infection, venom and poison.

Alexipharmic : Remedy for poison.

Allergenic/Allergy : A disorder in which the body becomes hypersensitive to particular antigens, which provide characteristic symptoms whenever they are subsequently encountered, whether inhaled ingested, injected or otherwise contacted.

Allograft : Transplantation of an organ from one person into another; term also used for skin graft, from a cadaver.

Alopecia : Absence of hair from areas where it normally grows.

Alterant/Alterative : Causing a favourable change in the function of the body or metabolism.

Ambylopia : Defective vision approaching blindness.

Amenorrhoea : The absence or stopping of the menstrual periods.

Amoebic dysentery : Infestation of large intestine by protozoan *Entamoeba histolytica,* causing passage of mucous and blood.

Anabolic : Promoting synthesis.

Anaesthetic : An agent that reduces or abolishes sensation, affecting either the whole body or a particular region.

Analgesic : An agent which relives pain without causing loss of consciousness.

Anaphylactic : An agent which causes allergy.

Anaplasty : Loss of distinctive characteristics of a cell or tissue due to proliferation, as in cancer.

Anasarca : Accumulation of great amount of fluids in all the body tissues.

Androgenic : A agent which produce masculine characters.

Angina pectoris : Severe but temporary attack of cardiac pain which may radiate to the arms; results from myocardial ischaemia.

Angiogenesis : The process of vascularization of a tissue during the development of new capillary blood vessel.

Anodyne : Any treatment or drug that soothes and caeses pain.

Anorexia : Loss of appetite.

Anthelmintic : Any drug or chemical agent used to destroy parasitic worm.

Antiacid/Antacid : A substance which neutralizes or counteracts acidity.

Antianginal : Counteracting the chest pains.

Antialdosteronic : An agent which inhibits the formation of aldosterone, a hormone.

Antibacillary : An agent which inhibits bacillary infection.

Antibacterial : An agent which destroys bacteria.

Antibilious : An agent which prevents excess bile production.

Antibiotic : A substance produced by or derived from a micro-organism that destroys or inhibits the growth of micro-organisms.

Anticephalagic : An agent which reduces or prevents headache.

Anticholinergic : An agent inhibiting the action of cholineric nerve by interfering with the action of acetylcholine, a chemical which transmits its impulses at neural or myoneural junctions.

Anticoagulant : An agent that prevents the clotting of blood.

Anticonvulsant : A drug that prevents or reduces the severity of fits (convulsions) in various types of epilespy.

Antidepressant : A drug that alleviates the symptoms of depression.

Antifebrile : Any agent that reduces or allays fever.

Antifertility : Any drug or agent that prevents fertilization.

Antifungal : An agent that destroys fungi.

Antigonadotropic : A drug that prevents or reduces the synthesis and release of hormones by the pituitary gland that acts on the testes or ovaries to promote production of sex hormones.

Antihistamine/Antihistaminic : A drug that suppresses some of the effects of released histamine.

Antilithic : An agent that prevents the formation of calculi or promote their dilution.

Antimicrobial : An agent that destroys microbes.

Antimycobacterial : An agent that destroys fungi and bacteria.

Antinematodal : An agent that destroys nematodes.

Antinociceptive : The process of pain transmission.

Antioxidant : Any substance that delays the process of oxidation by neutralizing free radicals.

Antiperiodic : Preventing the regular recurrence of a disease.

Antiphlogistic : Acting against heat or inflammation.

Antiprotozoal : An agent that prevents or destroys protozoans.

Antipyretic : A drug that reduces fever by lowering the body temperature.

Antiradiation : A drug that counteracts the effect of radiation.

Antiscorbutic : An agent acting against scurvy.

Antisecretory : An agent that inhibits secretion.

Antiseptic : A chemical that destroys or inhibit the growth of disease-causing bacteria and other micro-organisms.

Antispasmodic : Preventing or relieving spasms.

Antispermatogenic : A drug that retards the production of sperms.

Antistress : An agent that prevents or reduces stress.

Antithyroid : An agent used to decrease the activity of thyroid gland.

Antitussive : A drug that suppresses coughing possibly by reducing the activity of the cough centre in the brain.

Antivenemous : An antiserum containing antibodies against specific poisons in the venom of snake, spider or scorpion.

Antiviral : An agent that destroys virus.

Antiyeast : An agent that destroys yeast.

Anuresis : Retention of urine in the bladder.

Aperient : A mild laxative.

Aphrodisiac : An agent that enhances sexual excitement.

Aphthae/Aphthous : Small grey ulcerated area in the mouth.

Apoplexy : Sudden loss of consciousness caused by haemorrhage or thrombosis.

Apoptotic : Causing cell death.

Appendicitis : Inflammation of the vermiform appendix.

Arrhythmia : Any variation from the normal rhythm of the heart beat.

Arteriosclerosis : Degenerative arterial change associated with advancing age.

Arthralgia : Pain in a joint.

Arthritis : Inflammation of one or more joints, characterized by swelling, warmth, redness, of the overlying skin, pain and restriction of motion.

Articular : Pertaining to a joint or articulation.

Ascites : The accumulation of fluid in the peritoneal cavity, causing abdominal swelling.

Asphyxia : An inability to breath (suffocation).

Astigmatism : A defect of vision in which the image of an object is distorted, usually in either the vertical or the horizontal axis.

Astringent : An agent that contracts organic tissue thus lessening secretion.

Ataxy : The shaky movements or unsteady gait that result from the brains failure to regulate the body's posture and the strength and direction of limb movements

Athelete's foot : Ringworm of the foot caused by *Tinea pedis.*

Atherogenic : An agent that has the capacity to accelerate the process of formation of encysted tumour.

Atherosclerosis : A disease of the arteries in which fatty plaques develop on their inner walls, with eventual obstruction of blood flow.

Atony : Weak.

Attenuant : An agent that reduces the diseases producing ability (virulence) of a bacterium or virus.

Auscultation : A technique used to listen to the body sounds for diagnostic purposes (particularly the heart and lungs).

Bactericide : Any agent that destroys bacteria.

Bacteriostat : Agent that arrests bacterial growth.

Basal cell carcinoma : Malignant growth of basal cell.

Bechic : Any thing that relieves or cures cough.

Beri Beri : A disease due to thiamine (Vitamin B) deficiency, marked by polyneuritis, cardiac pathology and edema.

Biliary colic : Pain resulting from obstruction of the gall bladder or common bile duct, usually by a stone.

Biliary hyperplasia : Excessive proliferation of cells in the bile duct.

Black water fever : A malignant form of malaria occurring in the tropics, especially Africa resulting in great destruction of red blood cells and causing dark coloured urine.

Blastogenesis : Multiplication or increase in the blast cells.

Blenorrhagia : A copious discharge of mucus, particularly from the urethra.

Blisters : A swelling containing watery fluid and sometimes also blood (blood blister) or pus, within or just beneath the skin.

Bradycardia : Slow rate of heart contraction resulting in slow pulse rate.

Bright's disease : Inflammatory affection of the kidney.

Bronchial carcinoma : Malignant growth of bronchi.

Bronchitis : Inflammation in the bronchi.

Bronchodilator : An agent that causes widening of the air passages by relaxing bronchial smooth muscle.

Buboes : An enlarged and inflamed lymph node particularly in the groin, due to such infections as plague or syphilis, etc.

Bursa : A fibrous sac lined with synovial membrane and containing a small quantity of synovial fluid.

Bursitis : Inflammation of bursa.

Cachexia : A term denoting a state of constitutional disorder such as malnutrition and general ill health.

Calculi/Calculus : An abnormal concretion usually composed of minerals salts, occurring within the body.

Cancer : Any malignant tumor including carcinoma and sarcoma.

Cancerous ulcers : An ulcer that transforms itself into a cancerous form.

Carbuncles : An acute inflammation involving several hair follicles sloughing with discharging sinuses.

Carcinogenic : A substance that, when exposed to living tissue may cause cancer.

Carcinoma : A malignant new growth made up of epithelial cells tending to infiltrate surrounding tissues and to give rise to metastatis.

Cardiodepressant : An agent that depresses the effective pumping action of the heart.

Cardiopathy : Relating to the diseases of the heart.

Cardioprotective : Agent that protects the heart.
Cardiospasms : Spasms of the cardiac sphincter.
Carminative : An agent that stimulates or improves the digestive system.
Catabolic : An agent that breaks down the complex substances in the food into simpler ones.
Cataract : Opacity in the lens of eye, resulting in blurred vision.
Catarrh : Inflammation of mucous membrane with constant flow of mucus.
Catatonia : An abnormal condition usually associated with schizophrenia.
Cathartic : A drug used to stimulate or increase the frequency of the bowel evacuation or to encourage a softer or bulkier stool.
Cauterization : Destruction of tissue by direct application of a heated instrument.
Cephalagia/Cephalalgia : Pain in the head.
Cephalic : Relating to the head.
Cerebro tonic : Providing strength to the brain.
Cervical adenitis : Inflammation of gland or a group of glands in the neck region.
Chancre : The primary syphilitic ulcer associated with swelling of local lymph glands; it is painless, indurated and highly infectious.
Chicken pox : A mild highly infectious diseases caused by a virus that is transmitted by air borne droplets.
Chilblain : Itching and burning sensation in the fingers and toes in reaction to cold.
Cholagogue : An agent that stimulates gall bladder contraction to promote bile flow.
Cholecystitis : Inflammation of the gall bladder.
Choleretic : An agent that stimulates the secretion of bile.
Cholinergic : Liberation of acetylcholine by parasympathetic nerves.
Cholinesterase inhibition : Inhibition of the enzyme that breaks down choline ester into its choline and acid components.
Chorea : The ceaseless occurrence of rapid, jerky involuntary movements.
Chronic : Describing a disease of long duration involving very slow changes.
Chylous urine : The presence of chyle in the urine (chyle = digested alkaline fat).
Cicatrizant : Drug which heals and encourages the growth of new tissues over a wound.
Cirrhosis : Interstitial inflammation of the liver.
Clastogenic : Any substance or process causing breaks in chromosomes.
CNS depressant : An agent that depresses the central nervous system.
Colic : Severe pain resulting from periodic spasm in abdominal region.
Colitis : Inflammation of the colon.
Collyrium : An eye salve or eyes wash.
Colonic adenocarcinoma : Malignant tumour of the lymphoid tissue of the colon.
Colonitis : Inflammation of the colon.
Colonorrhoea : Mucous discharge from the colon.
Colporrhagia : Haemorrhage from the vagina
Condyloma/Condylomata : Simple tumour arising from a non-glandular epithelial surface.
Conjunctivitis : Inflammation of the conjunctiva (the delicate membrane lining the eye lids covering the eye ball).
Consumption : Any disease causing wasting of tissues, especially pulmonary tuberculosis.
Contusions : Bruising and bleeding into the tissue while the skin remains unbroken.

Convulsions : Involuntary contractions of muscles resulting from abnormal cerebral stimulation with or without loss of consciousness.

Corn : An area of hard thickened skin on or between the toes.

Corneal opacity : Opacity of the transparent circular part of the front of the eyeball.

Corneal ulcers : Ulcer of the transparent circular part of the front of the eye ball.

Coronary dilator : Dilation of the heart by a crown like nerve or artery supplying blood to the heart.

Coryza : Profuse discharge from the mucous membrane of the nose.

Cosmology : The science of universc as a whole; a treatise on the structure and part of creation.

Cramp : A prolonged painful contraction of a muscle.

Curare : Any of a wide variety of highly toxic extracts from various botanical sources, employed to control muscle spasms in tetanus and as a muscle relaxant in surgical operations.

Cutaneous diseases : Diseases relating to the skin.

Cutaneous melanoma : A tumour arising from the pigment-producing cells of the cutaneous layer.

Cyanosis : A bluish tinge manifested by hypoxic tissue observed under the nail, lip and skin, due to lack of oxygen.

Cystic ameloblastoma : Malignant jaw tumour in the form of cyst.

Cystic lesions : Pathological change in a cyst.

Cystitis : Inflammation of the urinary bladder.

Cytoprotective : An agent that protects the cells.

Cytotoxic : A drug that damages or destroys cells; used to treat various types of cancer.

Debility : Lack or loss of strength; weakness.

Decongestant : An agent that reduces congestion.

Delirium : An acute disorder of the mental process accompanying organic brain diseases.

Demulcent : A mucilagenous fluid that allays irritation and smoothens inflammation, especially of mucous membrane.

Dental caries : Demineralization of the tooth surface caused by bacteria.

Dentifrice : A powder or paste for cleaning the teeth.

Deobstruent : Medicine that removes obstruction.

Depilatory : An agent applied to the skin to remove hair.

Depressant : An agent that reduces the normal activity of body system or function.

Depurative : An agent that removes impurities.

Dermatitis : Inflammation of the skin.

Dermatopathy : Any skin disorder.

Dermatosis : Any diseases of skin, particularly one without inflammation.

Detoxicant : Any agent that neutralizes the effect of toxins.

Dhobi's itch : A skin disease, a type of allergic dermatitis, caused by wearing clothes laundry marked with ink used in India by native laundry men.

Diabetes mellitus : A disorder of carbohydrate metabolism in which sugars in the body are not oxidized to produce energy due to deficiency of pancreatic hormone insulin.

Diaphoretic : An agent that causes profuse perspiration.

Dietetics : Interpretation and application of the scientific principles of nutrition in health and disease.

Diphtheria : An acute infectious disease caused by *Corynebacterium diphtheriae* and its toxin, affecting the membranes of the nose, throat, or larynx causing fever, pain, and respiratory obstruction.

Discutient : Causing the dispersal or disappearance of a pathologic condition, such as a tumour.

Diuretic : Increasing the amount of urine.

Down's syndrome : A human condition where chromosome No. 21 is represented thrice, instead of twice, making for 47 chromosomes in the individual; also known as trisomy 21; congenital disorder; mentally subnormal with facial features resembling the Mangolian races.

Douche : A stream of fluid directed against the body externally or into the body.

Dropsy : Abnormal accumulation of serous fluids in cellular tissues or in body tissue.

Duodenal ulcer : An ulcer in the duodenum, caused by the action of acid and pepsin on the duodenal lining (mucosa).

Duodenitis : Inflammation of the duodenum.

Dypsia : Thirst.

Dysentery : Inflammation of the bowel with evacuation of blood and mucus.

Dysmenorrhoea : Painful menstruation.

Dyspepsia : Impairment of the power or function of digestion; usually applied to epigastric discomfort after meals.

Dysphagia : Difficulty in swallowing.

Dyspnoea : Difficult or laboured breathing.

Dysrhythmia : An abnormality in an otherwise normal rhythmic pattern, as of brain waves.

Dysuria : Painful or difficult urination.

Ecbolic : An agent that hastens childbirth by stimulating contractions of the womb.

Eczema : A superficial inflammatory process involving primarily the epidermis, marked early by redness, itching, weeping, oozing, and crusting, and later by scaling, lichenification and often pigmentation.

Edema/Oedema : Abnormal infiltration of tissues with fluids.

Edward's syndrome : A human condition with autosomal trisomy of chromosome No. 18, associated with mental subnormality.

Elephantiasis : A chronic filarial disease due to infection with *Brugia malayia* or *Wachereria bancrofti*, marked by inflammation and obstruction of the lymphatic system, hypertrophy of the skin, affecting the legs and genitals.

Emaciation : Wasting of the body, caused by such conditions as malnutrition.

Embrocation : A lotion that is rubbed onto the body for the treatment of sprains and strains.

Emetic : An agent that cause vomiting.

Emetocathartic : Any agent that promotes vomiting and motion.

Emmenogogue : An agent or measure that promotes menstruation.

Emollient : An agent that softens and soothens mucous membrane.

Encephalitis : Inflammation of the brain.

Encephalopathy : Disease of the brain.

Endometrial adenocarcinoma : A malignant growth of endometrium.

Endothelioma : A malignant tumour of endothelial cells.

Enteritis : Inflammation of the intestine.

Enterokinase : The former name of enteropeptidase (an enzyme).

Enterorrhagia : Bleeding within the intestinal tract.

Epistasis : Bleeding from the nose, caused by physical injury or associated with fever, high blood pressure or blood disorder.

Epithelioma : A tumour of epithelium (a covering of internal and external surfaces of the body): a former term for carcinoma but now also used to describe benign tumours.

Erectile dysfunction : Abnormal functioning of erectile tissue.

Eruptions : Any lesion that appears on the surface of the skin and is characterised by its prominence and redness.

Erysipelas : An infection of the skin and underlying tissues by the bacterium *Streptococcus pyogenes*.

Erythema : A normal flushing of the skin caused by dilation of blood capillaries.

Erythropoietic porphyria : A condition where the skin is photosensitive due to disturbed porphyrin metabolism resulting in itching, burning and swelling of the skin.

Estrogenic : One of a group of steroid hormones (including estriol, estrone and estradiol) which control female sexual development, prompting the growth and function of female sex organs and female secondary sexual characteristics.

Eugenics : The science dealing with factors which improve successive generations of the human race.

Evacuant : Emptying, purgative or cathartic.

Excoriation : The destruction and the removal of the surface of the skin or the covering of an organ by scraping or by application of the chemical.

Exhilarant : Filled with high spirits; joyful and lively.

Expectorant : A drug that enhances the secretion of sputum by the air passage.

Febrifuge : A treatment or a drug that reduces or prevents fever.

Fibrinolysis/ Fibrinolytic : The dissolution of fibrin in blood clots (fibrin is the matrix on which blood clot is formed).

Fibroids : A benign tumour of fibrous and muscular tissues, one or more of which may develop in the muscular wall of the uterus.

Fibrosarcoma : A malignant tumour of the fibroblastic cells.

Fibrosis : Formation of excessive fibrous tissue in a organ.

Fibrositis : Pain and stiffness of uncertain origin affecting soft tissues of limbs and trunk.

Fibrous histiocytoma : Benign tumour of histiocytes.

Filariasis : Disease caused by the presence of filariae in the tissues of the body, often resulting in occlusion of the lymphatic channels.

Fistula : An abnormal passage between two organs or between an organ and the exterior.

Flatulence : The expulsion of gas or air from the stomach through the mouth by belching or from the bowel by passing gas.

Fluxes : An abnormally copious flow from an organ or cavity.

Follicular carcinoma : Malignant growth in the follicle (follicle = a small secreting sac).

Follicular lymphoma : Any of several types of non-Hodgkin's lymphoma in which the lymphomatous cells are clustered into nodules or follicles.

Freckles : A brown spot of a skin commonly found on the arms and face of young people with fair complexion.

Fungistatic : An agent that inhibits the growth of fungi.

Fungitoxic : An agent toxic to fungi.

Furunculoid sores	: Sores resembling a boil.
Galactagogue	: An agent inducing or increasing the flow of milk.
Ganglion	: Group of nerve cells from which nerve fibres radiate.
Gangrene	: Death and decay of any part of the body due to deficiency or cessation of blood supply.
Gastralgia	: Pain in the stomach.
Gastric ulcer	: An ulcer in the stomach lining caused by the action of acid, pepsin, and bile.
Gastritis	: Inflammation of the stomach.
Gastroenteritis	: Inflammation of the mucous coat of the stomach and intestine due to bacterial infection.
Gastropathy	: Any disease of the stomach.
Genotoxic	: A poisonous substance that harms an organism by affecting its DNA.
Geriatrics	: Branch of medical science dealing with old age and its diseases.
Germicide	: Destructive of micro-organisms.
Gingivitis	: Inflammation of the gum.
Glaucoma	: A condition in which loss of vision occurs because of an abnormally high pressure in the eye.
Glycosuria	: Presence of sugar in the urine.
Glycosylation	: The process of adding sugar units to protein.
Goitre	: Enlargement of the thyroid gland causing a swelling in the front part of the neck.
Gonadotrophic	: Having an affinity for or influence on the gonads.
Gonorrhea/Gonorrhoea	: An infectious disease of venereal origin caused by bacterium *Neisseria gonorrhoeae*, affecting the genital mucous membranes of either sex.
Gout	: A form of metabolic disorder in which sodium biurate is deposited in the cartilages of the joints.
Granular lids	: Small rounded out growth, made up of small blood vessels and connective tissue on the eyelid.
Gravel	: A sandy deposit, present in the bladder, which may pass with the urine.
Gripe	: A sharp pain in the stomach.
Haemangioendothelioma	: A tumour of endothelial cells lining the inner surface of blood vessels.
Haematemesis	: Vomiting of blood.
Haematinic	: Pubstance that is required for the production of the red blood cells and its constituents.
Haematuria	: Presence of blood in the urine.
Haemoglobinuria	: Presence of free haemoglobin in the urine.
Haemolytic	: Causing destruction of red blood cells.
Haemophilia	: Hereditary disorder in which blood clots very slowly due to deficiency of one of the coagulation factor.
Haemopoietic progenitor	: Stem cell within the bone marrow, responsible for formation of cellular elements in the blood.
Haemoptysis	: Spitting of the blood.
Haemorrhage	: The escape of blood from a vessel.
Haemorrhoids	: Enlarged veins in the wall /around the anus usually a consequence of prolonged constipation or occasionally diarrhoea.
Haemorrhoidal diathesis	: Condition where in tissue of the haemorrhoids reacts to stimuli.

Haemostat/ Haemostatic : Arrest of bleeding by the physiological properties of vasoconstriction and coagulation or by artificial means.

Hallucinogenic : Agent that cause false perception without a sensory stimulus.

Helminthiasis : The condition resulting from infestation with worms.

Hemicrania : A headache affecting only one side of the head as in migraine.

Hemiplegia : Paralysis of one side of the body due to a cerebrovascular accident on the opposite side.

Hepatic disorders : Disorder of the liver.

Hepatic neoplasm : Growth of a tumour in the liver.

Hepatitis : Inflammation of the liver due to virus infection or such disease as amoebic dysentery and lupus.

Hepatoblastoma : Malignancy in the abdomen.

Hepatocellular carcinoma : Malignant growth of liver cells.

Hepatoma : Primary carcinoma of the liver.

Hepatomegaly : Enlargement of the liver to such an extent that it can be felt below the rib margin.

Hepatopathy : Relating to diseases of liver.

Hepatoprotective : An agent or drug that protects the liver.

Hepatosplenomegaly : Enlargement of liver and spleen.

Hepatotonic : A liver tonic.

Hepatotoxic : Damaging or destroying liver cells.

Hernia : The protrusion of an organ or tissue out of the body cavity in which it normally lies.

Herpes : Any inflammatory skin disease marked by the formation of small vesicles in cluster; the term is usually restricted to such diseases caused by Herpes virus and is used alone to refer to *Herpes simplex* or to *Herpes zoster.*

Hiccup/Hiccough : An involuntary spasm of the respiratory organ ending in closure of the glottis and production of sound.

Histiocytes : Cells derived from reticuloendothelium which acts as scavengers.

Histiocytoma : Tumour containing histiocytes.

Hydrocele : The accumulation of watery liquid in a sac, usually the sac surrounding the testes.

Hydrogogue : Producing watery discharge especially from the bowels.

Hydrophobia : Aversion to water.

Hyperacidity : Excessive acidity.

Hyperaemia : Excess of blood in an organ of body part

Hyperalbuminuria : Excessive albumin.

Hyperbilirubinemia : Excessive bilirubin in the blood, found in physiological jaundice in newborn.

Hypercholesterolemia : Excessive reduction of the enzyme cholinesterase.

Hyperdipsia : Intense thirst.

Hyperglycaemia : Excess of glucose in bloodstream.

Hyperlipaemia/-lipidaemia : Excessive fat in the blood.

Hyperlipoproteinaemia : Higher level of lipid and proteins.

Hyperprolactinaemia : Excessive production of the hormone prolactin.

Hypertensive/ Hypertension : Persistently high arterial blood pressure.

Hyperthermia	: Exceptionally high blood temperature about 41°C or above.
Hyperthyroid	: Over activity of the thyroid gland.
Hypertriglyceridemia	: Excessive triglyceride level in the blood.
Hypertrophy	: Increase in the size of tissue or structure, independent of natural growth.
Hypnotic	: A state resembling sleep brought about by utilizing the mental mechanism of suggestion.
Hypocholesterolemic	: Deficiency in production of the enzyme cholinesterase.
Hypoglycaemia	: Reduced glucose level in blood stream.
Hypokalaemic/Hypokalaemia	: Abnormally low potassium level of the blood.
Hypolipidaemic	: Deficiency of lipids in the body.
Hypotensive	: Low blood pressure.
Hypothermia	: Greatly decreased temperature.
Hypoxia	: Reduction of oxygen level to the tissues.
Hysteria	: A psychoneurosis with symptoms based on convulsion characterized by lack of control over acts and emotions by morbid self consciousness, by anxiety, exaggeration of the effect of sensory impressions etc.
Immunomodulator	: An agent that modulates the immune system.
Immunostimulant	: An agent which stimulates immune reaction.
Immunosuppressant	: An agent which prevents the occurrence of an immune reaction.
Impetigo	: A streptococcal or staphylococcal skin infection marked by vesicles or bullae that become pustular, rupture and form yellow crusts.
Impotency	: Inability in a man to have sexual intercourse.
Indolent	: Describing a diseases process that fails to heal or has persisted.
Induration	: The hardening of tissue.
Infantile diarrhoea	: Diarrhoea of a child or baby less than one year old.
Infantile paralysis	: An epidemic viral infection of the motor neurons of the interior horns in brain stem and spinal cord in children.
Insanity	: A degree of mental illness such that the affected individual is not responsible for his actions.
Insecticide	: A preparation used to kill destructive or disease carrying insects.
Insomnia	: Inability to fall asleep or to remain asleep for an adequate length of time.
Insulinogenic	: A agent which enhances the production of insulin.
Intellect promoting	: That which promotes thinking faculty or reasoning power.
Intermittent fevers	: A fever that rises subsides and returns again.
Invigoration	: To impart vigour, strength or vitality.
Ischaemia	: Deficient blood supply to the heart.
Jaundice	: A condition characterized by a raised bilirubin level in the blood (hyperbilirubinaemia).
Kapha	: Phlegm.
Kaposi's sarcoma	: A cancer that causes patches of abnormal tissue to grow under the skin, in the lining of the mouth, nose and throat.
Keloid	: An over growth of scar tissue, which may produce a contraction deformity.
Keratitis	: Inflammation of the cornea of the eye.
Keratolytic	: Having the property of breaking down keratinized epidermis.

Klinfelter's syndrome : A human condition with 44 autosomes and XXY sex chromosomes making a total complement of 47; the individual though a male, has large breasts and atrophied testes and is sterile; a pragmatic male but a genetic female.

Lactagogue : An agent that promotes secretion of milk by the mammary glands of the breasts, which usually being at the end of the pregnancy.

Larvicide : An agent that destroys larvae.

Laryngitis : Inflammation of the larynx and vocal cords, due to bacteria or viruses or irritation by gases, chemicals, etc.

Laryngopathy : Relating to diseases of larynx.

Laxative : One promoting bowels evacuation by increasing faecal volume.

Leishmaniasis : A disease caused by a protozoa *Leishmania donovani*.

Leucoderma : An acquired condition with localized loss of pigmentation of the skin.

Leucorrhoea : An abnormal mucous discharge from the vagina.

Leukaemia/leukemia : A malignant proliferation of the leucopoietic tissues usually producing an abnormal increase in leukocytes in the blood, with immature cells among them (blood cancer).

Leukopenia : A reduction in the number of white blood cells in the blood.

Leukoplakia : White thickened patch occuring on mucous membranes of lips, mouth and genitalia due to syphilis or pre-cancerous changes.

Ligature : Material used for tying blood vessels or sewing the tissues.

Liniment : A medicinal preparation that is rubbed onto the skin or applied on a surgical dressing.

Lithiasis : Formation of calculi.

Lithnotriptic/Lithotriptic : Dissolving vesicular calculi.

Lumbago : Incapacitating pain low down in the back; a symptom of fibrositis of lumbar muscles, spondylitis and prolapsed intervertebral disc.

Lupus : Any of several chronic skin diseases.

Lymphoblastic leukaemia : Malignant lymphoma in which single or multiple tumours arise from lymphoblasts in lymph nodes.

Lymphoid leukaemia : A malignant proliferation of leucocytes in the lymph.

Lymphoma : A malignant tumour of the lymphatic tissue.

Mania : Denoting obsession, compulsion or exaggerated feeling.

Mastalgia : Pain in the breast.

Mastication : The act of chewing.

Mastitis : Inflammation of the breast, usually caused by bacterial infection through damaged nipples.

Maturant : To bring to maturity.

Measles : Highly infectious viral disease that mainly affects children.

Melancholia : Term in psychiatry to mean severe forms of depression.

Melanoma : A malignant condition of plasma cells, usually in the bone marrow.

Meningitis : An inflammation of the meninges due to infection by viruses or by the bacteria responsible for pneumonia, syphilis or tuberculosis.

Menorrhagia : Painful menstruation.

Menstruation : The discharge of blood and fragments of womb lining from the vagina at intervals of about one month in women of child bearing age.

Metastatic/Metastasis : Transference of disease from one part of the body to another.
Metropathy : Any uterine disease.
Metroptosis : The downward displacement of the uterus.
Metrorrhagia : Bleeding from the womb when menstruation is not due.
Micturition : The act of passing urine.
Migraine : A recurrent throbbing head ache that characteristically affects one side of the head.
Mitogenic : Any substance that induces mitosis.
Molluscicidal : An agent or drug used to kill molluscs.
Monoclonal antibodies : Identical antibodies
Mumps : A common viral infection mainly affecting children.
Muscular atrophy : Diminution in size and function of the muscle.
Mutagenic : An agent that induces mutation.
Myalgia : Pain in the muscles.
Mycosis : Disease caused by any fungus.
Mydriatic : Drug which causes dilation of the pupil.
Myeloma : Malignant condition of the plasma cells in the bone marrow.
Myelosis : Disease of the spinal cord.
Myocardia : Middle layer of the heart wall.
Myocarditis : Inflammation of the myocardium.
Myopia : Short sightedness.
Myotic : Excessive contraction of the pupil of the eye.
Myasthenia gravis : Muscular weakness of voluntary muscle of the eye.
Myxoma : Tumour of the connective tissue.
Narcotic : A drug that produces abnormally deep sleep.
Nasal catarrh : Inflammation of the nose.
Nematicide : Agent used to destroy nematodes.
Neoplasm/neoplasia : Any new growth; refers to pathological process in tumour formation.
Nephritis : Inflammation of the kidney.
Nephropathy : Diseases of the kidney.
Nephrosis : Degenerative, non inflammatory change in the kidney.
Nephrotoxic : Any substance that prevents or inhibits the functions of kidney cells or cause their destruction.
Neuralgia : Pain extending along the course of one or more nerves.
Neurasthenia : Nervous condition consisting of lassitude, inertia, fatigue, loss of initiative, restless fidgeting, over-sensitivity, irritability and an asthenic physique.
Neuroblastoma : Malignant tumour of adrenal medulla from tissue of sympathetic origin.
Neurolathyrism : A disease of the nerves.
Neuroparalytic : Any agent that paralyze the neurons of the nervous system.
Neuropathy : Relating to disease of the nervous system.
Neurosis : Functional disorder consisting anxiety, depression and hysteria arising as a result of anxiety and stress.
Nocturnal emissions : Emission of semen during the night.

Nyctalopia : Night blindness due to vitamin A deficiency.

Nymphomania : An extreme degree of sexual promiscuity in a female.

Odontalgia : Tooth ache.

Odontopathy : Any diseases of the teeth.

Oligodendroglioma : Neuro ectodermal tumour of the neuroglial cell of central nervous system.

Opthalmic/Opthalmia : Pertaining to the eye.

Opthalmopathy : Relating to disease of the eye.

Oral ulcers : Ulcer of the mouth.

Orchitis : Inflammation of the testis.

Osteoarthritis : Degenerative arthritis (affecting the synovial joints).

Osteogenic : Agent that helps in bone formation.

Osteolathyrism : A disease of the bones.

Osteoporosis : A disease in which the bones become extremely porous due to loss of bone density.

Osteosarcoma : Malignant tumour of the bone.

Otalgia : Pain in the ear.

Otitis : Inflammation of the ear.

Otorrhoea : A discharge from the external auditory meatus.

Ovaritis : Inflammation of the ovary.

Oxytocic : A drug that induces or accelerates labor by stimulating the muscles of womb to contract.

Palliative : Anything which serves to alleviate but cannot cure a disease.

Palsy : Paralysis with involuntary tremors.

Paludism : A type of malarial disease.

Pancreatitis : Inflammation of the pancreas.

Papillary carcinoma : Malignant growth of a papilla.

Papillary meningioma : Benign growth in the papillary meninges.

Papilloma : A tumour arising from a nonglandular epithelial surface.

Parapsychology : The study of phenomena such as telepathy and clairvoyance.

Paraplegia : Paralysis of the lower limbs, bladder and rectum.

Parasiticide : An agent that destroys parasites.

Parkinsonism : A disorder of middle-aged and elderly people characterized by tremor, rigidity, and a poverty of spontaneous movements.

Parotitis : Inflammation of the parotid salivary glands.

Parturition : The act or process of giving birth to a child.

Patau's syndrome : A human condition of trisomy of chromosome number 13, associated with mental abnormality making for 47 chromosomes.

Pectoral : Of or pertaining to the chest or breast, relieving disorders of the respiratory tract, as an expectorant.

Pectoralgia : Pain in the chest.

Pediculosis : Infestation with lice.

Pericarditis : Acute or chronic inflammation of the membranous sac (pericardium) surrounding the heart.

Periodontal : Treatment of tissues associated with teeth and gum diseases.

Pertussis : Whooping cough.

Phagocytosis : The engulfment and digestion of bacteria and other foreign particles by a cell.

Phantom tumours : Tumour of the abdomen.

Pharmacognosy : The study of drugs of plant and animal origin.

Pharmacoepia : A book or list of drugs with directions for their preparation.

Pharyngitis : Inflammation of the part of the throat behind the soft palate.

Pharyngodynia : Pain in the pharynx.

Pharyngopathy : Relating to the disease of the pharynx.

Phlegmatic : Emotionally stable; not easily excited.

Phlogistic : Of, relating to, or inducing inflammation.

Photophobia : An abnormal intolerance of light in which exposure to light produces intense discomfort of the eyes with tight contraction of the eye lids and other reactions aimed at avoiding the light.

Photosensitization : Sensitive to light.

Phthisis : Wasting of the body.

Piles : Varicosity of the veins around the anus.

Piscicide : An agent which kills fishes.

Pitta : One of the three bioentities recognized as syndrome and responsible for digestion, body warmth, thermostatic balance, appetite, thirst, taste, complexion, body texture and higher mental functions.

Pityriasis : Any of a group of skin diseases typified by the development of fine bran like scales.

Pleurisy : Inflammation of the pleura (serous membrane investing the lungs and lining the walls of the thoracic cavity).

Pneumonia : Inflammation of the lungs caused by bacterial infection.

Pneumonitis : Inflammation of lung tissue.

Polio : Infantile paralysis; an epidemic viral infection that attacks the motor neurons of the anterior horns of the brain stem and spinal cord, in children.

Polyneuritis : Any disorder involving all the peripheral nerves.

Polyp : A pedunculated tumour arising from any mucous surface, such as cervical, uterine, nasal, etc., usually benign but may become malignant.

Polyuria : The production of large volume of urine, which is dilute and of a pale colour.

Post partum : After the birth.

Post prandial : Following a meal.

Proctoptosis : Prolapse of the rectum.

Prophylactic : Pertaining to the prevention of the development of diseases.

Prostatic hypertrophy : The hypertrophy of prostate glands.

Prostatic neoplasia : Cancer of prostate gland.

Prurigo : A chronic itchy skin disease of unknown cause.

Pruritus : Itching, caused by local irritation of the skin or sometimes nervous disorders.

Psoriasis : A chronic skin disease in which itchy scaly red patches formed on the elbows, forearms, knees, legs, scalp and other parts of the body.

Psychosis : A severe mental illness in which the sufferer looses contact with reality.

Psychotropic : That which exerts specific effect upon the brain cells.

Pterygium : A triangular overgrowth of the cornea, usually on the inner side, by thickened and degenerative conjunctiva.

Ptosis : Drooping of the eyelid.

Ptyalism : The excessive production of saliva; a symptom of certain nervous disorders, poisoning (by mercury, mushrooms or organophosphates), or infection.

Pulmonary tuberculosis : Any infectious disease caused by the bacillus *Mycobacterium tuberculosis* and characterised by the formation of nodular lesions (tubercles) in the lung tissues.

Purgative : A drug causing evacuation of fluid faeces.

Purpura : A disorder characterized by spontaneous escape of blood from the capillaries into the skin, manifested by red spots or oozing.

Purulent : Consisting of pus.

Pustules : A small blister on the skin containing pus.

Pyelitis : Inflammation of the pelvis or the kidney.

Pyorrhoea : A flow of pus, usually referring to that from teeth sockets.

Pyrexia : Rise in the body temperature above normal.

Pyrosis : Heart burn.

Quartan fever : The term applied to intermittent fever with paroxysms (sudden, temporary attack) occurring every 72 hours.

Radicular : From the root of any source.

Radioprotective : An agent that protects against radiation.

Rectalgia : Pain in the rectal region.

Rectitis : Inflammation of the rectum.

Refrigerant : An agent that relieving fever and thirst.

Refsum's disease : A genetic disorder of fatty acid accumulation; a progressive disease.

Regurgitation : The bringing up of undigested material from the stomach to the mouth.

Rejuvenating : An agent that restores youth full appearance and vigour.

Renal calculi : Stones in the kidney.

Renal cell carcinoma : Malignant growth of renal cell.

Renal colic : Spasm of ureter due to stones, resulting in pain.

Resolvent : Promoting resolution or the dissipation of a pathologic growth.

Restorative : Having the power to restore and renew health.

Retinopathy : Any non inflammatory disease of the retina.

Rheumatalgia : Pain of muscles and joints.

Rheumatism : Disorder or diseases of connective tissue associated with pain, stiffness and swelling of muscles and joints.

Rheumatoid arthritis : A form of arthritis that is the second most common rheumatic disease, after osteoarthirtis.

Rhinitis : Inflammation of the mucous membrane of the nose.

Ringworm : A fungal skin infection spreading in a circular manner.

Roachicide : An agent that kills cockroaches.

Roborant : A strengthening agent.

Rubefacient : An agent that causes reddening or warming of the skin.

Salve	: An ointment.
Sarcoma	: Malignant growth of mesodermal tissue (connective tissue, muscle and bone).
Scabies	: A parasitic contagious skin diseases caused by itch mite.
Scalding of urine	: Burn produced by urine.
Schizophrenia	: Mental illness characterized by disorganization of a patient's personality.
Sciatica	: Pain in the line of distribution of the sciatic nerve in the leg.
Scirrhus/Cirrhous	: A carcinoma that provokes a considerable growth of hard, connective tissue.
Scleroderma	: Localized edema of the skin, followed by hardening, atrophy, deformity and ulceration.
Sclerosis	: Abnormal hardening or fibrosis of a tissue.
Scrofula	: Tuberculosis of lymph nodes, usually those in the neck, causing the formation of abscesses.
Scurvy	: A disease caused by deficiency of vitamin C (ascorbic acid) .
Sedative	: An agent that gives a calming effect.
Semen coagulant	: An agent that coagulates semen.
Semenoma	: A malignant tumour of the testes.
Septicaemia	: The persistence and multiplication of living bacteria in the blood stream.
Sialogogue	: An agent that increases the flow of saliva.
Sinusitis	: Inflammation of the sinuses.
Soporific	: An agent that induces profound sleep.
Spasmolytic	: A drug that relieves spasm of smooth muscles.
Spasms	: Convulsive, involuntary muscular contraction.
Spermatogenesis	: The process by which mature spermatozoa are produced in the testes.
Spermatorrhoea	: The involuntary discharge of semen without orgasm.
Spermicidal	: An agent that kills spermatozoa.
Spinal sclerosis	: Degenerative changes of nerve sheaths in the spinal cord.
Splenitis	: Inflammation of the spleen.
Splenomegaly	: Enlargement of the spleen.
Splenopathy	: Any disease of the spleen.
Sprue	: Deficient absorption of food due to disease of the small intestine.
Stomachalgia	: Pain in the stomach.
Stomachic	: Agent that strengthens or stimulates the stomach.
Stomatitis	: Inflammation of the mouth.
Stomatopathy	: Any disorder of the stomach.
Strangury	: Slow and painful urine flow.
Stroke	: A sudden attack of weakness or paralysis, usually affecting one side of the body.
Stupor	: A state of marked impairment of, but not complete loss of consciousness.
Stuttering	: Repeating consonants.
Stye	: An abscess in the follicle of eye lash.
Styptic	: An astringent applied to stop bleeding.
Sudorific	: An agent that causes an increase in sweating.
Suppositories	: Medicament that is inserted inside an orifice which melts at body temperature.

Suppuration	: The formation of pus.
Synovia	: The fluid secreted by the membrane lining the joint cavity.
Syphilis	: A chronic venereal disease caused by the *Treponema pallidum*, resulting in the formation of lesions throughout the body.
Tachycardia	: Excessively rapid action of the heart.
Tactile	: Pertaining to the sense of touch.
Taenifuge	: An agent that expels tape worms.
Teratogenic	: Any substances, agent or process that induces the formation of developmental abnormalities in a foetus.
Tetanus	: An acute infectious disease, affecting the nervous system, caused by the bacterium *Clostridium tetani,* in which muscles contract and stiffen.
Thalassaemia	: Genetically transmitted haemoglobin abnormality.
Thermogenic	: The production of heat.
Thomson-Friedenreich antigen	: Is a cryptic disaccharide structure on human erythrocytes and expressed on carcinoma cell.
Thrombosis	: Intravascular formation of a blood clot.
Thrush	: Disease caused by infection with species of candida.
Tourniquet	: An apparatus for the compression of the blood vessels of a limb.
Toxaemia	: A generalized poisoning of the body due to bacteria or damaged tissue.
Toxicosis	: The deleterious effects of a toxin; poisoning; includes any disease.
Trachoma	: Contagious inflammation affecting conjunctiva, cornea and eye lids due to infection by *Bedsonia*.
Transitional cell tumour	: Benign growth in the lining of the renal pelvis.
Tremor	: A rhythmic alternating movement that may affect any part of the body.
Trichogenous	: Promoter of hair growth.
Tridosha	: Disorders or disturbance of *Vata*, *Pitta* and *Kapha*.
Tubercular glands	: Having small rounded swellings or nodules not necessarily caused by tuberculosis.
Tuberculosis	: A specific, infective disease of the lungs caused by *Mycobacterium tuberculosis.*
Tubular necrosis	: Death of tubular cells.
Tumours	: A new growth of tissue due to uncontrolled and progressive multiplication of cells.
Turner's syndrome	: A human condition with 45 chromosomes, with one sex chromosome missing; a genetic male with characters of female such as genitalia, atrophic ovaries and webbed neck.
Tympanites	: Abdominal distension due to accumulation of gas in the intestine.
Tymphanitis/ Tymphinitis	: Inflammation of the tymphanum (the cavity of the middle ear).
Ulcer	: A local defect or excavation of the surface of an organ, produced by sloughing of necrotic inflammatory tissue.
Unguents	: Ointment
Urecemia	: An excess of uric acid in the blood.
Urethritis	: Inflammation of the urethra.
Urethrorrhagia	: Flow of blood from the urethra.
Urethrorrhoea	: Discharge from the urethra.

Uropathy : Any diseases of the urinary tract.

Urorrhagia : Excessive secretion of urine.

Urorrhoea : Involuntary flow of urine.

Urothelia : Tumour of the urethra.

Urticaria : A vascular reaction of the upper epidermis marked by transient appearance of slightly elevated patches which are redder than the surrounding skin and often attended by severe itching.

Uteritis : Inflammation of the uterus.

Uterorelaxant : An agent that relaxes the muscles of the uterus.

Vaginitis : Inflammation of the vagina characterized by intensely irritating discharge due to ciliated protozoans.

Vaginopathy : Any diseases of the vagina.

Varicose veins : Dilated veins, the valves of which become incompetent so that the blood flow may be reversed; most commonly found in the lower limbs.

Vascular tumour : Benign growth in the circulatory system.

Vasoconstrictor : An agent that causes a narrowing of the lumen of blood vessels.

Vasodilator : An agent that causes a widening of the lumen of blood vessels.

Vasomotor depressant : An agent that depresses the vasomotor system.

Vata : One of the three bioentities recognized in Ayurveda and responsible for respiration, locomotion, excretion, circulation and suppleness of the sensory organs.

Venereal buboes : A venereal infectious disease transmitted by sexual intercourse.

Venereal sores : Sores of venereal origin.

Veno-occlusive disease : Disease due to occluded hepatic veins.

Vermifuge : Any drug or chemical agent used to expel worms from the intestine.

Verminosis : A disorder caused by parasitic worms.

Vertigo : A sensation of rotation or movement of one's self or one's surroundings in any plane sometimes used erroneously to mean any form of dizziness.

Vesicular calculi : Accumulation of calcium in vesicles.

Vesicant : A blistering substance.

Visceromegaly : Abnormal enlargement of the viscera.

Vitiligo : Leucoderma.

Vulnerary/Vulnerant : Drugs used for healing wounds.

Warts : A small (often hard) benign growth in the skin.

Whitlow : A purulent infection involving the pulp of the distal phalange of the finger.

Whooping cough : An acute contagious disease, primarily affecting children, due to infection of the mucous membranes lining the air passage by the bacterium *Hemophaelus pertusis.*

Yaws : A tropical disease resembling syphilis caused by spirocheates.

Yellow fever : An acute febrile illness of tropics caused by group B Arbo virus spread by mosquitoes.

Zellweger's syndrome : A rare autosomal recessive congenital syndrome of multiple abnormalities.

Bibliography

The information provided in the Species Profiles was sourced from the publications in this bibliography, but were not individually cited in the profiles. The literature cited in the text of the Chapters is included here.

Abbiw, D.K.1990. *Useful Plants of Ghana*. Intermediate Technology Publications Ltd., and The Royal Botanic Gardens, Kew, U.K.

Adamo, P.D. 1998. *The Eat Right Diet.* Century Books Ltd.

Adams, S. 2000. *Legumes: Medicine Chest of the Future*. NH News, USDA. Agricultural Research Services, Washington.

Adiga, P.R., Padmanabhan, G., Rao, S.L.N. and Sarma, P.S. 1962. The isolation of a toxic principle from *Lathyrus sativus* seeds. *J. Sci. Ind. Res.* C., 21: 284.

Agarwal, V.S. 1970. *Wood Yielding Plants of India*. Indian Museum, Calcutta, India.

Agarwal, V.S. and Ghosh, B.1985. *Drug Plants of India (Root drugs)*. Kalyani Publishers, New Delhi, India.

Ahuja, B.S. 1965. *Medicinal Plants of Saharanpur*. Central Council of Ayurvedic Research, Gurukula Kangri Vishwavidyalaya, Haridwar, India.

Aida, K., Tawata, M., Shindo, H., Onaya, T., Sasaki, H., Nishimura, H., Chin, M. and Mitsuhashi, H. 1989. The existence of aldose reductase inhibitors in some kampo medicines (oriental herbal prescriptions). *Planta Medica,* 55: 22-26.

Aida, K., Tawata, M., Shindo, H., Onaya, T., Sasaki, H., Yamaguchi, T., Chin, M. and Mitsuhashi, H. 1990. Isoliquiritigenin: a new aldose reductase inhibitor from *Glycyrrhizae radix. Planta Medica*, 56: 254-258.

Airy Shaw, H.K. (Ed.) 1973. Willis, J.C., *A dictionary of flowering plants and ferns.* 8th Edition., Cambridge University Press, London, U.K.

Akamatsu, H., Komura, H., Asada, Y. and Niwa, Y. 1991. Mechanism of anti-inflammatory action of glycyrrhizin: effect on neutrophil functions including reactive oxygen species generation. *Planta Medica,* 57: 119-121.

Alam, M. K. 1992. Medical ethnobotany of the Marma tribe of Bangladesh. *Economic Botany*, 46: 293-298.

Alcaraz, J.M and Jimenez, H.J. 1988. Protective effects of hypolactin-8-glucoside on the rat gastric mucosa. In: *Plant Flavonoids in Biology and Medicine II: Biochemical, Cellular and Medicinal Properties.* (Eds.) Cody, V., Middleton, E., Jr., Jeffrey, B., Harborne, J.B. and Beretz, A., Alan, R. Liss, Inc., New York, U.S.A.

Ali, L., Khan, A.K.A., Hassan, Z., Mosihuzzaman, M., Nahar, N., Nasreen, T., Nur-e-Alam, M. and Rokeya, B. 1995. Characterization of the hypoglycemic effects of *Trigonella foenum-graecum* seed extracts on normal and diabetic model rats. *Planta Medica*, 61: 358-360.

Ali, M. 1994. *Textbook of Pharmacognosy*. CBS Publishers & Distributors, Delhi, India.

*Alroy, J., Ucci, A. A. and Pereira, M. E. A. 1984. Lectins - histochemical probes for specific carbohydrate residues. In: *Diagnostic Immunochemistry*, Vol 2. (Ed.) De Lellis, R. A., Masson Inc, New York, U.S.A.

Altern, J. 1997. A comparative survey of leguminous plants as sources of the isoflavones, genistein and daidzein-implications for human nutrition and health. *Complement. Med.,* 208: 7-12.

*Alzueta, C., Trevino, J. and Ortiz, L. 1992. *Journal of the Science of Food and Agriculture*, 59: 551-553.

Amagaya, S., Sugishita, E., Ogihara, Y., Ogawa, S., Okada, K. and Aizawa, T. 1984. Comparative studies of the stereoisomers of glycyrrhetinic acid on anti-inflammatory activities. *J. Pharm. Dyn.,* 7: 923-928.

Ambasta, S.P. 1986. *The Useful Plants of India*. Publication & Information Directorate, CSIR, New Delhi, India.

Anonymous, 1948. *The Wealth of India - Raw Materials*. Vol. I. Publication & Information Directorate, CSIR, New Delhi, India.

Anonymous, 1950. *The Wealth of India - Raw Materials*. Vol. II. Publication & Information Directorate, CSIR, New Delhi, India.

Anonymous, 1952. *The Wealth of India - Raw Materials*. Vol. III. Publication & Information Directorate, CSIR, New Delhi, India.

Anonymous, 1956. *The Wealth of India - Raw Materials*. Vol. IV. Publication & Information Directorate, CSIR, New Delhi, India.

Anonymous, 1959. *The Wealth of India - Raw Materials*. Vol. V Publication & Information Directorate, CSIR, New Delhi, India.

Anonymous, 1962. *The Wealth of India - Raw Materials*. Vol. VI. Publication & Information Directorate, CSIR, New Delhi, India.

Anonymous, 1966. *The Wealth of India - Raw Materials*. Vol. VII. Publication & Information Directorate, CSIR, New Delhi, India.

Anonymous, 1969. *The Wealth of India - Raw Materials*. Vol. VIII. Publication & Information Directorate, CSIR, New Delhi, India.

Anonymous, 1971. *Chaemotaxonomy of the Leguminosae*. Academic Press, London and New York.

Anonymous, 1972a. *Formulary of Siddha Medicines*. The Indian Medical Practitioners Co-operative Pharmacy and Stores Limited, (IMPCOPS), Madras, India.

Anonymous, 1972b. *The Wealth of India - Raw Materials*. Vol. IX. Publication & Information Directorate, CSIR, New Delhi, India.

Anonymous, 1975a. *Nutritional Improvement of Food Legumes by Breeding*. A Wiley Interscience Publication, New York, London.

Anonymous, 1975b. Research Group Shan Dong Province. *Res. Chin. Mat. Med.*, 6:95.

Anonymous, 1976a. *British Herbal Pharmacopoeia*. Part 1. British Herbal Medicine Association, West Yorks.

Anonymous, 1976b. *The Wealth of India - Raw Materials*. Vol. XI. Publication and Information Directorate, CSIR, New Delhi, India.

Anonymous, 1976c. *Medicinal plants of India*. Vol 1. Indian Council for Medicinal Plant Research (ICMR), New Delhi, India.

Anonymous, 1976d. *Medicinal plants of Nepal*. Ministry of forests, Department of Medicinal Plants, Thapathi, Katmandu, Nepal.

Anonymous, 1976e. *The Wealth of India - Raw Materials*. Vol. X. Publication & Information Directorate, CSIR, New Delhi, India.

Anonymous, 1976f. *The Wealth of India - Raw Materials*. Vol. XI. Publication & Information Directorate, CSIR, New Delhi, India.

Anonymous, 1977a. Health *Plants of the World. Atlas of Medicinal Plants.* Newsweek Books, New York.

*Anonymous, 1977b. Anticancer Res. Group, Nun Ji Li, Shan Dong Province. *Zhong Cao Tong Sin.,* 11:5.

Anonymous, 1977c. Herb teas: how safe? *Consumers Res. Mag.*, 3:35.

Anonymous, 1978. *The Ayurvedic Formulary of India*. Government of India, Ministry of Health & Family Welfare, India.

Anonymous, 1979a. *Tropical Legumes: Resources for the Future*. National Academy of Sciences, Washington, D.C., USA.

Anonymous, 1979b. *British Herbal Pharmacopoeia*. Part II. British Herbal Medicine Association, West York.

Anonymous, 1980. Effect of legume seeds on serum cholesterol. *Nutrition Reviews,* 38:159.

Anonymous, 1981. *Medicinal Plants of India*, Vol.2, ICMR, New Delhi.

Anonymous, 1983. *Glossary of Medicinal Plants of Karnataka*. Panchabhootha Bhagirathi Association, Gangenahalli, Bangalore, India.

Anonymous, 1988a. *Plant Flavonoids in Biology and Medicine II: Biochemical, Cellular and Medicinal Properties*. (Eds.) Cody, V., Middleton, E, Jr., Jeffrey, B., Harborne, J.B. and Beretz, A., Alan. R. Liss, Inc., New York, U.S.A.

Anonymous, 1988b. *Observations on Medico-Botany of Andaman-Nicobar Islands*. Central Council for Research in Ayurveda & Siddha, Government of India, New Delhi, India.

Anonymous, 1988c. *The Wealth of India - Raw Materials*. Vol. XII. Publication & Information Directorate, CSIR, New Delhi, India.

Anonymous, 1989. *Formulary of Siddha Medicines.* The Indian Medical Practioners Co-operative Pharmacy and Stores Ltd., Chennai, India.

Anonymous, 1991. *Uses of Tropical Grain Legumes*. ICRISAT., India.

Anonymous, 1992a. *Contribution to the Unani Medicinal Plants from North Arcot District of Tamilnadu*. Central Council for Research in Unani Medicine, Government of India, New Delhi, India.

Anonymous. 1992b. *Standardisation of Single Drugs of Unani Medicine*. Part I. Central Council for Research in Unani Medicine. Ministry of Health and Family Welfare, Government of India, New Delhi, India.

Anonymous, 1992c. *Standardisation of Single Drugs of Unani Medicine*. Part II. Central Council for Research in Unani Medicine. Ministry of Health and Family Welfare, Government of India, New Delhi, India.

Anonymous, 1995. *The Quintessene Tantras of Tibetaan Medicine*. Snow lion publication, Ithara, New York, USA.

*Anonymous, 1997. Anticancer Research Group, Nun ji L, Shan Dong Province. *Zhong Cao Tong Sin*, 11:5

Anonymous, 2002. *Handbook of Agriculture.* Indian Council of Agricultural Research, New Delhi, India.

Anton, J., Alldrick., Lake, B.G., Anthony, K., Mallet and Rowland, I.R.1988. Modulation of hepatic-phase I and gut-microflora xenobiotic metabolizing enzymes by dietary flavonoids. In: *Plant Flavonoids in Biology and Medicine II: Biochemical, Cellular and Medicinal Properties.* (Eds.) Cody, V., Middleton, E, Jr., Jeffrey, B., Harborne, J.B. and Beretz, A. Alan, R.Liss, Inc., New York, U.S.A.

Anton, R.1988. Flavonoids and Traditional Medicine In: *Plant Flavonoids in Biology and Medicine, II: Biochemical, Cellular and Medicinal Properties.* (Eds.) Cody, V., Middleton, E., Jr, Jeffrey, B., Harborne, J.B. and Beretz, A. Alan, R.Liss, Inc., New York, U.S.A.

Aragona, F., Ostardo, E., Prayer-Galetti, J., Piazza, R. and Capitanio, G. 1991. Angiosarcoma of the bladder: a case report with regard to histologic & immunohistohemical findings. *European Journal of Urology,* 20:161-163.

Arao, T., Kinjo, J., Nohara, T. and Isobe, R. 1995. Oleanene-type triterpene glycosides from *Puerariae radix.* II. Isolation of saponins and the application of tandem mass spectrometry to their structure determination. *Chem. Pharm. Bull.,* 43: 1176-1179.

*Arao, T., Kinjo, J., Nohara, T. and Isobe, R. 1997. Oleanene-type triterpene glycosides from *Puerariae radix.* IV. Six new saponins from *Pueraria lobata. Chem. Pharm. Bull.,* 45: 362-366.

Arao, T., Udayama, M., Kinjo, J. and Nohara, T. 1998. Preventive effects of saponins from *Pueraria lobata* root, *in vitro* immunological liver injury of rat primary hepatocyte culture. *Planta Medica,* 64: 413-416.

Arnold and Krochmal, C. 1984. *A Field Guide to Medicinal Plants.* Times Books, A Division of Random House Inc., New York, USA.

Arora, R.K. and Pandel, A. 1996. *Wild Edible Plants of India: Diversity, Conservation and Use.* ICAR, National Bureau of Plant Genetic Resources, New Delhi, India.

*Arshad, M., Beg, A. and Shaikh, A.R. 1965. Buffer action of saponins. 1. Analytical studies. *Pakistan J. Sci. Ind. Res.,* 8: 99-102.

Asolkar, L.V., Kakkar, K.K. and Charkre, O.J. 1992. *Second Supplement to Glossary of Indian Medicinal Plants with Active Principles.* Part 1 (A-K). CSIR, Publication & Information Directorate, New Delhi, India.

Astaldi, G., Massimo, L., Airo, R. and Mori, P.G. 1966. Phytohaemagglutinin and lymphocytes from acute lymphocytic leukaemia. *Lancet,* 4: 1265.

Atkinson, E.T. 1980. *Economic Botany of the Himalayan region.* Cosmos Publications, New Delhi, India.

Aub, J.C., Tieslau, C., and Lankester, A. 1963. Reactions of normal and tumor cell surfaces to enzymes, I. Wheat-germ lipase and associated mucopolysaccharides. *Proc. Natl. Acad. Sci.* U.S.A., 50:613-619.

Aubery, M., Font, J. and Bourrillon, R. 1972. Action of phytohaemagglutinins on normal and tumour cell growth. *Experimental cell research,* 71:59-64.

*Augustine, J., Beck, G., Kalbfleish, G. and Kagel, L.C. 1981. *Fd. Technol.*, 35:75-76.

Austin, D. F. and Bourne, G. R. 1992. Notes of Guyana's Medical Ethnobotany. *Economic Botany*, 46: 293-298.

Aykrayd, W.R. and Doughty, J. 1964. *Legumes in Human Nutrition.* FAO, Rome.

Baba, M. and Shigeta, S. 1987. Antiviral activity of glycyrrhizin against Varicella-Zoster virus *in-vitro. Antiviral Res.,* 7: 99-107.

Bacchi, E.M., Sertie, J.A.A., Villa, N. and Katz, H. 1995. *Styrax camporum* and *Caesalpinia ferrea* in chronic ulcer. *Planta Medica,* 61: 204-207.

*Bachofen-Echt, A. 1949. *"Der Bernstein und seine Einschlusse"*. Springer Verlag, Germany.

Badhwar, R.L., Rao, P.S. and Sethi, W. 1964. *Some Useful Aromatic Plants*. Forest Research Institute, Dehradun, India.

Bailey, L. H. 1949. *Manual of Cultivated Plants*. The Macmillan Company, New York, U.S.A.

Bakhru, H.K. 1997. *Herbs that Heal. Natural Remedies for Good Health.* Orient Paperbacks, New Delhi, India.

*Balasubramanian, S.C., Narasingarao, B.S., Ramanamurthy, P.S. and Shenolikar, I.S. 1962. *Indian J. Med. Res.*, 50: 779-793.

Baldev, B., Ramanujan, S. and Jain, H.K. 1988. *Pulse Crops (Grain legumes)*. Oxford and IBM Publishing Co. Pvt. Ltd., New Delhi, India.

Banchonglikitkul, C., Smart, J.D., Gibbs, R.V. and Cook, D.G. 2002. Lectins as targeting agents—the *in vitro* binding of lectins to lesions in the eye and mouth. *J. Biomed. Sci.,* 59:115-118.

*Banerjee, S., Rohtagi, K. and Lahiri, S. 1954. *Fd. Res.*, 19: 134-137.

Banks, P.M. and Mumby, W.H.A. 1963. Lightweight Concrete. *Brit. Patent,* 943: 631.

Banwell, J.G., Howard, R., Kabir, I., Adriab, T.E., Dimond, R.H. and Abramowsky, C. 1993. Small intestinal growth caused by feeding red kidney bean phytohaemagglutinin lectin to rats. *Gastroenterology,* 104: 1669 – 1677.

Baquai, F.U. 1977. *Traditional Medicine in Pakistan.* Humdard Foundation, Karachi, Pakistan.

Bardocz, S., Grant, G., Ewen, S. W. B., Duiguid, T. J., Brown, D. S., Englyst, K. and Pusztai, A. 1995. Reversible effect of phytohaemagglutinin on the growth and metabolism of rat gastrointestinal tract. *Gut,* 37: 353-360.

Barnet, R.E., Scott, R.E., Furcht, L.T. and Kersey, J.H. 1974. *Nature* (London), U.K., 249: 465.

Barret, B. 1994. Medicinal plants of Nicaragua's Atlantic coast. *Economic Botany,* 48: 8-20.

Basu, B.D. 1918. *Indian Medicinal Plants*. Indian Press Allahabad, India.

Beal, J.C. and Reinhard, E. 1981. *Natural Products as Medicinal Agents*. Hippokrates Verlag, Stuttgart, Germany.

Beer, T.W. 1992. Malignant thyroid haemangioendothelioma in a non-endemic goitrous region, with immunohistochemical evidence of a vascular region. *Histopathol.,* 20: 539-541.

*Benbessat, N. 1993. Ph.D. Thesis, Faculty of Pharmacy, HMS, Sofia, UK.

Bennett, S.S.R., Gupta, P.C. and Vijendra Rao, R. 1992. *Venerated Plants.* Indian Council of Forestry Research and Education, Dehra Dun, India.

Bentham, G. 1865. 'Leguminosae.' In *Genera Plantarum*, (Eds.) Bentham, G. and Hooker, J.D. Lovell Reeve & Co., London.

Bentley, R. and Trimen, H. 1980. *Medicinal Plants*. Vol. II. Allied Book Centre, Dehradun, India.

Beretz, A. and Cazenave, J.P. 1988. The effect of flavonoids on blood vessel wall interactions. In: *Plant Flavonoids in Biology and Medicine II: Biochemical, Cellular and Medicinal Properties.* (Eds.) Cody, V., Middleton, E. Jr., Jeffrey, B., Harborne, J.B. and Beretz, A. Alan, R.Liss, Inc., New York, U.S.A.

Berg, P.A. and Daniel, P.T. 1988. Effects of flavonoid compounds on the immune response. In: *Plant Flavonoids in Biology and Medicine II: Biochemical, Cellular and Medicinal Properties.* (Eds.) Cody, V., Middleton, E, Jr., Jeffrey, B., Harborne, J.B. and Beretz, A. Alan, R.Liss, Inc., New York, U.S.A.

Bergey, D.R., Orozco-Cardenas, M., DeMoura, D.S. and Ryan, C.A.1999. *Proceedings of the Natural Academy of Sciences.*U.S.A. 96:1756-1760.

Berry, E.W. 1937. Tertiary floras of Eastern North America. *Bot. Rev.,* 3: 31.

Bhattacharjee, S.K. 1988. *Hand Book of Medicinal Plants.* Pointer Publishers, Jaipur, India.

Bianchini, F. and Corbelta, F. 1977. *Health Plants of the World: Atlas of Medicinal Plants.* Newsweek Books, New York, U.S.A.

Bibby, M.C., Double, J.A., Philips, R.M., Loadman, P.M. and Gummer, J.A. 1988. Experimental anti-tumor effects of flavon acetic acid (LM975). In: *Plant Flavonoids in Biology and Medicine II: Biochemical, Cellular and Medicinal Properties.* (Eds.) Cody, V., Middleton, E, Jr., Jeffrey, B., Harborne, J.B. and Beretz, A. Alan, R.Liss, Inc., New York, U.S.A.

*Bidet, D., Gaignault, J.C., Girard, P. and Potier, P. 1987. *L'actualite Chimique,* pp 89-97.

Birk, Y. 1969. Saponins. In: *Toxic Constituents in Plant Food Stuffs.* (Ed.) Liener, I.E. Academic Press, New York, U.S.A., pp 169-210.

*Birkhill, I.H. 1935. *Dictionary of the Economic Products of the Malay Peninsula.* Crown agents for the colonies, London, U.K.

Bisset, N.G.1995. Arrow poisons and their role in developing medicinal agents. In: *Ethnobotany: Evolution of a Discipline.* (Eds.) Schultes, R.E. and.Von, R.S. Chapmann and Hall, London, U.K.

Biswas, K. and Chopra, R.N. 1982. *Common Medicinal Plants of Darjeeling and the Sikkim Himalayas.* Periodical Experts Book Agency, Delhi, India.

Blumler, M. A. 1991. Modelling the origins of legume domestication and cultivation. *Economic Botany,* 45: 243-250.

Boericke, W. 1991. *Homoeopathic Materia Medica.* Jain Publishers Private Ltd., Paharganj, New Delhi, India.

Bogdan, A.V. 1977. *Tropical Pasture and Fodder Plants.* Longman Group Ltd., London, U.K.

Boldt, D.H. and Banwell, G. 1985. Binding of isolectins from red kidney bean (*Phaseolus vulgaris*) to purified rat brush border membranes. *Biochem. Biophys. Acta.,* 843: 23-37.

Bonshek, R. E., Marcyniuk, B., Torgersen, E., Mann, D. M. N. and Stoddart, R. W. 1988. In: *Lectins - Biology, Biochemistry, Clinical Biochemistry,* Vol. 6. (Eds.) Freed D. and Hansen, T. C. Sigma Library, St.Louis, U.S.A.

*Booth, A.N., Robbins, D.J., Ribelin, W.E., De Eds, F., Smith, A.K. and Rackis, J.J. 1960. *Proc. Soc. Exptl. Biol. Med.,* 116:1067.

*Bos, K.D. and Jetten, J. 1989. Determination of tannin in Faba beans. In: *Recent Advances of Research in Anti-Nutritional Factors in Legume Seeds.* (Eds.) Huisman, T.F.B., van der Poel and Liener, I.E., The Netherlands. Pudoc, Wageningen.

Bose, T.K. and Som, M.G. 1986. *Vegetable Crops in India.* Naya Prokash, Calcutta, India.

Boulter, D. 1993. Insect pest control by copying nature using genetically engineered crops. *Biochemistry, 34*: 1453-1466.

Bourdillat, B., Delautier, D., Labat, C., Benveniste, J., Potier, P. and Brink, C. 1988. Mechanism of action of hispiduline, a natural flavon on human platelets. In: *Plant Flavonoids in Biology and Medicine II: Biochemical, Cellular and Medicinal Properties.* (Eds.) Cody, V., Middleton, E, Jr., Jeffrey, B., Harborne, J.B. and Beretz, A. Alan, R. Liss, Inc., New York, U.S.A., 211-214.

Bracke, M.E., Guido De Pestel., Vincent Castronova., Barabara Vyncke., Jean-Michel Foidart., Lucien, C.A., Vakaet and Mareel, M.M.1988. Flavonoids inhibit malignant tumor invasion in *in-vitro*. In: *Plant Flavonoids in Biology and Medicine II: Biochemical, Cellular and Medicinal Properties.* (Eds.) Cody, V., Middleton, E. Jr., Jeffrey, B., Harborne, J.B. and Beretz, A. Alan, R.Liss, Inc., New York, U.S.A.

Bradly, T.P. and Bonavide, B. 1980. *Journal of Immunology,* 127:208.

Brinck, U., Bosbach, R., Korabiowska, M., Schauer, A. and Gabius, H. J. 1995. Lectin binding sites in the epithelium of normal appendix vermiformis and in acute appendicitis. *Histology and Histopathology,* 10: 61-70.

Britton, G. 1983. *The Biochemistry of Natural Pigments.* Cambridge University Press, Cambridge, UK.

Brooks, S.A., Leathem, A.J.C. and Schumacher, U. (Eds.) 1997. *Lectin Histochemistry.* Bios. Scientific Publishers Limited.

Brown, J.C. and Hunt, R.C.1978. Lectins. *International Review of Cytology,* 52:277-349.

Bruhn, J.G. and Helmstead, B. 1981. Ethnopharmacology: Objectives, principles and perspectives. In: *Natural Products as Medicinal Agents.* (Eds.) Beal, J.L. and Reinhard, E. Hippokrates Verlag, Stuttgart, Germany.

Busse, W.W., Kopp, D.E. and Middleton, E. Jr.1984. Flavonoid modulation of human neutrophil function. *The Journal for Allergy and Clinical Immunology,* 73:801.

Bychkov, V. and Toto, P. D., 1987. Application of lectin binding in gynaecology, pathology. In: *Lectins - Biology, Biochemistry, Clinical Biochemistry,* Vol. 6. (Eds.) Freed, D. and Hansen, T. C., Sigma Library.

Cabrejas, M.A.A., Esteben, R.M., Waldron, K.W., Mina, G., Grant, G., Bardocz, S. and Pusztai, A. 1995. *J. Sci. Food & Agric.*, 69:429-435.

Caius, 1989. *The Medicinal and Poisonous Legumes of India.* Scientific Publishers, Jodhpur, India.

Campos, J. E., Martinez-gallardo, N., Mendiola-olaya, E. and Blanco-Labra, A. 1997. Purification and partial characterization of a protein inhibitor from tepary bean (*Phaseolus acutifolius*) seeds. *Journal of Food Biochemistry*, 21: 203-218.

Canary, J. J. 1983. Modern allopathic medicine and public health. In: *Traditional Medicine and Health Coverage.* (Eds.) Bannerman, R. H., Burton, J. and Wen-Chieh, C. WHO, Geneva.

Carvalho, M. R. B. and Sgarbieri, V. C. 1997. Heat treatment and inactivation of trypsin-chymotrypsin inhibitors and lectins from beans (*Phaseolus vulgaris*). *Journal of Food Biochemistry*, 21: 219-233.

Cassady, J.M. and Douros, J.D. (Eds.) 1980. *Anticancer Agents Based on Natural Product Models.* Academic Press, New York.

*Chattopadhyaya, H. and Banerjee, S. 1951a. *Science,* 113: 600-601.

*Chattopadhyaya, H. and Banerjee, S. 1951b. *Fd. Res.,* 16: 230-232.

*Chattopadhyaya, H. and Banerjee, S. 1952. *Fd Res.,* 17: 402-403.

*Chattopadhyaya, H., Nandi, N, and Banerjee, S. 1950. *Indian Pharmacist*, 5: 121-122.

Chauhan, D.V.S. 1968. *Vegetable Productions in India.* Ramprasad and Sons, Agra, India.

Chauhan, L. and Vijendra Rao, R. 2003. *Wood Anatomy of Legumes of India.* Bishen Singh Mahendra Pal Singh, Dehra Dun.

Cheeke, P.R. (Ed.) 1989. *Toxicants of Plant Origin.* Volume I, II, III & IV. CRC Press, Inc., Baca Raton, Florida, U.S.A.

Cheema, P.S., Malathi, K., Padmanaban, G. and Sarma, P.S. 1969. The neurotoxicity of β-N-oxalyl-α-β-diaminopropionic acid, the neurotoxin from the pulse, *Lathyrus sativus. Biochem. J.,* 112, 29.

*Chen, L.H., Wells, C.E. and Fordham, J.R. 1975. *J. Fd. Sci.,* 40: 1290-1294.

*Chitre, R.G., Desai, D.B. and Raut, V.S. 1955. *Indian J. Med. Res.,* 43: 575-583.

Choi, J.S., Lee, H.J., Park, K., Ha, J. and Kang, S.S. 1997. *In vitro* antimutagenic effects of anthraquinone aglycones and naphthopyrone glycosides from *Cassia tora. Planta Medica,* 63: 11-14.

Chopra, I.C. and Handa, K.L. 1961. *Review of Research on Indian Medicinal and Allied Plants.* ICAR, New Delhi, India.

Chopra, R.N. and Chopra, I.C. 1955. *A Review of Work on Indian Medicinal Plants (including indigenous drugs and poisonous plants).* ICMR, New Delhi, India.

Chopra, R.N., Badhwar, R.L. and Ghosh, S. 1965. *Poisonous Plants of India.* Vol. I. ICAR, New Delhi, India.

Chopra, R.N., Nayar, S.L. and Chopra, I.C. 1986a. *Glossary of Indian Medicinal Plants.* CSIR, New Delhi, India.

Chopra, R.N., Nayar, S.L. and Chopra, I.C. 1986b. *Supplement to Glossary of Indian Medicinal Plants.* CSIR, New Delhi, India.

Choudhury, B. 1990. *Vegetables.* National Book Trust, India.

Chung, K., Wong, T.Y., Wei, C., Huang, Y. and Lin, Y. 1998. *Critical Reviews in Food Science and Nutrition,* 88(6): 421- 464.

Clarke, J.H. 1982. *A Dictionary of Practical Materia Medica (Homoeopathy).* Vol. I, II and III. Jain Publishing Company, New Delhi, India.

Clay, H.F. and Hubbard, J.C. 1977. *The Hawaii Garden Tropical Shrubs.* The University Press of Hawaii, Honolulu.

*Codington, J.F., Copper, A.G., Brown, M.C. and Jealoz, R.R. 1975. *Biochemistry,* 14:855-859.

Cody, V., Middleton, E. Jr., Jeffrey, B., Harborne, J.B. and Beretz, A. (Eds.) 1988. *Plant Flavonoids in Biology & Medicine II: Biochemical, Cellular & Medicinal Properties.* Alan, R.Liss, Inc., New York, U.S.A.

Colucci, G., Moore, J.G., Feldman, M. and Chrispeels, M.J. 1999. *Proceedings of the National Academy of Sciences of the United States of America,* 96: 646-650.

Conerfore, S. C. 1996. Medicinal plants of two Mayan healers from San Andres, Peten, Guatemala. *Economic Botany,* 50: 327-336.

Conn, J.W., Rovner, D.R. and Cohen.E.L. 1968. Licorice induced pseudo-aldosteronism. *The Journal of the American Medical Association,* 205:492.

Conway, D. 1973. *The Magic of Herbs.* Jonathan Cape, London, U.K.

Cooke, T. 1967. *The Flora of Presidency of Bombay.* Vol. 1. Botanical Survey of India, Calcutta, India.

Cooper, H.L. 1972. Studies on RNA metabolism during lymphocyte activation. *Transplant Rev.,* 11:3.

*Cooper, H.S.1984. Lectins as probes in histochemistry and immunochemistry– The peanut lectin. *Human Pathology,* 15:904-908.

Corner, E.J.H. 1949. The durian theory of the origin of the modern tree. *Ann. Bot., N.S.,* 13: 367.

Cronquist, A. 1988. *The Evolution and Classification of Flowering Plants.* 2nd Edition. New York Botanic Garden, New York.

Cronquist, A. 1992. *An Integrated System of Classification of Flowering Plants.* Columbia Univ. Press, New York, U.S.A.

Cruz – Sanchez, F.F., Rossi, M.L., Butler, J.R., Carboni, P. Fineron, P.W. and Cokham, H.B. 1991. Oligodendrogliomas: A clinical, histological, inmunocytochemical & lectin binding study. *Histopathology,* 19:361-367.

*Cruz, G.L. and Cruz, B.H. 1965a. Livro verde das plantas medicinais e industriasis do brasil, V-1. G.L.

*Cruz, G.L. and Cruz, B.H. 1965b. Livro verde das plantas medicinais e industriasis do brasil, V-2. G.L.

Daljithsimha, K. 1974. *Unani Dravyagunadasthana.* Ayurvedic and Tibbi Academy, Lucknow, India.

Daly, D. 1997. Alternative medicine courses taught at the United States medical schools: An ongoing list. *The Journal of Alternative and Complementary Medicine,* 3: 405-410.

Danguy, A., Akif, F., Pajak, B. and Gabius, H.J. 1994. Contribution of carbohydrate histochemistry to glycobiology. *Histology and Histopathology*, 9:155-171.

*Daniel, P.T., Falcioni, F., Bery, A.U.J. and Bery, P.A.1986. Influence of cianidanol on specific and non-specific immune mechanisms. *Methods and Findings in Experimental Clinical Pharmacology,* 8:139.

Daniel, P.T., Holzschuh, J., Guo-Jun Diao. and Berg, P.A. 1988. Interference of the flavonoid compound Cianidanol with macrophage function and lymphocyte activating mechanisms. In: *Plant Flavonoids in Biology and Medicine II: Biochemical, Cellular and Medicinal Properties.* Cody, V., Middleton, E. Jr., Jeffrey, B., Harborne, J.B. and Beretz, A. (Eds.) Alan, R. Liss, Inc., New York, U.S.A.

Dastur, J.E 1962. *Medicinal Plants of India and Pakistan.* Taraporewala Sons and Co. Ltd., Bombay, India.

Dastur, J.E 1964. *Useful Plants of India and Pakistan.* Taraporewala Sons and Co. Ltd., Bombay, India. de Candolle, A.P. 1882. Origins of cultivated species. Reprint 1967. Hafner Publishing Co., London.

Depierreux, C., Midoux, P., Monsigny, M. and Roche, A.C. 1992. Drug targeting: synthesis and endocytosis of oligonucleotide-neoglycoprotein conjugate. *Nucleic acid Res.,* 20:4621-4629.

Deshpande, S.S., Sathe, S.K., Salunkhe, D.K. and Conforth, D.P. 1982. *Journal of Food Science*, 47: 1846-1850.

*Deshpande, V.H., Pendse, A.D. and Pendse, R. 1977. *Indian J. Chem., 15B*: 205-207.

Devlin, T.M. (Ed)., 1997. *Textbook of Biochemistry with Clinical Correlations.* Wiley-Liss, New York, U.S.A.

Dewalt, S. J., Bourdy, G., Chavez, L. R., de Michel, and Quenevo, C. 1999. Ethnobotany of Tacara: quantitative inventories of two permanent plots of North Western Bolivia. *Economic Botany*, 53(3): 237-260.

Dey, K. L. and Bahadur, R. 1972. The indigeneous drugs of India. *The Chronica Botanica*, New Delhi, India.

Dhungat, S.B. 1972. *Khair tree (Acacia catechu).* Chandra Katha Industries Limited, Najibabad, India.

Dickey, W. 1994. Lewis phenotype secretor status and coeliac diseases. *Gut,* 35: 769- 770.

Doll, R., Hill, I.D., Hutton, C. and Underwood, D.J. 1962. Clinical trial of a triterpenoid liquorice compound in gastric and duodenal ulcer. *Lancet*, 2: 793-796.

Donas, B., Mandoki, J., Kormendi, C. and Kernoczi, Z.S. 1993. *Planta Medica,* 59: 612.

Dranchenberg, C.B. and Papadimitriou, J.C. 1995. Aberrant pattern of lectin binding in low and high grade prostatic intraepithelial neoplasia. *Cancer,* 75: 2539-2544.

Drury, C.H. 1878. *The Useful Plants of India.* William H. Allen & Co., London.

Duke, A.K. 1981. *Handbook of Legumes of World Importance.* Plenum Press, New York and London.

Dummer, T. 1994. *Tibetan Medicine and Other Holistic Health Systems.* Paljor Publication, India.

Dunaevskii, Y.E., Gladysheva, I.P., Pavlukova, E.B., Beliakova, G.A., Gladyschev, D.P., Papisova, A.I., Larionova, N.I. and Belozersky, M.S. 1997. The anionic protease inhibitor BBWI – 1 from buckweed seeds. Kinetic properties and possible biological role. *Physiologia Plantarum,* 100: 483-488.

Dymock, W., Warden, C.J.H. and Hooper, D. 1980, *Pharmacographia Indica.* Vol. I. Periodical experts, Delhi, India.

Eagleton, G. 1999. Winged bean in Myanmar revisited. *Economic Botany,* 53: 342-353.

Eaton, S.B. and Konner, M. 1985. Paleolithic nutrition: A consideration of its nature and current application. *New England J. Med.,* 312: 283-289.

*Ekeke, G.I. and Shode, F.O. 1985. *Planta Medica,* 51: 504-507.

Ekeke, G.I. and Shode, F.O. 1990. Phenylalanine is the predominant antisickling agent in *Cajanus cajan* seed extract. *Planta Medica,* 56: 41-43.

Enomoto, M. 1987. Tannins (Tannic acids). In: *Naturally Occurring Carcinogens of Plant Origin: Toxicity, Pathology and Biochemistry.* (Ed.) Hirono, I. Kodansha Limited, Tokyo. pp. 161-166.

Ernst and Lehner, J. 1973. *Folklore and Odysseys of Food and Medicinal Plants.* George G. Harper & Co. Ltd., London, U.K.

Erskine, W., Smartt, J. and Muehebaver, F. J. 1994. Mimicry of lentil and the domestication of common vetch and green pea. *Economic Botany,* 48(3): 326-332.

*Essman, E.J.1984. The medicinal use of herbs. *Fitoterapia,* 55:279.

Etzler, M.E. 1985a. Distribution and function of plant lectins In: *The Chemistry and Biology of Lectins.* (Eds.) Liener., Sharon, N. and Goldstain, I.J. Academic Press, New York, U.S.A.

Etzler, M.E. 1985b. Plant lectins: molecular and biological aspects. *Annual Review of Plant Physiology,* 36: 209-234.

Fantz, P. R. 1991. Ethnobotany of *Clitoria* (Leguminosae). *Economic Botany,* 45: 511-520.

FAO, 1968. Q.f. Tindall, H.D. 1983. Vegetables in the tropics. Avi Publishing Company, Inc., West Port, Connecticut.

FAO, 1972. Q.f. Tindall, H.D. 1983. Vegetables in the tropics. Avi Publishing Company, Inc., West Port, Connecticut.

Farel, A., Steinmetz, M.D., Ollivier, E., Elias, R., Regli, P. and Balansard, G. 1992. Screening of triterpenoid saponins for antifungal activity. *Planta Medica, Suppl.,* A635.

Farooqi, A.A. and Sreeran, B.S. 2001. *Cultivation of Medicinal and Aromatic Crops.* Universities Press (India) Limited, Hyderabad, India.

Fathima, T. 1994. *Glossary of Medicinal Plants in Unani Medicine.* M.Sc., (Applied Botany) Dissertation. Bangalore University, Bangalore, India.

*Finney, S.H. and Sommers, G.F. 1958. *J. Pharmacol.,* 10: 613-620.

*Fitzpatrick, F.L 1964. *Our plant resources: Plants and their Economic Importance.* Holt, Rinehart and Winston, Inc., New York, U.S.A.

Fluck, H. 1988. *Medicinal Plants and their Uses.* W. Foulsham and Company Limited, London.

*Fordham, J.R., Wells, C.E. and Chen, L.H. 1975. Sprouting of seeds and nutrient composition of seeds and sprouts. *J. Fd. Sci.,* 40: 552-556.

Freed, D. 1991. Lectins in food: their importance in health and disease. *Jour.Nutr Med.,* 2: 45-64.

Fugita, T., Sezik, E., Tadata, M., Yesilada, E., Honda, G., Takeda, Y. and Tanaka. 1995. Traditional medicine in Turkey VII. Folk medicine in middle and west black sea. *Economic Botany,* 49: 406-422.

Fulla, F., Choi, Y., Kinghorn, A.D. and Bunyapraphatsara, N. 1990. *Planta Medica,* 56: 332-333.

Gabor, M. and Szent-Gyorgyi. 1988. The bioflavonoids: New results and perspectives of pharmacological research into benzo-pyrone derivatives. In: *Plant Flavonoids in Biology and Medicine II: Biochemical, Cellular and Medicinal Properties.* (Eds.) Cody, V., Middleton, E. Jr., Jeffrey, B., Harborne, J.B. and Beretz, A. Alan R. Liss, Inc., New York.

*Gaind, K.N., Budhiraja, R.N. and Kaul, R.N. 1966. *Indian J. Pharm.,* 28: 248-249.

Gamble, J.S. 1979. *Flora of the Presidency of Madras.* Vol 1. Botanical Survey of India.

George, A.J. 1965. Legal status and toxicity of saponins, *Food Cosmet. Toxicol.,* 3, 85-91.

Ghazanfar, S. A. and Sabahi, A. M. A. A. 1993. Medicinal Plants of Northern and Central Oman (Arabia). *Economic Botany*, 47: 89-98

Gibbs, R.D. 1974. *Chaemotaxonomy of Flowering Plants.* Vols. I, II, III and IV. McQueen's University Press, Montreal, Canada.

Glick, Z. and Joslyn, M.A. 1970a. Food intake depression and other metabolic effects of tannic acid in rats. *Journal of Nutrition,* 100:509.

Glick, Z. and Joslyn, M.A. 1970b. Effects of tannic acid and related compounds on the absorption and utilization of proteins in rats. *J. Nutr.,* 100:516-520.

Gomathy, V. and Gnanamanickam, S.S. 2004. Polygalacturonase-inhibiting proteins in plant defense. *Curr. Sci.,* 87: 1211-1217.

Gonzales-Campora, R., Montero, C., Martin-Lacave, I. and Galera, H. 1986. Demonstration of vascular and endothelium in thyroid carcinoma using *Ulex europaeus* agglutinin. *Histopathol.*, 10:261-266.

Goodman, M.P. 2004. Natural and personalized hormonal products. *Health and age,* (www.healthandage.com)

Goodwin, T.W. 1980. *The Biochemistry of Carotenoids* Vol I. Chapmann and Hall, London, U.K.

Gopalan, C., Rama Sastri, B.Y. and Balasubramanian, S. 1985. *Nutritive Value of Indian foods.* ICMR Offset Press, New Delhi, India.

Graves G. 1990. *Medicinal Plants.* Bracken Books, London, U.K.

Greenwald, P. 1996. Chemoprevention of cancer. *Sci. Amer.,* 275:96-99.

Greer, F.A.C. and Pusztai, A.1985. Toxicity of kidney bean (*Phaseolus vulgaris*) in rats: Changes in intestinal permeability. *Digestion*, 32:42-46.

Greer, M.A. 1962. The natural occurrence of goitrogenic agents. *Recent Prog. Horm. Res.,* 18:187.

Greer. F.A.C., Brewer, A. and Pusztai.A. 1985. Effect of kidney bean (*Phaseolus vulgaris*) toxin on tissue weight and composition and some metabolic functions of rats. *British Journal of Nutrition,* 54: 95-103.

Grieve M. 1967. *A Modern Herbal*. Hafner Publishing Company, New York and London.

Griffiths, D.W. 1981. *Journal of the Science of Food and Agriculture,* 28: 983-989.

Griminger, P. and Fisher, H. 1958. Dietary saponin and plasma cholesterol in chicken. *Proc. Soc. Exp. Biol. Med.,* 99: 424.

Gumbinger, H.G., Winterhoff, H., Sosa, A. and Wylde, R.1988. Investigations of the structure-function, relationships of flavonoid compounds with anti-gonadotrophic activity. In: *Plant Flavonoids in Biology and Medicine II: Biochemical, Cellular and Medicinal Properties.* (Eds.) Cody, V., Middleton, E. Jr., Jeffrey, B., Harborne, J.B. and Beretz, A. Alan, R.Liss, Inc., New York, U.S.A.

*Gupta, Y.P. and Das, N.B. 1956. *Proc. 43rd Indian Science Congress, Part III.*

*Gupta, Y.P. and Das, N.B. 1959. *Indian J. Agric. Sci.,* 29: 27-33.

Gurudeva, M.R. 2001. *Botanical and Vernacular Names of South Indian Plants.* Divyachandra Prakasana, Bangalore.

Haas, H., Falcone, F.H., Schramm, F., Haisch, k., Gibbs, B.F., Kaluchje, J., Poppelmanmn, M., Becker, W.M., Gabius, H. and Schlaak, M. 1999. *Eur. J. Immunol.,* 29:918-927.

Hamilton, A. 1995. The people and plant initiative In: *Ethnobotany.* (Ed) Martin, G.J. Chapmann and Hall, London, U.K.

Handa, S.S. and Kapoor, V.K. 1995. *Pharmacognosy*. Vallabh Prakashan, Delhi, India.

Hangerman, A.E. and Butler, L.G. 1980. *Journal of Agricultural and Food Chemistr.,* 28: 947-952.

Haraguchi, H., Inoue, J., Tamura, Y. and Mizutani, K. 2000. *Planta Medica,* 66: 569-571.

Harborne, J.B.1971a. Distribution of flavonoids in the leguminoseae. In: *Chaemotaxonomy of the Leguminosae.* (Eds.) Harborne, J.B, Boulter, D. and Turner, B.L. Academic Press, London.

Harborne, J.B. 1971b. Terpenoid and other low molecular weight substances of systematic interest in the leguminoseae In: *Chaemotaxonomy of the Leguminosae.* (Eds.) Harborne., Boulter, D. and Turner, B.L., Academic Press, London.

Harborne, J.B. 1973. *Phytochemical Methods.* Chapmann and Hall, London, New York, U.S.A.

Harborne, J.B. 1984. *Phytochemical Methods: A Guide to Modern Techniques of Plant Analysis.* Chapman and Hall, New York, U.S.A.

Harborne, J.B. and Baxter, H. 1996. *Dictionary of Plant Toxins.* John Wiley and sons, New York, Toronto, Singapore.

Harborne, J.B., Boulter D. and Turner. B.L. (Eds.). 1971. *Chaemotaxonomy of the Leguminosae.* Academic Press, New York, U.S.A.

Harder, D. K. and Smartt, J. 1992. Further evidence on the origin of the cultivated winged bean, *Psophocarpus tetragonolobus* (L.) DC. (Fabaceae): Chromosome numbers and the presence of host-specific fungus. *Economic Botany,* 46: 187-191.

Hartwell, J.L. 1967. Plants used against cancer, a survey. *Loydia*, 30: 379

Hartwell, J, L. 1982. *Plants Used Against Cancer*. Quarterman Publication Inc., Lawrence, Massachusetts, U.S.A.

Hasan, H.M.Z. 1989. *Selected Herbo-Mineral Remedies*. Ministry of Public Health, Government of Kuwait.

*Hatano, T., Yasuhara, T., Miyamoto, K. and Okuda, T. 1988a. *Chem. Pharm. Bull.*, 36: 2286.

*Hatano, T., Kagawa, H., Yasuhara, T., and Okuda, T. 1988b. *Chem. Pharm. Bull.,* 36: 2090.

*Hatano, T., Yasuhara, T., Fukuda, T., Noro, T. and Okuda, T. 1989. *Chem. Pharm. Bull.*, 37: 3005.

*Hattori, M., Fukuda, Y., Imoto, M., Koyama, Y., Nakano, S. and Urano, F. 1991. Histochemical properties of vascular and sinusoid endothelial cells in liver diseases. *Gastroenterol. J.,* 20:336-343.

Hayashi, K., Hayashi, H., Hiraoka, N. and Ikeshiro, Y. 1997. *Planta Medica*, 63: 102-105.

Heaney, R.K. and Fenick, G.R. 1987. Indentifying toxins and their effects: glucosinolates. In: *Natural Toxicants in Food: Progress and Prospects.* (Ed.) Watson, D.H., Ellis Horwood, Chichester, England, U.K.

Hedge, V.R., Maiti, T.K. and Podder, S.K. 1991. Purification and characterization of three toxins and two agglutinins from *Abrus precatorius* by using lactamyl-sepharose affinity chromatography. *Analytical Biochemistr.,* 194:101-109.

*Hedge, V.R., Dai, P., Patel, M.G., Puar, M.S., Das, P. and Pai, J. 1997. *J. Nat. Prod.,* 60: 537-539.

Hedrick, U.P. 1972. *Sturtevant's Edible Plants of the World.* Dover Publications, Inc., U.K.

Hein, H.O. 1992. The Lewis bloodgroup-a new genetic marker of ischaemic heart disease. *J. Inter. Med.,* 232: 481-487.

Heng, M. C., Friedlander, F. S. and Bennet, R, 1992. Expresion of *Ulex europaeus* agglutinin I lectin binding site in squamous cell carcinomas and their absence in basal cell carcinomas. Indicator of tumour type and differentiation. *Am. J. Dermatopathol.,* 14: 216-9

Herklots, G.A.C. 1972. *Vegetables in South-East Asia*. George Allon & Unwin Limited, London, U.K.

Hermann, M. 1973. *Marvellous World of Herbs and Medicinal Flowers*. Abbey Library, London, U.K.

Hersh-martinez, P. 1997. Medicinal plants and regional traders in Mexico: Physiographic differences and conservational challenge. *Economic Botany*, 51: 107-120.

Heywang, B.W. and Bird, H.R.1954. The effect of alfa-alfa saponin on the growth, diet, consumption and efficiency of diet utilization of chicks. *Poultry Science,* 33:239-241.

Heywood, V.H. 1978. *Flowering Plants of The World*. Croom Helm Publishers Limited, England, U.K.

Hiai, S. 1985. Chinese medicinal material and secretion of ACTH and corticosteroid. In: *Advances in Chinese Medicinal Materials Research.* (Eds.) Chang, H.W., Yeung, H.W., Tso, W.W. and Koo, A. World Scientific Singapore.

*Hikokichi, O., Shoichi, N., Akita, K. and Masaru, T. 1974. Immunosuppressive effect of licorice root extracts, *Chem., Abstr.,* 81: 76115.

Hilder, V.A., Gatehouse, A.M.R., Sheerman, S.E., Barker, R.F. and Boulter, D. 1987. A novel mechanism of insect resistance engineered into tobacco. *Nature,* 300: 160-163.

Hilgert, I., Horejsi, V., Angelisova, C. and Kristofova, H. 1980. Lentil lectin effectively induces allotransplantation tolerance in mice. *Nature,* 284:273-275.

Hill, A.F. 1989. *Economic Botany*. Tata McGraw-Hill Publishing Company Limited, NewDelhi, India.

*Hlodversson, R. 1987. *Animal Food Science and Technology*, 17: 245-255.

Holthofer, H., Virtannen, I., Kariniemi, A. L., Hormia, M., Linder, E. and Miettinen, A. 1982. *Lab. Invest.*, 47: 60-66.

*Hoppe, H.A. 1975. *Drogenkunde*, Vol. 1. De Gruyter, Berlin and New York. pp 139.

Hostettmann, K., Hostettmann, M. and Marston, A. 1991. Saponins. *Methods in Plant Biochemistry,* 7:435-471.

*Huang, K.C. 1993. *The Pharmacology of Chinese Herbs*. CRC Press, Boca Raton, Florida, U.S.A. pp. 86-87.

*Hunt, R.C., Bullis, C.M. and Brown, J.C. 1975. *Biochemistry,* 14:109-115.

Husain, A., Virmani, O.P., Popli, S.P., Misra, L.N., Gupta, M.M., Srivastava, G.N., Abraham, Z. and Singh, A.K. 1992. *Dictionary of Indian Medicinal Plants*. Central Institute of Medicinal and Aromatic Plants, Lucknow, India.

Hutchinson, J. 1964. *The Genera of Flowering Plants*. Academic Press, London and New York.

*Ikeda, T., Udayama, M., Okawa, M., Arao, T. and Kinjo, J. 1998. *Chem. Pharm. Bull.*, 46: 359-361.

*Inoue, H., Mori, T., Shibata, S. and Saito, H. 1987. *Chem. Pharm. Bull.,* 35: 3888-3893.

*Irvine, F.R. 1961. *Woody Plants of Ghana*. Oxford University Press, London, U.K.

Ishaaya, I. and Birk, Y. 1968. Unpublished results. Q.f. *Toxic constituents of Plant Food Stuffs* 1969 – (Ed.) Liener, I.E. Academic Press, New York and London.

*Ishida, H., Umino, T., Tsuji, K. and Kosuge, T. 1989. *Chem. Pharm. Bull.,* 37: 1616-1618.

Itzkowitz, S., Kjeldsen, T., Fiera, A., Hakomori, S.R., Yang, V.S. and Kim, Y.S. 1991. Expression of Sialosyl Tn and T-antigens in human pancreas. *Gasteroenterology,* 100:1691-1700.

Iwu, M. M.1986. Bioflavanones of *Garcinia*: Pharmacological and biological activities. In: *Plant Flavonoids in Biology and Medicine: Biochemical, Pharmacological and Structure-Activity Relationships*. (Eds.) Cody, V., Middleton, E. Jr., Jeffrey, B., Harborne, J.B. and Beretz, A. Alan, R.Liss, Inc., New York, U.S.A.

Jackson, B.D. 1893. *Index Kewensis. An Enumeration of the Genera and Species of Flowering Plants*. Vol 1- 4. The Calarendor Press, Oxford, U.S.

Jadhav, S.J., Reddy, N.R. and Deshpande, S.S. 1989. Polyphenols. In: *Handbook of World Food Legumes: Nutritonal Chemistry, Processing, Technology and Utilization*. Vol. I. (Eds.) Salunkhe, D.K. and Kadam, S.S. CRC Press Inc., Florida. pp 145-161.

Jain, S .K. 1968. *Medicinal Plants*. National Book Trust, New Delhi, India.

Jain, S.K 1981. *Glimpses of Indian Ethnobotany*. Oxford & IBH Publishing Company, New Delhi, India.

Jain, S.K. 1991. *Dictionary of Indian Folk Medicine and Ethnobotany*. Deep Publications, New Delhi, India.

Jain, S.K. 1996. *Ethnobiology in Human Welfare*. Deep Publication, New Delhi, India.

Jain, S.K., Sinha, B.K. and Gupta, R.C. 1991. *Notable Plants in Ethnomedicine of India*. Deep Publications, New Delhi, India.

Jamil, T.1997. *Complementary Medicine*. Butterworth-Heinemann, Oxford. U.K.

Jansman, A.J.M., Huisman, J. and van der Poel, A.F.B. 1993. *Animal Food Science and Technology*, 71: 627-641.

Jasalavitch, C.A. and Anderson, A.J. 1981. Isolation from legume tissue of an agglutinin of saprophytic pseudomonads. *Can. J. Bot.,* 59: 264-271.

Jin, C., Yin-chun, S., Gui-qui, C. and Wen-Dun, W. 1999. Ethnobotanical studies on wild edible fruits in southern Yunnan: folk names, nutritive value and uses. *Economic Botany,* 53: 2-14.

*Jin, Y.T. and Lin, L.M. 1989. Lectin binding patterns in squamous epithelium in experimentally induced hamster buccal pouch carcinoma. *J. Oral. Pathol. Med.,* 18:446-450.

*Johns, T. and Kakwaro, J. O. 1991. Food plants of Luo of Siaya district-Kenya. *Economic Botany,* 45: 103-113.

Johns, T., Mhoro, E. B., Sanaya, P. and Kiranarri, E. K. 1994. Herbal remedies of the Batemi of Ngorangoro district, Tanzania. A quantitative appraisal. *Economic Botany.* 48: 90-95.

Johns, T., Mhoro, E. B. and Sanaya, P. 1996. Food Plants and masticants of Batemi of Ngorongoro district. Tanzania. *Economic Botany,* 50: 115-121.

Johnston, M. and Colquhoun, A. 1996. Preliminary ethnobotanical survey of Kurupukari: An amerindian settlement of Central Guyana. *Economic Botany.* 50: 182-194.

Johri, R.K., Zutushi, H., Kameshwran, L. and Atal, C.K. 1985. Effect of quercitin and *Albizia* saponins on rat mast cells. *Indian J. Physiol. Pharmacol.,* 29, 43.

Jones, A. and Earle, F.R. 1966. Chemical analysis of seeds. II. Oils and protein content of 759 species. *Econ. Botany,* 20: 127-155.

Jood, S., Bishnoi, S. and Sehgal, S. 1998. Effect of processing on nutritional and antinutritional factors of moong bean cultivars. *Journal of Food Biochemistry,* 22: 245-257.

Joshi, B., Sainani, M., Bastawadae, K., Gupta, V.S., and Ranjekar, P.K. 1998. Cystine protease inhibitor from pearl millet: a new class of antifungal protein. *Biochem. & Biophys. Res. Comm.,* 246: 382-387.

*Joshi, K.C., Singh, P.J., Cox, P.J., Howie, R.A. and Thompson, H. 1982. *Tetrahedron,* 38:2703-2708.

*Joshi, S. 1987. Cell surface properties associated with malignancy of metastatic large cell lymphoma cell line. *Cancer Res.,* 47: 3551-3557.

Joshi, S.G. 2000. *Medicinal Plants.* Oxford & IBH Publishing Co. Pvt. Ltd., New Delhi, India.

Juninger, H.E. 1990. Bioadhesive polymer system for peptide delivery. *Acta Pharmaceutica Technologica,* 36:110-126.

Kachroo, P. 1970. *Pulse Crops of India.* Central Arid Zone Research Institute, Jodhpur, India.

Kakade, M.L., Arnold, R.L., Liener, I.E. and Warbel, P.E. 1969. Unavailability of cystine from trypsin inhibitors as a factor contributing to the poor nutritive value of many beans. *Journal of Nutrition.* 99:34.

Kakeji, Y., Machara. Y., Tsujitani, S., Baba, H., Ohno, S., Watanabe, A., Sugimachi, K. 1994. *Helix pomatia* agglutinin binding activity and lymph node metastasis in patients with gastric cancer. *Semin. Surg. Oncol.,* 10:130-134.

Kameswara Rao, C. and Sangeetaa, W. 1993. A sensitive method to detect saponins in plant tissues using snake head fish (*Channa striatus* Bloch) erythrocytes. *J. Pl. Sci. Res.,* 9:27-29.

Kameswara Rao, C. and Sangeetaa, W. 1996. In: *Advances in Pollen Spore Research.* Today and Tomorrows Publication.

Kameswara Rao, C. and Sangeetaa, W. 1997. Implications of lectins and saponins in pollen in the diagnosis and treatment of allergy caused by inhaled pollen. In: *Aerobiology.* Agashe, S.N. (Chief Ed.), Oxford & IBH Publishing Co., Pvt., Ltd., New Delhi. pp 487-497.

Kameswara Rao, C. 2000. *Database of Medicinal Plants.* Karnataka State Council for Science & Technology, Bangalore, India.

Kameswara Rao, C., Geetha, B.L. and Suresh, G. 2003. *Red List of Threatened Vascular Plant Species in India.* Botanical Survey of India, Howrah, India.

Kannan, S., Balaram, P., Chandran, C.J., Pillai, M.R., Mathew, B. and Nair, M.K. 1993. Expression of lectin specific cellular glycoconjugates during oral carcinogenesis. *J. Cancer. Res. & Clin. Oncol.,* 119:689-94.

Kardono, L.B.S., Tsauri, S., Padmawinata, K., Pezzuto, J.M. and Kinghorn, A.D. 1990. *Planta Medica*, 56: 673-674.

*Kariniemi, A. l., Holthofer, H., Miettinen, A. and Virtanen, I. 1983. *British Journal of Dermatology*, 109: 523-529.

Kaufman, P.B., Cseke, L.J., Warber, S., Duke, J.A. and Brielmann, H.L.1999. *Natural Products from Plants.* CRC Press, Boca Raton, Boston, London, New York.

Kawa, S., Koto, M. and Ogavchi, H. 1991. Preparation of pancreatic carcinoma associated with mucin. *Scand. J. Gasteroenterol.,* 26:981-992.

Kawaguchi, K., Alves, S.M., Watanabe, T., Kikuchi, S., Satake, M. and Kumazawa, Y. 1998. *Planta Medica,* 64: 653-655.

Kawai, T. and Suzuki, M. 1994. Histochemical and biochemical aspects of lectin binding glyco protein in primary & metastatic adenocarcinoma in the lung. *Modern pathology,* 7:175-180.

Kay, 1979. Q.f. Tindall, H.D. 1983. *Vegetables in the tropics.* Avi Publishing company, Inc., West Port, Connecticut.

Kays, S. J. and Dias, J. C. S. 1995. Common names of commercially cultivated vegetables of the world in 15 languages. *Economic Botany,* 49: 115-152.

Kellokumpu, I., Kashi, K. and Anderson, L.C. 1986. Lectin binding sites in normal, hyperplastic, adenomatous and carcinomatous human colorectal mucosa. *Acta pathological Microbiologica et Immunologica Scandinavia.* 194: 271–280.

Kensil, C.R., Wu, J.Y. and Soltysik, S. 1995. Structural and immunological characterization of the vaccine adjuvant QS-21. In *Vaccine design: the Subunit and Adjuvant Approach.* (Eds.), Powell, M.F. and Newman. M.J. Plenum Press, New York.

Kent, J.T. 1975. *Repertory of the Homoeopathic Materia Medica with Word Index.* Jain Publishing Co., New Delhi, India.

*Kereselidge, E.V., Pkheidze, T.A., Kemertelidize, E.P., Khardziani, S.D., Dzhaparidze, T.N. and Makhardze, Sh. K. 1975. Fibrinolytic activity of *Ruscus ponticus* and *Ruscus hypophyllum* saponins. *Soobshch. Akad. Nauk Gruz., S.S.R.,* 78: 485.

Keshavamurthy, K.R. and Yoganarasimhan, S.N. 1990. *Flora of Coorg (Kodagu)*, Karnataka, India. Vimsat Publishers, Bangalore, India.

*Keung, W., Lazo, O., Kunze, L., Vallee, B.L. 1996. *Proc. Natl. Acad. Sci.,* 93: 4284-4288.

Keys, J.D. 1995. *Chinese Herbs.* Charles E. Turtle Company Inc., Rutland, Vermont & Tokyo, Japan.

Kienholz, E.W., Jensen, L.S. and Mcgimis, J. 1962. *Poultry Sci.*, 41:367.

Kim, B.K., Cho, H.Y., Kim, J.S., Kim, H.W. and Choi, E.C. 1993. Studies on constituents of higher fungi of Korea. (LXVII). Antitumour components of the cultured mycelia of *Ganoderma lucidum. Korean J. Pharmacog.*, 24:203-212.

*Kimura, Y., Okuda, H., Okuda, T. and Arichi, S. 1988. *Phytother. Res.,* 2:140.

Kirthikar, K.R. and Basu, B.D. 1975. *Indian Medicinal Plants.* Vols. I & II Periodical Experts, New Delhi, India.

*Kitagawa, I., Wang, H. K., Saito, M., and Yoshikawa, M. 1983a. Saponin and Sapogenol. XXXI. Chemical constituents of the seeds of *Vigna angularis* (Willd.) O. Ohashi, (1) Triterpenoidal sapogenols and 3-furanmethanol-B-D-glycopyranoside, *Chem. Pharm. Bull.,* 31:644.

*Kitagawa, I., Wang, H. K., Saito, M., and Yoshikawa, M., 1983b. Saponin and Sapogenol. XXXII. Chemical constituents of the seeds of *Vigna angularis* (Willd.) O. Ohashi, (2) Azukisaponins I, II, III & IV. *Chem. Pharm. Bull.,* 31:674.

*Kitagawa, I., Wang, H. K., Saito, M., and Yoshikawa, M. 1983c. Saponin and sapogenol. XXXIII. Chemical constituents of the seeds of *Vigna angularis* (Willd.) O. Ohashi, (3) Azukisaponins V & VI. *Chem. Pharm. Bull.,* 31:683.

*Kitagawa, I., Wang, H.K., Katagi, A., Fuchida, M., Miura, I. and Yoshikawa, M. 1983d. *Chem. Pharm. Bull.,* 31:689.

Klen and McReady, 1975. Q.f.Yamaguchi, M. 1983. *World Vegetables: Principles, Production and Nutritive Values.* Avi Publishing Company Inc., West Port, Connecticut.

Knott and Deanon, 1967. Q.f. Tindall, H.D. 1983. *Vegetables in the Tropics.* Avi Publishing Company, Inc., West Port, Connecticut.

Kochaar, S.L.1981. *Economic Botany in the Tropics.* Macmillan India Limited, Delhi, India.

Kohrae, J., Matthias, S., Irmsches, K. and Rolf-Dieter, M. 1988. Flavonoids effect on transport, metabolism and action of thyroid hormones. In: *Plant Flavonoids in Biology and Medicine II: Biochemical, Cellular And Medicinal Properties.* (Eds.) Cody, V., Middleton, E. Jr., Jeffrey, B., Harborne, J.B. and Beretz, A. Alan, R.Liss, Inc., New York, U.S.A.

Koiwa, H., Bressan, R.A. and Hasegawa, P.M. 1997. Regulation of protease inhibitors and plant defense. *Trends in Plant Science,* 2: 379-384.

Koiwa, K., Shade, R.E., Zhu-Salzman, K., Subramanian, L., Murdock, L.L., Nielsen, S.S., Bressan, R.A. and Hasegawa, P.M. 1998. Phage display selection can differentiate insecticidal activity of soybean cystatins. *Plant Journal,* 14: 371-379.

*Konoshima, T., Kokumai, M., Kozuka, M., Tokuda, H., Nishino, H. and Iwashima, A. 1992. *J. Natl. Prod.,* 55: 1776-1778.

Konoshima, T., Kozuka, M., Haruna, M. and Ito, K. 1991. *J. Nat. Prod.,* 54: 830-836.

*Koyko, K., Kazuvo, Y., Satoshi, T., Toshiaki, O. and Tatsuro, I. 1994. *Japanese J. Cancer Res.,* 85:744-749.

Kreig, M.B.1965. *Green Medicine: the Search for Plants that Heal.* George G. Harrap & Co. Ltd., London, U.K.

Krishna Murthy, T. 1993. *Minor Forest Products of India.* Oxford and IBH Publishing Co. Pvt. Ltd., New Delhi, India.

Krochman, A., Walters, R.S. and Doughty, R.M. 1969. *A Guide to Medicinal Plants of Appalachia*. N.S.D.A. Forest Service Research Paper. NE 138.

Kroes, B.H., Beukelman, C.J., Van den Berg, A.J.J., Quarles van Ufford, H.C. and Labadie, R.P. 1993. Complement inhibitory activity of α-Glycyrrhetinic acid. *Planta Medica,* 59: 585.

Kulanthaivel, P., Janzen, W.P., Ballas, L.M., Jiang, J.B., Hu, C., Darger, J.W., Seldin, J.C., Cofield, D.J. and Adams, L.M. 1995. Naturally occurring protein kinase C inhibitors II. Isolation of oligomeric stilbenes from *Caragana sinica*. *Planta Medica,* 61: 41-44.

*Kuroyanagi, T. and Sato, M. 1966. *Allergy,* 15: 67-75.

Kurup, P.N.V. 1977. *Handbook of Medicinal Plants*. Vol. 1. Central Council for Research in Indian Medicine and Homoeopathy, New Delhi, India.

Kurup, P.N.V. 1983. Ayurveda. In: *Traditional Medicine and Health Care Coverage*. (Eds.) Bannerman, R.H., Burton, J. and Wen-Chieh, C. WHO, Geneva.

*Kylen, A.M. and McCready, R.M. 1975. *J. Fd. Sci.,* 40: 1008-1009.

*Lacaille-Dubois, M.A. and Wagner, H. 1996. *Phytomedicine,* 2: 363-386.

Langer, P. and Greer, M.A. 1977. *Antithyroid Substances and Naturally Occurring Goitrogens*. S. Krager, Basel.

Latha, P. 1992. *A Probe into Health Problems. Herbal Formulations and Quantitative Requirements of Raw Drugs. An Analysis through Computer Database, Part II.* M.Sc (Applied Botany) Dissertation, N.M.K.R.V. College for Women, Bangalore University, Bangalore, India.

Lawrence, P.K. and Koundal, K.R. 2002. Plant protease inhibitors in control of phytophagous insects. *Electronic J. Biotech.,* 5.

Lee, H.S., Ko, H.R., Ryu, S.Y., Oh, W.K., Kim, B.Y., Ahn, S.C., Mheen, T.I. and Ahn, J.S. 1997. Inhibition of phospholipase C gamma 1 by the prenylated flavonoids from *Sophora flavescens*. *Planta Medica,* 63: 266-268.

Lehninger, A.L. 1979. *Biochemistry.* Indian edition. Kalyani Publishers, Ludhiana and New Delhi, India.

Lewis, W.H. and Lewis, M.P.E. 1977. *Plants Affecting Man's Health.* A Wiley Interscience Publication, John Wiley & Sons, New York, U.S.A.

*Leyel. 1937. *Herbal Delights*. Faber and Faber Limited, London, U.K.

*Li, W., Asada, Y. and Yoshikawa, T. 1998. *Planta Medica*, 64: 746-747.

Liener, I. E. 1975a. In: *Protein Nutritional Quality of Foods and Feeds* (Ed.) Friedman, M., Marcel Dekker, Inc., New York, U.S.A.

Liener, I.E. 1975b. Anti-tryptic and other anti-nutritional factors in legumes. In: *Nutritional Improvement of Food Legumes by Breeding.* (Ed.) Maxmilner. Wiley-Interscience publication, New York.

*Limas, C. and Lange, P. 1986. T-antigen in normal and neoplastic urothelium. *Cancer,* 58:1236-1245.

Lindley, J. 1981. *Flora Medica: A Botanical Account of all the More Important Plants Used in Medicine.* Ajay Book Service, New Delhi, India.

Lipp. F.J. 1996. *Herbalism: The Healing Power of Plants*. Duncan Baird Publishers, London, U.K.

Lis, H. and Sharon, N. 1986. Lectins as molecules and as tools. *Ann. Rev. Biochem.,* 55:35-67.

Livingstone, C. 1978. *Pocket Medical Dictionary, B.I.* Churchill Livingstone Pvt. Ltd., London, U.K.

Loggia, D. R., Zilli, C., Del Negro, P., Redaelli, C. and Tubaro, A.1988. Isoflavones as spasmolytic principles of *Piscidia erythrina.* In: *Plant Flavonoids in Biology and Medicine II: Biochemical, Cellular and Medicinal Properties.* (Eds.) Cody, V., Middleton, E., Jr., Jeffrey, B., Harborne, J.B. and Alian Beretz. Alan, R. Liss, Inc., New York, U.S.A.

*Longstaff, M. and McNab, J.M. 1991. *British Journal of Nutrition,* 65: 199-216.

*Lovati, M.R. 1992. *Planta Medica,* 58: 580-581.

Lurz, G., Greinwald, R., Witte, L. and Czygan, F.C. 1990. Quinolizidine alkaloids in *Spartium junceum. Planta Medica.* 56:522.

Mabey, R. 1988. *The Complete New Herbal: A Practical Guide to Herbal Living.* Elm Tree Books, London, U.K.

*Maekawa, M. and Nishinune, Y. 1985. *Biol. Reprod.,* 32:419-425.

Mager, J., Chevion, M. and Glaser, G. 1980. Favism, In *Toxic Constituents of Plant Food stuffs.* Second Edition. Liener, I.E. (Ed.) Academic Press, New York.

Maheshwari, J.K., Singh, K.K. and Saha, S. 1981. *The Ethnobotany of the Tharus of Kheri District.* National Botanical Research Institute, Lucknow, India.

Maheshwari, J.K., Singh, K.K. and Saha, S. 1986. *Ethnobotany of Tribals of Mirzapur District, Uttar Pradesh.* National Botanical Research Institute, Lucknow, India.

Maheshwari, P. and Umarao singh. 1965. *Dictionary of Economic Plants in India.* Indian Council of Agricultural Research, New Delhi, India.

Mahindra, S.N. 1992. *Indian Plant Perfumes.* Metropolitan Book Company Private Limited, New Delhi, India.

Mahishi, P. 1993. *Health Problems, Herbal Formulations and Quantitative Requirements of Raw Drugs.* M.Sc (Applied Botany) Dissertation. N.M.K.R.V College for Women, Bangalore University, Bangalore, India.

Maikhuri, R. K. and Gangwar, A. K. 1993. Ethnobiological notes on the Khasi and Garo tribes of Meghalaya, North East India. *Economic Botany.* 47: 345-357.

Maiti, R. 1997. *Phaseolus* sp. In: *Bean science.* Oxford & IBH Publishing Co. Pvt. Ltd., New Delhi, India.

*Maldomado, G., Porras, F., Fernandes, L., Vazques, L. and Zenteno, E. 1994. *Immunol. Invest.,* 123:429-436.

Malik, C. P. (Ed). 1982. *Annual Review of Plant Sciences.* Vol. II. Kalyani Publishers, New Delhi, India.

Manandhar, N. P. 1995. An inventory of some herbal drugs of Nyagdi district, Nepal. *Economic Botany,* 49: 371-379.

Mansfield, J.W. 1983. Antimicrobial compounds. In: *Biochemical Plant Pathology.* (Ed.) Callow, J.A., John Wiley & Sons, New York, U.S.A.

Marafioti, T., Barresi, G. and Batolo, D. 1994. Lectin histochemistry of human meningiomas. *Histology and Histopathology,* 9: 535-540.

Martin, G.J. 1995. *Ethnobotany.* Chapman & Hall, London, U.K.

Martindale, W. 1972. *The Extra Pharmacopoeia.* 26th Edition. Pharmaceutical Press, London.

Mazumdar, S., Sengupta, S.K. and Parem, R. 1993. Binding pattern of eight different lectins in healthy subjects and patients with dysplastic and malignant lesions of the oral cavity. *Int. J. Oral Maxillofae. Surg.,* 22:301-5.

*Mazzuca, M., Lhermitte, M., Lafitte, J. J. and Roussel, P. 1982. *Journal of Histochemistry and Cytochemistry,* 30: 956-966.

Meijia, E. G. D., Hankins, C. N., Paredes lopez, O. and Shannlon, L. M. 1990. The lectins and lectin like proteins of tepary beans (*Phaseolus acutifolius*) and tepary common bean (*Phaseolus vulgaris*) hybrids. *Food Biochemistry,* 14: 117-126.

Messina, M. and Barnes, S.1991. The role of soy products in reducing risk of cancer. *Journal of the National Cancer Institute,* 83:541-546.

Metcalf, R.L. 1986. The ecology of insecticides and the chemical control of insects. In: *Ecological Theory and Integrated Pest Managment.* (Ed.) Kogan, M. John Wiley & Sons, New York, U.S.A.

*Metcalf, T.N., Villanueva, M.A., Schindler, M. and Wang, J.L. 1986a. *Journal of Cell Biology,* 102: 1350-1357.

*Metcalf, T.N., Villanueva, M.A., Schindler, M. and Wang, J.L. 1986b. *Proceedings of the National Academy of Sciences.*U.S.A., 83:95-99.

*Mickel, C.E. and Standish, J. 1947. Susceptibility of processed soy flour and soy grits in storage to attack by *Tribolium castaneum. University of Minnesota Agricultural Experimental Station Technical Bulletin*, 178: 1-20.

*Miettinen, M., Holthofer, F., Lehto, V.P., Miettinen, A. and Virtanen, I. 1983. *American J. Clin. Pathol.,* 79:32-36.

*Mikhailova, I.Y., Nikolova, M.P. and Stoyanov, D.P. 1965. Effect of some pharmacological means on experimental atherosclerosis. *Chem. Abstr.,* 64: 11735b.

Milliken, W. 1997. Special-purpose legume genetic resources conserved for agricultural, industrial and pharmaceutical use. *Economic Botany*, 51: 251-263.

*Minura, H., Nishida, H. and Lunuma, N. 1996. *Planta Medica,* 62:150-153.

*Mircevova, L. and Simonova, A. 1966. Effects of various substances on adenosine triphosphatase activity in the erythrocyte memberane. *Collection Czech. Chem. Commun.,* 31: 4145-53.

*Mirelman, D., Galun, E., Sharon, N. and Logan, R. 1975. Inhibition of fungal growth by wheat germ agglutinin. *Nature* (London), U.K., 256: 414-416.

Mitaru, B.N., Reichert, R.D. and Blair, R. 1984. The binding of dietary proteins by sorgum tannins in the digestive tracts of pigs. *Journal of Nutrition,* 114: 1787-1796.

*Mitscher, L.A., Drake, S., Gollapudi, S.R., Harris, J.A. and Shankel, D.M. 1986. In: *Antimutagenesis and Anticarcinogenesis Mechanisms.* (Eds.) Shanke, D.M., Hartman, P.E., Kada, T. and Hollaender, A. Plenum Press, New York and London.

*Mitscher, L.A., Park, Y.H., Clark, D. and Beal, J.L. 1980. *J. Nat. Prod.,* 43: 259-269.

Moon, A. and Kim, S.H. 1997. Effect of *Glycyrrhiza glabra* roots and glycyrrhizin on the glucuronidation in rats. *Planta Medica,* 62: 115-119.

Morris, C.J. and Thompson, J.F. 1956. The identification of (+) S-methyl–L–cysteine sulfoxide in plants. *J. Am. Chem. Soc.*, 70, 1605.

Morris, J.B. 1997. Special-purpose legume genetic resources conserved for agricultural, industrial and pharmaceutical use. *Economic Botany,* 51: 251-263.

Morton, J.F. 1977. *Major Medicinal Plants: Botany, Culture and Uses.* Charles C. Thomas Publishers, Bannerstone House, U.S.A.

Morton, J.F. 1981. In: *Atlas of Medicinal Plants of Middle America.* Charles C. Thomas Publishers, Springfield, U.S.

*Moseley, G. and Griffiths, D.W. 1980. The effect of diets containing field beans of high or low polyphenolic content on the activity of digestive enzymes in the intestines of rats.

Journal of the Science of Food and Agriculture, 31: 255-259.

Muduli, D.S., Marquardt, R.R. and Guenter, W. 1982. Effect of dietary vicine and Vitamin E supplementation on productive performance of growing and laying chickens. *Br. J. Nutr.,* 47: 53.

Mugedo, J. Z. A. and Waterman, P. G. 1992. Sources of tannin: alternatives to wattle (*Acacia mearnsii*) among indigenous Kenyan species. *Economic Botany,* 46(1): 55-63.

Mukherjee, B.(Ed.) 1992. *Traditional medicine.* Oxford & IBH Publishing Co., Pvt., Ltd., New Delhi, India.

Muller, J. 1981. Fossil pollen records of extinet angiosperms. *Bot. Rev.* 47: 1- 142

*Muller-Stoll, W.R. and Madel, E. 1967. Die fossilen Leguminosen-Holzer. Eine Revision der mit Leguminosen Verglichenen fossilen Holzer und Beschreibunger alterer und neuer Arten. *Palaeontographica, Abs. B.,* 119: 95-174.

*Munoz, M., Barrera, E. and Meza, I. 1981. *El uso medicinal y alimenticio de plantas nativas y naturalizadas in Chile.* Museo nacional de Historia Natural, Publicacion ocasional N 33, Santiago de chile.

*Muzquiz, M., Burbano, C., Ayet, G., Pedrosa, M.M. and Cuadrado, C. 1999. BASE: *Biotechnologie, agronomie, societee et environmnet,* 3: 210-216.

Nadasdy, T., Larizik, Z., Blick, K.E., Johnson, D.L., Burst-Singer., K. Nast, C., Cohen, A.H., Ormos, J. and Silva, F.G. 1995. Human acute tubular necrosis: A lectin and immunohistochemical study. *Human Pathology,* 26:230-239.

Nadkarni, A.K. 1976. *Indian Materia Medica.* Popular Prakashan Pvt. Ltd., Bombay, India.

*Nakashima, H., Okubo, K., Honda, Y., Tamura, T., Matsuda, S. and Yamamoto, N. 1989. Inhibitory effect of glycosides like saponin from soybean on the infectivity of HIV *in vitro. AIDS*, 3: 655-658.

*Namba, T. 1980. *Colored illustrations of Wakan-Yaku.* Vol. 1. Hoikush Publishing Company Limited.

*Nandi, N. and Banerjee, S. 1950a. *Indian Pharmacist,* 5: 63-68.

*Nandi, N. and Banerjee, S. 1950b. *Indian Pharmacist,* 5: 202-205.

Narayana, K. 1995. *Drugs.* Jayashri Publications, Bangalore, India.

Narayanan, T.R. and Dabadghao. P.M. 1982. *Forage Crops of India.* Indian Council of Agricultural Research, New Delhi, India.

Nasir, E and Ali, S.I., (Eds.) 1997. *Flora of West Pakistan.* University of Karachi, Karachi.

Nayar, M.P., Ramamurthy, K. and Agarwal, V.S. 1989. *Economic Plants of India.* Botanical Survey of India, Calcutta, India.

Neogrady, S., Galfi, P., Veresegyhazy, T., Bardocz, S. and Pusztai, A. 1994. Lectins as markers of rumen epithelial cell differentiation. *Histochemical Journal,* 26: 197-206.

Newdick, J. 1991. *The Magic of Herbs.* Salamander Books Limited, London, U.K.

Newman, H.A., Kummenow, F.A. and Scott, H.W. 1957. Dietary saponin, a factor which may reduce liver and serum cholesterol levels. *Poult, Sci.,* 37: 42.

Nicolson, G.L. 1974. The interaction of lectins with animal cell surfaces. *International Review of Cytology,* 39:89-190.

*Nishioka, I. 1983. *Yokugaku Zasshi*, 103: 125-142.

Nowell, P.C.1960. Phytohaemagglutinin: An initiator of mitosis in cultures of normal human leukocytes. *Cancer Research,* 20:462-466.

Oakenfull, D. and Sidhu, G.S. 1989. Saponins. In: *Toxicants of Plant Origin.* (Ed.) Cheeke, P.R., CRC Press, Florida, U.S.A. pp. 97-141.

Ochse, J.J. 1977. *Vegetables of the Dutch East Indies*. A. Asher & Co., Amsterdam. pp. 362-439.

Oh, S., Kinjo, J., Shii, Y., Ikeda, T., Nohara, T., Ahn, K.S., Kim, J.H. and Lee, H. 2000. *Planta Medica*, 66: 506-510.

Oliver-Bever. 1986. *Medicinal Plants in Tropical West Africa*. Cambridge University Press, Cambridge.

Olmado, S.F., Sanchez, M.G., Gomez, R.L., Rozo, J. and Carbonero, P.L. 1987. Plant proteinaceous inhibitors of proteinases and α-amylases. *Plant Mol. Biol. & Cell Biol.,* 4:275-334.

*Ominami, H., Kimura, Y., Okuda, H., Arichi, S., Hoshikava, M. and Kitagawa, I. 1984. Effects of saponins on liver injury induced by highly peroxidized fats in rats, *Planta Med.,* 50, 440.

*Oosterhout, V.A.J.M., Landenius, A.R.C., Savelkoul, H.F.J., Van Ark, I., Deshman, K.C. and Nijikamp, F.P. 1993. *Am. Rev. Resp. Dis.,* 147: 548-552.

Ortiz, L.T. Centeno, C. and Trevino, J. 1993. *Animal Feed Science and Technology,* 41: 271-278.

Osaka Y. and Kenkyusho, K.K. 1984. Cosmetics containing soyabean saponins. *Jpn. Kokai Tokkya Koho,* 59:106-410.

Osborn, A. 1996. Saponins and plant defence—a soap story. *Trends in Plant Sci.*, 1: 1360-1385.

*Oshima, Y., Okuyama, T., Takahashi, K., Takizawa, T. and Shibata, S. 1988. *Planta Med.,* 54: 250-254.

*Ottenjam, R. and Rosch, W. 1970.Therapie des ulcus ventriculi mit carbenoxolon-natrium. *Ergebnisse eins doppelblind versuchs.Med.Klin.,* 65:74.

Ozanne, B. and Sambrook, J. 1971. *Nature New Biol.,* (London), 232: 156–160.

Padgette, S.R., Taylor, N.B., Nida, D.L., Bailey, M.R., McDonald, J., Holden, L.R. and Fuchs, R.L. 1996. The composition of glyphosate-tolerant soybean seeds is equivalent to that of conventional soybeans. *Journal of Nutrition,* 126:702-716.

Painter, G. and Power, E. 1982. *A Garden of Old Fashioned and Unusual Herbs*. Hodder & Stoughton, Auckland, London, U.K.

*Pandey, D.K., Chandra, H. and Tripathi, N.N. 1982. Volatile fungitoxic activity of some higher plants with special reference to that of *Callistemon lanceolatus* DC. *Phytopath.,* 105: 175-182.

Pannetier, C., Giband, M., Couzi, P., Letan, V., Mazier, M., Tourneur, J. and Hau, B. 1997. Introduction of new traits into cotton through genetic engineering: insect resistance as an example. *Euphytica,* 96: 163-166.

Patnaik and Dhawan. 1986. Pharmacological studies on some Indian medicinal plants. In: *Current Research on Medicinal Plants of India.* (Ed.) Dhawan, B.N. Indian National Science Academy, New Delhi, India.

Pearce, P.T., Zois, I., Wynne, K.W. and Funder, J.W.1982. *Panax ginseng* and *Eleutherococcus senticosus* extracts *in-vitro* studies on binding to steroid receptors. *Endocrinology. J.*, 29:567-573.

Perry, L.M. 1980. *Medicinal Plants of East and South-East Asia.* The MIT Press, London, England.

Peary, W. and Peavey, W. 2000. Sproutnet. International Sprout Supply.

*Peterson, D.W. 1950. Effect of sterols on growth of chicks fed on high alfalfa diets or diets containing Quillaja. *J. Nutr.,* 42: 597.

Pignol, B., Etienne, A., Carstes de Paulet, A., Deby, C., Mencia-Heurta, J.M. and Braquet, P. 1988. Role of flavonoids in the oxygen free radical modulation of the immune response In: *Plant Flavonoids in Biology and Medicine II: Biochemical, Cellular and Medicinal Properties.* (Eds.) Cody, V., Middleton, E. Jr., Jeffrey, B., Harborne, J.B. and Beretz, A. Alan, R. Liss, Inc., New York, U.S.A.

Pineau, N., Aucouturier, P., Brugier, J.C. and Preud'Homme, J.L. 1990. Jacalin: a lectin mitognic for human CD4 T lymphocytes. *Clinical and Experimental Immunology*, 80:420-425.

Platt, 1962, Q.f. Tindall, H.D. 1983. *Vegetables in the Tropics.* Avi Publishing company, Inc., West Port, Connecticut.

Polhill, R.M. and Raven, P.H. 1981. (Eds.) *Advances in Legume Systematics.* Vols. 1 & 2. Royal Botanic Garden, Kew, U.K.

Pompei, R., Flore, O., Marccialis, M.A., Pani, A. and Loddo, B. 1979. Glycyrrhizic acid inhibits virus growth and inactivates virus particles. *Nature*, 281: 689-690.

Poretz, R.P., Riss, H., Timberlake, J.W. and Chien, S. 1974. Purification and properties of the hemagglutinin from *Sophora japonica* seeds. *Biochemistry.* 13: 250 – 256.

Premnath, 1976. *Vegetables for the Tropical Region.* ICAR, New Delhi, India.

Price, W.A. 1939. *Nutrition and Physical Degeneration. A Comparison of Primitive and Modern Diets and Their Effects.* P.B. Hoeber, New York, U.S.A.

Principe, P. 1991. *Monetizing the Pharmacological Benefits of Plants.* US Environmental Protection Agency, Washington, D. C., U.S.A.

Purkayastha, R.P. 1990. *Economic Plants and Microbes.* Today & Tomorrow's Printers & Publishers, New Delhi, India.

Purseglove, J.W. 1991. *Tropical Crops: Dicotyledons.* Longman Scientific & Technical Group, New York, U.S.A.

Pusztai. A. 1988. Transport of proteins through the membranes of adult gastrointestinal tract- A potential for drug delivery? *Advanced Drug Delivery Review,* 3:215-228.

Pusztai, A.,1991. *Plant Lectins.* Cambridge University Press, Cambridge.

Pusztai, A., Clarke, E.M.W., Grant, G. and King, T.P. 1981. The toxicity of *Phaseolus vulgaris* lectins. Nitrogen balance and immuno chemical studies. *Journal of the Science of Food and Agriculture.* 32: 1037–1046.

*Pusztai, A. and Bardocz, S. 1996. Biological effects of plant lectins on the gastrointestinal tract: metabolic consequences and applications. *Trends in Glycoscience and Glycotechnology*, 8:149-165.

Pusztai, A., Croy, R.R.D., Stewart, J.C. and Watt, W.B.1979. Protein body membranes of *Phaseolus vulgaris* cotyledons: isolation and preliminary characterization of constituent proteins. *New phytologist*, 83:371-378.

Pusztai, A., de Oliveira, J.T.A., Bardocz, S. and Grant, G. 1987. In: *Lectins: Biology, Biochemistry, Clinical biochemistry.* (Eds.) Bog-Hansen, T.C. and Freed, D.L.J. Sigma Chemical Company.

Pusztai, A., Ewen, S. W. B., Grant, G., Peumans, W. J., Van Damme, E. J. M., Coates, M. E., and Bardocz, S. 1995. Lectins and also bacteria modify the glycosylation of gut receptors in the rat. *Glycoconjugates. J.,* 12: 22-35.

Pusztai, A., Koninkx, J., Hendrix, H., Kok, W., Hulscher, S., Van Damme, E.J.M., Peumans, W.J., Grant, G. and Bardocz, S. 1996. Effect of the insecticidal *Galanthuse nivalis* agglutinin on metabolism and activities of brush border enzymes in the rat's small intestine. *Journal of Nutritional Biochemistry,* 7:677-682.

Qui, G., Elst, E.R., Bossch, M.W., Poel, A.F.B. and van der Poel, A.F.D. 1996. *Animal Feed Science and Technology,* 57:313-324.

*Race, K. and Sanger, S. 1975. *Arch. of Pharmacol.*, 5:110-112.

Rackis, J.J. and Gumbman, M.R. 1981. In: *Antinutrients and Natural Toxicants in Food.* (Ed.) Ory, R.L. Food and Nutrition Press, Connecticut, U.S.A.

Rademaker, T.W., Parekh, R.B. and Dwek, R.A. 1988. Glycobiology. *Ann. Rev. Biochem.,* 57:785-838.

Radha, B.1992. *A Probe into Health Problems, Herbal Formulations and Quantitative Requirements of Raw Drugs: An Analysis Through Computer Database,* Part 1. M.Sc (Applied Botany) Dissertation. Bangalore University, Bangalore, India.

Ramaswamy, S.V. and Razi, B.A. 1973. *Flora of Bangalore District.* University of Mysore Press, Mysore, India.

Rao, C.K. 1986. Seed lipid content of the Galegae (Fabaceae). *Acta. Bot. Indica.*, 14: 156-161.

Rao, C.K. and Subhashini, G. 1986. Saponins and Leucoanthocyanins in *Cassia* L. *Current science,* 55(6): 320-321.

Rastogi, P.R. and Mehrotra, B.N. 1990. *Compendium of Indian medicinal plants.* Vol. I. Central Drug Research Institute, Lucknow. Publication and Information Directorate, New Delhi, India.

Ratikanta Maiti, 1997. *Bean Science.* Oxford and IBH Publishing Co. Pvt. Ltd., New Delhi, India.

Raut, A.M. 1993. Concept of toxicology in *Ayurveda.* In: *Traditional Medicine,* (Ed. in chief) Mukherjee, B. Oxford & IBH Publishing Co., Pvt., Ltd., New Delhi.

Ray, P. and Gupta, H.N. 1980. *Caraka Samhita.* Indian National Science Academy, New Delhi, India.

Ray, P., Gupta, H.N. and Roy, M. 1980. *Susruta Samhita (Scientific synopsis).* Indian National Science Academy, New Delhi, India.

*Reddy, N.R. and Pierson, M.D. 1994. *Food Research International.* 27: 281.

Reddy, N.R., Pierson, M.D., Sath, S.K. and Salunkhe, D.K. 1985. Dry bean tannins – a review of nutritional implications. *J. An. Oil Chem. Soc.*, 62: 541.

Ree, H.J. 1988. PNA-a marker of Reed-Stenberg cells in paraffin sections. In *Lectins: Proc. of 9th Lectin meeting,* Cambridge. (Eds.) Bog-Hansen, T.C. and Freed, D.L.J.

Reisner, Y. 1983. Differential agglutination by soybean agglutinin of human leukemia and neuroblastoma cell lines: Potential application to autologous bone marrow transplantation. *Proceedings of the National Academy of Sciences,* U.S.A.

Remani, P., Nair, R.A., Kusumakumari, J.P., Bhattathri, V.N., Vijayakumar, T. and Krishnan Nair, M. 1998. Jackfruit lectin binding pattern in leukaemia. In: *Proceedings of 10th Kerala Science Congress,* Thiruvananthapuram.

*Revers, F.E. 1946. Has Licorice juice a healing action on gastric ulcer? *Ned.Tijdschr.Gencesk.*, 90: 135.

Rhodes, E.G.H., Baker, P., Rhodes, J.M., Davies, J.M. and Cawley, J.C. 1991. Peanut agglutinin in combination with CD19 monoclonal antibody has potential as a purging agent in myeloma. *Experimental Hematology,* 19:833-837.

Richardson, M.J. 1991. Seed storage proteins: the enzyme inhibitors. In: *Methods in Plant biochemistry.* (Ed) Richardson, M.J., Academic Press, New York.

Robbins, W. W. and Ramaley, F. 1937. *Plants Useful to Man.* P. Blakiston's Sons & Co. Inc., Philadelphia, U.S.A.

Robert, A., Nezamis, J.E., Lancaster, C. and Hanchar, A.J. 1979. Cytoprotection by prostaglandins in rats. *Gastroenterology* 77: 433-443.

Robinson, W.E., Montefiori, D.C. and Mitchell W,M., 1987. Evidence that mannosyl residues are involved in human immunodeficiency virus type I (HIV-I pathogens). *AIDS Research and Human Retrovirus*, 3: 265 – 282.

*Rocher, Y., 1965. Isolation of saponins from *Ficaria ranunculoides. Chem. Abstr.,* 64: 6418.

*Rodriguez-Burger, A.P., Manson, A. and Nielson, S.S. 1998. *Journal of Agricultual and Food Chemistry,* 46:4806-4813.

Rong, H., Stevens, J.F., Deinzer, M.L., Cooman, L.D. and Keukeleire, D.D. 1998. *Planta Medica*, 64: 620-627.

Ross, I.A. 1999. *Medicinal Plants of the World.* Vol. I. Humana Press, Totowa, New Jersey.

Ross, I.A. 2000. *Medicinal Plants of the World.* Vol II. Humana Press, Totowa, New Jersey.

Rusznyak St. and Szent – Gyorgyi, A. 1936. Vitamin: Flavonols as vitamins. *Nature* (London) 138: 27.

Ryan, C.A. 1989. Insect-induced chemical signals regulating natural plant protection responses. In: *Variable Plants and Herbivores in Natural and Managed Systems.* (Eds.) Denno, R.F. and McClure, M.S., New York, Academic Press.

Ryan, C.A. 1990. Protease inhibitors in plants: genes for improving defenses against insects and pathogens. *Annual Review of Phytopathology,* 28: 425-449.

Ryder ,S.D., Parker, N., Ecclestone, D., Haqqani, M.T and Rhodes, J.M. 1994. Peanut lectin stimulates proliferation in colonic explant from patients with inflammatory bowel disease and colon polyp. *Gastroenterology,* 106: pp 117–124.

*Ryu, S.Y., Choi, S.U., Lee, C.O. and Zee, O.K. 1992. Anti-tumor activity of *Psoralea corylifolia. Archives of Pharmacol Research,* 15:356-359.

*Ryu, S.Y., No, Z., Kim, S. and Ahn, J. 1995. *Arch. Pharm. Res.,*

Said, H.M. 1969. (Ed) *Humdard Pharmacopoeia of Eastern Medicine.* Institute of Health and Tibbi Research, Karachi, India.

Said, H.M. 1983. *The Unani System of Health and Medicare in Traditional Medicine and Health Care Coverage.* (Ed) Bannermann, R.H., Burton, J. and Wen-Cheih, C. WHO, Geneva.

Saito, H. 1993. Purification of active substances of *Aloe arborescens* and their biological and pharmacological activity. *Cytotherapy Research,* 7:S14-S19.

*Saito, K., Arai, N., Sekine, T., Ohmiya, S., Kubo, H., Otomasu, H. and Murakoshi, I. 1990. *Planta Medica,* 56: 487-488.

*Sakai, Y., Nagase, H., Ose, Y., Sato, T., Yamada, A., Hibi, M. and Yamada, F. 1986. *Mutation Research,* 174:1-4.

Saldanha, C.J. 1982. *Flora of Karnataka*. Vol. I. Amerind Private Ltd., New Delhi, India.

Saldanha, C.J. and Nicolson, D.H. (Eds.) 1976. *Flora of Hassan District: Karnataka, India*. Amerind Publishing Co. Pvt. Ltd., New Delhi, India.

*Saleh, M.M., El-Olemy, M.M., Metawie, H.M., Abou-Zied, A.H. and Abdel-Salam, M.S. 1990. *Planta Medica,* 56: 610.

Salunkhe, D.K. 1982. Legumes in human nutrition. *Curr. Sci.,* 51: 387-394\

Salunkhe, D. K., Jadhav, S. J., Kadam, S. S. and Chavan, J. K. 1982. Chemical, biochemical and biological significance of polyphenols in cereals and legumes. *Crit. Rev. Food Sci. Nutr.,* 17: 277.

Salunkhe, D. K., Chavan, J.K. and Kadam, S.S. 1990. *Dietary Tannins: Consequences and Remedies.* CRC Press.U.K.

Salvayre, R., Negre, A., Affany, A., Lenoble, M. and Douste-Blazy, L. 1988. Protective effect of plant flavonoids, analogs and Vitamin E against lipid peroxidation of membranes. In: *Plant Flavonoids in Biology and Medicine II: Biochemical, Cellular and Medicinal Properties.* (Eds.) Cody, V., Middleton, E. Jr., Jeffrey, B., Harborne, J.B. and Beretz, A. Alan, R.Liss, Inc., New York, U.S.A.

Sambamurthy, A.V.S.S. and Subrahmanyam, N.S. 1989. *A Textbook of Economic Botany*. Wiley Eastern Limited, India.

Sambamurthy, A.V.S.S. and Subrahmanyam, N.S. 1998. *A Textbook of Modern Economic Botany*. CBS Publishers & Distributor, New Delhi, India.

Sangeetaa, M. 1994. *Lectins and Saponins in Pollen and Stigmas.* Ph.D. thesis, Bangalore University, India.

Sangeetaa, W. and Kameswara Rao, C. 1993. Saponins in pollen. *Grana,* 32: 255-259.

Sanjappa, M. 1992. *Legumes of India*. Botanical Survey of India, Howrah, India.

Santapau, H. & Henry, A.N. 1973. *Dictionary of the Flowering Plants in India.* CSIR, New Delhi, India.

*Santidrian, A. and Marzo. 1989. *J. Sci. Food and Agric.,* 47:435-442.

Sardana, R.K, Ganz, P.R., Dudani, A.K., Tackaberry, E.S., Cheng, X. and Altosaar, I. 1998. Synthesis of recombinant human cytokine GM-CSF in the seeds of transgenic tobacco plants. In: *Recombinant Proteins from Plants. Production and Isolation of Clinically Useful Compounds.* (Eds.) Cunningham, C. and Proter, A., Humana Press Inc., Totowa, N.J.

Sarker, A.B., Akogi, T., Teremoto, N., Nose, S., Yoshino, T. and Kondo, E. 1994. *Bauhinia purpurea* agglutinin (BPA) binding to normal & neoplastic thyroid glands. *Pathological Research and Practice*, 190:1005-1011.

Sathyanarayana Bhat. 1993. *Plants with Antimicrobial Activity: Database and Lectin Distribution*. Ph.D. Thesis. Bangalore University, Bangalore, India.

Sathyanarayana Bhat and Kameswara Rao, C. 1993a. *Bharatheeya Visha Vaidya,* (in Kannada). Directorate of Indian Systems of Medicine and Homeopathy, Government of Karnataka.

Sathyanarayana Bhat and Kameswara Rao, C. 1993b. *Ayurveda Parichaya*: an introduction to *Ayurveda* (in Kannada). Directorate of Indian Systems of Medicine and Homeopathy, Government of Karnataka.

Sathyanarayana Bhat and Kameswara Rao, C. 1993c. *Siddha Vaidya Parichaya:* an introduction to *Siddha* system (in Kannada). Directorate of Indian Systems of Medicine and Homeopathy, Government of Karnataka.

Sathyanarayana Bhat and Kameswara Rao, C. 1993d. *Unani Vaidya Parichaya:* an introducton to *Unani* system (in Kannada). Directorate of Indian Systems of Medicine and Homeopathy, Government of Karnataka.

Satyananda, N. 1989. *Studies on Legume Lectins.* Ph.D thesis, Bangalore University, Bangalore, India.

Schenk, S.U., Lambein, F. and Werner, D. 1991. Broad antifungal activity of α- isoxazolinonylalanine, a non-protein amino acid from roots of pea (*Pisum sativum*) seedlings. *Biol. Fertil. Soils.,* 11: 203-209.

*Schoham, J., Inbar, M. and Sachs, L.1970. *Nature* (London), U.K., 227:1244-1246.

Schultes, R.E. and Reis, S.Von. 1995. (Eds.) *Ethnobotany: Evolution of a Discipline.* Chapmann and Hall, London, U.K.

*Schwartz, A. and Middleton, E. Jr. 1984. Comparison of the effects of quercetin with those of other flavonoids on the generation and effector function of cytotoxic T lymphocytes. *Immunopharmocology,* 7:115.

*Sehwechhimer, K., Weiss, G., Sehnable, P. and Moller, P. 1984. *Histochemistry*, 80:165.

Shah, C.S. and Qadry, J.S. 1985. *A Text Book of Pharmacognosy.* B.S. Shah Prakashan Ahmedabad, India.

Shalini, B.U. 2002. *Studies on Economically Important Legumes of India: Database and Distribution of Lectins and Saponins.* Ph.D., thesis, Bangalore Univrsity, India.

Shanmugavelu, K.G. 1989. *Production Technology of Vegetable Crops.* Oxford and IBH Publishing Co. Pvt. Ltd., New Delhi, India.

Sharma, A. and Sehgal, S. 1992. Effect of processing and cooking on the antinutritional factors of faba bean (*Vicia faba*). *Food Chemistry.* 43 (5): 383–385.

Sharma, P.C., Yelne, M.B. and Dennis, T.J. 2000. *Database on Medicinal Plants Used in Ayurveda.* Vol. I. Central Council for Research in Ayurveda and Siddha, New Delhi, India.

Sharma, P.C., Yelne, M.B. and Dennis, T.J. 2001. *Database on Medicinal Plants Used in Ayurveda.* Vol. II. Central Council for Research in Ayurveda and Siddha, New Delhi, India.

Sharma, P.C., Yelne, M.B. and Dennis, T.J. 2002. *Database on Medicinal Plants Used in Ayurveda.* Vol. III. Central Council for Research in Ayurveda and Siddha, New Delhi, India.

Sharma, P.C., Yelne, M.B. and Dennis, T.J. 2003. *Database on Medicinal Plants Used in Ayurveda.* Vol. IV. Central Council for Research in Ayurveda and Siddha, New Delhi, India.

Sharma, P.V. 1993. *Sodasangahrdayam, Essentials of Ayurveda.* Motilal Banarsidas Publishers Private Limited, Delhi, India.

Sharon, A. 1994. *Database of Plants Used in Gastro-Intestinal Disorders and Effects of Plant Extracts on Pathogenic Enteric Bacteria.* Ph.D Thesis, Bangalore University, Bangalore, India.

*Sharon, N. 1976. Lectins as mitogens In: *Mitogens in Immunobiology.* (Eds.) Oppenheim, J.J. and Rosenstreich, D.L. Academic Press, New York, U.S.A.

Sharon, N. 1983. Lectin receptors as lymphocyte surface markers. *Advances in Immunology,* 34:213-298.

Sharu Raj, M. 1990. *Studies on Lectins in Some Food Plants.* M. Phil., Dissertation. Bangalore University, Bangalore, India.

Sheng-Chukuo, Sheng-Chin Chen, Lin-Hui Chen, Jih-Pyang Wang and Che-Ming Teng. 1995. *Planta Medica,* 61: 307-312.

*Shet, M.S., Madaiah, M. and Ahmed, K.N.1989. *Indian Journal of Experimental Biology,* 27:58-61.

Shibata, S. 1972. Saponins with biological and pharmaceutical activity. In: *New Natural Product and Plant Drugs with Pharmocological, Biological and Therapeutical Activity.* (Eds.), Wagner, H. and Wulff, P. Springer Verlag, Berlin.

Shibata, S., Inove, H., Iwata, S., Ma, R., Yu, L., Ueyama, H., Takayasu, J., Hasegawa, T., Tokuda, H., Nishino, A., Nishino, H. and Iwashima, A. 1991. *Planta Medica,* 57: 221-224.

*Shinohara, M., Nakano, Y., Kaise, H., Izawa, T. and Miyazaki, W. 1980. *Chemical Abstracts,* 92: 153139b.

Shiraki, K., Takase, K., Tameda, Y., Hamada, M. Kosaka, Y. and Naakano, T. 1995. A clinical study of lectin reactive alpha-fetoprotein as an early indicator of hepatocellular carcinoma in the follow-up of cirrhotic patients. *Hepatology,* 22:802-807

*Shoham, J., Inbar, M.and Sachs, L. 1970. *Nature* (London). pp 227:1244.

*Shoshana, M., Shlomo, M. and Shimon, S. 1988. *Cancer Res.,* 48: 4573-4577.

Shubha Rani, R. 1995. *Plants Used in Dental Care in India.* Ph.D. Thesis, Bangalore University Bangalore, India.

Shue, G. L., Kawa, H., Kato, M., Oguachi, H., Kobayarhi, T., Koival, T., Tokoo, M; Furuta, S. and Homa, T. 1993. Expression of glycocongugates in pancreatic, gastric and colonic tissue by *Bauhinia purpurea, Vicia villosa* and Peanut lectins. *Scand. J. Gastroenterol.,* 28:599-604

Siess, M-H., , Christophe Broudrd, Marie-France Vernevaut and Marc Suschetet. 1988. Effects of feeding quercetin and flavon on hepatic drug metabolizing enzymes of rat. In: *Plant Flavonoids in Biology and Medicine II: Biochemical, Cellular and Medicinal Properties.* (Eds.) Cody, V., Middleton, E. Jr., Jeffrey, B., Harborne, J.B. and Alian Beretz. Alan, R.Liss, Inc., New York, U.S.A.

Silva, F.G., Nadasdy, T., Laszik, Z, 1993. Immuno histochemical and lectin detection of the human nephron in health and disease. *Arch. Pathol.Lab.Med.,* 117:1233-9.

Simpson, G. B. and Conner-Ogorzaly, M. 1986. *Economic Botany. Plants in Our World.* Mcgraw-Hill Book Company, Singapore.

Singh, H.R. and Arora, R.K. 1978. *Wild Edible Plants of India.* ICAR, New Delhi, India.

Singh, J., Kamboj, S.S., Sandhu, R.S., Shangary, S. and Kamboj, K.K. 1993. Purification and characterization of a tuber lectin from *Alocasia indica. Phytochemistry,* 33:979-983.

Singh, N.P and Karthikeyan, S. (Eds.) 2000. *Flora of Maharashtra state.* Vol. I. Botanical survey of India, Government of India.

Singh, N.P., Chauhan, A.S., Mondal, M.S., (Eds.) 2000. *Flora of Manipur* Vol. I, Botanical survey of India, Government of India.

Singh, N.P., Singh, K.P. and Singh, D.K., (Eds.) 2002. *Flora of Mizoram,* Vol. I. Botanical survey of India, Ministry of Environment and forests, Government of India.

*Singh, U. 1993. Protein quality of pigeon pea as influenced by seed polyphenols and cooking process. *Plant Foods for Human Nutrition.* 43: 171-179.

Singh, U.R., Wadhwani, A.M. and Johri, B.M. 1965. *Dictionary of Economic Plants of India.* Indian Council of Agricultural Research, New Delhi, India.

Sinha, R.K. 1998. *The Renaissance of Traditional Herbal Medicine.* INA Shree Publishers, Jaipur, India.

*Sitori, C.R., Agradi, E., and Conti, F. 1977. *Lancet,* 1: 257-277.

Sivaranjan, V.V. and Indira Balachandran. 1994. *Ayurvedic Drugs and Their Plant Sources.* Oxford & IBH Publishing Company Private Limited, Delhi, India.

Skerman, P.J. 1977. *Tropical Forage Legumes*. Food & Agricultural Organisation of United Nations, Rome.

Smartt, J. 1976. *Tropical Pulses*. Longman Group Limited.

Smith, P.M. 1976. *The Chaemotaxonomy of Plants*. Edward Arnold, London, U.K.

So, L.L. and Goldstein, I.J. 1967. Protein-carbohydrate interaction. Application of the quantitative hapten inhibition technique to polysaccharide-concanavalin A interaction. Some comments on the forces involved in concanavalin A-polysaccharide interaction. *Immunol.,* 99: 158-163.

*So, L.L. and Goldstein, I.J. 1968. *J. Biol. Chem.,* 243: 2003.

*Sokolchik, I.G., Sukolinskii, V.N., Morozkina, T.S., Kukhta, V.K. 1992. *Radiobiol.,* 32: 728-730.

*Soo, R. 1963. "*Fejlodestorteneti Novenyrendszartan Tankonyokiada*". Budapest.

Soto-Hernandez, M. and Jackson, A.H. 1994. *Erythrina* alkaloids: isolation and characterization of alkaloids from seven *Erythrina* species. *Planta Medica.* 60: 175-177.

*Sotomatsu, S., Takahashi, Y., Hiroi, J., Namikata, A. and Okano, N. 1959. *Skin and Urology,* 21: 138.

*Srivastava, R.C, Puri, V., Srimal, R.C. and Dhavan, B.N.1986. Prostanoid mediated effects of centchroman, a nonsteroidal oral contraceptive. *Agent Actions,* 18:596-599.

Srivatsava, R.C. 1989. *Drug Plant Resources of Central India (An inventory).* Today & Tomorrow's Printing & Publishers, New De1hi, India.

*Stastny, J. and Das Gupta, T. 1994. *Anticancer Res.,* 14 : 587-591.

Stoddart, R. W., Ball, R. Y., Jones, C. J. P. and Mitchinson. In *Lectins, Biology, Biochemistry, Clinical Biochemistry.* Vol. 6, (Eds.) Freed, D. and Hansen, E. C., Sigma Library, 1988.

Stuart, G.A. 1985. *Chinese Materia Medica. Vegetable Kingdom.* (International Book Distribution, Dehradun, India). Oriental Booksellers & Publishers, London.

Stuart, M. 1989, *The Encyclopedia of Herbs and Herbalism.* MacDonald & Co. Publishers Ltd., USA.

*Sun, C.O., Wang, S.R., Zou, Z. and Chun, Q. 1983. *Acta Pharm. Sin.,* 18: 81-85.

*Suter, M. M., Augustin-Voss, H. G., Pantano, D. M., Flanders, J. A. and Varvayanis, M. 1991. Differentiation-dependent expression of lectin binding sites on normal and neoplastic keratinocytes *in vivo* and *in vitro. Journal of Histochemistry and Cytochemistry,* 39: 1103-1112.

Suvachittanont, W. and Jaranchavan, P. 2000, Mitogenic effect of *Parkia speciosa* seed lectin on Human lymphocytes. *Planta Medica,* 66:699-704.

*Suvachittanont, W., Pothirakit, P. 1988 Protein in *Parkia speciosa* seeds. *Songela Med. J.,* 6: 23-30.

*Suzuki, H., Ohta, Y., Takino, T., Fujisawa, K., Hirayama, C., Shimizu, N. and Aso, Y. 1977. *Igakuno Aymi.,* 102: 562.

Swiader, J.M., Ware, G.W. and Mccollum, J.P. 1992. *Producing Vegetable Crop.* Central International Book Distribution Company, Lucknow, India.

*Takachi, M., Matsunami, S., Nishizama, J., Lino, C. and Tanaka, Y. 1999. Haemolytic and antifungal activity of saponins. *Planta Medica,* 65: 1585-86.

*Takashi, I., Keiichi, U. and Hisayuki, K. 1986. Gourd saponins as antioxidants in oils, in food, cosmetics and pharmaceuticals, *Japan Kokai Tokkyo Koho,* 61: 130-390; *Chem. Abstr.,* 105:51815.

Takhtajan, A. 1997. *Diversity and Classification of Flowering Plants.* Columbia Univ. Press, New York, U.S.A.

Talbot, C.F. and Etzler, M.E.1978a. Isolation and characterization of a protein from leaves and stems of *Dolichos biflorus* that cross reacts with antibodies of seed lectin. *Biochemistry,* 17:1474-1479.

Talbot, C.F. and Etzler, M.E.1978b. Development and distribution of *Dolichos biflorus* lectin as measured by radioimmunoassay. *Plant Physiology,* 61:847-850.

Tanaka, H., Etoh, H., Shimizu, H., Makita, T. and Tateishi, Y. 2000. *Planta Medica,* 66: 578-579.

Tang, W. and Eisenbrand, G. 1992. *Chinese Drugs of Plant Origin.* Springer-Verlag, Berlin, Heidelberg.

*Taubert, P. 1894. Leguminosae. In: *Die Naturlichen Pflanzenfamilien.* (Eds.) Engler and Prantl.

Tawata, M., Yoda, Y., Aida, K., Shindo, H., Sasaki, H., Chin, M. and Onaya, T. 1990. Anti-platelet action of GU-7, a 3-arylcoumarin derivative, purified from *Glycyrrhizae radix*. *Planta Medica*, 56: 259-263.

Telikepalli, H., Gollapudi, S.R., Keshavarz-Shokri, A., Velazquez, L., Sandmann, R.A. and Veliz, E.A. Jagannadha Rao K. V., Siva Madhavi A., Mitscher L.A.1990. Isoflavonoids and a cinnamyl phenol from root extracts of *Erythrina variegata. Phytochemistry,* 29: 2005-2007.

*Terpstra, A.H.M., West, C.E., Fennis, J.T.C.M. 1984. *Am. J. Clin. Nutr.,* 39: 1-7.

Thakur, R.S., Puri, H.S. and Akhtar Hussain. 1991. *Major Medicinal Plants of India.* Central Institute of Medicinal & Aromatic Plants. Lucknow, India.

Thomson, W.A.R. 1976. *Herbs that Heal.* Adam & Black Limited, London, U.K.

Timoshenko, A., Chrenkevich, S., and Gabius, H. 1995. *Viscum album* agglutinin induced aggregation of blood cells and the lectin effects on neutrophil function. *Biomed. Pharmacother.,* 49: 53-158.

Tindall, H.D. 1983. *Vegetables in the Tropics.* Avi Publishing Company, Inc., West port, Connecticut.

*Tomohisa, A. and Shintara, A. 1984. Topical formulations containing saponins and hydroquinone glycosides for cosmetics and pharmaceuticals, *Jpn Kokai Tokkyo Koho,* 61, 210, 004; *Chem. Abstr.,* 106,72705p.

Toms, G.C. 1981. Lectins in Leguminosae. In. *Advances in Legume Systematics.* Part 2. (Eds.) Polhill, R.M. and Raven, P.H. Royal Botanic Gardens, Kew.

Toms, G.C. and Western, A. 1971. Phytohaemagglutinins. In: *Chaemotaxonomy of the Leguminosae.* (Eds.) Harborne, J.B., Boulter, D. and Turner, B.L. Academic Press, New York.

Trease, G.E. and Evans, W.C. 1978. *Pharmacognosy.* Bailliere Tindall, London, U.K.

Trivedi, P.C. 2002. *Ethnobotany.* Avishkar Publishers and Distributors, India.

Troup, R.S. 1986. *Indian Woods and Their Uses.* Soni Reprints Agency, Delhi, India.

*Tu, Z., Yin, Z. and Wu, M. 1992, Prospective study on the diagnosis of hepatocellular carcinoma by using alpha-fetoprotein reactive to lentil lectin. *Chin.Med. Sci. J.,* 7:191-6.

*Tunis, M.1964. *Journal of Immunology,* 92:864.

Uma, K. 1993. *Documentation of Aromatic Plants with Reference to Ayurvedic Literature.* M.Sc (Applied Botany) Dissertation, Bangalore University, Bangalore, India.

Uphof, J.C.T.H. 1968. *Dictionary of Economic Plants.* Wheldon & Wesley Ltd., Codicote. Herts.

Urwin, P.E., Lilley, C.J., Mcpherson, M.J. and Atkinson, H.J. 1997. Resistance to both cyst and root knot nematodes conferred by transgenic *Arabidopsis* expressing a modified plant cystatin. *Plant Journal,* 12: 455-461.

Vain, P., Worland, B., Clarke, M.C., Richard, G., Beavis, M., Liu, H., Kohli, A., Leech, M., Snape, J. and Christou, P. 1998. Expression of an engineered cysteine proteinase inhibitor (Oryzacystatin-I delta D86) for nematode resistance in transgenic rice plants. *Theoretical and Applied Genetics,* 96: 266-271.

*Vallidevi, A., Ramanuja, M.N., Rao, N.A.N. and Nath, H. 1972. *Indian J. Nutri. Dietet.,* 9: 336-341.

Van Damme, E.J.M., Peumans, W.J., Pusztai, A. and Bardocz, S. 1998. *Handbook of Plant Lectins: Properties and Biomedical Applications.* John Wiley & Sons., New York.

*Van der Stoep, J. 1981. *Fd. Technol.,* 35: 83-85.

*Van Etten, C.H. and Tookey, H.L. 1983. Glusinolates. In: *Handbook of Naturally Occurring Food Toxicants.* (Ed.) Reichcigl, M. Jr., CRC Press, Boca Raton, FL.

Van Etten, C.H., Gagne, W.E., Robbins, D.J., Booth, A.N., Daxenbichler, M.E. and Wolff, I.A. 1969. Biological evaluation of Crambe seed meals and derived products in rat feeding. *Cereal Chem.*, 46: 145.

Variers, P.S. 1994. *Indian Medicinal Plants.* Vol. I and Vol. II Orient Longman Limited, New Delhi, India.

Variers, P.S. 1995. *Indian Medicinal Plants.* Vol. III and Vol. IV. Orient Longman Limited, New Delhi, India.

Variers, P.S. 1996. *Indian Medicinal Plants.* Vol V. Orient Longman Limited, New Delhi, India.

Vavilov, N.I. 1932. *Studies on the Origin of Cultivated Plants.* Leningrad.

Vedavathi, M. and Sudhakar. 1997. *Tribal Medicine of Chittor District, Andhra Pradesh (India.),* Herbal folklore Research Center, Tirupathi.

*Venugopal, K. and Rama Rao, G. 1978. *Indian. J. Nutr. Dietet.,* 15: 7-11.

Vered, Y., Grosskopf, I., Palevitch, D., Harsat, A., Charach, G., Weintraub, M.S. and Graff, E. 1997. *Planta Medica,* 63: 237-240.

Vijayajkumar, T., Leslie, M., Augustine, J., Nir, M.B., Remani, P and Aleykutty, M. 1992. Tissue binding pattern of plant lectins in benign and malignant lesions of thyroid. *J. Expt. Pathol.,* 6:11-12.

*Vijayakumar T. and Bhattathri V.N. 1995. *Oral Oncology,* 4:303.

*Vijayakumar, K., Sidharaju, P. and Janardhanan, K. 1996. *International Journal of Food Sciences and Nutrition,* 47:263-272.

Vijayakumar, T. 1998. Isolation, purification and characteriztion of plant lectns and their application in oncology. *Rheedea*, 72:75.

*Vijayakumar, T., Bhattathiri, V. N., Remani, P., Bindu, L., Chandralekha, B. and Nair, M. K. 1998. *British Journal of radiology,* 71: (Suppl.), 21.

Vlietinck, A.J., Dirk, A., Berghe, V. and Haemers.A.1988. Present status and prospects of flavonoids as anti-viral agents. In: *Plant Flavonoids in Biology and Medicine II: Biochemical, Cellular and Medicinal Properties.* (Eds.) Cody, V., Middleton, E. Jr., Jeffrey, B., Harborne, J.B. and Alian Beretz. Alan, R. Liss, Inc., New York, U.S.A. 280:283-99.

Vonreis Altschul, S. 1973. *Drugs and Foods from Little Known Plants.* Harvard University Press, Cambridge.

*Wagner, H. and Elbl, G. 1992. *Planta Medica,* 58: 297.

*Walker, R.A.J. 1984. *Pathol.*, 144:109-118.

Walker, R.A.J. 1985a. The binding of peroxidase labeled lectins to human breast epithelium. IV. The reactivity of breast carcinoma to peanut, soyabean & *Dolichos biflorus* agglutinin. *J. Pathol.*, 145:269-277

Walker, R.A.J. 1985b. *Ulex europaeus*. I-Peroxidase as a marker of vascular endothelium: its application in routine histopathology: *Path.Res. Pract.*, 185:826-35.

Walker, R. A. J., 1988. The use of lectins in histology and histopathology – review. In: *Lectins-Biology, Biochemistry, Clinical Biochemistry,* Vol. 6. (Eds.) Freed, D. and Hansen, T. C. Sigma Library

Wallis, T.E. 1995. *Text book of Pharmacognosy*. CBS publishers & distributors, Delhi, India.

Walter, K.S. and Gillett, H.J. (Eds.) 1998. 1997 *IUCN Red List of Threatened Plants*. Compiled by The World Conservation Monitoring Centre, IUCN-The World Conservation Union, Glands, Switzerland and Cambridge, U.K.

*Wang, Q., Zhao, L. and Mai, K. 1996. *Zhong eaoyao*, 27:229-230.

Watson, 1971. Q.f. Tindall, H.D. 1983. *Vegetables in the Tropics*. Avi Publishing Conmpany, Inc., West Port, Connecticut.

Watt, G. 1972. *A Dictionary of the Economic Products of India*. Vol. I - Vol. VI. Periodical Experts, Delhi, India.

Watts, L.R. and Watts, G.S. 1944. *The Vegetable Growing Business.* Orange Judd Publishing Company Inc.,

Weder, J.K.P. 1981. Protease inhibitors in the Leguminoseae In: *Advances in Legume Systematics* Part 2. (Eds.) Polhill, R.M. and Raven, P.H.Royal Botanic gardens, Kew, U.K..

Weder, J.K.P.1985. Chemistry of legume protease inhibitors and their use in taxonomy. *Qual. Plant.: Plant Foods Hum. Nutr.* 35, 183-194

Weenen, H., Nkunya, M.H.H., Bray, D.H., Mwasumbi, L.B., Kinabo, L.S. and Kilimali, V.A.E.B. 1990. *Planta Medica*, 56: 368-370.

*Wettstein, R. 1933-1935. *"Handbuch der systematischen Botanik"*. Auf. 4, Leipzig-Wien.

Willett, W.C. 1994. Diet and health. What should we eat? *Sci.Kakadeence.,* 264: 532-537.

Williamson, V.M.and Hussey, P.S.1996. Nematode pathogenesis and resistance in plants. *Plant Cell,* 8:1735-1745.

Wilson. K. and Goulding, K.H. (Eds.) 1992. *A Biologist's Guide to Principle & Techniques of Practical Biochemistry*. 3rd Edition. Cambridge University Press, Cambridge, U.K.

Wolff, I.A. and Kwolek, W.F. 1971. Lipids of Leguminosae. In: the *Chaemotaxonomy of Leguminosae.* (Eds.) Harbone, J.B., Boulter, D. and Turner, B.L. Academic Press, New York. pp 231-255.

Wu, J.Y., Gardner, B.H., Murphy, C.I., Seals, J.R., Kensil, C.R., Ecchia, J., Beltz, G.A., Newman, G.W. and Newman, M.J. 1992. Saponin adjuvant enhancement of antigen specific immune responses to an experimental HIV-1 vaccine. *J. Immunol.,* 148: 1519- 1525.

* Xie, C., Lin, R.C., Antony, V., Lumeg, L. 1994. *Alchol. Clin. Exp. Res.,* 18: 1443-1447.

Yamaguchi, M. 1983. *World Vegetables: Principles, Production and Nutritive Values*. Avi Publishing Company. Inc., West port, Connecticut.

*Yamamoto, S., Miyamoto, A., Awachi, Y. and Hirishima, Y. 1963. *J. Therap.,* 45: 157.

Yamawaki, M., Zurbriggen, A., Richard, A., Vandevele, M. 1993. Saponin treatment for *in situ* hybridization maintains good morphological preservation. *J. Histochem. Cytochem.*, 41:105-109.

Yang, Y., Hyun, J., Sung, M., Chung, H., Kim, B., Paik, W., Kang, S. and Park, J. 1996. *Planta Medica,* 62: 353-354.

Yen, Y., Lee, M.C. and Salzmann, M.1986, A comparison with secretory and proliferative endometrium. *Anat. Rec.*, 215:262-266

Yeshidanden. 1986. *Health through Balance: An Introduction to Tibetan Medicine*. Snow Lion Publication, Ithaca, USA.

Yoganarasimhan, S.N. 1996. *Medicinal Plants of India.* Vol. I–Karnataka. Interline Publishing Private Limited, Delhi, India.

Yoganarasimhan, S.N. 2000. *Medicinal Plants of India.* Vol II–Tamil Nadu. Cyber Media, Bangalore, India.

Yoshikoshi, M., Yoshiki, Y., Okubo, K., Seto, J. and Sasaki, Y. 1996. *Planta Medica,* 62: 252-255.

Yoshioka, H., Inada, M., Ogawa, K., Ohshio, G., Yamabe, H., Hamashima, Y. and Miyake, T,. 1989. Lectin histochemistry in ulcerative colitis and Crohn's disease. *Journal of Experimental Pathology*, 4: 69-78.

Zafar, R. 1994. *Medicinal Plants of India.* CBS Publishers & Distributors, Delhi, India.

Zang, S.M., Wu, M. and Chen, H. 1989. Expression of galactin–3 modulated T cell growth and apoptosis. *Histochemistry,* 92:171-5

*Zani, F., Cuzzoni, M.T., Daglia, M., Benvenuti, S., Vampa, G. and Mazza, P. 1993. *Planta Medica*, 59:502-507.

Zeven, A.C. and Zhukovsky, P.M. 1975. *Dictionary of Cultivated Plants and Their Centers of Diversity (excluding ornamentals, forest trees and lower plants).* Centre for Agricultural Publishing and Documentation, Wageningen.

Index

A

B

C

D

E

F

G

H

I

J

K

L

M

N

O

P

Q

R

S

T

U

V

W

X

Y

Z